An Engineer's Guide to Automated Testing of High-Speed Interfaces

Second Edition

For a complete listing of titles in the
Artech House Microwave Library,
turn to the back of this book.

An Engineer's Guide to Automated Testing of High-Speed Interfaces

Second Edition

José Moreira

Hubert Werkmann

ARTECH HOUSE

BOSTON | LONDON

artechhouse.com

Library of Congress Cataloging-in-Publication Data
A catalog record for this book is available from the U.S. Library of Congress.

British Library Cataloguing in Publication Data
A catalogue record for this book is available from the British Library.

Cover design by John Gomes

ISBN 13: 978-1-60807-985-8

685 Canton Street
Norwood, MA 02062

10 9 8 7 6 5 4 3 2 1

Para os meus paid Inês e o meu irmão Carlos.
—José Moreira

For everyone who supported me getting to the next levels
—whenever, wherever
—Hubert Werkmann

Contents

Preface to the Second Edition

Since the first edition of this book, the automated testing of high-speed digital interfaces has continued to evolve with data rates now exceeding 28 Gbps in volume production devices and 56 Gbps/Gbaud devices already on the horizon. This continuing increase in data rate generates new challenges, especially in the DUT test fixture area. Also new multilevel standards like PAM-4 or MIPI C-PHY are starting to become more common requiring new test strategies. In addition, some of the previously important data transmission schemes like at-cycle source synchronous interfaces lost traction somehow outside the main memory market segments.

There have been several updates in this second edition. The section on the Hypertransport interface has been removed in this edition and a section on MIPI C-PHY has been added. Although XDR memory is not widely present in the semiconductor markets anymore, we kept this section in the second edition because of its unique differential bidirectional nature. A new appendix on the relation between jitter and phase noise was added. Although no new chapters have been added or removed, there have been multiple updates on all chapters.

Acknowledgments

We would like to thank all colleagues and customers with whom and who we had the privilege to work in the last years since the first edition of this book and who have helped to improve our knowledge in this area.

If you have any questions or feedback to the authors, please direct them to joseanton.moreira@googlemail.com or hubert.werkmann@arcor.de.

Preface to the First Edition

The starting point for writing this book was a discussion between the two of us about our common experience in our ATE test application support work for high-speed digital interfaces and the challenges these interfaces pose on the test engineers we supported. Initially we just planned to write an extended application note about the nature of jitter and jitter testing. This was due to the fact that we identified this topic as the one that all engineers that in the one or other way have responsibility for getting high-quality high-speed devices out of a fabrication line had most questions about.

However, it became clear to us very soon that just another jitter document would not really satisfy the needs of these engineers because their main difficulty is not necessarily to understand the test theory around jitter. It rather is the transfer of this knowledge about the theory into a practical implementation of device test flows and the various cross dependencies between device building blocks, interface architectures, test system interfacing, integration of external instrumentation, and efficient test implementations for production testing.

Thus, we analyzed the real-world test implementation issues we came across in the past years, from the instable test that resulted in production yield loss because of insufficient device training, over the massive device failures that were caused because just a single pulse within millions of clock cycles on a PLL input was missing in the test pattern, to the marginal data eye test due to suboptimal test fixture design.

As a result of this analysis, we saw the need to provide a comprehensive introduction to the characterization and production testing of high-speed I/O digital interfaces using automated test equipment (ATE) that does not just list the test requirements for these interfaces and propose test implementations but also gives some background on why the test requirements are defined the way they are.

The target audience of this book are mainly test engineers working on design verification, characterization, and production testing of multigigabit I/O interfaces. However, the book provides a general overview over the challenges these test engineers face in their daily work with high-speed devices that can be helpful also for chip designers and product engineers to minimize these challenges using appropriate DfT and device test strategies to ensure sufficient device quality with less test implementation effort.

After introducing the challenges of high-speed digital testing on ATE in Chapter 1, we provide a general introduction to high-speed digital signals and related topics in Chapter 2 to set the basic framework to understand the following chapters.

The underlying basis for this book is the high-speed I/O digital standards that evolved in the past years. In Chapter 3, we discuss some of these standards. We selected the standards that are covered with regard to the general test requirements and challenges they represent for ATE implementations. The chosen standards reflect a cross section of the current high-speed I/O interfaces in the market with respect to test requirements. In this chapter we do not want to give a very detailed description and test implementation examples for the dedicated interface. The main purpose is to extract the general architectural characteristics for each interface that define the type of tests that need to be implemented and can be leveraged to other interfaces that use similar architectural implementations.

After this, we present the basics of ATE systems and instrumentation in Chapter 4 with emphasis on the critical instrumentation for high-speed digital applications like, for example, digital pin electronics cards, sampler/digitizer cards, and power supplies. One of the core chapters of the book is Chapter 5 that covers the details of key measurements for high-speed I/O interfaces and the options a test engineer has to implement these measurements on an ATE system.

Chapter 6 focuses on how high-speed I/O interfaces can be tested efficiently in a production test environment. The key discussion in this chapter deals with at-speed loopback testing. Characterization and test of high-speed digital I/Os often requires complementing an ATE with external measurement instruments for some dedicated tests. Chapter 7 presents some of the bench instrumentation that is relevant for engineers working on high-speed digital applications on ATE and also some of the related accessories.

One of the topics that commonly is recognized as a key bottleneck for the ability to test and characterize high-speed digital interfaces on an ATE is the electrical interfacing between the ATE pin electronics and the

DUT. Chapter 8 covers this bottleneck and discusses ATE test fixture design including associated topics like sockets and wafer probing.

Last but not least, we discuss some advanced topics on high-speed digital testing with ATE in Chapter 9. These topics include the definition of EPA/OTA, the importance of linearity, focus calibration, and the important topics of de-embedding and test fixture loss compensation through equalization.

In the appendices of the book, we cover several additional topics that are not directly related to the test of high-speed devices itself. However, these topics are helpful to understand the background of some of the test requirements and test implementations that are presented in the main chapters of the book.

Appendix A provides a brief overview of the Gaussian distribution and some examples of obtaining analytical results that are relevant for high-speed digital applications.

Appendix B presents a detailed description of the dual Dirac jitter model and the associated random and deterministic jitter separation algorithm.

Appendix C presents an overview of pseudo-random bit sequences and other important data patterns.

Appendix D discusses the topics of encoding, scrambling, disparity, and error correction that are required to understand the implementation of some of the high-speed digital standards.

Appendix E provides an introduction to the topic of time-domain reflectometry and transmission that is important for test fixture characterization.

Appendix F introduces the concept of S-parameters that are again important for test fixture characterization and modeling.

Appendix G provides an overview of the different simulation tools that are relevant for high-speed digital applications, especially for test fixture design.

Appendix H presents several approaches to characterize a manufactured test fixture electrical performance, not only regarding the high-speed signal paths but also the power distribution network.

Appendix I discusses the challenge of calibrating the amount of jitter injected by an ATE driver into the device under test (DUT) since receiver jitter tolerance is a key part in some of the high-speed digital standards.

Acknowledgments

We first thank Heidi Barnes, Hideo Okawara, and Bernhard Roth from Verigy and Michael Comai from AMD for contributing to the sections on test fixture design, sampler/digitizer ATE card, data eye profile, active equalization, and

HyperTransport. We also thank Verigy for allowing us to use several pictures we have obtained in the course of our work.

We thank Gert Haensel from Texas Instruments, Wolfgang Maichen from Teradyne, Christoph Zender from DHBW Horb, Istvan Novak from Sun Microsystems, Eric Bogatin from Bogatin Enterprises, Marcus Mueller from Agilent Technologies, and Takahiro J. Yamaguchi from Advantest Laboratories, Ltd., for reviewing the book or parts of it. We also thank all the colleagues from Verigy that have contributed to the review of this book, especially Shinji Fujita, Frank Hensel, Joe Kelly, Bernd Laquai, Clemens Leichtle, Jinlei Liu, Kosuke Miyao, Joerg-Walter Mohr, Jan-Ole Brandt, Roger Nettles, Claus Pfander, Robert Schneider, and Juergen Wolf. Thanks to Andreas Olenyi and Oliver Guhl for their support inside Verigy.

Orlando Bell from GigaTest Labs, and Bob Schaefer and Mike Resso from Agilent Technologies provided help in some of the measurements presented in this book.

The author José Moreira thanks his university mentor Professor Carlos Bispo from the Institute for Systems and Robotics in Lisbon for instilling in him the enjoyment of technical writing.

Finally we thank all the other colleagues that we had the privilege to interact with and who helped shape the authors' knowledge of high-speed digital testing with ATE.

1

Introduction

It is an amazing fact of modern technology that we are able to manufacture integrated circuits (IC) with hundreds of millions of transistors switching at gigahertz clock rates and containing hundreds of high-speed I/O terminals running at multigigabit data rates.

It even is more amazing that we are able to guarantee the correct behavior of such devices when delivered to the consumer.

The most amazing matter of fact, however, is that such devices can be delivered ensuring not only the device operation by itself, but also that their compliance with complex high-speed I/O specifications that guarantee error-free communication with other chips and allow the implementation of powerful yet highly integrated electronic systems we all enjoy (more or less) in our daily life.

These high-speed I/O standards that guarantee the interoperability of devices from different manufacturers and help to facilitate the widespread adoption of multigigabit I/O interfaces into applications such as wired communications, computation, and consumer electronics typically set directions in the industry and serve as a roadmap that pushes IC manufacturers to higher data rates. Figure 1.1 shows the data rates for some of the more common high-speed digital interfaces listing both what is available and what is under development as of the writing of this book.

One of the most critical steps in the manufacturing process of devices that employ such complex high-speed I/O standards is IC testing. While device testing does not usually get the same recognition as the design and front-end process disciplines of the semiconductor industry, IC testing is becoming more important and more complex with the continuous movement to enhanced process technologies and higher clock frequencies or data rates. Besides the technical necessity to guarantee correct device interoperation, the

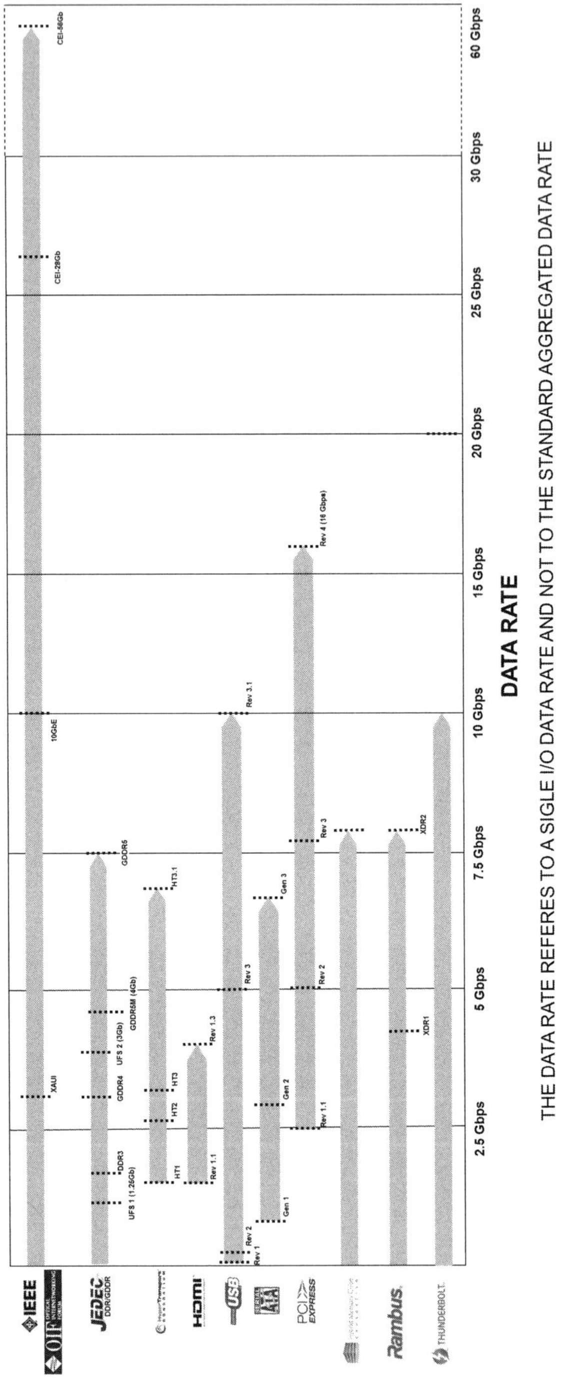

Figure 1.1 Some high-speed digital standards available and under development as of the writing of this book.

implications of IC testing on cost also have to be stressed. Very often device manufacturers only consider the liabilities that are associated with IC testing like capital investment and recurring device manufacturing cost but they neglect the asset that a thorough and reliable test implementation represents because it avoids the delivery of faulty parts to a consumer. It is important to acknowledge that the subject of IC testing is vast and spans areas in multiple dimensions, from IC design to the final application integration into a system. This is reflected in the test-related topics that have to be considered along the manufacturing life cycle of a device. Examples of these topics include design for test (DfT) strategies, design validation and characterization, production testing, system level test, test fixture design, and test cell integration.

The objective of this book is to cover the specific topic of testing high-speed digital interfaces with automated test equipment (ATE). When discussing the testing of high-speed digital interfaces with automated test equipment, there are four main areas a test engineer has to address. These are characterization and design validation of the initial silicon, production testing, accuracy and correlation between bench instrumentation and ATE or between characterization and production, and finally the test fixture used for the high-speed digital testing of the device under test (DUT).

1.1 Characterization and Design Verification

The high-speed digital interfaces embedded in multifunctional integrated circuits have achieved a significant degree of complexity. This complexity requires the test engineer to have a broad range of knowledge covering several disciplines such as analog design, digital design, probability and statistics, and signal integrity to be able to understand and solve the technical test challenges posed by these interfaces. Figure 1.2 shows a block diagram of a multigigabit I/O cell. Looking at the block diagram, one can estimate the complexity that state-of-the-art I/O cells contain. This complexity in combination with the data rates these cells are operating, both today and in the future, presents additional challenges for design validation, characterization, and production testing.

For the test engineer, the challenge of this movement to higher data rates is not as simple as just using more advanced measurement instruments or ATE systems that support the increase in data rates. To address the timing restrictions that these higher data rates bring, the complexity of modern standards is increasing. State-of-the-art high-speed I/O standards now require additional complex measurements for the test of the I/O cell such as jitter generation, and jitter tolerance. These new challenges require the test engineer to understand and implement a new set of measurements that were not

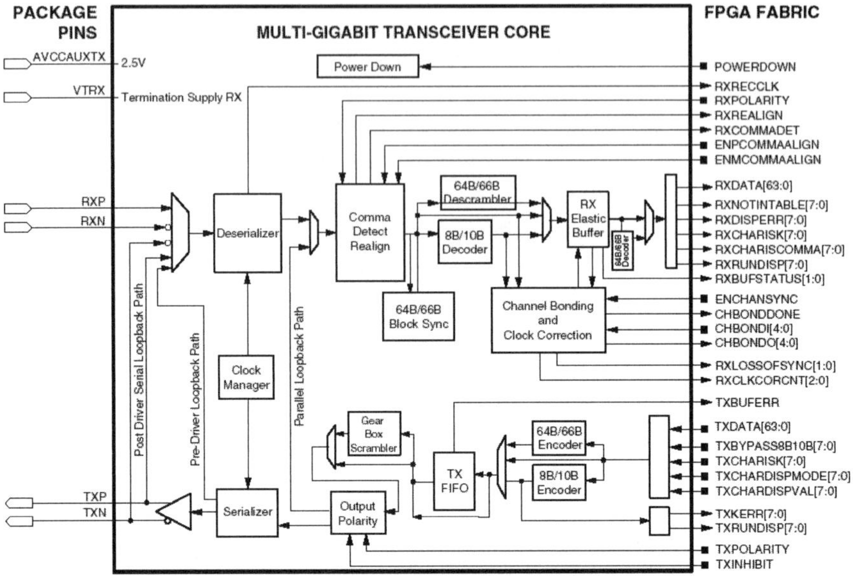

Figure 1.2 Xilinx Rocket I/O cell block diagram (reprinted with permission from [1]).

required in the past. Figure 1.3 shows the receiver requirements from the PCI Express standard [2].

Parameter	Description	Min	Max	Units	Notes
UI	Unit interval without including of SSC	199.94	200.06	ps	Over 10^6 UI
$T_{RX\text{-}HF\text{-}RMS}$	1.5 – 100 MHz RMS jitter		3.4	ps RMS	Spectrally flat, 3
$T_{RX\text{-}HF\text{-}DJ\text{-}DD}$	Max Dj impinging on Rx under test		88	ps	2,4
$T_{RX\text{-}SSC\text{-}RES}$	33 kHz Refclk residual	--	75	ps	
$T_{RX\text{-}LF\text{-}RMS}$	< 1.5 MHz RMS jitter	--	4.2	ps RMS	Spectrally flat
$T_{RX\text{-}MIN\text{-}PULSE}$	Minimum single pulse applied at Rx	120		ps	2
$V_{RX\text{-}MIN\text{-}MAX\text{-}RATIO}$	Min/max pulse voltage ratio seen over an time interval of 2 UI.	--	5		2
$V_{RX\text{-}EYE}$	Receive eye voltage opening	120		mVPP diff	1,3
$V_{RX\text{-}CM\text{-}CH\text{-}SRC}$	Common mode noise from Rx	--	300	mVPP	2

Figure 1.3 Example of part of the DUT receiver requirements from the PCI Express standard. (From: [2]. ©2009 PCI-SIG. Reprinted with permission.)

1.2 Production Testing

Production testing of high-speed digital interfaces presents a different set of challenges when compared to design validation and characterization. A high-speed digital standard might define a series of measurements that the interface must comply with, but in production testing the cost pressure associated with the target application makes it impractical to implement all of these measurements. The characterization and design verification tests ensure compliance with the standard or design requirements while the lower cost production test focuses on a minimal set of tests to verify that the process parameters for a given device have not shifted, to check basic functionality, and to screen for known failure mechanisms.

At a high level, one can assume that the cost of testing will be proportional to the test time for each device and the capital cost of the measurement instrumentation. This simplified look at cost of testing highlights the importance of faster measurement techniques, a lower cost per measurement ATE system, and the benefit of parallel or multi-site testing with a single insertion on an ATE system. Implementing these cost reduction methods can require significant changes in the test methodology and the types of tests compared to those found in high-speed I/O standards. For example, it is not uncommon to reduce cost of testing by using a less expensive ATE system where the measurement instrumentation does not have the data rate or bandwidth to support the device under test. This type of situation requires test methodology such as DfT approaches and at-speed loopback to ensure the full speed functionality of the high-speed I/O interface.

1.3 Accuracy and Correlation

The ability of an ATE system to measure the performance of a high-speed digital interface depends on the measurement accuracy/repeatability and correlation that can be obtained for a given application. This can be quite challenging with increasing data rates. Basic requirements like the timing accuracy for a state-of-the-art ATE system are now specified in the picosecond range with random jitter values in the hundreds of femtoseconds. This type of timing accuracy demands fast digital edge rates with multigigahertz bandwidth which increases the challenges when trying to maintain the same accuracies over multiple channels in an ATE system.

The narrow timing margins also create a challenge for correlation between ATE systems and bench instrumentation since slight timing variations and test fixture differences between measurement systems can have a significant impact. This correlation is complicated even further for measurements such as jitter separation where the methodology, hardware performance, and algorithms used for the measurement can vary dramatically between instrument manufacturers.

To address these challenges and the trade-offs between measurement systems, the test engineer needs to have a complete understanding of the capabilities and accuracy of the ATE system and how these parameters influence the measurements they need to apply. In addition, the test engineer has to have a full understanding of the instrumentation that the ATE system must correlate to, including any specific measurement algorithms these instruments apply.

1.4 The ATE Test Fixture

The design of the hardware and signal path for multigigabit instrumentation is of little use if the signal cannot be transmitted to the device under test (DUT) with at least the performance of the ATE instrument itself [3]. The test engineer now must have an increased understanding of the impact the signal path has on the characterization of multigigabit digital I/O interfaces. In order to assure that this understanding is transferred correctly into the overall test solution, he has to be involved in the design of the test fixture also known as DUT loadboard, device interface (DI), or device interface board (DIB), which connects the ATE instrumentation to the DUT. In the past, most test engineers were only responsible for providing a schematic or netlist of the needed DUT/ATE pin assignments, leaving most of the test interface design details (e.g., trace geometry, PCB dielectric material) to the PCB layout team. The success of the test fixture simply meant getting all of the point-to-point connections right and sometimes having additional circuitry on the test fixture to support a particular application test.

This simple pin assignment approach is no longer appropriate for test fixtures that are intended to test high-speed digital interfaces. In this case the actual signal integrity of the routing on the test fixture is critical for the success of the application, and must be engineered to meet the testing requirements of the high-speed I/O channel. This challenge is illustrated in Figure 1.4 where it is shown that the signal performance at the ATE interconnect where the system specifications are set is not the same as that found further away at the DUT. If this interface between the ATE interconnect and DUT is not engineered correctly, it can close the eye in a data eye diagram of a high-speed I/O completely. This in turn will have a direct impact on the measurement accuracy in a device characterization application or on the production yield in a manufacturing test environment.

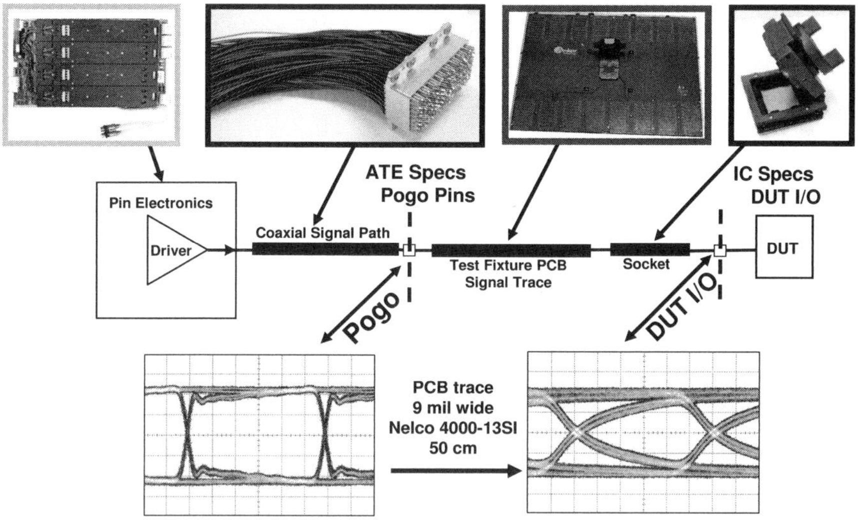

Figure 1.4 The signal integrity challenge in an ATE system. (From: [4]. ©2010 Advantest. Reprinted with permission.)

It is also important to take into account that the signal integrity of a test fixture is not restricted to the performance of the connection between the ATE channels and the I/O cells of the DUT. The generation and detection of the multigigabit signal requires a power distribution network that has low noise and fast switching current response characteristics to prevent voltage drop on the supply lines.

1.5 The Future

One indication from looking into the past is that pin counts and data rates will continue to increase in the future, pushing the limits of existing electrical interconnects. Technologists have often predicted the need to move to optical technology for short-range applications and the demise of the copper interconnect. Not long ago, 2.5 Gbps was considered the limit for copper-based electrical interconnects and that the next generation of chip-to-chip interconnect needs to use optical signal transmission [5]. Although copper-based electrical interconnects have fundamental limits [6], the fact that the majority of the electronics industry is copper-based makes staying with copper interconnects a significant advantage. Because of that, the electronics industry will continue to solve the challenges associated with higher data rates in copper interconnects by using new technologies and smarter approaches not only in the design of the IC I/O cells but also in the data transmission. But at the same time new technologies are creating new test challenges like through-silicon vias, silicon photonics and high data rate transmission using wireless links.

1.5.1 TSV Technology

New capabilities in semiconductor manufacturing technology will have an impact on required test approaches and test requirements in the future. One example for such evolving manufacturing technologies is through-silicon via (TSV) processes [7]. TSV processes allow the formation of via structures from the active surface of semiconductor dies to the back side of the die. In combination with micro-bumps on either side of a die and a potential back-side metallization, this allows vertical stacking of multiple dies with vertical signal routing paths through the stack using the TSV structures.

With TSV interconnects, new packaging technologies are enabled that make use of the vertical chip-to-chip signal paths as well as the higher density of micro-bump device signal interfaces. An example for true three-dimensional packaging (3D-packaging) using TSVs is the stacking of mobile application processors and memory devices which is the target application of the Wide I/O memory interface that was standardized by JEDEC in 2011 [8].

Another example for TSV-enabled packaging is 2.5D integration. With 2.5D integration two advantages of TSV connections are used. One is the ability to vertically stack devices to reduce real estate requirements of a system and the other is the high device interface signal density. An example for a 2.5D integrated system is graphics applications that use high-bandwidth memory (HBM). Such a system consists of one or multiple memory TSV stacks that are mounted side by side to a graphics processing unit (GPU) on an interposer substrate that allows high density signal routing. The memory stack in this application provides a large amount of memory capacity on a small footprint and the interposer substrate in combination with the memory TSV signal interface allows the routing of one or multiple 1,024-bit data buses between the memory stacks and the GPU [9]. The wide data bus provides data transmission bandwidth at moderate data rates that are far beyond narrow high-speed memory interfaces. Nevertheless, TSV technology also will be used to drive up data rates due to the short signal path lengths that are possible with both 2.5D integration and 3D integration.

In any case, TSV systems will impact test requirements and test methodologies [10, 11]. This spans from dedicated instrumentation for TSV process development that has to provide extreme parallelism over new scan-architectures for debugging TSV-based systems to the requirement for at-speed wafer test and known good die (KGD) testing due to the necessity to use bare dies for any TSV-based system.

1.5.2 Silicon Photonics

Optical networks have been the backbone for data transmission across middle and large distances. There has been usually a separation between traditional high-speed digital testing and the testing of the optical/electical components used on optical networks [12]. This was mainly because in an optical system there was somehow a separation between the devices traditionally tested by a high-speed digital test engineer (e.g., MUX/DEMUX digital components) from devices tested by the optical test engineer (e.g., laser drivers, transimpedance amplifiers, lasers, photodiodes) [13].

With the advent of silicon photonics this traditionally separate devices are being integrated into a single packaged 3D IC or even in a single die [14]. This means that both the high-speed digital side and the optical side need to be addressed by the test engineer. In the case of a high-speed digital test engineer, it does require understanding this new world of optoelectronics and photonics [15–17]. Also new ATE instrumentation and test methodologies will be required to handle the photonics side measurement requirements [18–20].

1.5.3 Millimeter-Wave Wireless Communications

Millimeter-wave communication systems (e.g., 60 GHz WLAN/WPAN) offer the ability to transmit extremely high data rates over short distances in a wireless medium [21, 22]. There are multiple standards in discussion. For the test engineer characterizing and testing these integrated circuits, the challenge is testing on one side the wireless signal that is running at millimeter-wave frequencies (e.g., 60 GHz) and on the other side the high-speed digital wire interface that might be running at a data rate of 4 Gbps.

References

[1] J. Moreira, B. Sadler, and D. Ferguson, "Parametric Testing of a 10Gbs I/O Cell in Production Through a Parametric Loopback Approach with Retiming," *IEEE International Design and Test Workshop*, 2006.

[2] PCI-SIG, *PCI Express Base Specification Revision 2.1*, Mar. 2009.

[3] J. Moreira, M. Tsai, J. Kenton, H. Barnes, and D. Faller, "PCB Loadboard Design Challenges for Multi-Gigabit Devices in Automated Test Applications," *DesignCon*, 2006.

[4] B. Roth and J. Moreira, "Equalization in the HSM6800 Pin-Electronics," *Verigy VOICE Users Conference*, 2010.

[5] N. Savage, "Linking with Light," *IEEE Spectrum*, Aug. 2002.

[6] D. A. B. Miller and H. M. Özaktas, "Limits the Bit-Rate Capacity of Electrical Interconnects from the Aspect Ratio of the System Architecture," *Journal of Parallel and Distributed Computing*, no. 41, 1997.

[7] P. Garrou, C. Bower, and P. Ramm, *Handbook of 3D Integration: Technology and Applications of 3D Integrated Circuits*. New York: Wiley-VCH, 2008.

[8] JEDEC Solid State Technology Association, *Wide I/O Single Data Rate*, Dec. 2011. JSED229.

[9] J. Kim, "Blazing A Trail to High Performance Graphics," *AMD 2011 Technical Forum and Exhibition*, 2011.

[10] H.-H. S. Lee and K. Chakrabarty, "Test Challenges for 3D Integrated Circuits," *IEEE Design & Test of Computers*, vol. 26, no. 5, 2009.

[11] B. Noia and K. Chakrabarty, *Design–for–Test and Test Optimization Techniques for TSV–based 3D Stacked ICs*. New York: Springer, 2014.

[12] D. Derickson, *Fiber Optic Test and Measurement*. Upper Saddle River, NJ: Prentice-Hall, 1998.

[13] B. Razavi, *Design of Integrated Circuits for Optical Communications, 2nd Edition*. New York: John Wiley & Sons, 2012.

[14] L. Chrostowski and K. Iniewski, *High-Speed Photonics Interconnects*. Boca Raton, FL: CRC Press, 2013.

[15] M. Hochberg, N. C. Harris, R. Ding, Y. Zhang, A. Novak, Z. Xuan, and T. Baehr-Jones, "Silicon Photonics: The Next Fabless Semiconductor Industry," *IEEE Solid-State Circuits Magazine*, 2013.

[16] S. O. Kasap, *Optoelectronics and Photonics: Principles and Practices, Second Edition*. Pearson, 2013.

[17] L. Chrostowski and M. Hochberg, *Silicon Photonics Design: From Devices to Systems*. Cambridge, UK: Cambridge Universtity Press, 2015.

[18] D. Watanabe, S. Masuda, H. Hara, T. Ataka, A. Seki, A. Ono, and T. Okayasu, "30-Gb/s Optical and Electrical Test Solution for High-Volume Testing," *IEEE International Test Conference*, 2013.

[19] T. Fujibe, "A 28Gbps, 16-lanes High Throughput Optical Transceiver Test System for 100Gbps Datacom Interconnection Devices," *Advantest VOICE Users Group Conference*, 2015.

[20] M. Galli, M. Rigamonti, G. Astone, R. Barbon, D. Sala, R. Aranzulla, P. Mooney, D. Adorni, J. L. Jeanneau, Y. Osuga, H. Shibata, J. Moreira, H. Werkmann, Z. Zhang, and F. Pizza, "A Silicon Photonics Wafer Probing Test Cell," *Burn-In and Test Strategies (BITS) Workshop*, 2016.

[21] J. Wells, *Multi-gigabit Microwave and Milimeter-Wave Wireless Communications*. Norwood, MA: Artech House, 2010.

[22] S. Emami, *UWB Communication Systems: Conventional and 60 GHz: Principles, Design and Standards*. Norwood, MA: Artech House, 2010.

2

High-Speed Digital Basics

This chapter reviews some important concepts that are necessary when analyzing high-speed digital signals and interfaces. These concepts provide a basis for the following chapters. General references on this topic for the interested reader are [1–3].

2.1 High-Speed Digital Signaling

The inputs/outputs of a high-speed digital interface are electrical waveforms (ignoring optical interfaces) that represent information encoded in a digital signal. Several options to encode a digital signal electrically exist as shown in Figure 2.1. Most high-speed digital standards use a non-return-to-zero (NRZ) approach for the signal being transmitted. Return-to-zero (RZ) signaling is usually not used on high-speed digital interfaces due to its bandwidth requirements. Another type of signaling is pulse amplitude modulation (PAM). In this approach the signal to be transmitted is encoded with different voltage levels thus, for example, a four-level PAM interface can transmit the same information as an NRZ single level interface at half of the clock frequency, as shown in Figure 2.1. Although PAM and RZ signaling are important for certain types of applications, in this chapter we will mainly discuss NRZ type signaling with the exception of Section 2.7 where we will briefly discuss PAM signaling in some more detail.

Figure 2.2 shows one example of analyzing a digital interface with a commercial ATE system. The figure shows the programmed time-domain waveforms for the driver channels on the ATE system that stimulates the respective receiver on the DUT and the digital waveforms measured by the ATE receiver from the appropriate DUT drivers. From these received

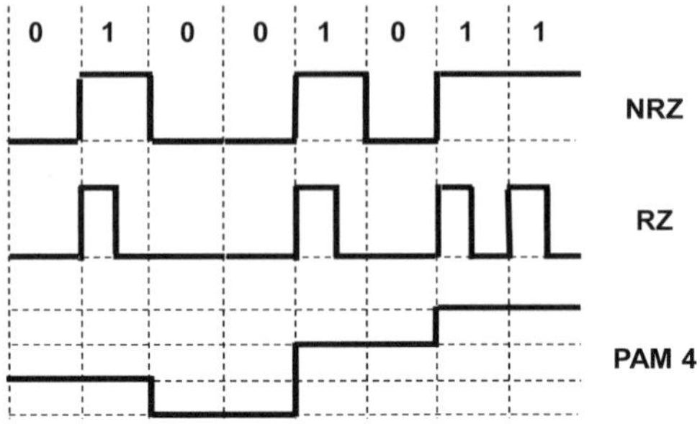

Figure 2.1 Comparison of the different types of signaling approaches for high-speed digital interfaces.

waveforms it is possible to infer the logic value based on the expected logic voltage thresholds.

Unfortunately this view is no longer sufficient for characterizing and debugging modern I/O digital interfaces [4]. The increase of the data rate on the DUT I/O data makes the timing and level margins more critical. The question is no longer only whether there are any errors on the data transmission through the digital interface but also how many bits can be transmitted on average before the first error occurs. The reason is that in multigigabit interfaces it is no longer possible to guarantee an error-free data transmission for an indefinite amount of time. This fact requires the use of new techniques to analyze the performance of digital interfaces like the data eye diagram, BER bathtub curve, and jitter analysis, which are discussed in this chapter and the following chapters.

2.1.1 Out-of-Band Signaling

Traditional electrical interfaces typically consist of signals that take care of the data transmission itself (e.g., address and data signals) and signals that are used for control and status information (e.g., reset signal, interrupt signals). These control and status signals are referred to as sideband signals. Besides increased data rates, high-speed I/O interfaces also tend to reduce the amount of sideband signals and integrate the status and control information that needs to be conveyed to partner devices into the regular communication protocol that runs on the signals which are also used for payload data exchange. The most extreme example of this are high-speed I/O interfaces that use serial links. These typically eliminate sideband signals completely to get maximum benefit from the serialization and an optimal ratio of signals that transfer payload data and signals that transfer sideband information.

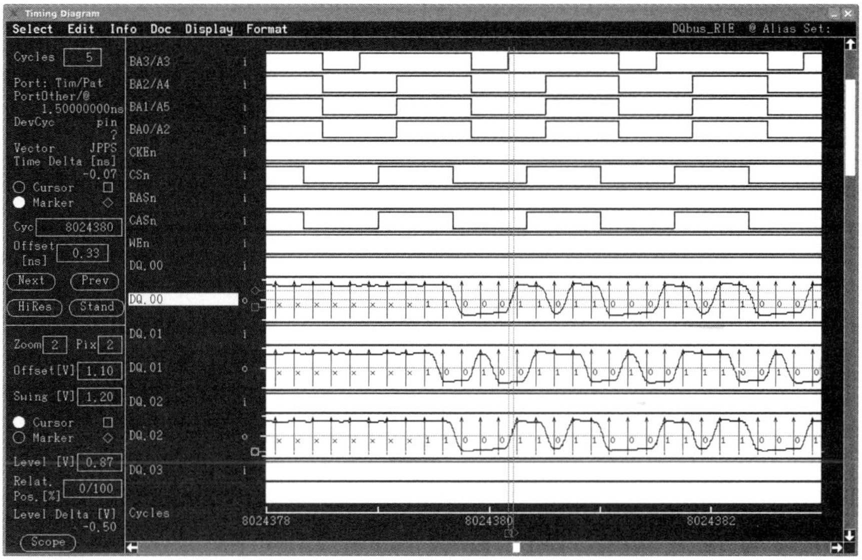

Figure 2.2 Classical view of a digital interface obtained with an ATE system (courtesy of Advantest).

However, for some of the control and status information that needs to be transferred, integration into the standard protocol that is applied for data exchange is not possible. This, for example, might be the case for serial links that are in power saving modes, but still need to transmit few sideband data, or during link (re)initialization. Often signaling mechanisms that do not match the signaling used for normal data exchange of the respective high-speed I/O standard are applied to convey this information in such a case. These non-standard signaling mechanisms are called out-of-band signaling. It has to be stressed that out-of-band signaling can be implemented very differently between the various high-speed I/O interfaces.

2.1.2 Data Eye Diagram

The data eye diagram is a methodology to display and characterize a high-speed digital signal in the time-domain. The data eye diagram is derived from the digital waveform by folding the parts of the waveform corresponding to each individual bit into a single graph with a one unit interval (UI)[1] width on the timing axis as shown in Figure 2.3 [2].

[1]One unit interval (UI) corresponds to the period of one bit and is sometimes also referred to as the bit period. For example, in a 10 Gbps data signal the UI is 100 ps.

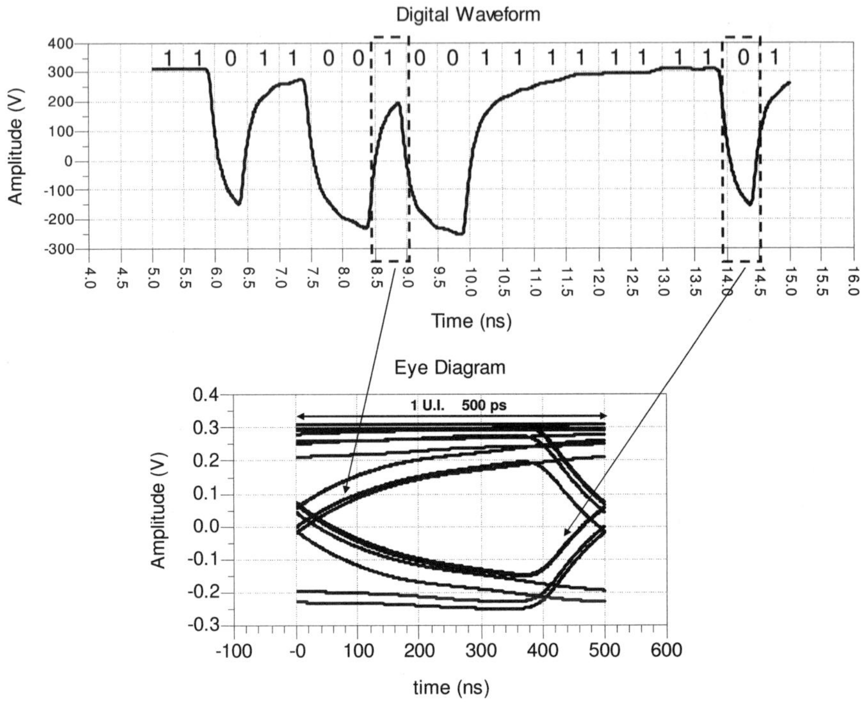

Figure 2.3 Construction of the data eye diagram from a digital waveform.

Contrary to the classical digital waveform representation of Figure 2.2, the data eye diagram does not allow the identification of the individual bits in the bit pattern. However, showing all the bits folded into a single UI allows the easy visualization of the quality of the digital signal. This is because both the best and the worst-case bit transitions on the waveform will be folded together as shown in Figure 2.4. In the figure two different waveforms are displayed, and although it is not easy to compare the performance of both digital signals when looking at the full waveforms, using the data eye diagram, it is obvious that waveform 2 is worse than waveform 1.

Figure 2.5 shows some of the typical nomenclature associated with an eye diagram. One key item is the optimal strobing point in the data eye diagram. This is the point with the maximum timing and level margin. Another important item is the amount of timing jitter (Section 2.4) and amplitude noise (Section 2.4.3) that are present on the data eye and that reduce the timing and level margin.

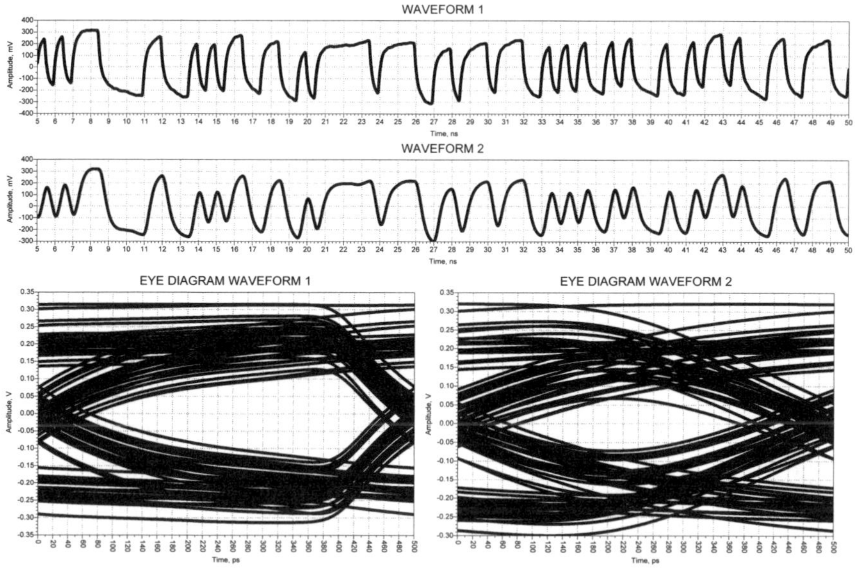

Figure 2.4 Comparing two waveforms on the time-domain showing the advantage of a data eye diagram display.

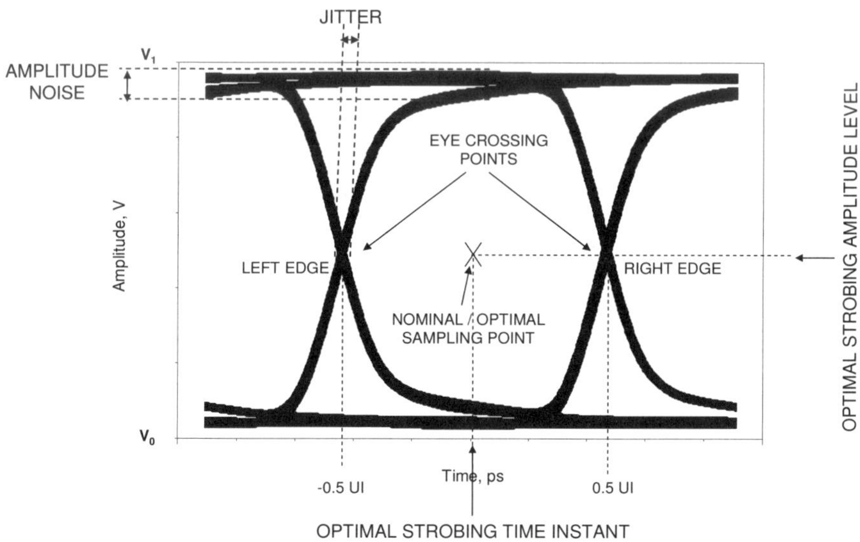

Figure 2.5 Nomenclature typically associated with a data eye diagram.

2.1.3 Differential Signaling

Most high-speed digital interfaces use differential signaling for electrical transmission. Differential signaling uses two complementary output drivers (differential legs) to drive two independent signal transmission lines [5]. Figure 2.6 shows a comparison between single-ended and differential signaling and how differential signaling is defined. While a single-ended signal is characterized by its high voltage level value V_{HIGH} and its low voltage level value V_{LOW}, a differential signal is usually characterized by its differential amplitude V_{DIF} and the common-mode value V_{CM} of the two differential legs.

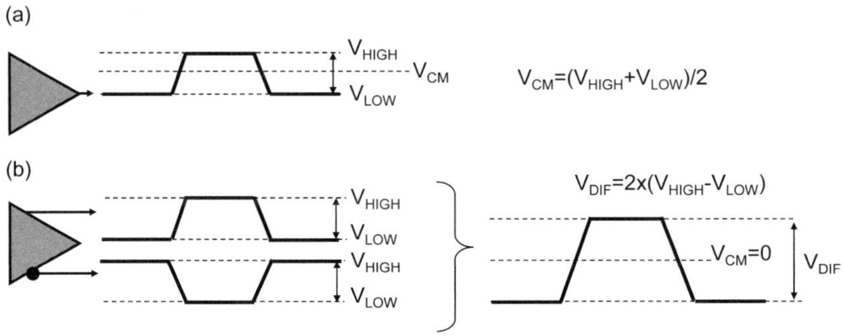

Figure 2.6 Comparison of (a) single-ended and (b) differential signaling.

Differential signaling has several properties that make its use advantageous for high-speed digital interfaces [5]. They include lower dI/dt, lower EMI, higher gain, and more immunity to crosstalk if transmitted through a coupled transmission line. Of course, the drawback is the need to use twice the number of pins and transmission lines compared to a single-ended signaling approach.

2.1.4 Transmission Line Termination

Electrical signals in the context of high-speed digital applications are typically transmitted through coaxial cables or controlled impedance printed circuit board traces with a characteristic impedance of 50 Ω for single-ended signaling and 100 Ω for differential signaling as shown in Figure 2.7 ([5, 6] are good starting points for the concepts of impedance and controlled impedance design).

The challenge of transmitting high-speed digital signals with fast rise times is that any impedance change seen by the signal while it travels through the signal path will create a reflection (e.g., due to relays and vias). Some of

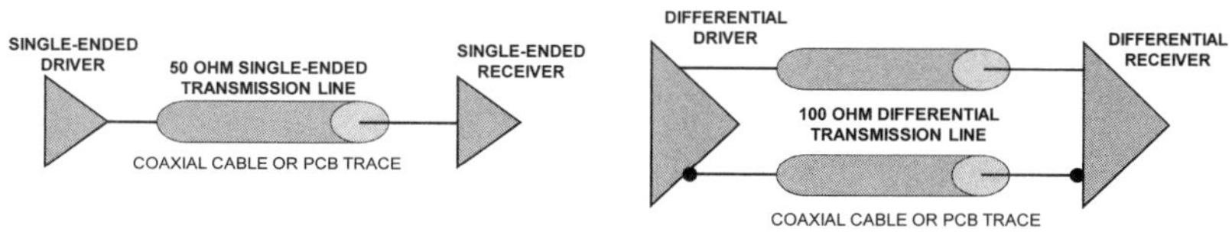

Figure 2.7 The 50 Ω single-ended and 100 Ω differential transmission system.

these points will be discussed in more detail in Chapter 8. In this section we are interested in the impedance discontinuity from a DUT driver/receiver or pin electronics driver/receiver. This is illustrated in Figure 2.8.

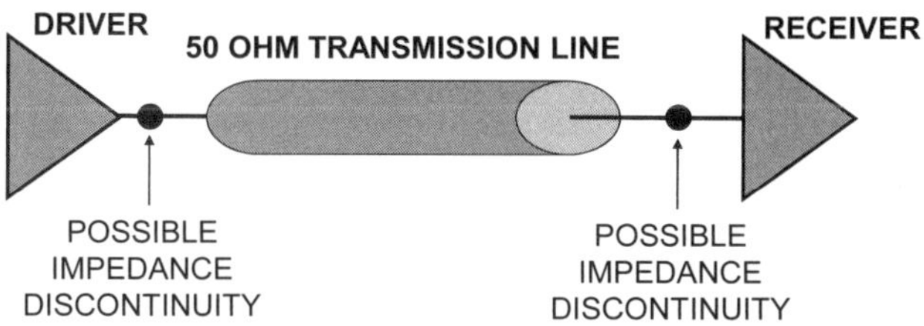

Figure 2.8 Impedance discontinuities at the pin electronics and DUT receiver.

Each time an electrical signal "sees" an impedance discontinuity, a portion of the electrical signal is reflected. The value of this reflected portion can be computed by (2.1) where Z_0 is the transmission line impedance before the discontinuity and Z_1 is the impedance of the transmission line right after the discontinuity that is encountered by the incident wave V_I and results in a reflected wave V_R:

$$V_R = \frac{Z_1 - Z_0}{Z_0 + Z_1} V_I \tag{2.1}$$

This equation shows that to prevent any reflections at the driver or receiver as shown in Figure 2.8, it is necessary that the impedance of the driver and receiver is equal to the impedance of the transmission line. This can be achieved by the use of proper termination resistors as shown in Figure 2.9 for the case of a single-ended 50 Ω transmission line.

Several methods for transmission line termination exist. In this section we will concentrate on the most common structure for high-speed digital applications which is shown in Figure 2.9 and is named series termination (for the ATE driver side) and parallel termination (for the DUT receiver side). Other possible termination methods are discussed in [6].

It is important to note that the 50 Ω resistor on the receiver parallel termination is usually connected to a termination voltage source. This is

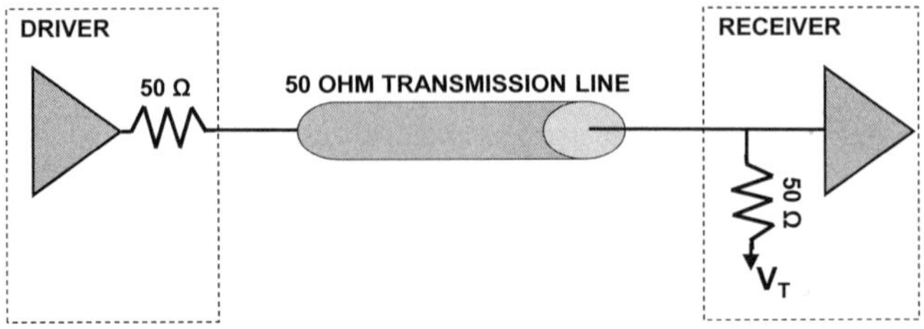

Figure 2.9 Properly terminated transmission line.

important to minimize any constant current flowing between the driver and receiver as shown in Figure 2.10. This is one of the values that typically needs to be set when programming the ATE pin electronics receiver.

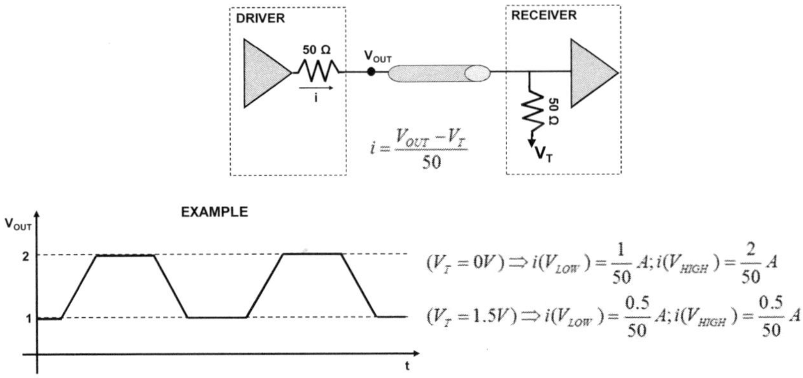

Figure 2.10 The importance of the 50 Ω voltage termination on a receiver and its impact on the current flowing between the driver and receiver.

There are some cases where the measurement instrument has a fixed 50 Ω to ground parallel termination that cannot be changed. Section 7.17.1 shows a possible approach to address this case when a termination voltage is required by the application. With differential signaling typically two types of termination schemes, named "center tap" and "cross termination," are used on the receiver as shown in Figure 2.11. From an electrical point of view there is no difference between the two configurations, although some standards or receiver implementations might prefer one of them.

2.2 Time and Frequency-Domains

Digital signals can be analyzed in both the time and frequency-domains. Digital signals are characterized by the fact that they ideally have a rectangular

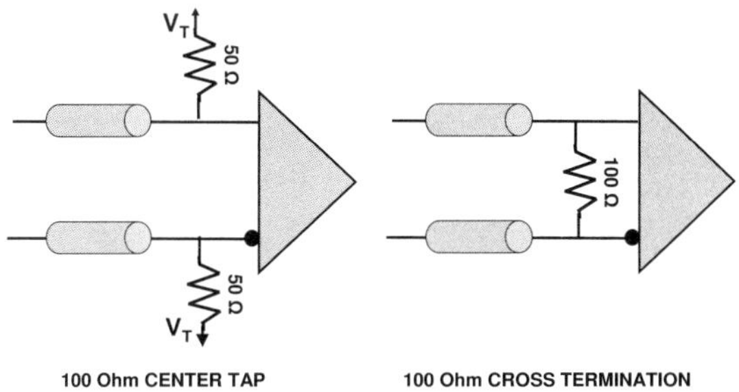

Figure 2.11 Differential signaling receiver termination schemes.

shape as shown in Figure 2.12 (top). For a random binary sequence with bit period T_B, it is possible to compute its power spectrum as shown in (2.2) [7, 8]. It results in the *sinc* function shape, which is defined in (2.3).

$$\text{Power Spectrum}(f) = \log\left[T_B\left(\frac{\sin(\pi f T_B)}{\pi f T_B}\right)^2\right] \tag{2.2}$$

$$sinc(x) = \begin{cases} 1 & (x = 0) \\ \frac{sin(x)}{x} & (\text{otherwise}) \end{cases} \tag{2.3}$$

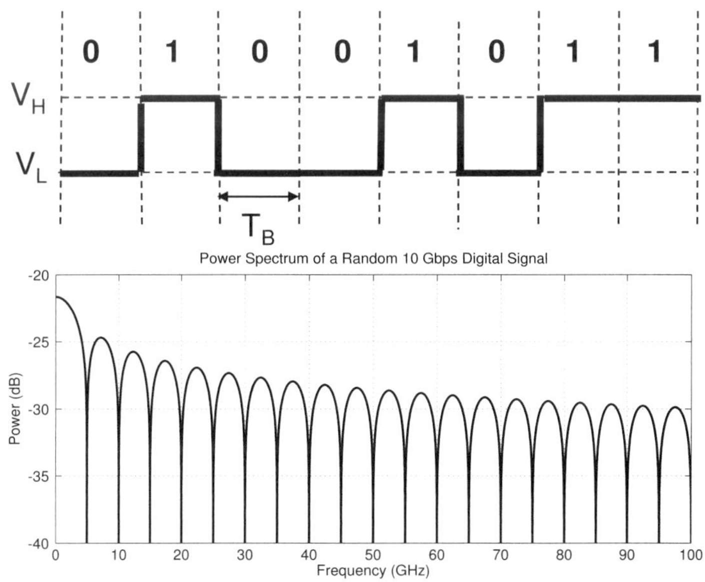

Figure 2.12 An ideal NRZ digital signal (top) and the power spectrum of a perfect NRZ random digital signal at 10 Gbps (bottom).

Figure 2.12 (bottom) shows the computed power spectrum for a 10 Gbps random binary sequence. Note that a null in the power spectrum exists at each multiple of the frequency corresponding to half of the bit period ($T_B/2$). Although Figure 2.12 shows the time-domain and frequency-domain views for a "perfect" digital signal, real signals are not perfect. Figure 2.13 shows a real 10 Gbps signal with a PRBS $2^7 - 1$ data pattern (see Appendix C) measured in the time-domain with an equivalent-time sampling oscilloscope and in the frequency-domain with a spectrum analyzer (see Chapter 7).

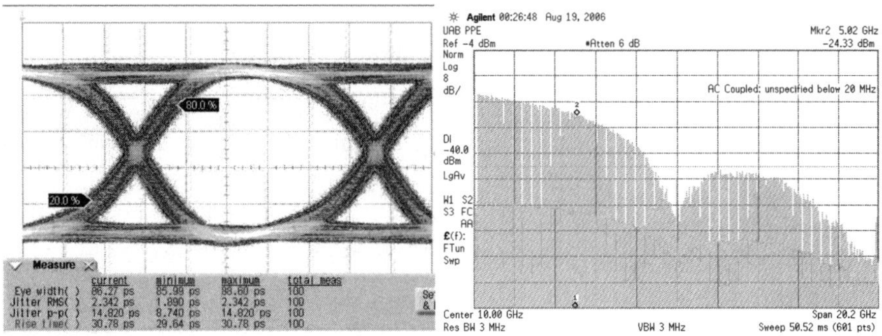

Figure 2.13 Viewing a high-speed digital signal in the time-domain (left) and frequency-domain (right).

The time-domain measurement is in the form of a data eye diagram showing a signal with a fast rise time (30 ps 20/80[2] with a good "open eye"). The frequency-domain measurement shows one important property of high-speed digital signals, which is the fact that they are broadband, having energy from a low frequency determined by the used data pattern until the highest frequency that is determined by the signal rise time. This is a very important factor that makes high-speed digital signals different from high-frequency RF signals that are typically narrowband around a specific frequency point.

The frequency content of the digital signal will depend on the data rate, on the pattern that is used, and on the rise time of the driver. Figure 2.14 shows the frequency spectrum of two data signals with the same data rate and data pattern but originating from different drivers. The figure shows that the driver with the slowest rise time (in this case the one with a maximum data rate of 3 Gbps) has almost no energy above 10 GHz while the 10 Gbps driver still has energy components beyond 10 GHz.

[2] 20/80 refers to the voltage levels used to compute the signal rise time meaning that the 20% and 80% point of the full signal amplitude are used to compute the rise time. 10/90 is also sometimes used for rise time measurements (see Section 5.4.1).

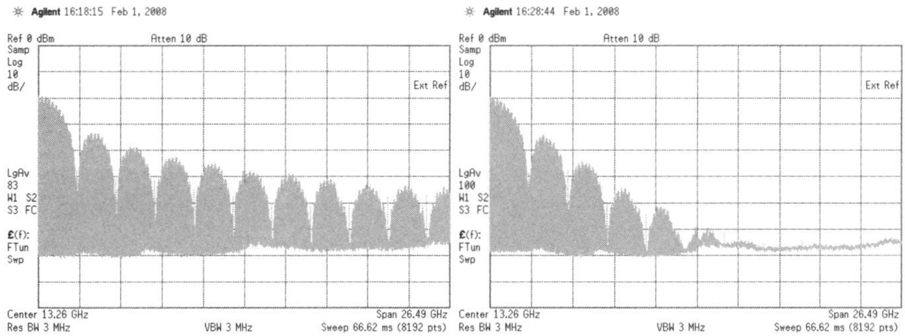

Figure 2.14 Comparison of the measured frequency spectrum of a PRBS $2^7 - 1$ data signal at 2.5 Gbps generated by two different drivers with different rise times (left: a 10 Gbps driver; right: a 3 Gbps driver).

2.2.1 The Concept of Bandwidth and Its Pitfalls

The concept of bandwidth for a measurement instrument or for a passive interconnect like a cable or a test fixture is very often used to define the required or expected performance of a measurement instrument or interconnect. Typically the term "bandwidth" refers to the frequency, where the frequency response of the measurement instrument or interconnect is -3 dB[3] below that of the DC point as shown in Figure 2.15. Note that for coaxial cables the term "bandwidth" is sometimes used with another definition in mind. This topic will be further discussed in Section 7.16.1.

The problem with this definition is that it is very limited in the sense that systems with the same bandwidth value might have very different frequency responses. Figure 2.16 shows the frequency response of two different systems with the same bandwidth and the simulated data eye diagram for each of the systems.

The resulting data eye diagrams show a significant performance difference between the different frequency responses demonstrating the pitfalls of relying on the bandwidth value for evaluating the time-domain performance of an instrument or interconnect. The main point is that knowing where the system frequency response crosses the -3 dB point is not as important as knowing how the system behaves in the frequency-domain (i.e., does it roll off very fast after the -3 dB point or does it roll off slowly and in this way not attenuating so strongly the frequencies after the -3 dB point?). Also note that the flatness of the frequency response before the roll-off is also very important.

[3]The -3 dB corresponds to a reduction of the signal amplitude by a factor of 1.41 and is also known as half power point or half power bandwidth (HPBW).

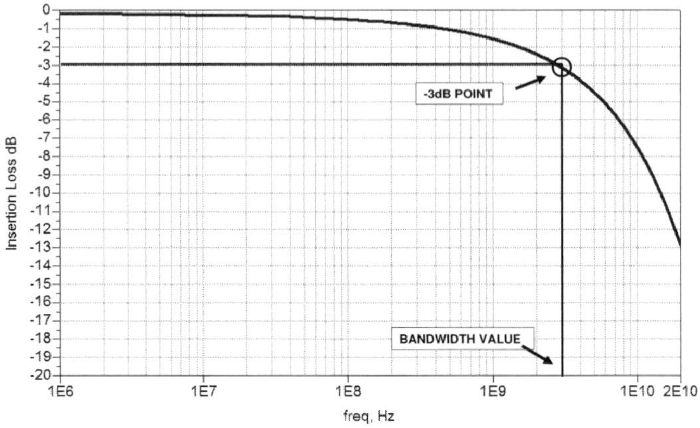

Figure 2.15 Example of the response of a measurement instrument or interconnect and the −3 dB definition of bandwidth.

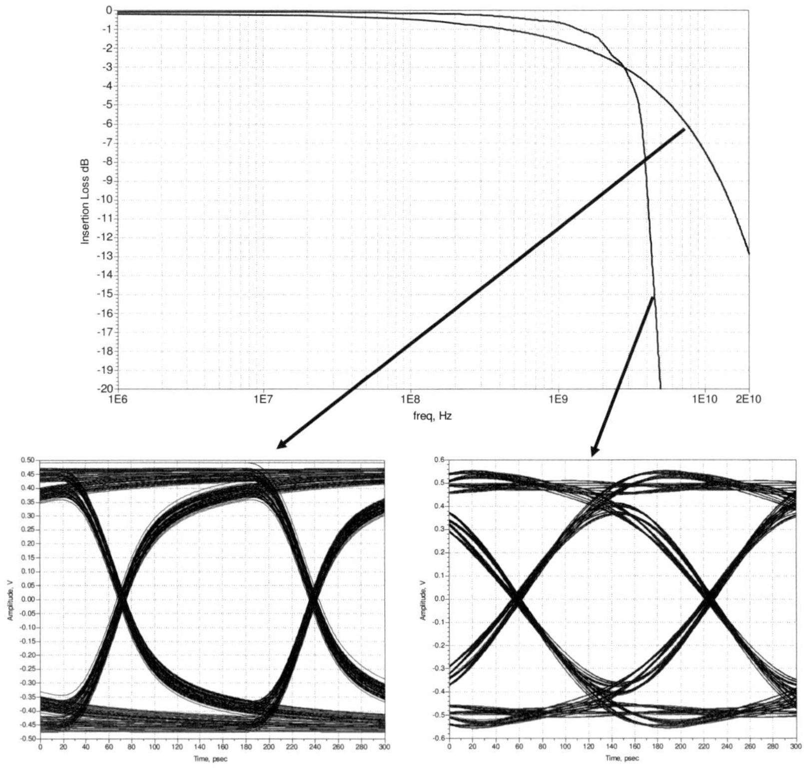

Figure 2.16 Frequency response of two different systems that have the same bandwidth value to a 6 Gbps PRBS data pattern.

Although the complete frequency response of a measurement instrument or interconnect might not be available in several situations, it is important to know the frequency response not only for the −3 dB point but also the −6 dB and −10 dB point.

2.3 Bit Error Rate

The concept of bit error rate (BER) is the cornerstone for analyzing the performance of a high-speed digital link. Figure 2.17 shows a basic block diagram of a unidirectional digital interface. Errors due to noise and signal degradation are generated in different parts of the digital interface by physical processes. Depending on the magnitude of the noise and the margins of the digital interface, it might happen that a given noise event will generate an error on the data transmission. The probability of a transmission error occurring is given by the bit error rate (BER).

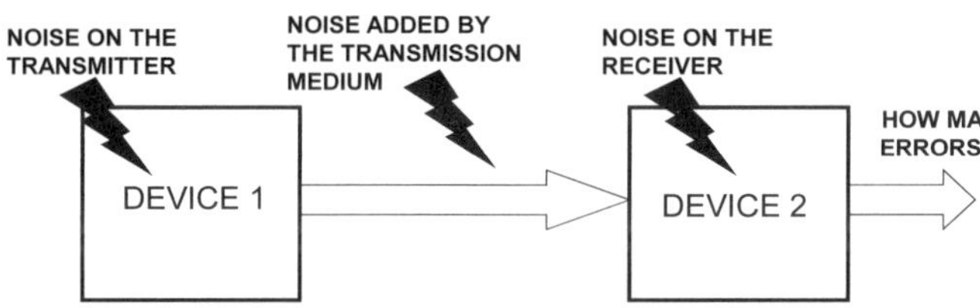

Figure 2.17 Simplified block diagram showing several sources of noise in a digital interface.

The bit error rate[4] is defined as the ratio of the number of failed bits to the total number of transmitted bits as shown in (2.4).

$$\text{BIT ERROR RATE (BER)} = \frac{\text{NUMBER OF FAILED BITS}}{\text{NUMBER OF TRANSMITTED BITS}} \quad (2.4)$$

A BER of 0.5 means that, on average, half of the transmitted bits will suffer from logical errors.

Note that the BER definition depends on the number of transmitted bits (i.e., typically when performing a BER measurement, only a limited number of bits are transmitted and measured). Since errors can be generated by

[4]Bit error rate is also referred to by some authors as bit error ratio. In fact, as described in [2], bit error rate refers to a timing related parameter (e.g., number of errors per second) while bit error ratio could be considered the most appropriate term for (2.4). But since most engineers and technical publications use the term bit error rate when referring to (2.4), we will also use it through this book, although the reader should be aware of this possible distinction.

random and deterministic events, for a perfect BER measurement an infinite number of measured bits would be required. Since this is not practical, a limited number of bits is used.

Another important point is the fact that a BER measurement result will also depend on the applied pattern, especially when using a limited size pattern. Figure 2.18 shows a very simplistic example that demonstrates this dependency. In the figure, the noise of the receiver moves some of the strobing edges to wrong positions that might generate bit errors. The measurement setup uses a 10-bit pattern to compute the BER of the system. With the first pattern a BER of 0.1 is measured, but with the second pattern, which is different only in a single bit, the measured BER jumps to 0.2.

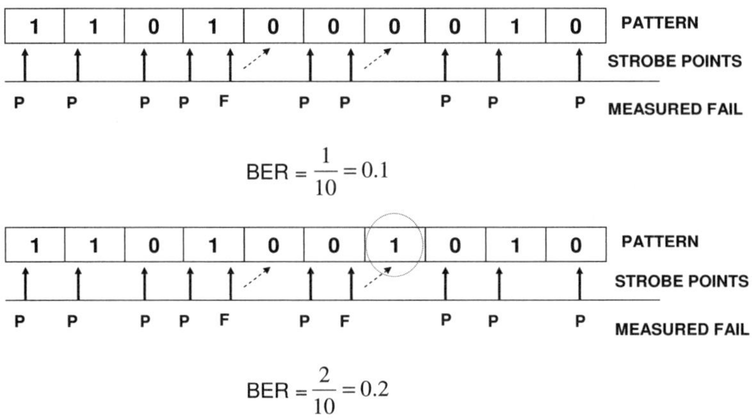

Figure 2.18 Simplistic example showing how the measured BER can be different from the "real BER" due to the used pattern.

For measuring a DUT BER, one needs to be able to transmit bits to the DUT, find out if the transmitted bit was correctly received by the DUT, receive bits from the DUT, and find out if the received bits correspond to the intended bits to be transmitted by the DUT. Figure 2.19 shows a block diagram of two possible measurement setups.

In the first setup (Figure 2.19, left), a loopback configuration is available between the receiver and driver in the I/O cell. This is a common DfT technique in modern I/O cells [9]. In this way a known pattern can be sent to the DUT I/O cell receiver, which in turn is transmitted by the I/O cell driver after going through some internal circuitry without any modification on the bit pattern (the exact circuitry will depend on how the driver/receiver loopback path was implemented). This approach is discussed in detail in Section 6.4.3.

Another more complex setup is shown in Figure 2.19 (right). In this case there is no DfT mechanism available to help with measuring the

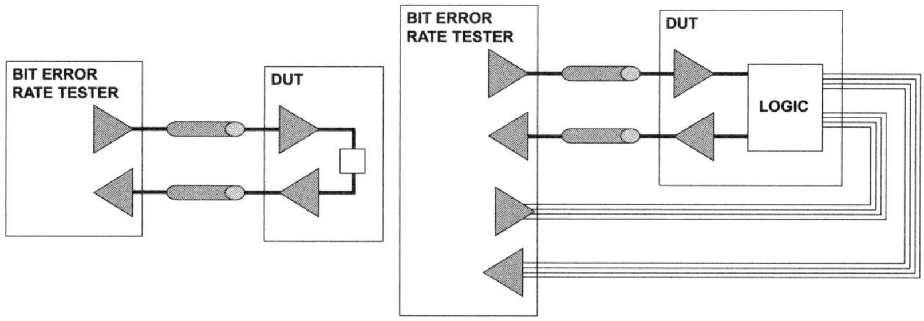

Figure 2.19 Block diagram of bit error rate measurement setup for a digital I/O cell in a DUT (left: The DUT I/O cell can be programmed to a loopback mode; right: the data must be obtained after the internal logic of the device in some lower-speed pins).

I/O cell BER. The BER must then be measured by analyzing all relevant inputs/outputs of the DUT. This can be a prohibitive task and that is why DfT is typically used on high-speed I/O cells, although simpler devices like multiplexers/demultiplexers have their BER rates sometimes measured in this fashion (e.g., [10]).

Although the previous examples only concentrated on measuring the BER of a single I/O cell in an IC, it is also possible to measure the BER of a complete digital link including the driver, receiver, and the link medium (e.g., a copper trace on a backplane printed circuit board) as shown in Figure 2.20. Typically this is more of a system test than an IC test performed with an ATE system.

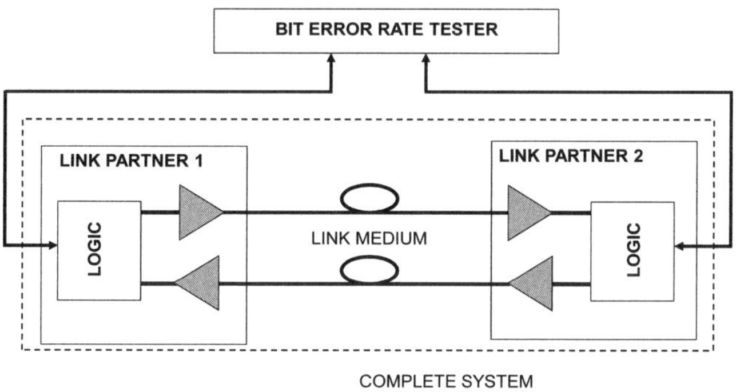

Figure 2.20 Block diagram of a bit error rate setup measurement for a digital link.

2.4 Jitter

Jitter and jitter testing are very important but sometimes complex subjects. There exists a myriad of resources spread through different books, articles, and application notes. Some of the available references on this topic include [1, 2, 11–14]. This section addresses some of the important points regarding jitter in high-speed digital interfaces.

The International Telegraph and Telephone Consultative Committee (CCITT) has defined jitter as "short-term non-cumulative variations of the significant instants of a digital signal from their ideal positions in time." A significant instant for a digital signal can be defined as the rising or falling edge of a bit transition crossing a voltage threshold level. To make this definition easier to understand, Figure 2.21 presents the most common jitter definition, which is the time interval error (TIE), also sometimes referred to as absolute jitter or timing jitter. There is also a relationship between jitter and phase noise for a bit clock type signal that can sometimes generate some confusion. This relationship is discussed in detail in Appendix J.

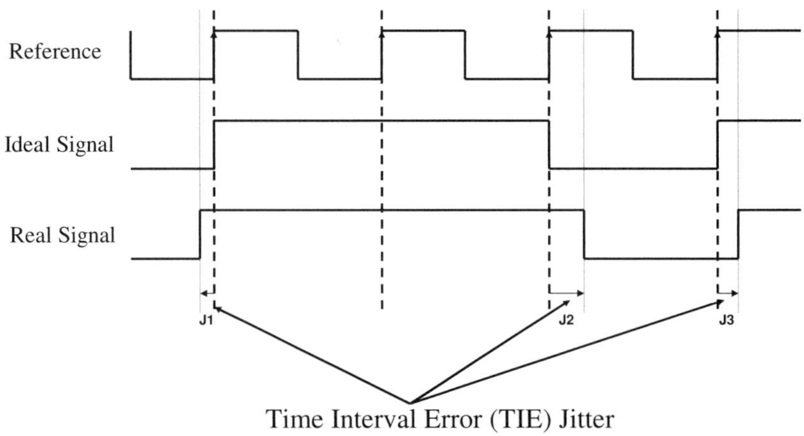

Figure 2.21 Standard definition of jitter on a digital signal, also referred to as time interval error (TIE), absolute jitter, or timing jitter.

In Figure 2.21, the correct timing (i.e., the ideal positions for the rising and falling edges of the signal) is defined by a perfect reference clock. An ideal signal should have the rising and falling edges timing aligned with this clock for each bit. A real signal in turn will have an imperfect timing and the timing difference between the real signal edge and the ideal position is the instant jitter value at that point or the time interval error. One important consequence is that jitter is only defined or measured when there is a transition (e.g., from a logic 1 to a logic 0). This means, for a digital signal, only one jitter value is measured at each transition resulting in an array of jitter values.

The TIE jitter definition, although the predominantly used definition of jitter, is not the only one. Another possible definition for jitter in a digital signal, especially for clock signals, is the cycle-to-cycle jitter definition as shown in Figure 2.22 [15]. In this case the jitter value is measured by looking at the variation of the adjacent periods such as $t_2 - t_1$.

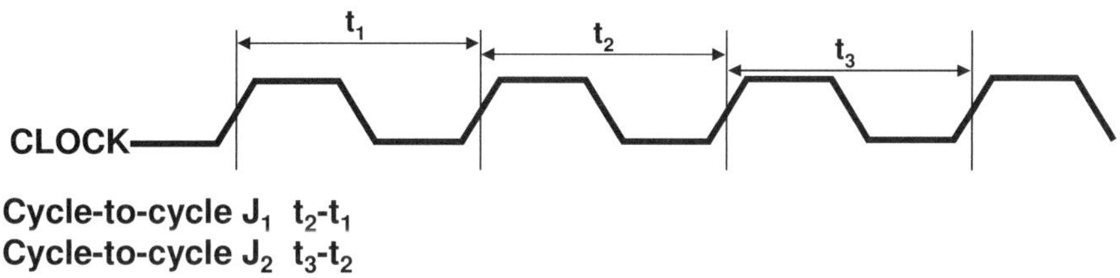

Figure 2.22 Cycle-to-cycle jitter definition.

For clock signals it is also typical to define the maximum deviation of each single period of the clock signal under measurement from that of the ideal clock as the period jitter [15, 16], which is illustrated as t_1, t_2, and t_3 in Figure 2.22. Note that some authors use the term period jitter with a jitter measurement definition equivalent to TIE jitter which can cause confusion.

2.4.1 Jitter Histogram

The jitter histogram is one of the most commonly used ways of analyzing the jitter of a digital signal. The idea is to measure the time interval error of each edge of the signal and generate a histogram of the measurements. The exact way that the measurement array of values required for the histogram is generated depends on the specifics of the measurement methodology and instrument (to be discussed with more detail in Chapter 5). Typically an array of values for the measured jitter at each transition is generated. From this array a histogram can be computed. Figure 2.23 shows an example of an array of jitter values.

To obtain the histogram from the array, a series of time interval buckets or bins (e.g., 1 ps wide buckets) are created and then the number of jitter measurements that are inside each bucket is counted. The histogram can then be normalized to a percentage by dividing the number of samples in each bucket by the total number of measurement samples used to generate the histogram.

26.84	15.59	17.50	5.44
-19.77	7.69	27.91	29.79
36.36	13.93	6.42	28.74
-7.49	-2.25	32.47	-0.55
-29.03	-0.78	21.25	18.48
-12.37	1.76	4.28	-6.43
18.69	-15.79	17.54	13.22
21.12	28.46	3.89	38.31
3.20	0.13	-8.30	3.14
5.75	13.73	7.17	-6.01
12.66	-17.10	0.72	-10.00
-29.18	-21.50	-7.29	14.33
-11.63	-1.82	35.42	26.75
-36.60	-5.06	4.43	42.51
-8.98	23.90	54.61	1.08
18.99	12.12	-5.92	3.26
14.35	10.81	11.29	-12.65
45.76	-28.90	31.65	32.24
3.33	-19.35	54.58	-1.51
-43.13	4.04	6.07	-9.46
33.79	-6.96	-15.81	43.68
25.65	25.80	16.07	16.20
-11.65	26.82	-26.40	14.33
4.45	-11.62	-5.48	-20.11
-4.52	-21.84	10.40	8.68

Figure 2.23 Example of a set of 100 measured TIE jitter values (in picoseconds).

Figure 2.24 shows the plotted histograms (nonnormalized and normalized) for the jitter measurement samples of Figure 2.23 where a resolution of 500 fs was used for each bucket with a total span of -100 ps to 100 ps.

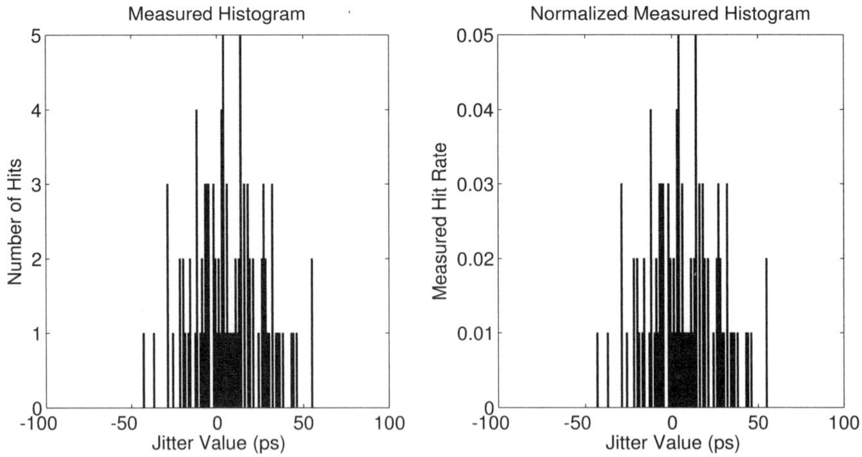

Figure 2.24 Jitter histogram plots (left: nonnormalized; right: normalized).

The jitter histogram provides a visualization on the amplitude of the jitter in the measured signal and might provide some clues on what type of jitter is present by analyzing its shape. In Section 5.5, two important measures (peak-to-peak and RMS jitter) that can be obtained from the jitter histogram are discussed in detail.

2.4.2 Jitter Categorization

TIE jitter is typically divided into several subcategories depending on its properties [12, 17, 18]. These properties sometimes provide clues to the origin of the jitter allowing the designer to more easily find the jitter source in the design. Figure 2.25 shows a typical categorization of jitter. Sometimes slightly different categorizations are used by different authors. The two main categories of jitter are random and deterministic jitter.

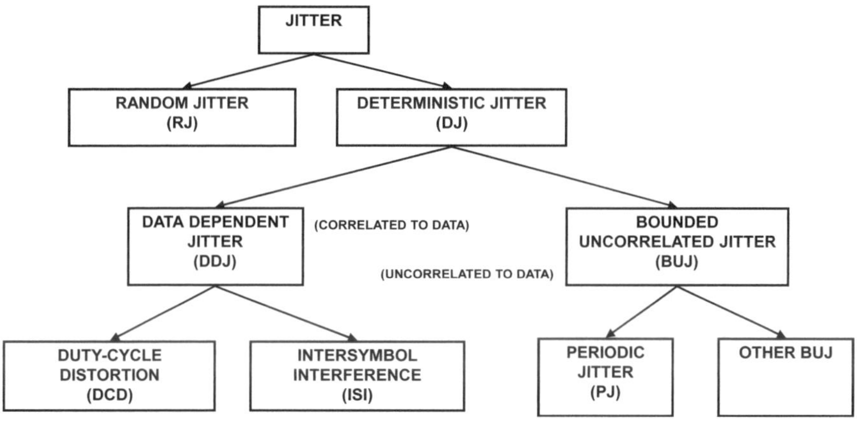

Figure 2.25 Categorization of jitter into different components.

2.4.2.1 Random Jitter

Random jitter (RJ) is defined as being unbounded in the sense that by definition there is always a probability (although it can be infinitesimal small) of the jitter value reaching any value. Random jitter is usually modeled using a Gaussian (also known as normal) probability distribution [19]. One important reason for using a Gaussian distribution to model random jitter is due to the central limit theorem (see Appendix A). The theorem states that the sum of an infinite number of arbitrary probability distributions is a Gaussian distribution. Since in any semiconductor device there are multiple noise sources that contribute to the generated random jitter (shot noise, flicker noise, and Johnson noise [20, 21]), using a Gaussian distribution to model the random jitter behavior is a reasonable approach. Figure 2.26 shows a simulation example of a digital data signal at a data rate of 5 Gbps that contains random jitter (with a 10 ps RMS[5] value).

In Figure 2.26 it is not easy to discern the jitter on the signal waveform. Only the representation in a data eye diagram makes the presence of jitter clearly obvious. The jitter histogram shows a distribution that is not exactly matching the expected Gaussian distribution due to the limited number of samples. If the number of samples to be acquired would be increased, the histogram would converge more closely to the expected Gaussian distribution.

2.4.2.2 Deterministic Jitter

Deterministic jitter (DJ) is mainly characterized by being bounded (i.e., the maximum and minimum jitter values are limited, unlike with random jitter). Deterministic jitter can be further divided into two main categories: data-dependent jitter (DDJ) that is correlated to the data pattern of the signal being measured and bounded uncorrelated jitter (BUJ) that is uncorrelated to the data pattern being measured. From these two categories several subcategories exist depending on the underlying physical cause: periodic jitter (PJ), other bounded uncorrelated jitter (other BUJ), duty-cycle distortion (DCD), and intersymbol interference (ISI).

2.4.2.3 Data-Dependent Jitter

Data-dependent jitter (DDJ), as the name indicates, is correlated with the data pattern. It is typically subdivided into two subclasses: duty-cycle distortion (DCD) and intersymbol interference (ISI).

In the context of data-dependent jitter, the metric data-dependent pulse width shrinkage (DDPWS) is sometimes used since it is defined in the Fibre Channel standard. Basically, it corresponds to the difference between one symbol period and the minimum pulse width of a repeating data sequence [22].

2.4.2.4 Duty-Cycle Distortion Jitter

Duty-cycle distortion (DCD) jitter refers to the jitter generated in a clock signal with different widths of the logical high and the logical low bit. Figure

---

[5]The root mean square (RMS) value of a distribution is defined in Section 5.5.2. For a Gaussian distribution the RMS values correspond to the standard deviation σ (see Appendix A).

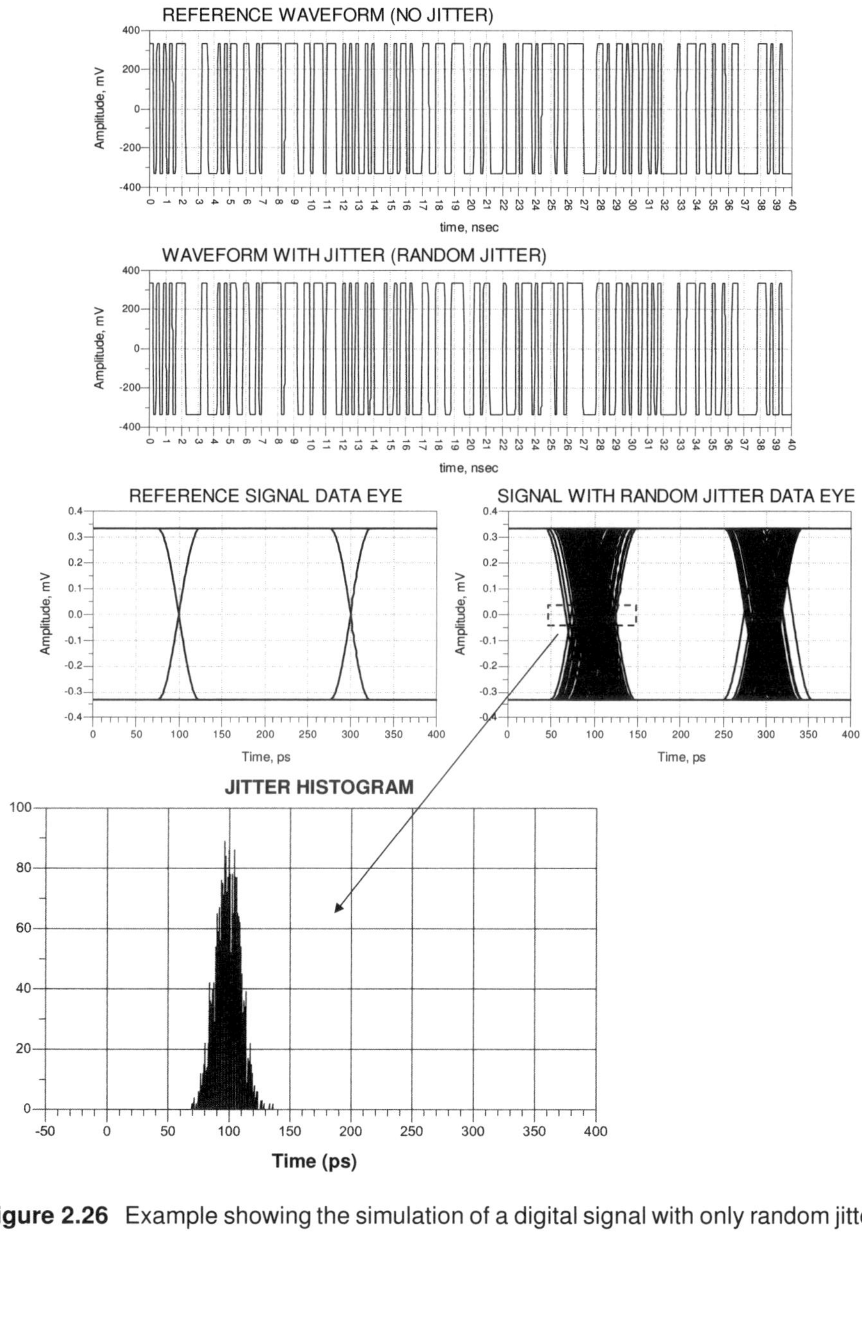

Figure 2.26 Example showing the simulation of a digital signal with only random jitter.

2.27 shows an example of the added DCD jitter for a bit clock with a duty cycle unequal to 50%.

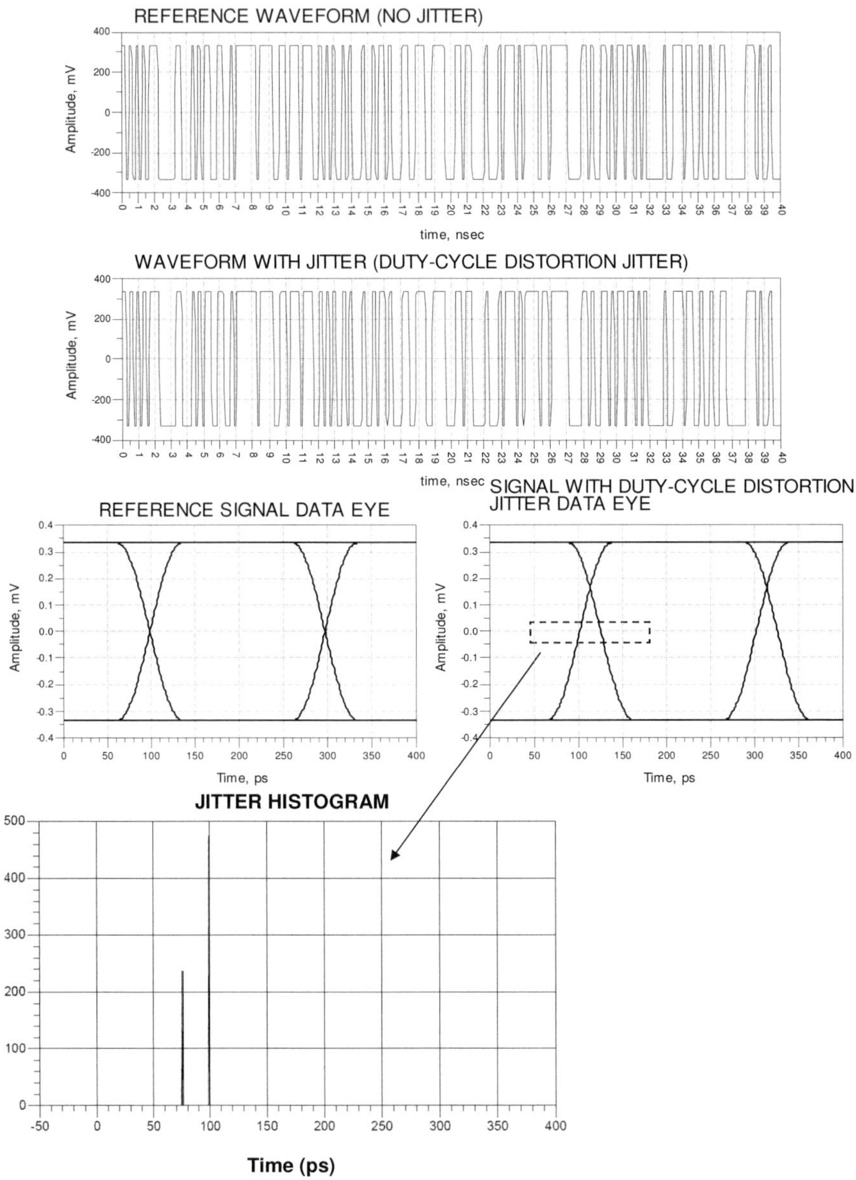

Figure 2.27 Example of data-dependent jitter in the form of duty-cycle distortion (DCD) jitter.

In the context of DCD jitter, there is another type of jitter referred to as F/2 jitter. In the case of F/2 jitter the width of even bits is different from the width of odd bits which also results in a similar jitter histogram distribution but in a different data eye diagram [23, 24].

Some possible causes of DCD are imbalance in the driver source and sink current, nonlinear loads, marginal timing of output drivers, and common-mode voltage in differential signals.

2.4.2.5 Intersymbol Interference Jitter

Intersymbol interference jitter (ISI) is based on the influence of the preceding bits on a transition. It can be observed when a digital waveform passes through a bandwidth limited channel like a lossy signal path in a printed circuit board trace. This is exemplified in Figure 2.28.

From the figure it is possible to see the effect of a lossy channel on the data signal waveform (more on this in Chapter 8) and how the previous bits have an effect on the current bit transition and in this way add jitter. In the case of a printed circuit board trace, the shape of the bit rising edge will depend on how the line was charged by the previous bits, and it is this dependence that creates the data-dependent jitter. Interestingly, this can be seen in the example of Figure 2.29 where a pattern consisting of two repetitive symbols is plotted in the frequency and time-domains showing the contribution of each symbol for the ISI. Some more analytical work on predicting DDJ is presented in [25].

2.4.2.6 Bounded Uncorrelated Jitter

Bounded uncorrelated jitter is defined as deterministic jitter that is uncorrelated with the data pattern. It is typically divided into two subclasses: periodic jitter and other types of bounded uncorrelated jitter.

2.4.2.7 Periodic Jitter

Periodic jitter (PJ) corresponds to jitter that has a periodic nature but is not correlated to the signal data pattern. One example is the crosstalk from an adjacent noncorrelated clock signal into the data signal being measured. One typical model for periodic jitter is a sinusoidal waveform, which will also be important for the jitter tolerance test discussed in Section 5.7.2. Other examples of periodic jitter waveforms, like triangular waveforms, are used, however with limited practical use. Figure 2.30 shows an example of the simulation of a 5 Gbps data signal with sinusoidal periodic jitter.

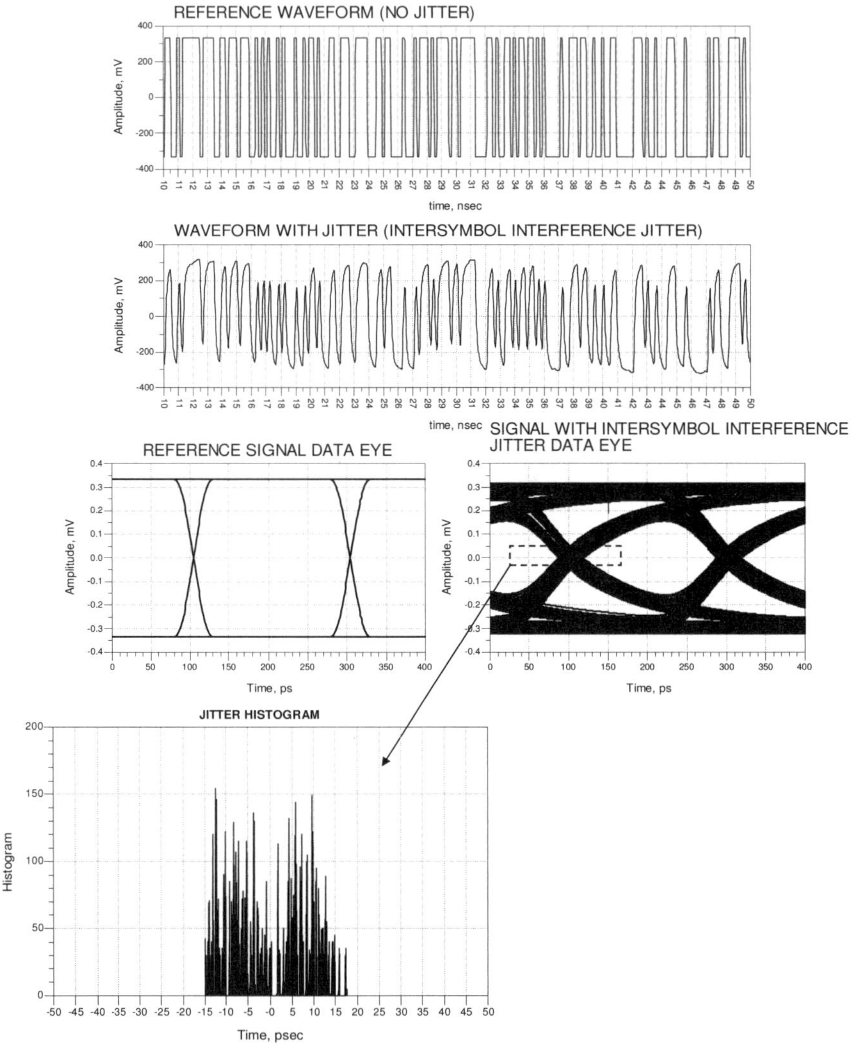

Figure 2.28 Example of data-dependent jitter in the form of intersymbol interference (ISI) jitter.

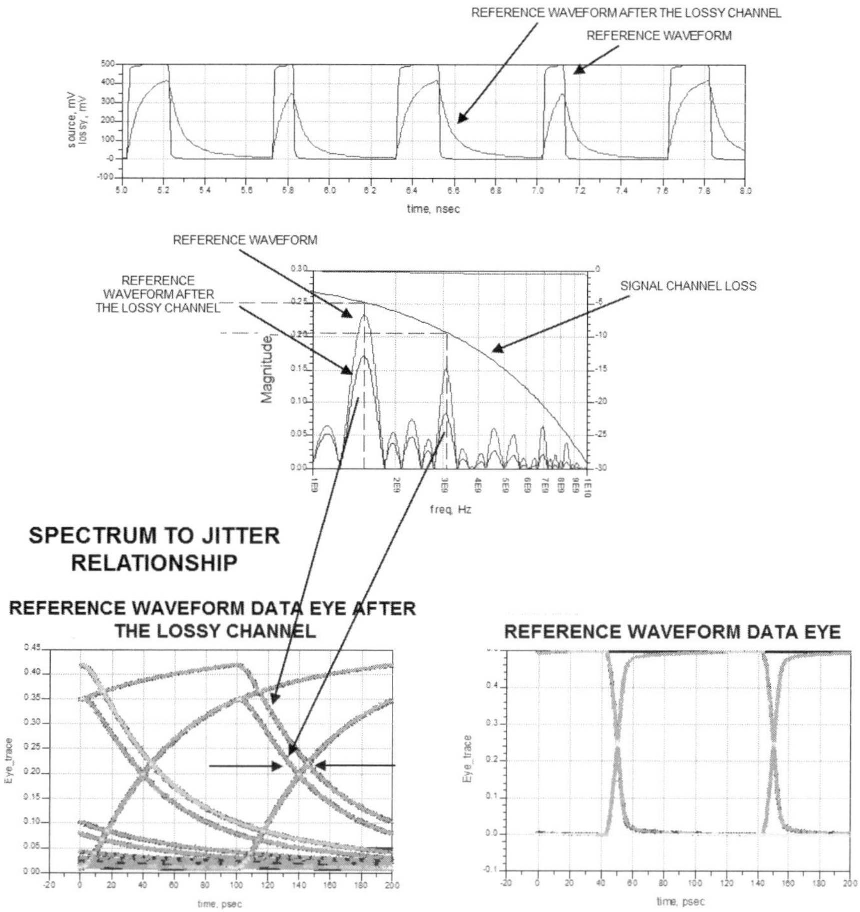

Figure 2.29 Signal path loss effect on the spectral power density and data eye for a two-symbol alternating pattern 110000010000001100000100000... [26].

The histogram shows the typical shape expected from a data signal with sinusoidal deterministic jitter with the time difference between the two peaks at the extremes corresponding to the periodic jitter amplitude.

Although periodic jitter could be considered just as bounded uncorrelated jitter, it has its own class since crosstalk from clock sources is always a critical point for designers and it will be shown in Chapter 5 that periodic jitter is also critical for jitter tolerance testing.

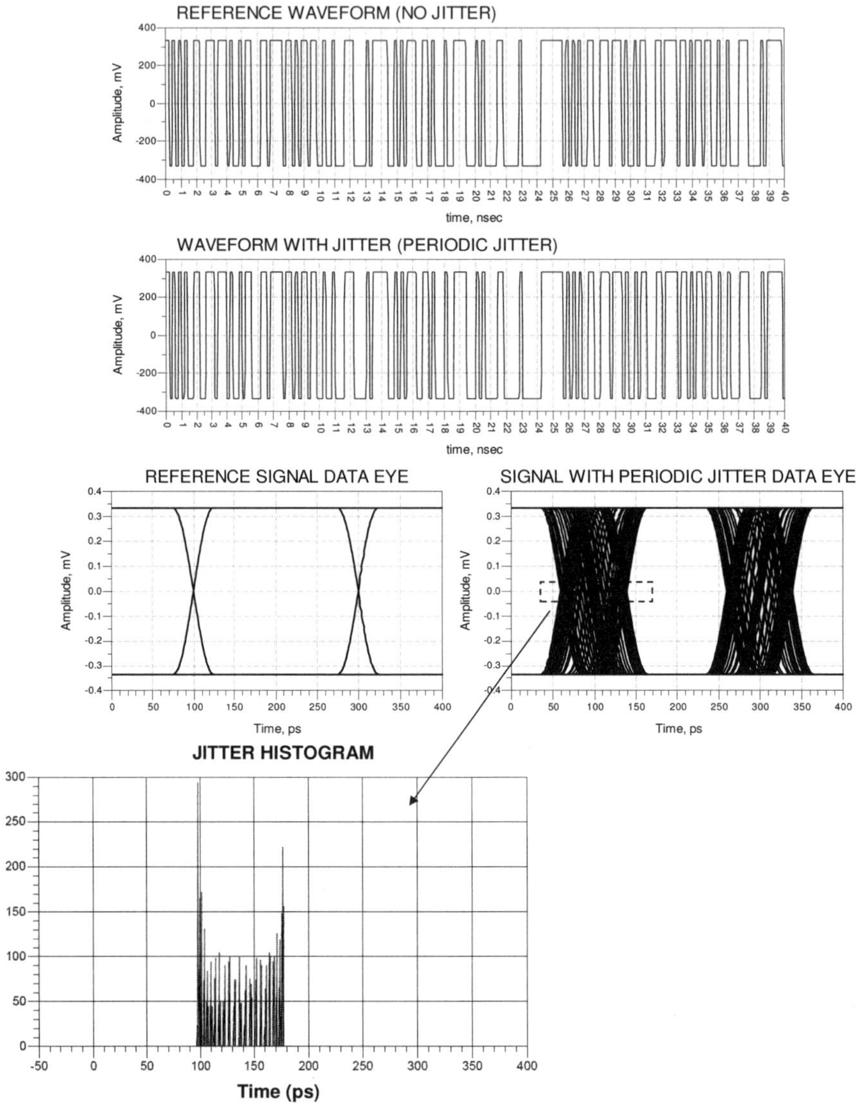

Figure 2.30 Example of periodic jitter (PJ).

2.4.2.8 Other Bounded Uncorrelated Jitter

One example of a mechanism that causes this type of jitter is the crosstalk from an adjacent data line belonging to a different bus. It will be different from periodic jitter since the jitter spectrum will not correspond to the case of crosstalk from an adjacent clock signal since the crosstalk source is not a periodic signal. To demonstrate how this type of BUJ looks on the time-domain, the simulation setup in Figure 2.31 was used where a 5 Gbps data line is coupled (i.e., physically very close in the PCB board) to a 3.2345 Gbps data line belonging to a different bus. The crosstalk from this lower-speed data line will create BUJ on the 5 Gbps data signal as shown in Figure 2.32.

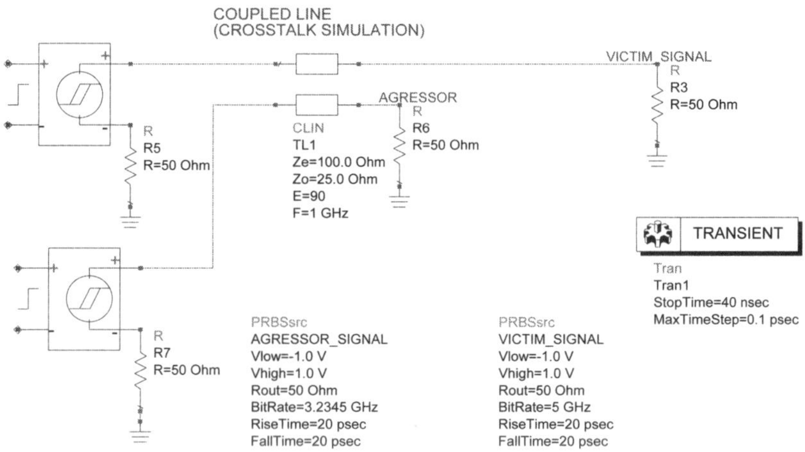

Figure 2.31 Simulation setup for demonstrating other type of bounded uncorrelated jitter.

In this example the crosstalk is very strong in order to clearly show its effect. It is easy to see the degradation of the signal on the time-domain waveforms and on the data eye diagram. Note also that the jitter histogram does not have a shape that would allow one to clearly conclude that the jitter was of the BUJ type and not random due to the fact that the crosstalk source is uncorrelated.

2.4.3 Amplitude Noise and Conversion to Timing Jitter

The previous sections have only discussed timing jitter, but a digital signal also contains noise in the voltage levels. While timing noise is referred to as jitter, voltage level noise is usually referred to as amplitude noise. One critical point is that any amplitude noise in a signal is translated into timing jitter as shown in Figure 2.33 [2].

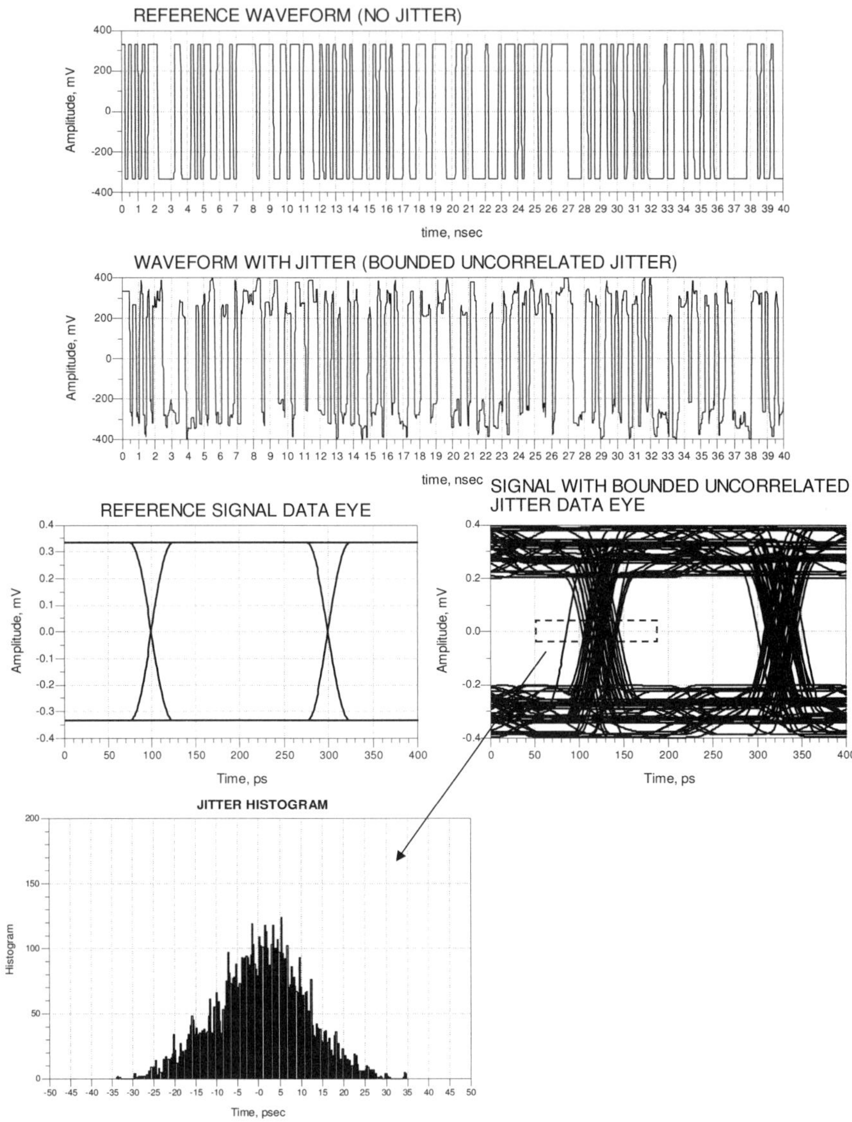

Figure 2.32 Example of other bounded uncorrelated jitter.

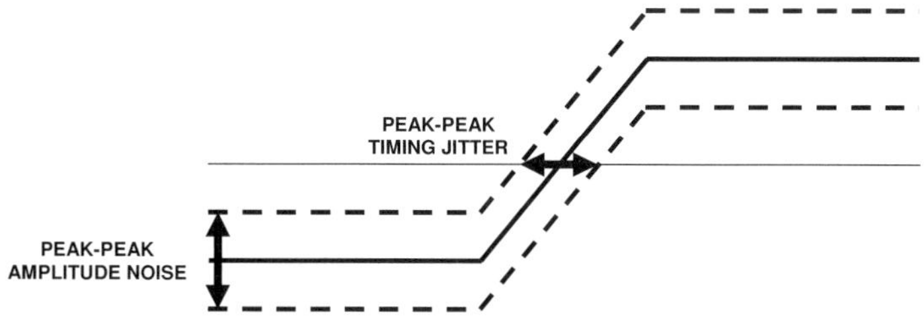

Figure 2.33 Amplitude noise to timing jitter conversion using a linear model.

Using a linear model, the timing jitter peak-to-peak value can be computed from the amplitude noise peak-to-peak value using the following equation:

$$\text{Timing Jitter (peak-to-peak)} = \left(\frac{dV}{dt}\right)^{-1} \times \text{Amplitude Noise (peak-to-peak)} \tag{2.5}$$

where dV/dt is the slew rate of the digital signal.

2.4.4 Jitter in the Frequency-Domain

The previous sections looked at jitter in the time-domain (e.g., the jitter histogram), but it is also possible to analyze jitter in the frequency-domain. This approach can provide significant insights on the causes of the measured jitter. How to compute the jitter spectrum is directly related to the methodology used to measure the jitter.

The straightforward way to analyze jitter in the frequency-domain is to compute its spectrum (i.e., the jitter frequency spectrum). This can be easily done by performing the discrete Fourier transform (DFT) on the time-domain jitter signal. The challenge is how to obtain the time-domain waveform of the jitter embedded on the data signal waveform. For example, if the measured data is in the form of a real-time sampled waveform, then the jitter waveform can also be obtained by comparing the transition times of the waveform with the ideal transition times. One challenge is that only during logical transitions the possibility is given to calculate a jitter value point for this time instant. A solution to obtain the missing jitter values during intervals without logic transitions is to apply an interpolation algorithm.

Figure 2.34 shows an example of a jitter spectrum comparison between a digital signal with only random jitter and a digital signal with a periodic sinusoidal jitter component at 100 MHz. Note that the data eye diagram and measured histogram in both cases are very similar. Only the analysis in the frequency-domain clearly reveals the presence of a periodic jitter component at 100 MHz. One reason why the frequency-domain view is so powerful is that the random jitter is spread across the frequency range allowing the deterministic jitter components to become more visible when compared to a time-domain approach like a histogram.

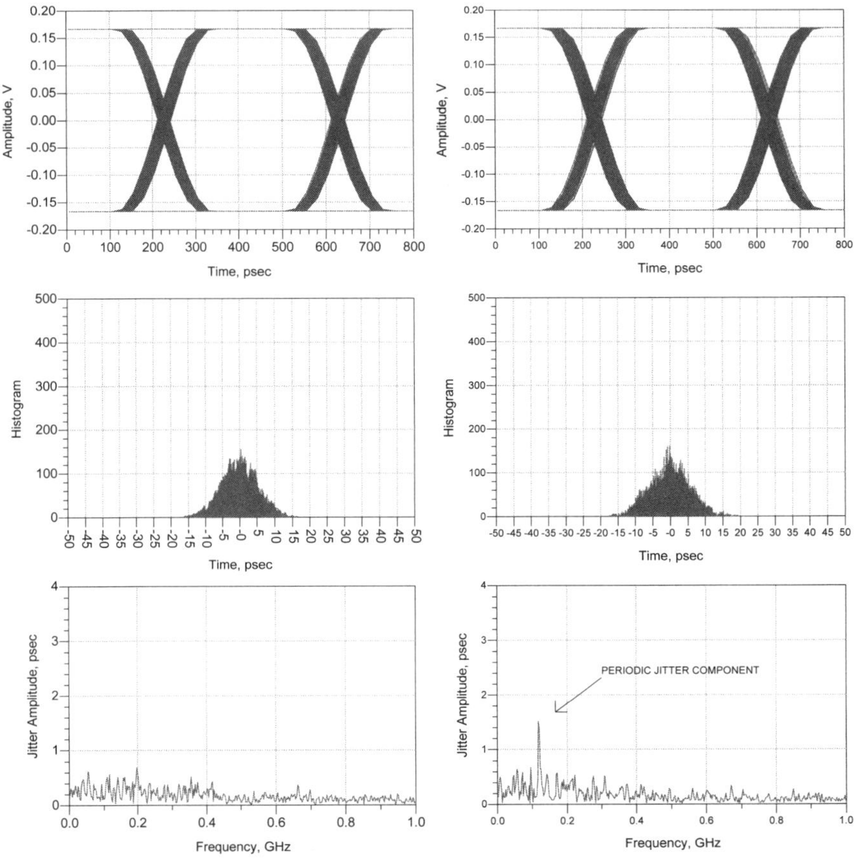

Figure 2.34 Comparison of the jitter spectrum obtained from a digital signal with only random jitter (left) and a signal with random jitter and a periodic jitter component in the form of sinusoidal jitter at 100 MHz (right).

Figure 2.35 shows an example of a jitter spectrum comparison between a digital signal with random and deterministic jitter in the form of ISI from a lossy signal trace and a digital signal with random jitter and deterministic jitter in the form of BUJ obtained through crosstalk from another digital signal using the approach presented in Figure 2.31.

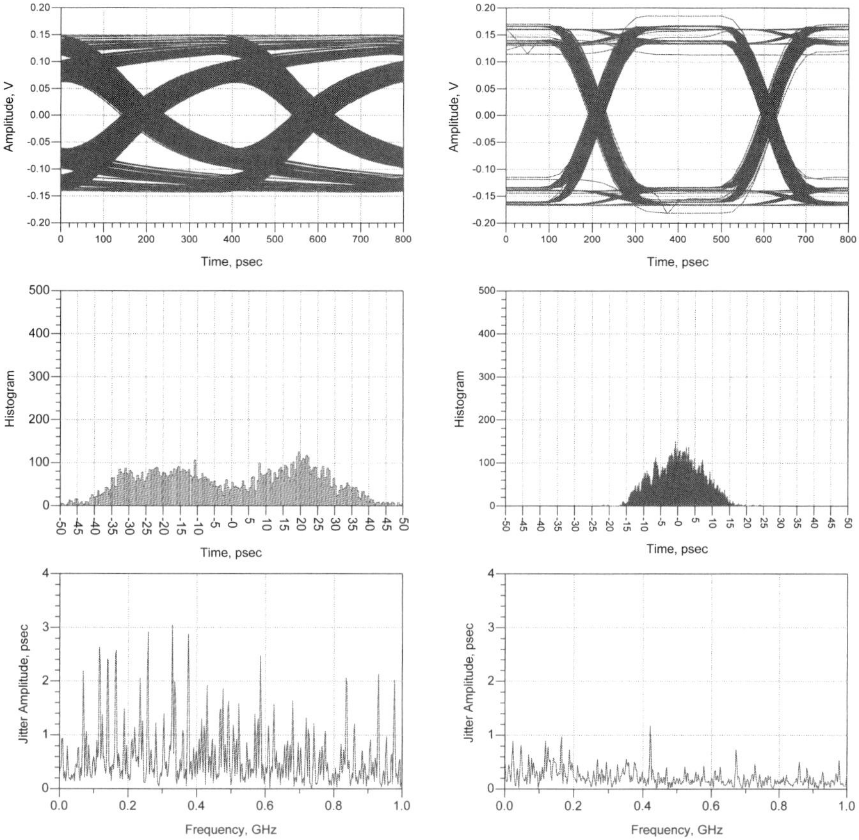

Figure 2.35 Comparison of the jitter spectrum obtained from a digital signal with random jitter and deterministic jitter (left) in the form of ISI and a signal with deterministic jitter in the form of BUJ (right).

In this case it is easy to observe that, for an ISI-type deterministic jitter, the jitter spectrum is composed of several discrete lines at multiple frequencies. In the case of BUJ jitter, the spectrum is not that different from a random jitter spectrum, making it more difficult to separate the BUJ from the random jitter noise floor. Note that jitter components having very low frequencies (e.g., below 10 Hz) are typically not called jitter but wander [2, 27].

As mentioned before, jitter frequency decomposition techniques depend on the measurement approach. In the case where the measurement instrument is digital pin electronics, the technique to obtain the frequency jitter decomposition is not straightforward, since we do not have a real-time sampled waveform available. One possible technique is using the error density as described in [28]. This technique is discussed in Section 5.5.5.

2.5 Classification of High-Speed I/O Interfaces

Digital data transmission interfaces can be classified according to the clocking architecture that they use. Lower-speed interfaces typically deploy common clock architectures, where devices communicating with each other derive their data generation and sampling timing from a centrally provided system clock. This approach has proven to not be feasible with high-speed I/O interfaces. The distribution of a central clock with the required timing accuracy, especially with regard to clock skew over a whole system easily results in high system design efforts and leads to high system cost. Thus, high-speed interfaces typically use source synchronous timing concepts where each device derives its own timing from a local or central system clock, but the data transmission timing reference is determined by the sending device that transmits this clock information together with the sent data. The receiving device derives the phase of this transmitted clock information and uses it to correctly latch the incoming data.

It is important to mention that the term "source synchronous" as used in this book has to be taken literally for any interface that transmits phase information about the transmitted signals together with the transmitted data, regardless if it is on the same or separate signals. Sometimes this term is misleadingly used to describe interfaces that explicitly transmit a clock signal together with data signals. Using our more general interpretation of source synchronous interfaces, all important high-speed I/O interface standards today use a source synchronous data transmission concept. Source synchronous interfaces can be subclassified in embedded clock interfaces where the transmission clock phase information is sent together with the transmitted data on the same physical signal trace. The other representatives of the source synchronous group are at-cycle source synchronous interfaces and forwarded clock interfaces, which lead to the overall classification shown in Figure 2.36. All source synchronous interfaces have the advantage that the clock information is originating from the same source as the data signals. Thus, transmitter timing variations such as thermal drifts of certain jitter components are present on both, the data content and the clock. This widens the margins to latch the data correctly at the receiving device. The architectural topologies for the different interface classes are shown in Figure 2.37.

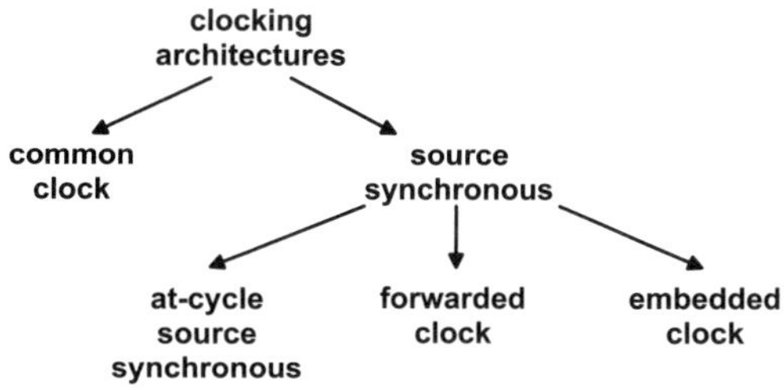

Figure 2.36 Clocking architectures for high-speed interfaces.

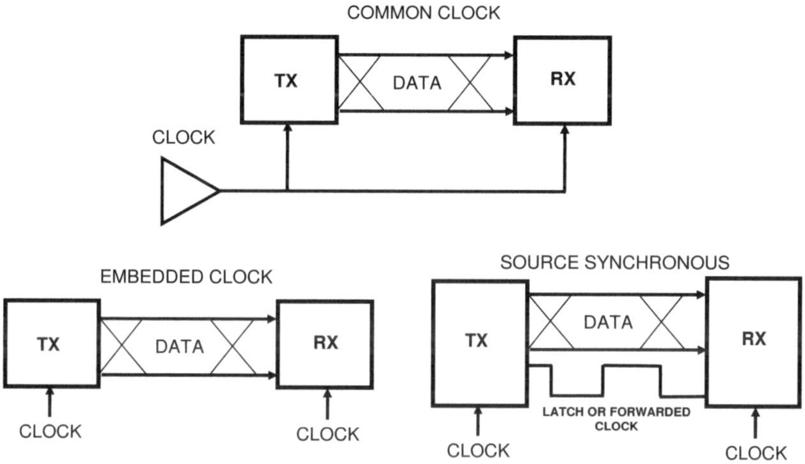

Figure 2.37 Different clocking models for high-speed digital transmission.

2.5.1 Common Clock Interfaces

As mentioned, although common clock interfaces are the classical clocking architecture for digital interfaces this interface type usually is not found in the high-speed I/O interface arena. Correct data transmission using a common clock interface relies on accurate clock distribution with well-defined relative phase conditions between clock and data for each device. System design parameters that impact this phase relationship are skew introduced by the clock routing paths, skew within the data connections for a bus, and skew between these data signals and the distributed clock signals. Since for high-speed I/O interfaces bit time intervals are small, the allowed margins for these skew components are so tight that high-speed I/O system design is not possible anymore in an economical way using common clock architectures.

2.5.2 At-Cycle Source Synchronous Interfaces

At-cycle source synchronous interfaces transmit a clock signal together with the signals of the data bus. For wide data buses, at-cycle source synchronous interfaces usually divide the bus into subbuses and one clock signal is transmitted together with the signals of each subbus. At the receiving device, the received clock signal is directly used to latch the associated data signals. For at-cycle source synchronous interfaces, there is a one-to-one phase relationship (isochronous timing) between the clock and the data signals. Thus, clock and signal connections have to keep the same electrical length from the transmitter up to the latching circuit in the receiver. The isochronous timing conditions for at-cycle source synchronous interfaces have the advantage that even high-frequency jitter components caused by a transmitter can be tracked by this kind of interface as long as they affect clock and data in the same way. However, for very high-speed interfaces with short bit times, the margins for the allowed setup/hold times at the receivers become small. This can cause very similar challenges for designing systems using this kind of interfaces as in the case for common clock based systems. Even small variations in the electrical length between data and clock signals or nonuniform timing variations between data and clock can cause a violation of the allowed setup/hold times.

2.5.3 Forwarded Clock Interfaces

In contrast to at-cycle source synchronous interfaces, the clock signal of forwarded clock interfaces serves as a relative phase reference only. The difference between the at-cycle source synchronous and the forwarded clock concept is shown in Figure 2.38. For at-cycle source synchronous interfaces, an isochronous timing concept is used where a fixed relative timing relationship between a clock edge and a corresponding data bit has to be kept for the complete transmission path. However, forwarded clock interfaces use a mesochronous concept where this fixed clock edge to data bit phase relationship is not existent anymore for the data transmission. Of course, a valid relative timing relationship that does not violate setup/hold time margins between clock edges and data bits needs to be achieved at the receiving end of the data transmission path. However, for forwarded clock interfaces it is not required that a distinct clock edge at the source keeps its phase relationship with the corresponding data bit while being transmitted. Valid setup/hold time conditions at the receiving end of the transmission path are typically set up using training steps during device initialization and potentially during device operation if this is required. The benefit of this training concept is that PCB designers do not need to keep phase matched trace lengths between the clock and data connections as well as within the data connections. Static phase mismatches on these connections are compensated by the device training.

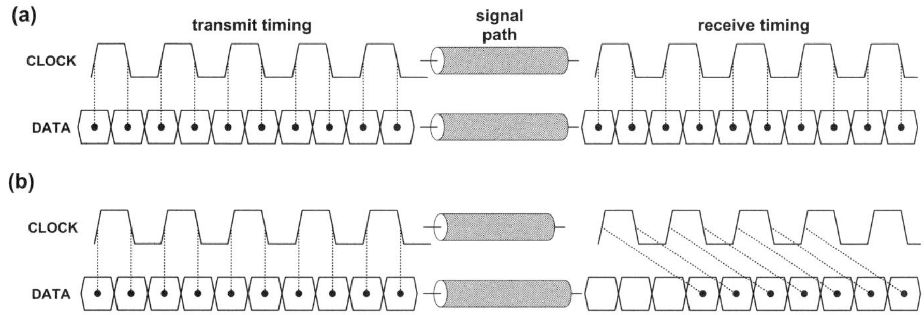

Figure 2.38 Comparison of (a) at-cycle source synchronous interface versus (b) forwarded clock interface.

2.5.4 Embedded Clock Interfaces

Embedded clock interfaces combine the clock information and the data content within the same signal. On the receiver side, a clock data recovery (CDR) circuit separates the clock and data information from this signal and reconstructs the data content. The benefit of combining the data and clock phase information on the same signal is that both are affected equally by signal disturbances. This allows embedded clock interfaces to transmit data at very high data rates on a PCB. In order to allow the extraction of clock phase information and the reconstruction of the data in the receiver, embedded clock data transmission has to fulfill requirements such as having enough data transitions on the transmitted signal. This usually is fulfilled by deploying appropriate coding or scrambling (see Appendix D). Also, the methodology used to extract data and clock information (see Section 2.6.3) has influence on which kind of timing variations can be tolerated on embedded clock interfaces and which kind is likely to cause data transmission problems.

Digital interfaces can be further divided into unidirectional (also referred to as simple duplex) and bidirectional interfaces (also referred to as full duplex). Unidirectional interfaces only allow transmission of data in one direction, meaning that each I/O cell requires a separate transmitter and receiver pin. A bidirectional interface uses the same media or pin for the transmitter and receiver parts of the I/O cell as shown in Figure 2.39.

Clearly full duplex interfaces have the advantage of requiring fewer pins since both the transmitter and receiver share the same pins. The challenge with this type of approach is the I/O cell design. Typically bidirectional interfaces

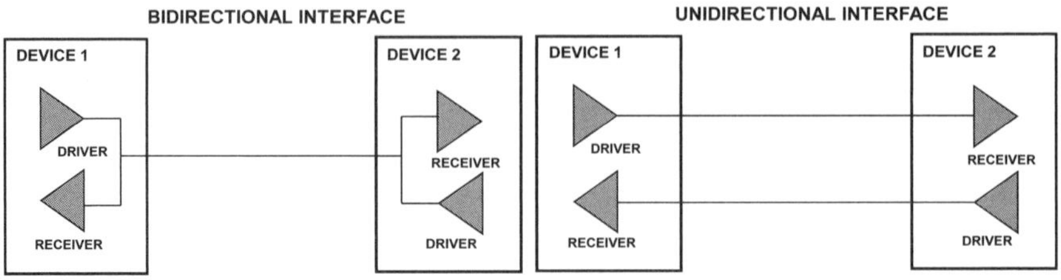

Figure 2.39 Comparison of unidirectional and bidirectional interfaces.

are behind unidirectional interfaces regarding the maximum achievable data rate.

2.6 Hardware Building Blocks and Concepts

This section provides a high-level description of the important building blocks of high-speed I/O cells. In a high-speed I/O interface, the most important building blocks are phase locked loops (PLL), delay locked loops (DLL), clock and data recovery (CDR), and equalization circuits [29]. These components of an I/O cell define the timing behavior of the high-speed signals under measurement and it is important for a test engineer to understand the characteristics of these circuits.

2.6.1 Phase Locked Loop (PLL)

A phase locked loop (PLL) is a circuit that aligns the signal phase of an oscillator contained in the PLL with the phase of a reference input signal [30]. Besides this phase alignment or clock synchronization functionality, another application area for PLLs is to derive clock signals that are synchronized to the provided reference signal but run at different frequencies than the reference signal. In high-speed I/O circuits, this functionality of a PLL is used to generate a high-speed bit clock from a lower-speed system clock. The derived high-speed clock is used to determine the final I/O data rate and among other things triggers the bit transitions of the high-speed transmitters of a device. Figure 2.40 shows a basic block diagram of a PLL.

PLLs are implemented as closed loop systems. The (potentially divided) input frequency serves as comparison frequency F_{comp} input for a phase detector with an integrated charge pump. This phase detector with the help of the integrated charge pump generates a current on its output that represents the instantaneous phase difference between the comparison frequency and the PLL output frequency that is fed back via a divider to the second input

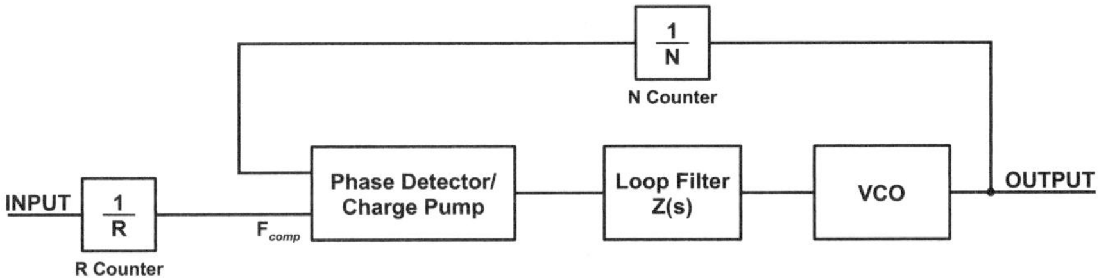

Figure 2.40 Basic block diagram of a phase locked loop.

of the phase detector. The generated current is converted via the loop filter impedance $Z(s)$ into an average voltage that is fed into a voltage controlled oscillator (VCO). The VCO converts the voltage at its input into a proportional output frequency that serves as output of the PLL as well as input to the feedback loop of the PLL. The N counter (divider) shown in this figure provides the capability to derive other frequencies than just the one defined by the reference signal. With a frequency division of N in the feedback loop, it is possible to create a high-frequency clock that is synchronized with a reference clock that has a $1/N$ frequency of the output clock. The loop filter is the most critical component in the PLL since it is the part that predominantly determines the performance and stability of the whole PLL circuit. It also is responsible for the PLL characteristics that are important to understand for test engineers measuring high-speed I/O signals. PLL loop filters are implemented as low-pass filters. Most PLLs integrated in consumer and computation high-speed I/O devices use second order low-pass filters that allow control of the natural frequency ω_n and loop dampening ζ with the design of the filter. The open loop transfer function of a typical second order PLL loop filter follows (2.6). This results in a transfer function for the closed loop PLL system as described by (2.7). Figure 2.41 shows these transfer functions for the open loop filter and for the overall PLL. A description of how these transfer functions are derived from the single PLL components can be found in [31].

$$H(s) = \frac{\omega_n^2}{s^2 + 2\zeta\omega_n s + \omega_n^2} \qquad (2.6)$$

$$H(s) = \frac{2\zeta\omega_n s + \omega_n^2}{s^2 + 2\zeta\omega_n s + \omega_n^2} \qquad (2.7)$$

The PLL transfer function describes the phase relationship of the PLL output relative to the PLL input with respect to the frequency at which the phase change happens. Since phase changes of a clock signal can be transferred into the time-domain as drifts or jitter, this transfer function also describes how jitter with a certain frequency is transferred through the PLL. At low frequencies all of the jitter that is present at the PLL input is transferred to the output of the PLL. As illustrated in Figure 2.41, there is a frequency-range where the jitter is amplified from the PLL input to the PLL output. The amount of amplification in this peaking range mainly is determined by the dampening factor ζ selected for the loop filter. The magnitude of the jitter attenuation at the high frequencies is mainly determined by the order of the loop filter that is used in the PLL. As shown in Figure 2.41, the second order loop filter that we considered in our discussion attenuates high-frequency jitter with a slope of −40 dB/decade.

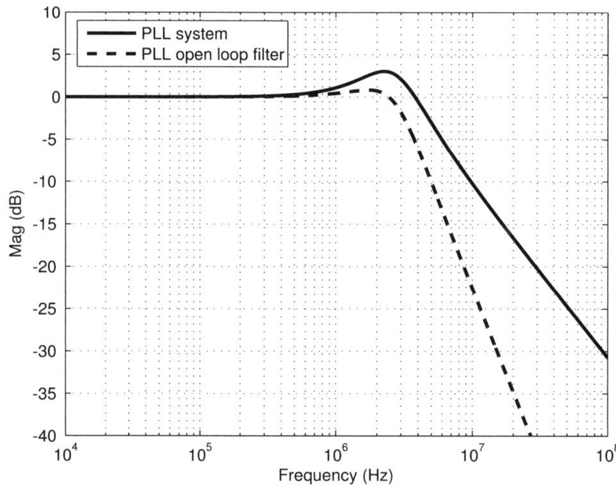

Figure 2.41 Transfer function of a second order PLL and open loop filter.

Another characteristic of PLLs that influences the test implementation for high-speed I/O interfaces is the fact that the phase relationship between the PLL input and its output is neither necessarily predictable nor repeatable. As a consequence of this behavior, the phase offset between the PLL output and its input might change if the input to a PLL is driven by a signal that has not enough signal transitions to allow the PLL keeping its lock state. The same effects can be observed when the PLL goes through a (potentially local) power cycle that might be triggered by a global reset or a reinitialization of the device that contains the PLL. This behavior causes timing phase uncertainties of a DUT in a test environment. As a consequence, the ATE timing usually has to be readjusted to the phase of the signals that are part of the clock domain defined by the PLL output if the PLL lost its lock state due to missing input data transitions or a PLL reset. This is also true if the PLL is restarted without changing the DUT.

2.6.2 Delay Locked Loop (DLL)

A delay locked loop (DLL) is a feedback system that tracks the phase of a reference signal. Compared to a PLL, a DLL does not contain a VCO as phase control element, but a delay line that is controlled by the phase detector. Also DLLs typically contain a first order loop filter. The usage of a first order filter simplifies the goal to achieve a stable servo loop for a DLL compared to the PLL. A DLL also does not exhibit jitter accumulation as is the case for a PLL and has the big advantage that its input can be shut down without causing phase jumps on its output after restarting the input signal. However, a DLL typically does not filter jitter of its reference input and cannot be used as M over N frequency synthesizers like PLLs. Figure 2.42 shows a typical block diagram of a DLL.

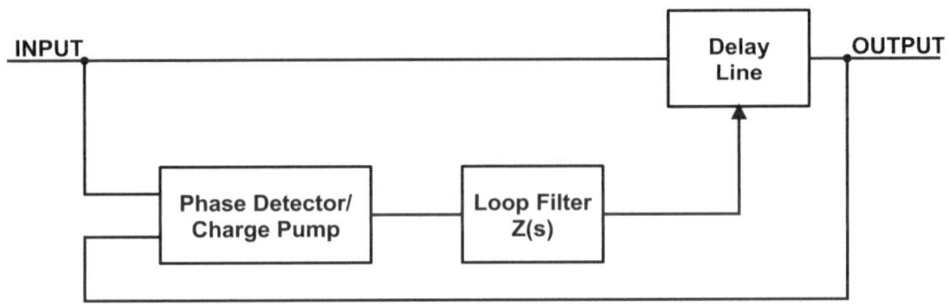

Figure 2.42 Basic block diagram of a delay locked loop.

DLLs are commonly used in high-speed I/O designs, especially to implement features like skew compensation between multiple interdependent input signals of a device. Figure 2.43 shows a basic block diagram for such a skew compensation circuit and how DLLs in such circuits are used to remove the skew between signals (e.g., clock and data). With this feature, IC designers can address one major challenge in high-speed design, which is the alignment between the different data and clock signals. Without the usage of DLLs, in the past system designers had to make sure that all connections for these inter-dependent signals were matched with respect to electrical length. With increased data rates, this task became more difficult. With the use of DLLs, the device can take care of the alignment internally, simplifying PCB routing for system designers significantly.

2.6.3 Clock and Data Recovery (CDR)

Clock and data recovery (CDR) is used for embedded clock signals to extract the clock information from the received signal and to use this extracted clock signal to recover the data content from the received signal. There are a variety of different architectures and approaches for clock and data recovery (CDR)

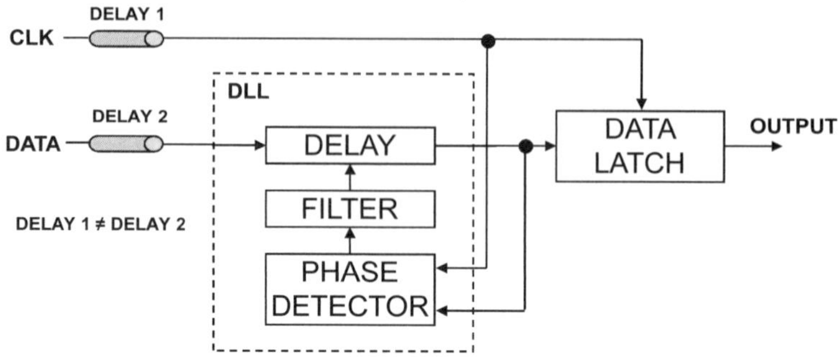

Figure 2.43 Basic block diagram of using a DLL to remove the skew between a clock and a data signal.

from high-speed digital signals and a large amount of published work in this area is available [7, 30, 32]. The most commonly used CDR circuit architecture is based on the usage of a PLL to extract a clock signal from the transitions that are present on the received signal. Figure 2.44 shows a simple block diagram of such a PLL-based CDR circuit. A requirement to guarantee the correct operation of a PLL-based CDR is that the received signal has to provide enough transitions so that the CDR phase cannot drift relative to the incoming signal. The required transition density usually is achieved by applying appropriate coding or scrambling (see Appendix D) to the transmitted data.

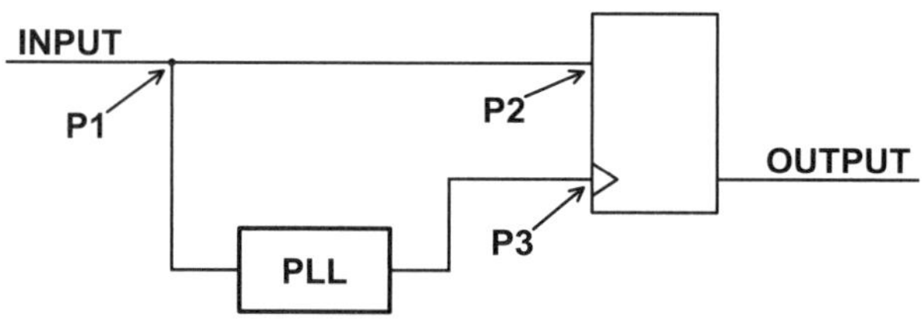

Figure 2.44 Basic block diagram of a PLL-based clock and data recovery circuit.

In a PLL-based CDR, the incoming signal is routed in parallel to the input of a PLL and the data input of the input latch circuitry for that pin. The PLL in the CDR generates a clock that is phase locked to the incoming signal. This extracted clock is used to sample the data content of the incoming signal in the center of the data bits. Since the sampling clock is derived from the incoming signal via a PLL, the transfer function as described in Section 2.6.1 also applies for the clock extraction. In particular, this means that low-frequency jitter in the received signal by the CDR will be passed into the extracted clock signal while high-frequency jitter in the received signal will be filtered according to the PLL transfer function. This low-pass filter characteristic of the PLL in the CDR has quite some impact on the functionality of the overall CDR and the way that jitter measurement results of high-speed I/O transmitters have to be interpreted.

Let's assume the signal at P1 of Figure 2.44 only contains low-frequency jitter that passes the loop filter of the PLL completely. Let's also assume that the PLL does not introduce a phase difference between the signal at P1 and the extracted clock at P3. In such a case the jitter of the signal at P2 and the jitter of the clock at P3 are identical in amplitude and frequency. Thus, at all instances the timing delta between the signal transitions at P2 and the clock transitions at P3 are identical and maximum setup/hold times can be kept for latching the data content of the received signal. If the jitter frequency of the signal at P1 is identical to the frequency at the −3 dB point of the transfer function for the PLL, then the amplitude of the jitter of the extracted clock at P3 only will be 50% of the jitter in the signal at P2. This difference will cause a relative movement of the data eye at P2 compared to the clock edges at P3. This relative movement will reduce the available setup/hold times for the data latch and thus will shrink the available data eye at the data input of the latch. If the jitter frequency increases further to values where the jitter transferred through the PLL is more attenuated, the relative movement between the signal at P2 and the clock edges at P3 increases and the available setup and hold times for extracting the data content of the received signal are reduced down to a level where no reliable data extraction is possible anymore if the jitter amplitude on the received signal is too large.

Due to this behavior, jitter on a received signal can be separated into harmless low-frequency jitter that has no influence on the data extraction and harmful high-frequency jitter that reduces device internal setup/hold times of the data extraction and can cause bit errors. The exact classification of which jitter frequencies are harmless and which jitter frequencies are harmful is determined by the transfer function of the PLL used in the CDR. This separation into harmful and harmless jitter is especially important if the jitter that is measured on a transmitter needs to be evaluated against a single maximum jitter amplitude allowed by a specification. This evaluation has to take into account whether the receiver CDR will be sensitive with regard to the measured jitter or will be able to tolerate it. In order to do this evaluation, the measured jitter has to be weighted with the inverse bandwidth characteristic of the used CDR.

An example for such a weighting function is shown in Figure 2.45. If the jitter measured on a transmitter is weighted with this function, the harmless low-frequency jitter components will contribute less to the overall jitter than the harmful high-frequency jitter components [12]. For jitter measurements using bench equipment, the jitter weighting is achieved by using a golden PLL with the desired bandwidth characteristic in the trigger path.

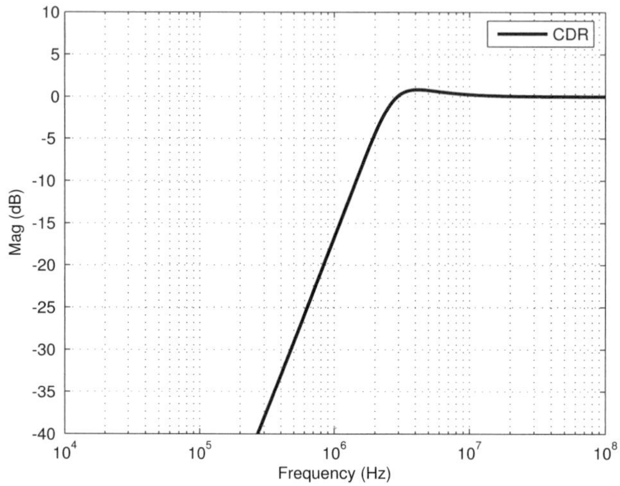

Figure 2.45 Example of an inverse CDR loop bandwidth plot.

One important point to note is that on a digital application the CDR loop bandwidth is not independent of the pattern. The pattern transition density has an effect on the CDR performance. In most designs halving the transition density can reduce the loop bandwidth by half. It is important to understand when looking at a given CDR loop bandwidth specification for which pattern density it was specified [33].

2.6.4 Pre-Emphasis/De-Emphasis and Equalization

The fact that high-speed digital signals are transmitted through a lossy medium like a printed circuit board (PCB) trace presents a challenge when moving to higher data rates. Since the transmission medium loss usually increases with frequency, it has a direct impact on the jitter added to the signal (ISI) and in this way on the BER of the transmission link (more on this topic in Section 8.2). Thus, it becomes critical to compensate for this loss when moving to higher data rates. One option is to improve the transmission medium loss by using lower loss materials and better design strategies but this option implies a higher cost and in some cases cannot solve the problem completely, especially for very high data rates. The other option is to compensate for the transmission medium loss by using equalization on the I/O cell driver and receiver. Basically equalization seeks to either emphasize the high-frequency components or de-emphasize the low-frequency components of the transmitted or received signals [13, 34]. Table 2.1 shows a list of different possible equalization approaches [35]. Note that for each of these equalization approaches there are multiple implementation options.

Table 2.1
List of Equalization Approaches

Implementation	Notes
Continuous time linear equalizer (CTLE)	Used in ATE pin electronics equalization implementations
Transversal FIR	
Rx sampled FIR	Most common approach for transmitter equalizer implementation and sometimes also referred as a feed-forward equalizer (FFE)
Tx sampled FIR	
Decision feedback equalizer (DFE)	Most common approach for receiver equalization

Equalization is an important and vast topic. References [3, 35, 36] provide more detailed information on equalization since the objective of this section is only to provide a very high-level overview. Figure 2.46 shows an implementation of a transmitter equalizer using a FIR filter [37]. In the filter each element T corresponds to a delay of one UI. The input signal propagates through the delay elements of each stage. The input samples are multiplied by the tap coefficients (c_K). For each transmitted bit, the output of the taps are summed to provide the output signal. One challenge with any equalization approach like the one in Figure 2.46 is the choice of the optimal values for each tap. The challenge is that the optimal values will depend on the loss characteristic of the transmission medium.

On the transmitter side, equalization is sometimes referred to as transmitter pre-emphasis/de-emphasis. The transmitter pre-emphasis and de-emphasis techniques try to compensate for the frequency-dependent loss of the transmission medium by amplifying the high-frequency components of the signal (pre-emphasis) or attenuating the low frequencies (de-emphasis). The shape of the frequency attenuation or amplification will depend on the exact implementation of the de-emphasis or pre-emphasis algorithm that is usually implemented using a sampled FIR approach. The most common method for pre-emphasis is to set the first bit of any series of equal bit values to a higher

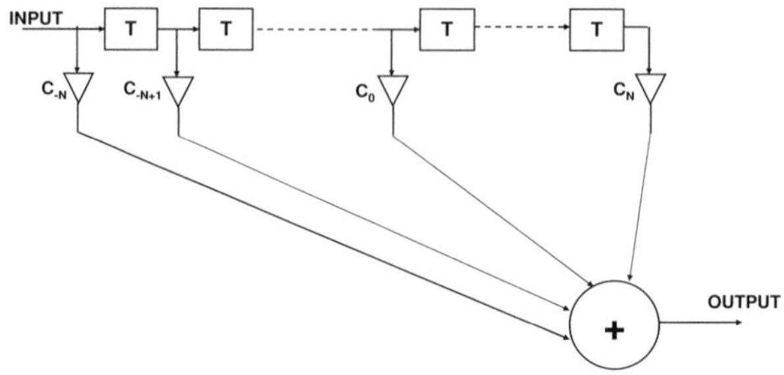

Figure 2.46 Example of a transmitter equalizer implementation using a finite impulse response (FIR) filter.

voltage level than the rest of the bits as shown in Figure 2.47 (top). In the case of de-emphasis, the typical approach is to set the rest of the bits in a sequence of equal bits to a lower voltage level with the exception of the first one as shown in Figure 2.47 (bottom). Although both methods might cause different design implementations on the driver, with regard to the waveform geometry aspect, they are exactly the same. Figure 2.48 shows a comparison of the data eye at a link partner receiver when the link partner transmitter uses pre-emphasis or not in a lossy signal path. Note that in the pre-emphasis example the signal at the receiver contains a smaller amount of ISI jitter providing a better timing margin for the receiver to correctly strobe the incoming bits.

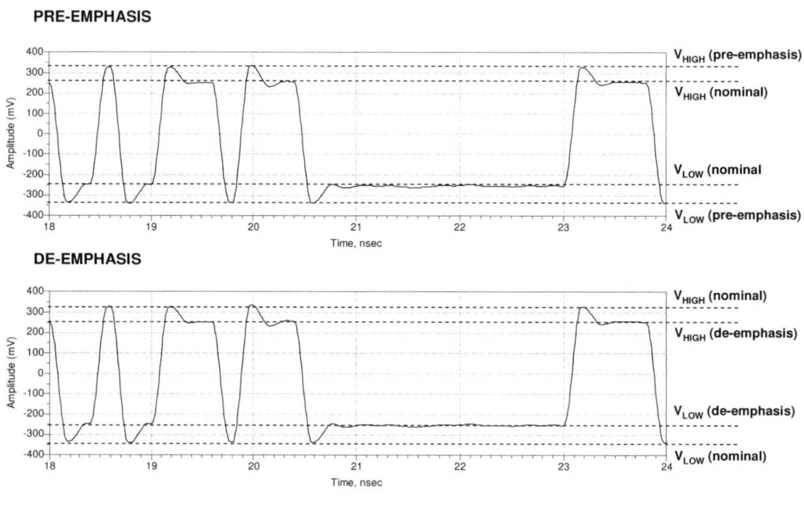

Figure 2.47 Examples of pre-emphasis (top) and de-emphasis (bottom) on a digital signal.

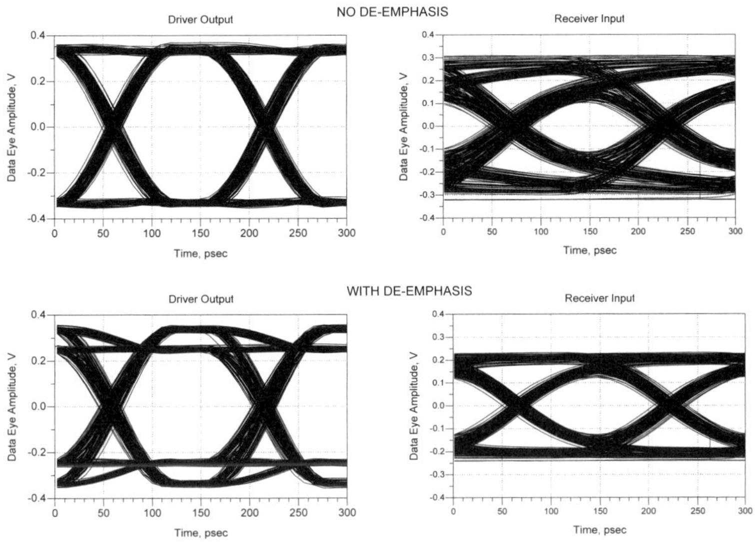

Figure 2.48 Example the improvement on the data eye at the link partner receiver when using pre-emphasis on the driver.

Figure 2.49 shows a high-level diagram of a decision feedback equalizer which is the most common type of receiver equalization implemented in high-speed digital receivers. The equalizer consists of a feed-forward and feedback equalizer implemented as FIR filters. This topology makes the DFE a nonlinear equalizer. Both the feed-forward and feedback equalizers on the DFE will consist of multiple taps that require appropriate values. Those values can be predefined or found automatically by the I/O cell through some training sequence and will depend on the transmission medium loss.

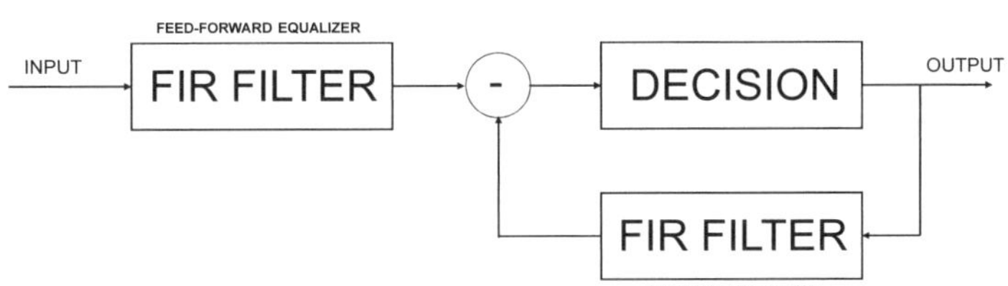

Figure 2.49 High-level block diagram of a decision feedback equalizer.

2.7 Multilevel Signaling

Multilevel signaling encoding like PAM-4 provides certain advantages for increasing the number of transmitted bits while using the same bandwidth [38]. For example, with PAM-4 encoding, four levels are used allowing to encode two data bits in a single timing interval while an NRZ signal will encode a single bit as show in Figure 2.50. Section 9.8 discusses testing multilevel interfaces using standard NRZ ATE pin electronics.

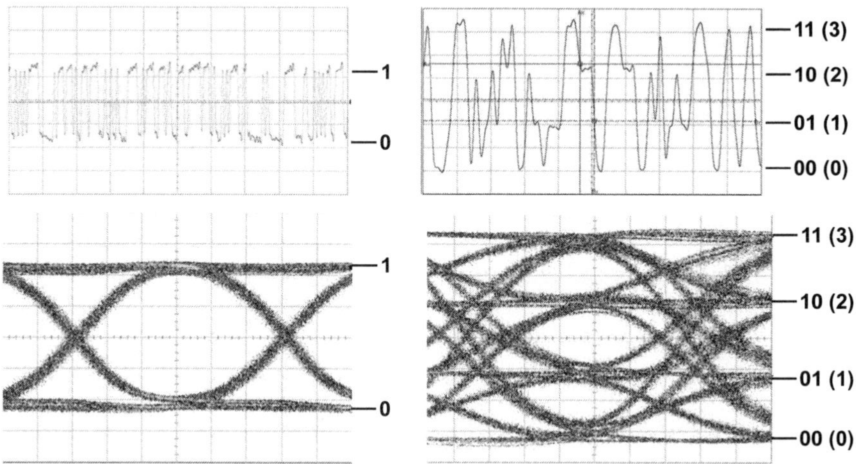

Figure 2.50 Comparison of NRZ (left) and PAM-4 (right) waveform and data eye.

Figure 2.51 shows a comparison of the power spectrum between a NRZ (PAM-2) signal and a PAM-4 signal for the same bit rate (e.g., 10 Gbps for NRZ compared with 5 Gbaud for PAM-4). It is possible to observe that the PAM-4 signal requires half the bandwidth of the NRZ signal for the same data rate. This fact allows by moving to a multilevel interface like PAM-4 to double the data rate but still use the same signal path bandwidth (e.g., the same backplane). The drawback is the reduction on the signal-to-noise ratio by having to make a decision between multiple levels. Also the design of the transmitter and receiver is more complicated.

Multilevel signaling presents new challenges for test and measurement since some of the standard concepts used for the analysis and measurement of NRZ signaling do not apply in a straightforward way to multilevel signaling [39–42]. One approach is to analyze the three data eyes corresponding to the three decisions levels as independent data eyes as shown in Figure 2.52. Correspondingly, you could also measure BER bathtub curves for each decision level data eye.

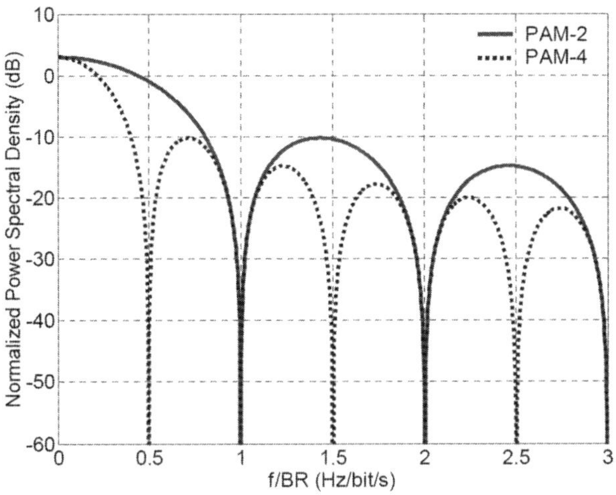

Figure 2.51 Power spectrum comparison between a NRZ (PAM-2) signal and a PAM-4 signal.

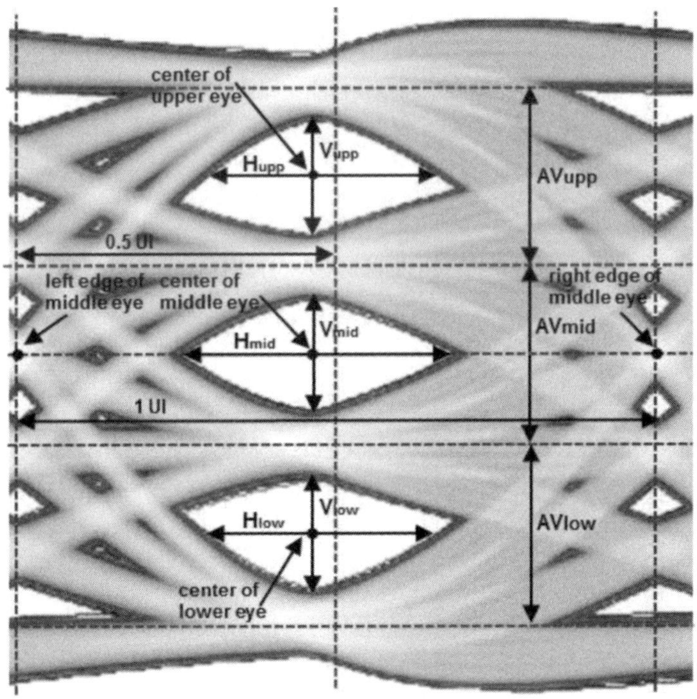

Figure 2.52 Analyzing each of the three decision levels data eyes on a PAM-4 data eye (reprinted with permission from [43]).

However analyzing a PAM-4 data eye as three independent NRZ data eyes is not the only way and it is based on the assumption that independent delays are available for each decision threshold. In other approaches current being evaluated in some standards related with PAM-4 interfaces special patterns are used for some measurements like jitter analysis. For example, the JP03A pattern, which consist of the symbols "03" or the pattern JP03B, which consists of 15 repetitions of the symbol sequence "03" followed by 16 repetitions of the symbol sequence "30," can be used for jitter analysis like jitter separation using current tools since the resulting data signal is NRZ like. It is also possible to perform a symbol error rate bathtub where at each timing offset we measure the number of symbols that failed.

References

[1] W. Maichen, *Digital Timing Measurements*. New York: Springer, 2006.

[2] D. Derickson and M. Mueller, *Digital Communications Test and Measurement: High-Speed Physical Layer Characterization*. Upper Saddle River, NJ: Prentice-Hall, 2007.

[3] S. H. Hall and H. L. Heck, *Advanced Signal Integrity for High-Speed Digital Designs*. New York: John Wiley & Sons, 2009.

[4] L.-T. Wang, C. E. Stround, and N. A. Touba, *System on Chip Test Architectures*. Morgan Kaufmann, 2007.

[5] E. Bogatin, *Signal and Power Integrity Simplified*. Upper Saddle River, NJ: Prentice-Hall, 2010.

[6] H. Johnson and M. Graham, *High-Speed Digital Design*. Addison-Wesley, 1993.

[7] B. Razavi, *Design of Integrated Circuits for Optical Communications, 2nd Edition*. New York: John Wiley & Sons, 2012.

[8] J. Redd and C. Lyon, "Spectral Content of NRZ Test Patterns," *EDN Magazine*, 2004.

[9] Y. Cai, B. Laquai, and K. Luehman, "Jitter Testing for Gigabit Serial Communication Transceivers," *IEEE Design and Test of Computers*, 2002.

[10] J. Moreira, G. Haensel, and F. Koban, "Addressing the Challenges of Implementing an At-Speed Production Test-Cell for 10Gb/s Wafer Probing," *DesignCon*, 2005.

[11] M. P. Li, *Design and Test for Multiple Gbps Communication Devices and System*. IEC, 2005.

[12] M. P. Li, *Jitter, Noise, and Signal Integrity at High-Speed*. Upper Saddle River, NJ: Prentice-Hall, 2007.

[13] Y. Takasaki, *Digital Transmission Design and Jitter Analysis*. Norwood, MA: Artech House, 1991.

[14] INCITS, *Fiber Channel – Methodology for Jitter and Signal Quality Specification – MJSQ*, 2004. Technical Report for Information Technology.

[15] "Understanding and Characterizing Timing Jitter," *Tektronix*, 2002.

[16] U.-K. Moon, K. Mayaram, and J. T. Stonik, "Spectral Analysis of Time-Domain Phase Jitter Measurements," *IEEE Transactions on Circuits and Systems - II: Analog and Digital Signal Procesing*, vol. 49, May 2002.

[17] N. Ou, T. Farahmand, A. Kuo, S. Tabatabaei, and A. Ivanov, "Jitter Model for the Design and Test of Gbps-Speed Serial Interconnects," *IEEE Design and Test of Computers*, July 2004.

[18] Maxim, "Jitter Specifications Made Easy: A Heuristic Discussion of Fibre Channel and Gigabit Ethernet Methods," *Application Note HFAN-4.3.0*, 2001.

[19] A. Papoulis and S. U. Pillai, *Probability, Random Variables and Stochastic Processes*. New York: McGraw-Hill, 2002.

[20] J. B. Johnson, "Electronic Noise: The First Two Decades," *IEEE Spectrum*, 1971.

[21] D. K. C. MacDonald, *Noise and Fluctuations*. Dover, 2006.

[22] International Committee for Information Technology Standards, *Information Technology - Fibre Channel - Methodologies for Signal Quality Specifications (FC-MSQS)*, Dec. 2011.

[23] G. Foster, "Comparing DCD and F/2 Jitter," *Synthesis Research SR-TN078 Measurement Brief*, 2008.

[24] D. Chow, S. Tian, Y. Ke, and K. Ren, "Analysis and Decomposition of Duty Cycle Distortion from Multiple Sources," *DesignCon*, 2013.

[25] B. Analui, J. F. Buckwalter, and A. Hajimiri, "Data-Dependent Jitter in Serial Communications," *IEEE Transactions on Microwave Theory and Techniques*, Nov. 2005.

[26] J. Moreira, H. Barnes, M. Howieson, and M. Broman, "Passive Equalization of DUT Loadboards for High-Speed Digital Applications," *Verigy Users Conference*, 2007.

[27] Silicon Labs, "A Primer on jitter, Jitter Measurement and Phase-Locked Loops," *AN687*, 2012.

[28] B. Laquai and R. Plitschka, "Testing High Speed Serial IO Interfaces Based on Spectral Jitter Decomposition," *IEC DesignCon 2004*, 2004.

[29] H. Zhang, S. Krooswyk, and J. Ou, *High Speed Digital Design: Design of High Speed Interconnects and Signaling*. Morgan Kaufmann, 2015.

[30] W. F. Egan, *Phase-Lock Basics*. Wiley-IEEE Press, 2007.

[31] D. Banerjee, *PLL Performance, Simulation, and Design*. Indianapolis, IN: Dog Ear Publishing, 2006.

[32] B. Razavi, "Challenges in the Design of High-Speed Clock and Data Recovery Circuits," *IEEE Communications Magazine*, Aug. 2002.

[33] G. Foster, "Clock Recovery Primer, Part I," *IEC DesignCon*, 2008.

[34] J. Liu and X. Lin, "Equalization in High-Speed Communication Systems," *IEEE Circuits and Systems Magazine*, 2004.

[35] B. Casper, P. Pupalaikis, and J. Zerbe, "Serial Data Equalization," *IEC DesignCon*, 2007.

[36] Agilent Technologies, "Equalization: The Correction and Analysis of Degraded Signals," *White Paper*, 2005.

[37] L. W. Couch, *Digital and Analog Communication Systems*. Upper Saddle River, NJ: Prentice-Hall, 2006.

[38] A. Healey and C. Morgan, "A Comparison of 25 Gbps NRZ & PAM-4 Modulation Used in Legacy & Premium Backplane Channels," *DesignCon*, 2012.

[39] Keysight Technologies, "PAM-4 Design Challenges and the Implications on Test," *Application Note 5992-0527EN*, 2015.

[40] M. Miller, "Exploring Analysis of Jitter, Noise and BER for PAM4," *DesignCon*, 2015.

[41] M. Agoston, M. Guenther, R. Poulo, K. Sepp, and P. Zivny, "Jitter, Noise Analysis and BER Eye Synthesis on PAM4 Signals on 400 Gbps Communication Links," *DesignCon*, 2016.

[42] Tektronix, "PAM4 Signaling in High Speed Serial Technology: Test, Analysis, and Debug," *Application Note*, 2015.

[43] M. P. Li, A. Healey, P. Fisher, C. Liu, E. Frlan, and D. Brown, "Baseline Proposal for CDAUI-8 Chip-to-Module (c2m) For IEEE 802.3bs," *IEEE P802.3bs 400 GbE Task Force*, 2015.

3

High-Speed Interface Standards

The objective of this chapter is to provide an overview of some representative high-speed digital interface standards, concentrating on the challenges for testing the physical layer of these interfaces and the functionality in the standards that drive these challenges. Since there are similarities on the electrical (physical) layer between the different high-speed digital standards, some of the them are described in detail in this chapter while others are not. Of course, the detailed standard descriptions in this book that go beyond the pure description of the physical layers cannot and should not be exhaustive in a sense that they cover all facets of the respective standard. The details that are discussed for these standards are selected with regard to their representative consequences for ATE-based testing that are also applicable in a similar form to other interface standards.

One important note regarding the data rate notation used in this chapter and through the entire book is the usage of bits per second versus transfers per second. Some standards define their data rate in terms of transfers per second. The reason is that encoding mechanisms used by those interfaces reduce the number of transmitted bits, which contain real data information. In some standards only 80% of the transmitted bits, for example, contain information since the rest of the bits are used by the encoding protocol. In this context, transfers per second represents the rate of the total number of transmitted bits including those of the encoding protocol and bits per second refers to the rate of transferred bits that contain real information. To avoid confusion, we use bits per second within this chapter to refer to the total number of transmitted bits.

3.1 PCI Express

3.1.1 Application Areas

The I/O bus system that connects the core components of a computation system has significant impact on the overall performance of such a system. When the predominantly used conventional PCI bus and its derivatives like the accelerated graphics port (AGP) became a performance bottleneck in computer systems and it turned out that the existing standard could not be enhanced in an economic way to address the immediate requirements with a solid path into the future, the PCI Special Interest Group (PCI-SIG association) started the development of the PCI Express (PCIe) technology.

In its initial implementations, PCIe mainly replaced AGP in computer systems. Over time, PCIe implementations became more pervasive and steadily pushed back conventional PCI in the PC area. In parallel, PCIe expanded into other form factors than just PC plug-in cards such as the ExpressCard or PCI Express Mini Card form factors. With the definition of the PCIe external cabling standard [1] in 2007, PCIe backplane expansion systems that mainly target industrial and enterprise applications became available.

3.1.2 PCI Express Fundamentals

PCI Express was specified by the PCI-SIG association in 2002. The guiding principle for the development of the PCI Express standard was to design an I/O interface that allowed scalability along the bandwidth needs of the next decade with system manufacturing costs at or below the I/O interfaces used at the time of PCIe standardization.

After the first incarnation of PCIe in 2002 [2, 3], a second generation of the interface was standardized in 2006 [4, 5], and the third generation was standardized in 2010 [6, 7]. The fourth generation of PCI Express is in the process of standardization at the writing of this book. Although the driver for a generation rollover was always increased performance requirements that resulted in data rate changes, each generation also included the adaptation of existing features and the introduction of new functionalities to allow operation at the higher data rates. An overview of the various PCIe generations with a selection of test-relevant features that changed over time is shown in Table 3.1. It is important to note that the column for the fourth generation of PCIe is preliminary and might change with the finalized specification.

In order to allow a smooth transition from conventional PCI to PCIe interfaces, software compatibility between the two I/O interfaces had to be ensured while considering software handles for the requirements of the future like improved hot-plug support, advanced power management, and

Table 3.1
PCI Express Generations

	Generation 1	Generation 2	Generation 3	Generation 4[1]
Release year	2002	2006	2010	2015
Data rate	2.5 Gbps	5 Gbps	8 Gbps	16 Gbps
Signal coding	8B/10B	8B/10B	128B/130B	TBD
Jitter filter	1-pole high-pass	Step bandpass	1st order CDR or 2nd order CDR	TBD
Tx equalization	Tx de-emphasis single preset	Tx de-emphasis two presets	Tx de-emphasis, preshoot and boost 11 presets	TBD
Rx equalization	None	None	Optional continuous time linear equalizer and 1-tap DFE	TBD

[1] All values stated here are expected only, since fourth generation of PCI Express was not finalized at the time of writing this book.

so on. Maintaining and emulating the existing operational and configuration capabilities of conventional PCI and enhancing these capabilities with the requirements of the future ensured software compatibility of PCIe with the existing conventional PCI standard. Since configuration parameters are passed from software into hardware on the transaction layer interface of the protocol stack, the levels from this layer down to the physical layer of PCI Express are shown in Figure 3.1.

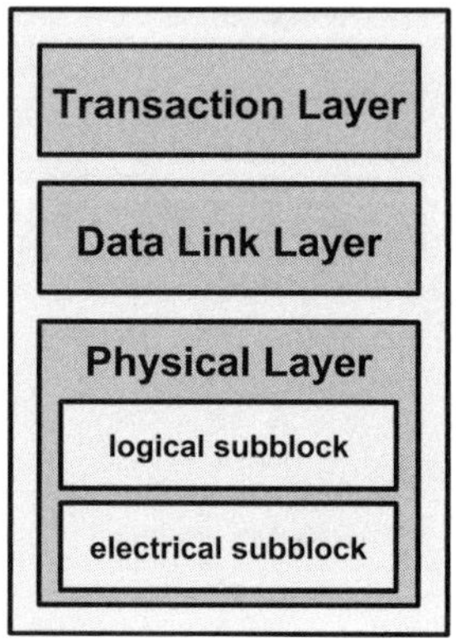

Figure 3.1 PCI Express protocol stack.

The interface from the upper layers to the transaction layer is defined in the PCIe architecture as a superset of conventional PCI. With this approach, software that was based on conventional PCI was also usable with PCIe hardware initially while additional functionality only usable with PCIe could be defined as well.

The basis for the PCIe architecture is a dual-simplex embedded clock interface that uses differential signaling on point-to-point connections without the need for physical sideband signal connections. An embedded clock architecture was selected for PCIe because it offers excellent data rate scalability that so far allowed the definition of multiple PCIe generations. All of these generations are based on the original PCIe architecture that provides data-transmission capability on standard low cost PCB material without the need to fundamentally change the underlying signal generation and data recovery mechanisms over a wide data rate range. A basic PCI Express link consists of two differential signal pairs, a transmit pair and a receive pair that establish the dual-simplex connection between two partner devices as shown in Figure 3.2. This combination of a single differential transmitter and a single differential receiver also is referred to as one lane. Data is exchanged via such a lane in a packet-based protocol. The packet structure of PCIe not only considers addressing information for the immediate link partner of a device, but also addressing information for bus segments and devices. With this, the implementation of complex PCIe bus topologies is possible that allow end-to-end communication between devices that do not have a common physical point-to-point connection but are physically separated from each other by one or more PCIe switch devices.

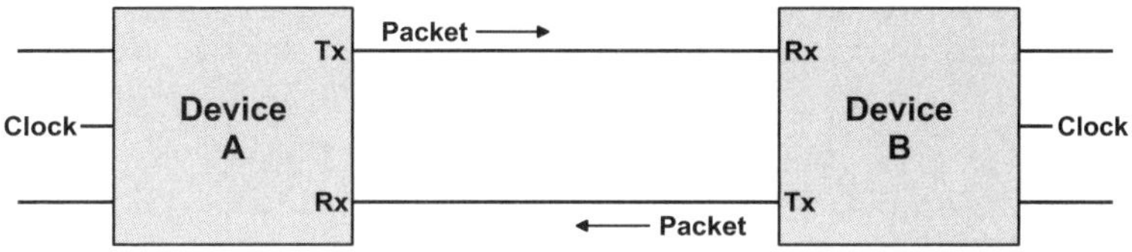

Figure 3.2 Fundamental PCI Express link.

In order to achieve scalability not only on the data rate axis, but also within one data rate class, PCIe makes use of a multilane concept. This means that a data link between devices can consist of multiple lanes. If there is more than one lane that makes up the link, a logical data stream in the transmitting device is split into several substreams, which are serialized separately as shown conceptually in Figure 3.3.

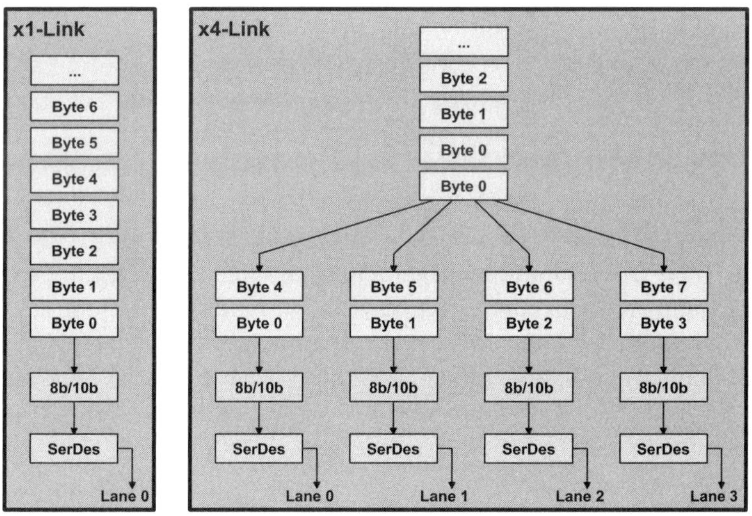

Figure 3.3 Multilane link concept.

Each of these substreams is then transmitted via a separate lane to the receiving device, which has to reconstruct the original logical data stream after deserializing the single substreams. Since each of the lanes carries its own clock information, multilane embedded clock interfaces can consider relative timing displacements of data packets on the single lanes after they passed the transmission media during the reconstruction of the original logical data stream. The relative displacements between the lanes are determined during link training in the protocol engine when a link is powered up. This allows system and PCB designers to have substantial differences in the physical trace length from differential pair to differential pair. Thus, system and PCB layout is simplified dramatically compared to centrally clocked bus architectures or at-cycle synchronous interfaces (see Section 2.5). Depending on the number of substreams or lanes, the PCI Express standard distinguishes x1, x2, x4, x8, x12, x16, and x32 links.

3.1.3 PCI Express Details

Before data alignment between the lanes of a link is possible, several other steps are executed during the power-up sequence of a PCIe link. The first of these steps is bit and symbol alignment per lane to enable the exchange of usable data between the link partner devices. After basic communication is possible on a per-lane basis, data rate and link width negotiation can take place. The initial training that covers all steps required to get the PCIe link into a fully operational mode is described in more detail in Section 3.1.4.3. Since the lane and link communication during the training is executed on a relatively low protocol level that does not yet include the protocol features that are used to assemble and disassemble the regular communication packets, PCIe contains a communication mechanism that uses low-level packet structures which are directly generated and disassembled on the physical layer. These low-level packets are called ordered sets and are transmitted simultaneously on all lanes of a link. An overview of the ordered sets defined for PCIe is shown in Table 3.2. Due to low level coding differences between Gen1/Gen2 and PCIe Gen3, the ordered set structures also are different between these two. While ordered sets for Gen1 and Gen2 are framed by a starting COM symbol and well-defined 8B/10B symbols at the end of the frame or even the entire ordered set, ordered set for Gen3 are only identified by the 128B/130B sync header and the first symbol of the ordered set.

Due to the embedded clock architecture that is used for PCIe, regular data transitions are required on the lanes of a link even if no payload data is transmitted. Thus, if no payload data is transmitted, PCIe requires a transmitter to generate idle data that is defined as the data byte $0x00$ for PCIe. Low level data coding together with scrambling used for PCIe ensures that even though the idle data is the zero byte, transitions are generated during the transmission of that data. While PCIe Gen1 and Gen2 used 8B/10B coding, this coding scheme was replaced by a 128B/130B coding scheme for Gen3

Table 3.2
PCI Express Ordered Sets

Ordered Set	Length (Gen1/Gen2) [symbols]	Length (Gen3) [bytes]	Description
TS1	16	16	Training ordered set used during link initialization
TS2	16	16	Training ordered set used during link initialization
FTS	4	16	Fast training ordered set used during link re-initialization from low-power L0s state
EIOS	4	16	Electrical idle ordered set is sent before entering electrical idle state
EIEOS	16	16	Electrical idle exit ordered set is sent for exiting electrical idle state for links working beyond 2.5 Gbps; serves as block alignment indicator for Gen3
SKIP-OS	2-6	8-24 (4 byte steps)	Skip ordered set is used to compensate link frequency offsets
SDS	NA	16	Start of data stream ordered set (only defined for Gen3)

communication in order to improve the ratio between payload data bits and bits that are stuffed into the data stream due to the coding scheme. Of course, due to the signal transitions, the transmission of idle data consumes power in the transmitter as well as the receiver of the link. In order to save power during phases of longer link inactivity, PCIe defines the electrical idle link state. In this state both legs of the differential signal are pulled to the common-mode voltage via a high impedance termination. The start of the electrical idle state is recognized by a receiver when it detects an EIOS ordered set. The receiver exits the electrical idle state when a certain differential voltage threshold is exceeded on its input. After electrical idle state, a (potentially shortened) link retraining is required.

In order to simplify interoperability between devices with PCIe communication interfaces, the standard requires AC-coupling of the PCIe lanes. AC-coupling simplifies system design, since device pairs can be used regardless of the common-mode voltage used on the legs of their differential signals. This allows combinations of devices that are manufactured in different processes, use different termination schemes, or use different supply voltages which is especially important for components that are connected via a PCIe cable but reside in subsystems that are physically separated from each other with the PCIe cable being the only connection between the subsystems.

Besides the required transition density of the electrical signal that is transmitted between link partners, this signal also has to be DC-balanced in order to avoid baseline wander on the AC-coupled transmission media. For Gen1 and Gen2 of PCIe the 8B/10B coding delivers DC-balanced signals with guaranteed transition density already due to the construction of the coding scheme. Thus, scrambling for the first two generations of PCIe was optional in a sense that it could be disabled by PCIe register settings. The scrambling (see Appendix D) that is used by PCIe Gen1 and Gen2 follows the polynomial $G(x) = x^{16} + x^5 + x^4 + x^3 + 1$. It is applied to the data bytes before 8B/10B coding for the transmitters and after 8B/10B decoding for the receivers. The 128B/130B coding that is used for PCIe Gen3 does not ensure the required transition density and DC-balance just by the coding. The reason for this is that in contrast to the 8B/10B coding scheme 128B/130B coding alone does not modify the payload data (data before coding) in the coding process, but only adds synchronization bits to the data stream (for details see Appendix D). For PCIe Gen3 scrambling is one contributor to guarantee transition density and DC-balance. Thus, scrambling is mandatory for PCIe Gen3 and cannot be disabled. The scrambling polynomial for PCIe Gen3 is $G(x) = x^{23} + x^{21} + x^{16} + x^8 + x^5 + x^2 + 1$. Scrambling rules for PCIe Gen3 are slightly more complex than for Gen1 and Gen2. Similarly to Gen1 and Gen2 scramblers are advanced depending on the content of the data stream and sections of the data stream might be excluded from scrambling on a per symbol basis. However, the rules for excluding symbols of bits from scrambling and advancing the scramblers are on a finer granularity than for Gen1 and Gen2. Also the scramblers are reinitialized periodically with Gen3 whenever a FTS or EIEOS ordered set occurs in the data stream. Although scrambling increases transition density, receivers for PCIe Gen3 have to be more tolerant to allow lower transition densities on the incoming data streams without loosing bit lock. In addition, the training ordered sets (TS1 and TS2) for Gen3 contain two symbols at the end of the ordered sets that are selected depending on running disparity counters to reduce the running disparity towards a more DC-balanced data stream.

In order to minimize electromagnetic interference (EMI) effects, PCIe requires support of spread spectrum clocking (SSC). SSC reduces the magnitude of frequency bins at the reference clock frequency and its derived frequencies in the power spectrum by applying a low-frequency phase modulation to the reference clock. More details on SSC are discussed in Section 5.8.3.

3.1.4 PCI Express Protocol

3.1.4.1 Transaction Layer

This layer of the PCI Express protocol stack is responsible for the formatting of the payload data it gets from the higher layers (usually the operating system interface) into transaction layer packets (TLP). A TLP consists of a header and a payload data section. Depending on the packet type, the header contains information about the packet format, address, and routing and a transaction descriptor that includes a transaction ID, traffic class selectors, and other attributes. The packet types supported by the PCI Express transaction layer are memory-, I/O-, and configuration-read/write transactions, as well as message transactions. Besides TLP assembly and disassembly, the transaction layer also is responsible for storing link configuration and link capability information. With regard to traffic prioritization, this layer implements virtual channels and traffic classes that allow isochronous data transfers via a PCI Express link and the usage of a single physical link as multiple logical links.

3.1.4.2 Data Link Layer

The main purpose of the data link layer is to ensure correct communication between two link partners. In the transmit data direction, the data link layer receives TLPs from the transaction layer. It complements these TLPs with a CRC code for error detection and a packet sequence number and hands this packet to the physical layer below which is responsible for physical data transmission. In the receive direction, the data link layer receives packets from the physical layer that were formatted in the same way from the data link layer of the partner device. On these packets, this layer performs an error check on the CRC code and in case no error occurred passes the received packets in TLP format to the transaction layer in the order given by the sequence number of the received packets. In order to recover from potential errors, the data link layer stores transmitted TLPs for potential retries if the data link layer of the partner device detected an error on a packet and requests retransmission of the data from the faulty packet on. Data retransmission is requested and acknowledged by sending packets that are generated solely within the data link layer. Besides indicating retransmission requests and acknowledgments, these data link layer packets (DLLP) also exchange flow control information like buffer size updates from link partner to link partner. In its function as a link between the transaction layer and the physical layer, the data link layer also conveys information about link states and power state requests between these two layers.

3.1.4.3 Physical Layer

The physical layer of PCI Express architecture consists of a logical subblock and an electrical subblock. The logical subblock defines the logical behavior of a data connection like signaling type, data scrambling, data coding, and the behavior of two devices connected via a PCI Express link over time, whereas the electrical subblock defines the low-level electrical details of a connection such as electrical parameters.

One core component of the logical subblock in the physical layer of the PCI Express architecture is the link training and status state machine (LTSSM). This finite state machine controls the state of a PCIe device from power-up to power-down with all training and configuration steps required to bring the PCIe link into a fully operational mode as well as the transitions between the different link operation modes.

A block diagram of a typical PCIe implementation for the physical layer without the LTSSM is shown in Figure 3.4.

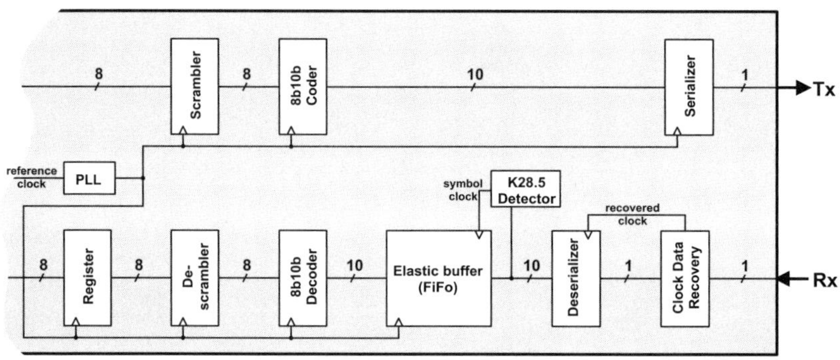

Figure 3.4 PCI Express PHY block diagram.

Link Training and Status State Machine

Embedded clock interfaces require that the receiving device synchronizes itself on the incoming data stream. This synchronization usually is done in multiple steps with the first step performing bit synchronization to allow the clock data recovery to sample the data bits with maximum setup/hold time margins. After bit synchronization is achieved, the receiving device identifies the word or symbol boundaries of the incoming data stream that is the basis for later identification of packet frames. With multilane interfaces, the electrical length differences between lane traces making up a link connection between two devices are compensated in the receiving device to allow correct reconstruction of the data distributed over the lanes for transmission. For PCI Express, one responsibility of the LTSSM is to perform these steps that are fundamental to start up a data link correctly. Besides these synchronization tasks that are common for all embedded clock interfaces, the LTSSM also performs PCI Express specific tasks such as receiver detection or link and lane configuration negotiation. Furthermore, the LTSSM contains states that implement PCI Express specific DfT features and power management. A high-level diagram of the LTSSM with its states and state transitions is shown in Figure 3.5. A short description of these high-level states follows.

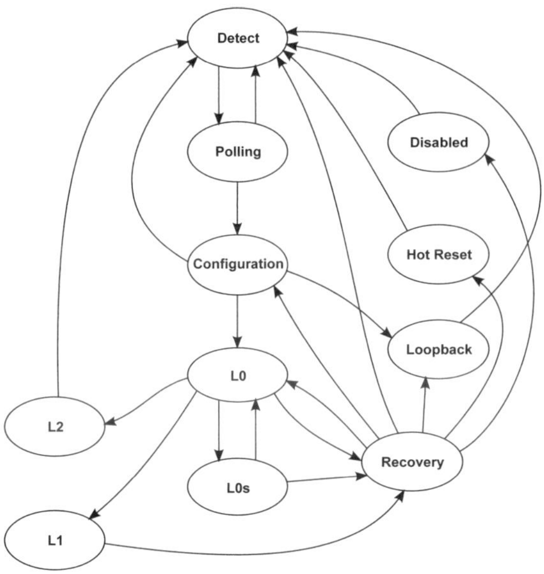

Figure 3.5 PCI Express link training and status state machine. (From: [5]. ©2002–2009 PCI-SIG. Reprinted with permission.)

Detect In the detect state of the LTSSM, a PCI Express device is checking whether a link partner is connected to its transmitters. Detection is done for each transmitter separately and is the basis for later link width negotiation between two link partners. From electrical idle signal state, a common-mode voltage step is applied to the media connected to the transmitters of a link. If a receiver is connected on the far end, the termination of that receiver prevents the media from following that step function quickly, but the media is charged slowly to the final voltage level. If no far end termination is connected, the media follows that step function quickly. A comparator integrated into the transmitter front-end senses the state of the media with some delay to the common-mode step applied to the media. The threshold voltage for that comparator is set in a way that it can distinguish the case of a connected receiver on the far end from the case of a floating media.

Polling The polling state of the LTSSM is responsible for establishing bit and word synchronization as well as the data rate negotiation between two connected link partners. The synchronization tasks are achieved by exchanging, analyzing, and adapting to training sequence ordered sets (TS1 and TS2) sent between two connected devices. In this state, a potential polarity inversion per differential connection is done if PCB layout restrictions force a system designer to connect the positive leg of a transmitter to the negative leg of a receiver.

Configuration This LTSSM state performs lane-to-lane deskew and evaluates the data content of the training sequences that are exchanged between link partners and branches into other LTSSM states (e.g., Disabled, Loopback) if the appropriate bits in the training sequences are set. Besides forced branch requests to other states, the training sequences also contain information about desired link configurations like scrambler control, the minimum number of training sequences required to establish bit and word lock, as well as link and lane numbering. To ensure a consistent link and lane numbering between link partners and to guarantee that an established link fulfills the link width capabilities of both link partners, link and lane numbers are negotiated between the two devices while in configuration state of the LTSSM.

L0, L0s, L1, and L2 After leaving the configuration state of the LTSSM to L0 state, a link between two partners is fully up and running and can be used for packet exchange. If no data exchange takes place, idle data characters are sent in L0 to keep bit and word lock on the receiver connected to a transmitter. L0s, L1, and L2 are different power saving states that implement three levels of power saving potential with the lowest power saving in L0s and the highest power saving in L2. Depending on the power saving level, resuming into data exchange mode L0 results in different return paths through the LTSSM to L0. These different return paths require different amounts of time to reestablish a fully operational link. The options to transit to L0 from one of the power saving states vary from just sending a certain amount of training sequences for L0s to a whole cycle through the LTSSM starting from detect state for L2.

Recovery Based on the negotiated link and lane numbers, recovery state is reestablishing bit and word synchronization and serves as a branching state to L0, Configuration, Loopback, Hot Reset, and Configuration.

Loopback The LTSSM loopback state establishes loopback paths between the receivers and transmitters of a PCI Express physical layer implementation (PHY). Data patterns stimulated on the Rx pins of a PCI Express device appear with some latency exactly as received on the associated Tx pins of the device.

Hot Reset In this state, transmitters signal their partner devices that a reset action was initiated from a higher protocol level by sending training sequences with the reset bit asserted. Upon receipt of training sequences with the reset bit asserted, a device sends training sequences with the reset bit asserted and enters detect state after a timeout period.

Disabled In disabled state, a link switches to electrical idle signaling after the partner devices have been informed about this by sending out training sequences with the disable link bit asserted.

3.1.5 Electrical Specifications

The electrical specifications of PCIe, like most other embedded clock signal interface standards, are oriented on the data eye openings a transmitter has to ensure and a receiver has to be able to identify in order to guarantee an error-free communication between two partner devices. The difference between the minimum transmitter eye and the minimum receiver eye defines the level and timing budget a system design can use for its connection losses.

The two data eyes nail down the most important level and timing parameters for PCI Express such as minimum differential peak-to-peak voltages, minimum data eye widths, and maximum timing jitter numbers. The importance of these data eyes is also indirectly contained in the detailed electrical parameters for PCI Express transmitters. For the transition time of a transmitter signal, only a minimum value is specified. There is no maximum specification. The only rule that limits the maximum transition time is the transmitter data eye compliance. Thus, for devices with low jitter, slower transition times are allowed than for devices with higher jitter numbers, since the transition time in combination with timing jitter on the transitions defines the data eye boundaries.

Besides the data eyes, another important parameter group for differential low swing signal interfaces are the DC parameters that define termination impedances of the transmitters and receivers. These impedances have substantial influence on the signal and data eye shape and the correct alignment of the positive and negative signal leg that form a differential signal. In addition to these DC impedance values, PCI Express also specifies parameters for the AC impedance matching in the form of return loss parameters for both transmitters and receivers.

Besides signal transition time, the electrical specification contains timing related parameters that define the allowed bit time or unit interval (UI) ranges as well as the lane-to-lane skew values that have to be met by the transmitters of a link and have to be tolerated by the receivers of a link. A UI width specification in fact is defining the allowed 300 ppm frequency offset between the nominal 100 MHz reference clock and the actual reference clock frequency. This is due to the fact that the reference clock input of a device directly translates to the same relative change in UI width on a transmitter

Another set of electrical parameters defines the level values that various common-mode voltage measurements have to fulfill. The most common of these parameters that also is found in most other differential signal interfaces is the AC peak common-mode voltage parameter. PCI Express defines this value for generation 1 devices as rms value that is measured over 10^6 UIs. This differs from the maximum absolute peak-peak values as they are used in most other standard specifications and for PCIe generation 2 devices. Another special parameter in the common-mode voltage parameter group of PCI Express is the absolute delta between the common-mode voltages of the positive and the negative leg of a differential signal. At a first glance, there is no obvious reason for this parameter because DC-balanced signal coding is used and the two legs of the differential signal usually are physically tied to the same current source in the transmitter. Thus, no difference in common-mode voltage between the legs should be expected. The underlying physical deviation for this parameter is duty-cycle distortion for the transmitting clock, which exhibits as a relative common-mode shift between the two legs of a differential signal even if DC-balanced data streams are used. Additional electrical parameters defined in the parameter lists of PCI Express are voltage levels for PCI Express specific operation modes like electrical idle and detect or time constants that apply for operation mode changes of the PCI Express front-ends.

Two parameters of the PCIe electrical specification list that require special consideration are the de-emphasis specification and the parameters around jitter, PLL bandwidths, and eye openings.

3.1.5.1 Equalization

Due to the positive effects more complex equalization has on the signal integrity of high-speed signals, it is no surprise that with the increasing data rate for every generation of PCIe also the equalization required by the standard got more sophisticated. The standard requires various forms of transmitter equalization depending on the PCIe generation. On the receiver side for the first two generations of PCIe, no equalization was mentioned in the standard. With the third generation receiver calibration is described in the standard, but no mandatory implementation guideline is given. The standard mentions, that receiver equalization is required for the longest PCIe channel that is allowed by the specification.

Transmitter Equalization Gen1/Gen2

In order to compensate expected losses caused by the bandwidth limitation of the PCB connection between two devices, only the first bit after a transition (transition bit) in the data stream is driven at the nominal voltage level for Gen1 and Gen2. The voltage level of subsequent bits that represent the same logical level as the first bit is attenuated by a specified ratio with respect to the nominal level. Thus, effects that are caused by the low-pass characteristic of a typical PCB interconnection are minimized. PCIe defines such a de-emphasis setting of −3.5 dB for 2.5 Gbps operation and two settings (−6.0 dB and −3.5 dB optionally) for 5.0 Gbps operation. Since the purpose of de-emphasis is to minimize timing errors at the far end of bandwidth limited connections, the presence of de-emphasized signals at the signal probe points has influence on quite some of the timing parameters that are specified and need to be checked. More details on de-emphasis can be found in Section 2.6.4. In order to have a common basis for the reported timing parameter numbers, the PCIe standard defines procedures to de-embed the de-emphasis effects from the measured signals for the parameters that are influenced by de-emphasis. Details on these procedures can be found in [5].

Transmitter Equalization Gen3

For the third generation of PCIe transmitter equalization changed significantly compared to the previous generations. Instead of a simple reduction of the levels for subsequent bits of the same polarity, Gen3 equalization takes into account the state of the current bit, the previous bit and the next bit in the data stream to derive the attenuation or amplification setting for the bit to be transmitted. The implementation of such an equalizer typically is done using a finite impulse response (FIR) filter with three inputs that sense the levels of the three bits that contribute to the equalization factor for a bit and apply dedicated weighting factors to each of these inputs to derive the final equalization factor as the sum of the weighted bit levels. These three inputs are referred to as taps and the positions of these taps at, before, or after the bit to be generated are called cursor, pre-cursor, and post-cursor. It has to be noted that this implementation of the transmitter equalization is compatible with the de-emphasis definitions for Gen1 and Gen2 because proper selection of the tap weighting factors allows to implement the required de-emphasis settings required for the previous generations of PCIe. In addition to the de-emphasis, the three-tap FIR filter equalization approach allows equalization control at a much finer granularity and thus can adapt much better to the actual signal transmission conditions. With the three taps and constant but different weighting factors for these, four distinct levels can be generated by the output of the filter. These levels also are referred to as normal, boost, de-emphasis, and pre-shoot. The meaning of these is described in Table 3.3. The PCIe standard defines 11 presets for 11 different tap factor combinations that implement 11 different combinations of normal, boost, de-emphasis, and pre-shoot levels. In order to adapt the equalization settings to the ones that are actually required for a given PCIe channel, the Recovery state of the PCIe LTSSM has been enhanced with Gen3 to include a state that implements the equalization negotiation between two link partners when the 8 Gbps mode of PCIe is entered.

Table 3.3
PCI Express Gen3 Equalization Levels

Level	Meaning
Normal	Signaling level for a bit after a transition and next bit at same logical level
Boost	Signaling level for a bit after a transition and next bit at opposite logical level
De-emphasis	Signaling level for a bit with previous and next bit at same logical level
Pre-shoot	Signaling level for a bit before a signal transition

Receiver Equalization

The PCI Express standard does not define how receiver equalization is to be implemented on the physical level, but gives recommendations on how to apply receiver equalization to measurements of stressed data eyes in

8 Gbps mode. Depending on the amount of stress applied to the data eye, this recommendation either is applying a first order CTLE only for low and medium stress and a first order CTLE plus a 1-tap DFE for high stress. The first order CTLE has fixed LF and HF poles and an adjustable DC gain with a minimum range from −6 dB to 12 dB in 1 dB steps.

3.1.5.2 Jitter

A central topic of discussion during the standardization of the second generation of PCIe was how the jitter specifications have to be defined to ensure proper device interoperability [8]. As a consequence of this discussion, transmitter phase jitter filtering functions were defined that need to be applied to the raw jitter that is measured on the transmitters. The filtering is required because PLLs providing the sampling clock for the CDR on the receiver side can track low jitter frequencies but not high jitter frequencies on the incoming data streams. Thus, high-frequency jitter is more likely to cause wrong data latching than low-frequency jitter and has to contribute to the specification boundaries to a higher degree than low-frequency jitter (also see Section 9.5). The filtering functions were selected as a single-pole high-pass filter for the 2.5 Gbps operation and a step bandpass filter for 5 Gbps operation. In combination with the jitter filtering functions, PLL bandwidth and peaking parameters for the transmitter PLLs were introduced because the selected filtering functions only are valid in combination with certain transmitter PLL characteristics.

As an analogy to the bandpass transmitter phase jitter filtering, the jitter parameters that are specified for the 5 Gbps receiver operation are separated for two phase jitter frequency ranges. For each frequency range a separate random jitter component and deterministic jitter component is specified. For the 2.5 Gbps operation, the allowed Rx jitter inherently is defined via a minimum eye width only. Also for the receiver side, PLL bandwidth and peaking values are specified for the PLLs used in the CDR circuit of the receiver.

3.1.6 ATE Test Requirements

If the life cycle of a device is analyzed regarding the usage of ATE equipment in the various test steps that apply, one can see that there are mainly three sections that usually require the involvement of ATE systems with their ability to integrate into an environment that allows gathering a substantial amount of device data efficiently. The first of these steps is the transition from design to manufacturing where ATE equipment is used to support design verification and device characterization in combination with box or bench test equipment. The second life cycle step with ATE equipment usually involved is the production and technology ramp, and the third step is high-volume manufacturing of mature devices. All three of these phases pose different requirements on the ATE instrumentation and test methodologies used.

For the design verification and characterization phase, most accurate tests on the complete parameter list are required. In this phase even parameters that can be guaranteed by design are verified directly to ensure that design implementation is correct. For design verification and characterization, the ATE-based test is complemented by box and bench equipment measurements or system level tests for test items that once verified can be guaranteed by design and which are more efficiently tested in a bench of system environment. For the parameters that are tested on the ATE, test instrumentation has to offer optimum performance regarding DC accuracy using integrated parametric measurement units (PMUs) to be able to accurately characterize DC parameters such as termination impedances and DC common-mode voltages. Best AC level and timing accuracy is important to be able to characterize data eye parameters such as differential swing and jitter parameters. For characterizing the de-emphasis behavior of PCI Express, the most critical instrumentation characteristic is analog bandwidth. Since de-emphasis is used to compensate bandwidth limitations of device connections, test instrumentation must not exhibit similar bandwidth limitations since the effect to be measured would not be visible any more in that case.

Another important instrument requirement is given by the multilane architecture of PCI Express. Multilane links on the one hand require accurate relative timing measurements among ATE tester channels in the picoseconds resolution range for Tx lane-to-lane skew measurements. On the other hand they also require wide relative timing shifts between these channels in the tens of nanoseconds range for the Rx lane-to-lane skew tolerance characterization.

For the production and technology ramp phase of the device cycle, ATE-based test flows usually do not test for parameters that are guaranteed by design and that have proven to be stable with enough margin to the allowed value boundaries during characterization. The most important task for PCI Express testing in this phase is to ensure data eye compliance and DC parameter accuracy. Data eye compliance in this regard also means specifically the requirements for jitter measurements and jitter tolerance. Also, tracking of other parameters that have exhibited narrow test margins during characterization is important during technology ramp. Since test time is already under tight control in this test phase, only the values of the most critical parameters that were identified during characterization are tracked. The compliance of all other parameters is ensured by guard-banded pass-fail tests. Since the measurement accuracy has to be in the same range as for characterization, requirements for the test equipment do not change substantially compared to the characterization phase. In fact, some of the requirements might even be tougher since test time is of more importance due to the higher volume that needs to be tested in this ramping phase. One example of this is the flexibility in termination mode switching that ATE instrumentation has to provide. Whereas multiple test programs with different hardware configurations are acceptable for characterization to change ATE termination modes from, for example, differential termination to high impedance termination, ATE equipment has to be capable of switching between these modes within one program on the fly for tests in the technology and device ramp phase.

In the volume-manufacturing phase, the main goal is to find the optimum balance between test cost and test coverage. In order to achieve this balance, DfT approaches such as Tx to Rx loopback configurations (far-end loopback; see Section 6.3.1.3) are used widely. Since most of these DfT approaches are either rather new or lack in parametric variability to really stress device components such as the PCI Express receivers, ATE equipment has to offer capabilities to support the DfT-based testing of PCI Express devices. One example for such a DfT support is to provide instrumentation that allows timing and level parameter variation in a loopback path between a Tx and Rx lane of the PCI Express interface.

3.1.7 Test Support

The LTSSM of PCI Express implements two states that are useful for testing the parametrics and a substantial portion of the logic inside the PCI Express physical front-end (PHY) implementation [9].

The first one is the Compliance substate of the Polling state. It is entered if a transmitter detects a far-end termination but does not receive a valid differential signal level on its associated receiver. This behavior indicates that the device is operated in a test environment and the transmitter starts to send a compliance pattern continuously. The content of the compliance pattern is defined to maximize ISI on the lanes and crosstalk between lanes of links with a width beyond x1.

3.1.7.1 Compliance Pattern for PCIe Gen1 and Gen2

For PCIe Gen1 and Gen2 maximum ISI is achieved by using a character with the maximum run length 8B/10B codes allowed in combination with a character containing the shortest run length. This leads to a pattern containing an alternating flow of a comma character (K28.5 with maximum run length 5) and a clock pattern character (D10.2 or D21.5 with maximum run length 1).

In order to obtain a well-defined compliance pattern, the standard specifies that the first K28.5 character of the K28.5-D21.5-K8.5-D10.2 compliance pattern building block has to have negative disparity. In order to achieve crosstalk maximization, it is required that not all of the lanes within a link generate this pattern synchronously, but that one victim lane stays stable for a certain number of UIs while all other lines operate as aggressors and perform maximum switching. This is achieved with the K28.5-D21.5-K8.5-D10.2 building block by inserting two additional delay characters in one of the lanes before and after this building block. If an overlay of a lane without delay character insertion is done with a lane that contains delay character insertion, one can see that the K28.5 characters with minimum switching activity on one lane falls together with the maximum switching activity characters D21.5 and D10.2 on the other lane. The delay characters are selected as K28.5 characters. A pair of two delay characters is required to keep the data insertion disparity-neutral. The delay character insertion starts at lanes 0, 8, 16, and 24 of a link, continues to shift through all lanes, and starts over on the initial lanes once each lane has inserted a building block with delay characters. With this insertion scheme, each of the lanes serves as victim once while all other lanes operate as aggressors.

The standard also defines a modified compliance pattern sequence that basically consists of the basic compliance pattern sequence that is appended by two error status symbols plus two K28.5 characters. In addition for the modified compliance pattern instad of inserting two delay characters, four delay characters are inserted into the sequences for multilane interfaces. The error state characters contain data for a 7-bit error counter value and one bit that signalizes that a lane locked to an incoming modified compliance pattern sequence.

3.1.7.2 Compliance Pattern for PCIe Gen3

The compliance pattern for PCIe Gen3 has to be defined differently from Gen1 and Gen2 because Gen3 uses 128b/130b coding instead of 8b/10b coding. As for Gen1 and Gen2, Gen3 also defines a standard compliance pattern and a modified compliance pattern. The standard compliance pattern consists of five sections. The first section is a sync header followed by a sequence of 64 ones and a sequence of 64 zeros. The second and third sections are sync headers followed by lane-specific 128-bit payload data sequences that are not scrambled. Section four is one EIEOS block and the last section of the standard compliance pattern is a sequence of 32 scrambled data blocks each consisting of 16 IDL data symbols. The modified compliance pattern for Gen3 is implemented differently depending whether a link is operated in separate reference clock independent SSC (SRIS) mode or not. If SRIS is not active, the modified compliance pattern starts with on EIEOS block followed by 256 scrambled data blocks each consisting of 16 IDL data symbols. This part then is followed by 255 sequences consisting of one SKP ordered set and again 256 scrambled data blocks with 16 IDL data symbols each. If SRIS is active, the modified compliance pattern also starts with one EIEOS block. This block then is immediately followed by 2,048 sequences with one SKP ordered set and 32 scrambled data block with 16 IDL data symbols.

3.1.7.3 Loopback State

The second LTSSM state supporting the test of PCI Express devices is the Loopback state that implements a far-end loopback path between the receivers of a PCI Express PHY with their associated transmitters. This mode of operation is activated by sending a training sequence ordered set (TS1 or TS2) with the loopback bit of the sequence enabled to a device receiver.

3.1.8 Test Challenges

Embedded clock interfaces in general and PCI Express technology in particular exhibit multiple challenges for ATE-based testing using general-purpose test instrumentation. One key challenge with this kind of interface that ATE systems have to be capable of handling is nondeterministic data generation of the DUT. With PCI Express there are two notions of non-determinism in the data streams a device generates [10].

The first notion of nondeterminism that is inherent for all embedded clock interfaces is the variable latency of the transmit data stream. This means that the exact position where a bit/word starts and where a bit/word ends is not predictable with respect to an external frequency reference (which the ATE runs on) and that the timing relationship between an embedded clock data stream and an external reference frequency might change if the reference clock feeding the PCI Express device's PLL is shut down and restarted. In order to deal with this challenge, ATE equipment has to be able to provide a continuously running reference clock and needs to have the capability to adapt its timing system to the timing dictated by the DUT.

One way to address this challenge with an ATE system is to use output-dependent timing approaches that identify bit and word boundaries using timing searches and match loop implementations as discussed in Section 5.1. The timing of the ATE is adapted to the timing values found during the searches to align the compare strobe positions to the correct positions with regard to the DUT data stream. A search function on the compare strobe position for the Rx pins is done to identify the center of the data eyes and the loopback latency. After these two variables have been identified, the strobe positions of the ATE comparators are programmed to the center of the data eye with the appropriate latency per lane.

The critical parameters of an ATE guaranteeing success with this approach are the maximum timing range in which the timing programming range, the at-speed match loop capabilities, and the possibility to keep the reference clock of the DUT running during all the searches and timing re-programming. Once the correct strobe positions are identified and set, correct functional tests can be performed on general purpose ATE hardware.

The second notion of nondeterminism is nondeterministic data that occurs on PCI Express devices. Nondeterministic data occurs because embedded clock devices like PCI Express components have to be able to handle frequency offsets between link partners and simultaneously rely on continuously running data streams. If two link partners communicate with a slight frequency offset, the slower partner is not able to process all the data at the pace it arrives. Since the data stream is continuously running, the receiving device has to skip some of the incoming data when its input elasticity buffers are full (or close to full). In order to prevent the device from skipping payload data, PCI Express foresees that transmitters have to insert skip ordered sets in specified interval boundaries into the data stream they send. Receiving devices can then skip or add dummy data (SKP symbols) to these skip ordered sets to adapt to a potential frequency offset without losing payload data.

The problem with skip ordered sets is that these are inserted into the data stream in a nondeterministic manner. It cannot be predicted with the help of simulations or other means when an ATE has to expect such a skip ordered set. Thus, tests requiring real-time comparison to prestored vector patterns are not possible in a normal operation mode of PCI Express if the ATE does not natively support skip ordered set handling. Moreover, undersampling techniques are not possible because the skip ordered sets occur at different positions within the repeated pattern section that is sampled. The only possibility to test a device with such a behavior is to do real-time sampling at the native data rate with postprocessing of the gathered data or to apply ATE hardware that can handle the interface protocol used by the DUT (see Section 9.7). Another methodology to test standard PCIe devices with ATE that does not have such capabilities is to configure the DUT in a far-end loopback configuration that allows the generation of data streams without skip ordered sets.

A better way to test the functional behavior of PCI Express, however, would be to control skip ordered set generation by means of DfT. There is either the possibility of making the occurrence of skip ordered sets predictable using DfT or to switch off the skip ordered set generation of the DUT completely. Since the ATE provides a well-controlled environment, it is ensured that no frequency offset will occur, and thus, no skip ordered set generation is necessary in the communication with ATE equipment.

The LTSSM poses another challenge on testing PCI Express devices. In order to walk through the state machine in a mission mode manner, handshaking between the DUT and the ATE is required since, in such a case, the ATE is considered as a partner device on the link. This, however, requires real-time response calculation of the ATE, which is not feasible with general-purpose ATE equipment. There are several ways to solve this problem.

The first of these solutions performs bit and word lock as described above on the LTSSM Polling.Compliance state. Once bit and word lock on the ATE is achieved, the LTSSM can be executed from Polling.Compliance to L0 by sending the data to the DUT that is expected within a certain time window. This data is independent of the DUT's response as long as the implementation of the LTSSM is correct.

Another approach to address this issue is to provide appropriate DfT that bypasses LTSSM states that are not required in an ATE environment. Since the DUT is well known by the test engineer, handshaking procedures like link width negotiation, link speed negotiation, link and lane numbering negotiation, and polarity inversion are not required in the ATE environment. Since all critical handshaking LTSSM states deal with the uncertainties a device might see in a system environment, it is feasible to bypass these states in the controlled ATE environment by means of DfT in the DUT.

A third option to address this issue would be to implement protocol-aware ATE hardware or use a golden device approach. Since protocol testing and the test of electrical low-level parameters, which is the target for ATE-based testing, access the device on very different levels, it might be difficult to get appropriate access to the low-level electrical parameters from a higher protocol level for measurement and debug purposes.

The last test challenges to be mentioned here are challenges on the lowest electrical level such as low swing capabilities in the range of less than 30 mV single-ended, analog bandwidth requirements in the range of 4 GHz for first generation PCI Express devices and raw data rate capabilities with sufficient signal quality. All of these low-level challenges have to be fulfilled on multiple parallel ATE channels to address the multilane architecture of PCI Express that can accommodate up to 32 differential lanes for a single link.

3.2 XDR DRAM

3.2.1 Application Areas

The target application areas for extreme data rate (XDR) DRAM devices are multicore processing systems, gaming, and graphics as well as consumer applications that require high memory bandwidth with limited data connection widths like HDTV devices. The usage of XDR DRAM devices in the mass market is closely linked to the Cell processor [11] and derivatives of the Cell architecture [12] that use XDR DRAM as main memory. Thus, XDR DRAM devices can be found in server blades that use Cell processor versions with XDR memory interface and consumer products that use Cell-based high-definition (HD) graphic processors. The most prominent representative of systems that use XDR DRAM components is the Sony PlayStation 3 game console with its Cell processor CPU.

3.2.2 XDR Fundamentals

The XDR technology has been developed by Rambus, Inc., and was introduced into the market in 2003. The XDR architecture consists of three core building blocks. These building blocks are XDR DRAM, XDR I/O (XIO), and the XDR memory controller (XMC). While XIO and XMC are a collection of hard and soft macros that can be licensed from Rambus, Inc., to be used in proprietary IC designs, XDR DRAM devices are complete DRAM components that use XIO as electrical interconnects and deploy a communication protocol defined by Rambus [13, 14]. XDR DRAM devices are manufactured under license by DRAM memory manufacturers.

The XDR technology represents a mixture between the traditional memory interface architecture with a parallel data bus and serial high-speed interfaces with a packet-based serial connection to transmit command and address data. The packets that contain command and address data are named requests in the XIO terminology. In contrast to traditional single-ended multi-drop memory data bus implementations, XDR uses differential point-to-point connections for its data bus. These point-to-point connections are not operated in a semiduplex manner to achieve a bidirectional data flow as is the case for the differential interfaces described so far. XDR in fact exchanges data on the data bus in a full-duplex way. This combination of a bidirectional differential point-to-point connection that uses a full-duplex data flow gives the XIO interface of XDR a unique position among other high-speed I/O interfaces and also opens up some interesting aspects from an ATE test point of view. In order to cover a wide application range with changing requirements for data bus width and memory depth, XDR memory devices have a programmable data width. This means that the number of data signals that form the data bus can be reduced from the maximum native width a device is designed for to a lower number. The data bus width reduction goes hand in hand with a column address range expansion which is reflecting an increase of memory depth. The additional column address bits that are used for this column address expansion are called subcolumn (SC) bits. The amount of decoded SC address bits depends on the programmed data width reduction. An example for the relationship between programmed data width and decoded SC bits is shown in Table 3.4 for devices with native data bus widths of 32 and 16 bits.

Table 3.4
XDR Subcolumn Addressing and Used Data Width (Source: [13])

Native Device Width	Programmed Device Width	Decoded SC Bits	Fraction of Selected Column Amplifiers
x32	x32	None	All
	x16	SC[4]	1/2
	x8	SC[4:3]	1/4
	x4	SC[4:2]	1/8
	x2	SC[4:1]	1/16
	x1	SC[4:0]	1/32
x16	x16	None	All
	x8	SC[3]	1/2
	x4	SC[3:2]	1/4
	x2	SC[3:1]	1/8
	x1	SC[3:0]	1/16

XDR DRAM devices transfer 8 data bits with one master clock cycle in an octal data rate (ODR) manner on each of their DQ signals. In order to simplify the electrical design of systems that use XDR memory devices, a mechanism for static skew compensation between the single data bus signals is provided by XDR. This deskew methodology called FlexPhase allows the adjustment of the device DQ data stimulus and capture times individually per pin to compensate for different electrical lengths of the DQ signal paths. The FlexPhase adjustment happens during a calibration step in the initialization sequence for a XDR DRAM device.

3.2.3 XDR DRAM Details

3.2.3.1 XDR Timing Architecture

XDR devices present themselves to the memory controller via two separate interfaces. Each of these interfaces comes with its own clock signal that is routed from the memory controller to the XDR device. The first of these interfaces is a low-speed serial interface that is used to initialize and configure the XDR devices. One part of the device initialization that is supported via this serial interface is the FlexPhase calibration. The other interface is the high-speed interface that transfers commands (or in XDR terminology request packets) from the memory controller to the XDR device and transfers data in both directions between memory controller and XDR memory device. The pins forming these interfaces are listed in Table 3.5.

Table 3.5
XDR Pin List

Signal Pins	Type	Group
CFM, CFMN	I	Clock
RST	I	Serial interface
CMD	I	Serial interface
SCK, SDI	I	Serial interface
SDO	O	Serial interface
RQ0..RQ11	I	Request
DQ0..DQ15	IO	Data
DQN0..DQN15	IO	Data

Clock Pin Group The differential clock from master (CFM) signal serves as timing reference for the request and data pins of the XIO interface of XDR memories. The active transition of the clock that serves as reference is the falling edge of the differential clock signal.

Serial Interface Pin Group The serial interface pins are used to control basic functions of the XDR memory such as initialization and control register programming. Whereas the SCK, RST, and CMD pins are routed in parallel to all memory devices that form a XDR DRAM system, the SDI and SDO signals connect the single memory devices in a daisy-chain manner.

Request Pin Group The XDR memory device receives its operation commands on the pins of the request pin group. The request pins run at the data rate of the clock pins. Request commands are transmitted as request packets consisting of two consecutive bits for each pin of the request pin group. The first of these bits is latched into the memory device with the falling edge of the differential CFM clock. The second bit is latched with the rising edge of that clock.

Data Pin Group The data on the differential data pins of XDR are transmitted in an octal data rate manner with eight data bits per CFM clock cycle.

A schematic block diagram describing the XDR clocking architecture is shown in Figure 3.6. For the low-speed serial interface, the RST, CMD, and SDI input signals are latched into the device with the falling edge of the SCK clock signal. The SDO signal is generated with a specified minimum setup and hold time around the falling edge of SCK.

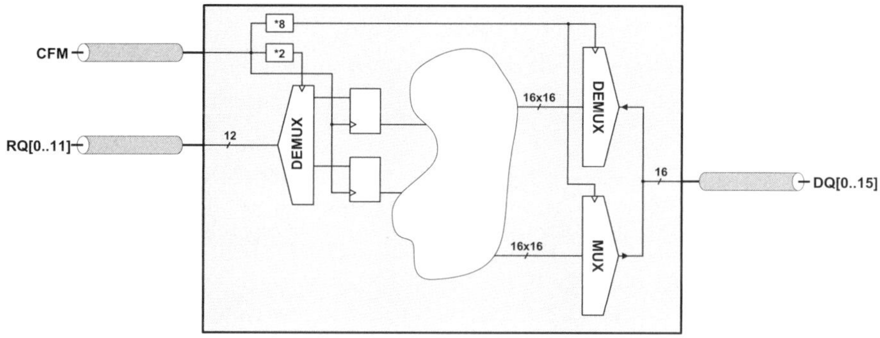

Figure 3.6 XDR timing architecture.

On the high-speed interface, the frequency of the differential CFM clock signal is multiplied in the XDR memory device by a factor of 2 and by a factor of 8. The multiplied internal clock signals are used to oversample the request-signals with two sampling points per CFM clock cycle. The DQ-signals are oversampled with 8 sampling points per CFM clock cycle for incoming data. For DQ data generated by the XDR memory, 8 data bits are sent out per CFM clock cycle. The oversampling of RQ and DQ signals happens in demultiplex circuits so that the nonmultiplied CFM clock signal can be used for further data processing after the demultiplexing. For the DQ signals that are generated by the XDR memory device, the data internally is clocked with the nonmultiplied CFM signal. Before the data is transmitted via the IO drivers, the data is multiplexed to eight times the internal data rate. As on the low-speed interface, also for the high-speed interface, the RQ and DQ signals are sampled and generated with the falling CFM clock edge.

3.2.3.2 XDR Specific Functionality

The mechanism of scalable bus width per XDR device is applied for XDR-based DIMM memory modules to provide flexibility for the population of

DIMM slots that are associated to a single x32 DQ channel. Due to the point-to-point topology of XDR, it is not possible to increase the storage amount served by such a DQ channel by just using a second DIMM that is daisy-chained to the already populated DIMM slots on the data lines. Instead, a topology called dynamic point-to-point (DPP) configuration was developed for XDR-based DIMMs. For a DPP base configuration that populates the first of two DIMM slots that are shared for one x32 DQ channel with an XDR DIMM, the second DIMM slot is occupied with a continuity module. This continuity module just loops back half of the DQ lines to the memory controller. In case a second DIMM is installed, the continuity module is replaced by that DIMM. The memory controller is able to recognize the second DIMM and configures the data width of the XDR components on both DIMMs into half of their native device width. The DQ lines that previously were looped back by the continuity module from the first DIMM to the memory controller now are directly connected to the DIMM in the second slot.

Another specific functionality of XDR memory devices to be discussed here is dynamic request scheduling. For some requests transmitted to XDR memory devices, the memory controller is allowed to specify a delay for the execution of that request. This delay control allows the memory controller to transmit the requests in a way that makes best use of the available bandwidth on the request bus while also keeping the amount of data transmitted within a certain period on the data bus at an optimum without avoidable gaps between data packets.

A feature that also targets optimized usage of the data bus bandwidth is the segmentation of the memory device into bank sets. The memory banks within different bank sets are served by internally separated data paths. This allows a quasi-parallel memory core access on the memory cells located in different bank sets. From an I/O perspective, this parallel access manifests itself in reduced gaps between data packets on the data bus if the data is associated to different bank sets. One specific XDR feature that uses this advantage of bank sets is Early Read After Write (ERAW). If the target bank set of a write request is different from the target bank set of a following read request, the write-to-read turnaround bubble on the data bus can be reduced compared to a single bank set operation. This is due to the fact that, internally to the XDR device, the read operation on one bank set can already start on the local data bus of that bank set while the write operation on the other bank set still occupies the local data bus of that bank set. Together with the bank set concept, simultaneous activation and simultaneous precharge for the different bank sets might be supported.

The last XDR-specific feature to be discussed here is the access to the XDR memory device via the low-speed serial interface. This interface is mainly used to initialize the XDR devices and to set the configuration registers of a device. The topology of the low-speed serial interface for a XDR memory system is shown in Figure 3.7. While the RST, CMD, and SCK signals are shared between all devices, the SDI and SDO pins form a daisy-chain that ends on the SRD input pin of the memory controller. The SDI input of the first device in this chain is connected to the VTERM supply through a termination component (as is the case also for SCK, RST, and CMD). Since all logic in the XDR environment is low-active, the VTERM voltage at the SDI input is interpreted as a logical low by the XDR device. During device initialization, each device in the daisy-chain is assigned a unique serial identification value. This ID value then allows access to specific devices within the chain via the shared control signals of the low-speed serial interface.

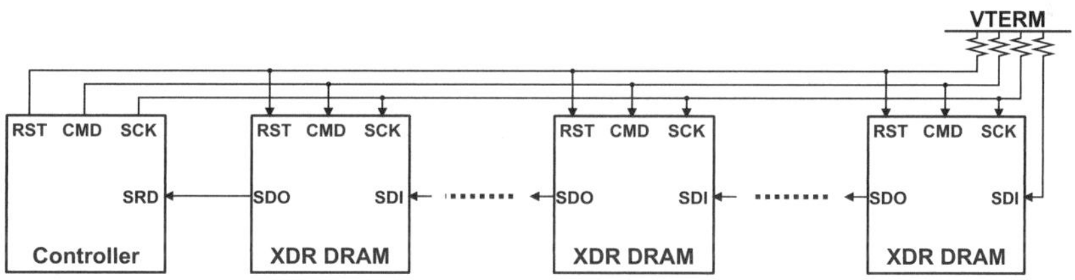

Figure 3.7 XDR serial interface topology.

3.2.4 XDR Protocol

Obviously, the low-speed serial interface and the high-speed IO interface of XDR deploy different protocols for the communication between memory controller and memory device. Although the low-speed interface should not pose any significant challenges from a testing perspective, we also will describe the data link layer of this interface because it plays a crucial role during the FlexPhase calibration that is described in the physical layer section.

3.2.4.1 Data Link Layer

For the XDR high-speed IO interface, a basic access is structured as shown in Figure 3.8. A request packet consisting of two consecutive bits on the request bus is transmitted with the first of the two bits center-aligned to the falling edge of the CFM clock. Depending on the command contained in the request packet, data on the data bus might occur with a defined latency to the start of the request. For some requests, no data transmission happens on the data bus. If the request triggers a data transmission, the direction of the data flow is determined by the request type (write or read request).

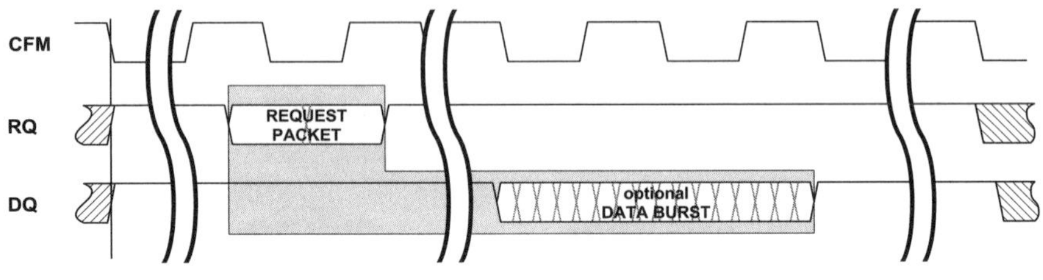

Figure 3.8 Basic access on the XDR high-speed interface.

A request packet contains operation code, address, and other control information for the memory device. XDR defines five types of request packets that are listed with the packet fields transmitted within each packet in Table 3.6.

The low-speed serial interface follows a serial protocol with a separately transmitted clock signal. During device initialization, each of the devices in the daisy-chain gets assigned a unique serial device ID (SID). Since the devices cannot be identified individually at power-up, the SID assignment cannot be done via command accesses on the shared CMD pin but has to follow a special protocol. After powering up the devices in the daisy-chain, the SCK pin starts clocking. Initially the RST pin is held low. The SDI input of the first device in the chain is connected to the termination voltage and thus sees a logical zero (negative logic) during the entire initialization sequence. The SDO outputs of all devices generate a logical one when the RST pin goes high after its initial low-state. With the transition of the RST pin to low, the first device in the chain samples the logical zero on its SDI pin and sets its SDO output pin to the state sampled on the SDI pin with a delay of one SCK clock cycle. Thus, the logical zero initially sampled by the first device in the chain with the first low-state of the RST pin ripples along the device daisy-chain with a delay of one SCK cycle per device. Each device contains a state machine that counts the SCK clock cycles between the RST pin transitioning to low and the SDI pin of that device sampling a zero for the first time. The value of this counter delivers a unique value for each device in the chain after the memory controller sees the low-state arriving on its SRD pin and is stored as SID in each device.

After this initialization, the memory controller can apply the standard protocol for the low-speed serial interface. This standard protocol initiates write and read transactions from the memory controller to the memory devices via the shared CMD pin and receives the data for read transactions from the

Table 3.6
XDR Request Packet Types

Request Packet	Packet Description	Packet Fields	Field Description
ROWA	Activation command	OP[3..0]	Operation code that defines the command in more detail
		DELA	Delay field for dynamic request scheduling
		BA[2..0]	Bank address
		R[11..0]	Row address
		SR[1..0]	Subrow address
COL	Read/write command	OP[3..0]	Operation code that defines the command in more detail
		WRX	Selects read or write command
		DELC	Delay field for dynamic request scheduling
		BC[2..0]	Bank address
		C[9..4]	Column address
		SC[3..0]	Subcolumn address for dynamic width
COLM	Write mask command	OP[3..0]	Operation code that defines the command in more detail
		BC[2..0]	Bank address
		C[9..4]	Column address
		SC[3..0]	Subcolumn address for dynamic width
		M[3..0]	Mask data
ROWP	Precharge/refresh command	OP[3..0]	Operation code that defines the command in more detail
		POP[2..0]	Delay field for dynamic request scheduling
		BP[2..0]	Bank address
		ROP[2..0]	Specifies refresh command in more detail
		RA[7..0]	Refresh address
COLX	Remaining commands (e.g., calibration, power-down ...)	OP[3..0]	Operation code that defines the command in more detail
		XOP[3..0]	Extended operation code

memory devices via the serial SDI/SDO daisy chain. The structure of such a basic access via the low-speed interface is shown in Figure 3.9.

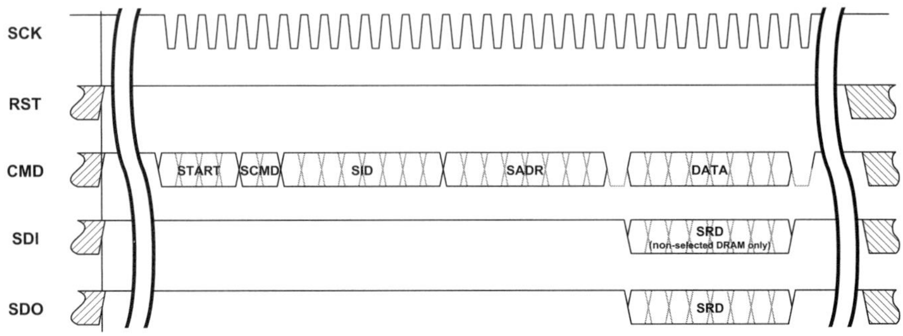

Figure 3.9 Basic access on the XDR low-speed interface.

After a START field that indicates the start of a transaction, the SCMD field defines the selected transaction type by a 2-bit opcode. The following serial byte defines the SID of the device in the chain the transaction is aimed at. Only the lower 6 bits of the byte contain the SID address. The upper 2 bits always are set to zero. The following serial byte SADR defines the address of the configuration register to be accessed with the transaction. After this address field, the serial data byte SWD is transmitted framed by two single zeros. For read transactions, the SWD byte is set to all zeros and the addressed device sends the requested SRD data byte on its SDO output. This data byte then ripples along the daisy chain toward the memory controller. The commands supported by this serial protocol are listed in Table 3.7. From a testing perspective, the serial forced read (SFR) and serial broadcast write (SBW) are important commands of the low-speed interface. These commands allow access to the DUT via the low-speed serial interface without knowing the serial ID of the device. Since in ATE test applications the devices typically are not configured in a daisy-chain manner, but are treated as single devices per test-site, the SID of the device does not have any functionality impact and just can be ignored by the ATE using the SFR and SBW commands.

3.2.4.2 Physical Layer

On its physical layer, XDR uses Rambus Signaling Levels (RSL) on unidirectional pins and Differential Rambus Signaling Levels (DRSL) on bidirectional pins. The only exceptions are the SDO pin, which uses CMOS signaling, and the differential clock inputs that use differential signaling with larger minimum swing than DRSL.

Table 3.7
XDR Serial Commands (Source: [15])

Opcode	Command	Description
00	SDW	Serial device write - device addressed by SID is written
01	SBW	Serial broadcast write - all devices are written
10	SDR	Serial device read - device addressed by SID is read
11	SFR	Serial forced read - all devices are read

RSL is the single-ended signaling implementation that already was used for RDRAM memory devices [16]. RSL is aimed at shared configurations with one pin (e.g., of the memory controller for the single request pins) driving the signal that is used by multiple receiving devices. After the last device in the sharing chain, the signal is terminated via a termination impedance to a termination voltage. Such a termination topology also is called fly-by termination. An example for a fly-by termination is shown in Figure 3.7 for the RST, CMD, and SCK pins of the low-speed serial interface. The termination impedance at the far end of the chain is defined to be 28 Ω for RSL signaling. The RSL input pins of the XDR devices are high impedance inputs. RSL signaling is based on the open drain methodology. RSL drivers only draw current if a low level (which is interpreted as logical high) is transmitted. If no data transmission takes place or a physical high level is transmitted, the signal line is pulled up to the termination voltage via the termination resistance at the end of the line. If XDR devices are operated in an ATE test environment with only a single RSL input connected to the ATE driver, a termination on the test fixture usually is not required because the wave that is reflected due to the impedance mismatch at the DUT input is fully absorbed by the 50 Ω source impedance of the ATE driver and does not impact other devices.

DRSL signaling that is used on the data pins of XDR memory devices is an enhancement of the RSL signaling that covers differential signaling. The use of differential signals improves noise immunity of the transmitted data and allows the use of low swing signals to reduce current consumption at increased data rates over RSL. Since DRSL is used for point-to-point connections only, on-die termination (ODT) is used instead of fly-by termination. The structure of a DRSL link and its signal transmission is shown in Figure 3.10.

DRSL uses bidirectional data exchange on the differential signal connection. Due to the bidirectional connection, signal transmission on a low level is fundamentally different from other unidirectional differential interfaces. The first difference is that the differential receivers deployed in a XIO front-end have to be implemented using high impedance inputs instead of a floating or center-tap termination [17]. The second difference is that

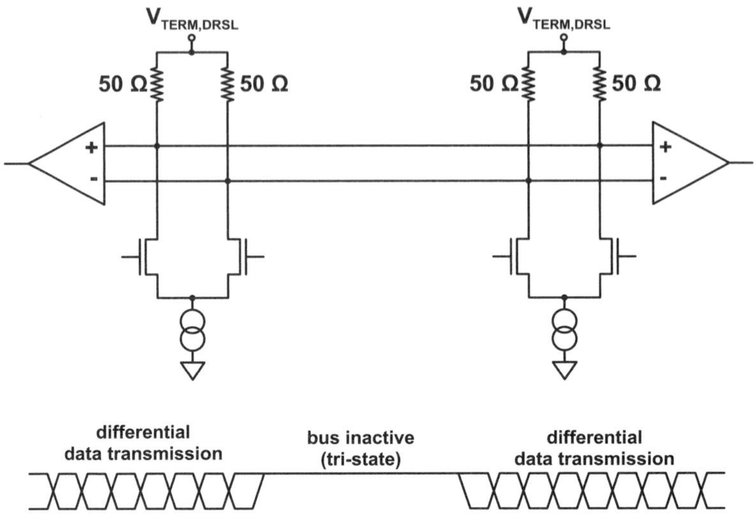

Figure 3.10 DRSL link structure and data transmission.

the termination on the receiving partner of a XIO link is provided by the termination resistances and termination voltage of the differential driver. As a consequence, the current that flows through the differential connection is not originating from the differential driver that stimulates the signal, but from the far-end termination. This is due to the fact that both link partners use the same termination voltage $V_{TERM,DRSL}$. Thus, there is no current flow on one of the differential legs and a current flow from the far-end termination into the current sink of the stimulating driver on the other differential leg. During periods of inactivity on the signal connection (e.g., for bus turnaround bubbles), the termination voltages pull both legs of the differential connection to $V_{TERM,DRSL}$.

3.2.4.3 Device Calibration

In order to guarantee the specifications required by DRSL, XDR DRAM components might support specific calibration transactions. These transactions allow the calibration of the on-chip termination impedances and the DQ drive currents to achieve optimal signal performance. The details of these calibrations usually are not disclosed by DRAM vendors. Thus, from a XDR testing perspective, it is sufficient to apply the required and supported calibration requests to calibrate impedance and current settings in the DUT.

The XDR architecture also defines calibration steps for the memory controller that apply in the same way to the ATE system in a XDR testing environment. Besides the impedance and current calibration, these steps also contain timing calibration. While current and impedance calibration are taken care of by the ATE system calibration, the timing calibration needs to be performed on a per-DUT basis because here the calibration results interact with the timing behavior of the DUT. In a XIO memory controller the timing calibration is covered by the FlexPhase feature.

FlexPhase allows the memory controller to adapt its timing on the DQ data pins separately for the drive data and the receive data. This timing adaptation compensates static skew components between the single DQ signal connections that for example are caused by PCB trace length differences, differences in package traces and bond wires between the pins and driver/receiver mismatches. Timing training in a first step adjusts the receiver sampling phase in the memory controller (RX TCAL) and in a second step the driver delay for each DQ pin of the memory controller. The timing adjustment happens in a way that after the training the receiver sampling phase and the driver delays are set to their optimum so that the DQ data in the memory controller and the memory device are sampled in the center of the respective data eye. In an ATE environment, the ATE system has to adjust its driver and compare strobe timing in a similar way as is the case for the memory controller.

Since there is no reliable data exchange possible via the DQ pins before the training, the data required for the RX TCAL is written to the XDR device via the low-speed serial interface using the write data serial load (WDSL) register. Reliable data communication on the low-speed serial interface is possible because the timing conditions required on this interface are relaxed enough to be guaranteed by design. Due to the fact that the amount of data required for the receiver phase calibration is relatively small, writing this data to the memory via the low-speed interface is feasible and does not have significant performance impact. After the data is loaded serially into the memory device, it is read at-speed via the data bus. Since the memory controller (or ATE) knows which data to expect, it can use this knowledge to adjust its timing as required to sample the single DQ signals in the center of their respective data eye.

After a successful RX TCAL calibration step, the driver timing calibration is done solely via the DQ bus because it is guaranteed that the read direction already has valid timing settings. Thus, mismatches between the data written from the memory controller to the memory device and the data read from the memory device stem from incorrect timing settings for the driver in the memory controller. During driver calibration, these timing settings then are varied to find the optimum setup.

3.2.5 Electrical Specifications

3.2.5.1 Level Specifications

XDR uses a supply voltage of 1.8 V. In contrast to true open collector or pseudo-open drain signals (see Section 3.3.4.2), the termination voltages $V_{TERM,RSL}$ for the RSL signals and $V_{TERM,DRSL}$ for the DRSL signals are not equal to the supply voltage, but are either provided via a separate pin for the RSL signals or generated at a lower level than VDD for the DRSL signals. Due to the nature of RSL and DRSL signal transmission, these termination voltages are identical to the high level of the respective signal voltages. While the minimum level swing for RSL, CMOS, and differential clock signals are specified well beyond 100 mV, the specified minimum swing of 50 mV for the DRSL signals is relatively small and requires special care for testing these device pins.

3.2.5.2 Timing Specifications

The timing specification parameters that are defined for XDR in general follow the known parameters of parallel bus interfaces. Of course, the most tight timing specifications for XDR are related to the high-speed DQ pins. Besides the skew between DQ pins, the most stringent parameters are the DQ setup and hold times for the DQ receive and the output delay variation over the consecutive bits of a single DQ pin. These parameters define the data eye width of the DQ bus with respect to the CFM clock.

3.2.6 ATE Test Requirements

The specific requirements of ATE for the test of XDR memory devices mainly stem from the new features and interface implementations XDR deploys. The first feature to be mentioned here is the timing calibration of the DRSL interface. To enable the economical test of XDR memory devices, the ATE system has to provide efficient means to support the timing calibration required by XDR. The ATE features used to perform timing calibration must provide a fast identification of the individual per-pin data eye timing and a per-pin readjustment of the ATE timing to the requirements of the DUT. Moreover, these timing features not only have to be fast if they are executed only on a few pins, but also in massive multi-site implementations where timing calibration needs to be done for hundreds of device pins.

Another requirement for the ATE is set by the bidirectional implementation of the differential DRSL signals. The inactive DRSL signals that pull both legs of the differential signal to the termination voltage are somehow untypical for conventional differential interfaces. The differential pin electronics of ATE systems might not support this specific signal condition, especially if true differential drivers are used. Thus, special care has to be taken that the ATE channels used for testing XDR have such a capability for their differential pins.

3.2.7 Test Support

XDR in its basic definition does not contain any dedicated features to support device testing besides the broadcast serial write and serial forced read instructions for the low-speed serial interface that simplify device initialization significantly. Since this basic definition specifies mandatory features only, this does not mean that specific XDR implementations do not have such test support features. In fact, XDR is prepared for additional manufacturer-specific features by a mandatory set of configuration registers. The content and functionality controlled by these registers are up to the specific XDR design. An example for the freedom that XIO/XDR gives to design engineers with regard to test features can be found in [17]. In the XIO design implementation described there, features like serial loopback paths, internal signal probing, and, of course, scan chains are used.

3.2.8 Test Challenges

Some of the test challenges with XDR were already mentioned in Section 3.2.6. We do not want to describe further the ones that are mainly on the ATE feature side. Here we want to mention two test challenges that are driving the parametric requirements for the ATE system to their limits. The first of these is the low voltage swing at which the DRSL pins of XDR operate. These swings that can go down to 50 mV are a challenge for drive and receive operations with the more challenging conditions for the ATE receivers that have to be able to reliably detect a difference of 25 mV to a fixed threshold voltage. The fact that XDR uses differential signals certainly helps for functional testing. However, if the differential signal implementation is analyzed on a low level, as in Section 3.2.4.2, it becomes obvious that it still has some relationships to single-ended implementations that might influence parametric measurements.

The other challenge is the accuracy required to do parametric timing measurements. Here especially, the setup/hold times defined for the DQ bus drive the requirements into limits that rarely can be fulfilled by ATE systems. Thus, parametric testing and characterization of XDR often require focused ATE calibration support (see Section 9.3).

3.3 GDDR SDRAM

3.3.1 Application Areas

As the name "graphics double data rate synchronous dynamic random access memory" (GDDR SDRAM) suggests, GDDR memory devices are used as high-speed memory for graphics processing units (GPU) to store texture and frame buffer information. Due to their tight linkage to GPUs, GDDR memory devices in general are used in a well-controlled system environment close to the GPU as, for example, on PC graphics cards or game console motherboards. Usually both of the link partners, GPU as well as GDDR memory devices, are directly soldered on the same piece of PCB avoiding connections that compromise signal integrity. This allows for substantially higher data rates for the GDDR interfaces than for the corresponding double data rate (DDR) main memory generations which are connected via long traces and socket connectors to their memory controllers in the chipset or the CPU.

3.3.2 GDDR Fundamentals

Like the DDR main memory standards, GDDR memory device standards are set by the Joint Electron Device Engineering Council (JEDEC). The most recent generation of the GDDR5 standard that was released by JEDEC in December 2009 is GDDR5 [18]. GDDR memories traditionally are ahead of main memory DDR devices with their I/O data rate, and they also often define feature subsets that are candidates for integration into DDR main memory standards. The interesting point about GDDR is that although the data rate goes up into the multigigabit data range (e.g., 6 to 7 Gbps for GDDR5), it still uses single-ended semiduplex data transmission. This is a fundamental difference to the interfaces discussed before that all use differential signaling and predominantly dual-simplex data transmission. In Table 3.8 a short comparison of the key characteristics between the GDDR generations since GDDR3 is shown. From this table it becomes obvious that keeping the single-ended semiduplex data transmission while increasing the data rate within the GDDR family is achieved by introducing device training steps and modifications of the device timing architecture to increase tolerance for timing parameter mismatches and their variations [19]. Data rate limitations due to the increased noise level caused by the relatively large signal swings required for single-ended signals are addressed by innovative features like data bus inversion (DBI) or address bus inversion (ABI). For the highest speed generation of GDDR, a cyclic redundancy check (CRC) based error code detection was introduced to address the increased likelihood for errors.

Table 3.8
GDDR Comparison

	GDDR3	GDDR4	GDDR5
Release year	2003	2006	2009
Data rate	1 .. 2.6 Gbps	2.2 .. 2.8 Gbps	3.2 .. 7.0 Gbps
Timing architecture	At-cycle source sync.	At-cycle source sync.	Forwarded clock
Device training	No	Data	Address, clock, and data
Bus inversion	No	Data	Data and address
Error detection	No	No	Yes

In the following sections we will focus on GDDR5 as the latest representative of the GDDR family that runs at the highest data rate and also poses new test requirements on ATE systems due to new functionality and clocking architecture changes that are unique for memory devices.

3.3.3 GDDR5 Details

3.3.3.1 GDDR5 Timing Architecture

The interface timing architecture of GDDR5 consists of three initially independent clock domains. The clock sources for all of these clock domains are not directly provided by a central system clock but are located in the memory controller. Thus, GDDR5 like the previous GDDR generations makes use of source-synchronous interfacing. However, in contrast to earlier GDDR generations, the standard GDDR5 operation mode no longer uses an at-cycle source-synchronous timing architecture for high-speed data transmission. Instead, a forwarded clock architecture is used. For lower-speed operation and better compatibility to earlier GDDR generations, GDDR5 offers the option to operate the devices in a classical at-cycle source-synchronous mode which will not be discussed in this section.

GDDR5 devices follow the traditional DRAM signal grouping into a command/control pin group, an address pin group, and a data pin group. An overview of the GDDR5 signal pins, their type, the clock domain to which they belong, and the groups with which they are associated is shown in Table 3.9.

Clock Pin Group In addition to the differential reference clock signal CK that was used in previous GDDR generations, GDDR5 introduces two new differential clocks: WCK01 and WCK23. These additional clock signals serve as references for the data pin group during write and read operations. The

Table 3.9
GDDR5 Pin List

Signal Pins	Type	Domain	Group
CK, CK#, WCK01, WCK01#, WCK23, WCK23#	I	-	Clock
RESET#, MF, SEN	I	-	Static
CKE#, CS#, RAS#, CAS#, WE#	I	CK	Command
A8/A7, A9/A1, A10/A0, A11/A6, A12/RFU, BA0/A2			
BA1/A5, BA2/A4, BA3/A3, ABI#	I	CK	Address
DQ0..DQ7, DBI#0, EDC0	IO	WCK01	Data
DQ8..DQ15, DBI#1, EDC1	IO	WCK01	Data
DQ16..DQ23, DBI#2, EDC2	IO	WCK23	Data
DQ24..DQ31, DBI#3, EDC3	IO	WCK23	Data

WCK signals run at a frequency that matches the specified device data rate in double data rate mode. The CK clock runs at half of this frequency. For the WCK signals, the JEDEC standard foresees an optional PLL in GDDR5 memory devices. This PLL predominantly will be required for the higher speed grades that are supported by GDDR5.

Command Pin Group For GDDR5, the data rate on the command pins is one quarter of the specified device data rate. This means that the command pin group represents the slowest group of pins of the device (besides the DC pin group that contains power supply and reference voltage pins). The command pins are the only group of pins for which no training step is specified.

Address Pin Group The GDDR5 address bus follows a time multiplexed data transmission concept. Over the nine address signals, 13 address bits and 4 bank bits are transmitted at half the specified device data rate.

Data Pin Group The GDDR5 data pin group that runs at the specified device data rate is divided into four DQ octets of eight bidirectional signal pins plus the two sideband signal pins DBI and EDC per DQ octet. While the DBI signals per DQ octet were already used for previous GDDR generations, the new EDC signals per DQ octet were introduced with GDDR5. These EDC pins serve as signals that transmit a CRC checksum per DQ burst from the memory device to the memory controller for read and write operations during normal operation.

The GDDR5 pin groups are organized along the three clock domains that are defined by the CK, WCK01, and WCK23 differential clock signals.

The CK signal represents the main reference clock and defines the domain command and address pins are referenced to. While the command data transmission follows a single data rate scheme with command bits sampled by the rising CK edges, addresses are transmitted in double data rate manner with their bits sampled at rising and falling CK clock edges. The WCK01 and WCK23 signals serve as clock references for the DQ data pins, the DBI pins and the EDC pins. Each of the two WCK signals is the reference for two of the DQ-octets and its associated DBI and EDC pins. The three clock domains of a GDDR5 device are synchronized by a dedicated training step during device training that will be described later. In order to facilitate this synchronization, the clocking architecture of GDDR5 allots for a circuitry that provides a feedback loop for CK to WCK phase comparison. A schematic block diagram of the GDDR5 timing architecture is shown in Figure 3.11. The figure does not represent an implementation example but a schematic view of signal-to-clock dependencies only.

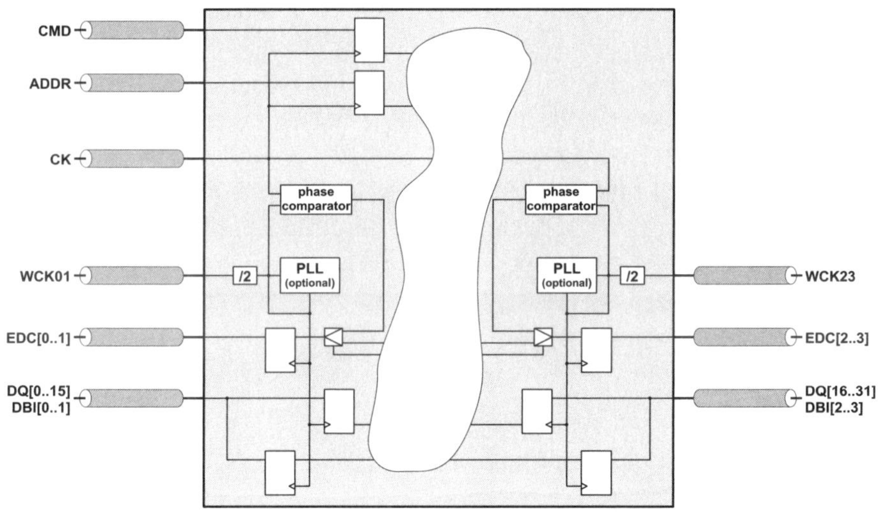

Figure 3.11 GDDR5 timing architecture. (From: [18]. ©2009 JEDEC. Reproduced with permission.)

3.3.3.2 GDDR5 Specific Functionality

JEDEC specifies 512 Mb, 1 Gb, and 2 Gb densities for GDDR5 devices. All devices are configurable during power-up to a x16 organization activating 16 of the 32 DQ signals or to a x32 organization that activates all 32 DQ signals. The activation of the x16 organization obviously implies an address space doubling compared to the x32 organization in order to be able to address all cells of the memory core. The 512-Mb devices support eight banks while 1- and 2-Gb devices support 16 banks. Like the previous GDDR generations and DDR3, GDDR5 supports a mirror functionality that is controlled by a dedicated device pin. If the mirror function is enabled, the signal-to-ball assignment of the device is reshuffled in a way that the ball-out is mirrored compared to a disabled mirror function. The benefit of the mirror function is that GDDR5 memory devices can be mounted on the front side and the back side of a printed circuit board in a way that the footprints for devices match not only geometrically but also from a signal point of view on both sides of the PCB. Thus, certain connections between the GPU and both memory devices like the ones for the command and address pins can be shared without generating additional routing overhead to achieve trace matching of the connections to the two different devices as shown in Figure 3.12.

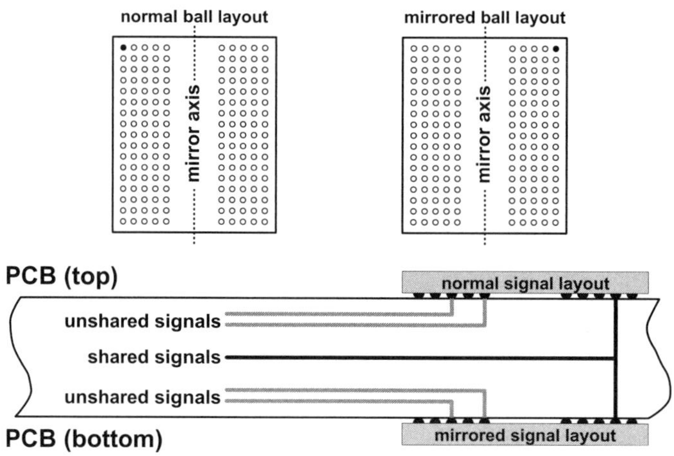

Figure 3.12 Mirrored device mounting to PCB. (From: [18]. ©2009 JEDEC. Reproduced with permission.)

Due to the congruent mounting of the devices to the PCB from both board sides, the mirror function is called the clamshell mode. From a GPU point of view, such a configuration of two memory devices with shared command and address pins looks the same as one single device with doubled capacity. This doubled capacity is represented either by a doubled address space if the devices are configured in x16 mode or by a doubled data word width if the devices are configured in x32 mode. In order to allow the command/address sharing usually used in clamshell mode, GDDR5 offers a dynamic adaption of the on-die-termination (ODT) values by programming the appropriate mode register (MR) bits of the device. By doubling the ODT values of the memory device for shared pins compared to non-shared operation, the GPU pins will see the same termination impedance for a dual-device shared configuration as for a single-device non-shared configuration. Separate MR impedance control for the WCK clock pins, the data pins, and the command/address pin group allows for the implementation of various sharing topologies.

As will be described in Section 3.3.4.2, GDDR5 uses its supply voltage VDDQ as pin termination voltage if ODT is turned on. As a consequence, each time a low level is applied to a terminated GDDR5 device pin, a current is flowing from the VDDQ supply via the termination impedance into the connecting trace. Thus, the amount of pins driven by a low level can contribute significantly to the dynamic power consumption and supply noise of a device. This is especially true if large signal swings are used, as in the case of single-ended interfaces.

With GDDR3 and GDDR4, the DBI technique was applied, which aimed to minimize the power consumption and supply noise caused by this effect on the data bus. DBI for GDDR3 and GDDR4 supported two implementation flavors, DBIdc and DBIac. Since DBIac was dropped for GDDR5, we will restrict the discussion to DBIdc and automatically refer to DBIdc if the term DBI is used.

With GDDR5, the DBI (DBIdc only) concept was extended to the address bus as address bus inversion (ABI). If bus inversion is enabled, the amount of address or data bits in low state on the respective bus is analyzed. This analysis happens over the complete bus width for the addresses and over single-byte sections for the data bus. If there are more signals in low state than in high state within an analyzed bus section, the logical levels of the pins in that section are inverted. This inversion is signalized on the DBI pin associated to a data byte for that byte or the ABI pin for the address bus.

Such a bus inversion implementation ensures that worst-case maximum half of the bits of a bus are drawing current via the termination impedances due to the applied low level while the other half is drawing no or only a minimal amount of current. The bus inversion for driving and receiving operations of course needs to work slightly differently. For receiving operations, depending on the status of the DBI/ABI pins, the received data needs to be inverted. For drive operations, the ABI/DBI signals are set according to the data/address bit analysis. Thus, both partners involved in a GDDR5 communication link, the memory device as well as the memory controller, have to support bus inversion if this feature is going to be used in the communication.

Another functional feature of GDDR5 to be mentioned here is the use of CRC codes (see Appendix D) for error detection during data transmission. The CRC polynomial used by GDDR5 is the CRC-8-ATM polynomial $x^8 + x^2 + x + 1$. The CRC code is calculated based on the 64 bits of one 8-bit burst per DQ byte plus the 8 burst bits of the associated DBI data. Thus, 72 bits are condensed via the CRC polynomial into one 8-bit CRC code as shown in Figure 3.13.

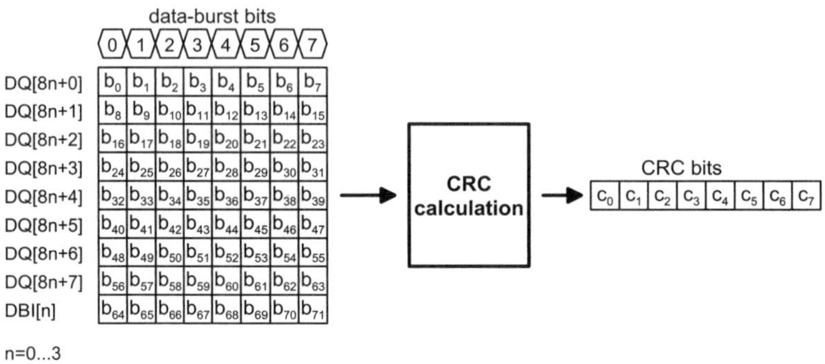

Figure 3.13 GDDR5 CRC code generation. (From: [18]. ©2009 JEDEC. Reproduced with permission.)

The CRC code data flow by GDDR5 is always from the memory device via the EDC pins to the memory controller. As long as no CRC codes are transmitted on the EDC pins, an EDC Hold Pattern consisting of 4 bits is transmitted repeatedly. This pattern is defined by the user with an appropriate mode-register setting. Thus, for normal memory operation, the EDC pins are unidirectional memory output pins.

For write operations, the memory device calculates the CRC code according to the data received on the DQ and DBI pins and returns this calculated CRC code to the memory controller. The memory controller compares the CRC code it receives after the write operation to the CRC code it has calculated on the write data it sent to the memory device earlier and thus can identify data transmission errors if both CRC codes do not match. For read operations, the memory device sends the locally calculated CRC data with some latency to the read data and the memory controller compares the received CRC code to the CRC code that it calculated based on the received data. Figure 3.14 shows a schematic overview of the data flow for read and write operations with enabled error detection.

In order to support the training of the DQ, EDC, and DBI pins that is described later in more detail, GDDR5 devices contain a FIFO that is written either via the address pins or via the DQ pins depending on the command used to send data to the FIFO. The data stored in the FIFO can be transmitted via the DQ, EDC, and DBI pins. With the ability to fill the FIFO via the address pins that run at only half of the DQ data rate, it is possible to provide known data on the pins that run at the highest data rate to adjust to their timing.

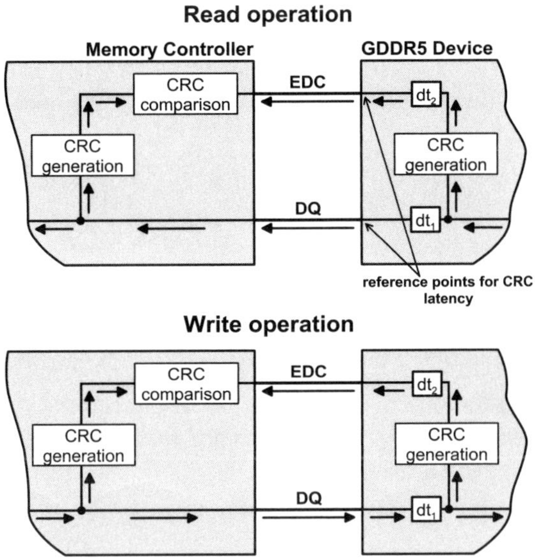

Figure 3.14 CRC data flow for read and write operations.

3.3.4 GDDR5 Protocol

Since the basic data access command sequences of GDDR5 devices are the same as for other synchronous DRAM components, we will not cover these in detail in this book. If the reader is interested in more detail of how DRAM devices are accessed and what boundary conditions need to be taken into account for these accesses, this can be found in [20]. In the following we will cover the GDDR5 specifics on the data link layer and the physical layer, which are the protocol layers with which test engineers primarily have to deal.

3.3.4.1 Data Link Layer

A typical GDDR5 access with address and data transfer is shown in Figure 3.15. In this figure one can see how a data transfer is initiated by an appropriate command that is applied on the command pins. If an address or any other data that needs to be supplied via the address pins is required with the command, the first slice of time multiplexed address data is supplied and latched into the device together with the command on the rising edge of the CK clock signal. The second slice of the time multiplexed address data is latched to the device with the following falling edge of the CK clock. Potential data is supplied or received with certain latency to the rising CK edge that latched the command. This latency is set by a user selected mode register (MR) setting.

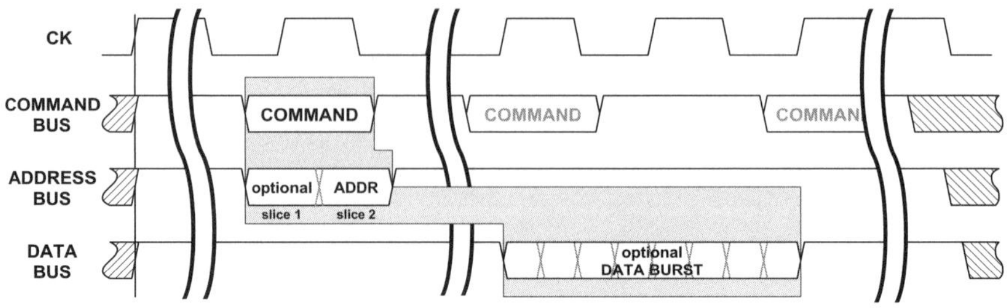

Figure 3.15 Basic GDDR5 access.

Another important detail to mention here is that for accesses that require DQ data in the write direction, the DQ bits are center aligned to the associated WCK clock signal. For accesses that expect DQ data in the read direction, the timing relation between DQ bits and the appropriate WCK clock signal is edge aligned. The same is true for the bits transferred on the DBI pins which are transmitted together with the associated data on the DQ pins. Potential CRC data is transmitted with a user selected latency to the start of a data burst that is set in the mode register.

As Figure 3.15 shows, GDDR5 operates with 8-bit bursts on its DQ signals. In contrast to previous GDDR generations, GDDR5 does not foresee an optional subburst addressing anymore. Thus, addresses that are supplied to GDDR5 devices address the memory cells on an 8-bit address grid with the LSB of the external address data supplied being the LSB+3 of the physical address space within the device.

GDDR5 CRC Operation

As already described in Section 3.3.3, the CRC code data flow for GDDR5 is always directed from the memory device to the memory controller. From the description above, it is obvious that the base data required to calculate the CRC code in the memory device is available with different latencies relative to the start of the operation for read and write operations. This latency difference is propagated to the EDC pins of the device and thus also is visible on the EDC pins for read and write operations as shown in Figure 3.16. The CRC latencies on the device interface are measured relative to the CK edge that aligns to the first DQ bit of a data burst.

3.3.4.2 Physical Layer

On the physical layer, GDDR5 uses electrical signaling according to the JEDEC 1.5V Pseudo Open Drain I/O (POD15) standard [21]. All device

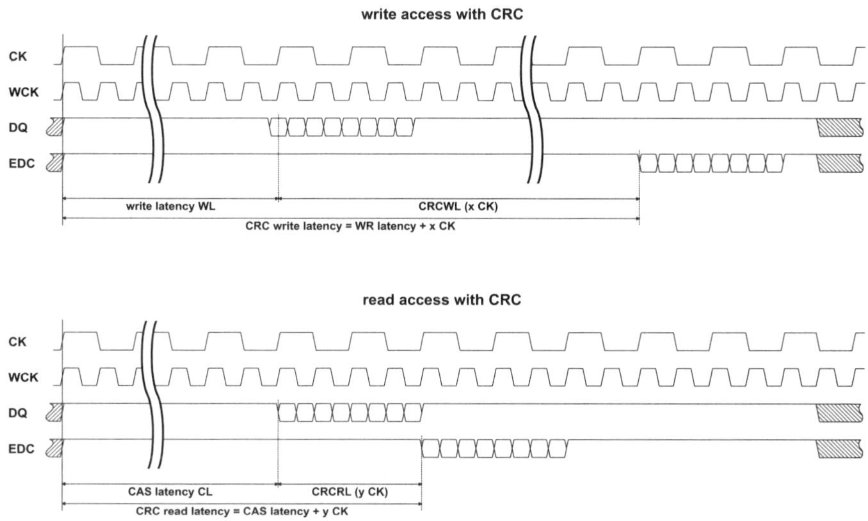

Figure 3.16 GDDR5 CRC timing for read and write accesses. (From: [18]. ©2009 JEDEC. Reproduced with permission.)

signals except the differential clocks CK, WCK01, and WCK23 are single-ended. The POD15 standard defines a flexible termination scheme that is able to maintain an equivalent 60 Ω termination load to the VDDQ supply on shared input pins with one, two, or four inputs operated in parallel as shown in Figure 3.17. The different receiver impedances required to achieve the effective 60 Ω impedance for the various configurations are programmed by the device according to user controllable register settings. GDDR5 allows separate impedance control for the Address/Command pin group, the Data pin group, and the WCK pins. Drive pins are defined with a 40 Ω impedance in the POD15 standard.

In order to allow the device to calibrate its on-die termination (ODT) impedance, the POD15 standard requires a device external reference resistor of 240 Ω to GND that is connected to the ZQ pin. GDDR5 deviates from the POD15 standard regarding this detail and requires a 120 Ω resistor to be connected between ZQ and GND. The threshold voltages used by POD15 I/O pins are derived from a reference voltage VREF that usually is provided from an external voltage source. With VREFD and VREFC, GDDR5 requires two of these external reference voltages. The threshold voltage for the data pin groups (DQ, DBI, and EDC pins) is derived from VREFD and the threshold voltage for the Address/Command pin group is derived from VREFC. GDDR5 also offers the option to provide a device internally generated VREFD voltage.

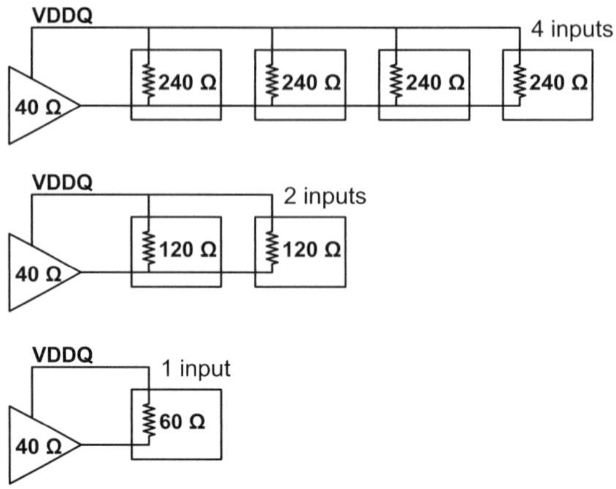

Figure 3.17 POD15 flexible termination scheme. (From: [21]. ©2006 JEDEC. Reproduced with permission.)

As mentioned already above, the differential WCK pins can be terminated using the on-die termination of the device. For the differential CK clock, this option is not available. Thus, external termination resistors with a value of 60 Ω each are required between the two CK signal legs and VDDQ. If the ODT option for the WCK pins is not used, also an external cross termination is required for the differential WCK pairs.

Although the GDDR5 training does not necessarily belong to the physical layer from a configuration and execution point of view, we cover it in this section because it has significant impact on the signal timing of the physical interface and requires timing control in the memory controller and the ATE on the lowest level that is accessible.

GDDR5 Training

In order to align the timing of the memory controller to the timing conditions a GDDR5 device expects, a training sequence as shown in Figure 3.18 is executed as a first step when powering up a GDDR5 device.

The only high-speed signal for which the timing is kept constant during all of the training steps is the differential reference clock CK since this clock serves as a timing reference for all the other signals in a GDDR5 device. After completion of all training steps, the timing relationship between the signals of the GDDR5 device is shown in Figure 3.19. It should be noted that this timing diagram shows the device internal timing relationship at the relevant sampling points inside the device.

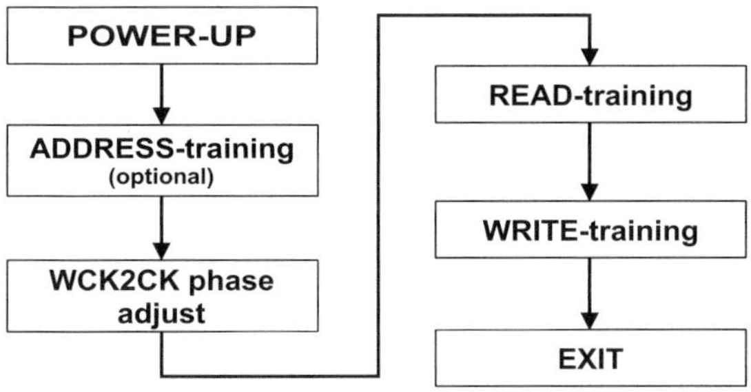

Figure 3.18 GDDR5 training sequence.

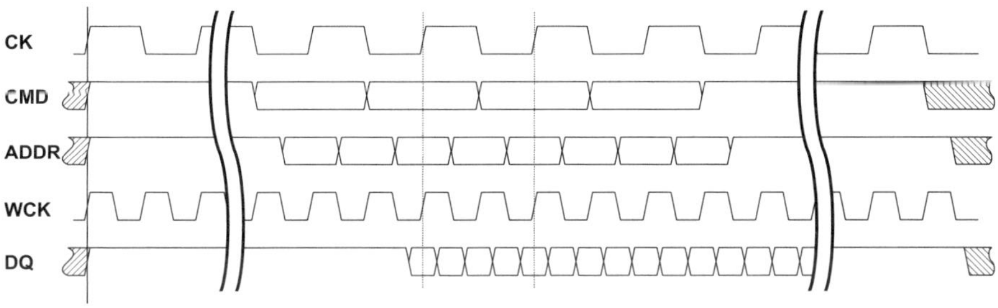

Figure 3.19 GDDR5 timing after training. (From: [18]. ©2009 JEDEC. Reproduced with permission.)

Address Training Address training is an optional step during GDDR5 training. This training step aligns the center of the address pin data eyes, which are generated by the memory controller/ATE, to the fixed CK-dependent latch timing that is valid within the GDDR5 device. For the address training, a feedback loop indicating the status of the training to the memory controller/ATE is created through selected DQ pins that transmit the data sampled on the address pins of the GDDR5 device back to the memory controller/ATE. The data rate on the DQ pins during the address training can be controlled by the memory controller/ATE through the rate at which new addresses are applied. This means the data transmitted by the DQ pins can be sampled safely by the memory controller/ATE even though no read training has taken place when the address training occurs.

Besides the training of the address pins, address training also contains a substep which adjusts the ABI timing to the timing location as required by the memory device. ABI training is performed after the training on the other address pins is completed. Instead of directly comparing to the transmitted address bits on the DQ pins, the memory controller/ATE tests for the inverted address data when performing ABI training.

Clock Training During WCK2CK phase adjustment, the rising edges of the WCK signals are aligned to the rising edges of the CK signal. GDDR5 memory devices contain phase detection circuitry with the functionality represented by the circuitry shown in Figure 3.20. The frequency for each of the two WCK clocks is divided by two in this circuitry and then sampled with the rising edges of the CK clock. The logical result of this sampling operation is transferred via the EDC pins to the memory controller/ATE.

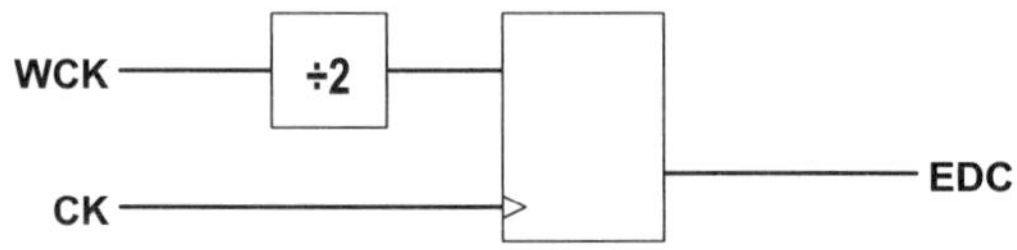

Figure 3.20 GDDR5 clock training phase comparator functionality. (From: [18]. ©2009 JEDEC. Reproduced with permission.)

Since the nominal frequency of the WCK clocks is twice the CK frequency, the divided WCK clocks and the CK clock deliver signals of the same frequency to both inputs of the sampling flip-flop. This means that the logical level captured by the flip-flop can be used as an indicator for the timing relationship between the rising edges of the respective WCK clock and the CK clock. As long as the output of the phase measurement flip-flop is a stable high or low, the rising edges of WCK and CK are far enough apart from each other to not cause setup/hold time violations of the sampling flip-flop. In the case where the WCK and CK edges are aligned to each other, the setup/hold time violations on the inputs of the flip-flop will cause instable data on its output, and thus on the associated EDC pin of the device.

The two possible states on the EDC pins are referred to as "early" or "late" states as shown in Figure 3.21. An early state is given if the divided WCK is already in the high state when the CK rising edge occurs. The device will indicate such an early state with a high level on the respective EDC output. The late state, however, is characterized by the rising edge of the divided WCK clock being positioned after the rising edge of the CK clock. Thus, a low state is captured into the phase identification flip-flop and transmitted via the respective EDC pin.

This behavior means the alignment point of the rising WCK and CK edges can be identified by simply moving the WCK edges versus the CK edges and scanning for a state change on the EDC pin(s).

Read Training During read training, the timing of the DQ sampling circuit in the memory controller is adjusted to match the center of the data eyes received from the memory device as shown in Figure 3.22. In order to perform the

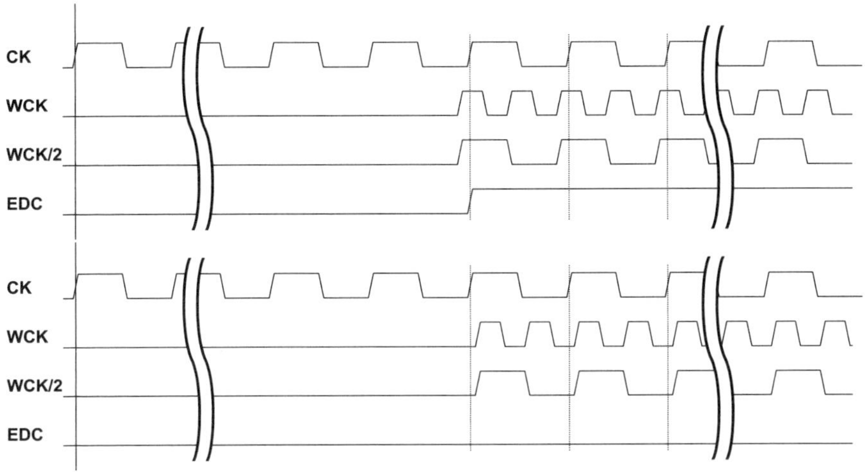

Figure 3.21 Early (top waveform) and late (bottom waveform) GDDR5 WCK to CK alignment states before clock training (reprinted with permission from [22]).

read training for GDDR5, the data used for the training is written to FIFOs in the device via the already trained address pins. A special load FIFO (LDFF) command is used to load this training data into the FIFOs. A read training command (RDTR) is then executed and one data burst is sent from the FIFOs to the DQ pins of the memory device. Since the memory controller knows which data to expect, it can align its sampling point per DQ pin by repeatedly executing read training commands and analyzing the data burst responses from the memory device.

Write Training In theory, write training could be done similarly to the read training by adjusting the DQ sampling point timing inside the memory device instead of the memory controller. However, in order to keep the logical complexity inside the memory device to a minimum, no timing adjustments are done by the GDDR5 device. This means the DQ transmitter timing in the memory controller must be adjusted in such a way that the memory device will receive the DQ data eyes at their optimum fixed sampling points per DQ, as shown in Figure 3.23.

As with the read training, no data transfers to and from the memory core take place during the write training. Special write training commands (WRTR) direct the memory to write the data it receives on its DQ inputs into the FIFOs that are also used for the read training. From there, the memory controller reads back the data the memory device sampled using the read training command. With the knowledge of the data that was sent to the

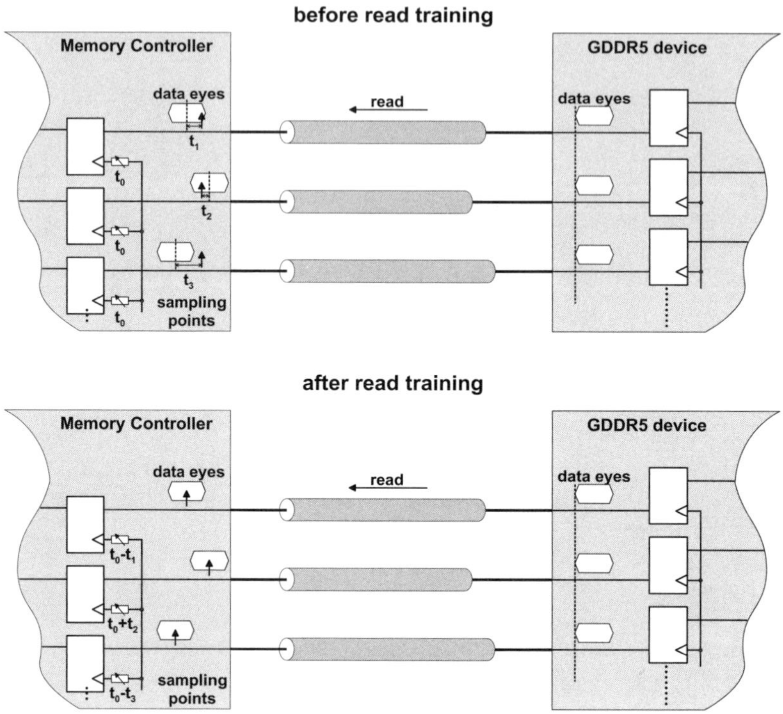

Figure 3.22 GDDR5 read training functionality (reprinted with permission from [22]).

memory device, and the results that the read training command delivers, the memory controller can adjust its transmitter timing for the single DQ pins until it reads back exactly the data that it originally sent into the FIFOs of the memory device before.

3.3.5 Electrical Specifications

3.3.5.1 Level Specifications

As already described, the level specifications for GDDR5 follow the POD15 standard. Since POD15 defines the electrical specifications for single-ended signals, the level swings for GDDR5 are relatively large compared to differential interfaces that run at similar data rates. The larger swings are required to provide enough noise margin around the threshold voltages used in the receivers of a GDDR5 interface. Due to the fact that POD15 mimics an open drain electrical interface, three drive voltage levels are required to operate a POD15 interface correctly. Besides the low and high input voltages that are usually set symmetrically around the respective reference voltage

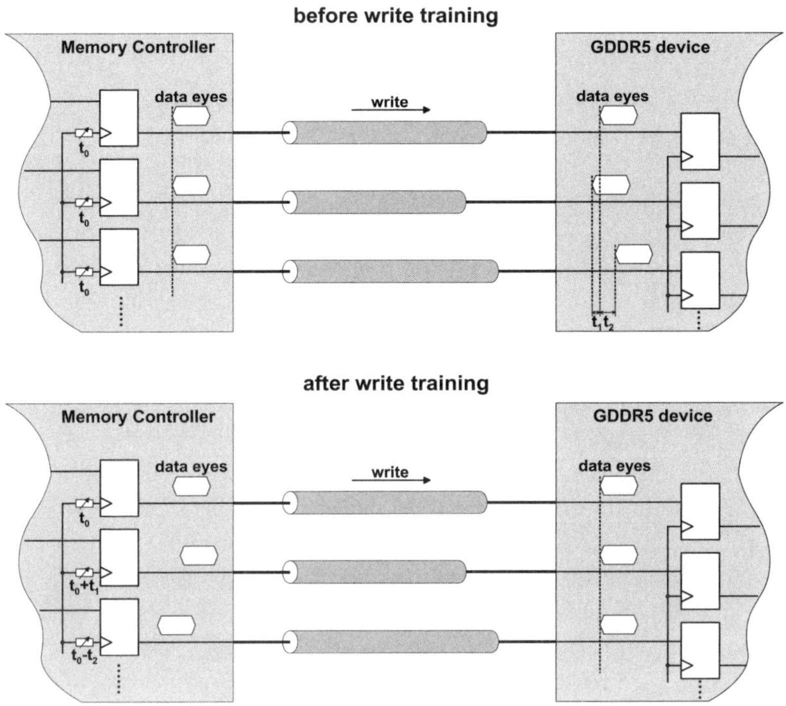

Figure 3.23 GDDR5 write training functionality.

(VREFC or VREFD), a termination voltage with the value of VDDQ is required that needs to be activated when the respective I/O pin is operated in receive mode.

3.3.5.2 Timing Specifications

As for other high-speed digital interfaces, the eye opening of the data signals is one of the most important timing related parameters that needs to be guaranteed for GDDR5 devices. Compared to previous GDDR generations and DDR memories up to DDR3, setup and hold times between the data signals (DQ) and their timing reference (WCK) have lower priority because the read and write training adjusts the timing in a way that optimum setup and hold times are achieved. Although GDDR5 can contain a PLL for its WCK input, no dedicated jitter specifications are available for GDDR5 as for the other standards discussed so far that contain PLLs as key building blocks. Since the application range and variety of components that communicate with GDDR5 devices is limited and a qualification of certain GPU/GDDR5 combinations can be done on a case-by-case basis, this is a feasible approach to ensure interoperability between components without a formal specification.

3.3.6 ATE Test Requirements

Besides the support for the high-speed device data rate, the primary requirement for testing GDDR5 is that the ATE is able to perform the device training steps because a successful training is a prerequisite for any other test. From a functional point of view, the ATE needs to support the generation of the CRC code patterns that are based on the DQ data used for the test. In order to be able to test the ABI and DBI functionality, the ATE also needs to support the generation of APG-based test patterns that stimulate and compare toggling ABI and DBI pin values.

For parametric measurements the timing accuracy of the ATE system is an important factor. In contrast to other memory interfaces the relative timing between ATE pins is of less importance because setup/hold times are only specified for the lower-speed command and address signals. Also the allowed skew for the DQ signals within the two double bytes does not require the ultimate skew accuracy of a few picoseconds between ATE pins, which, for example, is the case for accurate setup and hold time measurements between multigigabit signals.

The timing accuracy within a single pin like the timing delay line linearity is of much higher importance in order to do accurate data eye measurements. As mentioned above, although the GDDR5 specification does not contain any parametric jitter specification, it seems to be important to have the capability to do jitter measurements on the data interface of GDDR5 because jitter on the DQ and WCK pins will result in a reduction of the setup/hold time margins of the device internal data sampling flip-flop. In order to perform the required jitter tolerance and jitter transfer tests, jitter injection capabilities are required for the ATE channels serving the data pin group and the WCK pins and jitter measurement capabilities are required for the channels serving the data pin group.

3.3.7 Test Support

GDDR5 does not provide any dedicated features that are specified and implemented predominantly to support the component test or characterization besides a non-IEEE 1149.1 compatible boundary scan implementation which is mainly targeted to support interconnection tests for mounted GDDR5 components. GDDR5 devices, however, can be configured into a test mode via a mode-register setting. The device functionality that is enabled in this test mode is not further specified and thus is vendor-specific and varies from manufacturer to manufacturer.

Besides this test mode, some of the functionality required by the standard can be reused for test and characterization purposes. One component of the memory that is a perfect fit for such a reuse is the FIFO that is used for read and write training. The availability of this FIFO provides an easy way to access the memory with a user-defined pattern that is suited to do high-speed I/O characterization without the need to take care of memory refresh cycles during, for example, time consuming bit error rate measurements on the I/O cells for characterization purposes.

3.3.8 Test Challenges

The test challenges for GDDR5 cannot be simplified to just consider the high data rate although it is an important ATE hardware related factor. This is especially true if it comes to cost-efficient production solutions with maximized site-count. However, application-relevant challenges like the complex GDDR training sequence require appropriate tooling and ATE capabilities to adapt the timing of the test system to the device on a per-pin basis [22, 23].

Another application-related challenge is the variable read and write latency for the GDDR5 CRC generation. This variable latency might not be a significant issue if test patterns are derived from simulation. However, for patterns that are generated algorithmically, as is the case for memory patterns, this can be a challenge if the pattern generation hardware or the pattern compiler does not support the specific GDDR5 case. The last challenge to be discussed here is the impedance mismatch between the ATE pin electronics that usually are designed with a 50 Ω impedance and the GDDR5 device impedance which is 60 Ω for device inputs and 40 Ω for device outputs.

3.3.8.1 High-Speed Data Rate

Most of the test challenges presented by the high data rates that GDDR5 deploys do not differ from the other high-speed applications discussed so far. They are mainly created by bandwidth limitations of the signal paths between the ATE pin electronics and the DUT and the effects of these limitations on signal integrity. The fact that GDDR5—as XDR—transfers data in a semiduplex manner, of course, raises the same challenges that already were discussed in Section 3.2.8.

The difference with XDR, however, is that GDDR5 uses single-ended signaling. The large swings of this single-ended interface can have a significant impact on the requirements for the complete signal path that consists of the test fixture, the ATE internal signal routing, and the pin electronic front-end. In particular, critical items in this regard are, for example, crosstalk, power distribution, drive and threshold voltage offset accuracies, and the bandwidth of all signal path components.

3.3.8.2 Training Test Challenge

If the different training steps are analyzed with respect to their general requirements for an ATE-based test implementation, it becomes obvious very quickly that two timing alignment methodologies are required to address all GDDR5 training steps. These methodologies are eye center alignment and transition alignment, which are described in more detail in the following. The challenge with both of these training steps is that the ATE needs to provide an efficient way to adapt its timing individually per pin to the requirements of the device. The CDR functionality many ATE systems provide with their high-speed pin electronics only solves the challenge raised by the read training in this regard. The write training is not covered by this functionality because a CDR only works for device output pins, whereas the write training needs to adjust the timing on device input pins.

Of course, a potential CDR also is not usable for the WCK2CK phase adjust training. Here the rising edges of two signals need to be aligned relative to each other based on the feedback of the phase detection circuitry in the GDDR5 memory device. In order to achieve an accurate transition alignment between the WCK and CK signals, simply scanning for a transition of the state of the EDC feedback pin while moving the WCK timing might be sufficient for production-oriented functional testing, but not for device characterization. The reason for this is that in a real circuit, there is not one distinct time at which the early/late state transition on the EDC pin will occur. Instead there is a band or region in which that transition will occur with a certain probability for each distinct timing point within that band as shown in Figure 3.24.

This behavior is caused primarily by the jitter that is present on the CK clock and the WCK clocks. Because the jitter on these signals is typically uncorrelated, the early/late result on the EDC output might vary even if the timing settings for CK and WCK stay constant. The jitter on the clock signals and the relative timing variations between CK and WCK that are caused by the jitter follow statistical processes. The early/late measurement is influenced by this statistical behavior and needs to be analyzed by statistical means to get the most accurate training results.

Another contributor to the EDC early/late switching uncertainty is the setup and hold time behavior of the sampling flip-flop in the phase measurement circuitry. Since the goal for WCK2CK phase adjustment is

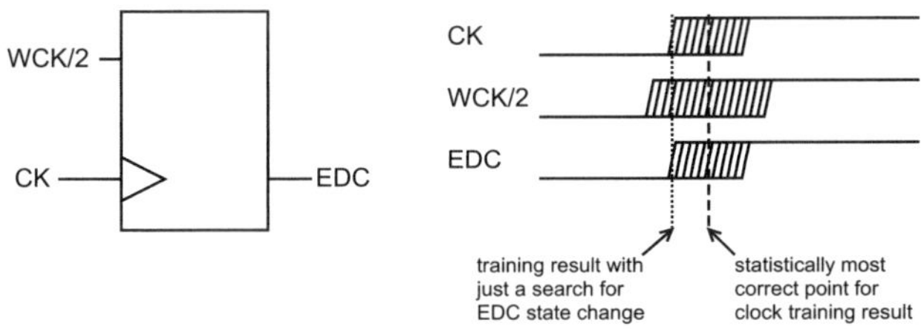

Figure 3.24 GDDR5 transition alignment uncertainty (reprinted with permission from [22]).

to align the transitions of the CK and WCK (and thus the divided WCK) clocks, the transition alignment point represents the worst-case setup/hold time situation for the sampling flip-flop. A statistical measurement approach for the EDC early/late indication will also cover the unpredictable behavior of the sampling flip-flop while it is operated in the setup/hold time violation band.

3.3.8.3 GDDR5 CRC Test Challenge

Testing the CRC functionality for GDDR5 represents a challenge because memory test patterns usually are generated algorithmically. While the pure generation of the CRC codes can be achieved through linear feedback shift registers on classical hardware-based APGs or XOR operations on compiler-based per-pin APGs, the common challenge is the handling of the varying latency between read and write operations.

The CRC compare data needs to be generated in the respective APG at the time the DQ data is valid. However, the generated CRC data has to be compared at a later point in time that corresponds with the different CRC latencies for write and read operations. For hardware-based APGs, this means that not only the very specific CRC polynomial used by GDDR5 needs to be implemented in hardware, but also the CRC checksum compare needs to be delayed, for example, by a variable length FIFO in the CRC signal path.

For compiler-based per-pin APGs the same requirements apply. The flexible adaptation to the CRC polynomial used by the application usually can be easily done via XOR operations in the pattern description itself. To address the variable CRC delay between CRC data generation and CRC compare, the APG compiler needs to support storage of the CRC data for a user-defined delay.

3.3.8.4 Termination Impedance Mismatch

As described in Section 3.3.4.2, POD15 drivers use a 40 Ω impedance that drives into 60 Ω terminated POD15 receivers. In contrast to this, ATE pin electronics usually provide a 50 Ω impedance for both drive and receive functionality.

The problem with this impedance mismatch is twofold. From an AC point of view, the impedance mismatch, of course, represents a discontinuity that causes reflections on the signal path. The mismatch, however, is small enough that reflections do not contain too much power and are absorbed to a large extent by the loss of the signal path. Also, an impedance matching on the test fixture usually would create even bigger AC issues due to mounted components as described in Section 8.9. The second aspect of the impedance mismatch is the DC effect that causes a shift of the device pin voltage levels compared to the ones programmed on the ATE drivers and receivers. In order to drive the desired voltage levels and measure the correct voltages at the device pins, an adaption of the ATE drive and receive levels to the impedance environment is required.

3.4 MIPI Standards

The Mobile Industry Processor Interface (MIPI) Alliance is an open membership nonprofit organization that acts as standardization body for the mobile device industry. Its purpose is to ensure reuse and compatibility of interfaces that are used in the mobile and mobile-influenced ecosystem. The different MIPI Alliance standards cover all possible interfaces that can be present in a mobile system from the physical layer to higher-level protocols. In the following sections we only will cover the MIPI Alliance Specification for C-PHY℠ Version 1.0 [24]. The reason for selecting MIPI C-PHY v1.0 is its unique data transmission scheme forming one lane out of three wires and using wire state transitions to encode data information. The other MIPI high-speed physical layer standard is MIPI Alliance Specification M-PHY®Version 4.0 [25]. It is widely used as classical embedded clock serial interface to implement various higher level protocols such as UFS [26] or M-PCIe.

3.4.1 MIPI C-PHY v1.0

3.4.1.1 Application Areas

MIPI C-PHY v1.0 was primarily but not exclusively developed as physical layer interface to connect camera and display devices to host processing elements in mobile systems. MIPI C-PHY v1.0 is used as a possible physical layer for the MIPI Camera Serial Interface (MIPI Alliance Specification for Camera Serial Interface 2 (CSI-2))[27] and Display Serial Interface (MIPI Alliance Specification for Display Serial Interface (DSI))[28] standards. The physical implementation of MIPI C-PHY v1.0 allows also to run the interface as MIPI D-PHY^{SM} [29] physical interface that was the physical layer interface MIPI CSI^{SM} and MIPI DSI^{SM} were based on exclusively before MIPI C-PHY v1.0 was developed.

3.4.1.2 MIPI C-PHY v1.0 Fundamentals

A MIPI C-PHY v1.0 link is a master-slave connection that consists of one or more lanes. One MIPI C-PHY v1.0 lane consists of three wires (wire trio). Typically these wires are denominated A, B, and C. MIPI C-PHY v1.0 lanes can operate and switch dynamically between a high-speed signaling mode for data transmission and a low-power mode that also is called escape mode. Besides high-speed mode and escape mode, the link also can be in control mode that controls transitions in the link state machine. In escape mode, a MIPI C-PHY v1.0 lane can switch to an ultralow-power (ULP) mode, initiate low-power data transmission (LPDT) or trigger predefined actions like link resets at its link partner. Receivers also can be directed into a wait state in escape mode. While ULP and trigger mode are mandatory features in escape mode, LPDT mode and receiver wait state are optional. In high-speed data mode, the data flow per lane is unidirectional. Bidirectional interfaces, which typically are not required for camera or display connections via CSI or DSI, are implemented using two lanes in dual-simplex configuration. In low-power control mode, MIPI C-PHY v1.0 allows a bus turn around on a link and thus allows a semi-duplex operation via LPDT mode. This bus turnaround, however, only applies to the low-power modes and not to the high-speed communication mode. While in high-speed data communication is implemented using embedded clock source synchronous data transmission, data exchange in low-power mode is done in a variation of at-cycle source synchronous operation with the clock being encoded into the data. Another difference in signal transmission between high-speed and low-power modes is the the receivers in high-speed mode are terminated 100 Ω differentially, whereas in low-power mode the receivers have unterminated high-impedance inputs. In both modes, the drivers have a single-ended line output impedance of 50 Ω. The low-power control and escape mode of C-PHY are implemented according to the equivalent modes defined for the MIPI D-PHY standard. Since the data rate of the low-power mode is limited to 10 Mbps, we will not consider this part of the standard extensively in the following sections.

3.4.1.3 MIPI C-PHY v1.0 Details

MIPI C-PHY v1.0 uses a unique data transmission scheme in its high-speed mode. Any of the three wires A, B, and C of a lane can take one of three states, high, low, or mid-level terminated. The mid-level termination can happen at either end of the wire or at both ends. For one symbol time always one wire drives a high or low level, another wire drives the opposite level and the third wire is mid-level terminated. On the receiver side of a lane, the three possible wire pairs A-B, B-C, and C-A of a lane are fed into differential receivers. Since only one of the considered differential pairs drives a full swing differential signal and the other differential pairs contain one signal that is mid-level terminated, for each symbol one of the differential receivers sees a strong one or zero and the other two receivers see weak ones or zeros. Since only the strong wire states are used for wire state detection, the three possible differential pairs produce six possible wire states. These are denoted HS_+X, HS_+Y, HS_+Z, HS_-X, HS_-Y, and HS_-Z. The transmitted symbol is coded by the transition from one wire state to the next wire state. A wire state change results in a change of at least one of three state attributes. These attribute changes are named polarity, rotation and flip. A polarity change is present if the wire pair that drives high and low changes and the direction of the differential signal changes. This means as soon as the direction sign of the wire state changes and the wire state character changes a polarity change happened. The rotation attribute signalizes a clockwise (CW) or counter-clockwise (CCW) change of the wire state character. The clockwise character state change direction has the circular sequence X, Y, Z. A flip attribute change is recognized if the wire state character stays constant, but only the direction sign of the wire state changes. A C-PHY symbol is a 3-bit value that codes the three wire state transition attributes. The polarity attribute is the LSB of the symbol, the flip bit is the MSB and the rotation attribute is the middle bit of each symbol.

The MIPI C-PHY v1.0 standard requires wire state changes for every symbol in order to reliably recover the clock information at the receive side of the lane. With six wire states, this allows transmitting one of five symbols with each wire state transition. With the possibility to select between five states for each state transition, each symbol carries the information equivalent of $log_2(5) = 2.321928$ bits. The MIPI C-PHY v1.0 high-speed data transmission symbol rate targets are in the range from 0.8 GBd (1 Bd = 1 symbol/second) to 3.0 Gbd, which reflects a raw data rate range of 1.8575 Gbps to 6.9657 Gbps.

In the high-speed mode MIPI C-PHY v1.0 transmits data in bursts that consist of 16 bit words as smallest entity. Each 16 bit word is coded into seven MIPI C-PHY v1.0 symbols. This results in a possible coding space of $5^7 = 78{,}125$ combinations with a combination space of $2^{16} = 65{,}536$ that has to be covered. This mismatch of combinations means that there are some coding combinations that do not reflect any 16 bit word. If a receiver detects such a combination, the behavior is implementation specific but has to be interpreted as invalid. With this, a MIPI C-PHY v1.0 link implements two layers of coding. In the transmit direction, the first layer codes the 16 bit words into five 3-bit MIPI C-PHY v1.0 symbols. These symbols are serialized with the least significant symbol being transmitted first. The serial symbol stream then is coded in the second coding layer to physical wire states of the wire trio that physically connects transmitter and receiver. In the receive direction, the decoding is applied with the two coding layers in reverse order. In order to distinguish the two coding layers, the MIPI C-PHY v1.0 standard refers to the 16 bit to seven-symbol coding as mapping operation and the reverse direction as de-mapping operation. The symbol to wire state translation is referred to as encoding layer and the reverse direction as decoding layer in the C-PHY standard. It is important to note that with the coding of the data based on state transitions, the actual code that is transmitted depends on the data history. This means that that identical 16 bit words might be represented by different wire state sequences depending on the data that was transmitted before.

For low-speed signaling, the three wires are considered and evaluated as single-ended signals. The state codes in low-speed mode are denoted LP-ABC with A, B, and C being replaced by the logical state (H or L), the respective wire transmits. Depending on the low-power mode that is active, control mode, or escape mode, the same state code has different meanings that are described in more detail in the next section. MIPI C-PHY v1.0 defines four valid low-power state codes, LP-000, LP-001, LP-100, and LP-111. In low-power escape mode, the wire states are low level coded in a return-to-zero (RZ) format. This means that a high state on a line is a high level followed by a low level within the same bit time. With such a coding and the fact that in escape mode the logical zero is coded by LP-001 and the logical one is coded by LP-100, the latch clock for the data in LPDT mode can be extracted from the wire states by an exor operation that is applied to the A and C wires. In low-power control mode, RZ coding is not applied but NRZ coding for which the wires keep their state constant over the duration of the state code is applied. This is possible because control mode is an asynchronous mode that does not require clock extraction from the wire states.

3.4.1.4 MIPI C-PHY v1.0 Protocol

A MIPI C-PHY v1.0 lane operates along a state diagram as shown in Figure 3.25. The state changes are initiated in control mode of the lane according to the low-power state codes or state code sequences described in the diagram.

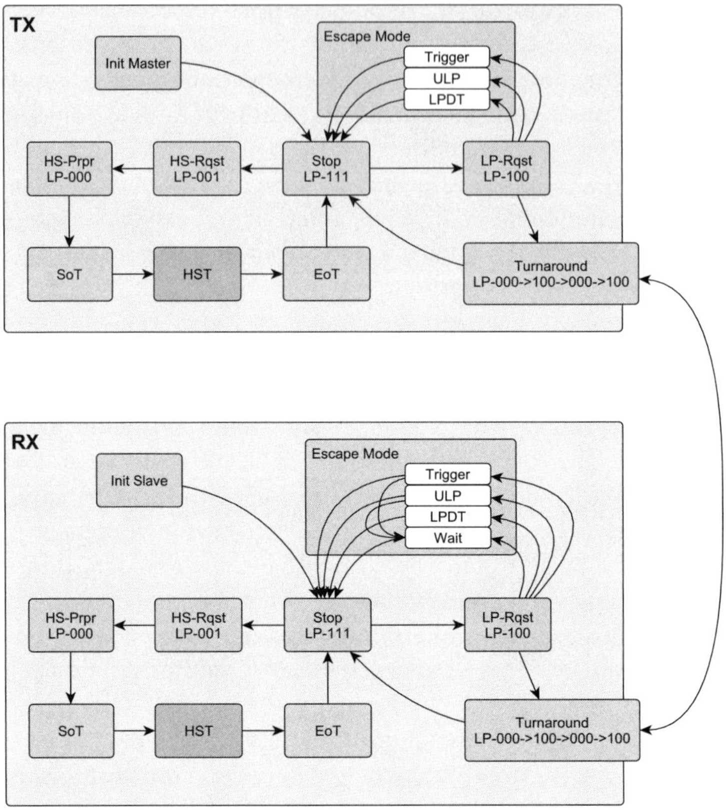

Figure 3.25 MIPI C-PHY v1.0 lane state diagram. Copyright (C) 2013-2014 by MIPI Alliance, Inc. and used by permission. All rights reserved.

The central state of a lane is the Stop state that is reached and kept with the LP-111 state code. In the state diagram the HST state represents the high-speed data transmission mode. During the sequence from Stop state to HST state, the receiver termination is turned on in HS-Prpr state. In this state also the low-power drivers are disabled and the high-speed drivers are enabled simultaneously for the transmit side of a lane. In the SoT (start of transmission) state the transmitter of a lane transmits a high-speed preamble followed by a sync word. The length of the preamble depends on the requirements of the receiver and has to be adjustable for a transmitter. The preamble also can contain an optional programmable symbol sequence. The preamble length is a multiple of seven symbols and besides the programmable part consists of "3" symbols. The sync word consists of a "3,4,4,4,4,4,3" symbol sequence. After the sync word the packet data to be transmitted starts.

At the end of packet data transmission a Post field consisting of multiple seven "4" symbol sequences is transmitted. The first of these seven symbol sequences identifies the end of packet data. The overall length of the Post field is adjustable. After the post field the state machine exits goes through an EoT (end of transmission) sequence that disables termination on the lane receivers. It also enables the low-power drivers and disables high-speed drivers on the lane drivers before. The low speed drivers force an LP-111 state code that brings the state machine to the Stop state.

From the Stop state also the Escape mode is reached with an escape entry procedure via the Lp-Rqst state. This entry procedure consists of the state code sequence LP-111, LP-100, LP-000, LP-001, and LP-000. After entering escape mode, the transmitter sends an 8-bit entry command that determines the substate of escape mode.

3.4.1.5 Electrical Specifications

The electrical specifications for MIPI C-PHY v1.0 is split into a set of parameters for high-speed data transmission and a set of parameters for low-power mode. While high-speed mode specifies a maximum differential voltage swing between a strong high and strong low level into 100 Ω of 300 mV around a common-mode voltage in the nominal range of 225 mV to 250 mV, the single-ended low-power voltage into an non-terminated line specifies a nominal low level of 0 V and a nominal high level of 1.2 V. Another difference between low-power and high-speed mode is that the nominal single-ended driver impedance for the high-speed drivers is 50 Ω. For a low-power driver, the single-ended driver impedance is at least 110 Ω.

3.4.1.6 Test Support

The MIPI C-PHY v1.0 standard does not require mandatory test support circuitry but recommends, for example, a PRBS generator and a programmable debug pattern generator for MIPI C-PHY v1.0 high-speed transmitters. On the receive side the corresponding pattern checkers with error detection and error counters are recommended. Another useful feature that some MIPI C-PHY v1.0 implementations deploy is a register controlled activation of high-speed mode. This simplifies testing of high-speed parameters independently of the lane state machine that would require control mode stepping through the state machine to enable the high-speed mode.

3.4.1.7 Test Challenges

The main test challenges for MIPI C-PHY v1.0, especially in ATE test environments, are coming from the fact that low-power mode and high-speed mode require dynamic termination, voltage level changes, and also the dynamic switching between single-ended and differential signaling modes. While dynamic voltage level and signaling mode changes sometimes can be accommodated by general-purpose ATE pin electronics, they typically are not designed to implement the very specific MIPI C-PHY v1.0 state machine based termination switches. Thus, as for a lot of bench tests, ATE-based test implementations might use active termination boards that are available for MIPI C-PHY v1.0.

Another test challenge is presented by the fact that MIPI C-PHY v1.0 receivers need to be stimulated by three-level signals and the fact that each wire of a wire trio has to be evaluated by two differential receivers. A similar challenge as for phase amplitude modulated (PAM) signals exists for the evaluation of data eye compliance and jitter measurements due to the effects of ISI. Since the start level and end level of a transition on the receive side can vary between four different levels, the ISI-induced effects manifest themselves differently depending on which of the four possible levels is the start level of a transition and which level is the end level of this transition. The MIPI C-PHY v1.0 standard contains a nonmandatory guideline on how to evaluate jitter numbers depending on the various transition possibilities.

3.5 Other High-Speed Digital Interface Standards

Apart from the high-speed digital standards presented in the previous sections, there are many other standards that are also important. One of these is the serial ATA standard (SATA) [30].

The SATA standard was developed to supersede the ATA and SCSI standards used mainly for connecting storage devices in computer platforms. The current standard (generation 3) runs at 6 Gbps with generation 2 at 3 Gbps and generation 1 at 1.5 Gbps. Two features of the SATA standard that might present some new challenges are the SATA implementation for out-of-band signaling (OOB) and spread spectrum clocking (SSC) requirements. In SATA, OOB is used for link initialization, recovery from low-power states, and for specific actions during test modes.

The signaling used for OOB applies low-speed differential data with gaps where both legs of the differential line will go to the same voltage as shown in Figure 3.26 [30]. In the gaps between data transmission response data from the link partner is expected. With this kind of signaling, a handshake protocol is implemented that allows communication between the link partners even if the link is not fully powered up and initialized. SSC is discussed in Section 5.8.3.

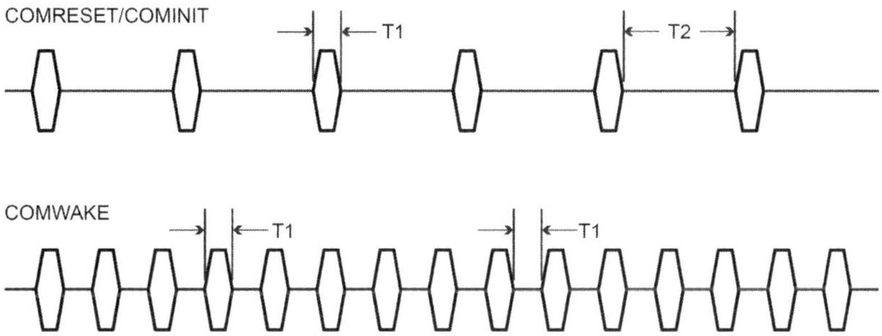

Figure 3.26 Out-of-band signaling in the SATA standard. (From: [30]. ©2002–2009 Serial ATA Organization. Reprinted with permission.)

Another important standard is the Universal Serial Bus (USB) [31]. USB was designed to allow many peripherals to be connected using a single standardized interface socket and to improve the plug-and-play capabilities by allowing devices to be connected and disconnected without rebooting the computer (hot swapping) and other important features. The design of USB is standardized by the USB Implementers Forum (USB-IF), an industry standards body incorporating several companies. The USB standard has several generations with different data rates. Generation 3 supports a maximum data rate of 5.0 Gbps.

The High-Definition Multimedia Interface (HDMI) standard is a compact audio/video connector interface for transmitting uncompressed digital streams. Like most standards, it has evolved over time in different generations [32]. HDMI/TMDS interface is a clock forwarding-based interface since a clock needs to be transmitted on a separate link with the data, but this clock provides only a reference to the link partner and is not used to directly latch the data at the receiver.

One of the major challenges of the HDMI/TMDS interface is its $AVcc$ voltage level of 3.3 V. This level might pose a challenge for testing HDMI especially for ATE pin electronics that is designed for high-speed digital applications that typically operate at lower voltage levels. While the specified swing value of HDMI should not be an issue for such a high-speed pin electronics, the common-mode voltage resulting from the 3.3 V requirement might be beyond the supported level ranges.

In situations where the measurement instrumentation cannot achieve this voltage level, two options are available. One option is to use an extra low-speed digital channel which has a larger voltage range or a power supply to help the high-speed digital channel to achieve the required common-mode voltage by using the appropriate connection circuit. Another option is to use a second power supply to create a virtual ground reference on the device and this way shift the required common-mode value inside the level range of the ATE pin electronics.

References

[1] PCI-SIG, *PCI Express External Cabling Specification Revision 1.0*, Jan. 2007.

[2] PCI-SIG, *PCI Express Base Specification Revision 1.0*, July 2002.

[3] PCI-SIG, *PCI Express Base Specification Revision 1.0a*, Apr. 2003.

[4] PCI-SIG, *PCI Express Base Specification Revision 2.0*, Dec. 2006.

[5] PCI-SIG, *PCI Express Base Specification Revision 2.1*, Mar. 2009.

[6] PCI-SIG, *PCI Express Base Specification Revision 3.0*, Nov. 2010.

[7] PCI-SIG, *PCI Express Base Specification Revision 3.1*, Oct. 2014.

[8] PCI-SIG, *PCI Express Jitter and BER Revision 1.0*, Feb. 2005.

[9] H. Werkmann, "PCI Express: ATE Requirements and DFT Features for Functional Test Optimization," *IEEE European Test Workshop*, May 2003.

[10] H. Werkmann, "Enabling the PCI Express Ramp – ATE Based Testing of PCI Express Architecture," *IEC EuroDesignCon*, Oct. 2004.

[11] M. W. Riley, J. D. Warnock, and D. F. Wendel, "Cell Broadband Engine Processor: Design and Implementation," *IBM Journal of Research and Development*, vol. 51, no. 5, 2007.

[12] H. Yasukawa, H. Goto, and K. Ueno, "LSI Architecture of SpursEngine™," *Toshiba Review*, vol. 63, no. 11, 2008.

[13] R. Perego, "XDR Architecture," *Rambus Advanced Information*, 2005.

[14] W. Wang and F. A. Ware, "XDR IO Cell," *Rambus Advanced Information*, 2006.

[15] Qimonda AG, *IDGV1G-05A1F1C-[40X/50X/55X] Internet Data Sheet*, Dec. 2008. http://www.qimonda.com/graphics-ram/gddr5/index.html.

[16] R. Crisp, "Direct Rambus Technology: The New Main Memory Standard," *IEEE Micro*, pp. 18–28, Nov. 1997.

[17] V. Khawshe, K. Vyas, R. Rangnekar, P. Goyal, V. Krishna, K. Prabhu, P. Kumar Venkatesan, L. Raghavan, R. Palwai, M. Thrivikraman, K. Desai, and A. Abhyankar, "A 2.4Gbps-4.8Gbps XDR-DRAM I/O (XIO) Link," *22nd International Conference on VLSI Design*, pp. 373–378, Jan. 2009.

[18] JEDEC Solid State Technology Association, *GDDR5 SGRAM*, Dec. 2009. JSED212.

[19] H. Werkmann and A. Olenyi, "High-Speed Memory Interfaces – Providing Bandwidth for the Future," *Verigy Users Group Conference (VOICE)*, Dec. 2007.

[20] D. T. Wang, *Modern Dram Memory Systems: Performance Analysis and Scheduling Algorithm.* Ph.D. thesis, University of Maryland at College Park, 2005.

[21] JEDEC Solid State Technology Association, *POD15 - 1.5V Open Drain I/O*, Dec. 2006. JSED8-20.

[22] H. Werkmann, D.-M. Kim, and S. Fujita, "GDDR5 Training – Challenges and Solutions for ATE-Based Test," *Asian Test Symposium*, Nov. 2008.

[23] H. Werkmann, H.-H. Jin, J.-W. Seo, and H.-W. Kang, "GDDR5 Test Challenges and Their Solutions on the Verigy 93000 HSM Series," *Verigy Users Group Conference (VOICE)*, Sept. 2008.

[24] MIPI Alliance, *MIPI Alliance Specification for C-PHY℠ Version 1.0*, Aug. 2014.

[25] MIPI Alliance, *MIPI Alliance Specification for M-PHY℠ Version 4.0*, Apr. 2015.

[26] JEDEC Solid State Technology Association, *Universal Flash Storage (UFS), Version 2.0*, Sept. 2013. JSED220B.

[27] MIPI Alliance, *MIPI Alliance Specification for Camera Serial Interface 2 (CSI-2)℠ Version 1.3*, May 2014.

[28] MIPI Alliance, *MIPI Alliance Specification for Display Serial Interface (DSI℠) Version 1.3*, Mar. 2015.

[29] MIPI Alliance, *MIPI Alliance Specification for D-PHY Version 1.2*, Aug. 2014.

[30] Serial ATE International Organization, *Serial ATA Revision 3.0*, June 2009. Gold Revision.

[31] *Universal Serial Bus 3.0 Specification*, Sept. 2008.

[32] Hitachi Ltd., Matsushita Electric Industrial Co. Ltd., Philips Consumer Electronics International B.V., Silicon Image Inc., Sony Corporation, Thomson Inc., Toshiba Corporation, *High-Definition Multimedia Interface Specification Version 1.3a*, Nov. 2006.

4

ATE Instrumentation for Digital Applications

The objective of this chapter is to provide a basic introduction to automated test equipment (ATE) for high-speed digital applications. For readers who have no ATE experience, [1, 2] can provide a starting point.

Like the semiconductor industry, ATE systems have evolved in the last decades with larger pin counts and higher data rates. Figure 4.1 shows three ATE systems from different manufacturers. From the hardware point of view, state-of-the-art ATE systems are typically composed of two parts: first, a testhead that can be filled by different pin electronics cards (digital, mixed-signal, power supplies, RF instruments) depending on the application one intends to address and second, a support rack that connects to the testhead and contains support infrastructure like workstation communication, power converters, and cooling. The exact split of functionality between the testhead and support rack components strongly depends on the ATE manufacturer and even on the application area an ATE system is targeted for. For example, some ATE systems that are targeted for specific applications like memory or mixed-signal do not have the option of testing other applications, even by adding or exchanging the pin electronics cards. Typically the pin electronics cards are provided by the manufacturer of the mainframe, although there have been efforts to allow different vendors to share the same testhead [3].

Typically the ATE system is not used on its own, especially in a production environment. It can be connected either to a handler for packaged parts testing or to a wafer prober for testing the parts at a bare die level as shown in Figure 4.2 [4].

Each pin electronics card typically contains several measurement channels. Figure 4.3 shows a few examples of digital ATE pin electronics

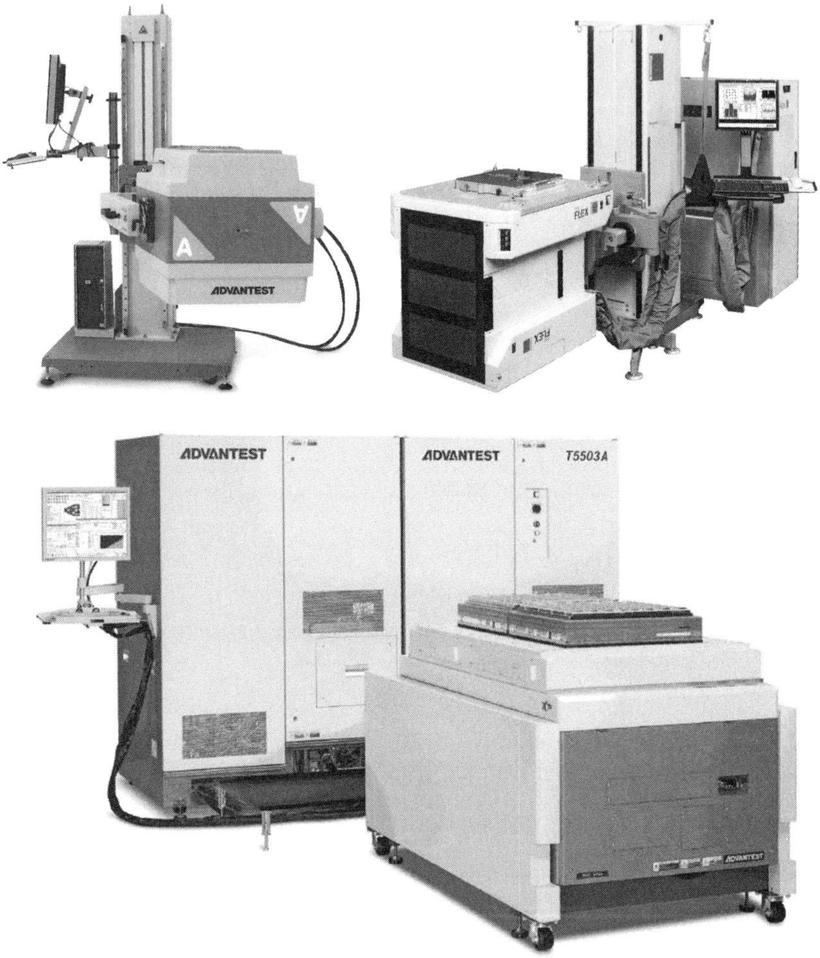

Figure 4.1 Example of automated test equipment (ATE) systems from different manufacturers (top left and bottom: courtesy of Advantest; top right: courtesy of Teradyne).

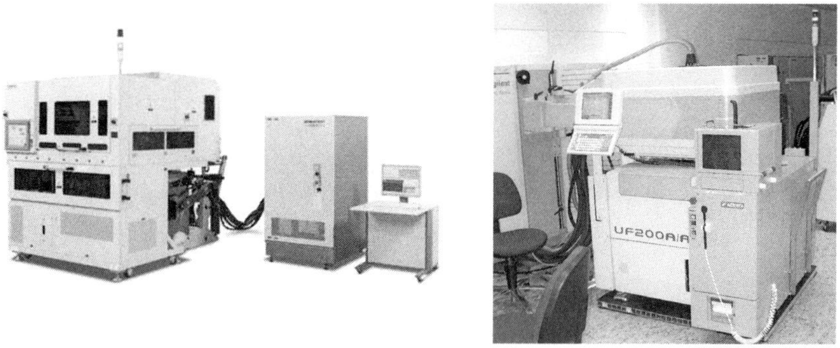

Figure 4.2 Example of an ATE system in a production environment: right: ATE system next to a handler for packaged parts testing (courtesy of Advantest); left: ATE system docked to a prober for wafer testing. (From: [5]. ©2005 Jose Moreira. Reprinted with permission.)

cards. Some pin electronics architectures divide the pin electronics card into channel assemblies containing a subset of the total number of channels. This architectural approach has significant advantages during tester maintenance where only the channel assembly containing the malfunctioning measurement channel needs to be substituted and not the entire pin electronics card. Figure 4.4 shows a picture of a channel assembly containing 16 single-ended bidirectional measurement channels that is a part of an ATE digital pin electronics card containing eight channel assemblies or a total of 128 single-ended measurement channels.

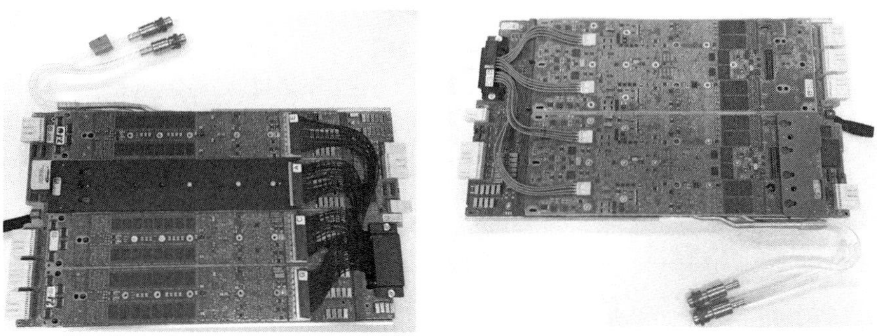

Figure 4.3 Photograph of a high density low-speed (128 single-ended measurement channels at 1.6 Gbps) ATE pin electronics card (left) and a low-density high-speed (8 differential driver and 8 differential comparators at 16 Gbps) ATE pin electronics card (right) (courtesy of Advantest).

Figure 4.4 Photograph of a pin electronics channel assembly containing sixteen 1.6 Gbps digital pins. Eight of these channel assemblies are then used on an ATE pin electronics card (courtesy of Advantest).

Figure 4.5 ATE to test fixture connection through a pogo pin assembly (top) and coaxial connector assembly (bottom) (courtesy of Advantest).

Another important hardware component is the connection of the ATE system to the test fixture where the DUT will reside. Figure 4.5 shows an example of the two main approaches to the ATE/test fixture high-performance interconnect. These are the pogo pin type assembly and the coaxial connector assembly. The figure also shows lower performance and lower cost connection approaches that are not required to be coaxial.

There are two common approaches in using ATE channels to measure and test high-speed digital I/O interfaces as shown in Figure 4.6. The first is based on using the driver/receiver in the ATE digital pin electronics card not only for performing functional tests but also for AC parametric measurements (transition time, jitter and so on). The second approach is based on using a sampler ATE card for characterizing the DUT transmitter and relying on the standard digital pin electronics for functional tests and the stimulus signal to the DUT receiver. Note that for this option the performance requirements on an ATE digital pin electronics are more relaxed than those on an ATE digital pin electronics card for complete I/O characterization.

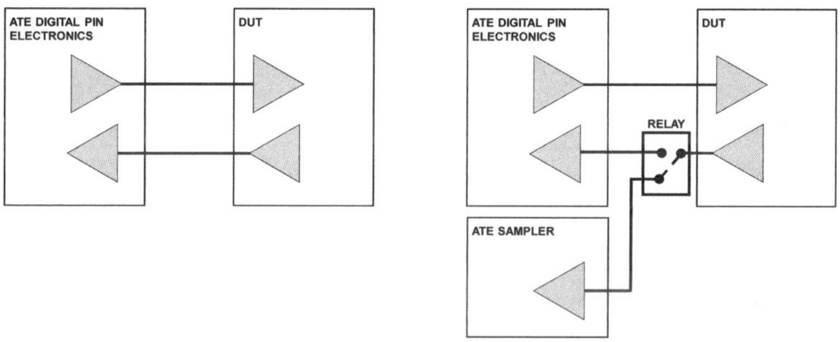

Figure 4.6 The two typical configurations for testing or characterizing a high-speed digital I/O cell with an ATE system using a pure digital pin electronics approach (left) and using a sampler/digitizer ATE instrument together with a digital pin electronics card (right).

Also note that it is technically possible to use other types of measurement instrumentation like high-performance time interval analyzers or a real-time sampling instrument. However, this type of instrumentation is typically not available as an integrated ATE card but as stand-alone instrumentation. It is feasible to integrate these instruments into an ATE system, although it does present some mechanical and signal integrity challenges (see Section 7.15).

It is important to note that specialized instruments, whether they are fully integrated or semi-integrated stand-alone, typically do not achieve the same amount of parallelism as is the case for digital pin electronics cards. Thus, usage of specialized instruments is only feasible for devices with only a few high-speed I/O pins without affecting the test time significantly.

The next sections discuss the digital pin electronics ATE card and the sampler ATE card in detail, but we will first discuss the different types of timing architectures for ATE systems. Digital power supply ATE cards are also briefly discussed.

4.1 ATE Timing Architectures[1]

ATE timing architectures are classified by the methodology they apply to derive the clock rate that they use to generate the waveforms or vectors that are applied to the DUT. Since this clock rate defines the tester period, the circuit generating the clock often also is referred to as period generator. There are three typical period generator implementations that have their unique advantages and disadvantages [6]:

- High-frequency clock timing architecture;
- Variable frequency clock timing architecture;
- Phase accumulator timing architecture (Flying Adder).

Two of these three implementations, the high-frequency clock architecture and the phase accumulator architecture use fixed frequency clocks and synthesize the required tester periods from these. The phase accumulator architecture also often is referred to as flying adder architecture. The third implementation is the variable frequency clock architecture that uses a PLL to derive the desired tester period. Due to the advantages and disadvantages of each of these architectures, it is important that the test engineer understands the differences. This especially is the case when converting applications between two different timing architectures. In some cases an ATE system might even have different ATE timing architectures for the digital and analog channels. In this case there are some synchronization challenges that need to be resolved [7].

Note that the discussion presented in this section is not extensive to the smallest bits and pieces since the objective is to convey the main concepts of the different ATE timing architectures. Real implementation of the ATE timing architectures are much more complex. Also note that the naming used might be different from the one used on a specific implementation linked to an ATE platform.

4.1.1 High-Frequency Clock Timing Architecture

In this architecture, a high-frequency master clock is used and the period of the digital pin electronics channel is derived by dividing this high-frequency clock

[1] This section was written in collaboration with Shusuke Kantake.

using a digital counter. It is a rather simple timing architecture that guarantees a very low timing jitter because the derived clock edge positions do not depend on any circuitry that adds jitter to the derived clock signal. This architecture has its limitations though because the ATE period resolution is determined by the fixed master clock frequency. In order to get fine period resolutions, very high-frequency master clock signals are required with this architecture. Thus, this architecture is not used in any of the major commercial high-speed ATE platforms.

4.1.2 Variable Frequency Clock Timing Architecture

In a variable frequency clock ATE timing architecture a programable fractional PLL is used to generate the ATE master clock based on a common reference clock. With a fractional PLL the desired target test cycle periods can be achieved accurately with a very fine resolution. Figure 4.7 shows a high-level block diagram description of how this ATE timing architecture works.

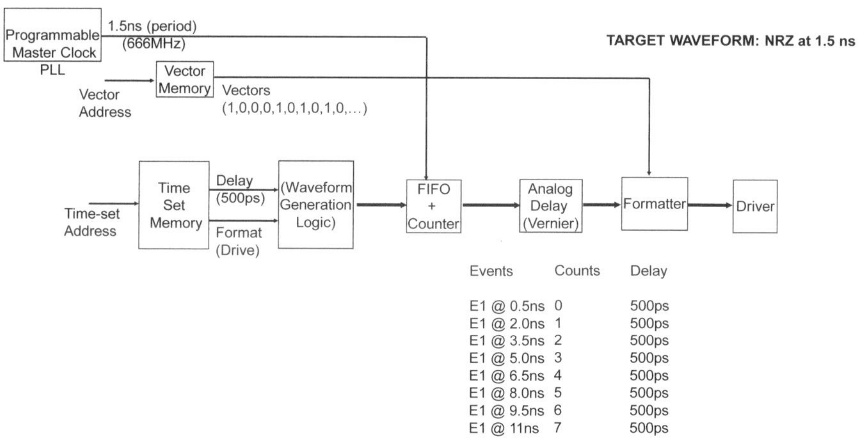

Figure 4.7 High-level block diagram of a variable frequency clock ATE timing architecture.

The drawback of the variable frequency clock timing architecture is that each time the master clock period is changed there is a latency until the master clock provides the new period due to PLL stabilization times. This latency usually corresponds to a large number of ATE cycles. The main advantage of this timing architecture is that by keeping the pattern generation only on the last stage of the timing generation chain and the direct usability of the synthesized clock for tester edge generation, it can allow for a better jitter performance (e.g., ISI).

4.1.3 Phase Accumulator Timing Architecture

The phase accumulator ATE timing architecture is also sometimes referred to as change time on the fly (CTOF) or change period on the fly (CPOF) timing architecture. The main characteristic of this ATE timing architecture is the use of a phase accumulator that implements a digitally controlled oscillator (DCO) to synthesize the clock frequencies that are used to generate the ATE timing.

4.1.3.1 Phase Accumulator

Phase accumulators are the key components in all digital frequency synthesizers like flying adder frequency synthesizers or direct digital synthesis (DDS) circuits. The purpose of the phase accumulator in these circuits is to keep track of the phase offset between a well-known reference frequency and a target frequency to be synthesized. With the knowledge of this phase offset, a signal with the target frequency can be synthesized from the known reference frequency.

4.1.3.2 Flying Adder

The principal idea of a phase accumulator architecture is the usage of a central reference clock that is fanned out via multiple delay lines to derive 2^m clock signals with the same frequency but with equidistant phase offsets Φ_i between the single delay line outputs that span over $180°$ of the original clock [8–10]. A multiplexer is used to select one of these phase offset clocks that then clocks a by-2 frequency divider circuit that generates a clock edge of the synthesized clock with every rising edge of the selected multiplexer input. With a static multiplexer control, this circuit would generate a phase offset clock at half the frequency of the central reference clock. Thus, the key to synthesize different frequencies based on the central reference clock is to control the multiplexer dynamically. This dynamic control selects a new multiplexer input with each positive edge on the clock divider input. The selection criteria is based on a target frequency in a way that the next edge of the selected multiplexer input is as close as possible to the next edge required by the target frequency to be synthesized. The target clock cycle time is specified as integer factor of the phase offset Φ_i. A functional block diagram of a flying adder frequency synthesizer is shown in Figure 4.8.

Due to the finite number of phase locations that can trigger a transition in the synthesized clock signal, for most frequencies to be synthesized that frequency cannot be achieved accurately but only in average [11]. The offset of the clock edges of the synthesized clock from their ideal positions have

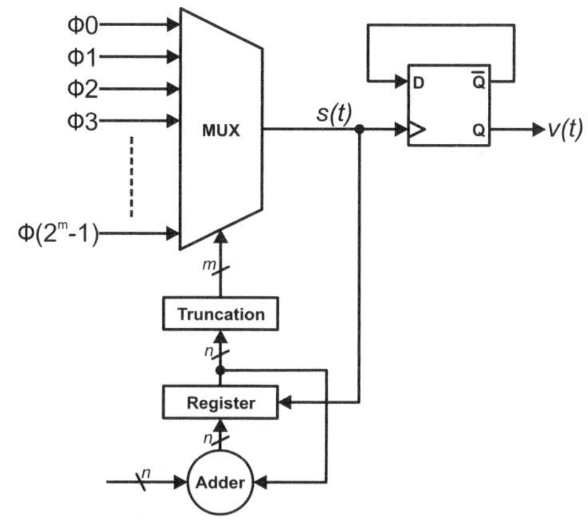

Figure 4.8 Block diagram of a flying adder frequency synthesizer.

to be taken into consideration when generating signal edges based on the synthesized clock and have to be corrected using a dedicated edge delay that is adjusted dynamically.

4.1.3.3 ATE Architecture

Figure 4.9 shows a high-level block diagram description of how the phase accumulator ATE timing architecture works. To make the description easier, let's assume we are using the ATE to generate a delayed non-return-to-zero (DNRZ) waveform with a tester cycle period of 1.5 ns and a delay of 500 ps. With the phase accumulator timing architecture, the master clock is running at a fixed period (1.25 ns in our example) that might be different from the tester cycle target period.

The timing set definitions for the phase accumulator architecture contain the information about the signal format (e.g., NRZ, DNRZ, RZ, RC ...) to be used as well as the timing offsets (relative to the cycle start) of the required data edges for the respective format and the desired tester cycle time for that timing set. In our example the data format to be used is DNRZ, the edge offset for the single edge required is 500 ps and the tester cycle time is 1.5 ns. The timing set information is combined on the fly with the test pattern data and the tester cycle time information in the phase accumulator and formatter stage of the timing architecture. The results of this combination are cycle counter and analog vernier delay settings for each data bit that are required to generate the desired edge positions based on the fixed central clock. Figure 4.9 shows the results of this combination in the phase accumulator and formatter stage for the data pattern and timing conditions of our example.

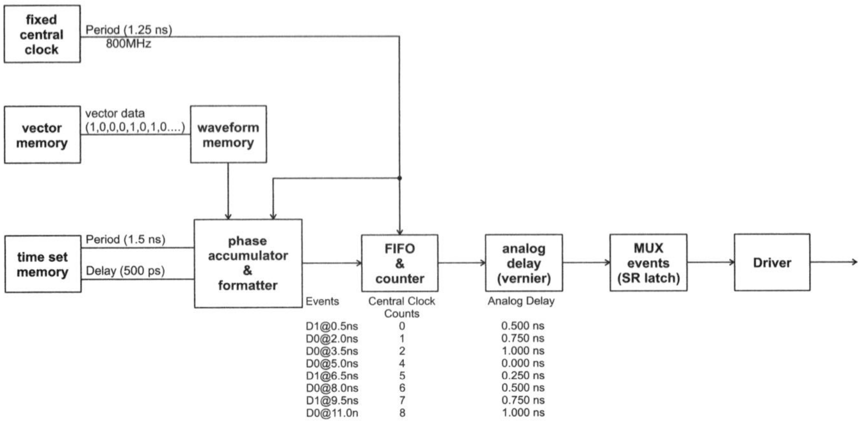

Figure 4.9 High-level block diagram of a flying adder ATE timing architecture.

Although the timing generation might look complex, this type of ATE timing architecture often is selected because it provides the ability to change the period of the tester cycles instantaneously from one cycle to the next (change timing on the fly) during the pattern execution.

One drawback of this architecture is the need to constantly reprogram the analog delay due to the mismatch between desired tester cycle period and the synthesized clock that runs at the desired frequency only in average. Thus, even if a timing edge is kept static relative to the start of a tester cycle, the different analog delay settings that are required to adjust to the fixed central clock result in increased intrinsic jitter of the generated edge positions due to the nonlinearities of the analog delay line.

4.2 Digital Pin Electronics ATE Card

The digital pin electronics ATE card is the main element of an ATE system for digital applications. It provides the stimulus in the form of a digital waveform with programmed levels and timing at a given data rate. It also compares an incoming digital waveform from the DUT to an expected pattern with a programmed voltage threshold and compare strobe timing.

The pattern data for stimulus and compare typically is stored in the vector memory of the ATE system. For certain applications or device building blocks (e.g., memory devices or embedded memory) some ATE systems generate the pattern data on-the-fly using dedicated algorithmic pattern generation (APG or ALPG) hardware [12]. Algorithmic pattern generation also is used by some digital pin electronics cards that not necessarily target memory test but offer DfT support (e.g., with PRBS data generation and comparison—see also Section 6.3.1.2). Figure 4.10 shows a high-level block diagram of the typical bidirectional driver/receiver pin electronics in an ATE card for high-speed digital applications. As further shown in the figure, digital pin electronics architectures usually include the possibility of performing DC measurements and sometimes even offer additional capabilities like an integrated time interval analyzer (TIA) for the measurement of timing events and jitter.

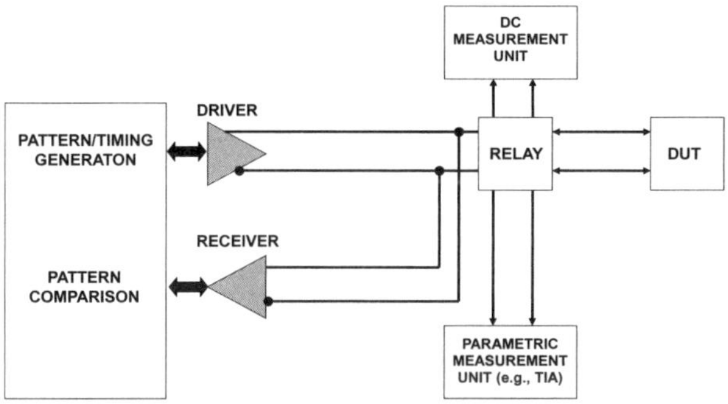

Figure 4.10 High-level diagram of a differential driver/receiver ATE pin electronics for high-speed digital applications.

The pin electronics driver/receiver topology can be divided into different categories as shown in Figure 4.11 with some ATE pin electronics implementations containing several of those topologies to allow the testing of a larger number of applications.

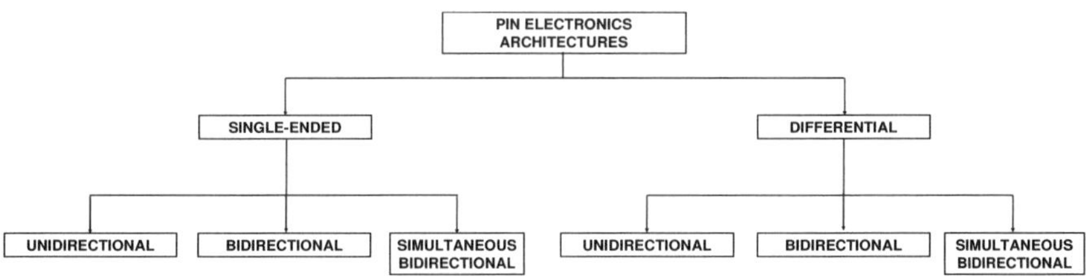

Figure 4.11 High-level categorization of the pin electronics architectures.

Figure 4.12 shows a block diagram of the three pin electronics topologies for differential high-speed digital applications.

In a unidirectional interface configuration, the driver and receiver are connected to separate pins. This is the typical topology that is required for serial interfaces like PCI Express [13]. This kind of topology allows the design of very high-performance pin electronics since one avoids the disadvantages

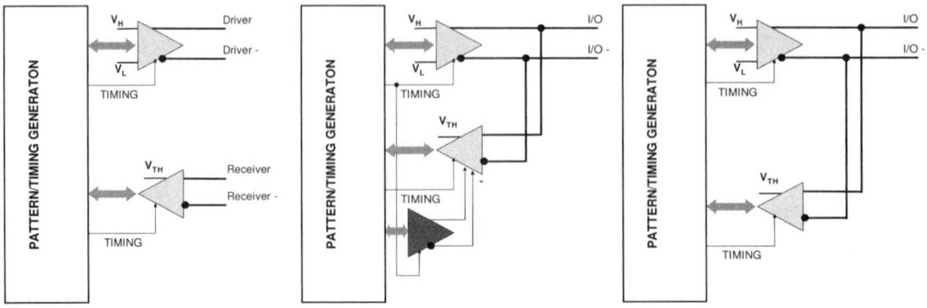

Figure 4.12 Different topologies for an ATE digital pin electronics driver/receiver assembly (left: differential unidirectional; center: differential bidirectional; right: simultaneous bidirectional).

of connecting the driver and receiver together, which are discussed later. The drawback is the lack of flexibility in assigning the ATE pins to the DUT and also the need to use techniques like dual transmission line (also referred to as fly-by) for testing bidirectional interfaces (see Section 9.4).

In a bidirectional configuration both the driver and receiver are connected together. This is the standard topology of most ATE digital pin electronics. This configuration allows any pin to be used as driver or receiver, increasing the flexibility in the ATE pin assignments with the DUT. From a performance point of view, the electrical connection of the driver and receiver does have its design challenges (e.g., the driver capacitance will be present when the receiver is measuring). Although this configuration can handle bidirectional interfaces, the round-trip delay between the pin electronics and the DUT might make bidirectional testing with a single pin impossible, again requiring the use of fly-by techniques.

In a simultaneous bidirectional configuration the challenge of the round-trip delay of a standard bidirectional configuration is solved by subtracting the ATE driver signal from the DUT signal being measured at the ATE receiver. This topic is further discussed in Section 4.2.6.

Apart from the standard capabilities of a digital pin electronics card and the possible driver/receiver configurations, there are several more advanced components that a digital pin electronics card might contain to provide specialized support for the characterization and testing of high-speed digital applications. Figure 4.13 shows an ATE pin electronics block diagram that includes several of these additional nonstandard components, which are described next.

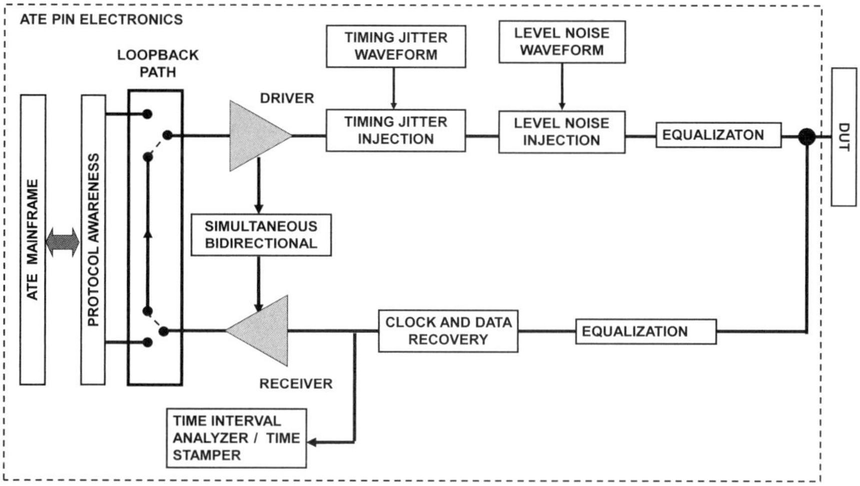

Figure 4.13 Block diagram of an ATE digital pin electronics including several nonstandard capabilities.

4.2.1 CDR and Phase Tracking

Providing clock and data recovery (CDR) or similar capabilities on the ATE pin electronics receiver has some significant advantages for high-speed digital applications. One major advantage is the ability to automatically set the receiver strobing point in the center of the data eye without having to resort to timing searches. Also, several high-speed standards require a CDR with a specific response to be included for measurements like jitter generation (more on this topic in Section 9.5).

However, it is important to note that implementing a CDR with a programmable loop filter is not a trivial task and will increase the pin electronics cost while a CDR with a fixed loop filter might not be appropriate for some high-speed standards depending on the loop filter characteristics. Figure 4.14 shows a block diagram of an ATE pin electronics receiver with an integrated phase detector block. The CDR functionality is achieved by using the phase detector in combination with a digital control loop that is connected to the ATE timing system.

Note that in the block diagram the control loop not only takes into account the phase information from the phase detector to set the strobing point in the middle of the data eye but also the timing shift of the edge that is required by the ATE timing system for measurements like a data eye diagram (Section 5.4.2).

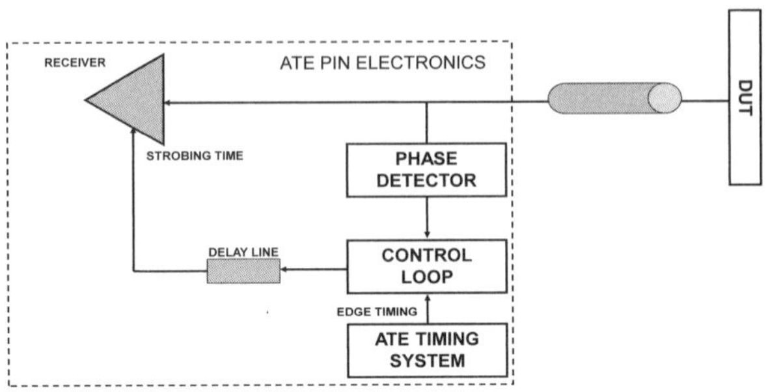

Figure 4.14 Example of a high-level block diagram of an ATE pin electronics receiver with an integrated phase detector.

4.2.2 Equalization

Equalization (also referred to as pre-emphasis/de-emphasis on the driver side) is one key item to address the challenge of waveform degradation due to the test fixture loss for high-speed digital measurements [14]. The challenge is that even when designing the test fixture using the best available materials and design techniques the loss might be too high for the measurement [15]. The reason is that the length of the signal traces on a typical ATE test fixture coupled with the number of pins makes it very difficult to create a solution that provides minimal degradation for a multigigabit digital signal. To address this problem, modern ATE pin electronics include an equalization block that compensates for the test fixture loss both on the drive and receive sides. Section 9.9.3 discusses equalization in more detail.

It is important to note that although equalization will compensate the test fixture loss, this does not mean that proper test fixture design is no longer important. Test fixture design is still critical, as will be discussed in Chapter 8.

4.2.3 Time Measurement Unit

Some digital pin electronics implementations include a time measurement unit (TMU) which can be used to record the time when a predetermined event occurs. This event is defined by the signal under test crossing a given level threshold with a given edge polarity (rising, falling, or either). By integrating this type of instrument on the pin electronics, measurements like the frequency of a clock signal can be performed in a very short time compared with other types of instruments [16, 17].

There are two approaches for implementing a TMU. The first is through a time interval analyzer (TIA), which performs relative measurements between two events, potentially synchronized to a pattern marker. The second approach is to use a time stamper that attaches a time stamp to each desired event with all events referenced to a common single time reference.

Modern ATE digital channels usually use a time stamper for implementing an integrated TMU unit with the time interval analyzer approach usually used for external instruments as discussed in Section 7.3 [18].

4.2.4 Timing Jitter Injection

The ability to support timing jitter injection on the pin electronics driver is critical for high-speed digital testing (see Section 5.7). The objective is to measure the ability of the DUT receiver to tolerate jitter that is present on the incoming data waveform. This requires that the ATE pin electronics driver is able to inject well-defined timing jitter on the stimulus data waveform to the DUT receiver. Several approaches are possible for jitter injection implementation on the ATE pin electronics.

4.2.4.1 Timing Jitter Injection Through a Voltage Controlled Delay Line

The most straightforward way to inject timing jitter into a waveform is to use a voltage controlled delay line as shown in Figure 4.15. Depending on the voltage level at the control input of the delay line, the time delay through it will vary. Consequently, this will delay the edges of the waveform. The voltage control signal can be generated by a waveform generator and can be completely arbitrary. The advantage of this approach is that it is continuous in time in the sense that the delay line is injecting jitter (delay) continuously on the waveform although for measurement purposes only the timing of the transition edges will be important.

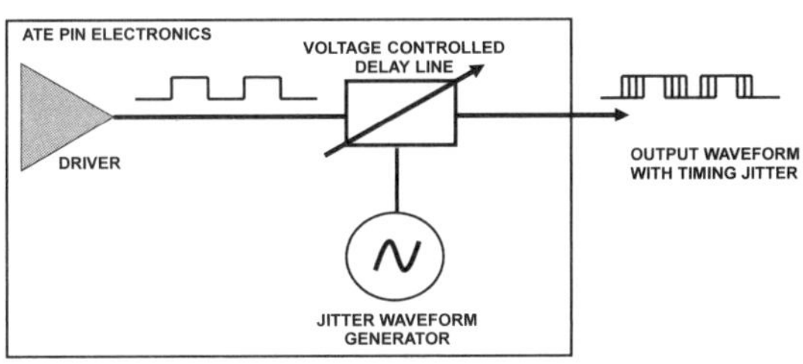

Figure 4.15 Jitter injection through a voltage controlled delay line.

One variation on the approach shown in Figure 4.15 is to inject the jitter into the reference clock of the ATE system which then is used to latch out the data from a FIFO as shown in Figure 4.16.

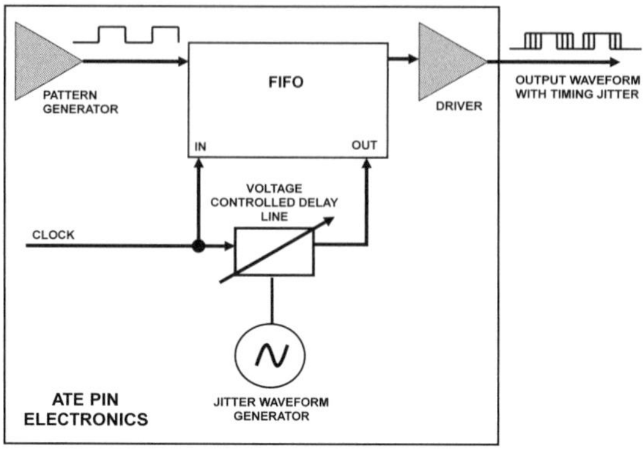

Figure 4.16 Jitter injection through a voltage controlled delay line in the reference clock and a FIFO.

4.2.4.2 Timing Jitter Injection Through Edge Time Reprogramming

Another possible approach for jitter injection is to use the timing programming capabilities of the driver pin electronics. Instead of correctly programming the timing for the driving edges of the data pattern, the timing is misaligned so that jitter is added to the pattern timing as shown in Figure 4.17 [19]. The disadvantage of this approach is that it depends on the timing programming capabilities of the pin electronics. Some ATE timing systems might be very restricted regarding reprogramming the driving edge timing. This can significantly restrict the ability to inject jitter using this approach. The other disadvantage is that this type of jitter injection is not easy to map to the typical jitter injection requirements from high-speed digital standards.

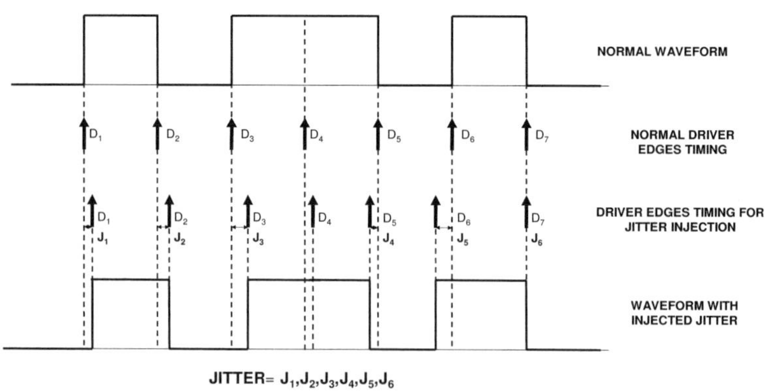

Figure 4.17 Jitter injection through edge timing programming.

4.2.4.3 Timing Jitter Injection Through Programmable Equalization/Pre-Emphasis

Since some ATE pin electronics cards for high-speed digital applications include programmable equalization (also referred to as pre-emphasis/de-emphasis) on the driver to compensate the test fixture loss, there is the possibility to use this programmable pre-emphasis to inject jitter in the form of data-dependent jitter (DDJ) to the DUT. This topic is discussed in more detail in Section 9.9.3.5.

4.2.5 Amplitude Noise and Common-Mode Voltage Injection

Another typical requirement on the pin electronics driver circuitry for high-speed digital applications is the ability to inject amplitude noise on the ATE driver signal to test the DUT receiver tolerance for this effect. While amplitude noise on a single-ended signal results in amplitude noise only on the DUT receiver, amplitude noise on the legs forming a differential signal can manifest itself in the form of two effects at the DUT receiver. These are differential amplitude noise and common-mode noise.

For differential interfaces typically both of these parameters are verified. If the amplitude noise on both legs of the differential interface has the same phase and magnitude at all instances (e.g., due to crosstalk on the differentially coupled connection), this translates into common-mode noise only. If the noise characteristics on the two legs differ, phase differences and magnitude differences translate into differential amplitude noise.

Two possible approaches that are used on pin electronics architectures to stimulate amplitude and/or common-mode noise are discussed below.

4.2.5.1 Continuous Time Voltage Source

The most straightforward way is to modulate the signal from the driver with a voltage waveform (e.g., a sinusoidal wave) as shown in Figure 4.18. If the two legs of a differential signal are modulated by separate modulation sources, nonuniform modulation for the differential legs is possible. This allows dedicated common-mode and differential amplitude noise injection. This approach could be implemented using a resistive network, for example.

4.2.5.2 Extra Switchable Level Range

Another less efficient method to inject amplitude noise also available in some ATE pin electronics is the ability to switch to another level setting during a functional test as shown in Figure 4.19. For differential interfaces, different

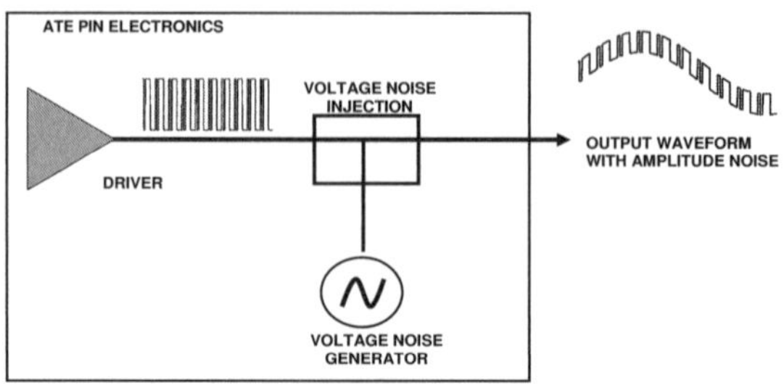

Figure 4.18 Amplitude noise injection through a voltage source.

levels on the two legs might be used to create differential amplitude noise injection. Timing edge offsets between corresponding edges of the differential legs can force common-mode variations. The alternative level control is typically implemented by additional data pattern states. The problem with this approach is that it typically only allows switching between two level ranges and this switch can take a significant amount of time. This makes this approach very limited for testing the DUT receiver for amplitude and common-mode noise tolerance.

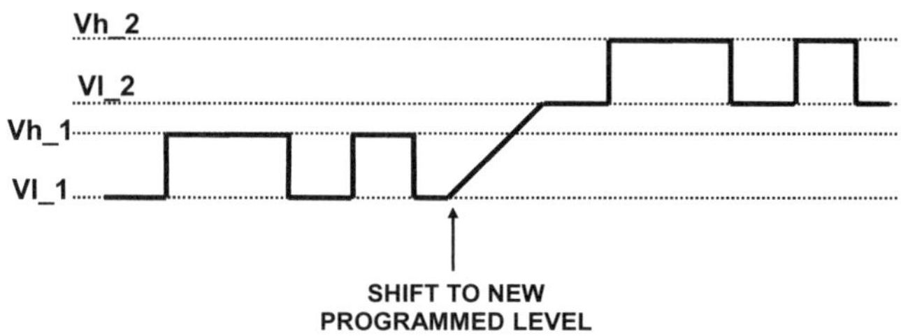

Figure 4.19 Amplitude noise injection through an additional level.

4.2.6 Bidirectional and Simultaneous Bidirectional Support

Support for bidirectional applications can depend significantly on the capabilities of the pin electronics. If the pin electronics architecture is unidirectional, then the only available solution to test a bidirectional interface is to use techniques like dual transmission lines [20, 21].

If the pin electronics contain a standard bidirectional architecture like in Figure 4.12 (center), then the pin electronics can natively test a bidirectional application [22]. The only challenge is that due to the long signal traces that are typical on an ATE test fixture, functional testing problems might arise when testing the application with a very short turnaround time between drive and receive. This is exemplified in Figure 4.20 where two examples of a read/write operation between the ATE pin electronics and the DUT are shown.

In the first example, illustrated in Figure 4.20(a), the test fixture delay is much smaller than the time between the read and write operations of the DUT. However, in the second example shown in Figure 4.20(b), the test fixture delay is larger than the first example, which creates a conflict between the driver and receive operation on the ATE pin electronics. In this case the pin electronics driver must start driving a signal while the receiver is still measuring the write signal from the DUT which will cause the driver signal to be superimposed to the write receive signal. In such a condition, which is described in more detail in Section 9.4, the pin electronics receiver will not be able to measure the write signal correctly.

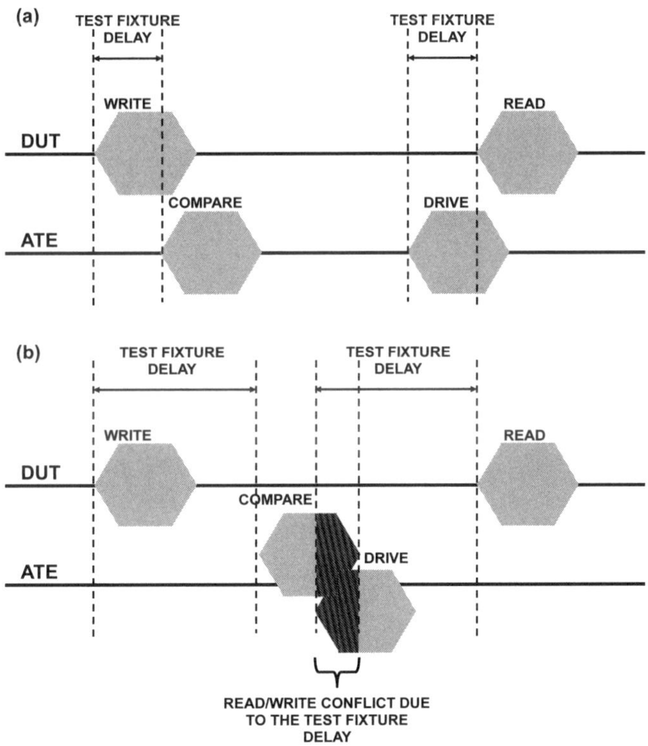

Figure 4.20 Demonstration of the impact of the test fixture delay on testing a bidirectional interface. In example (a) there is no conflict, but in example (b) where the test fixture delay is larger than (a), there is a read/write conflict on the ATE pin electronics.

One way to address this challenge is to use a simultaneous bidirectional (SBD) pin electronics architecture that is shown at a very high level in

Figure 4.12 (right). To completely avoid the round-trip time between the pin electronics and the DUT, the ATE pin electronics can drive and receive signals simultaneously because the receiver has a "copy" of the driver signal, which allows it to subtract the signal that is being driven from the signal at the input of the receiver and in this way figure out the signal that was driven from the DUT.

4.2.7 Protocol Engine

Some digital interfaces/standards have complex protocols that have a non-deterministic behavior in some situations; that is, the exact bit pattern to be expected cannot be determined a priori (e.g., through simulation). One way to address this challenge is to incorporate the capability to understand the protocol of the application being tested on the ATE pin electronics. Note that this protocol-aware approach is a significant change with regard to the classical functional test approach with a deterministic pattern. Section 9.7 discusses this topic in more detail.

4.2.8 ATE Loopback Path

An ATE loopback path provides a way for the data bits received by the ATE receiver to be sent to the ATE driver for transmission to the DUT. In this configuration the DUT driver generates a data pattern through some embedded DfT engine and the pattern is then looped back to the DUT receiver through the pin electronics (a more detailed discussion on loopback-based testing is done in Chapter 6). In this mode the pin electronics simply loop the signal from the DUT driver to the DUT receiver without having to generate any pattern. In some implementations it is even possible to recondition the electrical parameters of the signal (e.g., rise time, levels, jitter) to stimulate the receiver with controlled levels and/or timing parameters.

4.2.9 Parametric Measurements

Although digital channels historically are mainly seen as being responsible for performing functional tests with simple pass/fail results or for DC/scan type measurements, they can also perform detailed parametric measurements for level and timing parameters such as jitter histograms, device level measurements, or even full data eye diagrams. The capability of digital channels to perform this kind of parametric measurements for certain applications is mainly determined by their level and timing accuracy as well as the available methodologies for data post-processing. Besides the specified timing and level accuracy, the bandwidth of the pin electronics also has significant influence on the result of parametric measurements.

On a typical digital ATE pin electronics, the user has separate control of driver and receiver voltage and timing settings. Depending on the specific architecture of the ATE pin electronics, there might be some limitations (e.g., a true differential receiver might not have a variable threshold voltage).

For a standard functional test, the objective is to verify that the DUT reacts in the expected manner to a stimulus applied to its input pins. The stimulus and expected data patterns are typically derived from device simulations. For the execution of a functional test, the device input levels and timing parameters are set according to the requirements of the device specification. In order to check the DUT output pattern, the receiver voltage threshold is set to a value that allows distinguishing between the high and low output levels as defined by the DUT specifications. The compare strobe timing is set to occur in the center of the received DUT bits to guarantee the maximum possible timing margin.

In order to perform parametric measurements for level or timing parameters or both, the respective ATE control (drive levels, drive timing, compare level threshold, or compare strobe timing) is not set fixed to the optimum that guarantees maximum margins with respect to the DUT signal but is varied for iterative functional test executions. A change in the functional test result from one iteration to the next indicates that the test decision point that is controlled by the varied ATE resource has been reached. Using the knowledge of the values set for this ATE resource for both iterations that caused the test result transition allows the reconstruction of the respective signal value. An example for such a parametric level measurement is shown in Figure 4.21 for an output voltage measurement.

It is important to note that for output pins, such measurements often do not necessarily require a functionally passing test because it does not matter whether the test result transition for two iterations is from pass to fail or from fail to pass. One important condition for these kind of parametric measurements, however, is that the behavior of the DUT is repeatable and deterministic for a sequence of functional test executions. Due to the required test iterations to determine the parameters for a single measurement point, special care should be taken to the length of the test pattern executed per iteration since this will directly impact the required measurement time.

The extent to which this methodology can be used for parametric measurements strongly depends on other features that are available for the ATE receiver used. A basic functional test typically delivers a single pass or fail result for all bits that are compared within a test pattern. Thus, a parametric measurement that is performed as described before using such a typical single

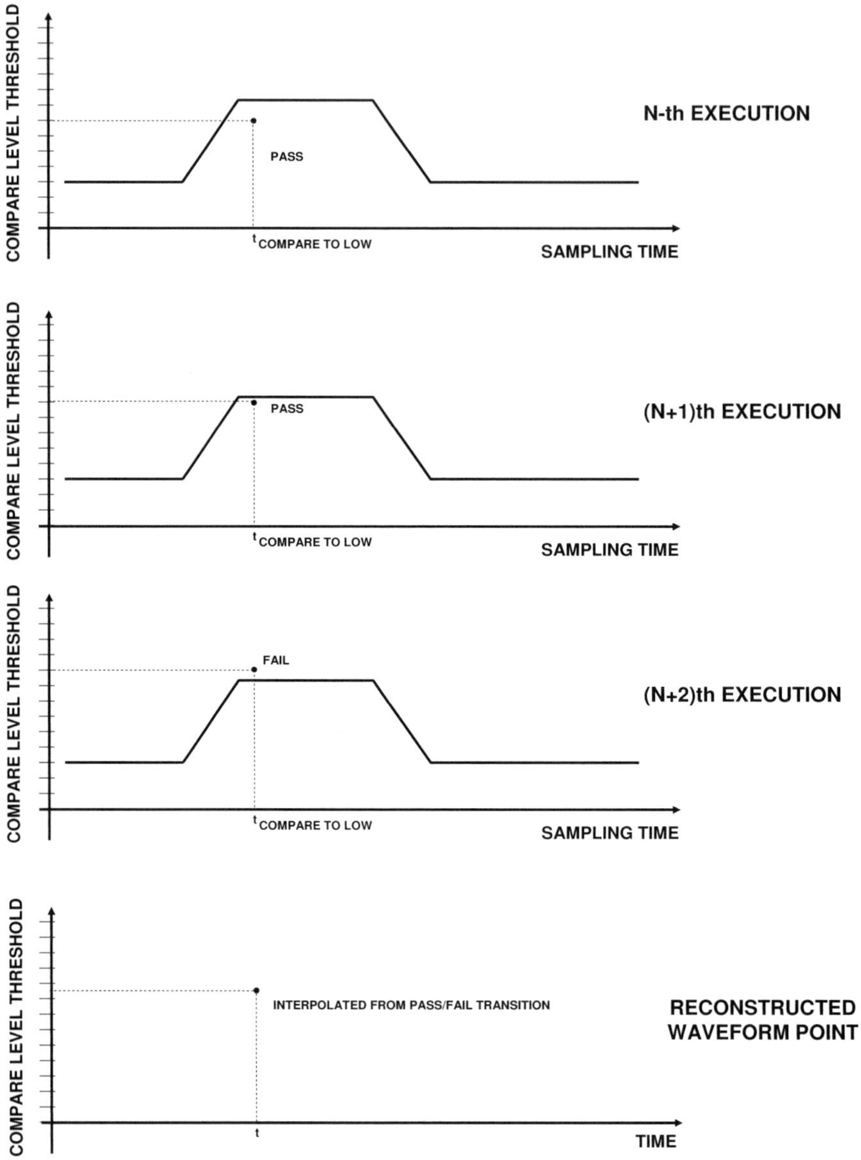

Figure 4.21 Acquiring the waveform using a strobe time/level threshold sweep approach.

pass-fail evaluation over all bits will deliver a result that represents the overlay of all bits of a pattern. In order to obtain the parametric values for one or more dedicated bits of the executed pattern, multiple executions of the parametric measurements would be required with only a single active compare at the desired bit positions if the ATE receiver only supports the classical single pass-fail reporting per pattern execution.

A significant improvement for parametric measurements is represented by ATE receivers that allow collection of and access to the pass-fail information per active compare strobe instead of the global pass-fail result only. Such a feature allows analysis of the parametric values on a per-bit basis as shown in Figure 4.22 for a per-bit level measurement without the need to repeat the measurement sequence per bit with an active compare at the bit of interest only.

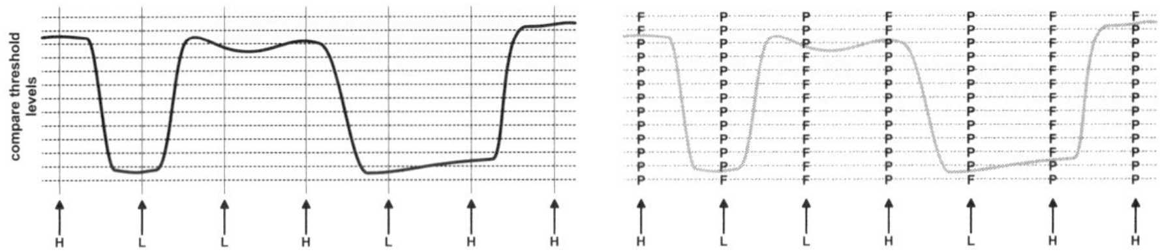

Figure 4.22 Using digital pin for parametric level measurements.

A further feature of ATE receivers that extends the described parametric measurement capabilities is a per-pin error counter, which delivers the number of errors for a pin over one complete pattern execution. If the number of errors per pin is evaluated for each test iteration with varying compare level or timing settings, the distribution of the received DUT signal over the iteration range of the level or timing is obtained. This distribution data can be used for statistical analysis of level or timing data.

Although the previous examples mainly concentrated on parametric level measurements, it has to be emphasized that the same approaches are also usable for parametric timing measurements if the ATE timing is varied instead of the level parameters.

An interesting point of this parametric measurement methodology is that if it is applied for level measurements it can be seen as being equal to using a 1-bit analog digital (A/D) converter. Thus, the same sampling methodologies as for sampler and digitizer instruments that will be described in Section 4.4.3 are applicable with digital receivers to sample analog signal waveforms. The accuracy and resolution of this approach depends on the bandwidth of the ATE receiver, the timing/level accuracy, and the timing resolution on the ATE edge control.

4.3 Sampler/Digitizer ATE Card[2]

In ATE systems for mixed signal applications, there are typically two kinds of analog measurement instruments available. One is called a digitizer and the other is called a sampler. Both of them use analog to digital converters (ADC) to convert the analog signal waveform at their inputs into a digital representation of that waveform. In order to understand the difference between both instruments and to select the right instrument for an application, it is important for a test engineer to understand how analog signals are acquired by mixed signal instruments and how the characteristics of these signals are influenced by the data acquisition.

The fundamental rule to obtain the correct waveform representation after sampling the analog waveform with fixed sampling intervals is Shannon's theorem that can be traced back to the sampling limit according to Nyquist [23]. Shannon's theorem states that if a signal over a given period of time does not contain any frequency component greater than f_x, all of the needed signal information can be captured with a sampling rate of $2f_x$. The sampling rate of $2f_x$ is referred to as the Nyquist rate. The corresponding time interval $(1/2)f_x$ is called the Nyquist interval. Strangely enough, the corresponding Nyquist frequency is f_x. If a signal is not sampled according to the rules of Shannon's theorem, an effect called aliasing falsifies the sampled signal.

4.3.1 Aliasing

In Figure 4.23 a periodic signal is sampled with timing sample intervals that are larger than half of the period of the signal itself. Thus, the used sampling interval is larger than the Nyquist interval and Shannon's theorem is violated by this fact. Such a kind of sampling also is called undersampling. If the data samples that should represent the original signal are analyzed without the knowledge of the original signal, the waveform reconstructed from the data samples has a lower frequency than the original signal. In the frequency-domain, aliasing folds the frequencies that are beyond the Nyquist frequency back into a frequency band below the Nyquist frequency with a distance to the Nyquist frequency that is the same as the distance of the original frequency to the closest integer multiple of the Nyquist frequency.

While aliasing is used sometimes on purpose to measure signals beyond the sampling capabilities of an instrument in an undersampling configuration, it is usually avoided if possible. If undersampling is used, special care has to be taken that the alias effects are taken into account during data analysis.

[2]This section was written in collaboration with Hideo Okawara.

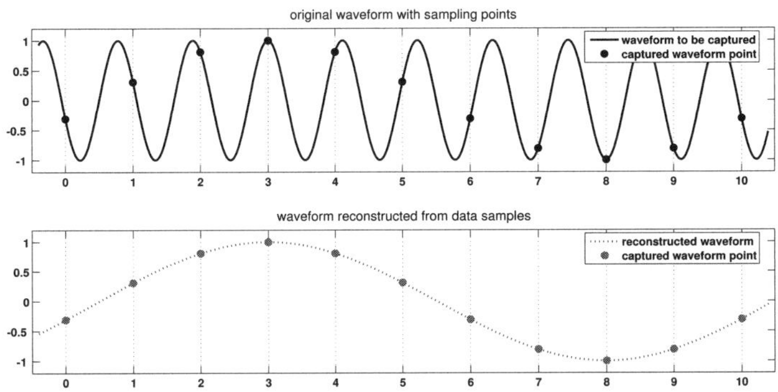

Figure 4.23 Signal aliasing for sampled data.

4.3.2 Digitizer

In order to avoid aliasing effects due to frequency components (e.g., noise components) beyond the Nyquist frequency, digitizers limit the bandwidth of their input signals. Figure 4.24 shows a typical block diagram of a digitizer.

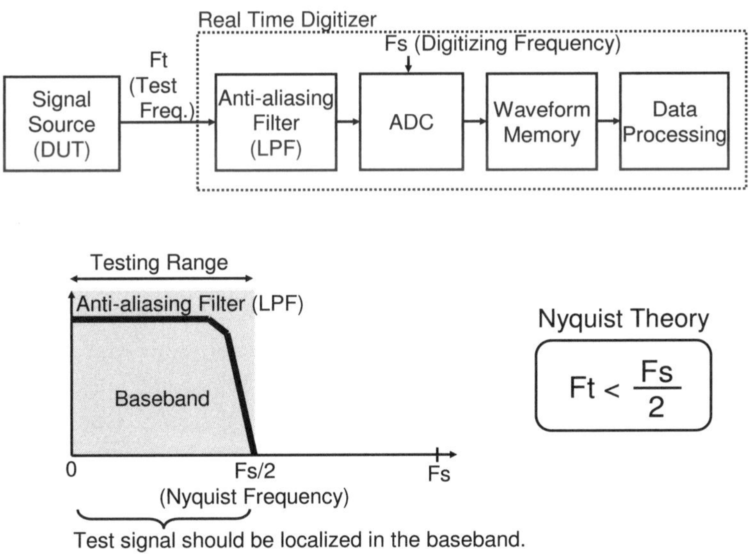

Figure 4.24 Digitizer block diagram.

The input signal first is fed through a low-pass filter (LPF) called the antialiasing filter. The band-limited signal after the antialiasing filter is digitized by the ADC. The sampling frequency of the digitizer that is controlled by the test engineer should be greater than twice the cutoff frequency of the antialiasing filter to conform with Shannon's theorem. The digitized data array is stored in the waveform memory of the ATE system. The data captured into the waveform memory is then processed using digital signal processing (DSP) methodologies to derive required test parameters. The analog input bandwidth of a digitizer is specified by the antialiasing filter and the real-time sampling frequency of the ADC.

In order to avoid aliasing within the full sampling rate range that a digitizer supports, the antialiasing filters that are used typically have a programmable bandwidth. This allows a user to adapt the antialiasing bandwidth to the sampling rate that was chosen for a certain measurement to avoid aliasing effects. The maximum antialiasing filter bandwidth usually is matched to the maximum sampling rates that are possible with the high resolution ADCs used in the digitizers in a way that Shannon's theorem cannot be violated. This, however, usually limits the bandwidth of digitizers in a way that they are not well suited for testing high-speed I/Os. As a general rule, a test engineer can assume that digitizers take care of the signals that conform to the Nyquist requirements.

4.3.3 Sampler

While the maximum frequency of a signal that the digitizer can measure is limited by its ADC sampling rate and the resulting maximum antialiasing filter bandwidth, a sampler does not have this constraint regarding analog input bandwidth. Figure 4.25 depicts a sampler block diagram and its frequency-domain coverage. The difference between the digitizer is that a sampler has no antialiasing filter. Instead, a sampling head with a track-and-hold stage is deployed at its front-end. The analog input bandwidth of the sampler is specified by the performance of the sampling head, which is usually several gigahertz. This enables sampler instruments to test high-speed I/Os that operate in the multigigabit range.

Of course, the ADCs used in samplers usually cannot implement real-time sampling rates that conform to Shannon's theorem. This is especially true if parametric measurements on the captured waveforms are in the focus where sampling intervals of few picoseconds might be required. Thus, high-speed I/O data capturing with samplers uses the knowledge about the expected data rate of the signal to be sampled and allows aliasing effects to happen for the sampled raw data. In a post-processing step, the aliasing effects are corrected and the original DUT signal is reconstructed.

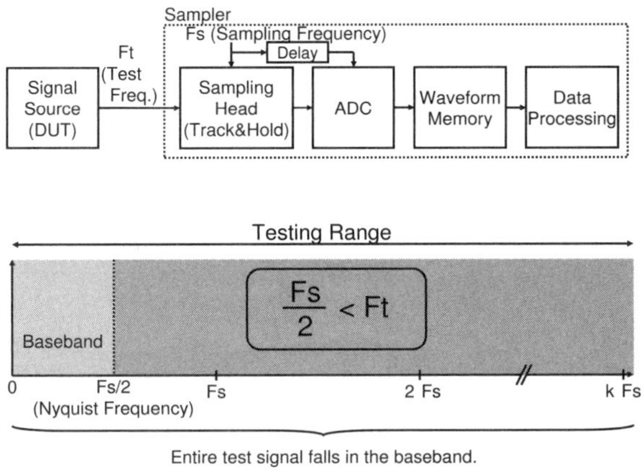

Figure 4.25 Sampler block diagram.

4.4 Parametric Measurements with Sampled Data[3]

The heart of a digitizer/sampler is a multibit resolution ADC that allows a direct transfer of the analog waveform voltage level into a digital representation within a single test pattern execution. As described in Section 4.2.9, digital pin electronics that provide a pass/fail result for certain level conditions can also be used to measure the same digital voltage level representation of a DUT signal with repeated test pattern executions and a different compare threshold setting for each test pattern iteration. Thus, although this section describes the fundamentals on how a digitizer or sampler can be used to acquire and measure a digital waveform, the same principles also can be applied on analog data captured by a digital channel into a digital waveform representation. Note that we will not discuss how sampler/digitizers are utilized in mixed-signal applications since it is not in the scope of this book. Nevertheless the reader should be aware that there is a significant additional body of theory related to sampler/digitizer usage [23, 24] that is not considered in this section.

4.4.1 Undersampling of High-Speed I/O Signals

As mentioned before, the maximum sampling rates of state-of-the-art sampler instruments are not sufficient to acquire multiple samples per UI of a high-speed I/O signal. In other words, sampler instruments cannot acquire high-speed I/O signals in real-time with the timing resolution required for

[3] This section was written in collaboration with Hideo Okawara.

parametric measurements. In order to be able to perform such measurements despite this limitation, samplers use undersampling.

Undersampling can be applied to periodic patterns only. The required samples of the signal that are taken during the different iterations of the periodic pattern are then folded back in a post-processing step into the unit test period (UTP) that represents the length of the fundamental pattern that is repeated. This post-processing step reconstructs one representation of the fundamental pattern. Since the sequence of the raw samples is not representing the sequence of the samples with regard to their phase location inside the UTP, the post-processing step that reconstructs the UTP has to reorder the samples along their appropriate phase within the UTP. The phase difference between subsequent samples in the reconstructed UTP defines the effective sampling rate that is much higher than the physical sampling rate and virtually can be increased to any value required within the accuracy limits of the sampler itself. The sample reshuffling operation is key for undersampling applications and usually is supported by ATE systems by a dedicated application programming interface (API).

For example, if a sampler is running at the sampling rate of 148.51 MHz and digitizes a 1.0 GHz sinusoidal waveform, the captured data and the phase offset of the samples with respect to the UTP look as in Figure 4.26. One UTP in this example matches exactly one cycle of the original sine wave. There are 15 sampling points in the UTP and exactly 101 cycles of the sine waveform are captured. Since 15 is not an aliquot[4] of 101, each of the sampling points hits a different phase for each cycle of the UTP. By reshuffling each sampling point according to its phase offset with respect to the UTP, a single cycle of precise sinewave can be reconstructed, as shown in Figure 4.27. The effective sampling rate achieved in this example would be 15 GHz.

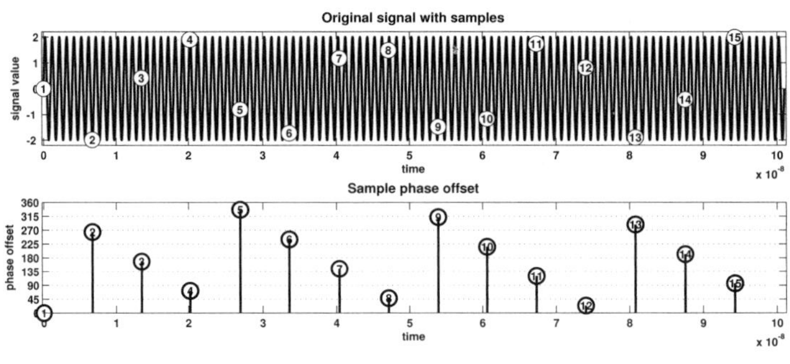

Figure 4.26 Original waveform with data samples and sample phase.

[4]An aliquot (or aliquot part) of an integer is any of its integer proper divisors. For instance, 2 is an aliquot of 12.

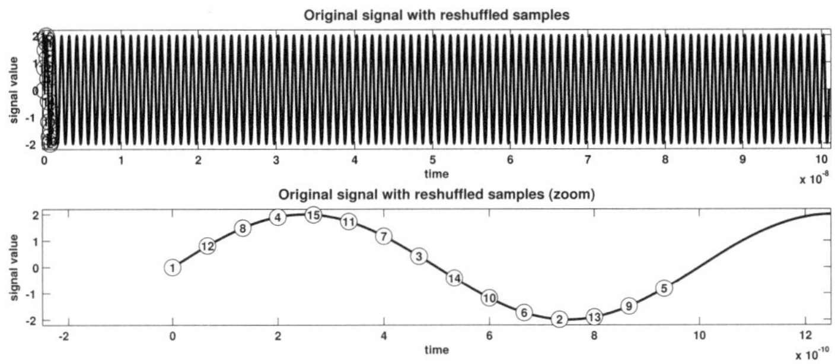

Figure 4.27 Reshuffled data and reconstructed UTP.

In the previous example the original signal was sampled to reconstruct a single UTP. Undersampling, of course, can also be used to sample the data for more than a single reconstructed UTP. In order to avoid phase jumps at the UTP boundaries of the reconstructed signal, the parameters used for sampling the data in this case have to fulfill the requirements of the coherency equation described in the following section.

4.4.2 Coherency Equation

Coherency is the key for successful undersampling measurements in digitizer/sampler applications. In order to have a coherent condition between the signal to be sampled and the sampling rate, the requirements defined by (4.1) have to be fulfilled [25].

$$\frac{F_t}{F_s} = \frac{M}{N} \tag{4.1}$$

In this equation, F_t is the frequency represented by the UTP of the signal to be measured and F_s is the sampling frequency of the sampler. M is the number of captured UTP cycles and N is the number of acquired samples per UTP. Thus, the effective sampling rate is obtained by the division of N with the length of one UTP. M and N must be integer numbers and should be mutually prime to get maximum signal information with a minimum of samples.

In order to provide more freedom for the selection of the sampling frequency F_s, the coherency condition can be extended according to (4.2). As with the standard coherency equation M_x is the aliased number of cycles into the baseband. For the extended coherency equation, M_x should be between $-N/2$ and $N/2$. Usually N is selected as a power of 2 to allow a direct FFT of the reconstructed signal. Thus, M_x usually is an odd number to maintain the guideline that N and M_x should be mutually prime. The new term K in the extended coherency equation determines how the frequencies of the signal to be measured are folded back into the baseband in dependency of the sampling frequency F_s for the selected coherency conditions. In the following we will call K the frequency zone number:

$$\frac{F_t}{F_s} = K + \frac{M_x}{N} \tag{4.2}$$

From the frequency-domain's point of view, the coherency (4.2) can be represented as the folded spectrum in Figure 4.28. If $M_x > 0$, the signal is located on the front page. If $M_x < 0$, the signal is located in the back page. Consequently $|M_x|$ indicates where the aliased signal would fall in the baseband. When the parameters for undersampling are selected, this information is used to set up the spectrum of the reconstructed signal in accordance with the application needs.

As one can see in Figure 4.28, a certain frequency range of the signal to be measured falls into a certain frequency zone for a given sampling frequency F_s. If the sampling frequency is changed, the frequency zones also change relative to the signal to be measured. In other words, the frequency range to be measured can be covered by multiple combinations of frequency zone numbers and sampling frequencies. This additional freedom allows a test engineer to adjust M_x in a way that assigns the favorite frequency bin location to the target signal by selecting the appropriate F_S and K combination that supports the desired M_x.

4.4.3 Capturing Digital Waveforms

Let's start by analyzing how a digitizer captures a waveform. Figure 4.29(a) shows a clock waveform with a frequency F_t. Three cycles of this clock waveform are digitized with 16 points of data so that the frequency condition is $F_t/F_s = 3/16$ and $M/N = 3/16$, where F_s is the sampling (digitizing) rate, M is the number of test signal cycles, and N is the number of data sampling points. The captured data consists of three cycles of the primitive waveform as Figure 4.29(b) shows. The samples are reshuffled as described earlier so that a single cycle of the precise waveform is obtained as shown in Figure 4.29(c).

Moving to a sampler in undersampling condition, the same waveform, for example, would be sampled over 35 cycles with 16 points of data, as shown in Figure 4.30(a). The frequency condition in this example can be described as $F_t/F_s = 35/16 = 2 + 3/16$, with $M_x = 3$. This setup samples three cycles of the primitive waveform as shown in Figure 4.30(b). Reshuffling the three-cycle waveform can replicate the single cycle of the original waveform shown in Figure 4.30(c).

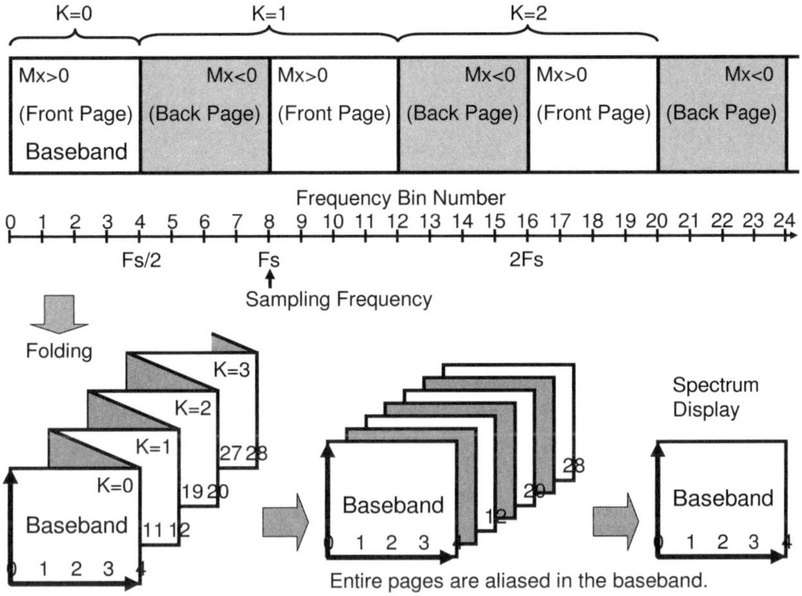

Figure 4.28 Coherency equation versus folded spectrum.

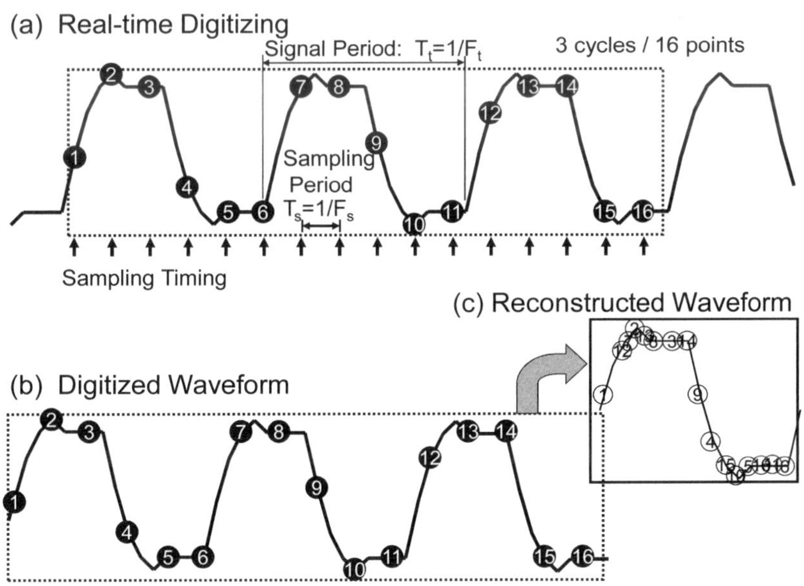

Figure 4.29 (a–c) Waveform sampled by a digitizer $F_t/F_s = 3/16$.

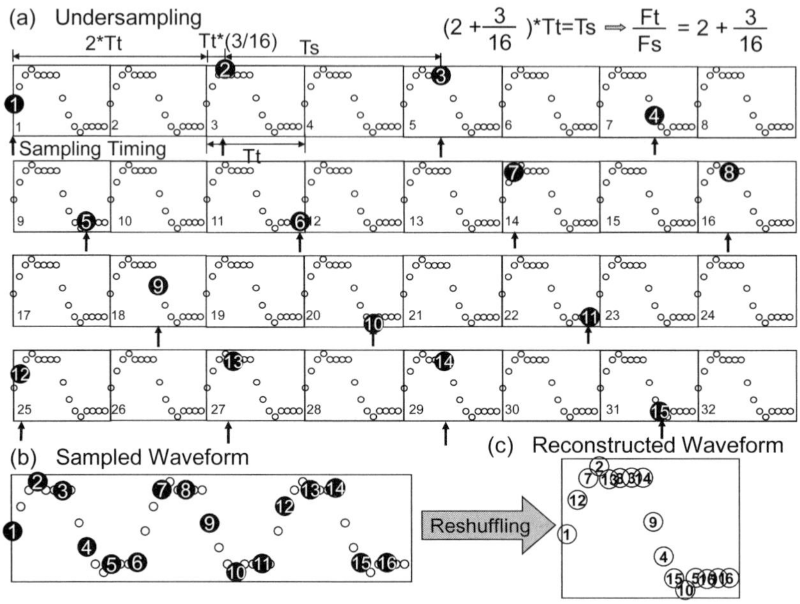

Figure 4.30 (a–c) Undersampling $F_t/F_s = 3/16$.

Figure 4.31 shows the case of a 31-cycle waveform that is sampled with 16 points, that is, $F_t/Fs = 31/16 = K + M_x/N = 2 - 1/16$ so that $M_x = -1$. The sampled data directly replicates a single cycle of the original waveform. However, it appears inverted. When M_x is negative, the waveform is reconstructed inverted. This does not pose a problem for analog signals that in most cases are analyzed in the frequency-domain using spectrum analysis methodologies. The signal reversion does not affect the signal spectrum. However, if such sampling conditions with negative M_x values are applied to digital signals, the reconstructed signal has to be corrected by changing the index of the waveform data array to obtain correct measurement results.

In real digital applications, usually longer pattern sequences than just the digital clock cycle considered in our previous description need to be captured for certain parameter measurements. Such real life sequences, for example, are PRBS patterns or the PCI Express compliance pattern. Coherent undersampling can be used also for these kinds of patterns as long as they are periodic and a DUT can be operated in a way that the desired pattern is generated repeatedly. In such a case F_t represents the frequency of the pattern repetition and the UTP represents the length of one complete pattern sequence that is repeated. Since coherent sampling increases the effective sampling rate at the cost of the number of required pattern repetitions, coherent sampling of

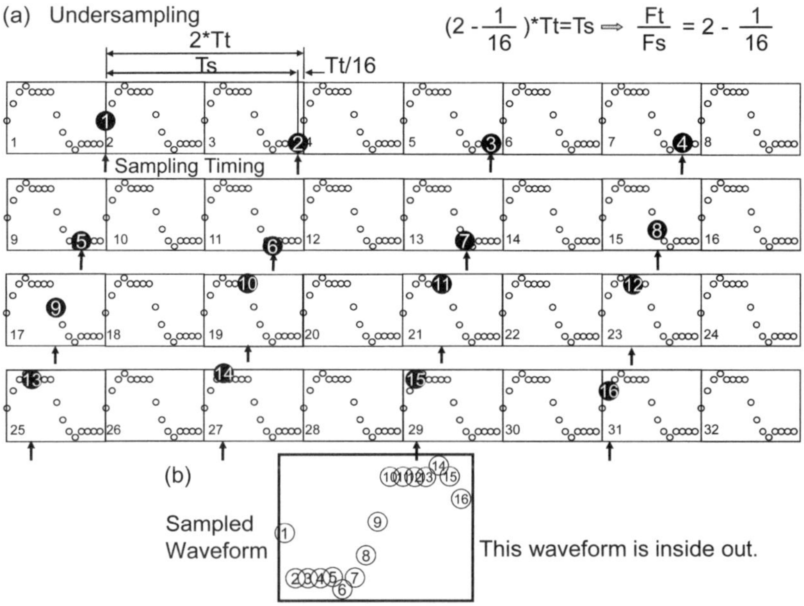

Figure 4.31 (a, b) Undersampling $F_t/F_s = 31/16$.

long pattern sequences with high resolution can increase the required pattern execution time significantly.

Example Let's assume that one wants to acquire one PRBS7 sequence for a 5 Gbps signal. The execution time for a single PRBS7 sequence consisting of 127 bits would be 25.4 ns. This time represents one UTP in our case and thus results in $F_t = 39.37$ MHz. If this PRBS sequence is to be sampled with a 10 ps resolution, $N = 25.4\ \text{ns}/10\ \text{ps} = 2{,}540$ sampling points are required. If M is selected to be 2,541, the pattern run time for the sampling execution would be about 64.5 μs. If more than one PRBS sequence is to be sampled, this pattern run time would increase accordingly.

4.4.4 Special Considerations for Coherent Sampling with Digital ATE Channels

The advantage of coherent sampling with digital ATE channels versus compare strobe movements in sampling resolution increments via the ATE timing system is the fact that a single functional test delivers data for all sampling points regardless of the sampling resolution. Of course, this comes at the cost of longer pattern execution times because the number of required waveform repetitions depends on the desired sampling resolution according to the coherence formula. However, executing a single functional test with a longer pattern typically is faster than executing multiple functional tests with timing changes in between. Also, coherent sampling allows high resolution sampling beyond the resolution of the timing system resolutions and without the negative influence of delay line nonlinearities.

Since the coherency requirement with M and N being prime to each other means that also F_t and F_s do not have friendly fractions, digital ATE channels that are used for coherent measurements need to run in different clock domains with a frequency offset to the channels that stimulate the DUT.

4.5 Power Supplies

ATE power supplies, also referred to as device power supplies (DPS), not only provide the needed voltage/current to the DUT but also typically offer the ability to measure important DUT parameters like the quiescent current (IDDQ) or the steady state current (IDD). In terms of requirements, power supplies are typically divided into subcategories like high pin-count, low noise, high current, and so on. Figure 4.32 shows pictures of a high pin count ATE DPS module (128 channels with a maximum current of 0.5 A per channel) and a high current ATE DPS module (4 channels with a maximum current of 40 A per channel).

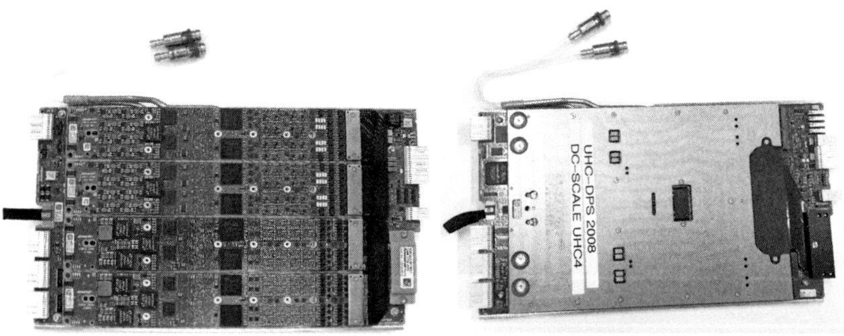

Figure 4.32 Pictures of a high pin count (128 channels), low current ATE power supply module (left), and a low pin count, high current ATE power supply module (right) (courtesy of Advantest).

The difference between a high current and a low current ATE DPS can also be seen on the interconnect between the ATE and the DUT test fixtures as shown in the example of Figure 4.33.

Figure 4.33 Pictures of a high pin count (128 channels) low current ATE power supply pogo pin block (left) and a low pin count, high current ATE power supply pogo pin block (right) (courtesy of Advantest).

High-current power supplies typically have to provide currents in the tens or hundreds of amperes since some applications like a microprocessor might require very high current values [26]. If a single power supply channel cannot provide the required amount of current, there is the option to gang multiple channels to obtain the required current value.

Another possible classification of DUT power supplies in ATE systems is between analog and digital power supplies: analog power supplies are intended for analog application or mixed signal application where a very low noise floor is needed, while digital power supplies are intended for digital applications where the noise floor requirements are not as high. This separation into analog and digital power supplies does not exist in some more recent ATE systems.

An ATE power supply contains two sets of connections for the DUT as shown in Figure 4.34. One is the power/ground pair (sometimes named force pair) that provides the power to the DUT. The other is the power and ground sense connections. These sense connections are used by the power supply control circuitry to measure the voltage at the DUT.

If the value at the DUT is different from the programmed value (note that the voltage at the DUT might be different from the voltage at the power supply output in certain time instants), the power supply control circuity will try to get the measured value equal to the programmed value by increasing/decreasing its own output voltage [27]. Because of this it is important that the power and sense lines are connected together as close as possible to the DUT power terminals. However, it is also important to note that this feedback circuit is slow and cannot compensate for the fast changes at the DUT power pins due to

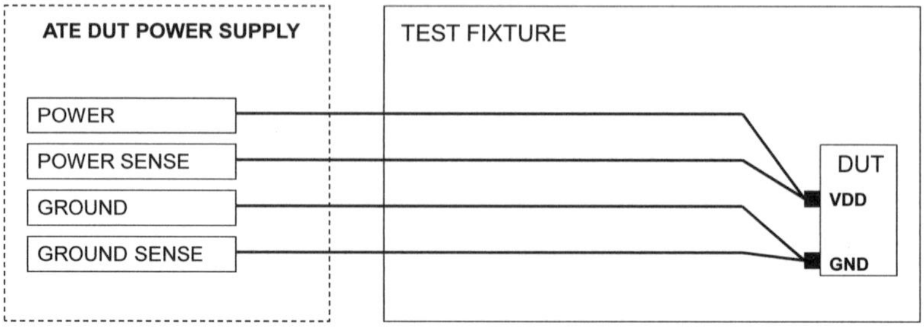

Figure 4.34 Block diagram describing the force/sense connections between an ATE DUT power supply and the DUT.

transistor switching. This is one of the design targets of the power distribution network (PDN) on the test fixture, which is discussed in Section 8.11.

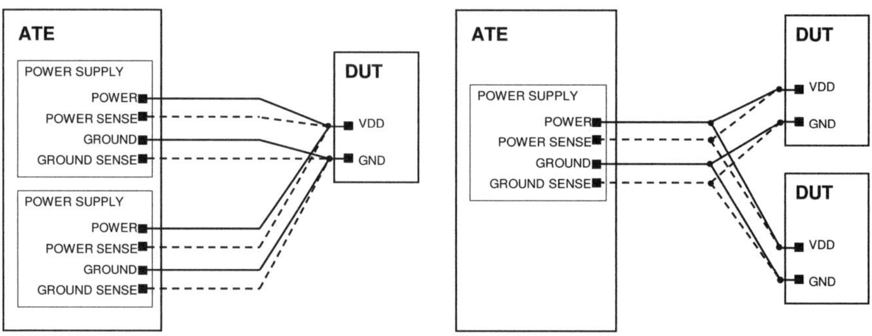

Figure 4.35 Block diagram describing the force/sense connections for two power supplies connected to a single DUT (left) and one power supply powering two DUTs (right).

In some situations it is necessary to use multiple power supplies to provide the required current to a DUT (power supply ganging) while in other situations like on a multi-site application one power supply is used to power multiple DUTs. Both of these options require proper connections of the power supplies' power and sense lines as illustrated on Figure 4.35. It is also important in both situations that the force lines ohmic loss is as balanced as possible. Note that when sharing one power supply between multiple DUTs, it is critical that all DUTs are running the exact same test program at the exact same time. This guarantees that the current required by each DUT is as similar as possible. The reason is that the sense value will be an average of the voltage at each DUT power pin.

References

[1] M. L. Bushnell and V. D. Agrawal, *Essentials of Electronic Testing for Digital, Memory, and Mixed-Signal VLSI Circuits*. New York: Springer, 2000.

[2] I. A. Grout, *Integrated Circuit Test Engineering*. New York: Springer, 2006.

[3] R. Rajsuman, N. Masuda, and K. Yamashita, "Architecture and Design of an Open ATE to Incubate the Development of Third-Party Instruments," *IEEE Transactions on Instrumentation and Measurement*, vol. 54, Oct. 2005.

[4] J. Kelly and M. Engelhardt, *Advanced Production Testing of RF, SoC and SiP Devices*. Norwood, MA: Artech House, 2007.

[5] J. Moreira, G. Haensel, and F. Koban, "Addressing the Challenges of Implementing an At-Speed Production Test-Cell for 10Gb/s Wafer Probing," *DesignCon*, 2005.

[6] M. G. Davis, "The Effect of Period Generation Techniques on Period Resolution and Waveform Jitter in VLSI Test Systems," *IEEE International Test Conference*, 1996.

[7] N. Zaman and A. Spilman, "Triggering and Clocking Architectures for Mixed Signal Testing," *IEEE International Test Conference*, 1998.

[8] L. Xiu and Z. You, "A "Flying-Adder" Frequency Synthesis Architecture of Reducing VCO Stages," *IEEE Transactions on Very Large Scale Integration (VLSI) Systems*, vol. 13, Feb. 2005.

[9] P. P. Sotiriadis, "Theory of Flying-Adder Frequency Synthesizers - Part I: Modeling, Signal Period and Output Average Frequency," *IEEE Transactions on Circuit and Systems I: Regular Papers*, vol. 57, Aug. 2010.

[10] P. P. Sotiriadis, "Theory of Flying-Adder Frequency Synthesizers - Part II: Time- and Frequency-Domain Properties of the Output Signal," *IEEE Transactions on Circuit and Systems I: Regular Papers*, vol. 57, Aug. 2010.

[11] P. P. Sotiriadis, "Intrinsic Jitter of Flying-Adderdder Frequency Synthesizers," *IEEE Sarnoff Symposium*, Mar. 2009.

[12] R. D. Adams, *High Performance Memory Testing: Design Principles, Fault Modeling and Self-Test*. Boston, MA: Kluwer Academic Publishers, 2003.

[13] PCI-SIG, *PCI Express Base Specification Revision 2.1*, Mar. 2009.

[14] B. Roth and J. Moreira, "Equalization in the HSM6800 Pin-Electronics," *Verigy VOICE Users Conference*, 2010.

[15] J. Moreira, M. Tsai, J. Kenton, H. Barnes, and D. Faller, "PCB Loadboard Design Challenges for Multi-Gigabit Devices in Automated Test Applications," *DesignCon*, 2006.

[16] J. Wolf, "Frequency and Jitter Testing in High-Volume Production Using the V93000 Pin Scale Per-Pin TIA," *Verigy VOICE Users Group Conference*, 2006.

[17] A. Roskin and P. Dayal, "Using the TMU for Accurate Jitter Measurements on Modern Phase Locked Loop (PLL) Devices," *Advantest VOICE Users Group Conference*, 2015.

[18] W. Maichen, *Digital Timing Measurements*. New York: Springer, 2006.

[19] U. Shankar, "Test Challenges for SONET/SDH Physical Layer OC3 Devices and Beyond," *IEEE International Test Conference*, 2001.

[20] D. C. Keezer and Q. Zhou, "Alternative Interface Methods for Testing High-Speed Bidirectional Signals," *IEEE International Test Conference*, 1998.

[21] J. Moreira, H. Barnes, and H. Werkmann, "DUT Loadboard Layout for Multi-Gigabit Bi-Directional Interfaces," *Verigy Users Conference*, 2007.

[22] S. Kojima, Y. Arai, T. Fujibe, T. Ataka, A. Ono, K. ichi Sawada, and D. Watanabe, "8Gbps CMOS Pin electronics Hardware Macro with Simultaneous Bi-directional Capability," *IEEE International Test Conference*, 2012.

[23] G. Roberts, F. Taenzler, and M. Burns, *An Introduction to Mixed-Signal IC Test and Measurement, 2nd Edition*. Oxford, UK: Oxford University Press, 2012.

[24] M. Baker, *Demystifying Mixed Signal Test Methods*. Amsterdam, NL: Elsevier Science, 2003.

[25] M. Mahoney, *DSP-Based Testing*. IEEE Computer Society Press, 1987.

[26] K. Aygün, M. J. Hill, K. Eilert, K. Radhakrishan, and A. Levin, "Power Delivery for High-Performance Microprocessors," *Intel Technology Journal*, vol. 9, Nov. 2005.

[27] D. Gizopoulos, *Advances in Electronic Testing: Challenges and Methodologies*. New York: Springer, 2006.

5

Tests and Measurements

This chapter presents a discussion on the typical tests associated with the driver and receiver parts of a high-speed digital interface. It is important to understand that all of these tests have one objective, to guarantee that the number of errors on a digital interface transmission will be below a certain threshold. Any digital interface has the possibility of having a transmission error at some point in time (independent of how good the physical layer design is). This is due to the physics associated with semiconductor devices and transmission over a nonideal medium. Note that at higher layers of the transmission protocol, some of these errors can be corrected by error correction methodologies embedded in the protocol [e.g., forward error correction (FEC)]. But there is always a price to be paid since more robust error correction methodologies imply a loss on bandwidth.

5.1 Bit and Pattern Alignment

As we have already described in Section 2.6, one of the key building blocks for high-speed I/O drivers are PLL circuits. One characteristic of high-speed I/Os that use PLLs to generate their high-speed clocks is that the phase offset between the reference input to the PLLs and the high-speed clocks generated by the PLLs that serve as timing reference for the high-speed I/O data of the device are typically not deterministic. Thus, the latencies between the reference clock or other ATE stimuli provided to the DUT and the high-speed signals generated by the DUT are not deterministic. The problem with this unknown phase and latency relationship is twofold.

The first aspect is the uncertainty where the compare strobe edges of the ATE system are located in relationship to the data bits the ATE comparator receives from the DUT (bit uncertainty). Without taking care of this bit uncertainty, it is possible that the ATE compares the received data bits in an unsafe area at or close to the bit transitions.

The other aspect is the uncertainty for the pattern offset between the data pattern generated by the DUT and the sequencer controlled compare vector data on the ATE system (pattern uncertainty) as shown in Figure 5.1.

The problem is further exacerbated by the fact that the latency between the stimuli of the DUT and the reaction of the DUT to these stimuli usually is not predictable exactly and might vary each time the DUT is powered up. The same effect might be observed in the case of a restart of the PLL input (e.g., the reference clock or the clock extracted from the incoming data), which also could change the phase and latency the device introduces between incoming clock/data and outgoing data to the ATE system.

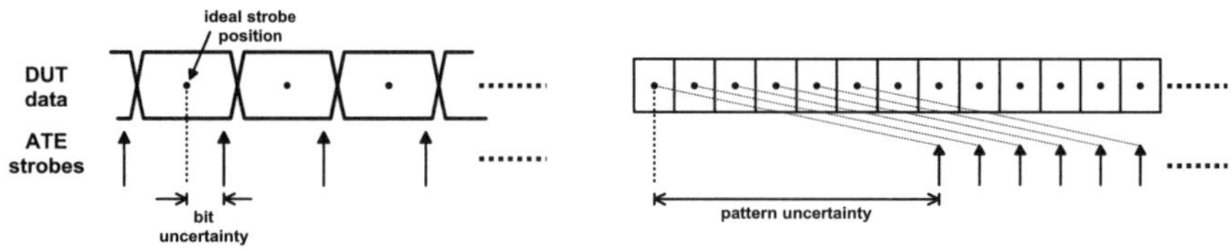

Figure 5.1 The alignment challenge for correctly performing a functional test.

The most straightforward way to solve this challenge is to perform a timing search on the ATE side until one finds the exact timing delay between the signal generated by the DUT and the ATE pin electronics receivers. Although this procedure is straightforward, it can be extremely time consuming because the pattern uncertainty that sets the requirements for the timing search range can be large while the bit uncertainty forces search steps that are significantly smaller than a single bit time (a single UI). In some situations the timing architecture of the ATE system might not even be able to handle the large delays associated with the pattern uncertainty.

Thus, a more general and faster way to overcome this alignment challenge is to compensate bit uncertainty and pattern uncertainty in two separate steps. In the first step, the bit uncertainty is compensated by bit alignment and in the second step, pattern alignment compensates for pattern uncertainty.

In a system environment, these uncertainties are usually handled by training sequences that usually have to be applied between two link partners as part of the power-up sequence or after the change of operating conditions. Although the details of this training vary between high-speed I/O standards (e.g., PCI Express, GDDR5), the fundamental target of the training sequences is to achieve bit and pattern alignment to guarantee correct link partner communication with maximum signal timing margins.

5.1.1 Bit Alignment

As mentioned before, bit alignment is the task of placing the ATE's strobe edge timing in the center of the data eye. The resulting values for this timing shift are less than one UI or bit time interval as the expression "bit alignment" implies. The first step in this regard is to identify the phase of the incoming data stream or, in other words, identify where the bit transitions in the incoming data stream occur relative to the ATE timing. Once this phase relationship is identified, the timing of the ATE comparators is offset by half a UI from the identified phase alignment point to set up the compare timing to the center of the received data eye.

ATE systems offer several ways to accomplish bit alignment. Timing search-based approaches search either for a single bit transition or for the transitions of multiple bits to identify the bit boundaries. A more accurate bit alignment is possible with an approach that scans the bit error rate (BER). Finally, some ATE platforms offer integrated hardware-based bit alignment.

5.1.1.1 Single-Bit Transition Search

The standard way to implement bit alignment on ATE is based on timing searches. Since a global fail-to-pass (or pass-to-fail) search with a compare to the full expected pattern only yields a result if potentially existing pattern offsets are also covered by the timing search range, this would lead to the problem of a large search range with a high timing resolution as mentioned above. In order to be able to use a timing search with a minimum search range, the compare pattern is set up with a single compare to high or low. With such a timing search setup, the closest bit transition in the direction of the search will be identified on the received pattern regardless of a potential pattern offset between received and expected data. For this search, a single compare threshold is set to the center of the expected level range in order to get the functional test result transition as good as possible at the center of the signal waveform transition as shown in Figure 5.2.

Taking into account that the used compare pattern only compares for a single bit within the received data stream, it is obvious that the search result obtained for the phase relationship between the DUT and the ATE system does not take into account the timing jitter and amplitude noise of the received signal. Thus, the accuracy achieved with this kind of phase identification is relatively low and might be acceptable for functional testing only.

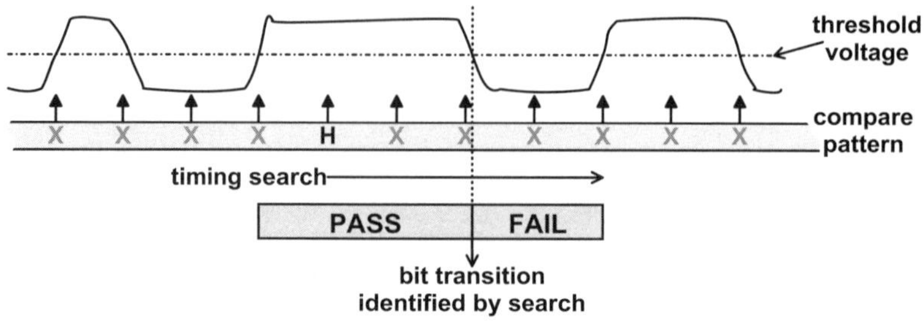

Figure 5.2 Bit alignment using a transition search.

5.1.1.2 Multiple-Bit Transition Search

One improvement of the result accuracy for a purely pass-fail based search would be to use a pattern that consists of a pattern sequence that is repeated by the DUT. Representatives for such patterns are, for example, the PCI Express compliance pattern or a PRBS pattern generated by the DUT. Such a pattern can hide the pattern uncertainty from the search to some extent because the compared pattern is reoccurring latest after the length of the repeated pattern sequence.

With this, a full compare to the pattern sequences is possible (with some mask margins at the start and the end of the pattern) for the search with a search range that corresponds to the length of one of the pattern sequences used. The extreme example for a pattern with these characteristics would be a bit clock pattern that consists of the pattern sequence 01. It is obvious that a pass-fail based search will yield a search result within a search range of two UIs even if the compare on the ATE is done for more than just a single bit. With this technique it is possible to take at least some of the jitter components into account that will be present with the functional pattern to improve the phase search accuracy.

5.1.1.3 Bit Error Rate (BER) Scan

A more sophisticated and accurate phase identification using a search approach can be accomplished, if the result is not just a simple fail-to-pass or pass-to-fail decision but collects the functional test errors observed for every search step. Such a measurement implementation delivers constant error count numbers for consecutive search steps inside the data eye. If the search steps are in the range of a data transition, the error count will vary from search step to search step.

Using error count results during the search has the advantage that the bit alignment can happen on a real functional pattern with a minimum search range or, more correctly, with a minimum error count scan range. Another advantage is that the full jitter characteristic of the received data pattern is obtained and can be analyzed. With an ATE channel that has the ability to measure the number of functional test errors for a single functional test, the compare timing is moved over a range of two UIs and for each timing step the number of errors for each pin to be analyzed is recorded.

The recorded error count signature (error density) per pin will show a variation of the error count over time regardless of to what data is compared. Note that we will not have a zero error region since we so far do not have a pattern alignment, but in the timing edges of the data eye the error density will vary due to jitter. Also the stable error count regions will vary in the number of errors observed due to the movement of the compare strobes for the same compare data bits from one data eye into the next data eye. The maximum values of the derivative of this error density provide the right and left edges of the data eye. The optimal compare strobe timing (with regard to the ATE timing period) will be in the center of these two points as shown in Figure 5.3.

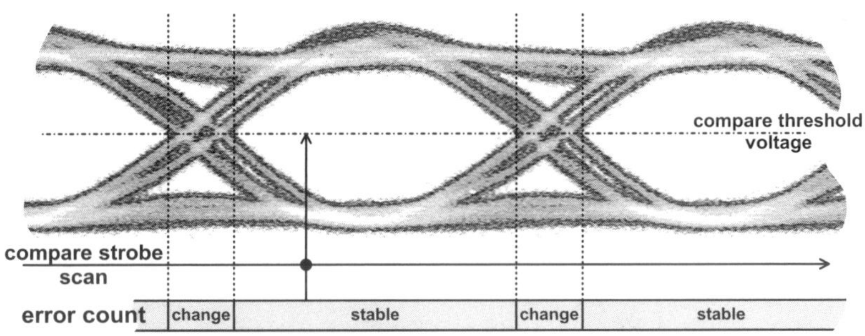

Figure 5.3 Bit alignment with measurement of the error density.

5.1.1.4 ATE Integrated

In order to simplify this task for the test engineer, the latest high-speed ATE pin electronics channels have embedded support (phase tracking and/or CDR) for the identification of the phase of an incoming data stream as described in Section 4.2.1. In addition to the identification of the phase, these channels usually also provide an automatic phase adjustment capability, which aligns the compare strobe edge to the center of the received data eye by setting an automatically calculated or user-defined timing offset to the identified data stream phase.

5.1.2 Pattern Alignment

After a successful bit alignment, it is necessary to perform a pattern alignment to obtain a functionally passing test. As for the bit alignment, some ATE cards have embedded support to achieve this task [1]. If such a feature is not available, there are several other possibilities to achieve pattern alignment. Pattern alignment using match loop constructs is one of these. Of course, also an alignment based on timing searches is possible in some cases. When timing search and programming ranges are not sufficient to cover the expected pattern offset, for some ATE systems an alignment using controlled ATE sequencer offsets is possible.

For pattern alignment, the main challenge is not necessarily the identification and compensation of the pattern offset between the DUT and the ATE system itself. A bigger challenge usually is to ensure that the relative timing offset between the DUT pattern start and the start of the sequencer(s) of the ATE stays constant for repetitive pattern executions that might be required to quantify the pattern uncertainty.

5.1.2.1 Match Loop

The classical way for ATE to adjust to unknown pattern latencies is the use of a match loop. A match loop is a feature of some ATE systems where the user defines a pattern that the ATE system loops around until it gets a functional pass. After this functional pass, the ATE will execute the target pattern that is used for the final pass-fail decision.

The advantage of such a loop construct is that the ATE sequencer(s) keep running and therefore no uncontrolled pattern offset change originating from sequencer restarts occurs. In order to achieve pattern alignment, a match loop is set up that compares to the pattern expected from the external pattern source (external instrument or DUT) plus one potentially masked compare bit as illustrated in Figure 5.4. Thus, a sliding compare window on the ATE is generated that advances by one bit relative to the pattern generated by the DUT with each match-loop iteration as shown in Figure 5.4.

It is obvious that at one of the loop iterations, the expected data matches the received data. If the match loop is left right after the matching loop iteration, pattern alignment is achieved [2]. It is important to note that a necessary requirement for the application of a match loop is the ability to implement a pattern offset between expected and received data in the match-loop body (in Figure 5.4 this offset is implemented by the additional compare bit). While this is not an issue when comparing to bit streams that run freely, for example, data originating from an external instrument or stimulus data generated autonomously by the DUT (e.g., via PRBS or other internal pattern generators), it can be a challenge if there is a one-to-one relationship between the stimulus data sent to the DUT and the data received from the DUT.

The problem in such a case is that both stimulus and compare data for the DUT are defined inside the match-loop body. If there is a fixed relationship between stimulus and compare data, the required pattern offset to implement the sliding compare window cannot be set up because either the stimulus or the compare data would have to be changed for each match-loop iteration for this purpose.

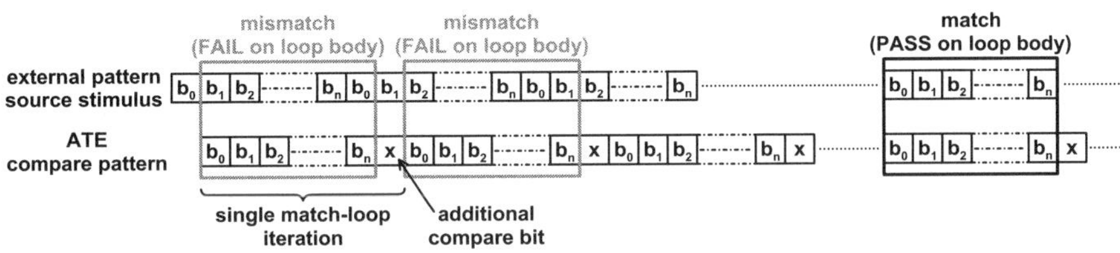

Figure 5.4 Sliding compare window implemented by a match loop.

5.1.2.2 Timing Search

Another way to achieve pattern alignment, is to do a timing search and to compensate the measured pattern offset by a timing adjustment on the ATE. Since bit alignment was done beforehand and compare strobes are positioned in the center of the data eyes, the search steps are set to one UI.

The problem with the timing search, however, in many cases is that the pattern offset can extend beyond the boundaries of the timing programming ranges of the ATE. If a timing search is possible (e.g., because the pattern repeated by DUT is shorter than the allowed timing programming range), one has to take care that the relative pattern offset between DUT and ATE stays constant for the repetitive sequencer starts that will be caused by this timing search. On ATE systems that support cycle times that are prime to each other and have synchronization mechanisms available to synchronize the different clock domains that are set up, a constant offset can be achieved by setting up a dummy clock domain.

This dummy clock domain runs at a cycle time that corresponds to the length of the pattern that the DUT loops on. The ATE synchronization mechanisms then take care that the sequencer(s) of the ATE channels that do the timing search always are restarted on a timing grid that has a spacing of the least common multiple of the two cycle times of the clock domains as shown in Figure 5.5. Thus, the sequencers of the searching pins always are restarted with the same relative pattern offset to the pattern generated by the DUT.

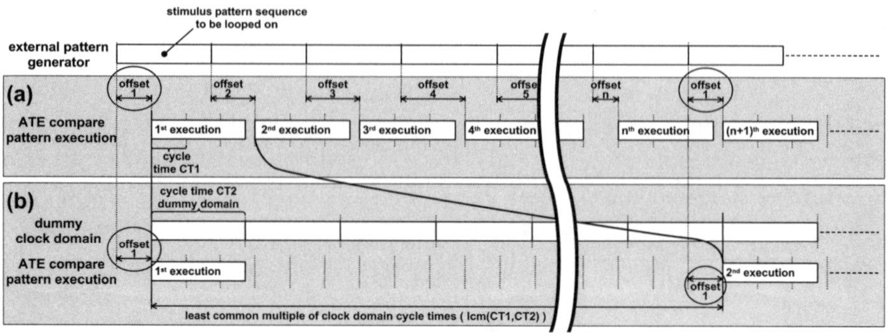

Figure 5.5 Relative ATE sequencer restart offset to external pattern (a) without and (b) with usage of clock domain synchronization.

5.1.2.3 Sequencer Offset

A final possibility for pattern alignment that even works with offsets bigger than the allowed timing programming range is to delay the generation of the compare pattern data on the ATE by inserting no operation (NOP) or dummy sequencer operations between the start of the sequencer(s) and the sequencer operations that generate the actual compare data. In order to calculate the necessary amount of NOP operations that need to be inserted, a portion of the received data is captured and its position within the expected data is analyzed.

The difference between the position of the captured data in the expected data and the position it actually was captured at represents the pattern offset that needs to be compensated by NOP operations to achieve pattern alignment. Of course, it is required that the captured data is a unique bit sequence within the expected data to be able to identify the pattern offset between the received and expected data. For PRBS patterns, this is achieved by capturing the number of bits that is used by the linear feedback shift register (LFSR) that generates the PRBS sequence. For other patterns the number of bits captured needs to be set in a way that guarantees that a unique bit sequence is captured in any case.

5.2 Functional Test

At-speed functional tests are not only at the core of ATE-based testing in general, but also are an important factor for testing high-speed interfaces. Some of the measurement algorithms that are used to characterize important high-speed parameters are based on functional tests. During a functional test, stimulus data is applied to the DUT by the ATE drivers, external instruments, or a combination thereof. The ATE receivers compare the data generated by the DUT as a reaction on the stimulating data to a data pattern defined inside the ATE system.

The pattern applied and compared by the ATE at a single timing instance also is called pattern vector or simply test vector. The minimum result a functional test delivers is whether the data pattern received from the DUT matches the ATE compare data or not (pattern pass or fail). The way that the data received from the DUT is interpreted with regard to the nomenclature used for the compare pattern is determined by the compare threshold voltage(s) used by the receiver pin electronics front-end. For high-speed pin electronics it is quite common that only a single threshold voltage exists, and thus, the data received can be distinguished between high and low only (no intermediate state is usually available for lower-speed dual-threshold comparators).

The generation of test vectors for functional tests can happen in various ways. Typically, test vectors are derived from logical device simulations or other means that result in predefined static stimulus and compare data stored in the vector memory of the ATE. In the memory test area, huge amounts of vectors are required to do exhaustive functional testing. Since it would not be affordable to keep these test vectors statically stored in a vector memory, memory test systems usually generate their test vectors dynamically with algorithmic pattern generators (APG) that allow a high-level definition of the test patterns using a programming language style [3].

The actual test patterns are generated from this description either directly on-the-fly using dedicated APG hardware on some ATE systems or during a compilation process that generates test patterns optimized for the sequencing capabilities on other ATE systems. Recently, similar approaches also found their way into nonmemory high-speed ATE systems and allow a mixture of the classical statically predefined vector pattern storage in vector memory and the usage of patterns that are generated algorithmically on-the-fly. One example for such algorithmically generated patterns are pseudo-random bit sequences (see Appendix C).

During a single functional test, there is only limited user control on the applied timing and level settings. Once the functional test is started, the predefined settings for these parameters are used. These may only be modified at predefined positions within the pattern to predefined values but not in a dynamic manner. Although some ATE systems offer the capability to dynamically adjust these parameters within their data pattern definitions depending on the test result of the previously executed pattern vectors, the execution of such a pattern is not considered a single functional test. The reason for this interpretation is the fact that the evaluation of the pass/fail result and the parameter adjustment depending on this result interrupt the at-speed pattern execution and thus introduce a pause in the at-speed pattern.

Besides just a single overall pass/fail result for the complete functional test, ATE systems have various other capabilities to deliver more information about the results of a functional test execution. Another level of information, for example, is the exact number of errors observed during a functional test or even the exact positions of these errors within the test pattern. All of this information might be available also on a per-pin basis depending on the ATE system. In general, the more information on a functional test execution that is made available, the better a functional test can serve as a starting point for parametric measurements that use this information in post-processing steps, for example, the measurement of a BER (see Section 2.3), which necessarily requires the exact error count information of a functional test.

Another aspect of functional testing that gained relevance with the increasing amount of packet-oriented, encoded high-speed interface standards is the topic of protocol-aware testing. In some sense this also can be interpreted as algorithmic pattern-based functional testing because the test engineer does not work on a physical layer presentation for the vector data anymore but on a layer that shows the payload data transferred from and to the DUT. The algorithmic part of the pattern generation in this case is the transfer of the payload data into actual physical data and vice versa. We will discuss protocol-aware testing in more detail in Section 9.7.

5.3 Shmoo Tests

Shmoo tests play a central role in the characterization and production ramp phase of a device. With the shmoo test methodology, test and device parameter margins can be identified and a test engineer can draw conclusions on the actual performance of a DUT and on the test repeatability and stability for the selected test conditions.

For a shmoo test, one or more of the test parameters controlled by the ATE (e.g., timing or level parameters) are swept over a range defined by the test engineer. The resolution of the single sweep steps also is defined as a shmoo test parameter by the test engineer. For each of the sweep steps, a functional test is executed and the result of the test is recorded. As already described in Section 5.2, this result might not just be the pass-or-fail information of the functional test but also the number of errors seen for the sweep point or any other result supported by the ATE. With this, a shmoo test allows identification of the test margin that a certain parameter has compared to the actual device performance.

While shmoo tests that sweep a single test parameter already deliver important insights into the DUT operation boundaries, the real power of shmoo tests is unleashed if multidimensional shmoos are applied to a DUT. For such a multidimensional shmoo, sweep settings are applied to several test parameters. During the shmoo test, a functional test is applied for all combinations of parameter settings that are defined by the single sweeps. The tests for all shmoo parameter combinations easily allow identification of parameter dependencies and parameter settings for a safe device operation. However, it has to be taken into account that the overall number of functional tests that are executed for such a multidimensional shmoo grows exponentially with the number of shmoo dimensions.

Thus, test engineers typically limit the number of dimensions in real test implementations to two or three to achieve reasonable test times. There might be cases, however, where it makes sense to tolerate long test times and use shmoo tests with more than three dimensions. This mainly is true for characterization measurements that have the target to gather data that allows distinguishing critical parameters that need to be considered during production testing from noncritical test parameters that might not even be checked in a production test flow.

Besides reasonable test times, shmoo tests with three or less dimensions also have the advantage that they can be displayed relatively easily in a graphical manner. Graphical representations of shmoo test results often are referred to as shmoo plots.

Shmoo plots usually simplify the interpretation of the shmoo test results for a test engineer. In order to speed up the execution of a multidimensional shmoo, shmoo algorithms like fast-shmoo are applied [4]. In a fast-shmoo, an initial parameter sweep is executed that does not use the minimum sweep resolution defined by the test engineer but a lower resolution to scan the complete sweep area. After this initial measurement, only the areas that are located between parameter settings that caused a change in the test result are scanned with the desired resolution. A typical example of such a fast-shmoo is shown in Figure 5.6 for a data eye measurement.

When setting up shmoo tests, one has to specifically take care of the fact that sweep parameters selected for one of the shmoo dimensions have to be independent of other test parameters. If this is not the case (e.g., a shmoo that sweeps the supply voltage typically affects DUT input voltages and compare thresholds), the test engineer has to use a shmoo setup that tracks these dependent parameters together with the parameter that is swept on one of the shmoo dimensions. Such a shmoo setup is often referred to as tracking-shmoo. The parameter tracking usually is done via a user-defined dependency equation that describes how the tracked parameter has to behave

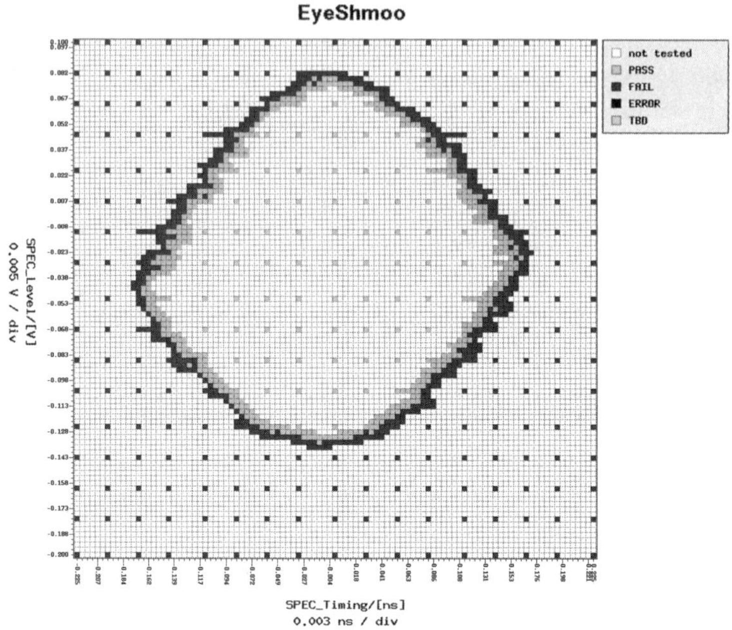

Figure 5.6 Fast-shmoo of device data eye.

in dependency of the parameter used in the shmoo setup. One example for such a tracked dependency would be the ATE drive and comparator threshold levels used during a shmoo test that sweeps the supply voltage of the DUT.

Another boundary condition that has to be taken into account is the potential dependency of the overall DUT functionality on one or more of the parameters used in the shmoo dimensions. This is especially true for high-speed interfaces that, for example, might require retraining after the change of parameters that have influence on the PLL phase locking of the high-speed I/O signals (e.g., reference clock frequency or supply voltage). If such a critical parameter is swept during a shmoo, the test engineer has to take care that the appropriate steps, for example, device retrainings, are performed at the correct instances during the shmoo execution.

A few examples for typical two-dimensional shmoo setups that are widely used for high-speed I/O devices are frequency shmoos that sweep the device operating frequency versus a data eye width measurement, data eye shmoos that sweep compare strobe timing versus compare level threshold, and VDD shmoos that show the dependency of the maximum operating speed of a device on the supply voltage level.

5.4 Fundamental Driver Tests

This section presents some of the typical measurements for testing an I/O cell driver and how these tests can be implemented on an ATE system.

5.4.1 Rise/Fall Time

The rise and fall time measurement delivers the amount of time the DUT driver takes to achieve a logic one voltage level starting at logic zero (for rise time) or starting at the logic one voltage level to achieve logic zero (for fall time). Typically rise or fall time is not measured from one voltage rail to another but from a given percentage from the rails voltage level, normally 10%–90% or 20%–80% as shown in Figure 5.7.

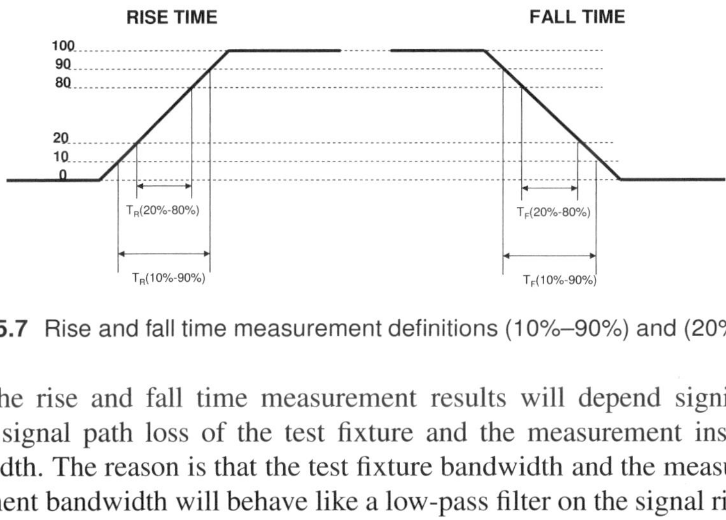

Figure 5.7 Rise and fall time measurement definitions (10%–90%) and (20%–80%).

The rise and fall time measurement results will depend significantly on the signal path loss of the test fixture and the measurement instrument bandwidth. The reason is that the test fixture bandwidth and the measurement instrument bandwidth will behave like a low-pass filter on the signal rise time, slowing it down. One way to address this issue is to use (5.1) to remove the effects of test fixture and instrument bandwidth. The formula requires the intrinsic rise time of the test fixture signal path and receiver bandwidth to be known so that it can be subtracted to the measured rise time to obtain the rise time at the DUT output. Note that this formula is only exact in special cases like Gaussian filters. In cases where the degradation is due to other causes like the skin effect in a coaxial cable, the formula is no longer exact [5].

$$Tr_{\text{MEASURED}} = \sqrt{(Tr_{\text{DUT}})^2 - (Tr_{\text{SIGNAL PATH AND RECEIVER}})^2} \quad (5.1)$$

Another important topic when measuring rise/fall times is the effect of jitter on the measurement. Jitter "spreads" the signal edge creating the problem of which points should be used for the time instants on the rise/fall time computation. The typical approach to address this challenge is to perform a histogram measurement on the two voltage levels intended to be used for measuring the rise time and then select the average value for the measurement as shown in Figure 5.8. This is equivalent to performing a waveform averaging in a real-time oscilloscope. This approach works very well in cases where the signal jitter is dominated by random jitter [6].

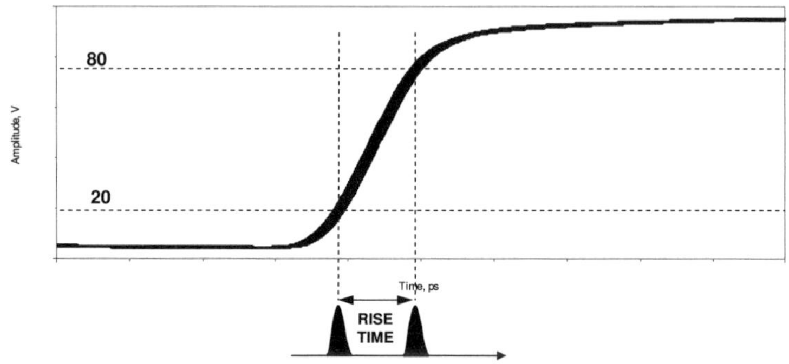

Figure 5.8 Using a histogram approach in a rise/fall time measurement.

5.4.2 Data Eye Diagram

The objective of the data eye diagram is to measure the degradation of a waveform from the DUT in comparison to the ideal one. In the ideal case, each bit in the data stream should have the time length at the optimal threshold voltage point equivalent to the but rate period T_{BIT} (e.g., 1 ns for a 1 Gbps NRZ signal) and the signal amplitude between the high and low levels should be equivalent to the expected ideal amplitude. Unfortunately a real signal will have jitter, amplitude noise, and limited rise and fall time as shown in Figure 5.9.

As already discussed in Section 2.1.2, a data eye diagram provides a visual representation of the parametric performance of the DUT driver. From a data eye diagram, several parametric performance values can be obtained like the rise/fall time, data eye width, data eye height, jitter, and the amplitude noise.

5.4.2.1 Measuring the Data Eye Diagram with a Digital Pin Electronics Receiver

To measure the data eye diagram with a digital pin electronics receiver, it is necessary to perform a shmoo test across the compare strobe timing and the

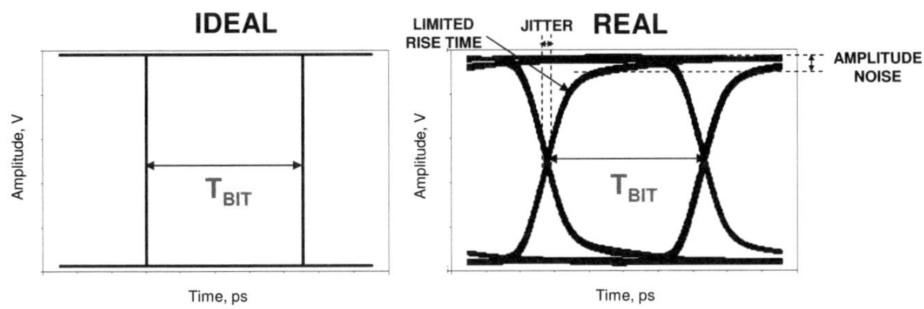

Figure 5.9 Ideal versus a real data eye.

voltage compare threshold. For this purpose the horizontal axis representing the strobe timing is divided into, for example, 100 steps. This is repeated for the vertical axis showing the compare voltage. For each combination of these steps (100 times 100 = 10,000), a functional test is executed by the digital pin electronics receiver.

The result of the functional test can either be a pass/fail or the number of failed bits if the pin electronics has the ability to count the number of failed bits. Figure 5.10 shows one example of a shmoo plot that corresponds to a data eye diagram measured with an ATE system. The data eye diagram in this case is plotted with different colors representing the different number of failed bits on the functional test. The test time will depend on the resolution used for the timing and voltage steps and the number of compared bits. The measured accuracy will depend on the digital pin electronics receiver bandwidth, the number of compared bits, the jitter and amplitude noise on the receiver compare strobe, and also on the compare strobe linearity associated with its movement along the time and voltage axis on the data eye diagram shmoo plot.

It is also possible to measure the data eye diagram with a digital pin electronics receiver using a coherent sampling approach as described in [7].

One of the major advantages of an ATE system is the ability to measure multiple I/O channels in parallel. This allows, for example, the overlay of multiple data eye diagrams corresponding to different pins in a I/O bus as shown in Figure 5.11. The 100% values correspond to a combination of strobe timing and compare voltages where the functional test for all pins in the interface bus results in a pass. Percentages below 100% indicate timing and voltage settings where at least one of the pins in the interface bus had a failure during the functional test. This type of overlay shmoo plot provides an overview of the performance difference between the different pins that compose the interface bus including any timing skew.

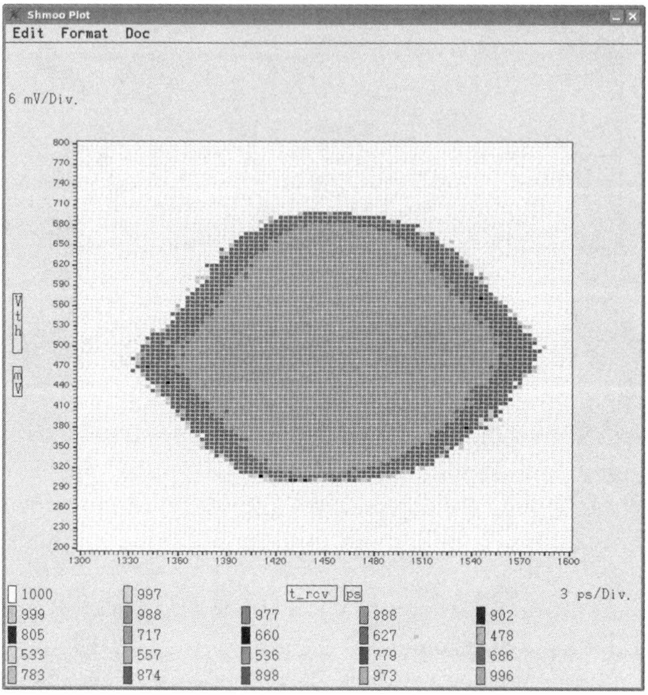

Figure 5.10 Example of measuring a data eye diagram with an ATE digital pin electronics receiver (courtesy of Advantest).

5.4.2.2 Measuring the Data Eye Diagram with a Sampler[1]

A sampler can also be used to generate a data eye diagram by applying the techniques described in Section 4.4. Figure 5.12 presents an example of a test configuration for performing a data eye diagram measurement with a sampler where the DUT drives a $2^7 - 1$ PRBS sequence at 2.5 Gbps.

In this example, the sampler is programmed to capture two sets of the entire PRBS pattern (i.e., a total 254 bits with $N = 65{,}536$ points of data). The sampler runs at the rate of 9.8423695 Msps by using the coherency configuration shown in Figure 5.12.

Figure 5.13(a) shows the captured waveform containing two sets of the PRBS sequence. By reshuffling the waveform with the key number 127 instead of 254, the data eye pattern can be reconstructed as shown in Figure 5.13(b) [8].

[1]This section was contributed by Hideo Okawara.

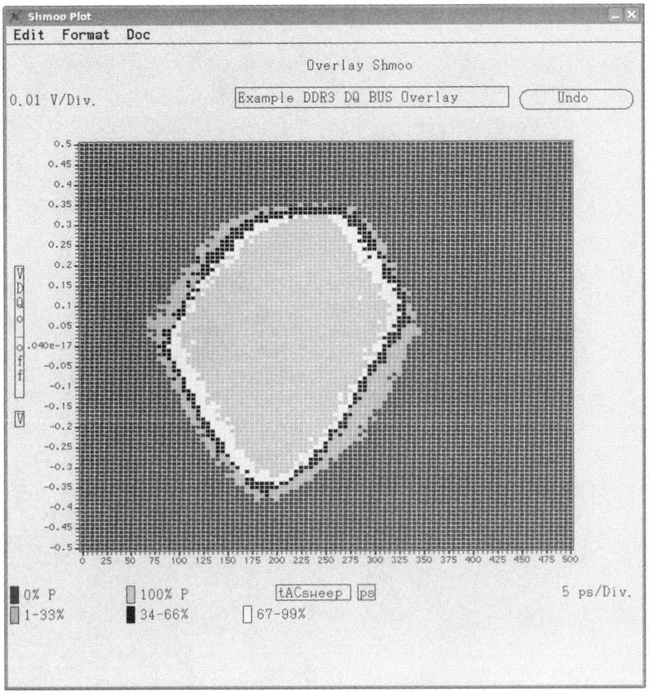

Figure 5.11 Overlay of multiple data eye diagrams from a DDR3 bus into a single data eye diagram (Courtesy of Advantest).

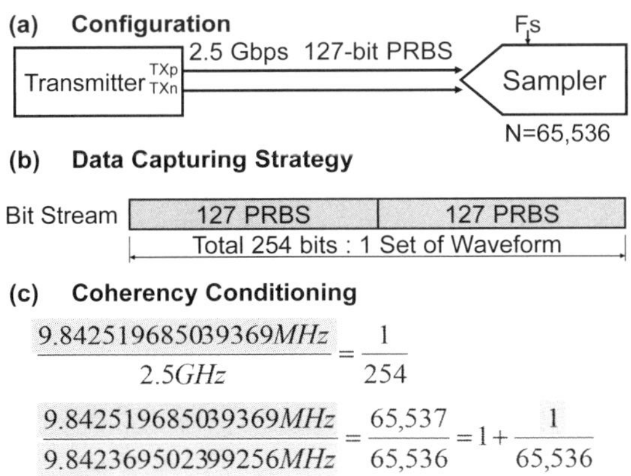

Figure 5.12 (a–c) Data eye measurement with a sampler configuration.

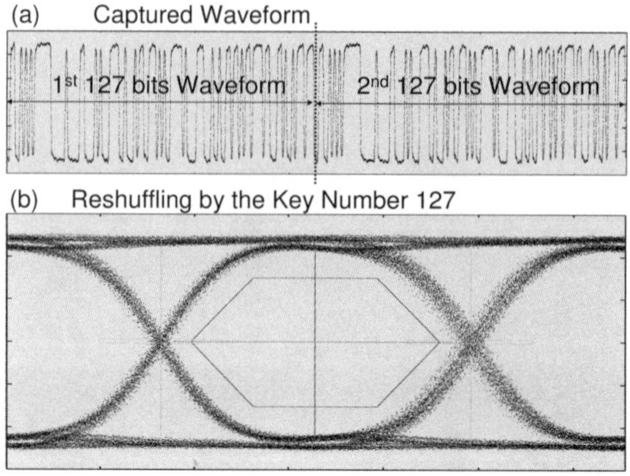

Figure 5.13 (a, b) Measured waveform and reconstructed data eye diagram.

5.4.2.3 Data Eye Width

The objective of the data eye width test is to measure the percentage of the ideal bit period where the setting of the strobe timing results in a functional pass (i.e., how much time margin is available for a device receiver connected to the DUT transmitter). This can be a complex task depending on the link implementation (e.g., some receiver designs have their own CDR circuit that is able to remove some of the timing jitter on the data). The question whether this circuit should be taken into account when measuring the eye width depends on the test requirements (e.g., some standards require a specific CDR to be used on all driver parametric measurements). This topic is addressed in Section 9.5 in more detail. In the remaining part of this chapter we assume no CDR circuit is used on the measurements.

Another issue is the fact that the peak-to-peak value of the random jitter increases over time (i.e., if the observation period is long enough, the data eye width will theoretically decrease to zero). This means that the measured data eye width depends on the number of acquired samples. This will be discussed more in detail in Section 5.4.3 with the BER bathtub curve.

Assuming that a data eye diagram is measured using a given amount of samples or bit compares, the data eye width is measured by computing the distance between the left and right edges of the data eye diagram at the optimum voltage threshold point as shown in Figure 5.14. In the figure the average values of the timing histograms at the optimal threshold point provide the data eye period but the data eye width will be the distance between the edges of the left and right histograms. Figure 5.15 shows an example of how the eye width is represented on an oscilloscope and the data eye shmoo on an ATE system.

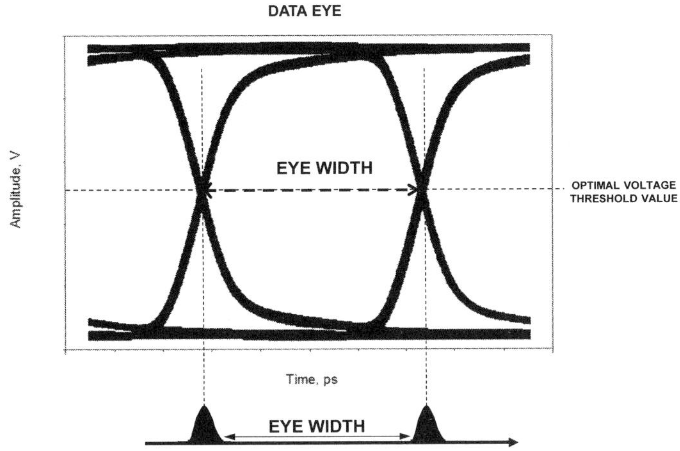

Figure 5.14 Measuring the data eye width.

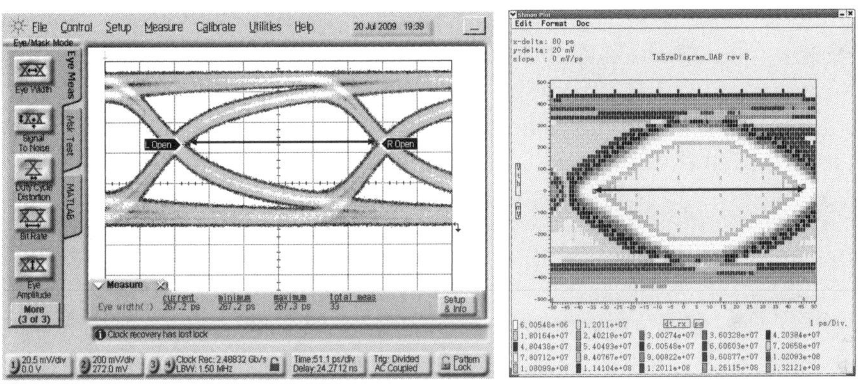

Figure 5.15 Examples of measuring the data eye width (left: Keysight Technologies DCA; right: shmoo with Advantest V93000 ATE).

5.4.2.4 Data Eye Height

The objective of the data eye height test is to measure the level margin available on the data eye from the DUT driver. This is important because the receiver on the link partner will receive a data eye that is degraded by the transmission channel. Therefore, it is critical to not only achieve enough timing margin but also ensure that the amplitude margin on the DUT driver data eye is sufficient for the link partner receiver to correctly identify the transmitted bits. The reasons why the data eye height does not correspond to the expected DC amplitude levels are the random noise inherent to the voltage levels and the bandwidth limitation on the DUT driver and the test fixture for a given transmitted bit, which depends on the preceding bits as shown in Figure 5.16.

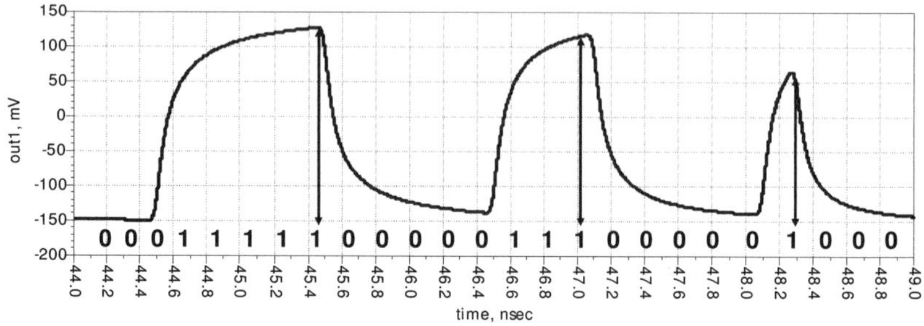

Figure 5.16 Example of a digital waveform after a lossy channel showing how the maximum amplitude reached by each bit depends on the previous bits.

When looking at a data eye diagram as a representation of a waveform, one can observe that the amplitudes of the various data bits on the data pattern will be superimposed. A data height measurement reveals the available voltage margin for the worst-case bit in a given data pattern. This can be accomplished by computing the difference between the top and bottom inner edges of the data eye diagram. There are several possible approaches as shown in Figure 5.17.

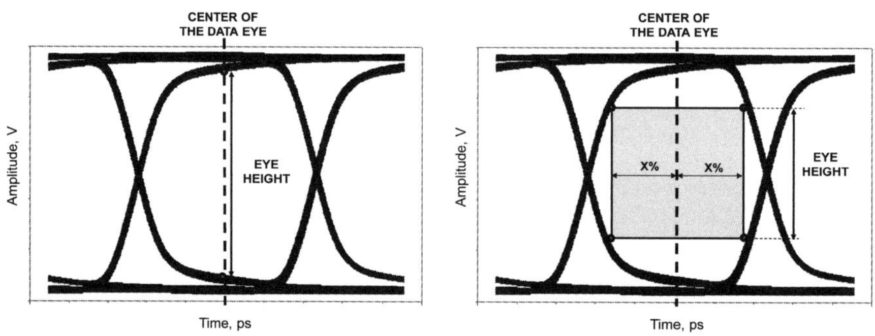

Figure 5.17 Possible approaches for measuring the data eye height.

One approach is to measure the voltage interval between the top and bottom edges of the pass/fail region taken at the center (optimal timing strobing point) of the data eye as shown in Figure 5.17 (left). Figure 5.18 (left) shows an example of this approach with an equivalent-time oscilloscope, and 5.18 (right) shows the same approach with an ATE-based data eye diagram.

Another option is to define a rectangular region around the center of the data eye where the top and bottom level rails must not cross as shown in Figure

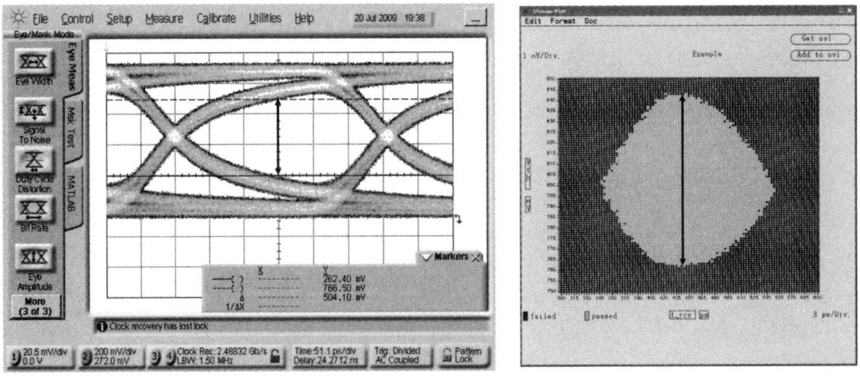

Figure 5.18 Examples of measuring the data eye height (left: Keysight Technologies DCA; right: shmoo with Advantest V93000 ATE).

5.17 (right). This approach takes into account that the receiver might not be ideal and that the strobing point might not be exactly at the center of the data eye. Figure 5.19 shows an example of this approach using an equivalent-time oscilloscope.

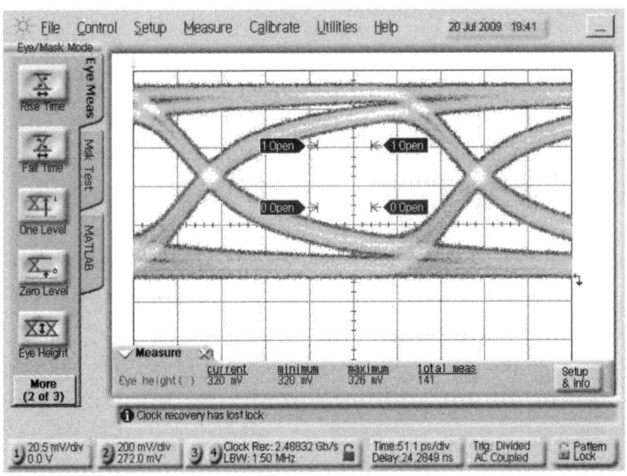

Figure 5.19 Example of measuring the data eye height with an equivalent-time oscilloscope using a pass region that is 20% of the UI around the data eye center.

5.4.2.5 Data Eye Mask Test

The data eye mask test is an approach to measure with a short test time if the data eye has enough performance for a given application or standard.

Instead of measuring the data eye parameters like the data eye height or width separately, a data eye mask is verified. Such a data eye mask consists of one or multiple geometric polygons (e.g., rectangles and hexagons) with a specific geometric configuration that is used to specify the minimum performance requirements on the DUT data eye. The data eye diagram must not "touch" any of the data eye mask polygons as shown in Figure 5.20. If a DUT I/O driver passes a data eye mask test defined in a standard, it means that the driver has the necessary level and timing margin to be compliant with the standard requirements.

With most measurement instruments (e.g., an equivalent-time sampler or an ATE sampler), the violations of the data eye mask are computed after the data eye diagram has been acquired. Applying digital pin electronics, the test time can be shortened by simply executing a functional test at the points specified by the data eye mask instead of acquiring the complete data eye through a shmoo plot. In Figure 5.20(b), for example, it is sufficient to check if any of the points for the data eye mask test polygon inside the data eye provide a fail during a functional test. For the polygon above the data eye also a simple functional test is performed which is exclusively compared to logical "high" and checks for errors. The test time of a data eye mask test can be further reduced by performing a fast eye mask test, which will be explained in Section 6.3.2.1.

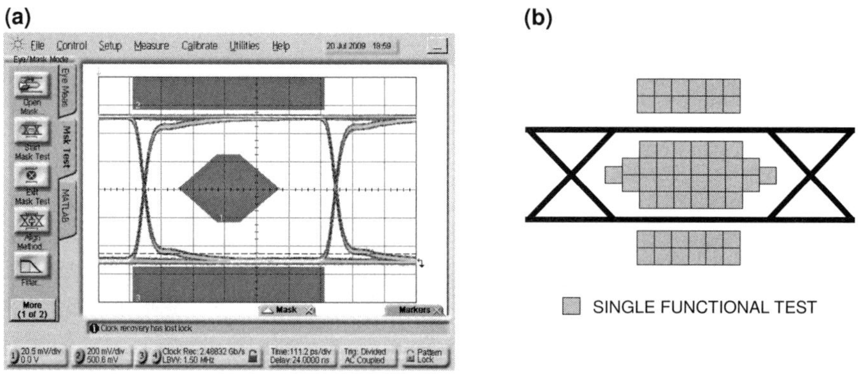

Figure 5.20 Examples of performing a data eye mask test. (a) Using an equivalent-time sampling oscilloscope. (b) Using a digital pin electronics receiver.

5.4.3 BER Bathtub Curve

One fundamental measure of the performance of a DUT driver is the BER bathtub curve[2] [9]. The concept of BER was already defined in Section 2.3. The BER bathtub curve measures the BER of the DUT driver by an ideal receiver at the different strobing instants across the data eye.

One way to think about it is to imagine that we have an ideal receiver (no intrinsic noise) where the strobing position of the receiver can be changed from its optimal position in the center of the data eye. If the DUT driver were also perfect without any intrinsic noise, then the "perfect" receiver should be able to set its strobe timing at any point of the bit period obtaining a zero BER at each point. In this case we would obtain the graph shown in Figure 5.21 where in every point inside of the data eye the BER is zero.

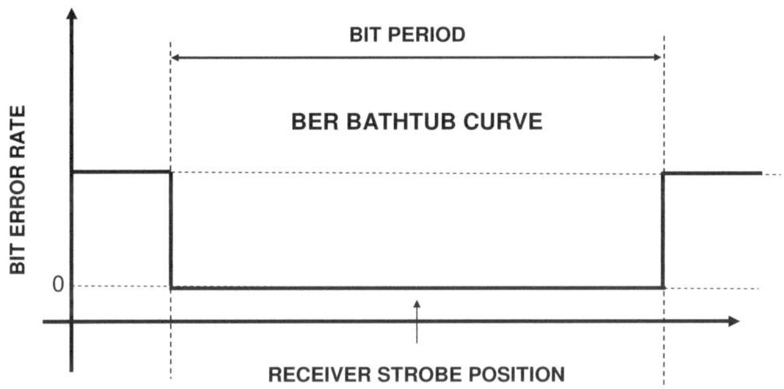

Figure 5.21 The BER bathtub curve assuming a "perfect" DUT driver.

Real DUT drivers have inherent noise; therefore, even a perfect receiver will not be able to correctly strobe the incoming data pattern for timing strobes at the edges of the data eye bit period. Figure 5.22 illustrates this fact by showing that the BER no longer goes abruptly to zero at the edges of the data eye. The most straightforward way to obtain a BER bathtub curve is to use a measurement instrument as the "ideal receiver" and then do a functional test with the receiver strobe covering the entire bit period in a given number of steps. In an ATE environment, a digital pin electronics receiver with the ability to record the number of errors in an at-speed functional test would suffice. It is important to note that the performance of the receiver in terms of bandwidth and jitter has a direct impact on the measurement accuracy since an ideal BER bathtub curve measurement would require an ideal receiver. Another important point is the linearity of the receiver strobing edge (see Section 9.1.3).

[2]Sometimes the measurement of the BER bathtub curve is referred to as a BERT scan.

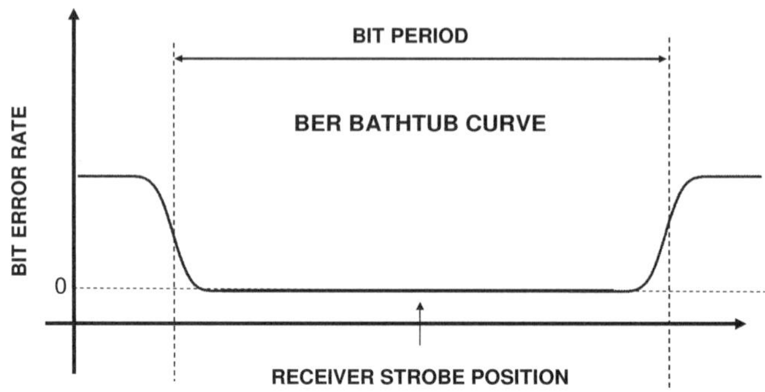

Figure 5.22 The BER bathtub curve for a realistic (not perfect) DUT driver.

The BER bathtub curve is usually plotted using a logarithmic scale for the BER instead of a linear scale as shown in Figure 5.23 to allow a more easy visualization of points with a very low BER value. The test time to obtain a BER bathtub curve measurement is given by (5.2).

$$\text{Test Time} = \text{Number of Compared Bits} \times \text{Bit Period} \times \frac{\text{Bit Period}}{\text{Step Resolution}} \tag{5.2}$$

For example, for a 10 Gbps pattern where 10^{12} bits are compared in each functional test and a resolution of 5 ps is used, the test time of just running the pattern for the complete BER bathtub curve measurement would be around 33 minutes.

Given the large time that it can take to acquire a BER bathtub curve, the test engineer is forced in some situations to reduce the test time by changing the two variables under his control that have a direct impact on the test time: the step resolution and the number of compared bits. It is important to understand that this change might have an impact on the measurement results depending on the jitter characteristics of the DUT. Figure 5.23 shows two BER bathtub curves from the same DUT where a different number of compared bits is used.

The figures show that by using a reduced number of points, the smallest BER value that is measured will be different. This is expected since to measure a BER of 10^{-10} we would need to measure at least 10^{10} bits and assume that one of those bits would be a bit error. A more realistic approach would be to make the compare bit pattern at least 10 times larger than the minimum BER value one intends to measure (Section 5.11 presents some more discussion on this topic).

The main objective of measuring a BER bathtub curve is to find out how much timing margin the DUT driver has. This is similar to the objective of the

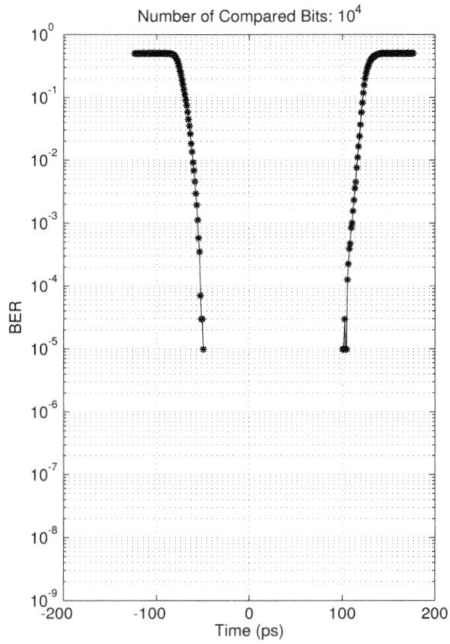

Figure 5.23 Influence of the number of compared bits on the BER bathtub curve (left: 10^4 compared bits; right: 10^9 compared bits).

data eye width measurement but the BER bathtub curve provides a more deep understanding of the effect of the random jitter through the compare pattern size on the data eye width. The reason is that the BER bathtub curve provides the timing margin for each possible BER value. For example, if the I/O design target is to have a maximum of one error per 10^8 transmitted bits, then the critical eye width value is the one obtained for a BER of 10^{-8} as shown in Figure 5.24.

Associated to the data eye width is the value of total jitter. Basically the total jitter at a certain BER threshold is equivalent to the bit period minus the eye width.

$$\text{Total Jitter (BER)} = UI - \text{Eye Width} \tag{5.3}$$

In most cases the design target is defined in terms of the amount of total jitter at a given BER threshold. The challenge is that most standards define the total jitter value for a BER of 10^{-12} which would imply a very large test time due to the required pattern size (e.g., 10^{13}). Section 5.5.6 discusses a different and faster approach to estimate the total jitter at a certain BER through random and deterministic jitter separation. In [10] an approach is proposed for a faster measurement of the total jitter through a BER measurement but avoiding the long test time of a standard BER bathtub curve.

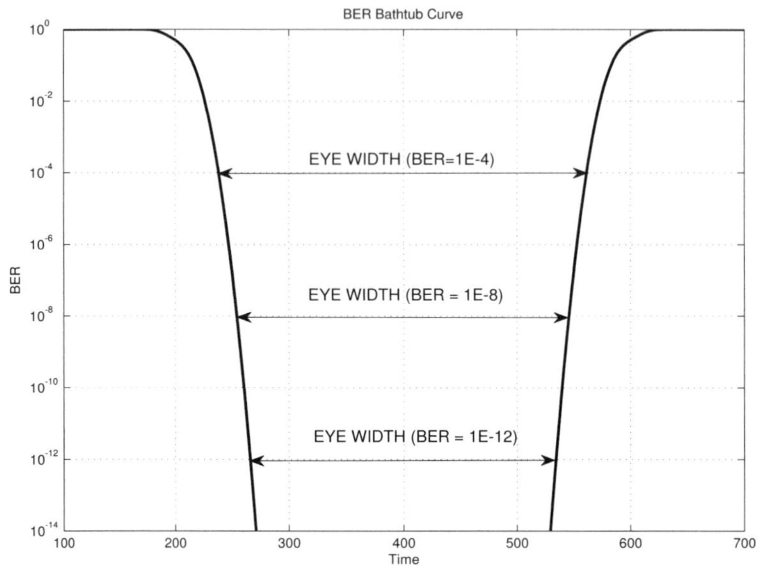

Figure 5.24 Measuring the data eye width value at different BER thresholds.

5.4.4 Skew

For at-cycle source-synchronous interfaces, transmitter or driver skew together with the receiver setup/hold time (described in Section 5.6.1) is one of the most critical test and characterization parameters. The skew between the transmitted clock and its associated data signals determines the margins that a receiver will have in the best case in regard to the required setup/hold times to correctly recognize the transmitted signal. The skew within the data signals determines the best case available valid data eye and thus the overall available margin for the sum of setup and hold time that is left for a receiver as shown in Figure 5.25.

It has to be noted that skew and level loss of the device interconnections will further reduce the margins an at-cycle source synchronous receiver will have available to recognize the transmitted data correctly. Thus, taking into account the overall timing budget of such a connection, the specified skew values for the transmit interface of a device are relatively small compared to the nominal data eye width at the specified data rate. This increases the sensitivity of data connections on even small deviations from the tight margins for the driver skew specifications and makes skew a must-test parameter.

Therefore, the skew between ATE channels becomes a critical factor for the accuracy of this measurement because the skew values to be guaranteed are in the range or even below the edge placement accuracies typical ATE

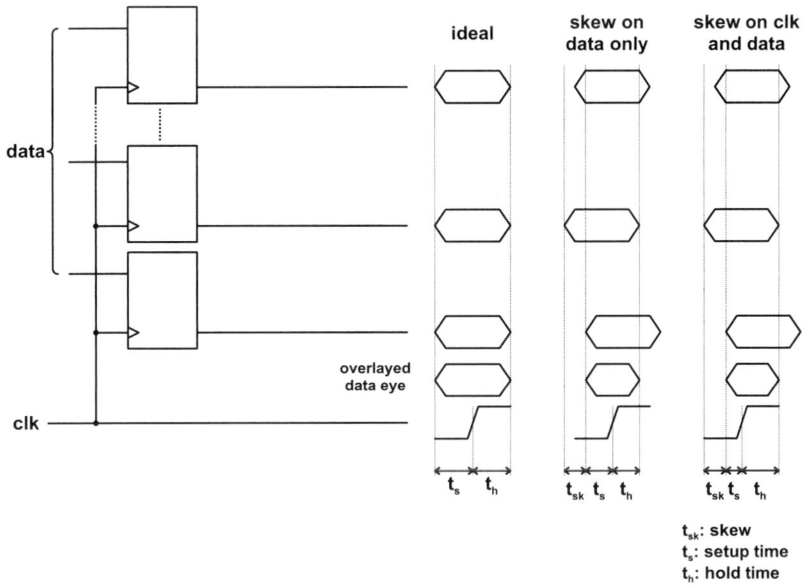

Figure 5.25 Transmitter or driver skew.

systems provide with their standard calibration procedure. As a consequence, focus calibration techniques as discussed in Section 9.3.1 might need to be applied to enable an ATE to test transmitter skew for at-cycle source-synchronous interfaces with sufficient accuracy.

For I/O interfaces that deploy device training mechanisms to compensate for static skew on the device interconnections (see Section 2.5), skew specifications usually are less tight and skew measurements are less important because the training ensures maximum margins for the receiver setup/hold regardless of potential transmitter and connection skew.

Characterization measurements for transmitter skew usually identify the time for corresponding signal transitions on two or more pins and calculate the skew as the difference between the times at which the transitions were found for the involved signals. Of course, an additional offset might need to be added to the measured skew if the skew specification is not related to signal transitions on the involved signals but, for example, the skew between a clock edge and the center of an associated data bit.

The instrumentation that can be used for skew measurements ranges from dedicated external support instrumentation like time interval analyzers (TIA; see Section 7.3) over digital pin electronics that have integrated TIA or time stamper functionality (Section 4.2.3) to purely digital pin electronics without this special functionality.

With the TIA or time stamper approach, it is important that the measurement resources to be used support relative timing measurements between at least two channels to be able to perform skew measurements. Another side effect that needs to be considered when TIA or time stamper circuitry is applied, is that these circuits usually introduce a latency (arming latency) between activating the measurement circuitry and performing the actual measurement. Typically the arming latency is specified with an uncertainty (variation) that is significantly larger than the skew to be measured for high-speed I/O interfaces. This can lead to the effect that corresponding items of the timing measurement tuples[3] for the signals to be analyzed are not referring to corresponding edges. A test engineer has to consider that such a case can be identified and the single measurement components are reshuffled in a way to get alignment of the measurement data along corresponding signal edges.

In case of digital pin electronics without TIA or time stamper functionality for skew measurements, the timing measurements to identify the signal transition locations are implemented by means of timing searches. The same search approaches as discussed in the bit alignment part of Section 5.1 are used to identify the timing of the signal transitions on the pins involved in the skew measurement. An important boundary condition a test engineer has to take into account when doing skew measurements is the fact that corresponding transitions between the signals have to be measured. If this requirement is not considered, singular effects that affect the transition timing at the measurement locations in the test pattern differently (e.g., ground bounce or crosstalk) might falsify the measurement. For a timing search that is based on functional tests, this means that not all bits of the pattern should be compared because the search in this case would identify the signal transition location as the overlay of all compared transitions in the pattern. Such an overlayed result per pin does not allow the extraction of the skew for associated signal transitions on the signals for which the skew should be measured. Thus, the pattern used for skew measurements has to only activate compare actions for either a single bit before or after the transition of interest or for the two bits separated by the transition of interest. With this, the associated transitions that are of interest for the skew measurement are marked and are the only ones that influence the search results.

As we have seen in Section 2.4, jitter is another factor that influences the timing of signal transitions and thus also has influence on the results of skew measurements. In order to minimize this influence, a test engineer should avoid using pattern data that introduces additional jitter (e.g., due to

[3]A set of timing values with one single timing value per measured signal.

bandwidth restrictions on the signal path connecting the ATE pin electronics with the DUT).

5.4.4.1 Differential Skew

In the context of skew, one important topic is differential skew (i.e., the skew in a differential signal pair). The question is not how to measure the differential skew but the impact of it on measurements like jitter when using a differential comparator. This is important because apart from the intrinsic skew of a DUT driver or receiver, there might also be differential skew added by the DUT test fixture signal path. A common misconception is that differential skew translates one to one into jitter which is not correct. In fact in the context of a data eye diagram or a jitter measurement, the differential skew impact should be seen more as an additional degradation of the signal path insertion loss as shown in Section 8.3.2 [11, 12]. There is also an increase in the common-mode amplitude.

Figure 5.26 shows an example of the impact of a differential skew on the data eye diagram for a 10 Gbps $2^7 - 1$ PRBS data pattern where 1E7 bits were compared. For a skew of 10 to 25 ps there is no visible impact on the data eye diagram when compared with the data eye diagram with no skew. For a differential skew of 50 ps the data eye width is slightly reduced and an increase on the rise time can be seen. For a differential skew of 75 to 90 ps the data eye width and data eye height are further reduced and the rise time also is degraded but still an open data eye can be observed even with a 90 ps differential skew. Note that all these measurements were done using a loopback setup between a differential ATE driver and comparator. The differential skew was added by using two trombone delay lines (see Section 7.17.6).

Figure 5.27 shows the differential and common-mode waveforms for the 10 Gbps $2^7 - 1$ PRBS data pattern with no differential skew and with a 50 ps differential skew measured with a real-time oscilloscope. The results show that with a 50 ps skew the rise time is degraded by 4 ps and the common-mode peak-to-peak amplitude more than doubles. Although the impact on the differential waveform is small, such an increase on the common-mode amplitude can create problems for the DUT receiver.

5.4.5 Pre-Emphasis and De-Emphasis Measurement

Modern I/O cells incorporate special techniques like pre-emphasis and de-emphasis to address the loss from PCB signal traces at high data rates as discussed in Section 2.6.4. Typically the objective is to measure the level of pre-emphasis or de-emphasis on the output waveform as shown in Figure 5.28 for a de-emphasis application.

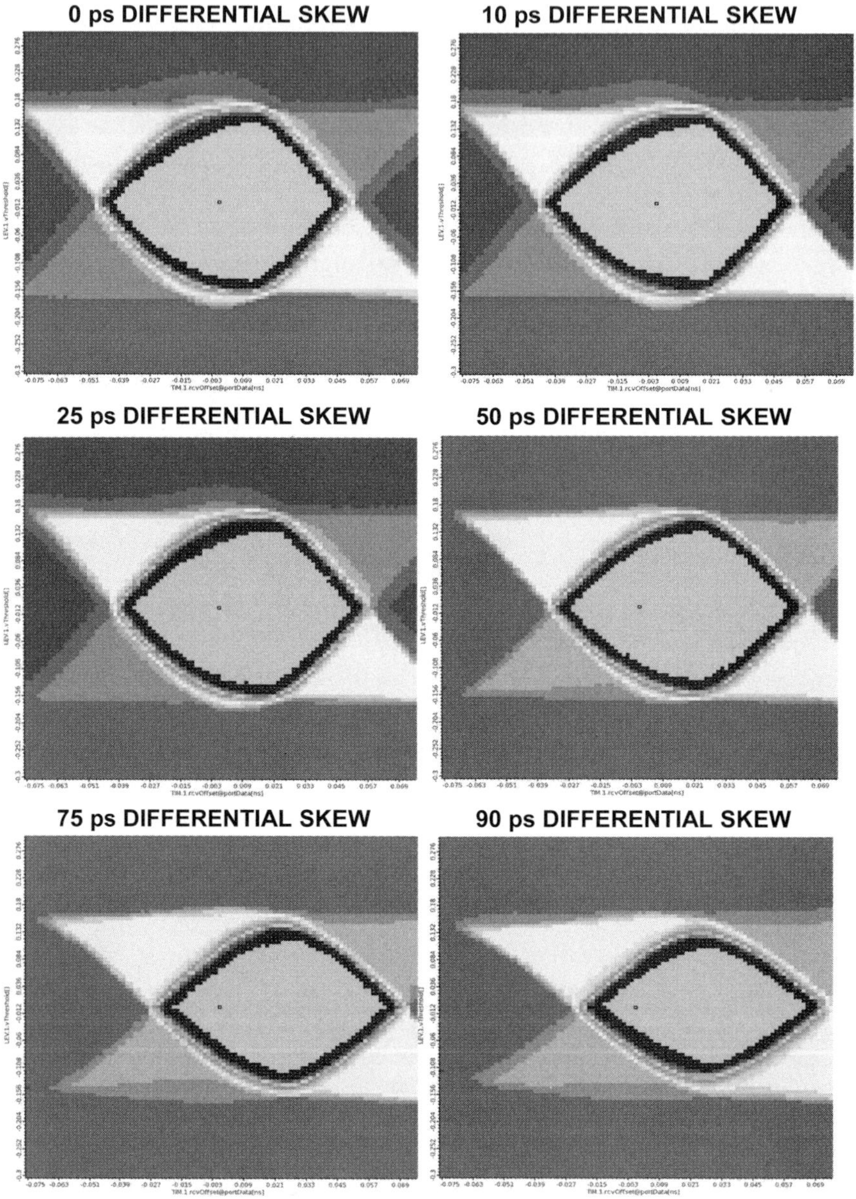

Figure 5.26 Example of the differential skew impact on the measured data eye diagram for a 10 Gbps $2^7 - 1$ PRBS data pattern with 1E7 compared bits.

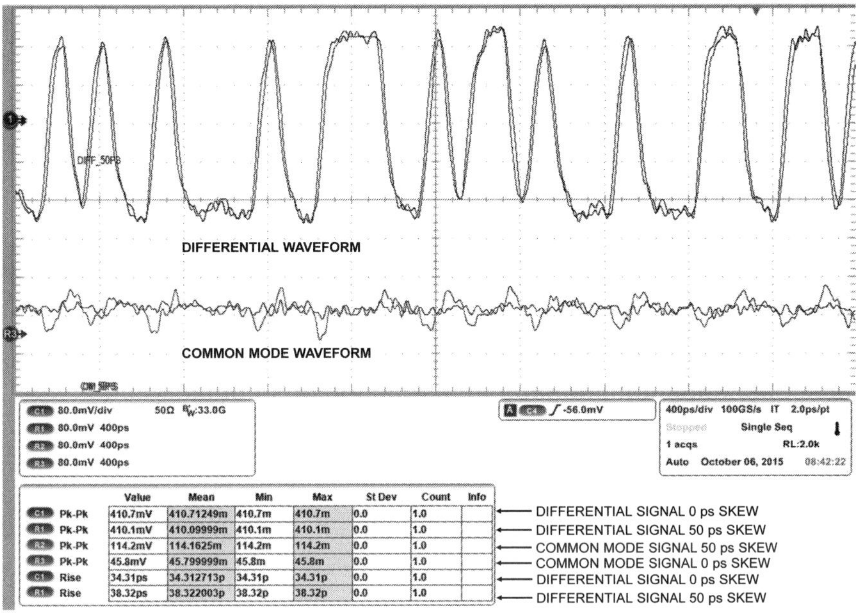

Figure 5.27 Measured differential and common-mode waveforms for a 10 Gbps, $2^7 - 1$ PRBS data pattern with no differential skew and with a 50 ps differential skew.

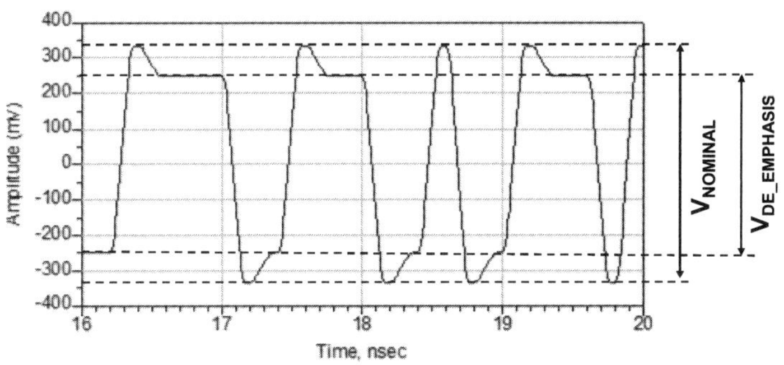

Figure 5.28 Measuring de-emphasis example.

The most straightforward option is to acquire the waveform with the digital pin electronics or with the sampler and measure the pre-emphasis or de-emphasis value on the acquired waveform though post-processing. Although this option is the most generic way to measure any type of pre-emphasis or de-emphasis implementation (there are several techniques with different complexity levels), it consumes a significant amount of test time, especially when using digital pin electronics.

A simpler approach with digital pin electronics is to measure the pre-emphasis/de-emphasis value through a two-step approach. First, a level shmoo test at the center of the data eye is performed only for the bits on the data pattern that are pre-emphasized or de-emphasized to obtain the eye height for the first bit after a transition. In a second step a level shmoo test is performed to obtain the eye height for the rest of the bits. The pre-emphasis/de-emphasis value will be the difference between these two shmoo values. The problem with this approach is that it can only address certain types of pre-emphasis/de-emphasis algorithms, for example, algorithms where only the first in a sequence of equal logic value bits is pre-emphasized.

5.5 Driver Jitter Tests

This section discusses the different types of jitter tests that can be performed on the driver of a DUT I/O cell.

5.5.1 Jitter Histogram

The jitter histogram is the most basic way of analyzing the jitter of a digital signal. The idea is to make a measurement of the timing of each edge of the signal compared to a reference signal [e.g., a low jitter clock source for measuring the time interval error (TIE) jitter] or even in some cases to the previous edge (e.g., when measuring cycle-to-cycle jitter) and generate a histogram of the measurements. The exact way the histogram is generated depends on the specifics of the measurement methodology and measurement instrument. For example, if the result of the measurement is an array of values for the measured jitter at each transition (e.g., using a time interval analyzer as the measurement instrument), then from this array a histogram can be generated as already discussed in Section 2.4.1.

In the case of a digital pin electronics receiver, the histogram can be obtained by the derivative of the shmoo plot of a functional test error count across the data eye edge transition area as shown in Figure 5.29. Basically one side of the BER bathtub curve can provide the jitter histogram for that transition by using the derivative on the BER bathtub curve.

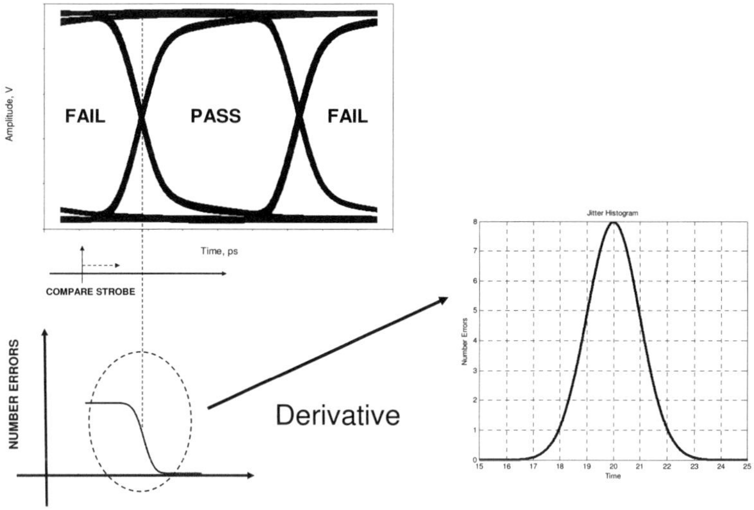

Figure 5.29 Obtaining the jitter histogram from an error count shmoo.

There are two typical values when measuring jitter that are associated with the histogram. They are the jitter RMS value and the jitter peak-to-peak value.

5.5.2 RMS Jitter

The jitter root mean square (RMS) value corresponds to the standard deviation of the jitter measurement (root mean square is typically used as a synonym for standard deviation in physical sciences). If one has access to the array of measured values the best estimate of the standard deviation (RMS jitter) can be computed in the following way:

$$\bar{\sigma} = \sqrt{\frac{1}{N-1}\sum_{i=1}^{N}(x_i - \bar{x})^2} \tag{5.4}$$

with the average $\bar{x}$ defined as

$$\bar{x} = \frac{1}{N}\sum_{i=1}^{N} x_i \tag{5.5}$$

In some cases the sample values might not be available and only the histogram is available. In this case the estimate for the standard deviation can be computed by the following formula:

$$\bar{\sigma} = \sqrt{\frac{1}{N-1} \sum_{i=1}^{N} (x_i - \bar{x})^2 \, H_p(x)} \tag{5.6}$$

where $H_p(x)$ is the value of the normalized histogram for the value x_i.

The RMS jitter value (standard deviation of the histogram) provides an idea of how much the measured data is "spread" as shown in Figure 5.30.

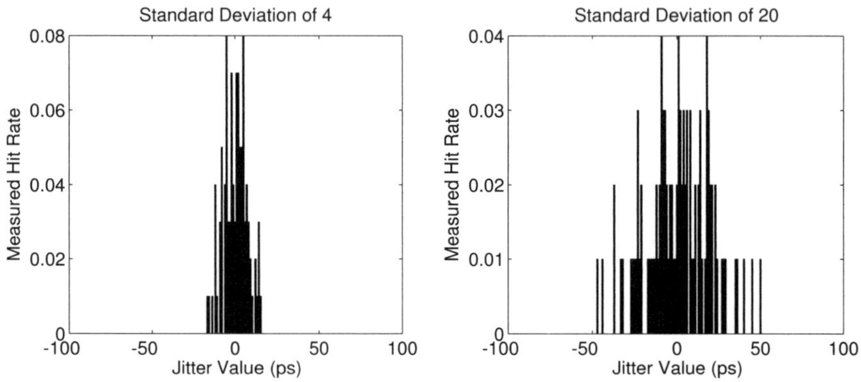

Figure 5.30 Comparing two histograms with different standard deviation (RMS jitter) values (left: $\sigma = 4$ ps; right: $\sigma = 20$ ps).

For the reader with RF background, there is a relationship between the phase noise measured on a sinusoidal clock signal and the RMS jitter value on that clock signal when only random jitter is present [13, 14]. Appendix J discusses this relationship in detail.

In the case of a data signal with only Gaussian jitter, the measured RMS jitter value of a jitter histogram is an estimate of the random jitter value, but in most cases even in the case of a clock signal there will be other jitter components apart from the random jitter. This means that the histogram RMS jitter value is in most cases a very poor estimate of the signal random jitter.

5.5.3 Peak-to-Peak Jitter

The peak-to-peak jitter value is a very commonly used measure of the jitter from a DUT. It corresponds to the difference between the two extremes of the jitter histogram (see Figure 5.31) or if looking to the array of measurement values it will correspond to the difference between the two extreme jitter values. In this example the peak-to-peak jitter value is 38 ps.

Although peak-to-peak jitter is a common measure used to evaluate the jitter performance of a DUT driver, it is also the most problematic one.

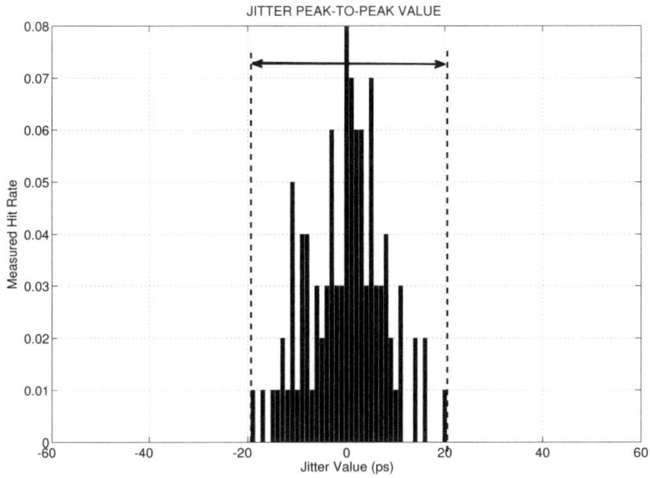

Figure 5.31 The jitter histogram peak-to-peak value.

The reason is the fact that the random jitter portion of the driver jitter is theoretically not limited (i.e., its peak-to-peak value will be infinite if one measures it for a sufficiently long period of time). This means that any peak-to-peak jitter value depends on the number of samples acquired. This can generate several issues, especially when correlating between different measurement setups. Figure 5.32 shows this fact by presenting four different independent measurements of the peak-to-peak jitter of a DUT with an equivalent-time oscilloscope using a different number of acquired samples. The results show that in the case of a small number of acquired samples (10 and 100) the jitter values vary significantly with the 100 samples case even showing a smaller peak-to-peak jitter value (each measurement was done independently of each other). For the 1,000 and 10,000 samples case, the peak-to-peak jitter value increases with the number of samples showing an increase of 49% in the peak-to-peak jitter value for the 10,000 samples case when compared to the 100 samples.

5.5.4 Drawbacks of the Error Count Approach for Jitter Measurements

One challenge with the approach presented in Section 5.5.1 for obtaining the jitter histogram arises when the error count curve is not monotonic. In this case the computed jitter histogram might even have negative values which do not have a real physical meaning. This is shown in Figure 5.33.

The reasons for this nonmonotonic behavior can be multiple. For example, the DUT might have a very low-frequency periodic jitter that is not

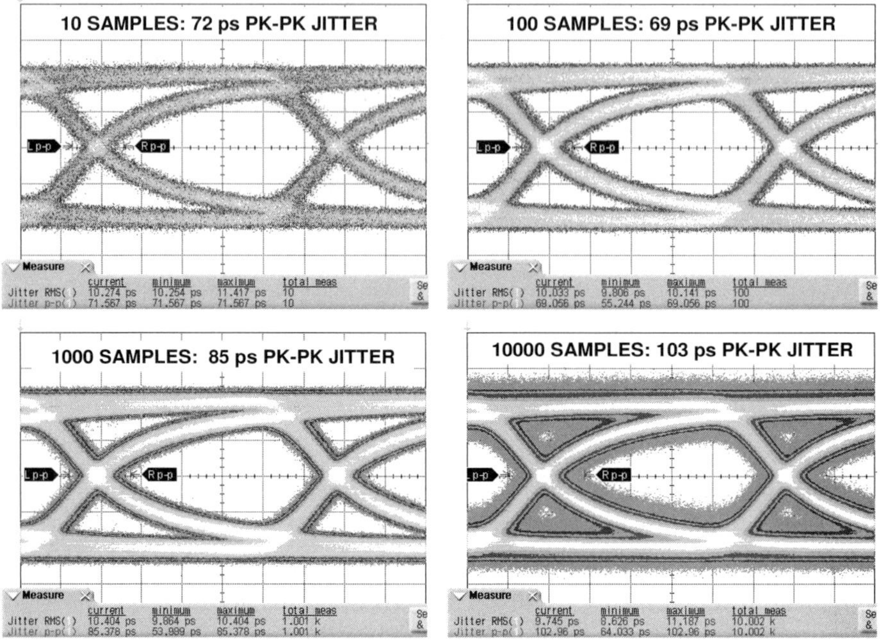

Figure 5.32 Variation on the measured peak-to-peak value with the number of acquired samples. The peak-to-peak jitter is measured with an equivalent-time oscilloscope and the measurement is restarted for each example.

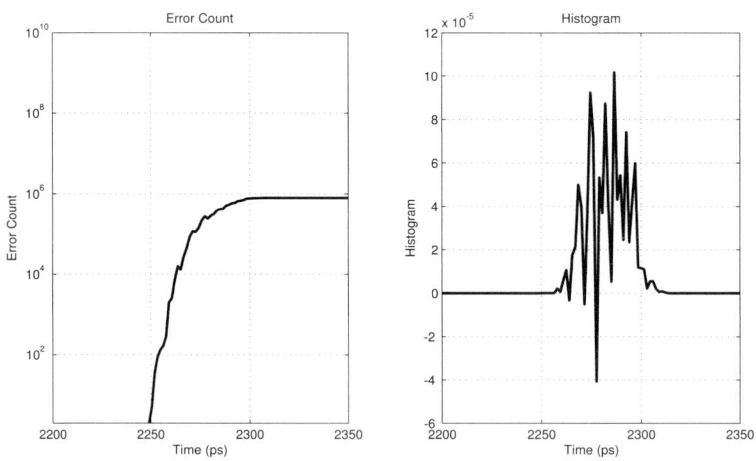

Figure 5.33 Obtaining an histogram with negative values from a nonmonotonic error count measurement.

properly averaged in the course of one functional test because of the pattern not being long enough, or the digital pin electronics timing delay line is nonmonotonic when using a very small time step resolution. Figure 5.34 shows one example where increasing the pattern size used on the functional test solved the negative histogram issue. However, this solution comes at the cost of a longer test time. Figure 5.35 shows another example where using a larger time step resolution solved the negative histogram problem but in this case at the expense of a lower measurement accuracy.

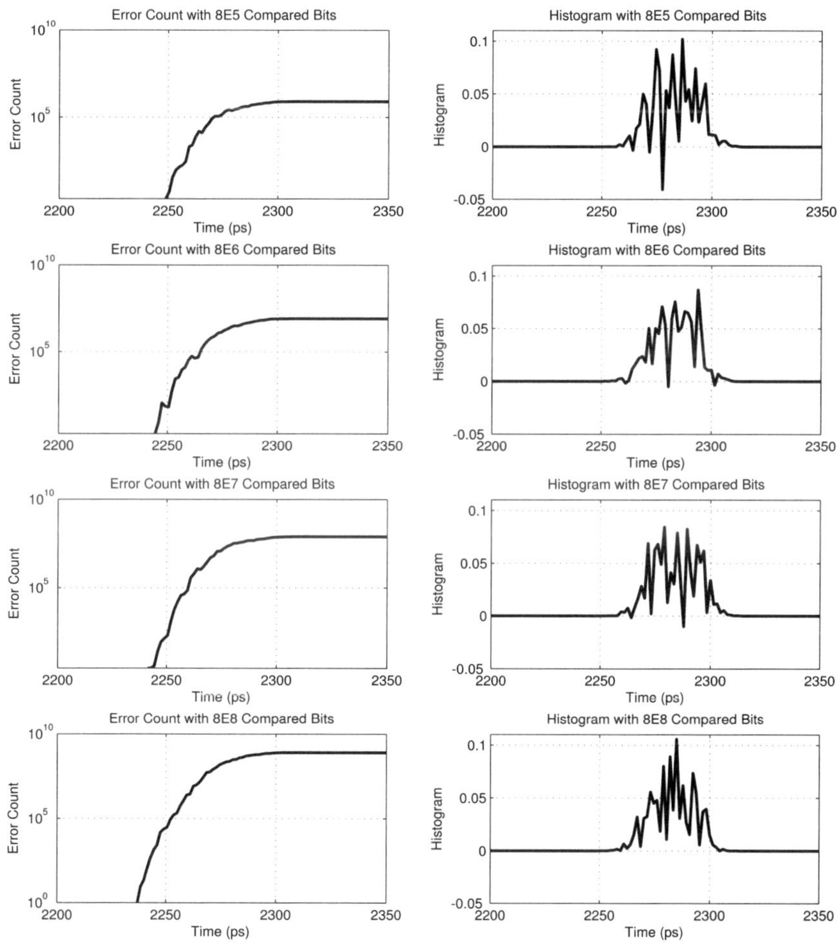

Figure 5.34 Pattern size influence on a nonmonotonic error count measurement and respective jitter histogram computation.

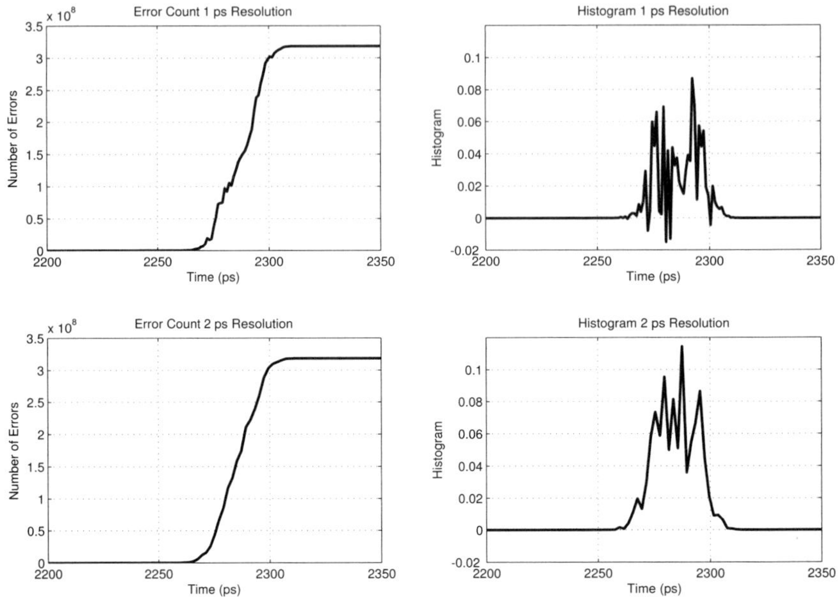

Figure 5.35 Time step resolution influence on a nonmonotonic error count measurement and respective jitter histogram computation.

Another option to address this issue is to force the error count curve to become monotonic by post-processing the measured error count curve [9]. One possible approach is shown in Figure 5.36 where a moving window averaging filter [15] described by (5.7) was used on the measured error count curve. It is important to note that by using a filter on the measured data to obtain a monotonic behavior on the error count curve, the measured data is changed and important features of the DUT or ATE in the measured data might be masked.

$$Y[n] = \frac{X[n]}{5} + \frac{X[n-1]}{5} + \frac{X[n-2]}{5} + \frac{X[n+1]}{5} + \frac{X[n+2]}{5} \tag{5.7}$$

A different but simpler and more accurate procedure for computing the jitter histogram when it is dominated by random or sinusoidal jitter is presented in [16]. This procedure is very effective in measuring jitter histograms avoiding the nonmonotonic issues described in this section.

5.5.5 Measuring the Jitter Spectrum

Although most standards define the jitter requirements for the DUT in terms of its random and deterministic jitter components, for design verification it is

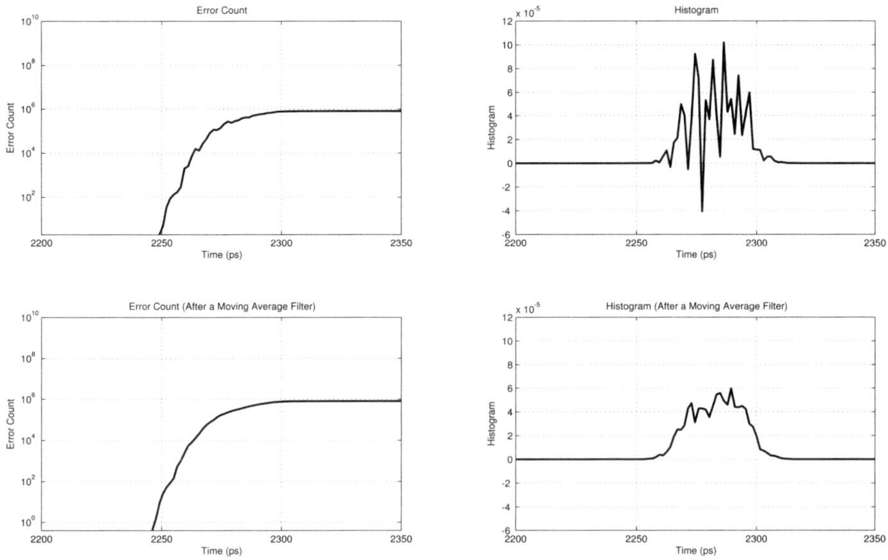

Figure 5.36 Post-processing of the error count curve to enforce a monotonic behavior through a moving average filter.

important to know the spectral distribution of the jitter generated by the DUT. This information could provide an indication of the root cause of the measured jitter (e.g., crosstalk from a system clock).

Although obtaining the jitter spectrum can be very simple, it does depend on the measurement approach. For example, with a real-time sampling oscilloscope, the waveform from the DUT driver can be acquired by sampling the signal in real-time and saving it for post-processing. It is then possible to perform a comparison of the pattern transition times with the transition times of an ideal nonjittered transition (since the data rate is known a priori). The resulting deviation values will be the jitter waveform sampled at the transition points of the pattern. It is now just a matter of applying the Fourier transform on the jitter waveform to obtain the jitter spectrum.

This procedure is shown in Figure 5.37. The maximum jitter frequency that can be measured corresponds to half of the data rate since it is only possible to sample a jitter value at a bit transition. However, this approach raises another issue due to the fact that, apart from a clock pattern, there is no guarantee that a transition will exist for each bit on the pattern. One way to address this issue is to interpolate the jitter waveform between transitions for the bits that had no transition.

It is important to note that obtaining the jitter spectrum from a clock signal or a data signal are two different challenges. An instrument such as a

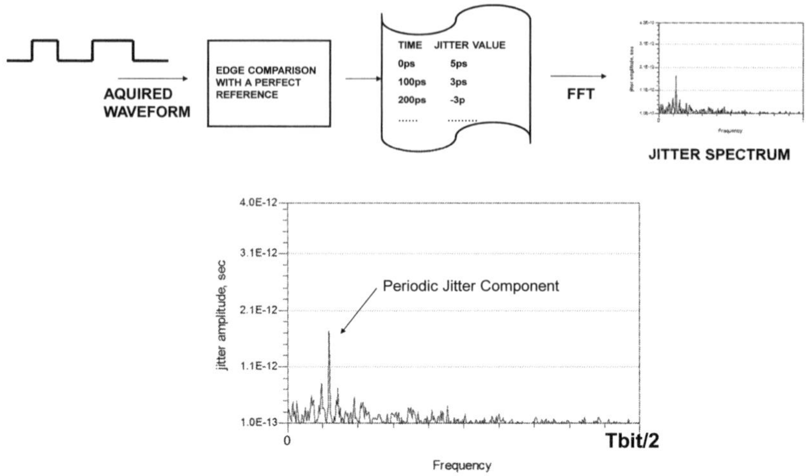

Figure 5.37 Measuring the jitter spectrum with a real-time sampling scope.

spectrum analyzer or algorithm that is able to measure the jitter spectrum from a bit clock signal might not be able to measure the jitter spectrum from a data pattern. The reverse is always true since measuring the jitter spectrum of a bit clock is a subset of measuring the jitter spectrum of a data pattern.

In [17] a technique is presented that allows the measurement of the jitter spectrum using a BERT or an ATE digital pin electronics. The main requirement for this technique is the ability to capture the pass/fail for each bit at-speed. The idea is to set the compare strobe timing not at the middle of the data eye as usual for a functional test but at the edge of the data eye. The jitter on the data signal will move the data eye left and right of the strobe point modulating in this way the pass/fail result for each bit on the data pattern as shown in Figure 5.38. The compare error signal can then be post-processed to obtain the jitter spectrum. The drawback is that although the jitter spectrum is obtained, the computed frequency amplitudes do not provide absolute jitter amplitude values but they can be displayed in relative values to the highest peak or to a calibration tone.

5.5.6 Random and Deterministic Jitter Separation

Several standards require the measurement of the random and deterministic jitter components of the jitter generated by the DUT driver (e.g., PCI Express [18]). The reasoning is that under certain conditions these two values allow the computation of the total jitter value for a given BER threshold that provides the timing margin or the data eye width from the DUT driver at that BER.

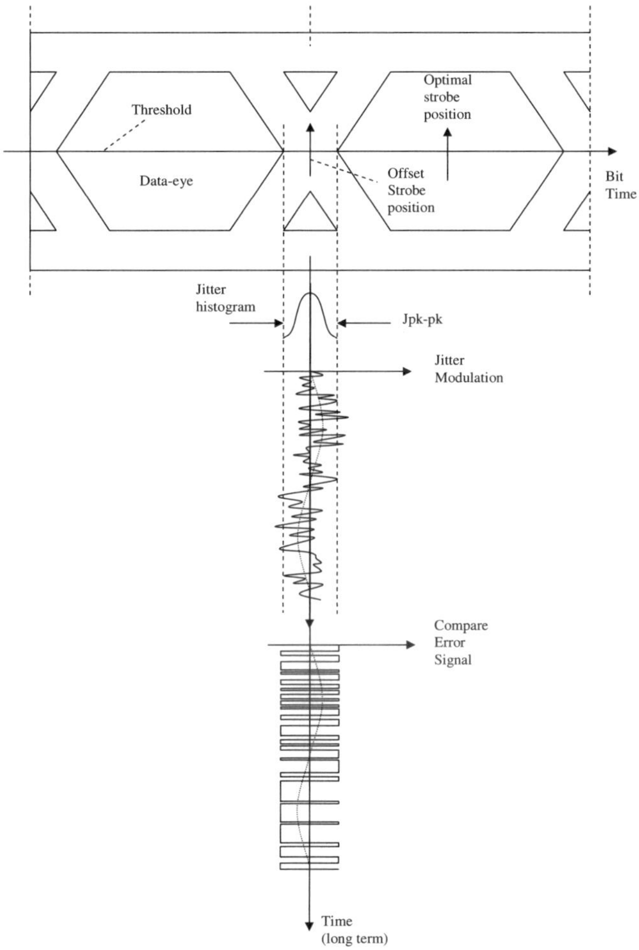

Figure 5.38 Measuring the jitter spectrum with a digital channel using at-speed capture. (From: [17]. ©2004 Bernd Laquai. Reprinted with permission.)

The first assumption is that the random jitter is modeled by a Gaussian distribution with variance σ (see Appendix A). The second assumption is that the deterministic jitter is bounded with a peak value (DJ_{PEAK}) and a peak-to-peak value (DJ_{PK-PK}). With these two assumptions, the total jitter (TJ) value at a given BER can be computed by the following equation:

$$TJ(BER) = DJ_{PK-PK} + \alpha(BER)\,\sigma \tag{5.8}$$

Equation (5.8) means that the total jitter $TJ(BER)$ at a given BER threshold is equal to the deterministic jitter peak-to-peak value plus the random jitter (RJ) variance (RMS jitter) value multiplied by an $\alpha(BER)$ value that depends on the BER threshold. Note that since the tails of the random jitter are theoretically infinite, if no BER target is defined then the total jitter value would theoretically be infinite. Table 5.1 presents some of the typical values used for $\alpha(BER)$ and Section A.3 shows how the values presented in the table are computed.

Table 5.1

Some Typical Used Values for $\alpha(BER)$

BER	α(BER)
10^{-9}	11.996
10^{-12}	14.069
10^{-16}	16.444

Equation (5.8) is of little value without the RJ and DJ values. Typically what is available is a jitter measurement in the form of a histogram, a BER bathtub curve, a jitter waveform, or a jitter spectrum. There are several possible algorithms to obtain the random and deterministic jitter values from jitter measurement data. The following subsections describe two algorithms: RJ/DJ separation based on the dual Dirac jitter model and RJ/DJ separation based on the jitter spectrum. In [19] a different RJ/DJ separation algorithm based on the identification of the best jitter model from the measured data is presented and compared to the RJ/DJ jitter separation algorithm based on the dual Dirac jitter model.

5.5.6.1 RJ/DJ Separation Based on the Dual Dirac Jitter Model

The dual Dirac jitter model assumes that the deterministic jitter is modeled by a dual Dirac distribution. The total jitter distribution is a convolution between the dual Dirac deterministic jitter model and the Gaussian distribution for the random jitter as shown in Figure 5.39. Appendix B discusses the dual Dirac model in detail.

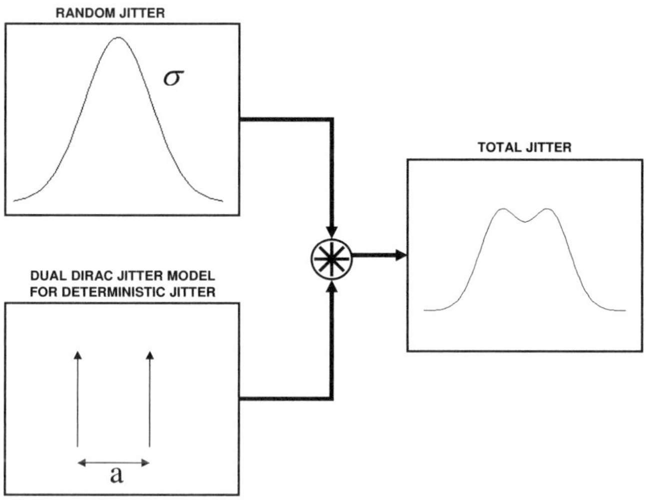

Figure 5.39 The dual Dirac jitter model.

With this model it is now possible to compute the RJ and DJ values from the measured histogram or bathtub curve. Since under this model, the deterministic jitter is bounded between two values, ($-DJ_{PK}$ to DJ_{PK}), it is possible to select a region outside these values (where the jitter is due only to random jitter) and fit a Gaussian curve to it. This is described graphically in Figure 5.40. Note that in the case of a BER bathtub curve the Gaussian distribution will have the form of an inverse erfc function (see Appendix A). Note also that there are two curves to be fitted on both sides of the jitter histogram or BER bathtub curve. Ideally they should be the same, but in reality they will usually be slightly different from each other.

Algorithmic details for the RJ/DJ separation algorithm using the bathtub curve can be found in [20] and Appendix B. For the histogram case, the RJ/DJ jitter separation based on the dual Dirac model was presented in [21] although it used an automatic procedure to compute the fitting range. The key point is that the σ value (standard deviation or RMS value) of the fitted Gaussian distribution will provide the random jitter value (typically one uses the average of the two fitted Gaussian curves) and the distance between the average of the Gaussian distributions will provide the deterministic jitter value. Figure 5.41 shows one example of an RJ/DJ separation based on the dual Dirac jitter model implemented on an ATE platform.

The choice of the fitting region is straightforward if one knows a priori the peak value of the deterministic jitter, but unfortunately this is one of the values we are trying to determine. The best strategy is to choose a fitting range that is far from the center in the case of a jitter histogram or with low

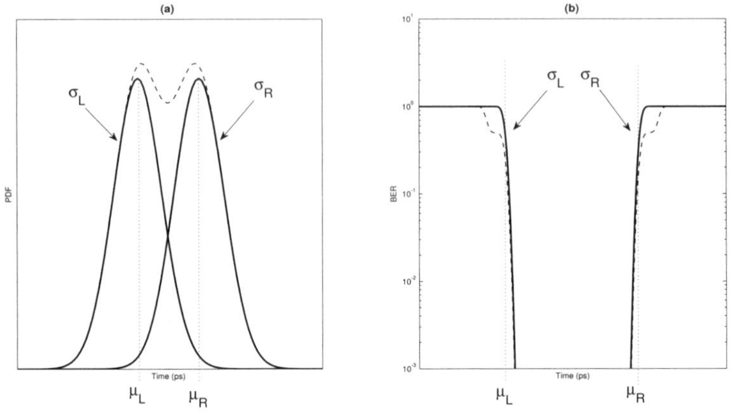

Figure 5.40 (a) Fitting a Gaussian curve to a jitter histogram or (b) fitting an inverse erfc function to a bathtub curve.

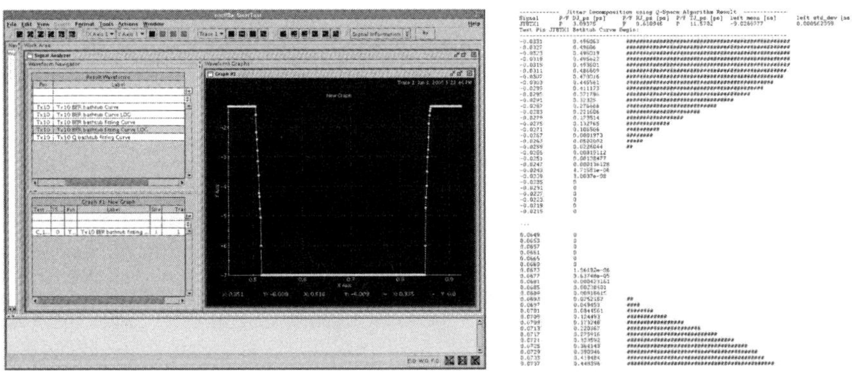

Figure 5.41 Jitter separation on an ATE platform. The left figure shows the GUI that is used during the test program debug and characterization phases to evaluate the jitter separation results, while the right figure shows the datalog that is used on a production environment where multiple pins and DUTs are measured (courtesy of Advantest).

BER values in the case of a BER bathtub curve. This approach guarantees that the fitting range is outside of the deterministic jitter area, but in turn this requires that the histogram or bathtub curve have measured values in this area which typically requires longer acquisition times which in turn means a longer test time. This is exemplified in Figure 5.42 for a jitter histogram and a BER bathtub curve.

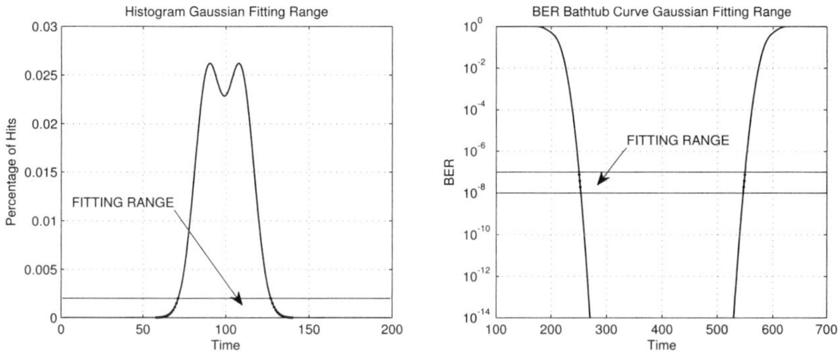

Figure 5.42 Fitting range of the histogram and bathtub curves.

The choice of the fitting range has an impact on the measurement results as shown in the example of Figure 5.43. In this example two different DUTs were measured. In the first DUT random jitter was injected so that the output jitter is dominated by random jitter, and on the second DUT both random and deterministic periodic jitter were injected. In both examples fitting ranges of $[10^{-3}, 10^{-5}]$ and $[10^{-7}, 10^{-9}]$ were used, respectively. The total jitter (TJ) was extrapolated to a BER threshold of 10^{-12}. The results show that in both examples the fitting range does have an impact on both the computed DJ and RJ values. Note that the values obtained with the $[10^{-7}, 10^{-9}]$ fitting range should provide a better accuracy than the ones obtained with the $[10^{-3}, 10^{-5}]$ fitting range.

It is important to note again that the RJ/DJ separation algorithm based on the dual Dirac jitter model is based on a very specific model for the deterministic jitter. Some authors have presented methods to further improve the accuracy of the algorithm by identifying the model in the distribution [22, 23] or using a different model from the dual Dirac [24].

For a total jitter measurement a full BER measurement at the target BER threshold is the closest one can get to a golden standard although it can take a significant amount of time. This means that when correlating to other bench instruments one needs to be aware of possible differences on the used algorithm before blindly comparing values (more on this topic in Section 5.5.9).

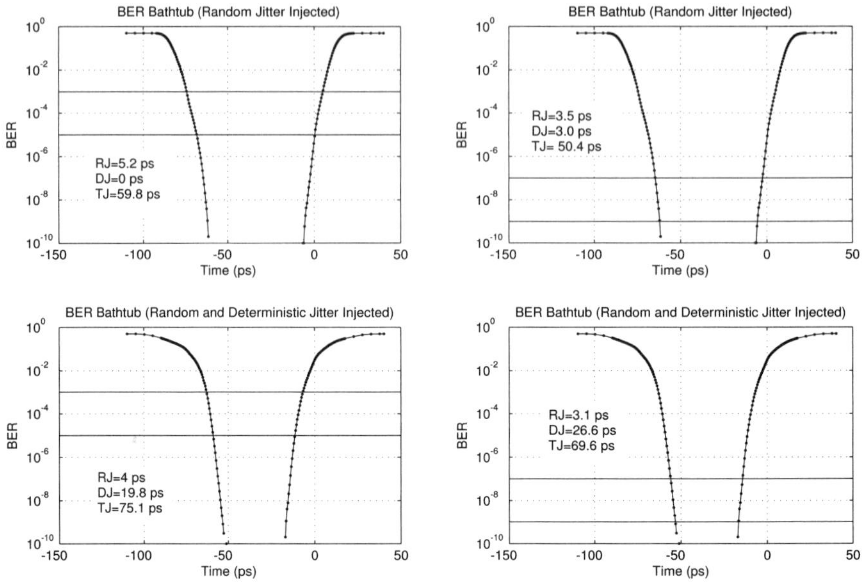

Figure 5.43 Demonstration of the impact of the fitting range in a signal dominated by random jitter (top) and by random and deterministic jitter (bottom). The two fitting ranges used in each example are $[10^{-3}, 10^{-5}]$ and $[10^{-7}, 10^{-9}]$.

One challenge when using a fitting range for the RJ/DJ separation algorithm is the duality between measurement accuracy and test time. Figure 5.44 shows two bathtub curves acquired from the same DUT but with a different number of compared bits. In this example a fitting range of $[10^{-3}, 10^{-7}]$ was chosen, but if a pattern with only 10^4 bits is used then no measurement values are available in the fitting range. In this case the test engineer needs to change the fitting range, which might imply a reduction on the measurement accuracy as already shown in Figure 5.43.

The best option is to start with a large pattern size and a safe fitting range (i.e., in the case of a BER bathtub curve with very low BER values) and then start to optimize the test time by reducing the pattern size and changing the fitting region in a series of small steps while keeping a close eye on the RJ/DJ separation values with each change.

Although the presented procedure for random/deterministic jitter separation and total jitter extrapolation shows limitations and accuracy issues due to the model assumptions, it is one of the few procedures that is used across the industry, and although when looking for the best accuracy it might not be the most appropriate one, for correlation between different measurement

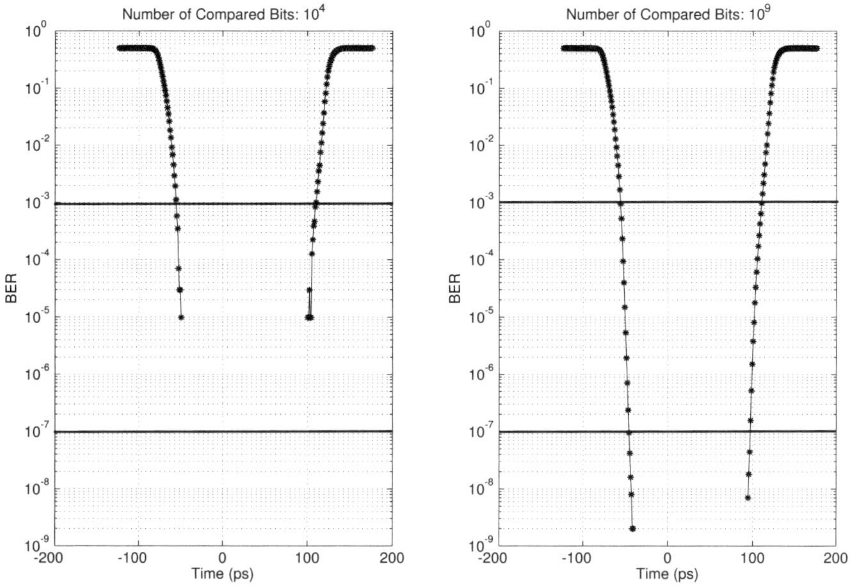

Figure 5.44 Bathtub curve acquired with a different number of compared bits for the same DUT (left: 10^4 bits; right: 10^9 bits).

instruments it might be the most appropriate one since it is implemented in the same way in various instruments.

5.5.6.2 RJ/DJ Separation Based on the Jitter Spectrum

Another possible RJ/DJ separation technique is to use the jitter spectrum [9, 25]. The idea is to assume that the random jitter corresponds to the noise floor of the measured jitter frequency spectrum and the deterministic jitter will correspond to the individual spectral lines that are above the jitter spectrum noise floor as shown in Figure 5.45.

One simple approach to accomplish this is to define a threshold value above which a spectral line is considered as deterministic jitter and not part of the random jitter noise floor. Although for a jitter spectrum dominated by random and periodic jitter as in the example of Figure 5.45, this approach seems straightforward, as it becomes much more challenging when trying to identify the jitter components for other types of deterministic jitter like ISI or BUJ as exemplified in Figure 5.46.

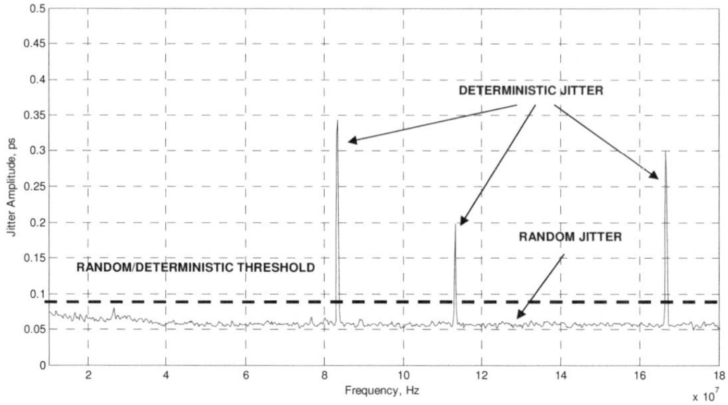

Figure 5.45 Jitter spectrum of a digital signal and the random deterministic jitter separation.

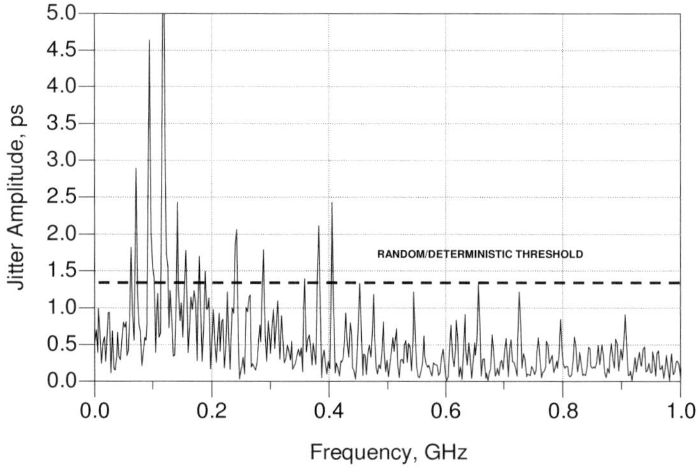

Figure 5.46 Challenge of setting a jitter separation threshold in the jitter spectrum results when multiple types of deterministic jitter are present (e.g., ISI and BUJ).

5.5.7 Measuring the Data-Dependent Jitter

In the context of deterministic jitter, data-dependent jitter (DDJ) is a critical component because it represents the portion of the deterministic jitter that is correlated with the pattern. Measuring the data-dependent jitter requires knowledge of the pattern under measurement; that is, where the pattern starts and its length. For some measurement instruments this is achieved by a pattern marker trigger capability (i.e., a trigger that indicates where the pattern starts since the DUT driver will be sending the pattern repetitively).

In the case of an ATE system, when using a digital pin electronics receiver, the knowledge of the pattern start and length is readily available in the functional pattern. For example, the receiver uses it for the functional test compare. One approach to measure the deterministic jitter would be to perform a jitter measurement for each bit of the pattern individually as shown in Figure 5.47 [26].

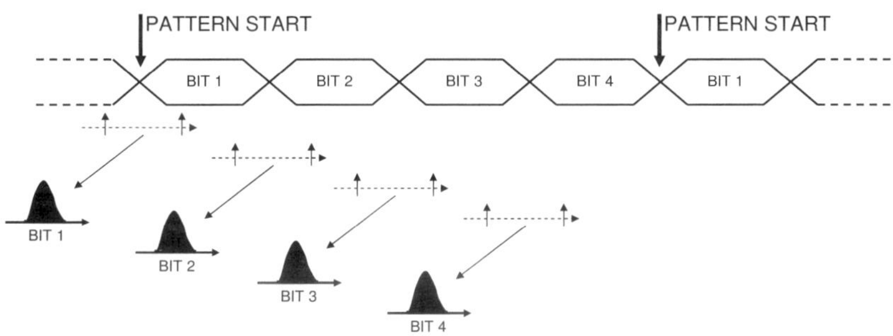

Figure 5.47 Measuring the jitter histogram of each individual bit in the data pattern.

The idea is to only measure the jitter histogram in the transition for bit X ignoring the contributions from the other bits in the pattern (i.e., if using an error count approach to obtain the histogram, only count the errors that happen at the particular bit X position). This procedure is repeated for the other bits in the pattern resulting in a jitter histogram for each bit position as shown in Figure 5.47. Assuming that the measurement setup consists of enough repetitions of the pattern, one would expect the jitter histograms for each bit to be equal if there is no data dependency on the jitter from the DUT driver. However, if there are jitter components that are data-dependent (e.g., ISI jitter from a lossy test fixture), then it should become visible by comparing the individual histograms for each bit (e.g., comparing the mean value of each histogram in regard to the expected bit transition instant).

5.5.8 Measuring Bounded Uncorrelated Jitter

In the context of deterministic jitter, bounded uncorrelated jitter (BUJ) represents the portion of the deterministic jitter that is not correlated with the pattern unlike DDJ. One way to perform this measurement with an ATE system using a digital pin electronics receiver is to measure the deterministic jitter in a single bit of the data pattern ignoring the other bits on the data pattern. In this way the measured jitter is uncorrelated with the data pattern. This is shown in Figure 5.48 where only the first bit in the data pattern is used in the jitter measurement [26]. To obtain the bounded uncorrelated jitter value, it is still necessary to perform an RJ/DJ separation on the measured jitter of the single bit in the pattern.

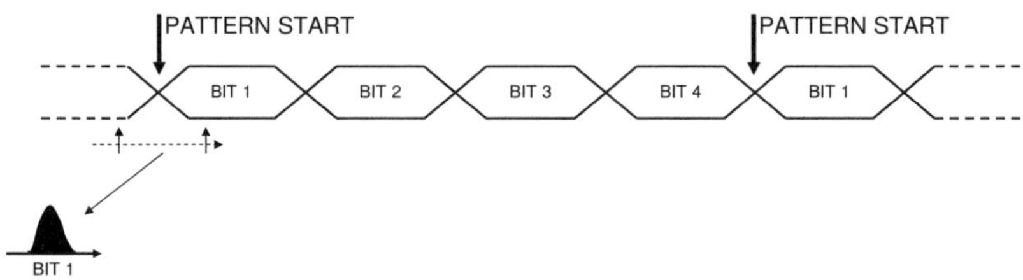

Figure 5.48 Measuring the jitter histogram of one individual bit in the data pattern.

5.5.9 Jitter Measurement Correlation

Correlation of jitter measurements can be a complex topic. The reason is that, unlike with measurement variables like frequency where there is a clear standard on how the measurement should be done and measurement differences between different measurement setups can be traced back to the instruments accuracy by a well-defined procedure, for jitter measurements this is not the case.

This creates a situation where a test engineer in certain cases might not be able to decide which instrument or technique to use and what is the expected error from a given setup. This is shown in Figure 5.49 where different bench instruments and one ATE pin electronics card were used to measure the deterministic jitter from a pattern generator with sinusoidal and random jitter that was added through a voltage controlled delay line. Note that not only are the instruments different but they also use different jitter separation algorithms or the same algorithm with different parameters.

References [27–32] present a comparison of selected instruments or algorithms, showing that significant differences can arise on the measured results for exactly the same DUT. This means that when correlating

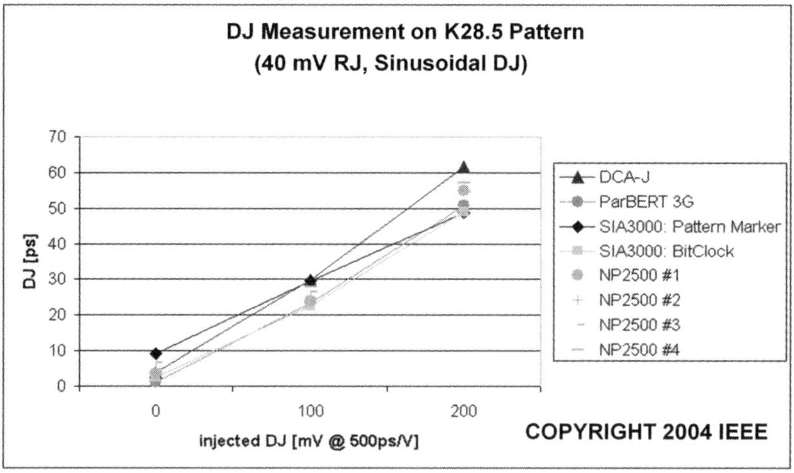

Figure 5.49 Deterministic jitter measurement on a K28.5 pattern from a pattern generator at 2.5 Gbps for different values of injected sinusoidal deterministic jitter (reprinted with permission from [20]).

between different types of instruments or algorithms one should expect differences on the measured values that cannot be easily traced back to a difference on the accuracy of the instruments. Some standards have tried to address this challenge with mixed results. This challenge is also present on ATE applications, not only when correlating between ATE and bench instrumentation but also between different ATE systems.

To better illustrate the jitter correlation difficulties, Figure 5.50 shows a jitter measurement of the same DUT using different measurement instruments. A PRBS $2^{15} - 1$ at 10 Gbps is measured with two real-time oscilloscopes from different vendors. The jitter is separated in its components by the jitter separation software of each instrument. The results are summarized in Table 5.2. They show that although the measured values are close, they are not equal and therefore show the difficulties of correlating values with a very high accuracy between two instruments with similar analog performance but with different jitter analysis software.

Another important point when correlating jitter data is to verify if a software of hardware clock and data recovery (CDR) is used. This can change the measured jitter values significantly as shown in the example of Figure 5.51 and Table 5.3.

This difficulty can also be seen in Figure 5.52 where a bit clock is measured with a real-time oscilloscope and a spectrum analyzer with phase noise measurement software. In this case the correlation is more complicated

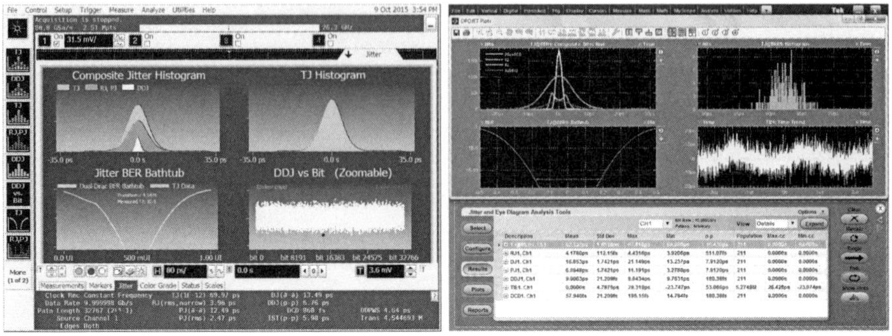

Figure 5.50 Comparison of a jitter separation measurement of a 10 Gbps PRBS $2^{15} - 1$ pattern between two equivalent-time oscilloscopes from different manufacturers. Both instruments are similar in terms of analog specifications.

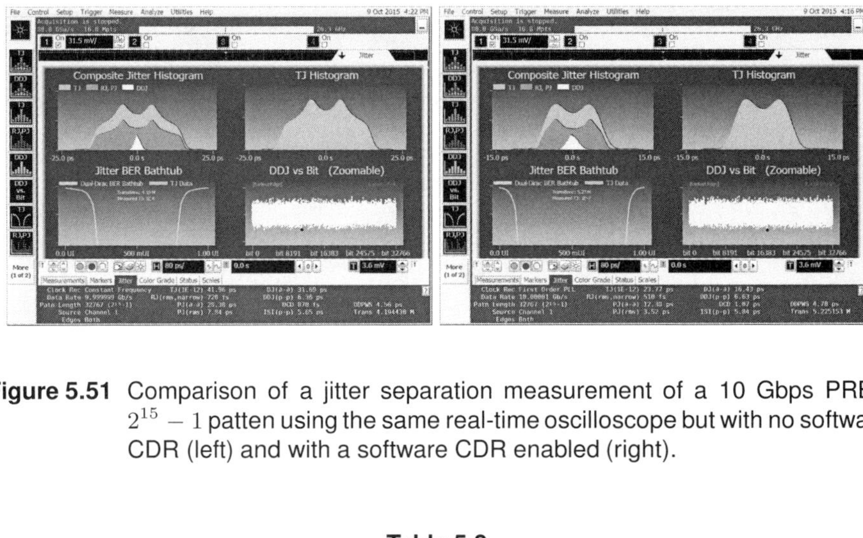

Figure 5.51 Comparison of a jitter separation measurement of a 10 Gbps PRBS $2^{15} - 1$ patten using the same real-time oscilloscope but with no software CDR (left) and with a software CDR enabled (right).

Table 5.2
Comparison of Jitter Measurements on PRBS $2^{15} - 1$ Pattern at 10 Gbps Between Two Real-Time Oscilloscopes

Jitter Value	Real-Time Oscilloscope 1	Real-Time Oscilloscope 2
RJ	3.96 ps	4.18 ps
DJ	13.49 ps	16.85 ps
TJ $(BER = 10^{-12})$	69.97 ps	62.14 ps

Table 5.3
Comparison of Jitter Measurements on PRBS $2^{15} - 1$ Pattern at 10 Gbps on a Real-Time Oscilloscope With and Without Software-Based CDR

Jitter Value	Real-Time Oscilloscope, no CDR	Real-Time Oscilloscope with CDR
RJ	0.72 ps	0.51 ps
DJ	31.69 ps	16.43 ps
TJ $(BER = 10^{-12})$	41.96 ps	23.77 ps

because the phase noise measurement includes not only the random jitter but also any periodic jitter in the measurement range while on the real-time oscilloscope the random jitter is separated from the periodic jitter. The real-time oscilloscope measured a value of 2.14 ps while the spectrum analyzer measured a value of 3.76 ps. However, if one analyzes the spectrum measured by both instruments (in the case of the real-time oscilloscope the jitter spectrum was computed by the jitter software) that is shown in Figure 5.53, it is possible to see that there is a good correlation on the periodic jitter component that is present on the jitter spectrum which also explains the measured random jitter value differences since this periodic jitter component will be present on the phase noise measurement range. This is a good example to demonstrate the possibility of inferring the reasons for the two jitter measurement differences by a deeper analysis of the jitter results of two instruments. Appendix J discusses phase noise measurements and correlation in more detail.

Unfortunately the solution to the jitter measurement correlation challenge is not as easy as choosing a "golden" jitter measurement instrument, since there is none. The best option is to compare the different measurement setups one intends to correlate using the same jittered signal and if possible with a calibrated amount of the different types of jitter. This topic is further discussed in Appendix I.

5.5.10 Driver Amplitude Noise

The driver performance is also defined by the amount of amplitude noise present. While noise in the time-domain is referred as jitter, in the voltage domain it is defined as amplitude noise. As already shown in Figure 2.33, amplitude noise will increase the amount of jitter and in this way not only decrease the timing margin on the driver signal but also the level margin in the form of a reduced data eye height. Like in the case of jitter, it is possible to analyze the amplitude noise using a histogram as shown in Figure 5.54.

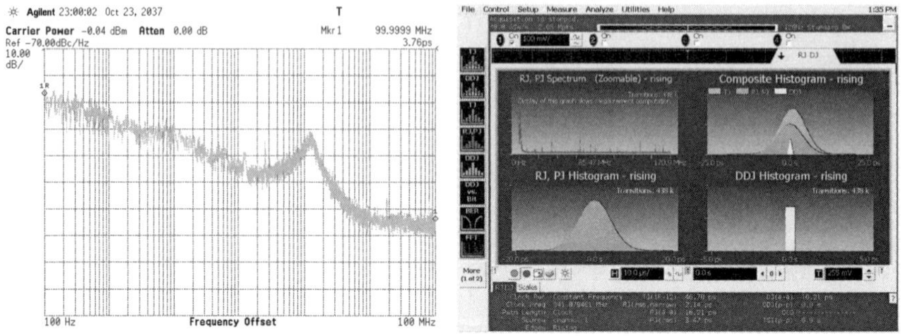

Figure 5.52 Comparison of a jitter measurement for a bit clock between a real-time oscilloscope with 13 GHz bandwidth and 40-Gsps sampling rate with jitter separation software and a spectrum analyzer with phase noise measurement software using a measurement band from 100 Hz to 100 MHz. Both instruments are from the same manufacturer. The real-time oscilloscope measured an RJ value of 2.14 ps and the spectrum analyzer measured a value of 3.76 ps.

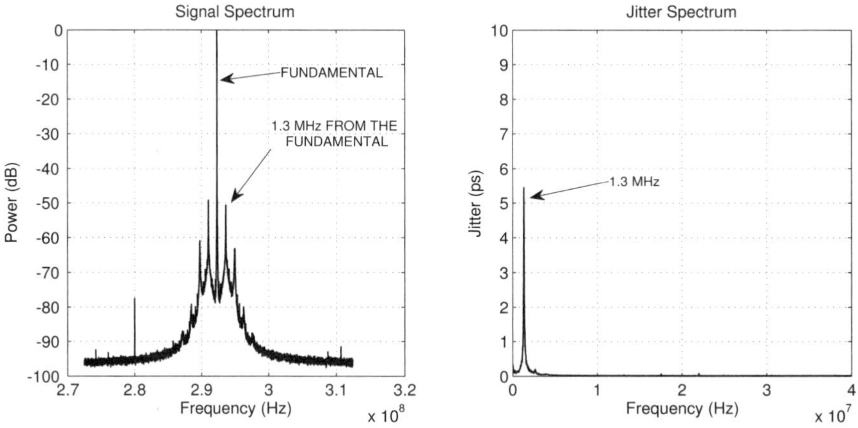

Figure 5.53 Comparing a measurement of a bit clock signal with regard to its jitter spectrum. The jitter spectrum obtained with real-time sampling oscilloscope (right) shows a periodic jitter component at 1.3 MHz that correlates with the phase modulation spur measured by the spectrum analyzer (left) that is located 1.3 MHz from the bit clock signal fundamental tone.

Amplitude noise also allows itself to categorization in different types like random and deterministic amplitude noise. In this case a lossy signal path not only adds deterministic jitter but also deterministic amplitude noise.

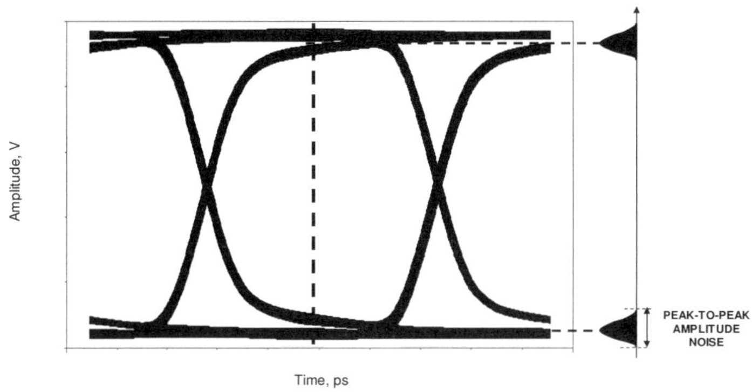

Figure 5.54 Histogram approach for the driver amplitude noise measurement.

Like for the timing case, it is also possible to generate a bathtub curve for the signal level. This fact allows one to generate a two-dimensional plot that provides a visualization of the effects of jitter and amplitude noise on the BER simultaneously as shown in Figure 5.55. The plot is also called a BER contour plot since each contour represents a collection of the same BER threshold.

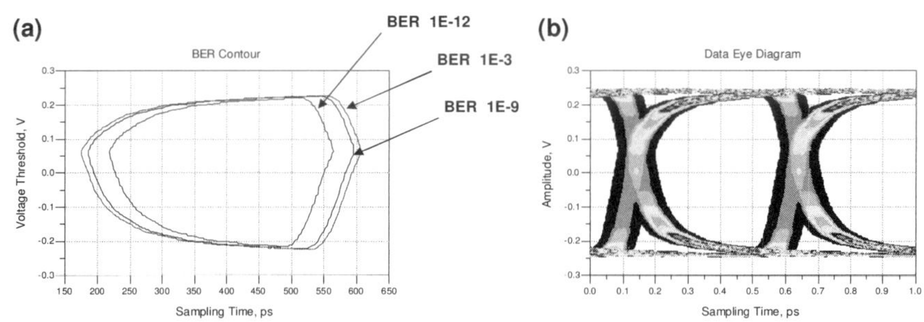

Figure 5.55 (a) Example of a two-dimensional BER plot (sampling time versus threshold voltage) also known as a BER contour plot and (b) the corresponding data eye diagram.

5.6 Fundamental Receiver Tests

This section presents some of the typical measurements for testing the receiver in a DUT I/O cell, and how these tests can be implemented in an ATE system.

5.6.1 Setup and Hold

The measurement on the DUT receiver side that corresponds to the driver skew measurement is the setup and hold measurement. Setup and hold time measurements determine the minimum time that a data signal has to be stable before (minimum setup time) or after (minimum hold time) the corresponding latching clock edge. Figure 5.56 shows this relationship graphically.

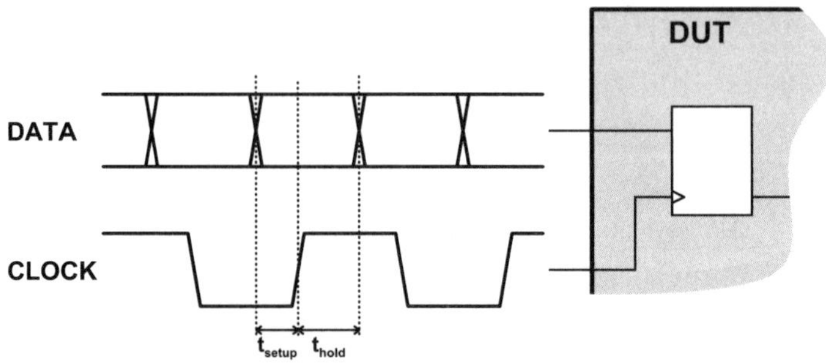

Figure 5.56 Receiver setup and hold measurements.

Like the skew measurements, setup/hold time measurements are implemented using the timing control and search capabilities of the digital pin electronics. Since the ATE drivers do not have the same timing deviations as might be the case for a DUT during the test pattern execution, in contrast to the skew measurement there is no need to restrict the pattern execution/compare on only a part of a full functional pattern set. However, test engineers should avoid the use of patterns that will generate significant data-dependent jitter on the signal path from the ATE driver to the DUT because this would cause varying timing offsets between the clock signal and the data signals.

In order not to cause device failure due to high jitter on the data signals and to avoid potential problems due to driver pulse widths below the specified limit, the setup/hold measurement is done with the bits set to their nominal UI width. For the setup/hold time measurement of DUT receivers, all ATE drive edges for each of the data signals are moved uniformly during a timing search. This way, the complete data eyes of the data signals are moved relative to the associated clock signal as shown in Figure 5.57. In order to measure the setup time t_{setup}, the search is started with the active clock edge and the left data bit edges aligned to each other ($t_{setup} = 0$). This results in a failing functional test. During the search, the data signal edges are moved into the negative direction relative to the active clock edges until a fail-to-pass transition for functional test results is detected.

The relative timing distance between clock and data signal at this fail-to-pass transition represents the minimum setup time that the DUT can handle.

For measuring the minimum hold time t_{hold}, the search is started with the right edges of the data bits aligned to the active clock edges ($t_{hold} = 0$). The search for the hold time is done by moving the data signal eyes into the positive direction relative to the active clock edges. Again, the relative timing offset between active clock edge and right data eye edge at this fail-to-pass transition represents the minimum hold time supported by the DUT.

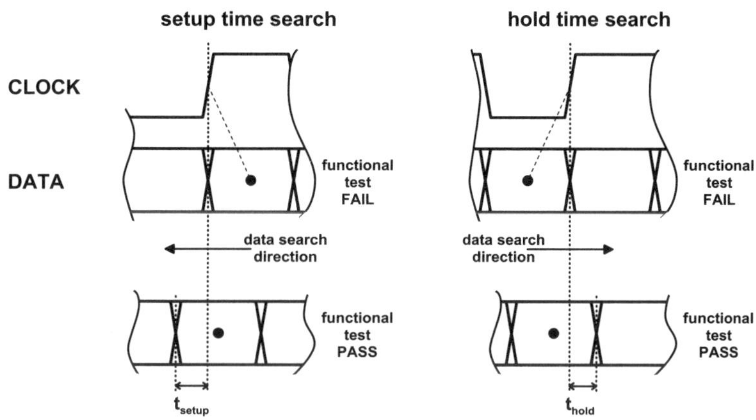

Figure 5.57 ATE search setup for receiver setup hold characterization.

5.6.1.1 Setup and Hold Derating

Sometimes relative AC timing parameters such as setup/hold times in data sheets are defined based on certain signal slew rate and level threshold assumptions. In a real-world environment, these assumptions typically do not hold because of unavoidable physical differences of signal driver implementations and losses of signal connections. Moreover, for certain signaling technologies setup and hold time specifications do not merely define required timing distances between signal reference level crossings, but also the time a signal stays between a certain signal threshold level and a different saturation level to accumulate the required charge for guaranteed signal latching [33]. Such a charge transfer model for setup/hold time specifications first was used by JEDEC for the DDR2 standard [34] and is applied since then for various memory standards. As a consequence of considering the integrated signal over time between two timing instances, the slew rates of the involved signals have an influence on the specified setup and hold values. Thus standards that deploy such a charge transfer model for setup and hold times define derating values that are used to adjust the setup/hold specification boundaries to the slew rates of the driver signals applied during the measurement. This adjustment assumes a base setup or hold time that is defined under the assumption of a defined slew rate of the involved signals. Derating values are specified in a two-dimensional matrix with the slew rates of each of the involved signals defining one dimension. These derating values are added to the measured setup and hold times according to (5.9) and (5.10) in order to get a base setup and hold time value that then can be compared to the specified minimum setup/hold limits. An example for a derating matrix is shown in Table 5.4.

$$tIS(total) = tIS(base) + \Delta tIS \tag{5.9}$$

$$tIH(total) = tIH(base) + \Delta tIH \tag{5.10}$$

Table 5.4
Setup/Hold Derating Value Table Example (values in [ps])

		clock slew rate							
		3.0 V/ns		2.0 V/ns		1.8 V/ns		1.6 V/ns	
		ΔtIS	ΔtIH	ΔtIS	ΔtIH	ΔtIS	ΔtIH	ΔtIS	ΔtIH
data slew rate	2.0V/ns	88	50	88	50	-	-	-	-
	1.5V/ns	59	34	59	34	67	42	-	-
	1.0V/ns	0	0	0	0	8	8	16	16
	0.9V/ns	-2	-4	-2	-4	6	4	14	12
	0.8V/ns	-	-	-6	-10	2	-2	10	6
	0.7V/ns	-	-	-	-	-3	-8	5	0
	0.6V/ns	-	-	-	-	-	-	-1	-10

In order to interpret the values in the derating matrix, one has to understand that for setup and hold time measurements, setup and hold time saturation voltage levels are defined with a symmetrical distance to the device threshold voltage V_{ref}. The charge that is accumulated over time to ensure data latching is defined by the area between the threshold voltage and the signal or the saturation level, whichever is closer to the threshold level. A specified setup and hold time means nothing else than a minimum charge that has to be accumulated by the clock and the data signals. This minimum charge is represented by minimum areas defined by the boundary of signal level, threshold crossing and saturation level. Standards assume certain slew rates for data and clock signals like DDR3 with nominal slew rates of 1.0 V/ns on data signals and 2.0 V/ns on data strobe signals that define the charge accumulation area boundary of the signal between threshold and saturation level crossings. With the assumed slew rates, the area between the device threshold voltage signal crossing and the saturation level signal crossing is well defined and is considered in the specified base setup and hold time values. The derating numbers in the slew rate matrix define how much the specified setup and hold times are influenced by the slew rate dependent charge transferred during the time the signal is in the area between device threshold voltage and the specified saturation voltage. The principle of this correction is shown in Figure 5.58 for a setup time correction with a data signal that has a slower slew rate than the one assumed for the base setup value. In this figure, the dashed data line represents a signal with the nominal slew rate assumed for the definition of the specified setup time. The solid data signal is the one with the slower slew rate. For figure simplification, we assumed the saturation voltages being the same as the signal levels used as setup/hold measurement levels. The accumulated charge required to ensure safe signal latching is represented by the marked areas separately for each of the two signals. In order to accumulate the same charge as the dashed reference signal, the solid data signal has to be integrated for a longer time due to its slower slew rate. The difference Δt of accumulation time represents the derating value that is specified in Table 5.4 for a data signal with the slew rate of the solid data signal. One can see that although both signals cross the measurement reference voltage V_h at the same time and thus result in the same measured setup time t_{meas}, the effective setup time (which is represented by the right boundary of the marked areas) of the signal with the slower slew rate is bigger than the effective setup time of the signal with nominal slew rate because it starts charge accumulation earlier. Thus, the signal with the slower slew rate can tolerate a measured setup time that is reduced by the the difference of the effective setup times of the two signals. This difference in effective setup time is considered by the derating factors of Table 5.4 and allows measured setup times that are shorter than the specified base setup times for signals that have smaller slew rates than the nominal signal and requires measured setup times that are longer than the specified base setup times for signals that have faster slew rates than the nominal signal. The same concept applies to the hold time derating as shown in Figure 5.59 for a faster than nominal slew rate. In our example we did not consider deviations from the nominal slew rate for the clock signal. In reality this also has to be considered and the combinations between data and clock slew rate variations are covered by the derating factor table.

5.6.2 Receiver Sensitivity

The objective of the receiver sensitivity test is to measure the minimum signal amplitude at the DUT's receiver input that still allows the receiver to correctly recover the data on the input signal. In design verification, the signal amplitude

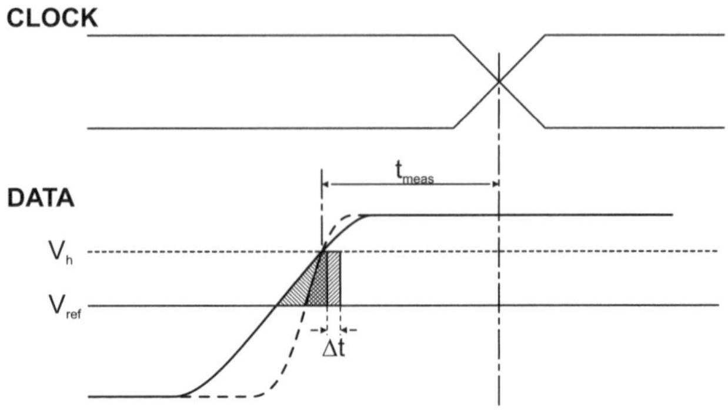

Figure 5.58 Setup time derating.

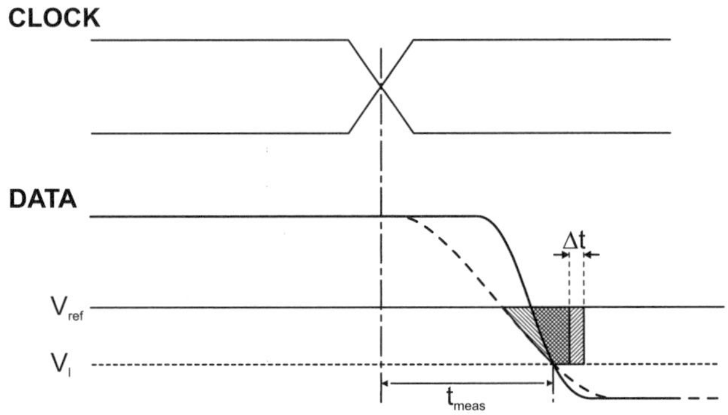

Figure 5.59 Hold time derating.

at the receiver input is reduced in a level search operation until the receiver starts to fail to recover the correct data.

In order to get the correct receiver sensitivity distribution for all pins and not only the worst-case minimum input amplitude voltage for a group of DUT receivers, this measurement has to be done serially for all DUT receivers to be measured. If each DUT receiver has an associated DUT dedicated output pin (e.g., in a far-end loopback configuration as described in Section 6.3.1.3) that allows a selective judgment whether a single receiver recognized its input levels correctly, the receiver sensitivity measurement can be done in parallel on all receiver pins as shown in Figure 5.60. Only the pass/fail evaluation for each search step needs to be done separately for each receiver/driver pair in this case.

In production, one simply programs the ATE driver with a specific minimum amplitude to be guaranteed and checks for a passing functional test with this setting. Although this test looks very simple from its definition point of view, it can create significant challenges for a high-speed application, especially if very accurate measurements are required.

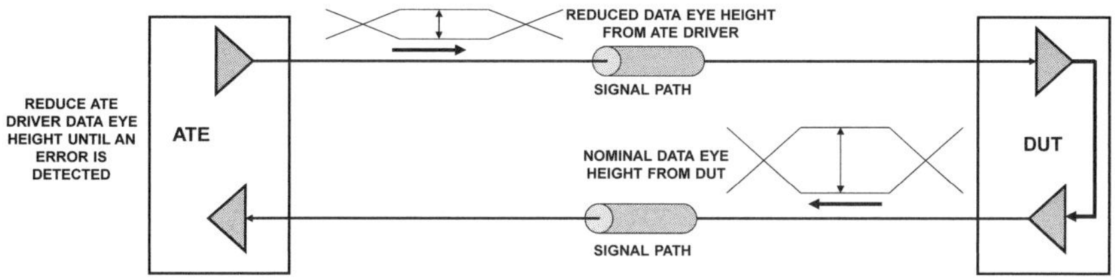

Figure 5.60 Example of a receiver sensitivity test where a far-end loopback is used on the DUT I/O cell.

The challenge starts with the fact that we are dealing with low signal amplitudes. Low amplitude waveforms with a good parametric performance are difficult to generate. Also receiver sensitivity is defined at the receiver input, but in a real setup a test fixture must be used where the signal will be degraded on its travel along the signal path from the ATE driver to the DUT due to the signal path loss and other effects described in Chapter 8. This signal degradation manifests itself in voltage loss, crosstalk from other signals, and amplitude noise from the test fixture (which is pattern-dependent).

All these factors make it clear that for receiver sensitivity, one needs to be very careful with the measurement setup. One approach to address the challenge of generating low amplitude signals with good parametric performance is to use high-bandwidth attenuators (Section 7.17.2). This allows the ATE driver to deliver an amplitude value lower than its specifications by programming the driver to an amplitude value within its specifications and then using the high-bandwidth attenuator to obtain an amplitude signal lower than the ATE specifications. Regarding the other challenges, appropriate design of the test fixture (Chapter 8) and focus calibration (Section 9.3) are some of the approaches to address them.

5.6.2.1 Crosstalk Effects on Receiver Sensitivity Testing

Although crosstalk is a topic that is important for any high-speed test, due to the fact that on the receiver sensitivity test, the measurement instrumentation will be driving very low amplitude signals, the effect of crosstalk from other signal lines might have a significant effect on the test result. Figure 5.61 shows a diagram describing how crosstalk might arise on a receiver sensitivity test. In this case, the ATE driver is driving a low amplitude signal to the DUT receiver while at the same time it is possible that the DUT driver is transmitting a signal with a normal amplitude to the ATE receiver. Crosstalk is proportional to the signal amplitude, rise time, and proximity between the aggressor and victim signals. Since higher data rates imply faster rise times, crosstalk is becoming a considerable effect.

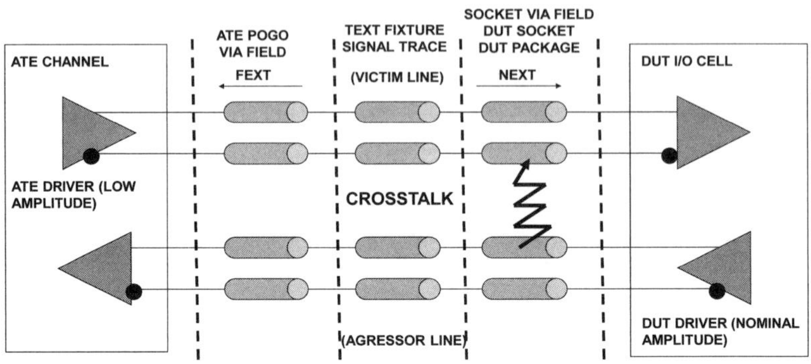

Figure 5.61 Diagram of the crosstalk effect on the receiver sensitivity test.

The fact that the ATE driver is sending a low amplitude signal makes this signal an ideal victim for crosstalk. Since the signal amplitude and rise time from the DUT are usually not under the test engineer's control (it is set by the IC design), the only variable over which the test engineer can have partial control is the proximity of the victim and aggressor signals, but this is restricted only to the test fixture since the DUT package and ballout are also predefined and outside the control of the test engineer. Another important point is that crosstalk only occurs in signal transitions, which means that depending on the skew between signals, crosstalk effects might have catastrophic effects or not. This is exemplified in Figure 5.62.

As mentioned before, minimizing crosstalk on the test fixture as well as keeping a tight control on the channel skew are important to prevent effects similar to the one shown in Figure 5.62.

5.7 Receiver Jitter Tolerance

The objective of a receiver jitter tolerance test is to verify the ability of the receiver in a DUT I/O cell to tolerate or compensate for the timing uncertainties (jitter) that can be expected for real signals in a real application. In Section 2.4 timing jitter was categorized in different categories depending on its characteristics and causes. In a jitter tolerance test the objective is to generate a waveform with some of these jitter categories and test if the receiver is still able to function with a given BER [35]. Figure 5.63 shows the jitter tolerance requirements from the PCI Express standard where it is possible to observe the different jitter categories that the DUT receiver needs to tolerate

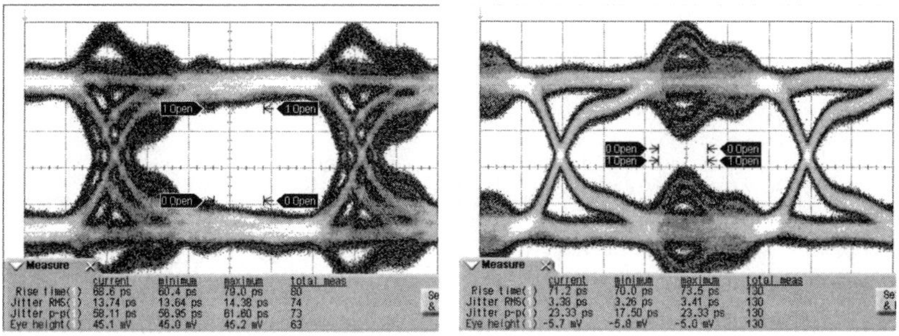

Figure 5.62 Effect of victim/aggressor skew on the crosstalk at the victim line. Note that the crosstalk amplitude is the same on both cases but in one it happens at the victim signal data eye edge with little effect on the eye center (left) and on the other it happens at the data eye center closing the data eye significantly (right).

to be standard-compliant. The next sections will discuss some of the jitter injection categories in detail.

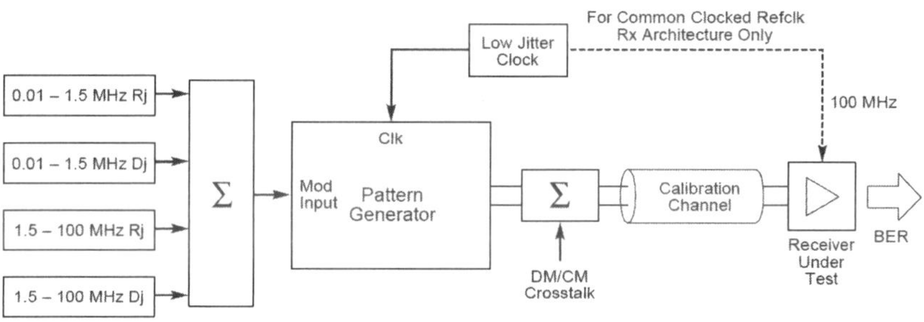

Figure 5.63 The PCI Express jitter tolerance compliance block diagram. (From: [18]. ©2002–2009 PCI-SIG. Reprinted with permission.)

5.7.1 Random Jitter Tolerance

Since any electronic system or part has some intrinsic noise, it is sometimes necessary to evaluate if a high-speed digital receiver is able to correctly detect the data pattern in the presence of a certain amount of random jitter. As discussed in Appendix A, random jitter is modeled using a Gaussian distribution. The Gaussian distribution probability density function is described by the following formula:

$$p(t) = \frac{1}{\sqrt{2\pi}\sigma} e^{-\frac{(t-\mu)^2}{2\sigma^2}} \tag{5.11}$$

The main challenge with random jitter tolerance is to provide a stimulus signal with the required random jitter content since on the receiver side it is only necessary to perform a standard functional test and measure if any bits failed. The typical technique is to use a noise source to inject random jitter on the stimulus driver through a voltage controlled delay line. Section 7.9 discusses noise sources in more detail. Figure 5.64 shows an example of a bit clock pattern with added random jitter in the time and frequency-domains. Section I.2 discusses in detail possible calibration approaches for random jitter injection.

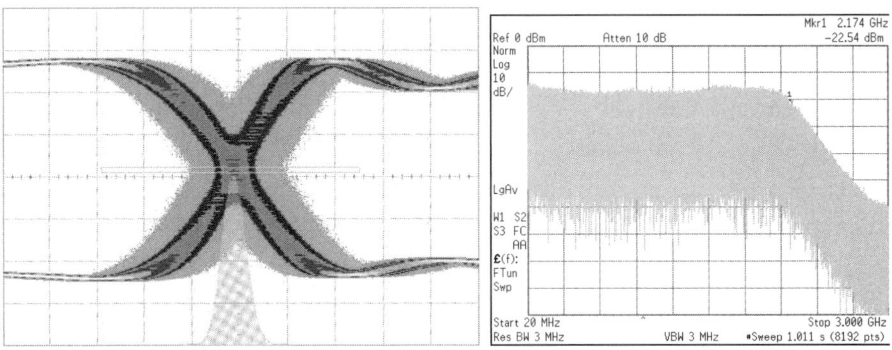

Figure 5.64 Example of a bit clock pattern with Gaussian random jitter injected through a voltage controlled delay line in the time-domain by measuring the jitter histogram with an equivalent-time oscilloscope (left) and on the frequency-domain with a spectrum analyzer (right).

Another approach is to use an AWG (see Section 7.8) for random jitter injection. Although a deterministic pattern is used to simulate random jitter, it can have some advantages especially when test time is critical as discussed in [36].

5.7.2 Sinusoidal Jitter Tolerance

Sinusoidal jitter tolerance is the oldest methodology for jitter tolerance testing [37]. One reason is that by changing the frequency of the injected periodic sinusoidal jitter it is possible to observe the behavior of the receiver in the frequency-domain as shown in Figure 5.65. The measurement approach is to increase the sinusoidal jitter amplitude for each frequency until a fail is detected during a functional test. Basically it is a shmoo plot of the periodic sinusoidal jitter frequency versus the sinusoidal jitter amplitude. It is important to note that depending on the pattern size used on the functional test, the measured jitter tolerance curve will correspond to a certain BER threshold. For BER thresholds like 10^{-12}, this test might take a considerable amount of time, although faster approaches are possible [38].

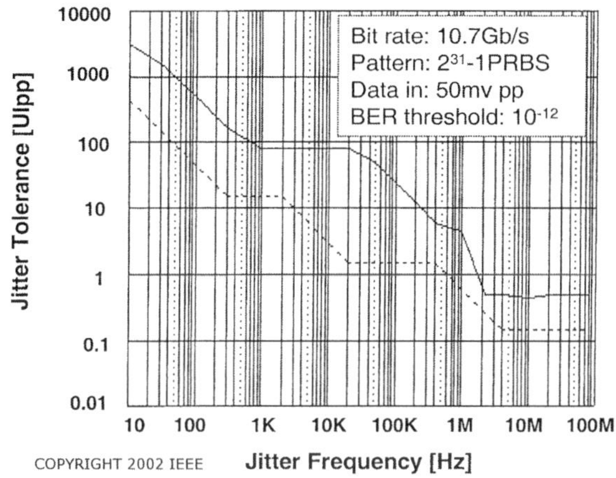

Figure 5.65 Example of a receiver jitter tolerance test with periodic sinusoidal jitter for a OC-192 application (reprinted with permission from [39]).

This type of test provides significant insight on the behavior of the receiver, and standards like OC-192 (SONET) [40] define specific jitter tolerance masks that the receiver must comply with.

Sinusoidal jitter with a peak-to-peak amplitude A is modeled by the following probability distribution function [41, 42]:

$$p_{\text{sinusoidal}}(x) = \begin{cases} \frac{1}{\pi\sqrt{(\frac{A}{2})^2 - x^2}} & |x| < \frac{A}{2} \\ 0 & \text{Otherwise} \end{cases} \tag{5.12}$$

Note that the probability distribution is independent of the frequency of the sinusoidal jitter. Figure 5.66 presents a plot of the probability distribution for a periodic sinusoidal jitter signal with a jitter peak-to-peak amplitude of 4 ps ($A = 4$).

Appendix I.1 discusses methodologies for sinusoidal jitter injection calibration. Figure 5.67 shows a measurement on the time and frequency-domain of the output of a pattern generator that has sinusoidal injected jitter.

Note that it is not possible to identify the sinusoidal jitter frequency in a histogram. This is only possible with a measurement in the frequency-domain. On the frequency-domain one can see the two side lobes appearing around

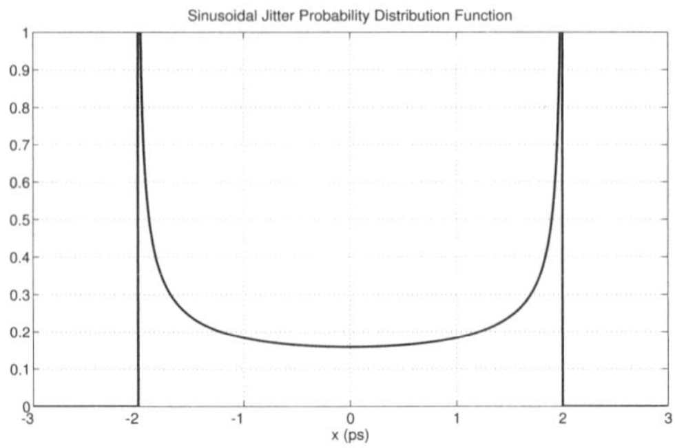

Figure 5.66 Sinusoidal jitter probability distribution ($A = 4$).

each harmonic spectrum of the bit clock signal as expected from modulation theory.

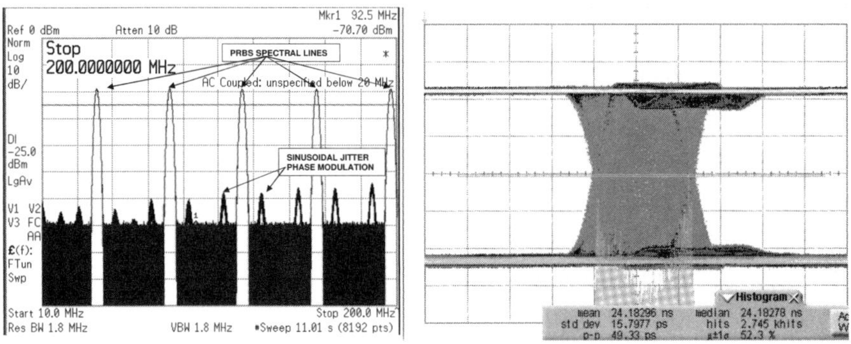

Figure 5.67 Sinusoidal jitter injection of a PRBS data pattern seen on the frequency-domain with a spectrum analyzer (left) and time-domain with an equivalent-time oscilloscope (right).

5.7.3 Data-Dependent Jitter (DDJ) Tolerance

The ability of a DUT receiver to tolerate the data-dependent jitter (DDJ) is a critical parameter for a high-speed I/O cell and is usually specified in modern high-speed standards. Unlike jitter tolerance to sinusoidal jitter that is typically in the low-frequency range, DDJ adds high-frequency components to the signal jitter spectrum.

The specifications for DDJ tolerance are sometimes defined using a certain trace length on a printed circuit board with a specific geometry and dielectric material type. Figure 5.68 shows an example of a printed circuit board implemented to address several DDJ standards.

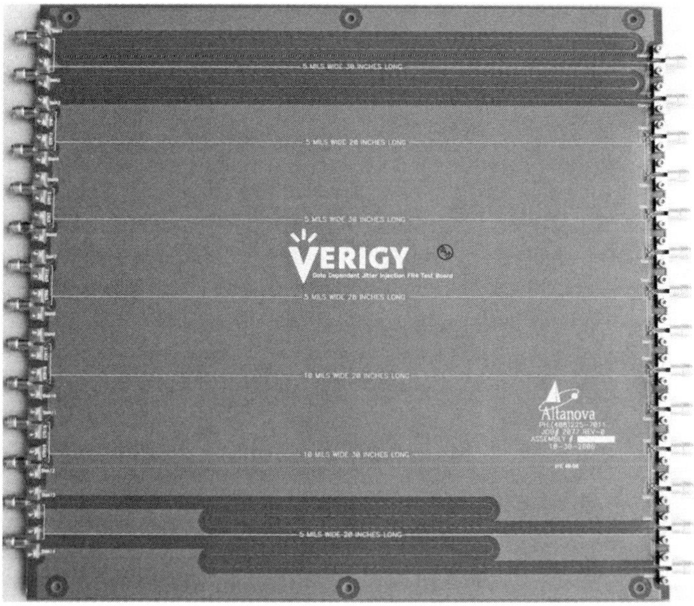

Figure 5.68 Printed circuit board on FR4 containing several traces compliant with different high-speed standards (Courtesy of Advantest).

Figure 5.69 shows the data eye diagram of a 5 Gbps digital signal after traveling through one of the DDJ injection signal traces of the PCB shown in Figure 5.68. The data eye diagram shows a large amount of DDJ added to it. It is the job of the receiver to correctly recover the signal among all that jitter.

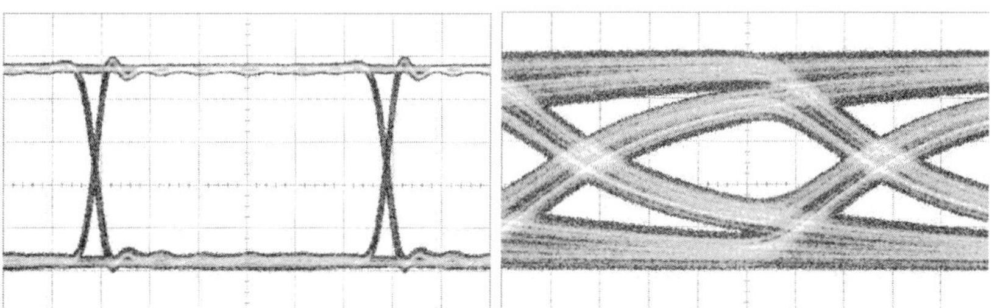

Figure 5.69 A 5 Gbps PCI Express compliance pattern before (left) and after (right) traveling through a DDJ injection PCB signal trace.

The approach of using a printed circuit board to inject the required DDJ, although usable in a bench setup, brings considerable challenges on an ATE environment when trying to test multiple I/O cells at the same time. One approach to address this challenge is to use a filter instead of a real printed circuit board [43]. This allows the filter to be directly assembled into the ATE test fixture. In [44] a filter design is presented that tries to mimic the frequency response of the printed circuit board trace. Another approach discussed in Section 9.9.3.5 is to use the equalization capabilities of some ATE pin electronics architectures to inject DDJ.

5.7.4 Bounded Uncorrelated Jitter (BUJ) Tolerance

In some situations it is necessary to test the tolerance of a DUT receiver to bounded uncorrelated jitter (BUJ). Injection of BUJ in a data stimulus signal can be achieved by different approaches. A typical approach when a delay line is used for jitter injection is to use a pattern generator running asynchronously to the data pattern. The pattern generator is programmed with a PRBS data pattern as shown in Figure 5.70 [26].

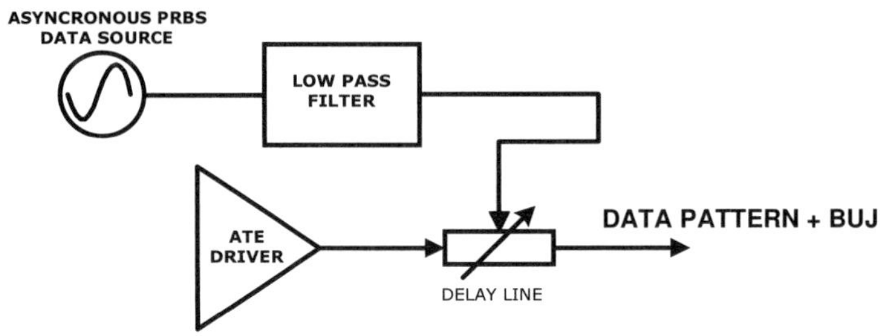

Figure 5.70 High-level block diagram of a bounded uncorrelated jitter (BUJ) injection setup using an asynchronous PRBS source and a delay line.

A low-pass filter is sometimes also used at the output of the BUJ pattern generator source to limit the maximum frequency of injected jitter. The setup in Figure 5.70 can present some challenges for implementation in an ATE system due to the asynchronous requirement. The BUJ peak-to-peak value will be determined by the BUJ pattern generator output amplitude which defines the maximum phase deviation of the delay line of Figure 5.70. Figure 5.71 shows the frequency spectrum from a BUJ pattern generator running at 2 Gbps with a PRBS $2^7 - 1$ and a PRBS $2^{31} - 1$ data pattern after a 100 MHz low-pass filter. This spectrum will be the injected BUJ jitter spectrum and should simulate the jitter spectrum that would arise from crosstalk from an adjacent bus running at 2 Gbps. Note that for the PRBS $2^{31} - 1$ case the resulting spectrum looks indistinguishable from a bandlimited noise spectrum. Figure 5.72 shows the measured jitter histogram from a data pattern running at 10.325 Gbps with and without the BUJ from Figure 5.71.

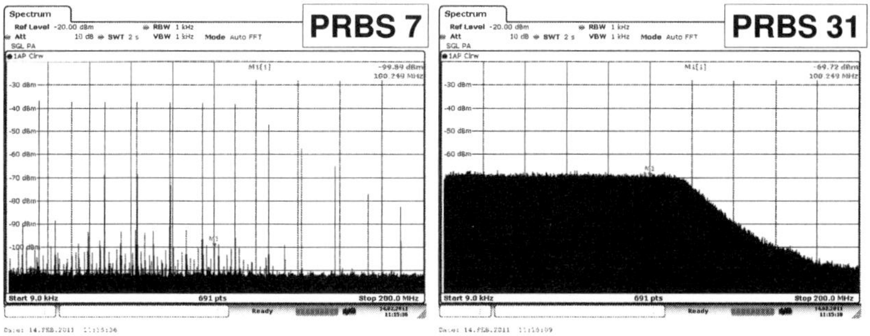

Figure 5.71 Frequency spectrum from a pattern generator source running at 2 Gbps with a PRBS $2^7 - 1$ (left) and a PRBS $2^{31} - 1$ (right) data pattern after a 100 MHz low-pass filter.

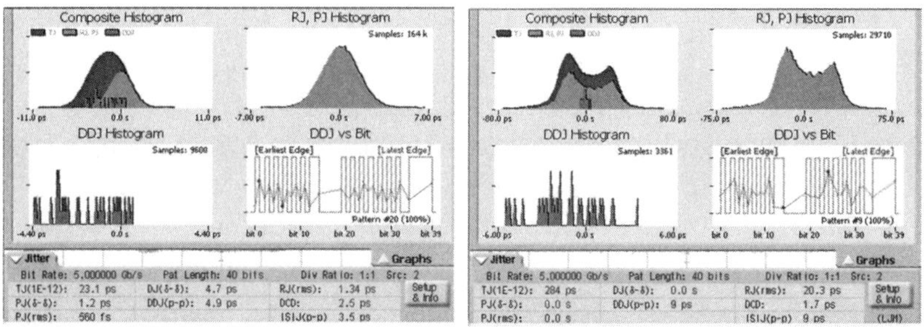

Figure 5.72 Measured jitter decomposition without injected BUJ (left) and with injected BUJ (right).

5.7.5 Testing the Receiver Equalizer

As already discussed in Section 2.6.4, modern high-speed digital receivers contain an equalization circuit that in most cases is controlled by a set of variables (e.g., the tap coefficients of a DFE receiver equalizer). These variables are usually set through registers or in some implementations automatically to an optimal value for the link signal path loss. This optimal value is typically determined by a link initialization phase.

The jitter tolerance test discussed in Section 5.7 assumes that the DUT receiver equalizer is already set to some optimal value. However, in the design verification or characterization phase, the objective might be to identify the optimal settings for the equalizer or verify that the settings that are automatically found by the link initialization algorithm are the correct ones as shown in Figure 5.73.

In the case of trying to identify the optimal values for the equalization settings, one option is to add the lossy signal path that the equalizer is intended

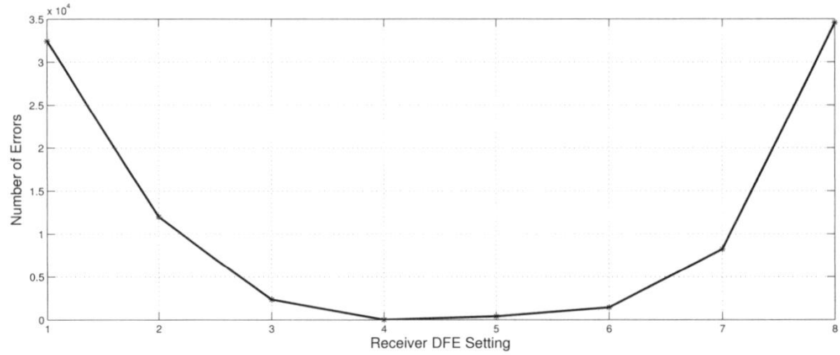

Figure 5.73 Measuring the ideal setting for a receiver DFE by using the number of errors on the received pattern. Note that this measurement is valid for a specific link loss.

to compensate on the test fixture. For testing the receiver equalizer, this signal path could be switched on and off between the ATE pin electronics and the DUT receiver through relays as shown in Figure 5.74. Although this option is straightforward and does not require any special features on the ATE pin electronics, it can result in a complex implementation for high pin-count devices. It also is limited to a single or a very limited number of lossy signal paths which might be insufficient for some applications that are designed for a large variety of signal path losses.

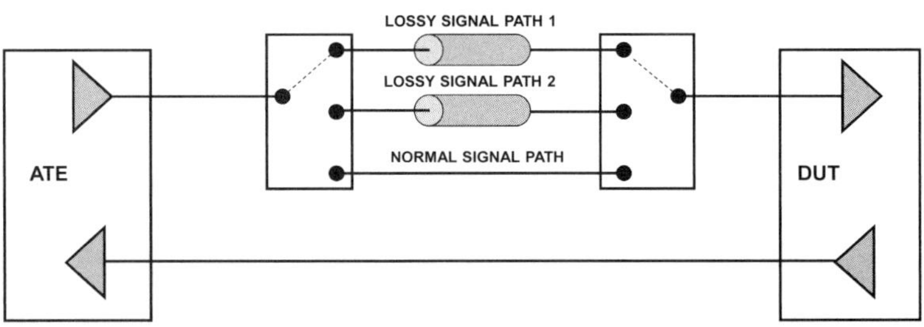

Figure 5.74 Receiver equalization testing by adding a lossy signal trace in the DUT test fixtures.

Another option is to inject the DDJ of the expected lossy signal path on the stimulus waveform to the DUT receiver using the ATE pin electronics (see Section 9.9.3.5). Note that the injected DDJ value and spectrum needs to have a correspondence to the PCB lossy signal path that is expected to be used in the link that the DUT is intended for. This approach provides the maximum flexibility since the DUT receiver equalizer can be tested for multiple different signal path losses without adding complexity to the test fixture. However, it does require the pin electronics driver to have this capability, which is not trivial.

In the case of a DUT I/O cell receiver that automatically sets the equalization parameters through a link initialization step, it is possible by implementing appropriate design for test (DfT) blocks on the DUT I/O cell to read the final settings of the equalizer and also overwrite those settings [45]. This would allow the test engineer to find the optimal equalization parameters for a given DDJ profile and compare them with the values obtained through the automatic link initialization process.

5.8 PLL Characterization

This section describes some of the tests that are targeted specifically to the PLL circuitry that is usually part of a high-speed I/O cell.

5.8.1 Jitter Transfer

Jitter transfer is defined as the amount of jitter in the input signal of a PLL circuit that is transferred to the output signal of the PLL. Figure 5.75 shows a diagram describing a possible jitter transfer measurement setup on the PLL of the DUT reference clock. Typically this measurement is displayed in the frequency-domain showing the ratio of measured jitter versus the injected jitter at a given frequency as shown in Figure 5.76.

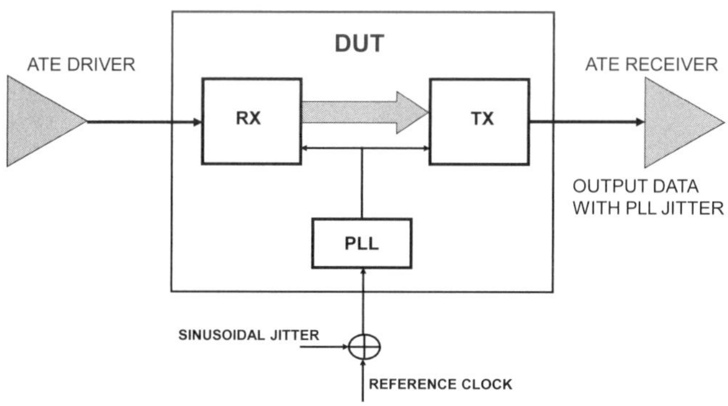

Figure 5.75 Jitter transfer measurement setup.

The idea is for each frequency to insert a certain amount of sinusoidal jitter at that frequency in the input signal and measure the sinusoidal jitter amplitude at that frequency on the output signal, displaying the ratio between them using a logarithmic scale as described in the following equation:

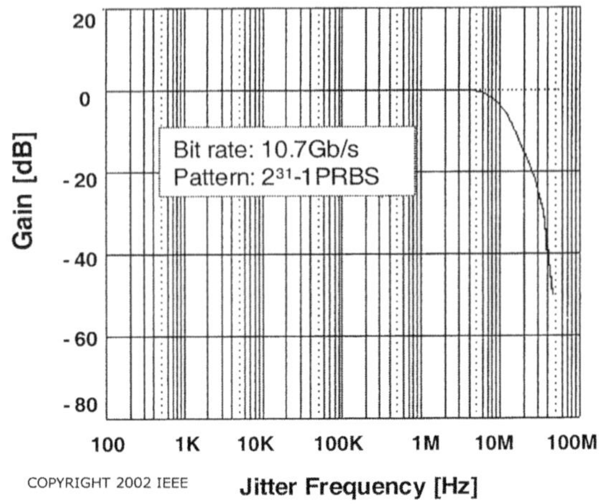

Figure 5.76 Jitter transfer measurement example (reprinted with permission from [39]).

$$\text{Jitter Transfer}(f) = 20\, log_{10}\left(\frac{\text{Output Jitter}(f)}{\text{Input Jitter}(f)}\right)\ \text{dB} \tag{5.13}$$

The challenge with this measurement is twofold. First, it is important to inject a specific amount of sinusoidal jitter at a specific frequency. This might require a calibration of the sinusoidal jitter amplitude (see Section I.1 for more details). The second challenge is to measure the sinusoidal jitter amplitude at the current measurement frequency step that is present on the output of the DUT. This measurement is not trivial, since the amplitude of the sinusoidal jitter at a specific frequency must be separated from all the other sources of jitter like random jitter or other deterministic jitter sources at different frequencies (e.g., crosstalk from another clock on the DUT). One simple approach when the device output jitter is dominated by random jitter is to use the RJ subtraction method that is discussed in Section I.1.2.

Other approaches have been developed for measuring the jitter transfer based on random jitter injection instead of using periodic sinusoidal jitter [46, 47]. The advantage of using random jitter to measure the jitter transfer is that it can be done with a single measurement instead of a shmoo test of different sinusoidal jitter frequencies. This is demonstrated in Figure 5.77 where a random noise source is used to inject random jitter on the reference clock of a DUT. A jitter decomposition is then performed on one of the high-speed outputs of the DUT that is running a clock pattern resulting in the PLL jitter transfer curve. For comparison the same measurement was done using a signal source analyzer instrument (see Section 7.6).

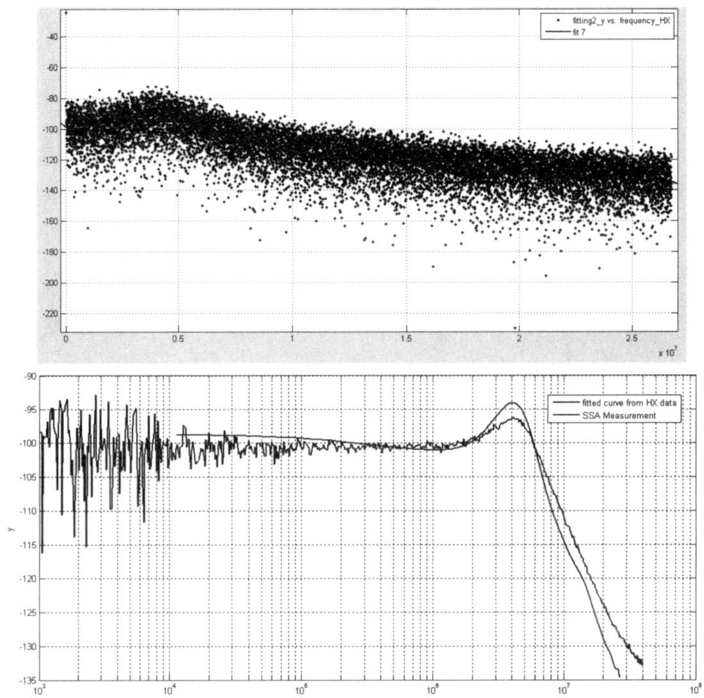

Figure 5.77 Measuring the jitter transfer characteristic of a PLL using random noise injected on the reference clock (top: jitter transfer measurement using an ATE digital pin electronics card, bottom: comparison with the measured result using a signal source analyzer instrument).

5.8.2 Frequency Offset

Frequency offset is a test designed to measure the ability of the DUT to work properly when the frequency on one of the DUT pins (e.g., the reference clock) does not correspond exactly to the nominal value. Figure 5.78 shows an example of a possible measurement setup for a frequency offset test.

A frequency offset is performed on the DUT reference clock or in another input like the incoming pattern data rate to the DUT I/O receiver, while keeping all the other inputs at the nominal frequency. A functional test is then performed to verify if the DUT is still working correctly with this setup. The challenge with this type of measurement is that typically the amount of frequency offset required is just some tens or hundreds of parts per

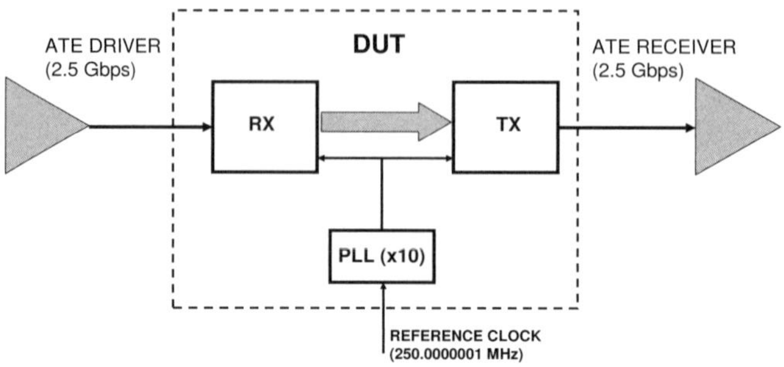

Figure 5.78 Example of a frequency offset measurement setup.

million (ppm) of the nominal frequency. This requires that the ATE timing architecture is able to offset the frequency of a driver channel by that amount while keeping the other ATE channels on the nominal frequency.

5.8.3 Spread Spectrum Clocking

Spread spectrum clocking (SSC) [48] is a requirement in several high-speed digital standards (e.g., FBDIMM and Serial ATA) and optional in others (e.g., PCI Express). The objective is to spread the energy of the clock (and therefore data) over a percentage of the frequency band. The reason is to reduce the power at the clock frequency (which is the highest) by spreading it using a specific modulation profile and in this way making compliance with emission standards easier. Figure 5.79 exemplifies how spread spectrum clocking works for the Serial ATA standard [49].

From the figure it is possible to see that spread spectrum clocking is a kind of frequency modulation (FM) of the reference clock of the device. The bit clock is modulated with an extremely low modulation frequency compared to the bit rate so that the bit rate can be very stable for a short period of time with no harmful effect on the BER.

5.8.3.1 Measuring SSC with a Sampler[4]

A sampler ATE instrument can be used to measure a spread spectrum clock [50]. Figure 5.80 shows an example of a measurement configuration and frequency-domain illustration of a spread spectrum clock. The data bit clock pattern is set to 1.5 Gbps to simulate a Serial-ATA (SATA) [49] application clock which is at 750 MHz. SSC for SATA is specified as 0.5% down spread so that the frequency deviation is 3.75 MHz at 750 MHz. Approximately 33 kHz

[4]This section was contributed by Hideo Okawara.

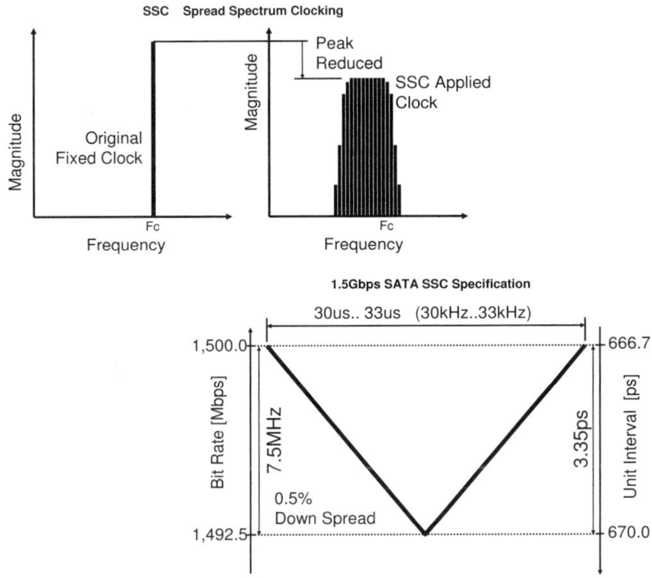

Figure 5.79 Description of how spread spectrum clocking (SSC) works (top: the spectrum peak reduction effect of SSC; bottom: the clock modulation profile).

sawtooth-shaped FM is applied. The signal is sampled with 4,096 points at 24.798 Msps so that the baseband is 12.399 MHz wide. The 750 MHz is aliased at around 6 MHz in the 12.399 MHz baseband, and the frequency deviation of 3.75 MHz can be appropriately captured in the baseband, as shown in Figure 5.80(b).

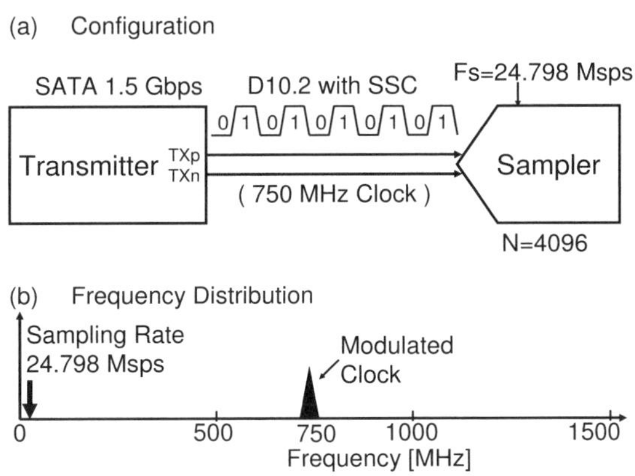

Figure 5.80 (a, b) Spread spectrum clock measurement configuration.

The test signal was measured with a spectrum analyzer and is shown in Figure 5.81 showing the frequency modulation on the clock signal.

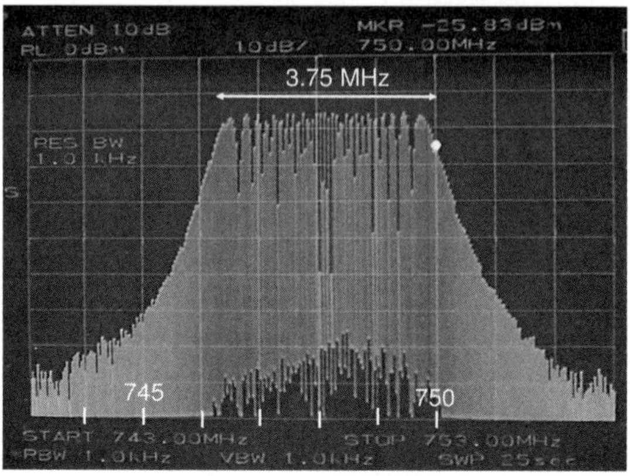

Figure 5.81 Clock with spread spectrum measured with a spectrum analyzer.

The block diagram of the data processing approach to obtain the modulation waveform of the spread spectrum clock is presented in Figure 5.82, which is the application of the orthogonal demodulation [51, 52].

The sampled signal from the measurement setup of Figure 5.80 is shown in Figure 5.83(a) which can also be seen in the frequency-domain as the spectrum shown in Figure 5.83(b).

To obtain the modulation signal, the cosine and sine reference (F_{ref}) waveform data are provided and multiplied with the sampled waveform data. The multiplied data (cosine and sine components) split the beat signal and the sum signal components, respectively. Then the data are low-pass filtered. The extracted beat signals of cosine and sine components are processed with the arc tangent mathematical operation reconstructing the instantaneous phase trend of the spread spectrum clock. The differential operation to the instantaneous phase trend derives the instantaneous frequency trend. Multiplying the reference frequency with the instantaneous frequency trend reveals the SSC modulation trend as shown in Figure 5.83(c).

It is also possible to measure a spread spectrum clock with a digital pin electronics ATE channel as described in [53].

5.9 Other Tests

5.9.1 Impedance Tests

In order to achieve the high data rates of high-speed I/O interfaces, the I/O driver and receiver circuitry of the devices usually has to provide a proper

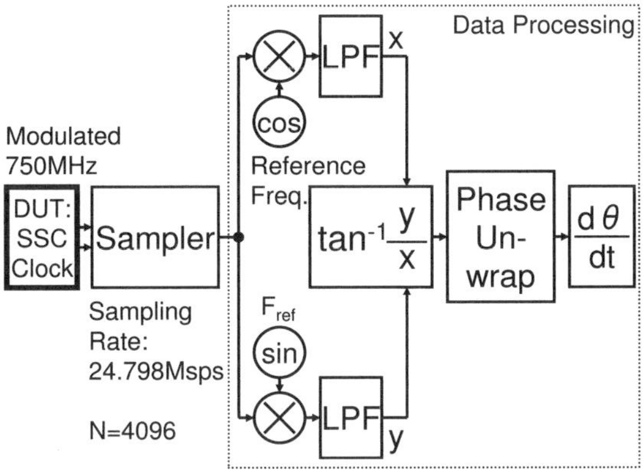

Figure 5.82 SSC analysis with the orthogonal demodulation method.

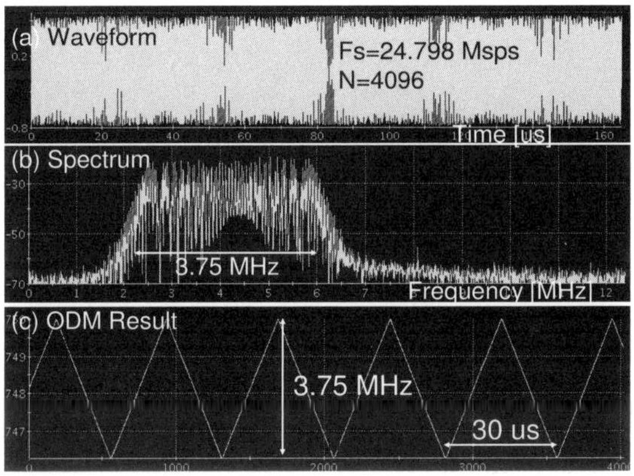

Figure 5.83 (a–c) SSC spectrum and reconstructed frequency trend.

signal termination. For differential interfaces, such a termination is mandatory because it transfers the constant current that flows in different directions for high and low states into a voltage difference that can be recognized by the differential comparator in the device I/O circuit. The impedance of a differential driver has to match the receiver impedance to avoid signal reflections on the signal path. For single-ended interfaces that operate at high data rates, proper termination is also important to avoid signal reflections that would disturb data transmission. The different termination schemes that typically are used in high-speed I/O circuits are shown in Figure 5.84.

Since the terminations in the I/O circuitry of a device can have a significant impact on the performance of the high-speed I/O signal path in a system, it is an important task to measure the I/O terminations at least in the

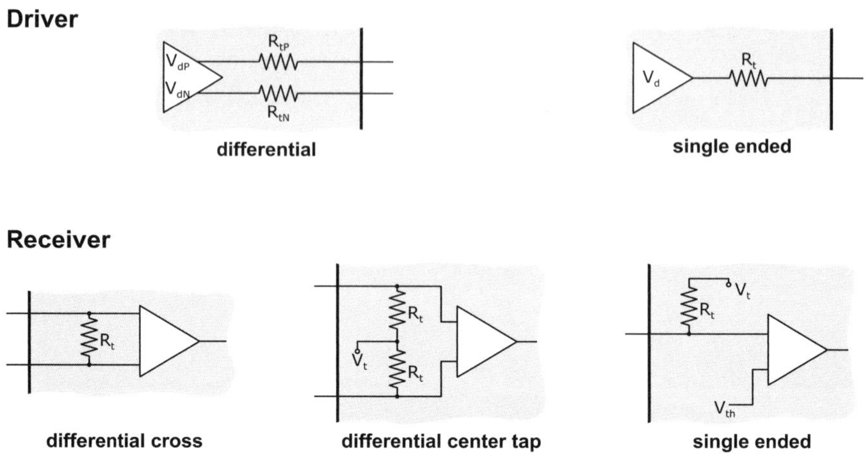

Figure 5.84 Typical termination schemes used in high-speed I/Os.

device verification phase. The measurement of the I/O terminations usually is referred to in a general way as impedance test, although it is important to note that in an ATE environment this is usually restricted to the measurement of the DC impedance (termination resistances) and not the AC impedance of a DUT that also takes parasitic inductance and capacitance into account. The AC impedance is measured through a return loss measurement that is described in Section 5.9.2.

The measurement of the termination impedances on the ATE is done using the PMU circuitry that is usually included in every ATE digital pin electronics card. With the PMU, the test engineer can force a current or voltage to the DUT pin and measure the voltage or current the DUT sets up as a reaction to this forced stimulus. For impedance measurements the PMU is used to determine current/voltage pairs that then are used to calculate the device impedance using Ohm's law. However, the boundary condition that a test engineer has to consider when measuring impedances of a real device is the fact that he does not know the exact values of internal device voltages that might have influence on the test results if the test setup is not selected in a correct way.

The device internal voltages V_t, V_d, V_{dP}, and V_{dN} as used in Figure 5.84, for example, are not known exactly and usually also cannot be measured directly in an accurate way. If these voltages, however, are not considered in impedance measurements that are based on a single voltage-current pair, they cause wrong impedance calculations. The reason for this is that when an impedance is calculated based on a single voltage-current pair, it inherently is assumed that if the force current I_f is equal to zero, then the measured voltage

V_m also is equal to zero as shown in the left graph of Figure 5.85. This is not the case, however, and the internal device voltages cause an offset of the voltage current relationship that is the base for our impedance calculation as shown in the right graph of Figure 5.85.

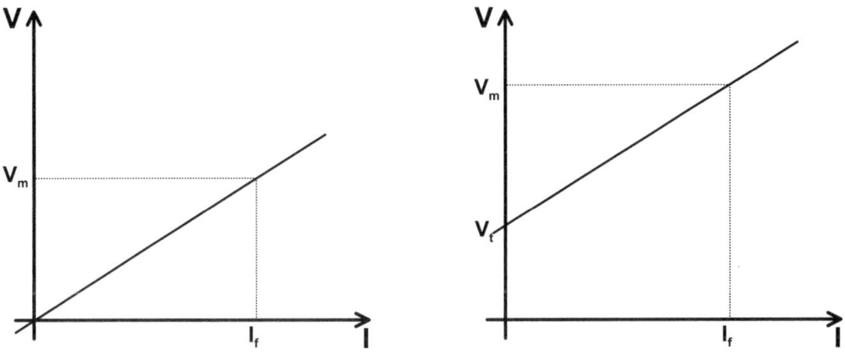

Figure 5.85 Difference of single point impedance measurements without consideration of DUT-internal voltages (left) and with consideration of DUT-internal voltages (right).

Since the impedance R is represented by the slope of the voltage current relationship, a simple and effective way to avoid this source of inaccuracies is to measure two voltage-current pairs and calculate the impedance as the relationship of the measured voltage difference for a forced current difference and this current difference as shown in Figure 5.86.

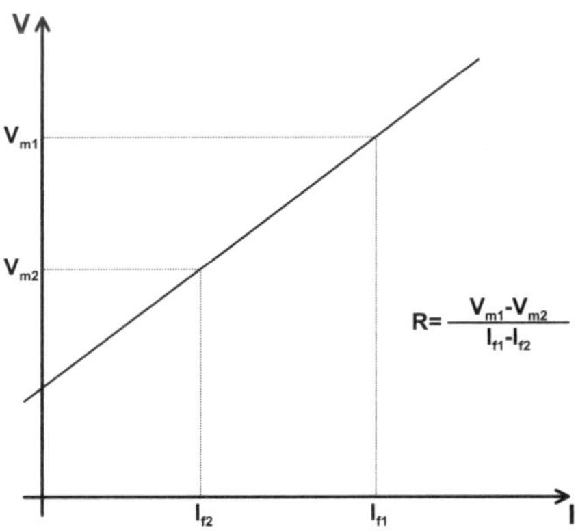

Figure 5.86 Two-point impedance measurement for voltage offset compensation.

All the test engineer has to ensure when performing such a two-point measurement is that the selected force currents I_{f1} and I_{f2} are in an area that has a linear characteristic for the I/O impedance. This is not always the case because I/O impedances often are realized with the help of biased transistors that might have a nonlinear behavior in certain regions.

The measurement setups required to determine the impedance values for single-ended and differential I/Os vary slightly and are described next. All measurement setups described support the required two-point measurement approach. Of course, as for all other DC measurements, and also for impedance measurements, the pins to be measured have to be configured into a static state.

5.9.1.1 Single-Ended Signals

For the single-ended impedance measurement, the setup is straightforward. The PMU of the ATE is connected to the input or output pin to be measured and for each of the two force currents selected by the test engineer, the voltage the device pin settles to is measured. The impedance is derived as the relationship between the measured voltage difference and the force current difference (see Figure 5.86).

5.9.1.2 Differential Signals

Impedance measurements for differential I/Os come in two flavors. The first is the common-mode impedance where the impedance between each leg of the differential interface and the common connection point of the two legs inside the DUT is measured. For differential receivers, this, of course, only makes sense if a center tap termination scheme is used.

The measurement configuration to do the common-mode impedance measurements is the same as for single-ended impedance measurements. The only additional item a test engineer has to take care of is the fact that the differential leg that is not being measured is kept at the same condition for both measurement executions of the two-point measurement on the opposite leg. The easiest way to ensure this is to disconnect the negative leg of a differential signal when the positive leg is measured and vice versa or to use the PMU on the leg that is not being measured and apply a constant voltage while the measurements on the other leg take place.

The other impedance measurement flavor for differential signals is the differential impedance. Here, the impedance between the two legs forming the differential signal is measured. The measurements for the differential impedance are independent of the differential termination scheme (floating cross termination or center tap termination) that is used for a differential signal. In order to measure the differential impedance, each of the two legs forming the differential signal is connected to a separate PMU. One of these PMUs forces the test current while the other one sinks the same test current. This ensures that there is no influence of other device internal current paths that might have an impact on the measurement result. In this PMU configuration, the voltage that settles on the differential legs is measured per leg and the voltage drop over the differential impedance is calculated as the difference of the two voltages that are measured on the two legs.

For the differential impedance measurement that is done in this form, it is not necessarily required to do a two-point measurement because it is ensured that no parasitic currents or voltages influence the measurement due to the selection of the same source and sink current on the opposite legs.

5.9.2 Return Loss

Return loss is an important parameter that is specified in several high-speed digital standards. The return loss specifies the amount of the signal energy that is reflected at the package terminals of either a driver or receiver pin due to impedance mismatches between the test fixture 50 Ω signal trace and the transition to the package, the package substrate signal traces, and the DUT driver output impedance or the receiver termination. It is easy to see that one would like to have all the signal energy delivered to the receiver internal circuit with no energy reflected. If a significant amount of the input signal energy is reflected before arriving at the internal receiver circuitry, it will have a smaller amount of the input signal to work with, making the probability of an error higher. Figure 5.87 shows the return loss requirements on the receiver from the PCI Express standard [18].

Return loss (RL) is defined by the following equation:

$$RL = 20 \log \left(\frac{Z_L - Z_0}{Z_L + Z_0} \right) \text{ dB} \tag{5.14}$$

where Z_L is the impedance of the receiver input and Z_0 the impedance of the transmission line feeding the input signal to the receiver. Detailed presentation on basic transmission line theory and return loss can be found in [54].

Note that on the requirements presented in Figure 5.87 the common-mode (CM) and differential (DIFF) return loss are distinguished. This is needed when defining return loss for a differential interface like PCI Express.

The standard approach to measure return loss is to use a vector network analyzer as shown in Figure 5.88. It is important to note that since the DUT is embedded on a test fixture as shown in Figure 5.88, it is necessary to de-embed the test fixture to obtain accurate measurements. Reference [55] provides a discussion on de-embedding techniques.

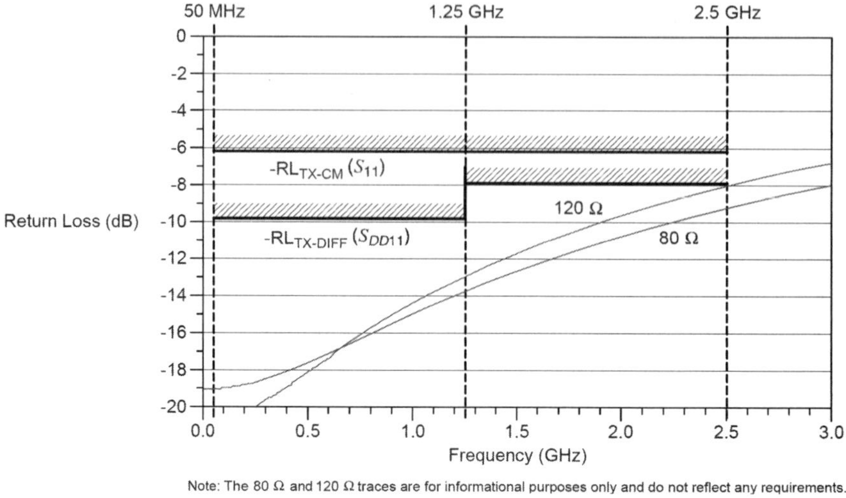

Figure 5.87 Return loss receiver requirements from the PCI Express standard (From: [18]. ©2002–2009 PCI-SIG. Reprinted with permission.)

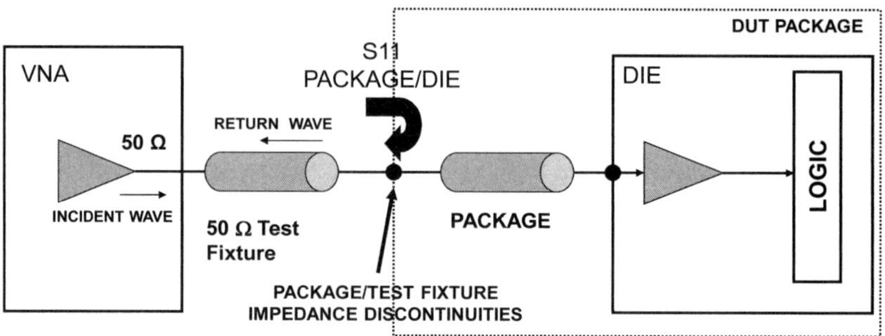

Figure 5.88 Measuring return loss with a network analyzer.

Return loss is usually measured on a bench setup with a vector network analyzer, although some ATE systems provide this capability also.

Figure 5.89 shows an example of a return loss measurement on a unidirectional interface (requiring a return loss measurement on the driver and receiver) and on a bidirectional interface where only one measurement needs to be done.

5.10 Power Consumption During IC Testing

It is important to be aware that IC power consumption during ATE IC testing can be significantly higher than during normal operation on its final

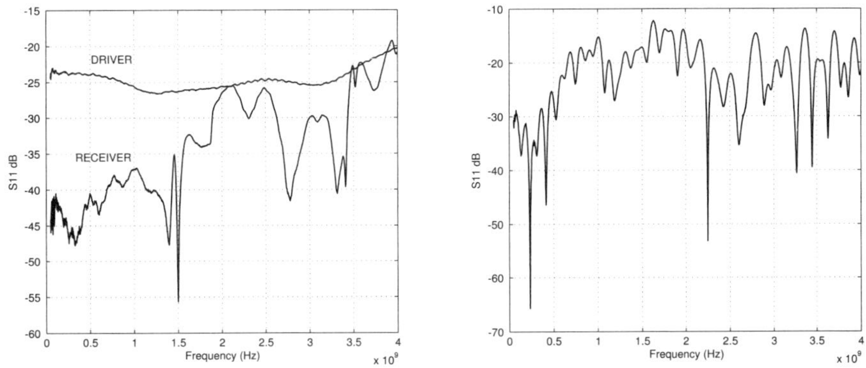

Figure 5.89 Example of the return loss measurements on a bidirectional interface (left) and unidirectional interface (right).

application. In some published cases power under test mode was up to 3.8 times higher than during functional mode [56].

The increased power consumption during ATE testing not only has an impact on the ATE test fixture PDN design (discussed in Chapter 8) but also on the DUT itself where this excessive power can have a negative impact on yield especially for at-speed scan [57].

5.11 Measurement Errors

One important question when performing any measurement is what is the associated error to the measured value. Errors can be classified in two categories: systematic and random [58–60]. There is little a test engineer can do in regard to systematic errors. The measurement instrument specifications should provide guidance for the systematic error that one can expect. For example, Figure 5.90 shows one example of the specification for an ATE pin electronics receiver. If the test engineer wants to measure the high level of a digital signal by moving the receiver level threshold up until he gets a fail, the associated error to the measured value will be +/−10 mV based on the specification of Figure 5.90.

Random errors are another category. In this case, the test engineer can address this type of error by making several measurements of the same variable and averaging the samples to obtain a more accurate result. This is a very important concept since it explains how the number of samples acquired has a direct impact on the accuracy of the measurement.

Mathematically if we intend to measure a variable X that is contaminated by Gaussian random noise (e.g., measuring frequency on a clock with

DC Performance

Threshold range	-2 to 7 V [1)]
Threshold resolution	2.5 mV
Threshold accuracy	±10 mV [2)]
Minimum overdrive	50 mV
Input leakage current	±10 μA
	(Characteristics: ±2 μA between 0 V to 5 V level)

1. Minimum high-low threshold difference: 100 mV.
2. Valid for threshold voltage range -2 V to 6 V and a DC Update period of 2 weeks. Characteristics: ±15 mV within a DC update period of 1 month

Figure 5.90 Specifications of the Advantest V93000 P1000 ATE pin electronics receiver (courtesy of Advantest).

random jitter or the BER of a DUT driver) with a variance σ by taking N samples of X, the obtained average value $\bar{X}$ will have a variance of $\sigma/\sqrt{N}$ [61, 62]. This means that with an increase in the number of measured samples of the variable X, the average value $\bar{X}$ gets closer to the real value of X.

References

[1] H. Shiotsuka and N. Masuda, "Measurement Technology for 6.5 Gbps High-Speed Serial Interface Using T2000," *Probo*, no. 25, 2005.

[2] H. Werkmann, "X-Mode Synchronization to Free Running Bitstreams," *Verigy Technical Note*, 2006.

[3] R. D. Adams, *High Performance Memory Testing: Design Principles, Fault Modeling and Self-Test*. Boston, MA: Kluwer Academic Publishers, 2003.

[4] P. Patten, "Divide and Conquer Based Fast Shmoo Algorithms," *Test Conference, International*, pp. 197–202, 2004.

[5] W. Maichen, *Digital Timing Measurements*. New York: Springer, 2006.

[6] J. R. Andrews, "Removing Jitter from Picosecond Pulse Measurements," *Picosecond Labs Note AN-23*, 2009.

[7] T. Ito, H. Okawara, and J. Liu, "RNA: Advanced Phase Tracking Method for Digital Waveform Reconstruction," *IEEE International Test Conference*, 2012.

[8] H. Okawara, "Novel Eye Pattern Test Method by Waveform Sampler," *Verigy Users Conference (VOICE)*, 2007.

[9] D. Derickson and M. Mueller, *Digital Communications Test and Measurement: High-Speed Physical Layer Characterization*. Upper Saddle River, NJ: Prentice-Hall, 2007.

[10] M. Mueller, R. Stephens, and R. McHugh, "Total Jitter Measurement at Low Probability Level, Using Optimizer BERT Scan Method," *IEC DesignCon*, 2005.

[11] M. Ahmad, M. Bugg, G. Fitzgerald, and M. Rost, "Skew in Twin-axial Cables and its Significance in Next Generation Differential Signaling," *DesignCon*, 2013.

[12] S. Farrahi, V. Kunda, Y. Li, X. Zhang, G. Blando, and I. Novak, "Does Skew Really Degrade SERDES Performance?," *DesignCon*, 2015.

[13] B. Brannon, "Sampled Systems and the Effects of Clock Phase Noise Jitter," *Analog Devices AN-756 Application Note*, 2004.

[14] K. Feher, *Telecommunications Measurements, Analysis and Instrumentation.* Atlanta, GA: Noble Publishing, 1996.

[15] W. H. Press, S. A. Teukolsky, W. T. Vetterling, and B. P. Flannery, *Numerical Recipes, The Art of Scientific Computing.* Cambridge, UK: Cambridge University Press, 2007.

[16] T. Yamaguchi, K. Asada, K. Niitsu, M. Abbas, S. Komatsu, H. Kobayashi, and J. Moreira, "A New Procedure for Measuring High-Accuracy Probability Density Functions," *IEEE Asian Test Symposium*, 2012.

[17] B. Laquai and R. Plitschka, "Testing High Speed Serial IO Interfaces Based on Spectral Jitter Decomposition," *IEC DesignCon 2004*, 2004.

[18] PCI-SIG, *PCI Express Base Specification Revision 2.1*, Mar. 2009.

[19] H. Werkmann, J. Moreira, and T. Yamaguchi, "State of the Art RJ/DJ Separation on the V93000: A Comparison between the Dual Dirac and the Blind PDF Jitter Separation Algorithms," *Advantest VOICE Users Group Conference*, 2015.

[20] G. Haensel, K. Stieglbauer, G. Schulze, and J. Moreira, "Implementation of an Economic Jitter Compliance Test for Multi-Gigabit Device on ATE," *IEEE International Test Conference*, 2004.

[21] M. P. Li, J. Wilstrup, R. Jessen, and D. Petrich, "A New Method for Jitter Decomposition Through Its Distribution Tail Fitting," *IEEE International Test Conference*, 1999.

[22] T. J. Yamaguchi, K. Ichiyama, H. X. Hou, and M. Ishida, "A Robust Method for Identifying a Deterministic Jitter Model in a Total Jitter Distribution," *IEEE International Test Conference*, 2009.

[23] F. Nan, Y. Wang, F. Li, W. Yang, and X. Ma, "A Better Method Than Tail-Fitting Algorithm for Jitter Separation Based on Gaussian Mixture Model," *Journal of Electronic Testing*, Sept. 2009.

[24] M. Aleksić, "Extraction of Jitter Parameters from BER Measurements," *IEEE Conference on Electrical Performance of Electronic Packaging and Systems*, 2011.

[25] B. A. Ward, K. Tan, and M. L. Guenther, "Apparatus and Method for Spectrum Analysis-Based Serial Data Jitter Measurement," *U.S. Patent 6,832,172 B2*, 2004.

[26] M. Lu and J. Moreira, "Implementation of Deterministic Jitter Characterization on the V93000," *Verigy Users Group Conference (VOICE)*, 2011.

[27] Y. Cai, S. A. Werner, G. J. Zhang, M. J. Olsen, and R. D. Brink, "Jitter Testing for Multi-Gigabits Backplane SerDes – Techniques to Decompose and Combine Various Types of Jitter," *IEEE International Test Conference*, 2002.

[28] J. Stimple and R. Stephens, "Precision Jitter Transmitter," *IEC DesignCon*, 2005.

[29] G. Foster, "Why Do Different Instruments Give Different Jitter Answers," *Synthesis Research SR-TN083 Measurement Brief*, 2009.

[30] M. Miller and M. Schnecker, "A Comparison of Methods for Estimating Total Jitter Concerning Precision, Accuracy and Robustness," *IEC DesignCon*, 2007.

[31] I. Zamek and S. Zamek, "Jitter Transformations in Measurement Instruments and Discrepancies Between Measurement Results," *IEEE International Test Conference*, 2005.

[32] W. Dalal and D. Rosenthal, "Measuring Jitter of High Speed Data Channels Using Undersampling Techniques," *IEEE International Test Conference*, 1998.

[33] B. Gervasi, "Memory Design Considerations That Affect Price and Performance," *Platform Conference*, 2002.

[34] JEDEC Solid State Technology Association, *DDR2 SRAM specification*. 79-2E.

[35] T. Lions, "Complete Testing of Receiver Jitter Tolerance," *IEEE International Test Conference*, 2010.

[36] R. Stephens and J. Calvin, "A New Method for Receiver Tolerance Testing Using Crest Factor Emulation," *IEC DesignCon*, 2010.

[37] P. R. Trischitta and E. L. Varma, *Jitter in Digital Transmission Systems*. Norwood, MA: Artech House, 1989.

[38] T. Yamaguchi, M. Soma, M. Ishida, H. Musha, and L. Malarsie, "A New Method for Testing Jitter Tolerance of SerDes Devices Using Sinusoidal Jitter," *IEEE International Test Conference*, 2002.

[39] J. Cao, M. Green, A. Momtaz, K. Vakilian, D. Chung, K.-C. Jen, M. Caresosa, X. Wang, W.-G. Tan, Y. Cai, I. Fujimori, and A. Hairapetian, "OC-192 Transmitter and Receiver in Standard 0.18-μm CMOS," *IEEE Journal of Solid-State Circuits*, vol. 37, Dec. 2002.

[40] Bellcore, *SONET OC-192 Transport System Generic Criteria*, 1998. GR-1377.

[41] N. Ou, T. Farahmand, A. Kuo, S. Tabatabaei, and A. Ivanov, "Jitter Model for the Design and Test of Gbps-Speed Serial Interconnects," *IEEE Design and Test of Computers*, July 2004.

[42] A. Papoulis and S. U. Pillai, *Probability, Random Variables and Stochastic Processes*. New York: McGraw-Hill, 2002.

[43] Y. Cai, B. Laquai, and K. Luehman, "Jitter Testing for Gigabit Serial Communication Transceivers," *IEEE Design and Test of Computers*, 2002.

[44] F. Zhang and W. Necoechea, "ISI Jitter Injection Filter Designs Using PIN and Varactor Diodes for SerDes Testing on ATE," *IEEE International Test Conference*, 2006.

[45] K.-T. Cheng and H.-M. Chang, "Test Strategies for Adaptive Equalizers," *IEEE Custom Integrated Circuits Conference*, 2009.

[46] B. Laquai, "A Model-Based Test Approach for Testing High Speed PLLs and Phase Regulation Circuitry in SOC Devices," *IEEE International Test Conference*, 2004.

[47] M. Ishida, T. J. Yamaguchi, and M. Soma, "A Method for Testing Jitter Tolerance of SerDes Receivers Using Random Jitter," *IEC DesignCon*, 2007.

[48] R. C. Dixon, *Spread Spectrum Systems: With Commercial Applications*. New York: John Wiley & Sons, 1994.

[49] Serial ATE International Organization, *Serial ATA Revision 3.0*, June 2009. Gold Revision.

[50] H. Okawara, "SSC Applied Serial ATA Signal Generation and Analysis by Analog Tester Resources," *IEEE International Test Conference*, 2009.

[51] H. Okawara, "Frequency/Phase Movement Analysis by Orthogonal Demodulation," *IEEE International Test Conference*, 2002.

[52] T. J. Yamaguchi, M. Soma, M. Ishida, T. Watanabe, and T. Ohmi, "Extraction of Instantaneous and RMS Sinusoidal Jitter Using an Analytic Signal Method," *IEEE Transactions on Circuit and Systems II: Analog and Digital Signal Processing*, vol. 50, June 2003.

[53] J. Liu, Y. Cai, L. Fang, and R. Ratemo, "Spread Spectrum Clocking Measurement—A Digital Approach," *IEEE IC Test Workshop*, 2004.

[54] D. M. Pozar, *Microwave Engineering*. New York: John Wiley & Sons, 2005.

[55] Agilent Technologies, "Agilent Signal Integrity Analysis Series, Part 3: The ABCs of De-Embedding," *Application Note 5989-5765EN*, 2007.

[56] S. Sde-Paz and E. Salomon, "Frequency and Power Correlation between At-Speed Scan and Functional Tests," *IEEE International Test Conference*, 2008.

[57] P. Girard, N. Nicolici, and X. Wen, *Power-Aware Testing and Test Strategies for Low Power Devices*. New York: Springer, 2009.

[58] N. Kularatna, *Digital and Analog Instrumentation: Testing and Measurements*. New York: The Institution of Electrical Engineers, 2003.

[59] *Calibration Philosophy in Practice, 2nd Edition*. Everett, WA: Fluke Corporation, 1994.

[60] L. Kirkup and B. Frenkel, *An Introduction to Uncertainty in Measurement*. Cambridge, UK: Cambridge University Press, 2006.

[61] J. R. Taylor, *An Introduction to Error Analysis*. Sausalito, CA: University Science Books, 1997.

[62] P. R. Bevington and D. K. Robinson, *Data Reduction and Error Analysis for the Physical Sciences*. New York: McGraw-Hill, 2003.

6

Production Testing

This chapter discusses testing of high-speed digital I/O interfaces with a production focus. In contrast to the measurement methodologies described previously, the purpose of production testing is not the retrieval of measurement values that allow an exact assessment of the individual performance of a device or design implementation. For production-oriented test implementations, the main goal is to guarantee the performance of a device according to a given device specification or datasheet while striving for minimum cost of test (COT) in a given test environment. The variable that a test engineer typically has under control to contribute to the COT is the test execution time per device. The performance of a device beyond the specification, datasheet requirements, or a certain speed bin usually is of minor interest for a production test run. As a consequence, not only can the measurement approaches used in production test implementations differ substantially from the ones we described previously, but also the test philosophy deployed for production tests. Figure 6.1 shows a diagram of the different possible approaches available to test a high-speed digital I/O cell. Note that this list is by no means exhaustive and other authors might use a different classification scheme.

Although classical ATE-based testing certainly dominates high-volume production, high-speed digital I/O interfaces have contributed to the fact that alternative or complementing test approaches like loopback testing have gained traction in recent years. The next sections discuss the different approaches presented in Figure 6.1 in detail. In the scope of this classification, our understanding of at-speed ATE is that the ATE instruments that are used to test the high-speed I/Os provide pattern and full timing generation at the native speed of the DUT's highest data rate signals. Thus, at-speed loopback instruments that rely on DfT circuitry in the DUT and/or low-speed ATE instruments to provide the required at-speed data generated by the DUT to their inputs are considered as low-speed ATE instruments, although they need to support at-speed data rates on their loopback signal paths. Another important facet of production test implementations is the topic of multi-site testing, which will be covered at the end of this chapter.

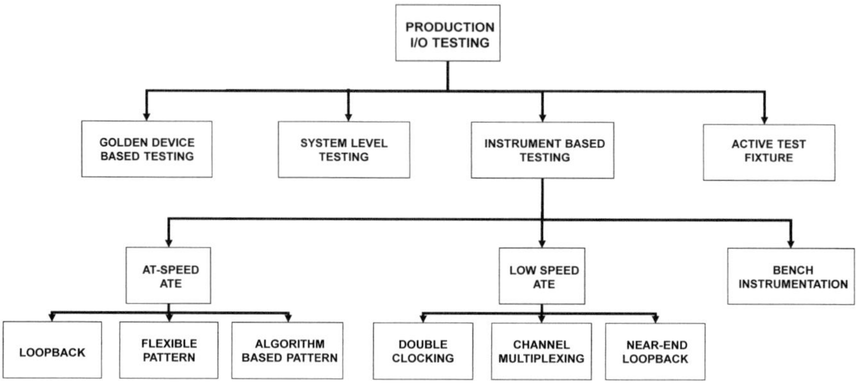

Figure 6.1 Diagram of possible approaches for testing a high-speed digital I/O cell in production testing.

6.1 Golden Device

A quite common alternative for production testing of selected devices is the use of a golden device with a proven good performance mounted on the test fixture. This is a feasible approach for symmetrical applications like a transceiver as shown in Figure 6.2 or when the golden device is the system partner of the DUT.

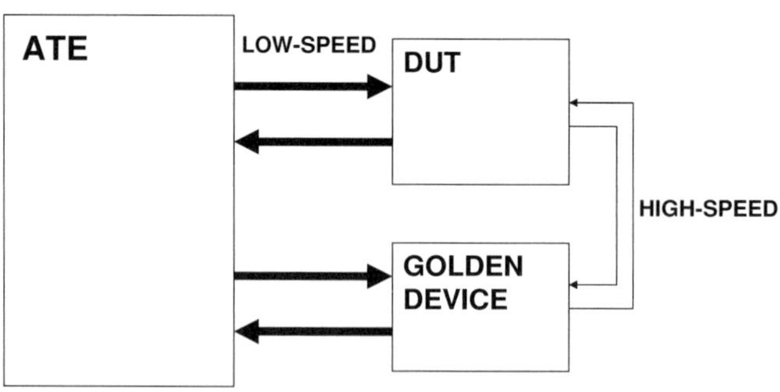

Figure 6.2 Golden device based testing of a transceiver application.

However, this solution has various challenges for the test engineer. Choosing the golden device and verifying its proper function in case of production issues (e.g., yield drop) might result in a large effort due to the restricted insight into the golden device or the lack of debugging possibilities. Furthermore, variations on multiple golden devices might cause correlation issues between different ATE systems on the production floor.

6.2 System-Level Test

Another option is the system-level test, where the DUT is tested in its target application (e.g., a microprocessor on a motherboard running special test software). From a cost point of view, this approach is very appealing but it can prove difficult to use for testing some physical parameters of the I/O cell. Typically system-level tests are used in conjunction with other approaches (e.g., as a second production step after a classical ATE-based testing of the DUT). Figure 6.3 shows one example of an automatic system for system-level testing. In this case the ATE system tests the DUT using a mission mode environment.

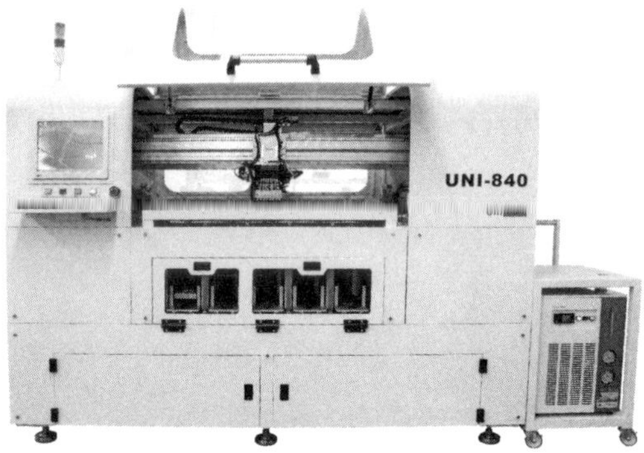

Figure 6.3 Example of an automatic PC-based system-level tester for GDDR memory applications (courtesy of UniTest).

6.3 Instrument-Based Testing: At-Speed ATE

As shown in Figure 6.1, there are two major options for production testing of I/O cells using an ATE system. One is to use an ATE system that is able to run at the same data rate as the DUT or using an ATE system that runs at a lower speed. The decision of which type of system to use can have a significant impact on the cost of test and also on the failure coverage. This section concentrates on the methodologies that can exclusively be performed by an at-speed ATE system. These methodologies are either the same type of tests that are done during the characterization phase or substitutes of these with shorter test execution times. A further classification of at-speed ATE systems is possible on the kind of ATE pattern generator, which is either flexible or algorithm-based. Finally, we distinguish between test solutions with and without far-end loopback. All these aspects are discussed next.

6.3.1 Physical Implementation

6.3.1.1 Flexible Pattern

The flexible-pattern ATE hardware corresponds to the classical at-speed testing approach where digital pin electronics (or sampler for the DUT driver parametric measurements) is used to test the DUT I/O cell by performing measurements using any kind of pattern, within the memory limits of the ATE system as shown in Figure 6.4.

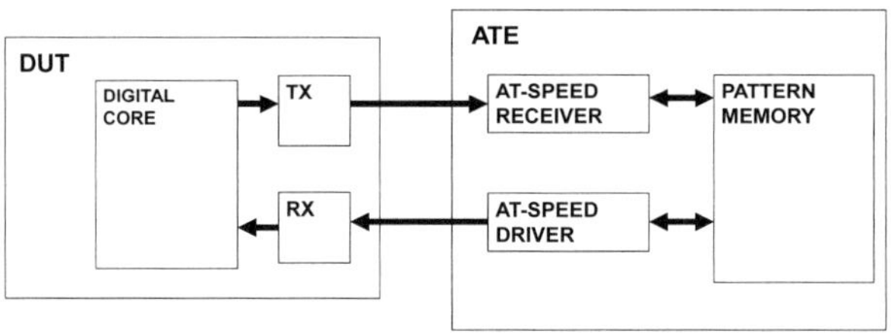

Figure 6.4 Block diagram of an ATE at-speed flexible-pattern approach that relies on the capabilities of the classical at-speed ATE system.

6.3.1.2 Algorithm-Based Pattern

The algorithm-based pattern approach differs from the flexible-pattern approach in the requirements on the ATE pin electronics. The ability to provide full pattern flexibility at high data rates increases the complexity and cost of the ATE pin electronics. One possible approach to reduce this cost while maintaining the capability to fully test an I/O cell at-speed is to restrict the used pattern to algorithmic patterns that can be generated and compared automatically from hardware (e.g., a PRBS pattern generator/analyzer). If the DUT contains DfT capabilities to create and evaluate the same algorithmic patterns, then it is possible to successfully test the DUT. This approach is shown in Figure 6.5.

In the algorithm-based pattern approach, the ATE driver stimulates a data pattern (e.g., a PRBS sequence) to the DUT receiver. A DfT engine

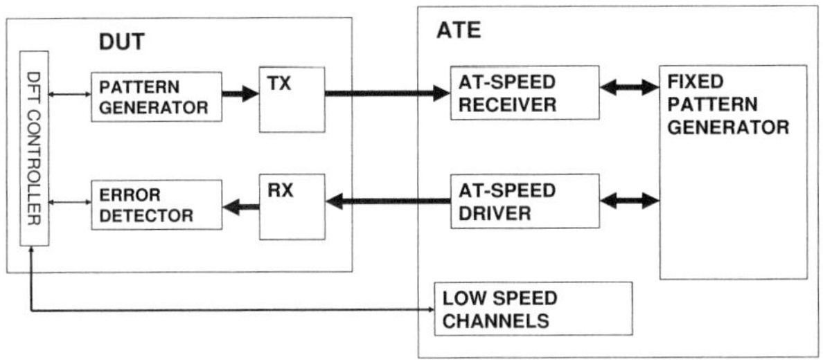

Figure 6.5 Block diagram of an ATE at-speed algorithm-based pattern approach.

contained in the DUT receiver checks the received bit stream for errors. Of course, both the pattern that is sent and pattern that is received have to be fully identical bit by bit. On the other side, the DUT driver sends a data sequence generated by the DfT engine to the ATE receiver that in turn checks it for errors. Note that the ATE system must create the compare pattern on-the-fly according to the same algorithms used on the DUT DfT engine.

6.3.1.3 Far-End Loopback

In a far-end loopback testing approach, a loopback between the receiver and driver of the DUT I/O cell is created by DfT paths inside the DUT. Two possible far-end loopback configurations are shown in Figure 6.6. These two configurations are called analog far-end and digital far-end loopback. The important feature of these configurations is that the loopback is generated inside the DUT, whereas the high-speed digital stimulus and comparison is performed by the ATE test system. In this case the device's DfT only needs to create the internal loopback paths. The pin electronics of the ATE are responsible for generating the high-speed digital stimulus signal and analyzing the output signal of the DUT driver.

The far-end analog loopback is performed inside or right after the analog I/O cell, while the far-end digital loopback is done much further inside the DUT in the core digital logic. The name "analog" is used because the loopback is done inside the high-speed digital I/O cell that contain elements that are considered analog (e.g., PLLs, amplifiers, and so forth) and because of that it can be considered an analog circuit, while the far-digital loopback is performed inside the digital core that is dominated by digital elements and for that reason is considered a pure digital circuit.

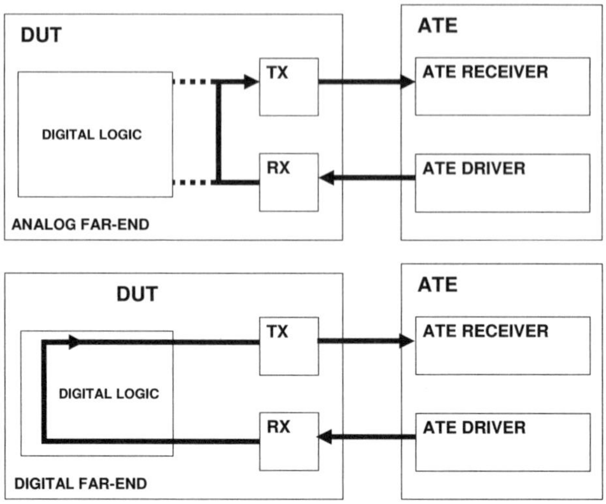

Figure 6.6 Block diagram of the possible far-end loopback approaches: analog far-end (top) and digital far-end loopback (bottom).

6.3.2 Parametric Testing

With an at-speed ATE system all the previously discussed methods for characterization and design verification of an I/O cell can be applied in production with a flexible or algorithm-based pattern approach with the DUT configured in a standard mission mode or using a far-end loopback approach. The problem is that, in production, test time is critical for the cost of test of each DUT and some of the tests might not be necessary to guarantee the needed failure coverage. Due to this, several test methodologies have been developed specifically for production testing. The next sections discuss some of these methodologies.

6.3.2.1 Fast Data Eye Mask

The data eye mask test described in Section 5.4.2.5 is an effective way to measure the timing and level performance of a DUT driver and is required by several standards.

In production testing it is common to use a small set of points instead of the full polygon of the data eye mask. This is typically called a fast data eye mask test. The idea is to do a set of functional tests for a selection of the data eye mask points instead of capturing the full data eye and verifying if there are any violations of the data eye mask polygon. This approach is significantly faster than the standard data eye mask test, and with an appropriate choice of the fast data eye mask points, the correlation with the data eye mask test can be very good. Figure 6.7 compares the fast data eye mask test with the standard data eye mask test. The selection of the number of points and their location is device specific and needs to take into account the expected data eye performance and the critical parts of the data eye to assure the needed failure coverage for the device.

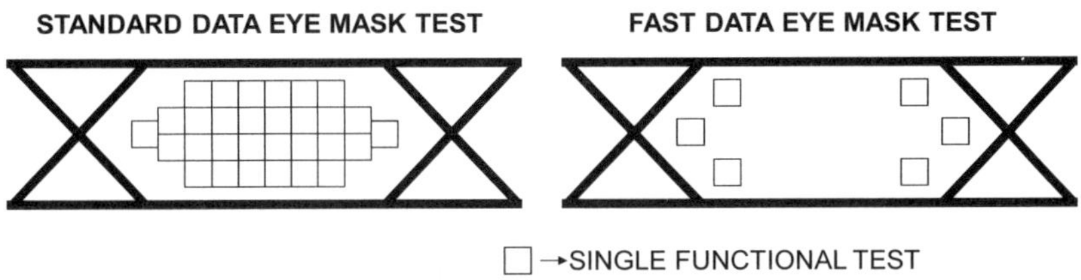

Figure 6.7 Comparison of the standard data eye mask test with the fast data eye mask test.

The drawback of this approach is that it is possible that the data eye mask test is violated at some point but the fast data eye mask test might not detect those violations, since only a reduced number of points are used for the data eye mask test. This risk will depend on the choice of the number of points to be used on the fast eye mask test and their location as shown in Figure 6.8.

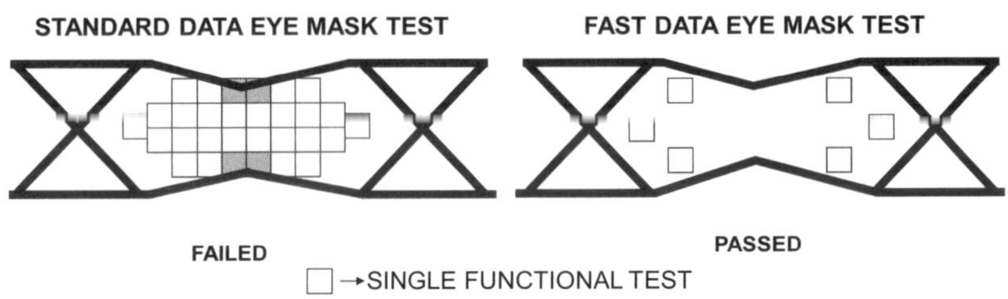

Figure 6.8 Example of how the number and location of the fast data eye mask points affect the fault coverage when compared to a standard data eye mask test.

6.3.2.2 Setup and Hold Time Measurement

Setup and hold time measurement are performed in production for several applications, but unlike the techniques used in Sections 5.4.4 and 5.6.1, in a production test the objective is not to measure the maximum setup and hold values for the DUT but instead ensure that the DUT is able to function with the worst-case setup and hold times stated by the DUT specifications. This means that in a production setup and hold test, the timing of the ATE driver or receiver is set with the worst-case setup and hold values plus potential guardbands and a functional test is performed with these settings.

For high-speed digital interfaces it is important to note that jitter can have a significant impact on setup and hold measurements, meaning that the result of the functional test will depend on the number of compared bits.

6.3.2.3 Transition-Time Measurement

Transition-time measurement as discussed in Section 5.4.1 can be too time consuming in a production test flow. One approach to reduce the test time when using a digital pin electronics card is to define a priori the transition-time threshold levels VTH (e.g., 20% and 80% of the signal amplitude), and move the compare strobe across the transition region starting first from the right and then from the left, and use the pass-to-fail transition-time instants to compute the transition time as shown in Figure 6.9. Like for setup and hold time measurements, jitter can have a significant influence on the result of this measurement, especially if all transitions of a pattern are actively compared with the functional tests executed during the timing search. In such a case, it is not guaranteed that the timing results for the two threshold levels originate from the same transition but from two separate transitions within the pattern that are displaced as a whole from the timing grid defined by the signal's data rate. As long as the jitter is small and the device's transition times have enough margin compared to the test limits, this effect can be ignored for a go/no-go test.

In order to keep the jitter that limits the usability of this measurement method to a minimum, a clock pattern should be used for this measurement. This can help to reduce the amount of data-dependent jitter (DDJ) that disturbs the measurement significantly. If the jitter of the signal to be measured plus the real transition time of the signal is close to or larger than the test limit for the transition time, it could happen that although all single transitions within the pattern fulfill the transition-time specifications, the test will fail because the jitter spreads the overlayed transitions too far apart from each other. In such a case and in the case that true value measurements are desired, it is required that only a single transition within the pattern is actively compared during the tests executed for the transition-time measurement.

A further test time reduction for go/no-go transition-time tests is achieved by executing the timing search just for one of the two threshold voltage levels. The compare strobe position for the test with the other threshold voltage then is set with an offset of the required test limit plus a potential guardband from the found timing position. With this setting, a single additional functional test after the first search will deliver the required pass/fail information at the cost of not getting the actual transition-time value but only guarantee that the transition time is below the allowed test limits. If the ATE receiver used to verify the transition time supports nonequidistant compare strobe and dual-threshold settings, another go/no-go test implementation is possible. For this implementation the two compare strobes checking for the desired transition are placed with a distance of the transition time that needs to be guaranteed (plus potential guardband). The two threshold levels are set to the high and low levels that are defined for the transition-time measurement (e.g., 20%/80% levels). With such a setting a timing search will yield a result if the required transition time is achieved by the DUT as shown in Figure 6.10.

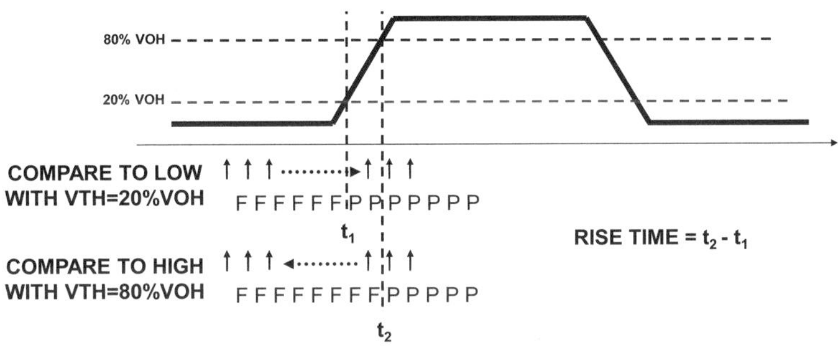

Figure 6.9 Measuring transition time in production with a digital pin electronics card.

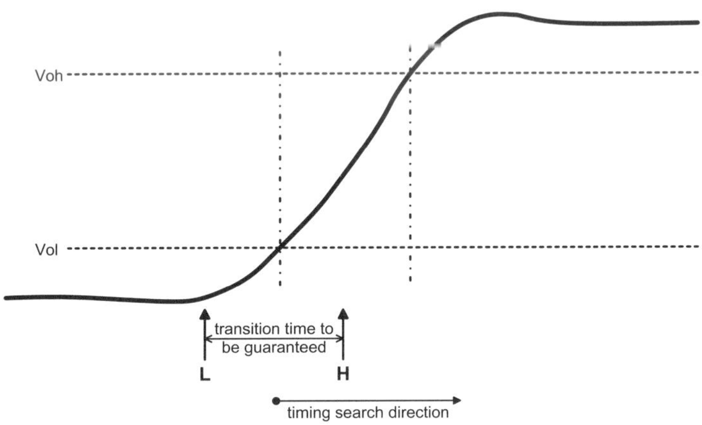

Figure 6.10 Transition-time go/no-go test for dual-threshold comparators.

6.4 Instrument-Based Testing: Low-Speed ATE

In this section we will discuss the options available for production testing of an I/O cell using a low-speed ATE system. They are divided into techniques like double data clocking and ATE channel multiplexing that try to perform classical at-speed testing with low-speed ATE system, and near-end loopback testing which uses DfT techniques to enable a self-test of the I/O cell.

6.4.1 Double Data Clocking

Performing a functional at-speed test in production is critical for some high-speed digital applications. This is especially true for devices that are manufactured in very advanced processes with minimum geometries because failure mechanisms in these processes might only exhibit in reduced maximum performance but not in hard functional fails at lower speeds. Double data clocking (DDC) is a methodology to test at full operating speed with reduced I/O data rates on the external signal connections of the DUT. Although DDC predominantly suits DRAM devices, and we focus here on these, it also is usable for other applications. With DDC, lower-speed ATE instrumentation can be used to test devices at-speed internally which helps to reduce capital investment for the test solution or extend the lifetime of already available ATE.

Some DRAM devices operate at quad- or even octal data rates (QDR/ODR). For these devices, only the data signals run at the data rate at which the devices are specified. The other device signals operate at lower speed at half, one-quarter, or one-eighth of that data rate. With DDC, instead of writing and reading single bits, the bits on the data signals of the memory device are doubled. Accessing the memory only with bit pairs of the same logical level reduces the effective I/O data rate on the high-speed memory signals by a factor of 2. The nondata signals can be kept running at their native speed in a DDC setup. In order to force transitions on any bit boundary, a dual pass test is executed for DDC setups where transitions are present for even bits in a first test run and these transitions then are moved to the odd bits in a second run as shown in Figure 6.11. The result overlay of both of these test executions then corresponds to single bit results.

It is important to note that with DDC only the high-speed data pins of the device are running at reduced speed. This means that all the device internal circuitry like multiplexers, PLLs, and so on operate at-speed. Since adjacent bits on an external data connection are usually not written to memory cells that are geometrically located next to each other, the two bits forming a lower data rate pulse on the external signal are separated in the device and travel on different signal paths inside the device as single bit pulses. Thus, from a

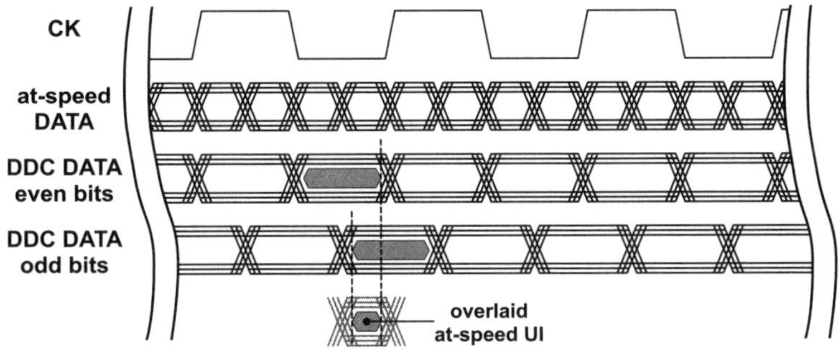

Figure 6.11 The double data clocking approach.

device operation point of view, most of the at-speed failures that are detected with an entire at-speed test are also detectable using the DDC approach. Figure 6.12 shows an example of a 4.8 Gbps XDR application comparing the device operation frequency dependent on the power supply voltage for an at-speed capable ATE system and a lower-speed ATE system using DDC. It has to be noted that the ATE systems used for this comparison have significant differences in their timing accuracy. As one can see, the difference between the two shmoo results is visible but not very significant. The device test operation margins are slightly reduced with DDC due to the parametric differences between the lower-speed ATE used for the DDC measurements and the at-speed ATE. However, if the device would not already be tested in a marginal condition on an at-speed ATE, there usually is still enough margin to test it on a lower-speed ATE in DDC mode instead.

One exception from the described DDC concept in the memory area is represented by any kind of DDR device where not only the data signals are operated at the specified data rate but also the clock signal(s). In order to keep the device operation at-speed internally for these devices, the clock signals need to be running at-speed while the data signals still can be operated in DDC manner. Using DDC for this kind of device can still be attractive for production testing because the number of required high-speed ATE channels can be significantly reduced compared to at-speed coverage of all device signals.

When using DDC setups, one of the most critical parameters is the bandwidth of the ATE instruments used to stimulate and check the device. Since these are designed for lower data rates than the ones at which the DUT natively runs, they obviously do not provide enough bandwidth for the DDC data rate to be pushed to its extreme (meaning double the data rate that the ATE hardware is designed and specified for). The weakest point of a DDC

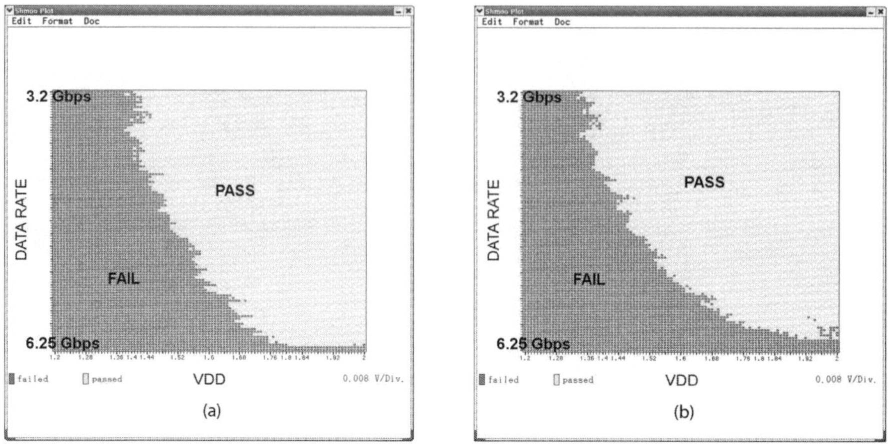

Figure 6.12 Comparison of shmoo results (a) using at-speed functional testing on a high-speed ATE system with (b) a double data clocking approach on a lower-speed ATE-system. (From: [1]. ©2008 Advantest. Reprinted with permission.)

setup in this regard is the fact that the ATE receiver reliably has to recognize a single at-speed data bit coming from the device. Such a single bit can occur if one of the double bits written to the memory flips its state due to a defect in the memory or parametric deficiencies of the device internal data path. The worst-case scenario would be that in a long sequence of double clocked zero bit pairs, one of these zero bits flipped to one (or in a long sequence of ones a single bit is flipped to zero). The ATE receiver has to be able to recognize this single bit that is sent by the DUT not with the reduced DDC data rate but in nominal speed. Since an ATE receiver used in DDC setups certainly will not be designed to handle that data rate, the signal that the receiver sees will only reach a reduced signal level due to the increased transition times caused by the receiver bandwidth limitation. If the signal levels seen by the receiver become marginal with regard to the threshold voltage set for this receiver, a reliable detection of such a single bit failure is no longer possible. Typically, ATE receivers that operate in the required speed ranges are implemented either as differential receivers with an inherent single threshold at the crossing of the positive and negative leg voltages or support a single threshold voltage only. An adaption of this threshold voltage to regain former margins is not possible because a threshold adaption to recognize single bit failures on a sequence of expected ones and also on a sequence of expected zeros is mutually exclusive as shown in Figure 6.13.

Thus, the maximum device data rate that can be achieved with a DDC setup is usually not just double the data rate that is supported by the

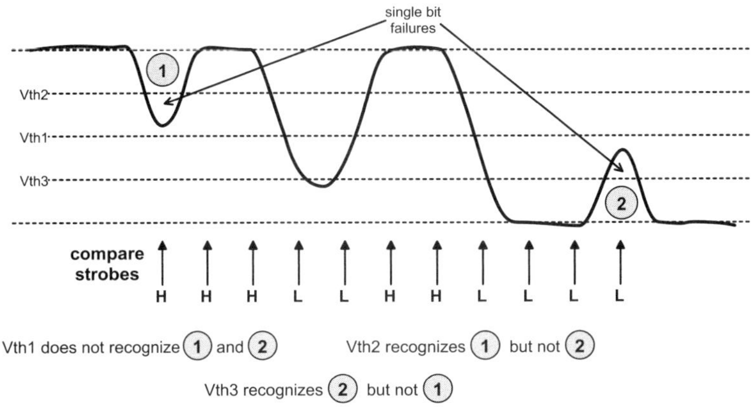

Figure 6.13 Mutual exclusive threshold for DDC single bit failures on bandwidth limited receivers.

ATE instrument used, but strongly depends on the actual bandwidth of that instrument.

6.4.2 Channel Multiplexing

ATE channel multiplexing tries to overcome the fact that the ATE system is not able to test the DUT I/O cell at-speed by multiplexing multiple channels to achieve the needed data rate [2]. This multiplexing can be done on the test fixture using, for example, a XOR logic gate or a multiplexing (MUX) IC. However, these approaches can have performance challenges for high-speed digital application and requires the use of a larger number of channels. Section 9.2 discusses this technique in more detail.

6.4.3 Near-End Loopback Testing

Near-end loopback testing is the preferred approach to implement built-in self-test (BIST) for high-speed digital I/O cells [3, 4]. The idea is to avoid the added cost of using ATE instrumentation to test the I/O cells by connecting the I/O driver to the receiver through an internal or external loop and using the loop to test if the DUT I/O cell is working.

Although the idea is simple, generating an implementation that has the needed fault coverage is not. In fact, there are several varieties of loopback testing with different types of complexity as shown in Figure 6.14. The classification/naming presented on the figure is not standard and other authors might use a different classification/naming for the same loopback methodology. It is important to note that loopback testing might have some limitations; for example, defect mechanisms that affect both the transmitter and receiver could get masked. Also in some loopback approaches the I/O cell transmitter and receiver work in a synchronous mode while in normal operation they work asynchronously and hence the clock and data recovery logic does not get exercised [5].

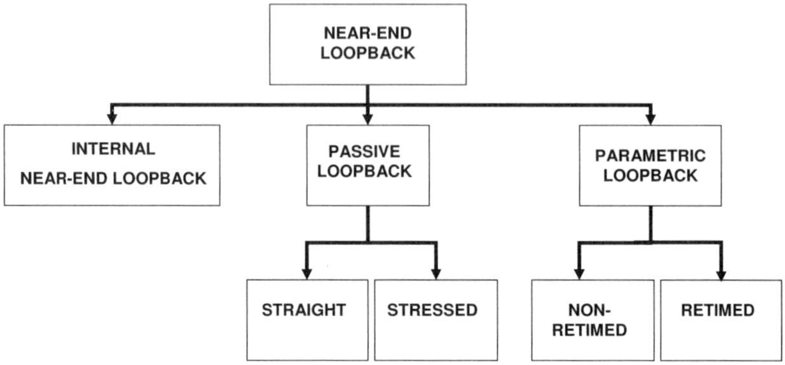

Figure 6.14 Loopback testing approaches with a low-speed ATE system.

Although some engineers see loopback testing as only intended for high-volume production, loopback testing can also provide important characterization data with the appropriate measurement methodology as will be discussed later in this chapter [6].

Loopback testing might also not be appropriate for all types of high-speed digital applications. The ability to create a loopback configuration implies a symmetry between the number of drivers and receivers on the DUT. A high-speed multiplexer cannot be tested using a loopback methodology since it lacks a receiver. One strategy to address this challenge is to add the missing drivers and receivers on a nonsymmetric interface to allow for a loopback testing approach even if they are of no use in the end application (they are only added for DfT purposes) [5]. Although this approach implies adding additional circuitry on the DUT and maybe even more pins on the package for the missing drivers or receivers, in some applications the cost of test advantages from being able to use a loopback approach surpasses the additional costs of implementing these extra drivers or receivers.

6.4.3.1 DfT Requirements for Loopback Testing

Loopback testing requires that the DUT contains some DfT capabilities implemented in the silicon. This is due to the fact that loopback testing requires the device to operate in a mode that is different from its normal operation mode (i.e., the device must generate the stimulus signal and also be able to receive this stimulus signal and compare it to the expected value as shown in Figure 6.15).

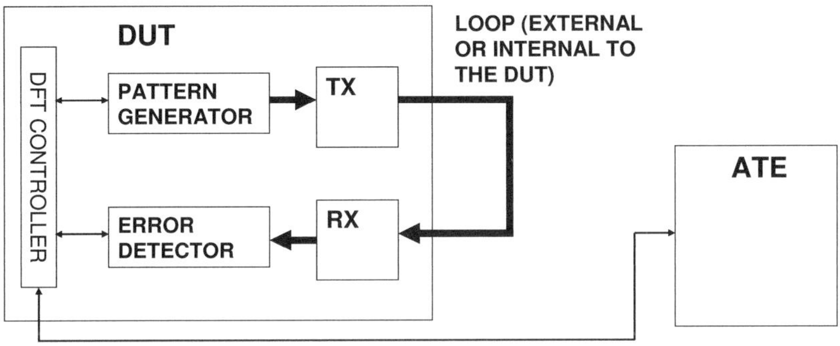

Figure 6.15 Block diagram of basic DfT requirements for loopback testing.

This DfT engine can be very simple or complicated depending on the application and how much test coverage one wants to achieve with the loopback test. Figure 6.16 shows an example of a loopback DfT on a DUT for a wired communications application [7]. Notice that in this example, the DUT is able to use different patterns on the loopback test. A loopback testing approach is not restricted to embedded clock type serial interfaces. Source-synchronous interfaces can also benefit from loopback testing techniques by using the appropriate DfT [8].

More complex DfT techniques for testing high-speed interfaces are available. For example, it is possible to include in the DUT receiver the capability to measure the received eye diagram and even more (see, for example, [9–11]) but this topic goes beyond the scope of this book.

6.4.3.2 Loop Delay

One important point when thinking about a near-end loopback approach is the loop delay between the transmitter and receiver. The DfT engine in the DUT generates the transmitted pattern on the driver and checks for errors in the received pattern by the receiver. There are some boundaries on the maximum delay the DfT engine can tolerate between the driver and receiver. The DfT engine is usually designed to tolerate a certain amount of maximum delay assuming a straight loopback between the driver and receiver. When one is using a long wired loopback signal path or a parametric loopback approach that might include several active components, it is important to check that the expected loop delay is still acceptable for the I/O cell DfT engine.

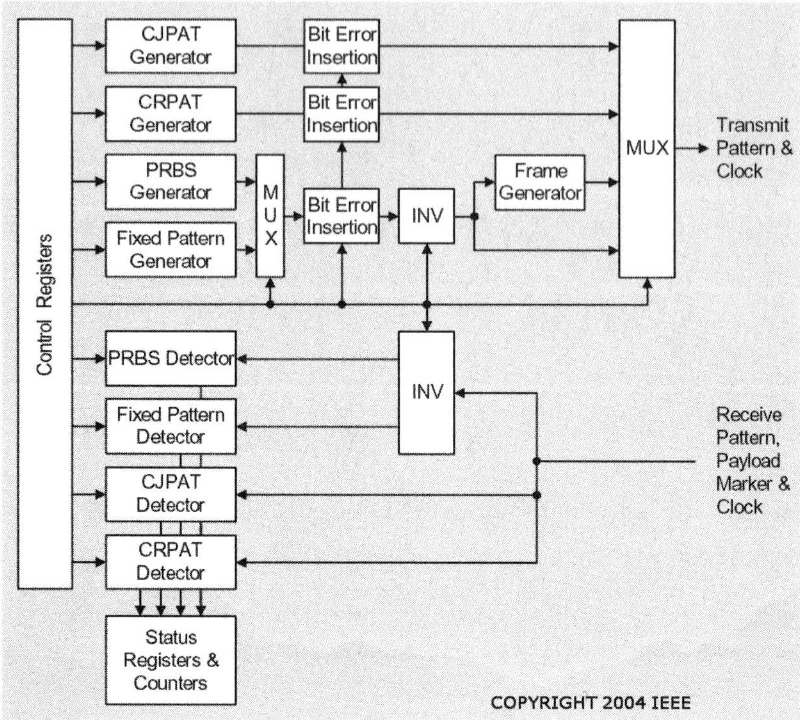

Figure 6.16 Example of a DfT implementation for loopback testing on a DUT (reprinted with permission from [7]).

6.4.3.3 Internal Near-End Loopback

Internal near-end loopback is implemented internally inside the silicon without going off the DUT die. Typically most DfT implementations to support loopback testing provide different types of internal loopback configurations together with external loopback support. Internal loopback has the advantage that it requires no support of the test fixture in terms of loopback wiring. In near-end loopback two configurations are possible: near-end analog loopback and near-end digital loopback. These configurations are shown in Figure 6.17.

The near-end digital loopback is realized at the digital logic level before the I/O cell, while the near-end analog loopback is done at the I/O cell level. The reasons for the analog/digital naming are the same as explained in Section 6.3.1.3. As shown in Figure 6.17, a DfT controller in the DUT is implemented to start and evaluate the results of the loopback test. It is even possible for some ICs to evaluate the results of the loopback test by functional testing for other I/O pins (typically at lower speed) as shown in Figure 6.18. Some

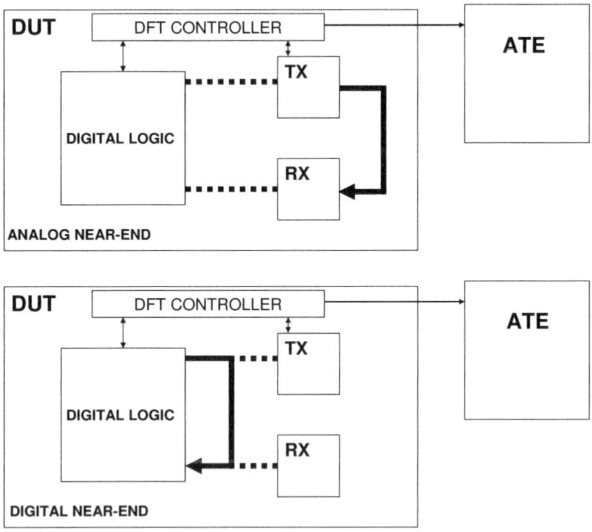

Figure 6.17 Block diagram of possible near-end loopback approaches: analog near-end (top) and digital near-end (bottom).

authors refer to the digital near-end loopback done before the serialization stage as a shallow looback and to the analog near-end loopback done after the serialization stage as a deep loopback [12].

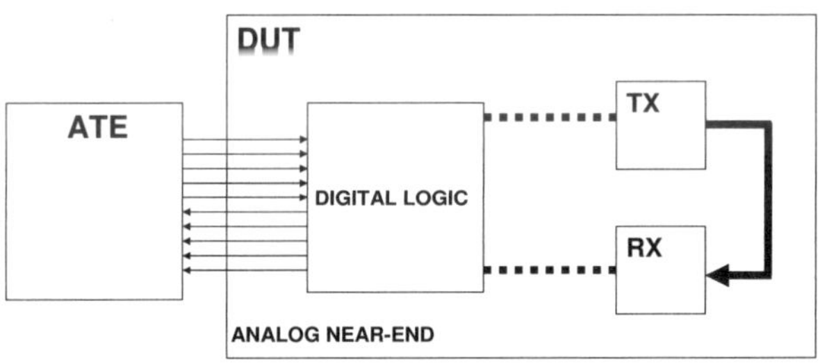

Figure 6.18 Block diagram of an analog near-end loopback approach without a DfT controller to evaluate the loopback results.

6.4.3.4 Passive Straight Loopback

Passive straight loopback (also referred to as wire loopback) is the most simple form of an external near-end loopback testing. As the name indicates, it is simply a wired connection on the test fixture of the DUT driver to its receiver. Figure 6.19 shows two examples of implementing this loop. Typically, little attention is given to the loop performance (this is different for a passive stressed loopback, as will be discussed later), and usually it is implemented as short as possible. The main idea is that if either the driver or the receiver are not working correctly, the functional test through the loop will show some type of problems (e.g., a nonzero BER). However, modern I/O cells and standards require compliance to more complex parametric measurements (e.g., jitter). This means that although a wire loopback can catch severe failures in the I/O cell, it might be unable to catch more complex parametric failures. It is important to note that this also depends on the complexity of the used DfT implementation, since it might support more parametric measurements with a loopback methodology (e.g., an integrated level shifter).

Figure 6.19 Implementation of a wired loopback on the test interface board (left: using a short microstrip loop; right: using a coaxial cable connected to a surface mounted connector).

6.4.3.5 DC and Scan Testing on Wire Loopback

Although a wire loopback testing approach does not require ATE channels to be connected to the DUT I/O cell for at-speed testing, there is typically the need to perform DC tests on the I/O cell pins (Rx and Tx) and even sometimes use these pins for scan chain access. There are several options to address this need. The most straightforward is to use relays as shown in Figure 6.20. For high pin count devices, however, it might be very difficult to have so many relays on the test fixture. Some relay manufacturers have developed relay configuration especially for these type of applications (see Section 8.8).

Another option that does not require relays is to use a combination of coils and a blocking capacitor as shown in Figure 6.21. The idea is to create a path from the pin electronics to the DUT Tx ad Rx pins at low frequencies where the coils behave like a short circuit and the capacitor as an open circuit. At high frequencies the capacitor becomes a short circuit creating the loopback path while the coils behave like an open circuit isolating the

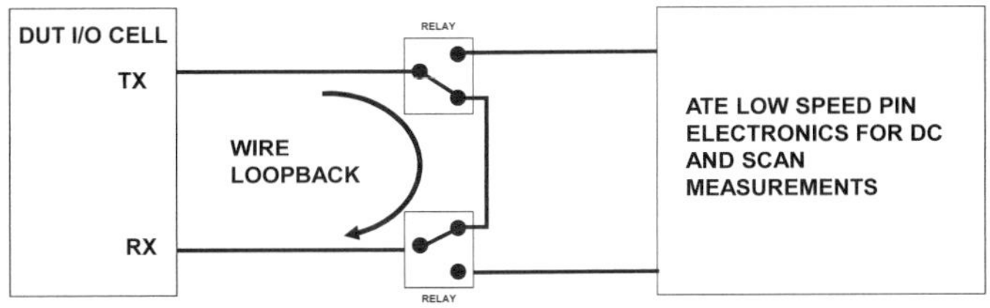

Figure 6.20 Using relays for DC and SCAN measurements on a wire loopback application.

loopback path from the pin electronics. This type of approach only works for I/O cells that support AC-coupled links. Figure 6.22 shows the simulated insertion loss of the TX to RX loopback path through the capacitor and the simulated insertion loss between the RX or TX pin to the DC measurement pin through the coil. A value of 1 pF was used for the capacitor and of 1 nH for the coils.

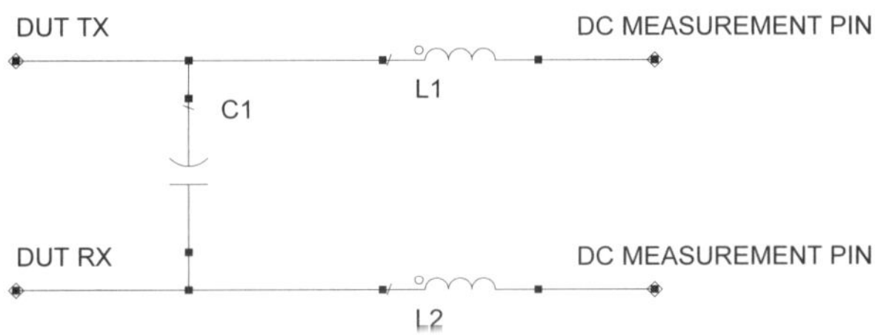

Figure 6.21 Using a coil and a blocking capacitor to provide DC measurement capabilities on a wired loopback application without using relays (the I/O cell must support AC-coupled links).

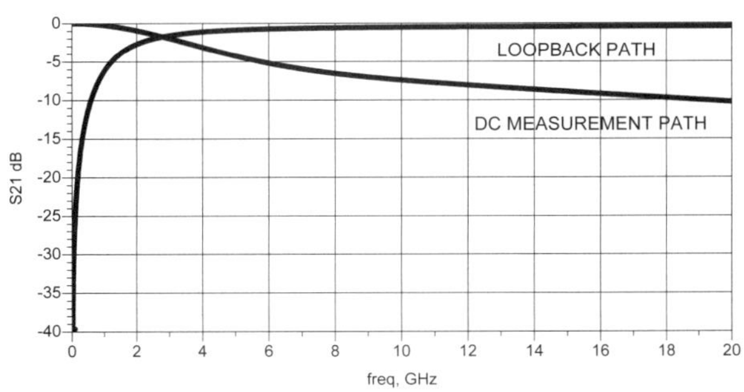

Figure 6.22 Frequency-domain response of the TX to RX loopback path and the TX/RX pin to the DC measurement pin for C=1 pF and L=1 nH.

From the results in Figure 6.22, the loss at low frequencies in the RX/TX path to the DC measurement pin is very low due to the LC circuit, which is the precondition for DC measurements and limited access for scan testing. Note that the simulation presented in Figure 6.22 is based on perfect components without any parasitics. For high data rates one possible implementation approach is to use two surface mounted bias network components (see Section 7.17.3) as shown in Figure 6.23.

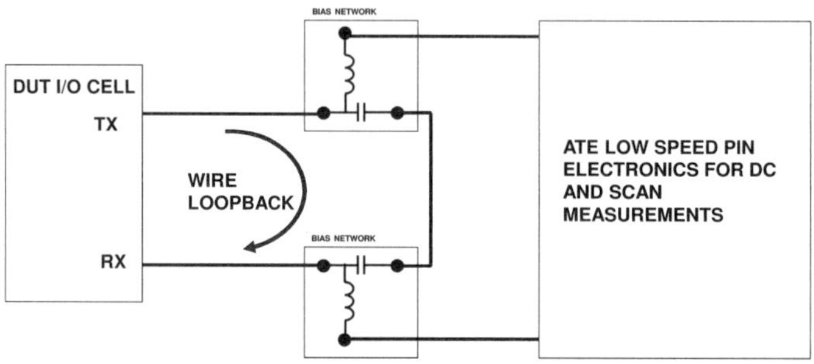

Figure 6.23 Using two bias network components to create a wired loopback configuration with ATE channel access without relays.

6.4.3.6 Passive Stressed Loopback

Passive stressed loopback tries to overcome some of the drawbacks of the passive straight loopback approach by trying to increase the fault coverage without a significant adder on the cost of test. The basic principle is to tune the loopback signal path on the test fixture toward parameters creating more stress on the receiver. The objective of this stress is twofold: the margins of the receiver are checked, but the loopback-generated stress is added to the driver characteristics and can cause excess stress to the receiver due to marginal driver performance.

One possible approach is to make the loopback signal path correspond to the worst-case signal path the application must support (e.g., some LAN devices' test fixtures are connected to a large roll with several meters of LAN cable for testing the device in loopback). Another option is to add a filter to the loopback signal path (e.g., in [13] a filter that generates data-dependent jitter is used on the loopback path). Another option is to implement the loopback signal path using a lossy signal trace on the test fixture. Figure 6.24 shows an example of this approach.

Since the passive stressed loopback always tests a driver and receiver combination, it is very challenging to design a passive stressed loopback characteristic that addresses the full fault coverage for both driver and receiver

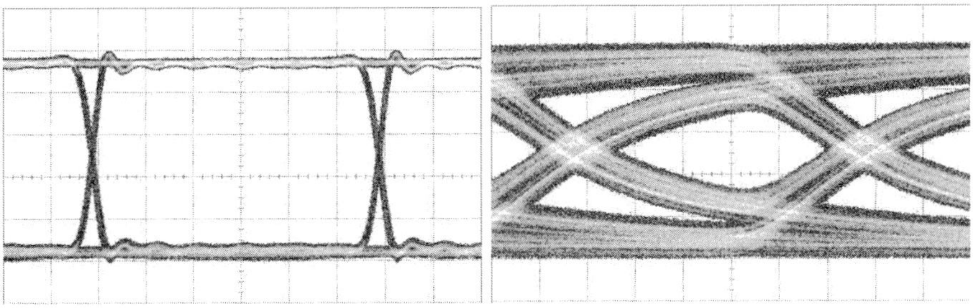

(a) (b)

Figure 6.24 Data eye (a) at the DUT driver output and (b) at the DUT receiver after a lossy signal path with a 50.8 cm (20 in) length and a trace width of 127 μm (5 mil) on an FR4 type dielectric.

on the DUT. The retimed parametric loopback approach that will be discussed in the following section tries to address this challenge.

6.4.3.7 Parametric Loopback

Parametric loopback (also sometimes called active loopback) tries to extend the fault coverage capability of the loopback approach by integrating active components on the loopback signal path [14–17]. The objective is to modify the waveform from the DUT driver in a controllable manner so that it increases the stress for the DUT receiver.

Figure 6.25 shows a block diagram of a simple parametric loopback implementation. In a parametric loopback the DUT output signal is fed into the loop receiver with a programmable level threshold. Besides some issues, which will be discussed later, this offers the ability to measure the data eye height of the waveform under test. The receiver output is then fed into a jitter injection unit, which adds a variable amount of jitter to the test signal. Finally, a driver unit with configurable values for amplitude and common-mode voltages transfers the modulated waveform to the DUT receiver.

This approach has significant advantages compared to the straight or stressed loopback. It is possible with this setup to change the shape of the waveform passing through the parametric loopback in a continuous way. This allows controlled sweeping of parameters to obtain parametric information to assess the DUT I/O performance. Figure 6.26 is a shmoo plot with parametric information for jitter tolerance created with a parametric loopback card of an ATE system. The vertical axis shows the peak-to-peak amplitude of the injected jitter and the horizontal axis is the sinusoidal input jitter frequency. For every pair of these two values, the BER measured by the DUT is plotted.

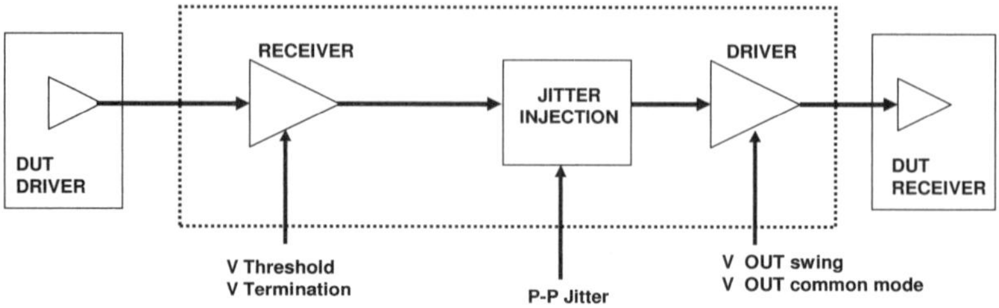

Figure 6.25 Block diagram of a parametric loopback.

As a precondition, the DUT (a 10 Gbps FPGA) contains DfT circuitry that can measure the BER of a functional test pattern [18].

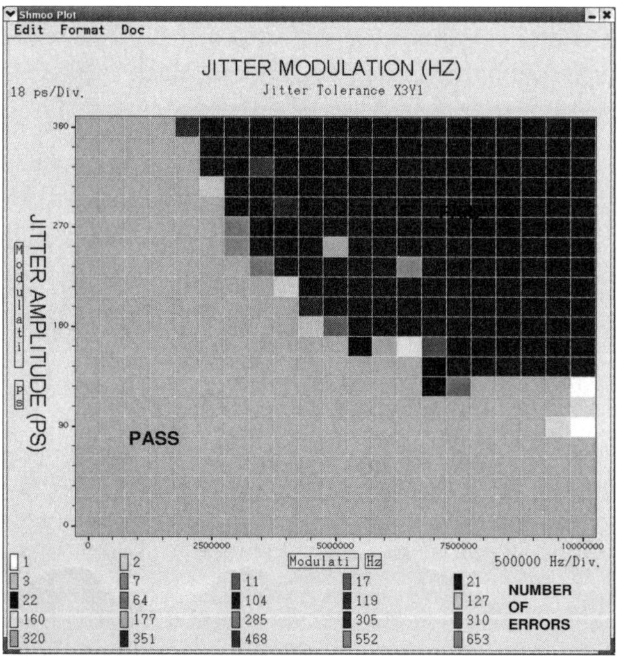

Figure 6.26 Jitter tolerance measurement example with a parametric loopback (courtesy of Advantest).

Figure 6.26 clearly demonstrates the capability of a parametric loopback to provide important characterization information on a DUT. This advantage is accompanied by the additional effort and cost for the integration of the loop components on the test fixture or the dedicated ATE module with appropriate control software.

Another challenge with a parametric loopback approach is that some interfaces/standards use out-of-band (OOB) signaling in the interface initialization (e.g., an electrical idle where the positive and negative legs have the same voltage level). A wire loopback approach would allow OOB signaling since the DUT Rx and Tx can understand this type of signaling while a parametric loopback might not be able to handle it properly. The reason is that circuitry to recognize and propagate this special protocol state would be required in the parametric loopback implementation. One solution is to include the capability on the parametric loopback to interpret the out-of-band type signals of the DUT Tx and send the corresponding signals to the DUT Rx. Another possible approach is to include the possibility of a direct wire loopback path on the parametric loopback loop (e.g., through the use of relays) to provide a way to handle this type of signaling during the initialization phases of the interface.

It is possible to increase the capability of the parametric loopback implementation of Figure 6.25. One possibility is to add a measurement unit to allow for additional parametric measurements of the DUT driver as shown in Figure 6.27.

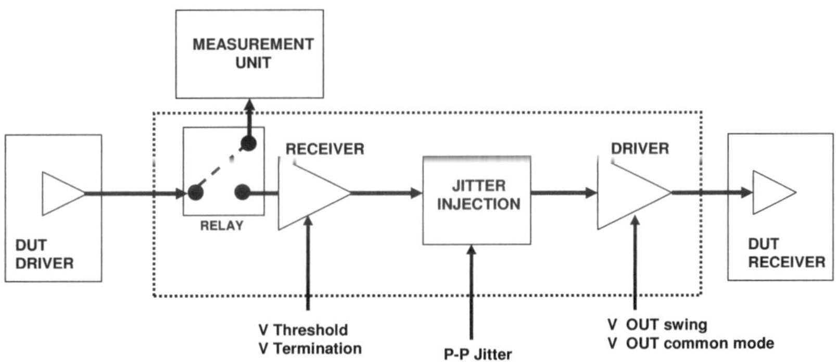

Figure 6.27 Block diagram of a parametric loopback with a measurement unit (e.g., TIA).

Again, this increases the cost of the solution but adds additional capabilities for increasing the test coverage. One possibility for the measurement unit would be a TIA that would allow parametric timing measurements, for example, rise/fall time or jitter. For reducing the cost of this approach, the measurement unit could be shared between different loops. Further improvement can be achieved by adding a DC measurement unit on both sides of the parametric loop or the integration of a retiming stage as shown in Figure 6.28.

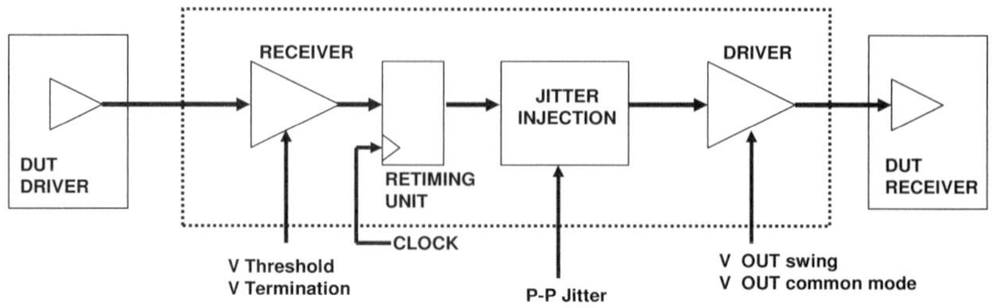

Figure 6.28 Block diagram of a parametric loopback with retiming.

The advantage of a retiming unit in the loop is the ability to precisely control the timing of the waveform fed into the DUT receiver. This is achieved because the retiming unit removes the jitter from the DUT output signal and thus the jitter of the waveform going into the DUT receiver will only be determined by the jitter of the retiming and the jitter injection unit as shown in Figure 6.29.

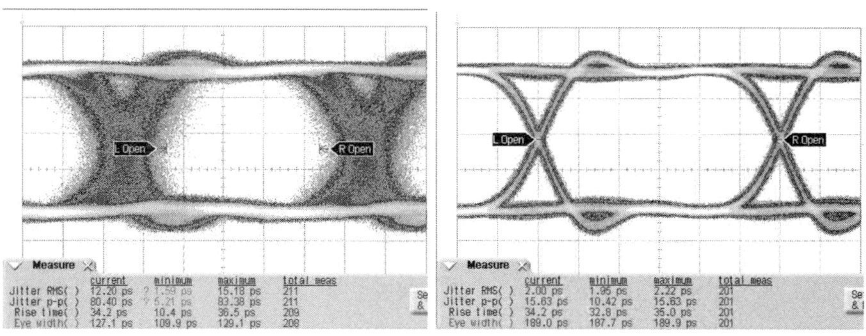

Figure 6.29 Comparison of the DUT Tx data eye after the ATE parametric loopback card but before the input of the DUT Rx (left: no retiming is used; right: retiming is used).

The essential feature of the shown configuration is the independency of the waveform into the DUT receiver from the one originating from the DUT driver. This completely excludes the influence of the DUT driver performance during the receiver test and therefore provides more accurate results as shown in the two shmoo plots in Figure 6.30 (i.e., it gets closer to the real performance of the DUT).

In Figure 6.30, the jitter tolerance results obtained with a retimed parametric loopback test are more accurate than with a nonretimed loopback test for the exact same I/O cell. The difference corresponds approximately to the jitter from the DUT driver (which might change from cell to cell) that is not removed in the nonretimed loopback approach.

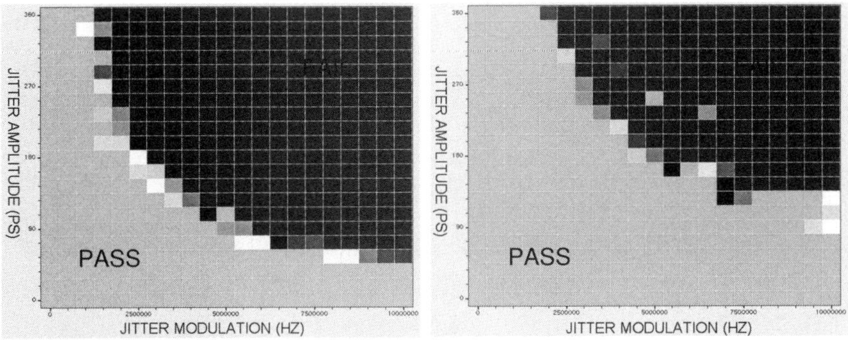

Figure 6.30 Comparison of a receiver jitter tolerance test at 10 Gbps with an ATE parametric loopback card (left: no retiming is used; right: retiming is used).

Another advantage of a retimed loop is its higher measurement accuracy for the DUT driver data eye height. Figure 6.31 shows timing diagrams during such a data eye height measurement, where the parametric loopback receiver threshold is varied over a predefined voltage range. When the programmed threshold reaches the driver voltage rails, the nonretimed loop on the left clearly shows a smaller duty cycle than the one using retiming on the right. This makes the task of the receiver under test more difficult than in normal operation mode since it is stimulated by a data stream with significant duty-cycle distortion compared with a retiming approach.

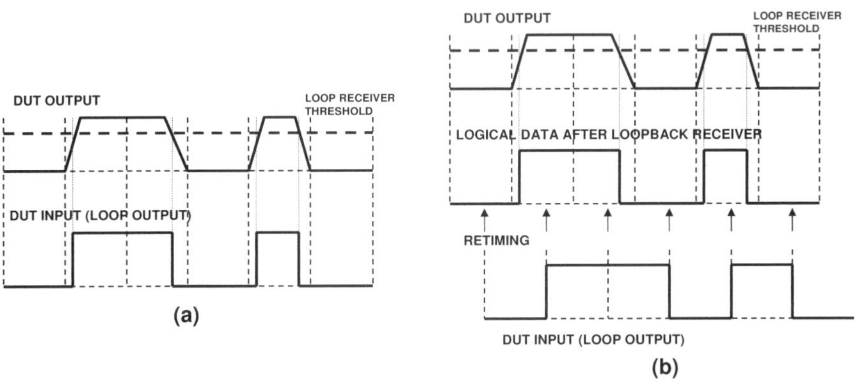

Figure 6.31 The duty-cycle issue (a) with nonretimed loopback versus (b) retimed loopback.

6.5 Instrument-Based Testing: Bench Instrumentation

Another option for production testing of a high-speed digital I/O interface with a low-speed ATE system is to use bench type measurement instrumentation on a rack and stack configuration integrated with the ATE system as discussed in Section 7.15 and show in Figure 6.32.

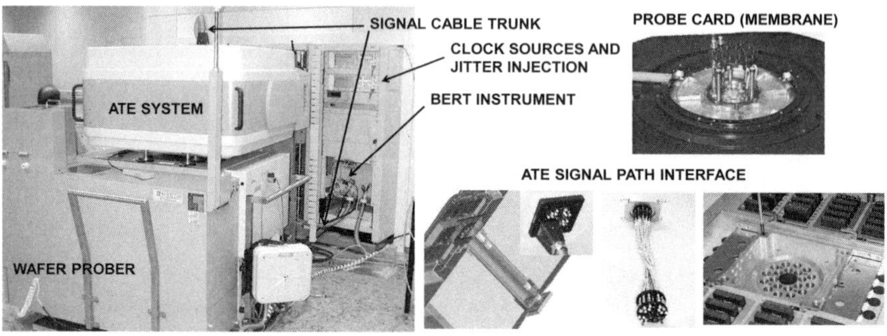

Figure 6.32 Example of at-speed 10 Gbps wafer probing test cell using a low cost ATE system and a 10 Gbps BERT instrument (from [19]).

Although such a test setup has significant drawbacks compared to a fully integrated ATE system, especially the signal path loss of long cable interconnects and test time. However, this approach might be a good option for low volume applications.

6.6 Active Test Fixture

Another option for production testing is to add instrumentation on the test fixture to test the high-speed IO interfaces at-speed while using a low cost ATE system. Since this implies adding active components to the test fixture, this approach is sometimes referred to as using an active test fixture or active loadboard. In [20] an example is shown of using this type of approach to test a 10 Gbps device. In [21] a compact module for testing high-speed I/O interfaces that can be integrated on the test fixture is proposed and in [22] an undersampling circuit for measuring 25 Gbps data eye diagrams is described that can also be integrated on the test fixture. ATE channel multiplexing on the test fixture as described in [2, 15, 23] and also in Section 9.2 are also valid approaches. Figure 6.33 shows an example of an active test fixture for high-speed digital applications consisting of multiplexing modules using low-speed ATE channels to obtain a high-speed data rate [24].

Another approach is to integrate active components into a removable measurement module that is located between the ATE test fixture and the ATE

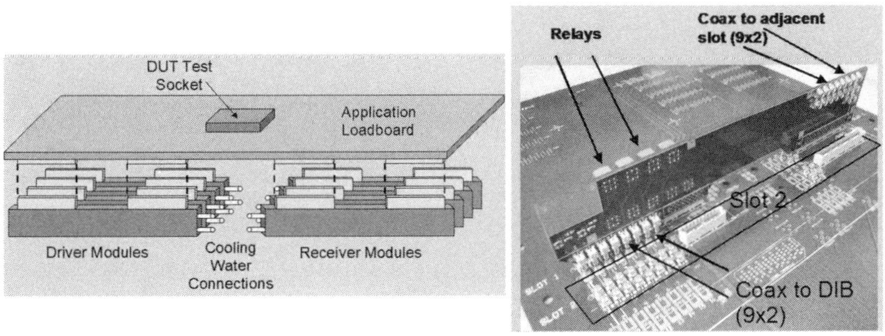

Figure 6.33 Example of an active test fixture approach for high-speed digital testing (reprinted with permission from [24]).

system. This allows not only to change the DUT test fixture depending on the DUT requirements but also remove the measurement module and use the ATE system for other applications as shown in Figure 6.34. Because the test fixture interface on top of the measurement module is mechanically equivalent to the ATE system test fixture interface, it allows for standard docking to handlers and probers [25–29].

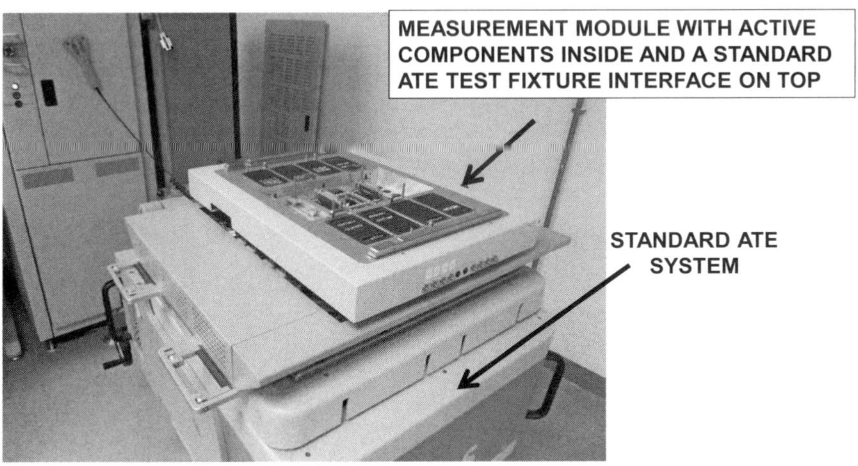

Figure 6.34 Example of a removable measurement module that is integrated between the ATE system and the ATE test fixture for at-speed digital testing of 32 Gbps applications (courtesy of Advantest).

Figure 6.35 shows an example from [30] where an active test fixture was implemented to allow the testing of DDR2/DDR3 applications at-speed using a lower data rate ATE system that would only be able to test a DDR1 application at-speed.

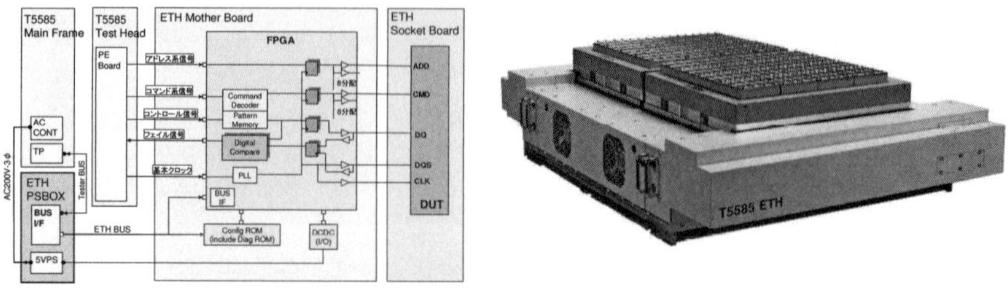

Figure 6.35 Example of an active test fixture approach DDR2/DDR3 testing based on a lower data rate ATE system (reprinted with permission from [30]).

Another possible approach is to leverage the current technology available in modern FPGAs and integrate such an FPGA on the test fixture to test the high-speed I/O interface [31, 32].

Challenges associated with the active test fixture approach are the calibration of the measurement components on the test fixture and correlation between different test fixtures used for production testing of the same device. The engineering effort required to develop this type of solution, especially in the high-speed digital area, can be significant.

6.7 Multi-site Testing

One popular concept that helps to reduce the cost of production testing is multi-site testing [33]. In a multi-site setup multiple devices are tested simultaneously on a single ATE system. The number of sites ranges from dual-site configurations for high pin count devices like a microprocessor up to 512 sites and higher. One key metric to describe such configurations is the multi-site efficiency, described by the following equation [34]:

$$\text{Multi-site Efficiency} = \left(1 - \frac{T_N - T_1}{(N-1) \times T_1}\right) \times 100\% \tag{6.1}$$

where T_N is the test time of an N-site configuration and T_1 is the test time of a single-site configuration. When performing a production test using N sites instead of a single-site, we would ideally expect that the test time of the N site configuration is the same as the test time of a single-site resulting in a value of 100% for the multi-site efficiency in (6.1). If the test time of an N-site configuration is N times the test time of a single-site, then the multi-site efficiency would be 0%. In reality, multi-site efficiency will be below 100%. One reason is that some tasks, like downloading data to the workstation for post-processing, will take more time with an increased number of sites. Also, in high-speed digital applications some tests might require the serial use of shared resources between sites like a high-speed measurement instrument.

6.7.1 Driver Sharing for Multi-site Applications

For production testing, especially in multi-site applications like high-speed memory, cost of test (COT) is very critical. One way to reduce COT is to have a large number of sites. In some situations it is advantageous to use a single channel to stimulate multiple DUTs and reduce the number of required ATE channels [35–37].

This approach is called driver sharing and is usually applied to the unidirectional input pins of the DUT (e.g., the address or control lines on a DDR application). There are several different topologies that can be used to implement a driver sharing methodology as shown in Figure 6.36. The most common approaches in this area are:

- **Active driver sharing:** In this approach, active circuitry is used for driver sharing.
- **Y-sharing:** This is the most traditional approach. With the right choice of trace impedance this structure does not require any termination at the end of the transmission lines. The DC levels remain unchanged compared to a non-shared topology. However, the signal trace impedance will have an impact on signal integrity because of the DUT input capacitance [37]. Another variant of this topology is to use the same signal trace impedance across the entire sharing topology by including a termination at the appropriate location.
- **Daisy chain sharing:** This topology is used for example in the address sharing inside a DDR3 DIMM module. In this context it is often referred to as a fly-by topology. The main advantage is that the DC amplitude stays the same independent of the number of shared DUTs. The challenge is to keep a good and equal signal integrity for all DUTs.
- **Modified star sharing:** This sharing topology is similar to a star topology allows for example the usage of any trace impedance on a Y-sharing type topology without termination after each DUT by placing an appropriate resistor on the trace-splitting point [38]. The possible drawback of such a configuration is the DC attenuation.
- **Hybrid sharing:** This sharing topology combines any features of the Y, Daisy chain, and modified star sharing topologies.

Figure 6.37 presents a block diagram of the major sharing topologies. Y2 means it is a Y-sharing topology with two shared DUTs, D8 means it is

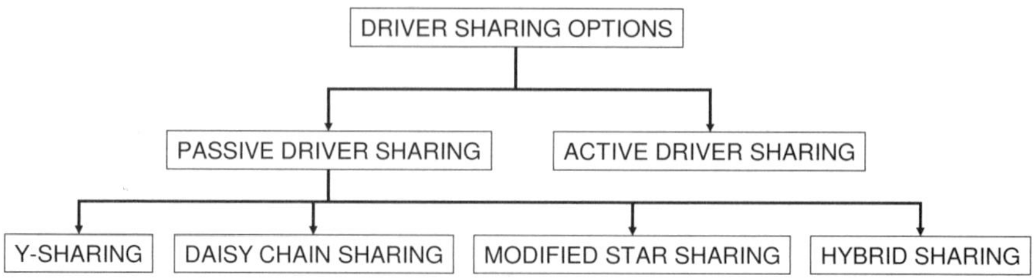

Figure 6.36 Possible driver sharing options.

a daisy chain sharing topology with eight shared DUTs, MS4 means it is a modified star sharing topology with four shared DUTs and Y2D4 means it is a hybrid sharing topology composed of Y-sharing of two DUTs followed by daisy chain sharing of four DUTs (this topology is also sometimes called mid-driven). So far, the nomenclature used for the sharing topologies is not standardized and other authors and companies might use a different naming scheme. Figure 6.38 shows an example of an active driver sharing architecture where extra drivers are used on the test fixture to support the sharing of the ATE driver between different DUTs.

It is important to note that several variations are possible in the driver sharing architectures presented in Figures 6.37 and 6.38. The challenge with driver sharing is to achieve the required signal performance at each DUT. Therefore, the signal has to be equal at each of the served DUTs even when one or multiple DUTs are deactivated from testing. To better understand this challenge, Figure 6.39 presents the results of a daisy chain driver sharing topology simulation with four DUTs. Note that each DUT input is modeled as a high impedance input with a parasitic capacitance. The timing delay caused by unequal run times to each DUT can be observed, but more critical is that the signal degradation at each DUT is different due to the presence of the multiple taps of the daisy chain configuration that create reflections that are dependent of the specific DUT position on the daisy chain.

For high-speed digital applications, driver sharing presents significant signal integrity challenges that require a very careful design of the sharing circuitry and signal traces [37]. Another important challenge resulting from driver sharing is how to perform DC measurements on the shared pins. To address this requirement, it is necessary to add a relay after the termination circuit as shown in Figure 6.40 for a D4 type driver sharing topology. With this setup it is not possible to do DC measurements on each shared DUT pin individually and therefore DC measurement results generally correspond to the resulting parallel circuit of all shared pins in the driver sharing topology [37].

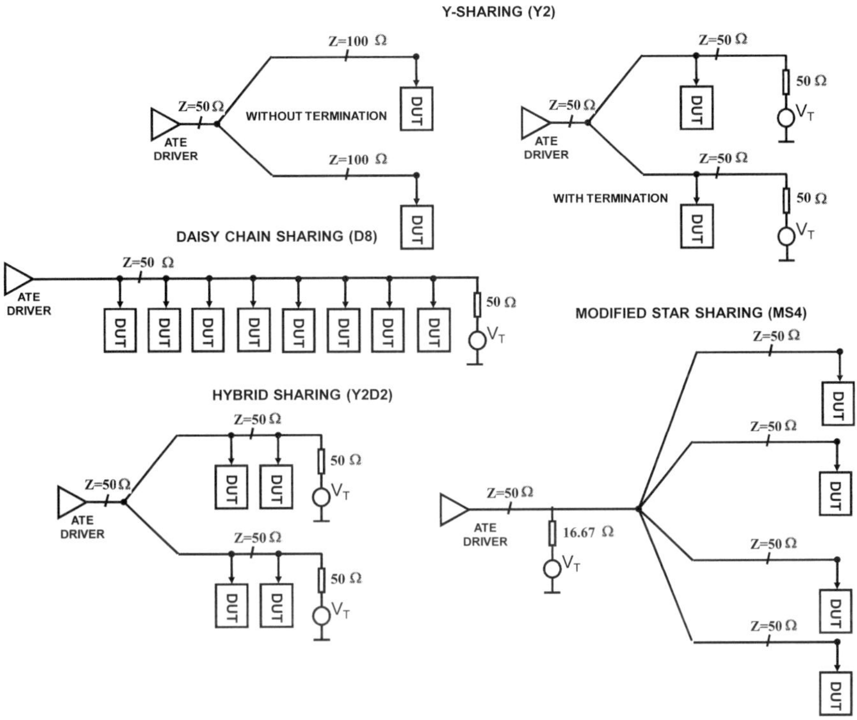

Figure 6.37 Major types of passive driver sharing topologies.

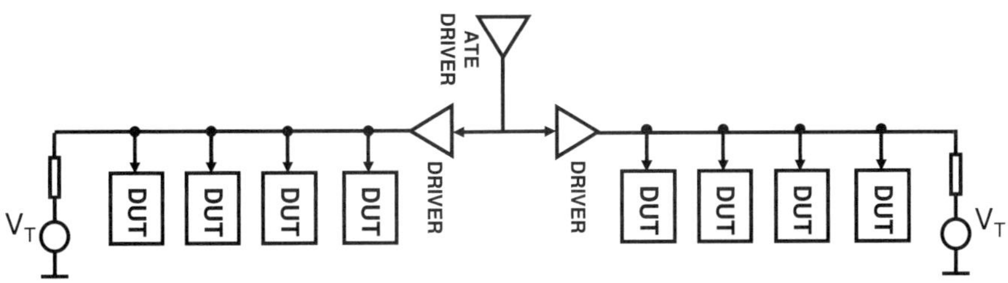

Figure 6.38 Example of an active driver sharing topology.

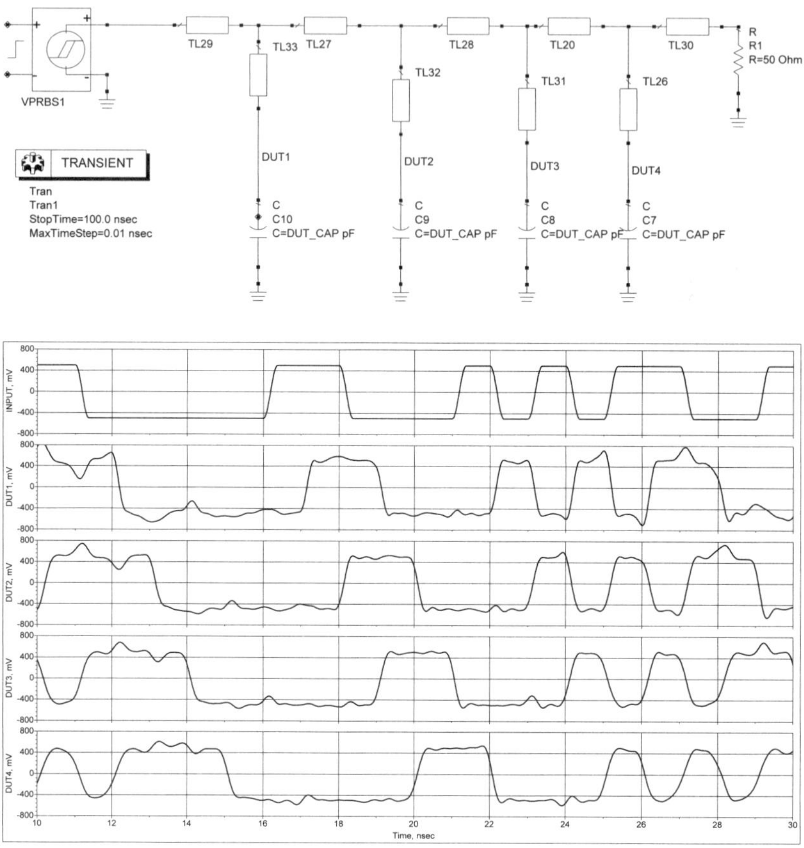

Figure 6.39 Simulation of a passive daisy chain topology for driver sharing with four DUTs.

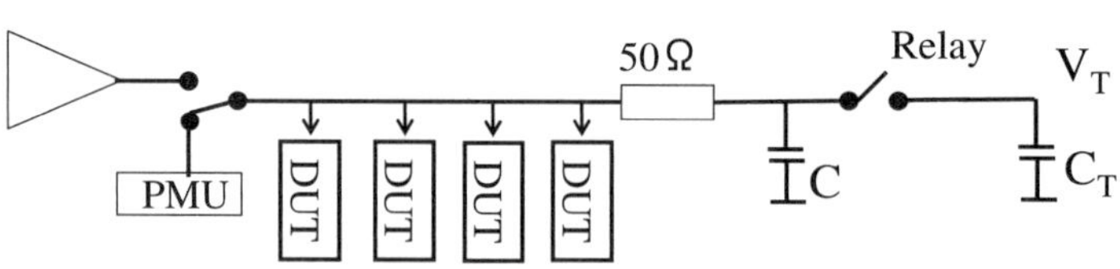

Figure 6.40 DC measurement challenges on a driver sharing topology.

On a final note, it is also possible to do sharing on bidirectional I/O DUT pins. Although the sharing topologies might be similar to driver sharing, I/O sharing is more complex, requiring significant support on the application side.

References

[1] K. Kato and T. Sonoda, "Mass Production Test Solution for XDR 4.8Gbps on the Leading Edge of DRAM," *Verigy Users Group Conference (VOICE)*, 2008.

[2] D. C. Keezer, D. Minier, and M.-C. Caron, "Multiplexing ATE Channels for Production Testing at 2.5 Gbps," *IEEE Design and Test of Computers*, 2004.

[3] A. Meixner, A. Kakizawa, B. Provost, and S. Bedwani, "External Loopback Testing Experiences with High Speed Serial Interfaces," *IEEE International Test Conference*, 2008.

[4] T. Pham, B. Koehler, D. Young, and L. Bushard, "Test Structure and Testing of the Microsoft XBOX360 Processor High Speed Front Side Bus," *IEEE International Test Conference*, 2006.

[5] I. Robertson, G. Hetherington, T. Leslie, I. Parulkar, and R. Lesnikoski, "Testing High-Speed, Large Scale Implementation of SerDes I/Os on Chips Used in Throughput Computing Systems," *IEEE International Test Conference*, 2005.

[6] J. Moreira, J.-W. Mohr, and R. Nettles, "At-Speed Loopback Testing: Strategies and Techniques," *2nd IEEE International Workshop on Automated Test Equipment (ATE VISION 2020)*, 2008.

[7] G. Haensel, K. Stieglbauer, G. Schulze, and J. Moreira, "Implementation of an Economic Jitter Compliance Test for Multi-Gigabit Device on ATE," *IEEE International Test Conference*, 2004.

[8] T. M. Mak, M. Tripp, and A. Meixner, "Testing Gbps Interfaces Without a Gigahertz Tester," *IEEE Design & Test of Computers*, vol. 21, no. 4, 2004.

[9] K. Krishna, "A Built-In Mixed Signal Test Architecture for a PCIe Phy in 90nm and 130nm CMOS," *PCI-SIG 2006*, 2006.

[10] B. Casper, A. Marti, J. E. Jaussi, J. Kennedy, and R. Mooney, "An 8-Gb/s Simultaneous Bidirectional Link with On-Die Waveform Capture," *IEEE Journal of Solid-State Circuits*, vol. 38, Dec. 2003.

[11] S. Sunter and A. Roy, "Structural Tests for Jitter Tolerance in SerDes Receivers," *IEEE International Test Conference*, 2005.

[12] A. Khare, P. Kishore, S. Reddy, A. Sanghani, and K. Rajan, "Methodology for Fault Grading High-Speed I/O Interfaces Used in Complex Graphics Processing Unit," *IEEE International Test Conference*, 2012.

[13] Y. Cai, B. Laquai, and K. Luehman, "Jitter Testing for Gigabit Serial Communication Transceivers," *IEEE Design and Test of Computers*, 2002.

[14] B. Laquai and U. Schoettmer, "A Low-Cost Vectorless ATE-Channel Architecture for Testing High-Speed IO Signal Integrity in High Volume Manufacturing," *IEEE ITC Future of ATE Workshop*, 2003.

[15] D. Keezer, D. Minier, P. Ducharme, and A. Majid, "An Electronic Module for 12.8 Gbps Multiplexing and Loopback Test," *IEEE International Test Conference*, 2008.

[16] W. A. Fritzsche and A. E. Haque, "Low Cost Testing of Multi-GBit Device Pins with ATE Assisted Loopback Instrument," *IEEE International Test Conference*, 2008.

[17] Y. Fan and Z. Zilic, *Accelerating Test, Validation and Debug of High Speed Serial Interfaces*. New York: Springer, 2011.

[18] J. Moreira, B. Sadler, and D. Ferguson, "Parametric Testing of a 10Gbs I/O Cell in Production Through a Parametric Loopback Approach with Retiming," *IEEE International Design and Test Workshop*, 2006.

[19] J. Moreira, G. Haensel, and F. Koban, "Addressing the Challenges of Implementing an At-Speed Production Test-Cell for 10Gb/s Wafer Probing," *DesignCon*, 2005.

[20] S. C. Ong, "Enabling Test-Time Optimized Pseudorandom Bit Stream (PRBS) 2E31 BER Testing on Automated Test Equipment for 10Gbps Device," *3rd IEEE International Workshop on Electronic Design, Test and Applications*, 2006.

[21] M. Hafed, D. Watkins, C. Tam, and B. Pishdad, "Massively Parallel Validation of High-Speed Serial Interfaces Using Compact Instrument Modules," *IEEE International Test Conference*, 2006.

[22] M. Lin, "Capture 25-Gbps Serdes Eye Diagram in ATE Production Environment," *IEEE International Test Conference*, 2012.

[23] D. C. K. D. Minier, M. Paradis, and F. Binette, "Modular Extension of ATE to 5 Gbps," *IEEE International Test Conference*, 2004.

[24] D. Keezer, D. Minier, P. Ducharme, D. Viens, G. Flynn, and J. McKillop, "Multi-GHz Loopback Testing Using MEMs Switches and SiGe Logic," *IEEE International Test Conference*, 2007.

[25] J. Moreira, F. Pizza, C. Borelli, F. Corneo, H. Werkmann, S.-X. Yang, D. Lam, and B. Roth, "A Pragmatic Approach for At-Speed Characterization and Loopback Correlation at 28 Gbps," *Advantest VOICE Users Group Conference*, 2014.

[26] J. Moreira, B. Roth, and C. McCowan, "An Active Test Fixture Approach for Testing 28 Gbps Applications Using a Lower Data Rate ATE System," *IEEE Asian Test Symposium*, 2012.

[27] J. Moreira, H. Werkmann, V. Filsinger, and B. Roth, "At-Speed Testing of 32 Gbaud PAM-4 Interfaces Using Automated Test Equipment," *DesignCon*, 2015.

[28] J. Moreira, H. Werkmann, I. Masahiro, B. Roth, V. Filsinger, and S.-X. Yang, "An ATE Based 32 Gbaud PAM-4 At-Speed Characterization and Testing Solution," *IEEE Asian Test Symposium*, 2014.

[29] J. Moreira, B. Roth, H. Werkmann, L. Klapproth, M. Howieson, M. Broman, W. Ouedraogo, and M. Lin, "An Active Test Fixture Approach for 40 Gbps and Above At-Speed Testing Using a Standard ATE System," *IEEE Asian Test Symposium*, 2013.

[30] S. Igarashi and S. Takeshita, "ETH Solution for Effective Use of Existing Testers," *Probo, Advantest Technical Report*, no. 36, 2011.

[31] S. Sunter and A. Roy, "A Self-Testing BOST for High-Frequency PLLs, DLLs, and SerDes," *IEEE International Test Conference*, 2007.

[32] A. M. Majid and D. Keezer, "Multi-Function Multi-GHz ATE Extension using State-of-the-Art FPGAs," *IEEE International Test Conference*, 2011.

[33] J. Rivoir, "Parallel Test Reduces Cost of Test More Effectively Than Just a Cheap Tester," *IEEE/SEMI International Manufacturing Technology Symposium*, 2004.

[34] J. Kelly, "Multi-Site Efficiency and Throughput," *Verigy Technical Note*, 2008.

[35] S. West, "Redefining Memory Test," *Evaluation Engineering*, Jan. 2009.

[36] Y. Jeon, S.-I. Kwon, C. Mack, and G. Westendorf, "Parallel Logic Test Interface Solutions," *Burn-In and Test Socket Workshop*, 2009.

[37] J. Moreira, M. Moessinger, K. Sasaki, and T. Nakamura, "Driver Sharing Challenges for DDR4 High-Volume Testing with ATE," *IEEE International Test Conference*, 2012.

[38] B. Laquai, "Signal Distribution Structure and Method for Distributing a Signal," *EPO Patent Application W02010/031418A1*.

7

Support Instrumentation

This chapter presents a discussion on bench measurement instrumentation and its relation to high-speed digital measurements with ATE. References [1–4] provide more information on the topic of measurement instrumentation. Also, the Web sites of the different measurement instruments manufacturers are good sources of information.

The measurement instruments presented in this chapter represent a set of tools to which ideally every engineer working on high-speed digital testing should have access. It is important to understand the capabilities of each instrument and how they can be used in a high-speed digital application.

7.1 Oscilloscopes

7.1.1 Real-Time Oscilloscopes

The real-time oscilloscope is probably the bench instrument with which most engineers will feel comfortable, since it resembles the analog oscilloscopes most electrical engineers encounter as their first bench instrument. Figure 7.1 shows a picture of a real-time oscilloscope and Figure 7.2 shows a high-level block diagram.

The challenge for the real-time oscilloscope is the acquisition of both time and level information in real-time using a sampling frequency that is higher than the data rate of the digital signal under measurement. It is obvious that for high-speed digital signals this is not a trivial task and is one of the reasons for the high cost of state-of-the-art real-time oscilloscopes.

Figure 7.3 shows the data gathering process. Modern real-time oscilloscopes are able to collect and store a very large number of samples in their memory. This allows the user to acquire very long non-repetitive waveforms

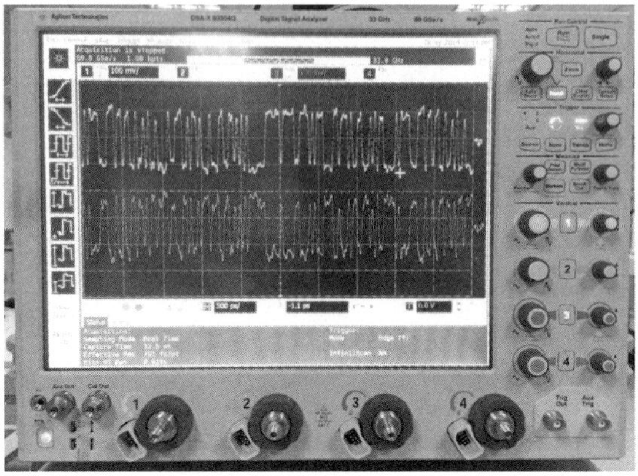

Figure 7.1 Picture of a real-time oscilloscope.

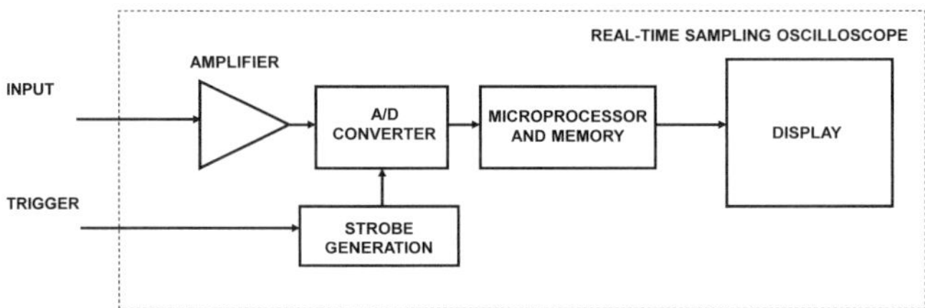

Figure 7.2 High-level block diagram of a real-time oscilloscope.

and transfer this data to a post-processing software package (e.g., MATLAB) where the user can process the data using any imaginable algorithm. This feature makes the real-time oscilloscope a very versatile tool, especially if non-repetitive single event effects have to be analyzed. The main challenge with the real-time oscilloscope is its signal performance; it is typically inferior to other instruments like an equivalent-time sampling oscilloscope. This means that for very accurate parametric measurements of a high-speed digital signal, the real-time oscilloscope might not provide the most accurate results possible. For example, at the time of this writing, although the best real-time and equivalent-time oscilloscopes have specified bandwidths of 100 GHz, due to other factors like timing resolution and noise, equivalent-time oscilloscopes provide more accurate measurements. Clearly, for versatility the real-time oscilloscope is without rival, but one must always be aware of its limitations.

7.1.2 Equivalent-Time Sampling Oscilloscopes

The equivalent-time sampling oscilloscope (sometimes also referred to as a digital communication analyzer or DCA and communication signal analyzer

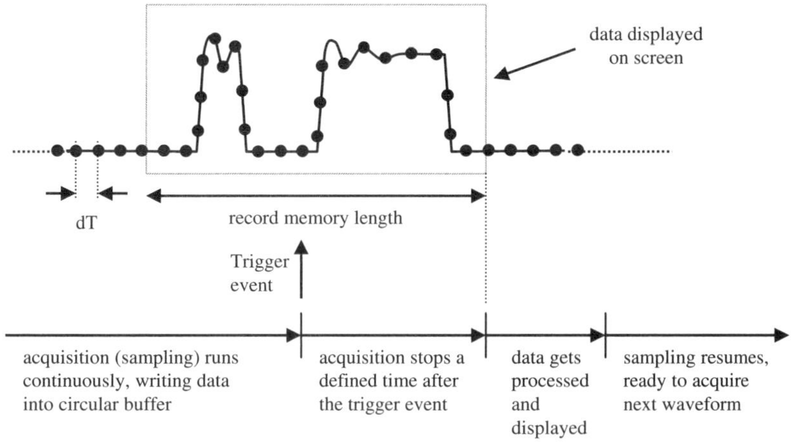

Figure 7.3 Data gathering process of a real-time oscilloscope. (From: [4]. ©2006 Springer. Reprinted with permission.)

or CSA) is intended to measure the parametric performance of a digital signal (e.g., jitter, rise time, period) with a very high timing accuracy and bandwidth by using an undersampling approach. Figure 7.4 shows a photograph of a commercially available equivalent-time oscilloscope.

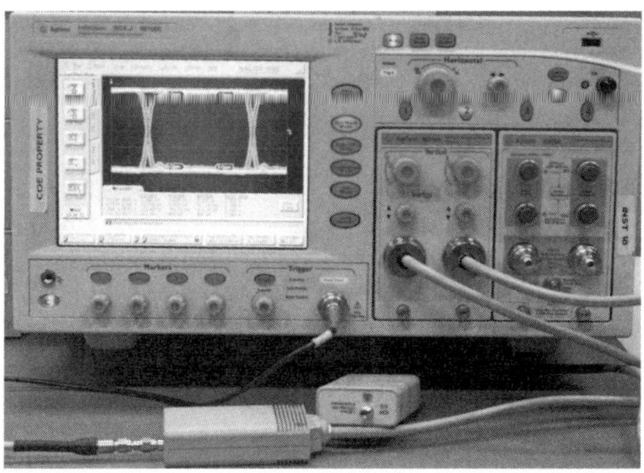

Figure 7.4 Picture of an equivalent-time oscilloscope.

Reference [2] provides an excellent introduction to equivalent-time oscilloscopes and Figure 7.5 shows the typical block diagram of an equivalent-time oscilloscope. The main idea is to use a trigger signal coupled with a high-precision delay line to sample the signal. This means that the signal to be measured has to be repetitive (i.e., a digital pattern must consist of a repetitive pattern). However, if a data eye diagram is the only measurement to be done, the pattern does not need to be repetitive since we are not interested in acquiring the entire pattern waveform but only measure the data eye diagram. Figure 7.6 shows a diagram explaining how the equivalent-time oscilloscope undersampling approach generates the measured waveform.

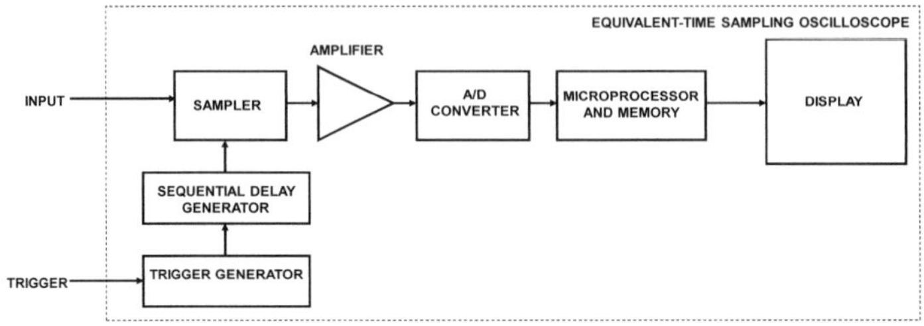

Figure 7.5 Typical block diagram of an equivalent-time sampling oscilloscope.

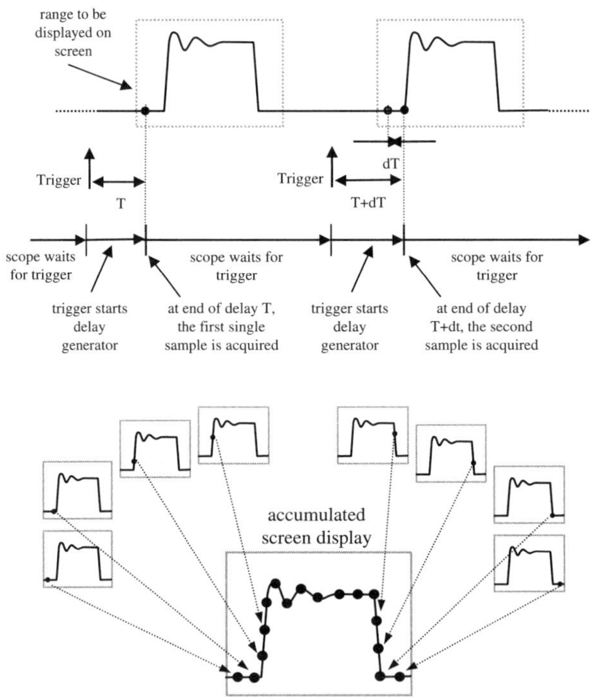

Figure 7.6 Data gathering process of an equivalent-time sampling oscilloscope. (From: [4]. ©2006 Springer. Reprinted with permission.)

An equivalent-time oscilloscope will display the full waveform of a repetitive digital signal as long as the trigger corresponds to the signal data rate divided by the pattern length in UI. Figure 7.7 shows an example for this type of measurement. Of course, the waveform will correspond to an overlay of several samples of the measured pattern at different points on the endless measurement loop similar to the waveform reconstruction discussed in Section 4.4.

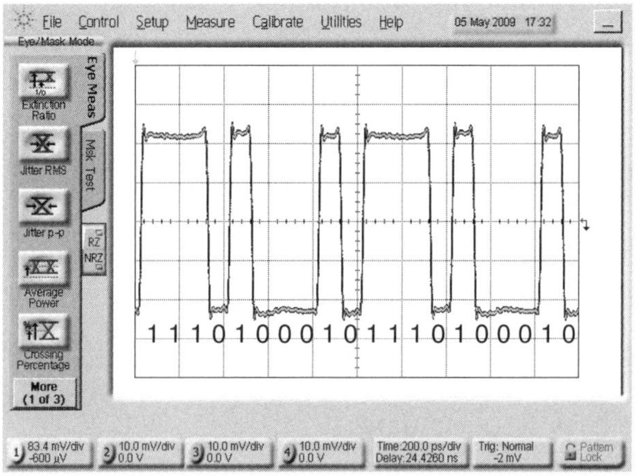

Figure 7.7 Ten-bit waveform (1011101000 repeated endlessly) at 10 Gbps measured by an equivalent-time oscilloscope using a trigger signal with 1 GHz frequency.

The best-known feature of the equivalent-time oscilloscope is the data eye diagram as shown in Figure 7.8. On a data eye all the bits in the pattern are merged into an overlay view (see Section 2.1.2). In this case there is no need for a repetitive patten, only for a trigger that is synchronous to the bits in the pattern. Such a trigger signal could even be obtained from a clock and data recovery (CDR) module. Of course, the data eye diagram does not provide any information on the measured pattern bit sequence but does provide information on the parametric performance of the signal under measurement like the average rise/fall time, the average signal amplitude, and the timing and level noise (jitter).

The equivalent-time sampling oscilloscope is an excellent tool for analyzing high-speed digital signals due to its accuracy and price (compared to a real-time oscilloscope). However, there are some areas where it cannot substitute for a real-time oscilloscope, especially when debugging non-repetitive problems or issues that require the waveform to be sampled in real-time. Ideally one would like to have both types of oscilloscopes, real-time and equivalent-time, available.

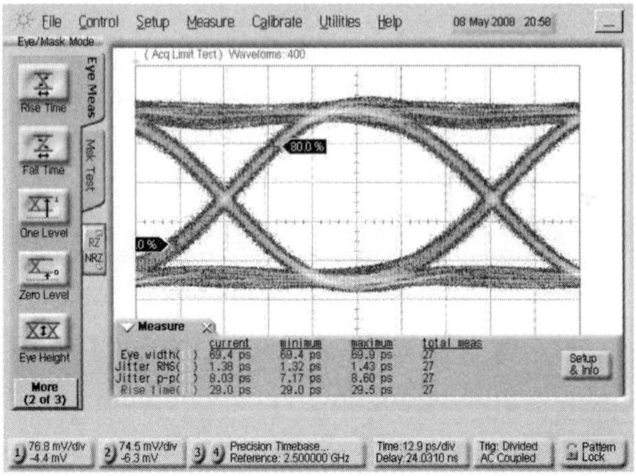

Figure 7.8 Data eye measurement with an equivalent-time oscilloscope for a 12.8 Gbps data signal.

7.2 Bit Error Rate Tester

The bit error rate tester (BERT) is the bench measurement instrument that is most closely related to a standard ATE digital pin electronics. It is able to stimulate and receive digital signals and compare the received signal to an expected pattern to identify bit failures. Current state-of-the-art BERTs are able to achieve data rates above 100 Gbps. Figure 7.9 shows a picture of a BERT instrument.

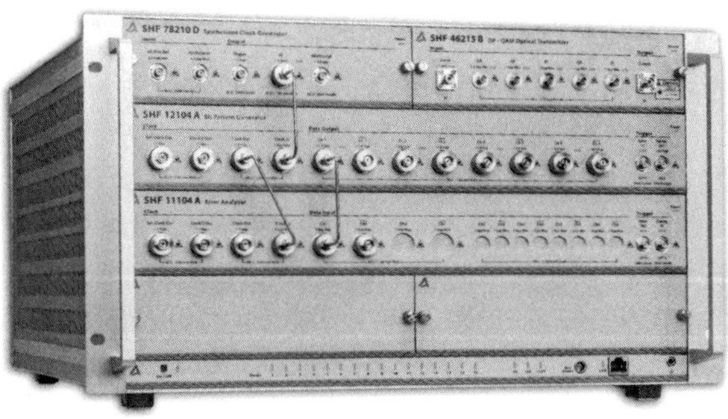

Figure 7.9 Picture of a bit error rate tester (BERT). Courtesy of SHF Communication Technologies AG.

Traditionally the BERT has been used to measure the BER of a DUT by running it for several hours. Fortunately, modern BERT instruments have significantly more capabilities [2]. A BERT is able to do data eye measurements, and jitter measurements including jitter tolerance. The main drawback of the BERT is that it is not as intuitive to use as an oscilloscope for some engineers. However, one should not forget that the BERT makes a pass/fail comparison at each bit generated by the DUT output and normally it is able to store the results of that comparison in memory. This allows the BERT to identify non-repetitive or sporadic failures. This fact, coupled with the ability to change the compare strobe with a resolution that is a fraction of the DUT output period, makes the BERT a formidable tool.

7.3 Time Interval Analyzer

The time interval analyzer (TIA) is an instrument that is able to measure the time intervals between consecutive events. These events are defined by the signal being measured crossing a certain level threshold [5]. TIAs are not only available as bench instruments but have also been integrated into ATE pin electronics. Figure 7.10 shows a picture of a TIA instrument.

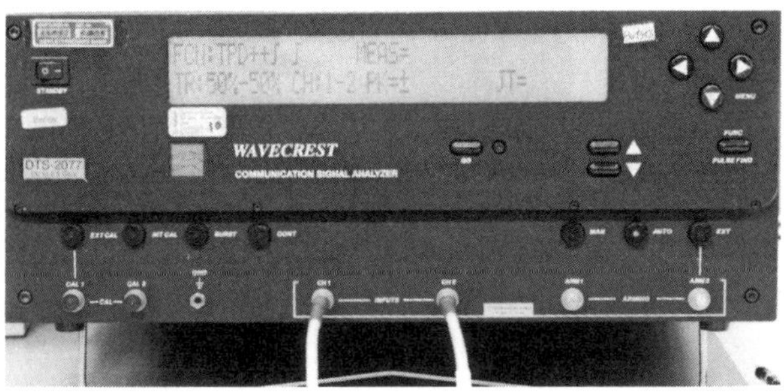

Figure 7.10 Picture of a time interval analyzer (TIA).

TIA measurement instruments can vary significantly in their capabilities, from very simple and low cost implementations intended to measure simple parametric values like frequency to highly sophisticated implementations that are able to characterize a high-speed digital DUT. Figure 7.11 shows a typical high-level block diagram of a TIA.

As stated before, a TIA measures time intervals between events defined as the signal under measurement crossing a certain voltage threshold. One possible differentiator between TIAs is how many thresholds one is able to use at the same time (e.g., having two thresholds allows the measurement of a signal transition time with a single measurement by setting the thresholds to the 20% and 80% points of the expected signal amplitude and measuring the time difference between the two events).

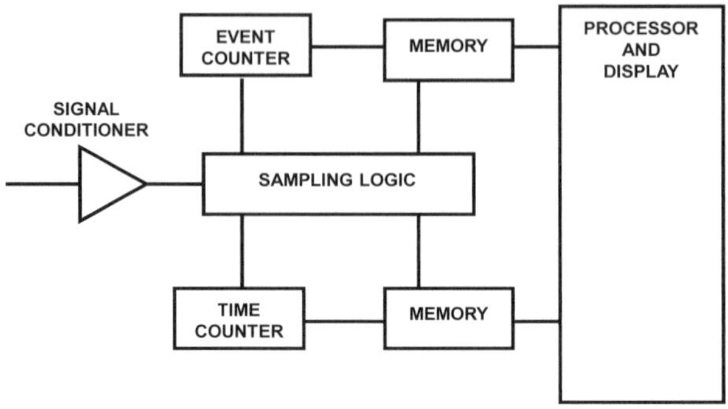

Figure 7.11 Simplified block diagram of a TIA.

Another important point is the number of consecutive events that a TIA is able to acquire before it has to prepare for another measurement run (rearming time). This rearming time is typically very large compared to the period of the signal under measurement and can generate some measurement challenges [6].

Depending on the capabilities of the TIA instrument, it is possible to measure several of the typical parameters of an I/O interface. This is achieved by using some options available in advanced TIAs (voltage threshold, random rearming time, pattern marker, multiple events) and with the appropriate data processing software.

7.4 Time-Domain Reflectrometry/Transmission (TDR/TDT)

A Time-Domain Reflectometry/Transmission (TDR/TDT) instrument is an ideal instrument for debug and validation of a PCB test fixture and complementary to a vector network analyzer. In most cases a TDR/TDT instrument is a module that is plugged into an oscilloscope as shown in the Figure 7.12 example. The critical parameters of a TDR/TDT module are the rise time of the stimulus step and the bandwidth of the comparator. The better these values are, the smaller the discontinuity on a PCB signal path that is possible to accurately measure. Appendix E discusses TDR/TDT measurements in detail.

7.5 Spectrum Analyzer

A spectrum analyzer is an instrument that allows the measurement of the spectral (i.e., frequency-domain) composition of an electrical signal.

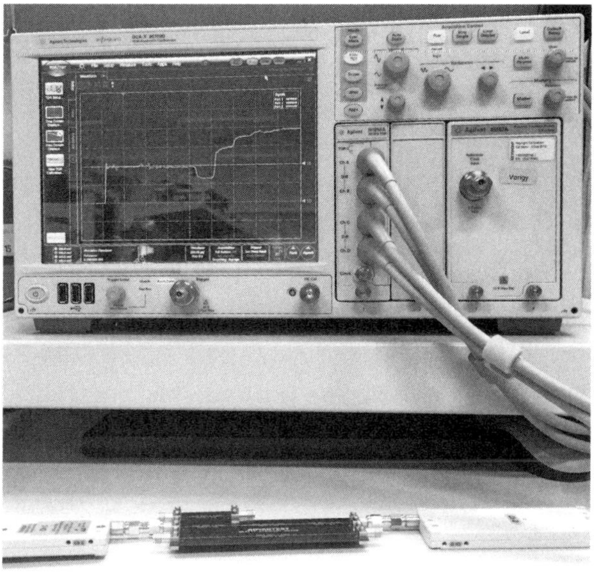

Figure 7.12 Example of a TDR/TDT module from Keysight Technologies.

Spectrum analyzers are typically divided into two large families (analog and digital) regarding the way the spectral information is measured, although some instruments also use a hybrid approach. Apart from the Web sites of spectrum analyzer manufacturers, [7, 8] provide a good introduction to this instrument. Figure 7.13 shows a picture of a spectrum analyzer.

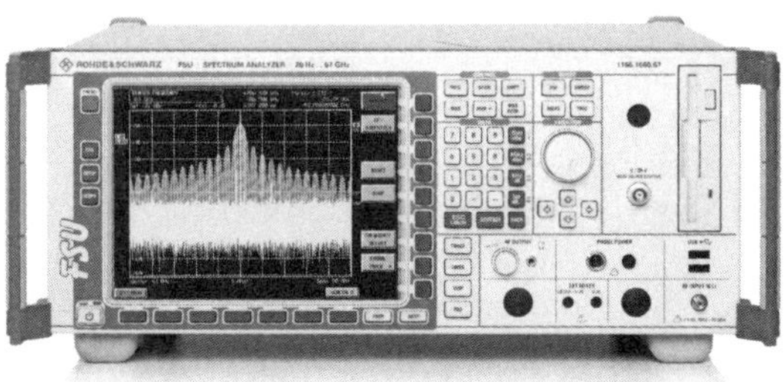

Figure 7.13 Picture of a spectrum analyzer (courtesy of Rohde & Schwarz).

It is important to note that although most real-time oscilloscopes provide a way to measure the spectral components of a signal by applying an FFT to the acquired signal, their measurement accuracy is inferior to a dedicated spectrum analyzer instrument.

For the test engineer working on the characterization and testing of high-speed digital interfaces, a spectrum analyzer can be used for the analysis of clock signals and the associated jitter. Section I.1 describes the calibration of sinusoidal jitter injection using a spectrum analyzer.

7.6 Signal Source Analyzer

A signal source analyzer (SSA) is an instrument dedicated to frequency-domain measurements like phase noise. Although a spectrum analyzer with the appropriate software can also measure the phase noise of a signal, an SSA can perform this measurement with more accuracy due to the use of techniques like the cross-correlation method. This is especially important for measuring clock signals with very low phase noise (random jitter). Figure 7.14 shows a picture of a signal source analyzer.

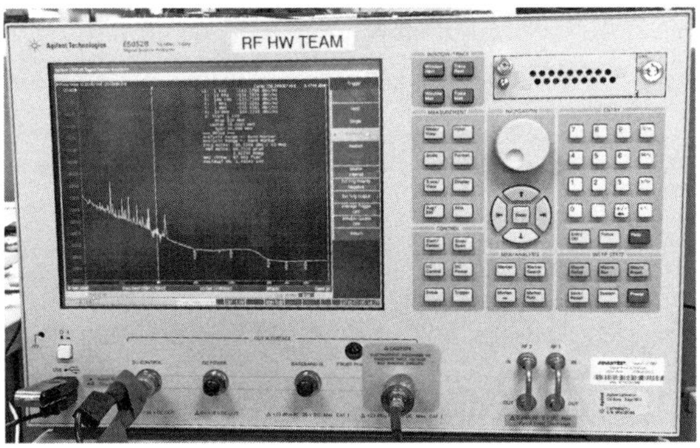

Figure 7.14 Picture of a Keysight Technologies E5052B signal source analyzer.

7.7 Vector Network Analyzer

A vector network analyzer (VNA) measures the properties of an electrical network, especially properties associated with the reflection and transmission of electrical signals. A network analyzer uses a controlled sinusoidal generator and measures the amplitude and phase of the transmitted and reflected waveforms at each frequency. In this way it characterizes an electrical network completely. The measured data is typically represented in the form of scattering parameters or S-parameters which are further discussed in Appendix F. Current network analyzers can achieve very high measurement frequencies (e.g., 110 GHz).

Apart from the Web sites of VNA manufacturers, [7, 9] provide a good introduction to this type of instrument. One important characteristic of a VNA instrument apart from its frequency and dynamic range is the number of ports it can measure. The ability to measure multiple ports (e.g., four) allows the user to more easily characterize multiport structures for items like crosstalk. Figure 7.15 shows a picture of a four-port VNA.

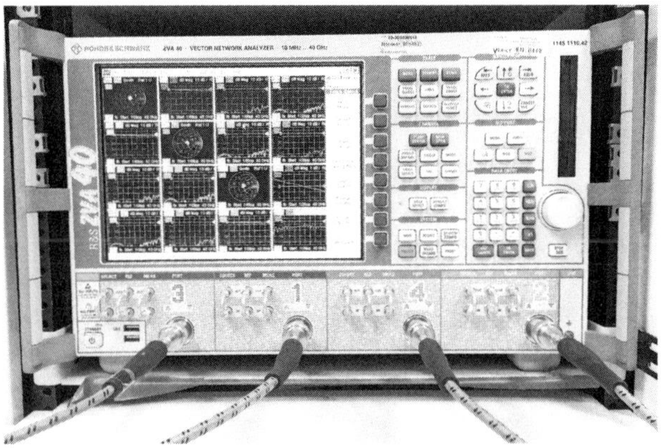

Figure 7.15 Picture of a four-port vector network analyzer (VNA).

In the area of high-speed digital interface characterization, the main applications of a VNA are on the return loss measurement of an I/O cell (Section 5.9.2) and the characterization of the test fixture elements. The fact that a VNA is able to completely characterize an electrical network through the measurement of its S-parameters makes it an ideal instrument to obtain measurement-based models for simulation (Appendix H). Vector network analyzers are also important for power integrity measurements as discussed in Section H.2. However, for power integrity measurements the requirements on the VNA are different from the requirements for signal integrity measurements. The VNA lowest frequency becomes more critical (e.g., 5 Hz), and the highest frequency only needs to be 1 to 3 GHz. Because of this, special VNA models have been developed for power integrity measurements [10].

7.8 Arbitrary Waveform and Function Generators

Arbitrary waveform generators (AWGs) and function generators (AFGs) are instruments that can be used to generate arbitrary types of signals other than the standard digital NRZ type that is the standard in pattern generators and bit error rate testers.

An arbitrary function generator provides different waveforms from a predefined set (e.g., sine wave, square wave). The user cannot define a completely generic waveform beyond this predefined set. However, an AWG is capable of generating a waveform based on a series of voltage points at a given sampling rate. This means that the user can specify any type of waveform. Note that although an AFG seems more limited, for its specific set of waveforms it might have a higher performance than an AWG. Figure 7.16 shows a picture of an AWG.

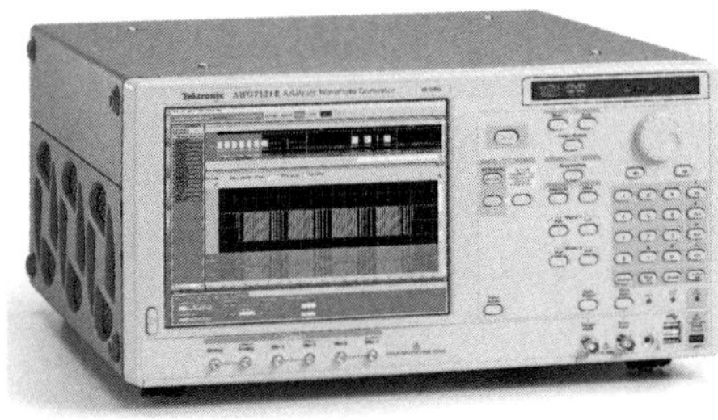

Figure 7.16 Picture of an arbitrary waveform generator (AWG). (Copyright Tektronix. Printed with permission. All rights reserved.)

Clearly an AWG is a very powerful instrument in the sense that any arbitrary waveform can be generated within the limits of the instrument sampling rate. For digital applications this type of instrument is especially interesting not only for generating modulation signals (e.g., for jitter injection) but also to generate digital waveforms at very high data rates including multilevel signal like PAM-4. References [1, 3] are a good starting point on this topic.

7.9 Noise Generators

Noise generators are instruments designed to generate voltage noise with a Gaussian distribution. Figure 7.17 shows a picture of a commercial noise generator instrument. Typically, this type of instrument uses a reverse-biased diode to generate the noise distribution with different selectable attenuation values available to set the correct noise level.

Characterizing the quality of a noise source is important [11], since in theory a true Gaussian distribution should have an infinite tail on the output voltage levels which would imply the capability to achieve infinite voltage levels (although with a very low probability, i.e., millions of years).

When looking at noise generators, one of the most important parameters is its crest factor. The crest factor of a Gaussian noise voltage source is defined

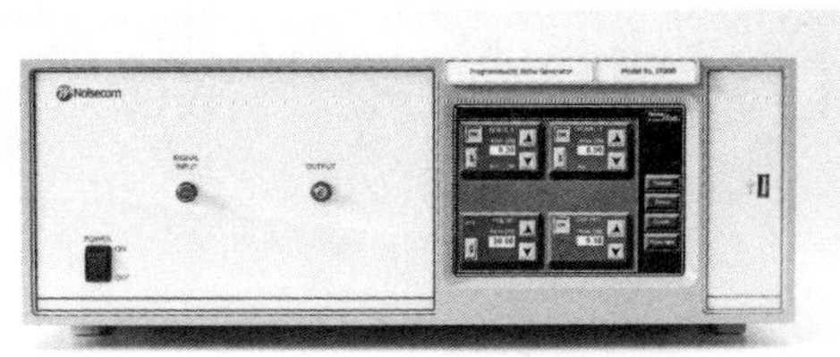

Figure 7.17 Picture of a voltage noise source instrument (Courtesy of Noisecom).

as the ratio of the peak voltage to the RMS voltage value (standard deviation) of the noise generator as shown in Figure 7.18. Sometimes the crest factor is defined as the ratio of the peak-to-peak voltage value to the RMS voltage value. One needs to verify which definition is being used when analyzing the performance of a noise source.

$$\text{CREST FACTOR} = \frac{V_P}{V_{RMS}} \tag{7.1}$$

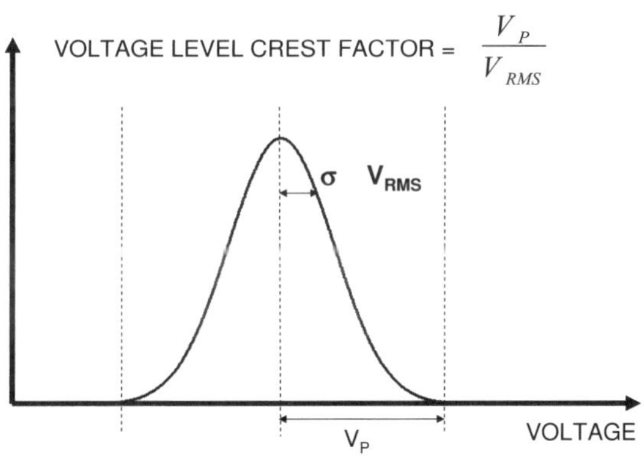

Figure 7.18 Crest factor definition for a voltage noise source.

Some arbitrary waveform and function generators provide the capability to generate a random waveform also. It is important to understand that there can be a significant difference on the generated Gaussian distribution when compared to a noise generator. Figure 7.19 shows a measurement on the time and frequency-domains of a pattern generator output that had jitter injected from an AWG type source set for random noise and the same measurements using a dedicated noise source instrument.

The two top graphs on the figure show a comparison of the jitter histogram taken with an equivalent-time oscilloscope (the pattern used was a bit clock pattern). Both histograms appear to have a Gaussian shape but

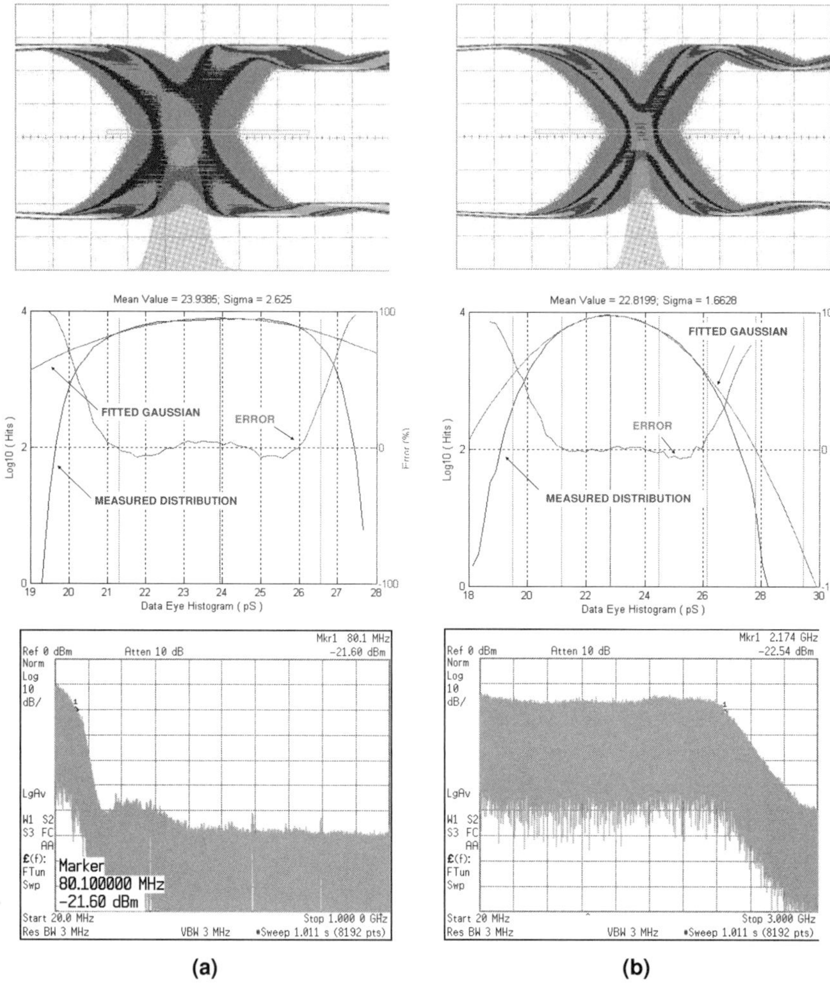

Figure 7.19 Comparison of (a) using the random function of an arbitrary waveform generator for random jitter injection with (b) using a dedicated noise source.

when analyzing the histograms with a post-processing software [12] that compares the measured histogram to the expected Gaussian fitted distribution in a logarithmic scale, one observes a significant difference between the AWG source and the noise source. Another area where this difference is visible is on the frequency-domain. In the bottom two graphs of the figure, the measured frequency spectrum is shown for both sources. Ideally, one would expect an infinite white noise spectrum, but on the measurements one observes that the AWG source only has spectrum bandwidth to approximately 80 MHz (this was the maximum rated frequency of the used AWG instrument) while the noise generator is able to achieve a noise bandwidth of approximately 2 GHz.

Complementary to the noise generation bench instruments, there are also noise generation modules that provide a fixed noise level as shown in Figure 7.20. These modules, due to their size, are better suited for integration into a test fixture. They also provide a more cost-effective way of combining several noise sources and low-pass filters to generate more complicated random jitter tolerance profiles that are required by some standards (see Section I.2).

Figure 7.20 Picture of a voltage noise source (Courtesy of Noisewave).

7.10 Sinusoidal Clock Sources

Sinusoidal clock sources, or signal generators, are an important part of any measurement setup since they generate a high-precision, continuous sinusoidal waveform that can be used as a high-precision time reference for triggering and allow the synchronization of several instruments. Figure 7.21 shows a photograph of a commercially available signal generator.

Note that a sinusoidal clock source will not only provide a "sinusoidal clock signal" with a given frequency and low jitter, but it might also provide the ability to modulate the output signal (important for low-frequency jitter injection) depending on the instrument capabilities. Also of importance is the 10 MHz synchronization input/output. The 10 MHz output is locked to the source PLL and is typically very stable (stability is defined by the parts per million that the 10 MHz signal might drift in time) and can be used to force other instruments to have their internal PLLs synchronized to this 10 MHz output. Section 7.15.2 discusses the synchronization topic in more detail.

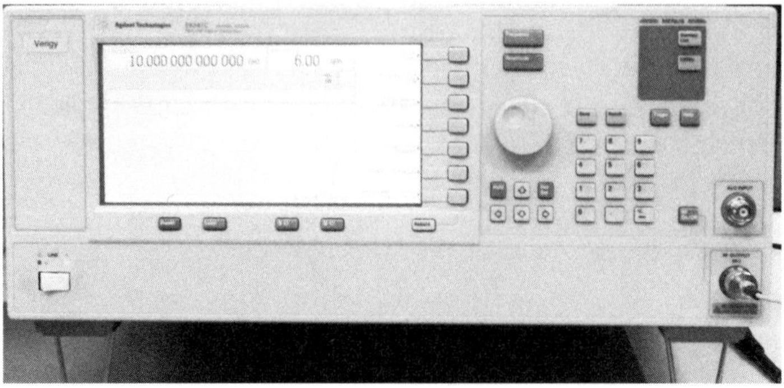

Figure 7.21 Picture of a signal generator that can be used as a sinusoidal clock source.

7.11 Clock and Data Recovery

Clock and data recovery units can be very helpful in several measurement situations. Figure 7.22 shows a picture of a CDR unit that is capable of clock and data recovery from data signal with a maximum data rate of 32 Gbps and with a programable loop filter.

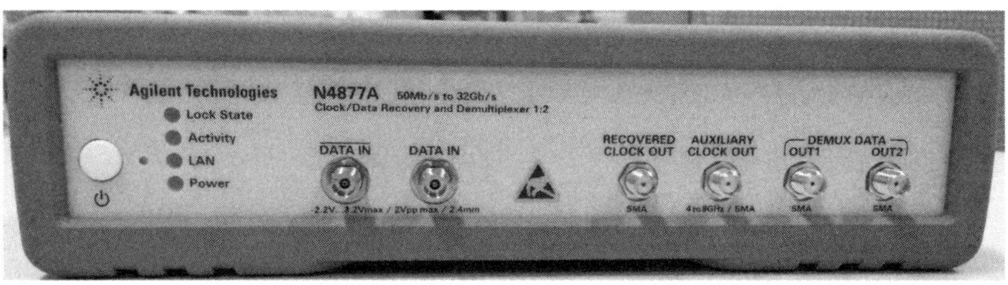

Figure 7.22 Picture of a hardware clock and data recovery unit (CDR) from Keysight Technologies.

7.12 Protocol Analyzer

Protocol analyzers are typically used during design validation to verify protocol engine implementations for correctness and completeness as well as during debugging of systems that use certain protocols to control their components. Protocol analyzers do not support parametric measurements of electrical interfaces like oscilloscopes but work strictly on the logic data that they observe on a protocol link. This logic data is translated by the protocol analyzer from the purely digital 0-1 sequences of the physical layer to the packets and packet payload data on higher protocol layers. In order to capture the desired data sections of the communication between link partners in an efficient manner, protocol analyzers offer complex trigger options and trigger sequences based on payload data, packet types, or link states.

Since protocol analyzers have to interpret the data based on higher protocol layers, the hardware of these instruments is partially protocol-specific and thus dedicated instruments are usually required for each protocol. An example of a PCI Express protocol analyzer is shown in Figure 7.23.

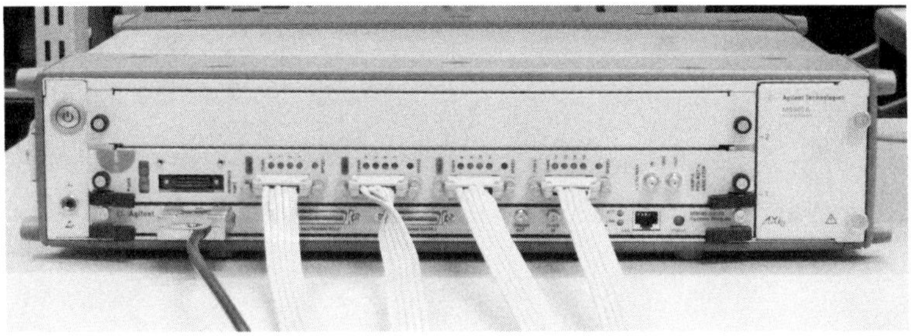

Figure 7.23 Picture of a protocol analyzer for PCI Express 3 from Keysight Technologies.

7.13 Switch Matrix

In some measurement setups it is necessary to switch a signal between multiple measurement instruments. Section 8.8 discusses relays in more detail in the context of test fixture design, but in a bench measurement setup it is sometimes necessary to have a switch matrix as an external instrument [13]. This allows changing automatically between different measurement instruments or between different DUT I/O cells. Figure 7.24 shows a photograph of a switch matrix that includes a GPIB controller and several high-performance coaxial RF relays.

7.14 Isolation Transformer

An isolation transformer can be very helpful in avoiding ground loops in measurements setups [14, 15], especially when using external measurement instrumentation with an ATE system. Figure 7.25 shows how a ground loop might arise between an external measurement instrument and an ATE system. To avoid this possibility, an isolation transformer can be added to the external instrumentation power line avoiding in this way any ground loop.

Figure 7.26 shows a high-level diagram of the needed isolation transformer. The critical feature is that the protective earth of the instrument

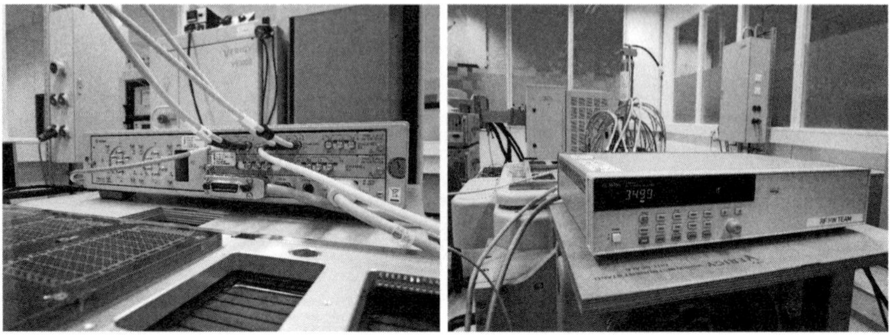

Figure 7.24 Picture of a GPIB controlled switch matrix from Keysight Technologies with different types of high-frequency coaxial relays (left: instrument back, right: instrument front).

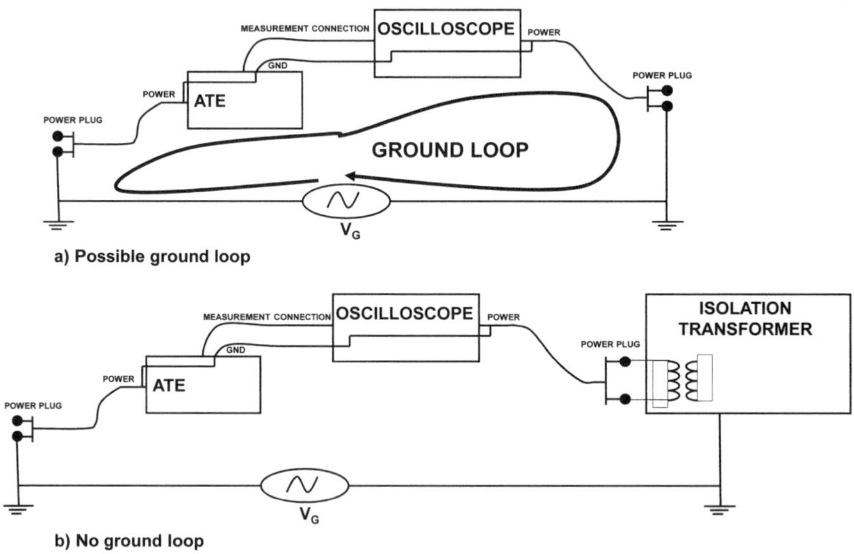

Figure 7.25 (a, b) Possible ground loop when connecting an external instrument to an ATE system.

is connected to the midpoint of the secondary winding of the transformer and not to the isolation transformer protective earth. In case one of the power lines gets connected to the instrument protective earth (PE), the fuse on that line in the isolation transformer will break the connection. Figure 7.27 shows a picture of an isolation transformer.

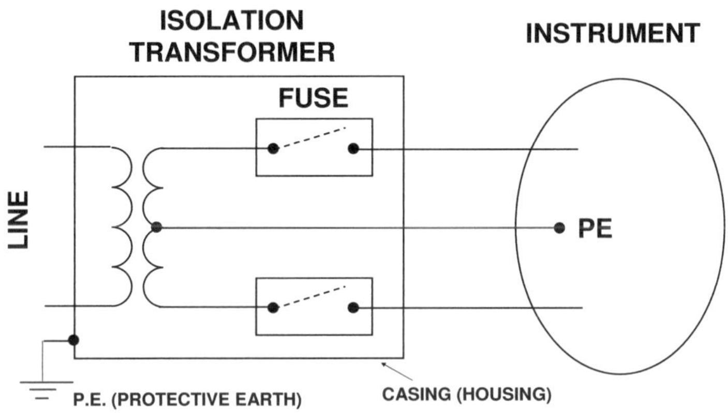

Figure 7.26 High-level diagram of an isolation transformer from the Toroid Corporation [16].

Figure 7.27 Picture of an isolation transformer from the Toroid Corporation [16].

7.15 Connecting Bench Instrumentation to an ATE System

Very often there is the need to connect external instruments to an ATE system. This can be as simple as connecting a voltmeter to an ATE DUT power supply to measure its voltage. Sometimes, however, it means to connect multiple external instruments like a parallel bit error rate tester to help testing a multigigabit I/O DUT on a low-speed ATE system. Figure 7.28 shows two examples of integrating external instruments with an ATE system. The problems that arise from connecting external instrument to an ATE system can be divided in two categories: signal integrity and synchronization. Software integration is also a significant challenge that might require a large effort, but it is usually not a bottleneck for integrating external instruments for high-speed digital applications.

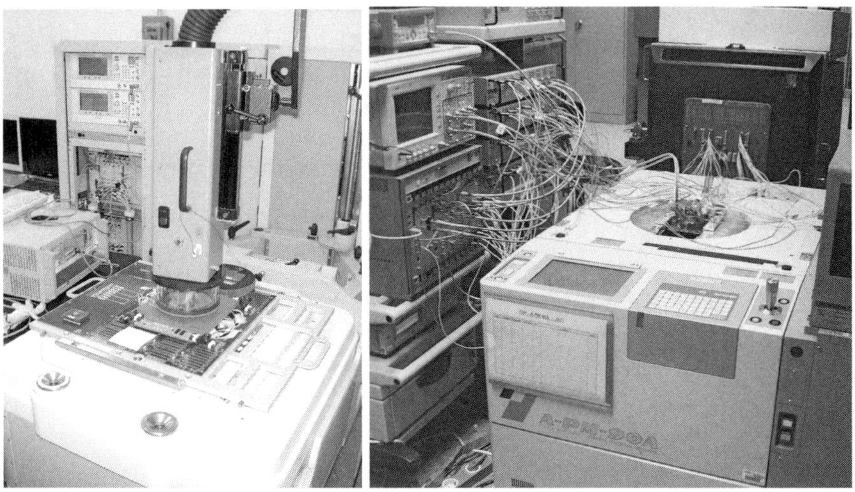

Figure 7.28 Examples of integrating bench instrumentation with ATE systems [17, 18].

7.15.1 Signal Integrity

The first challenge when connecting external instruments to an ATE system, especially when dealing with high-speed digital signals, is to provide a connection between the external instruments and the test fixture (i.e., connectors and cables) with the required performance or signal integrity. The challenge is easier if one knows up front that there is a need to connect external instrumentation to the ATE system when developing the test fixture.

For example, in Figure 7.29 there is an example of how knowledge of the need to use external instrumentation was taken into account on the test fixture design.

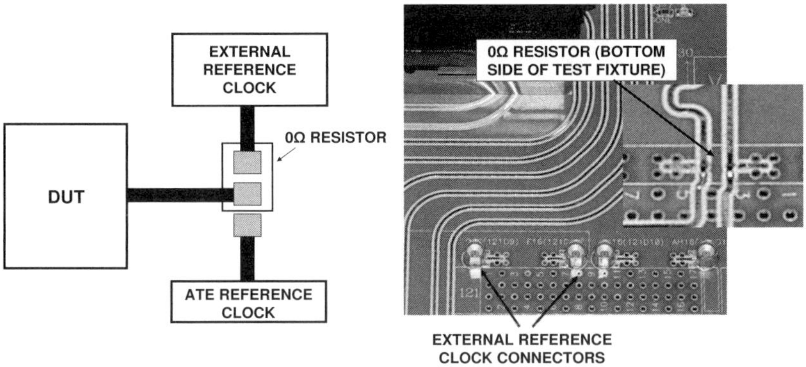

Figure 7.29 Test fixture layout allowing the DUT reference clock to be connected to an ATE pin channel or an external low jitter reference clock by moving a 0 Ω resistor.

In this application the initial problem was whether the ATE pin electronics could provide a reference clock with low jitter. The question was not on the amount of the pin electronics jitter (its value is known from the ATE specification), but if a lower jitter clock could improve the production yield even further. Having this knowledge, the test fixture was designed with the possibility of using the ATE pin electronics or an external clock source (connected through an SMP connector) as the reference clock for the DUT. This was accomplished by simply moving a 0 Ω resistor on the board. In this specific application, the discontinuity generated by the 0 Ω resistor was negligible.

On other occasions, the need for connecting external equipment arises from the need to debug problems on an application or test fixture that were nor foreseen. In this case, one needs to connect the external instrument to the test fixture or DUT socket using an appropriate probe. Appendix H discusses the topic of probing a test fixture in more detail. Note that probing a test fixture is not a trivial task. Not only should the appropriate probing techniques and equipment be used, but also appropriate electrostatic discharge (ESD) precautions should be taken, so as not to damage the ATE system and measurement equipment [19]. Note also when connecting bench instrumentation to an ATE system it is important to guarantee that no ground loops are created (see Section 7.14) and that any extra required shielding is implemented [14, 20].

7.15.2 Synchronization

Another separate challenge is the synchronization of external instruments to an ATE system. Sometimes when measuring signals from an ATE system with external instrumentation (e.g., an equivalent-time oscilloscope), the user might need a trigger signal from the ATE system to be provided to the external measurement instrument. This can easily be obtained from an unused ATE channel as shown on Figure 7.30.

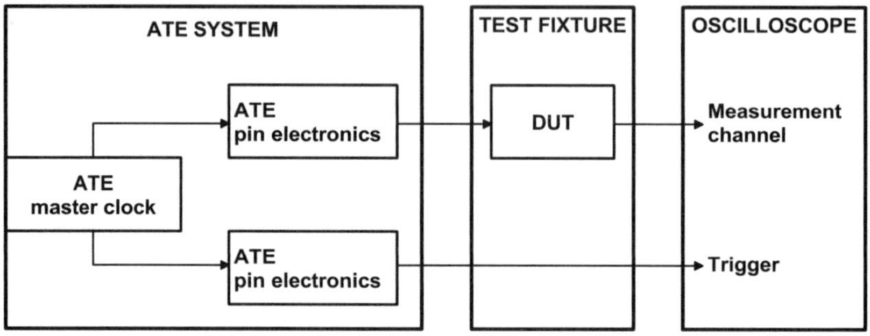

Figure 7.30 Measuring an ATE channel driver signal with a scope using an extra ATE driver channel as trigger.

In this setup the ATE driver channel is used as the reference for all timing measurements. There are some situations where using an extra ATE channel as the trigger is not good enough, especially when one wants to measure timing performance of the ATE or DUT against a high-accuracy external timing reference. Figure 7.31 shows four typical setups that can be used depending on the ATE system capabilities.

The main challenge is to ensure frequency synchronization between the reference timing source and the ATE system, guaranteeing that there are no frequency drifts or frequency offsets between the two systems. This can be accomplished by using the external timing reference as the master clock for the ATE system or locking the ATE master clock to the external timing reference with a synchronization signal (typically a 10 MHz frequency signal). Notice that these setups depend on the specifics of the ATE hardware architecture and in some ATE architectures this might not even be possible.

This type of setup can also be used for another type of synchronization challenge: using an external pattern generator as a signal source. Figure 7.32 shows an example of this type of setup for a demultiplexer DUT. In this case the external pattern generator provides a high-speed digital stimulus signal (e.g., 10 Gbps) to the DUT, which, in turn, demultiplexes this signal into four 2.5 Gbps data lines measured by the ATE pin electronics (assuming the ATE pin electronics maximum data rate in this example is 2.5 Gbps).

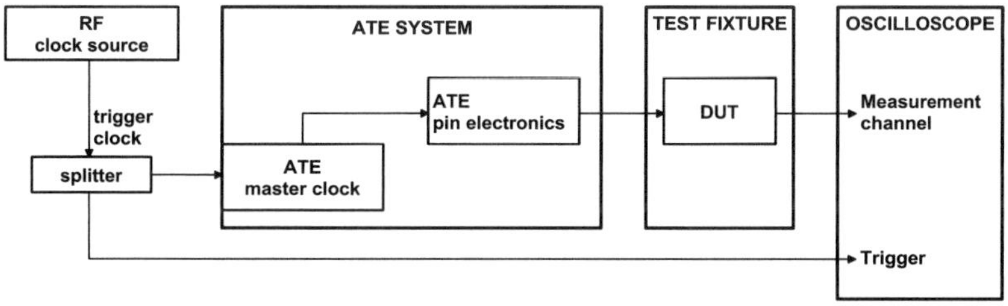

(a) The external RF source provides the ATE master clock directly.

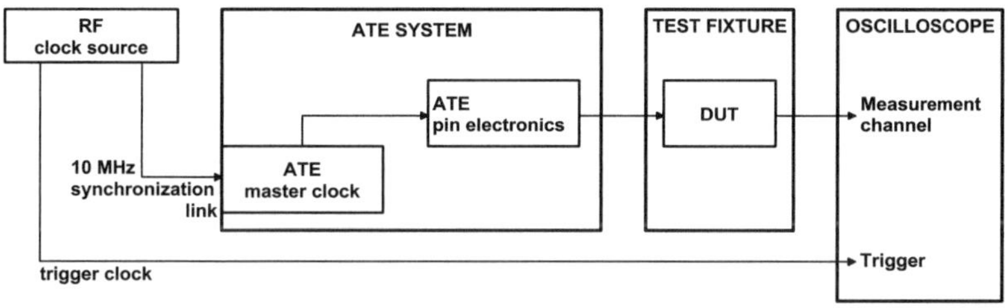

(b) The external RF source provides a 10 MHz reference signal to the ATE timing generation system.

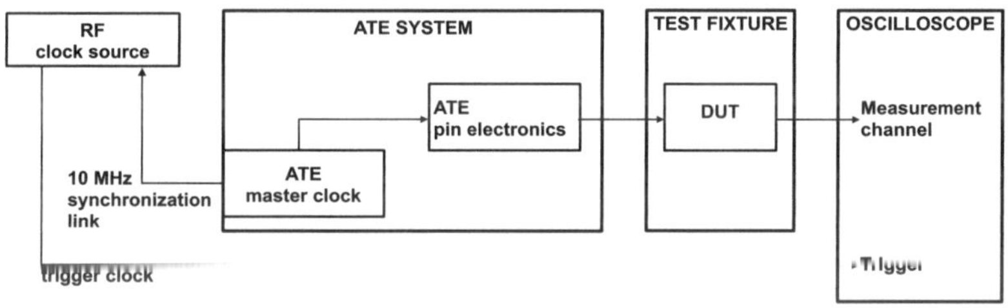

(c) The ATE timing system provides a 10 MHz reference signal to the external RF source.

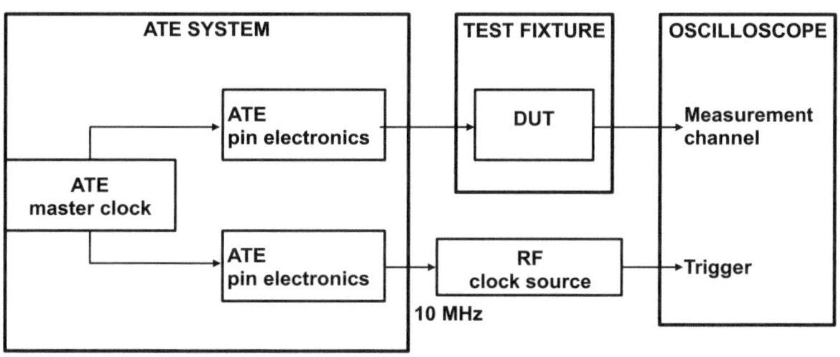

(d) The 10 MHz reference signal to the external RF source is supplied from one ATE digital channel.

Figure 7.31 (a—d) Possible options for synchronizing an external RF clock source to serve as a master timing reference.

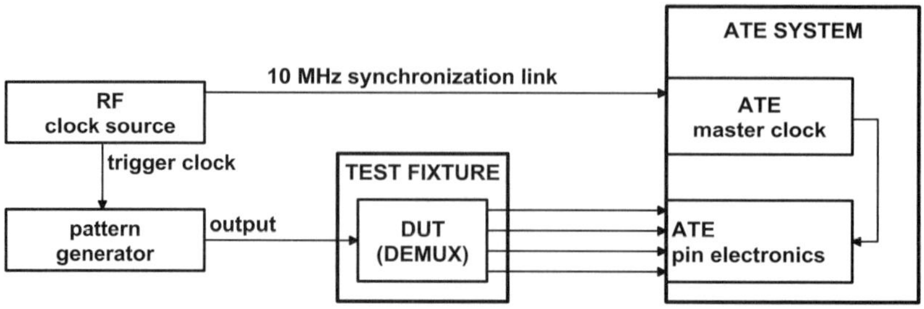

Figure 7.32 Connecting an external pattern generator to an ATE system.

As previously explained, with this setup it is guaranteed that there is no frequency offset or frequency drift between the ATE system and the external pattern generator. However, the fixed phase relationship between external instrument and ATE does not solve another problem, which is to know and to control the exact timing or phase relationship between the ATE system and the external pattern generator and thus the data received by the DUT [21]. The challenges with this pattern synchronization between ATE and external instruments is identical to the bit and pattern alignment discussed in Section 5.1. Thus, the solutions are also the same as presented in that section. The advantage for the synchronization to data provided by external instruments or DUT data derived from external instrument stimuli is the fact that the compare data on the ATE is well decoupled from the DUT stimulus data that is transformed to the DUT data stream to be compared. Thus, solution mechanisms that are based on dynamic pattern offsets between the incoming data stream and the compare data, like the match-loop approach described in Section 5.1.2.1, are easier to implement with external instruments than with DUTs that are completely stimulated by ATE drivers. Another benefit of external instruments stimulating the DUT is the fact that these run on their own clock domain and keep generating data to the DUT even if the ATE sequencers are stopped. For devices that potentially change the phase and/or latency of the data they generate if their stimulus data is stopped and restarted, this simplifies search-based synchronization methods that repeatedly stop and restart the ATE sequencers.

7.15.3 External Reference Clock Impact on Jitter Measurements

It is important to note that using external instruments like a clock source for oscilloscope triggering or to provide the reference clock to the DUT can have an impact on jitter measurement results. When discussing jitter measurements, the key question arises as to which reference is used for the jitter measurements. Not only has the intrinsic jitter of the reference created a baseline for the achievable measurement accuracy, even the measurement setup itself might have a significant influence. Figure 7.33 shows two different measurement setups for a jitter measurement of an ATE channel. The first setup uses an external reference clock for the oscilloscope triggering that also provides a 10 MHz reference signal to the ATE master clock. The second setup uses an additional ATE channel for triggering the oscilloscope.

If we assume only Gaussian random jitter being present on the setup, the amount of measured random jitter will depend not only on the random jitter present on the signal to be measured but also on the trigger signal.[1] Assuming also that the external reference clock has an intrinsic random jitter that is lower than the random jitter of the ATE channels, then the setup with the external reference clock for triggering provides a more accurate measurement of the ATE channel random jitter. In the presence of other jitter components on the ATE system (e.g., a periodic jitter component on the ATE master clock), when using an ATE channel for oscilloscope triggering, these jitter components might not be measured since they are present on both the signal to be measured and the trigger. If the external clock source is used instead, then this periodic jitter component will be measured because it is not present on the trigger signal from the external reference clock source.

Figure 7.34 shows an example where a 16 Gbps PRBS15 signal from an ATE channel is measured using both conditions described in Figure 7.33. The example shows that, as expected, setup (b) shows less measured jitter than setup (a), although the measured signal is exactly the same.

7.16 Coaxial Cables and Connectors

7.16.1 Coaxial Cables

Understanding and correctly using coaxial cables is very important for high-speed digital applications. Spending a significant amount of time optimizing the test fixture and then measuring its performance with the first "good-looking cable" in the laboratory is not advisable. It is important not only to understand the limitations of the applied coaxial cables, but also how to correctly select one in a given application.

Figure 7.35 presents a diagram of a typical coaxial cable and the equation for its characteristic impedance. Like on a PCB trace, the loss of a coaxial cable is mainly determined by the dielectric type (dielectric loss) and the inner core diameter (skin effect loss). Good quality coaxial

[1] Assuming that only Gaussian random jitter is present on the signal to be measured and on the trigger, the measured random jitter is given by: $\sqrt{Jitter_{signal}^2 + Jitter_{trigger}^2}$.

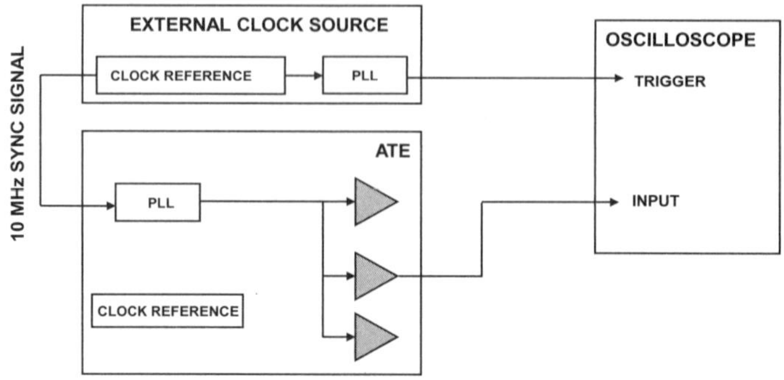

a) Using an external clock source as trigger

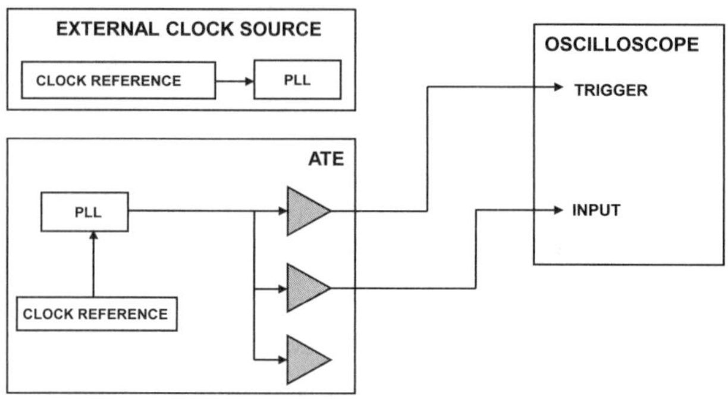

b) Using another ATE channel as trigger

Figure 7.33 (a, b) Comparison of two different jitter measurement setups.

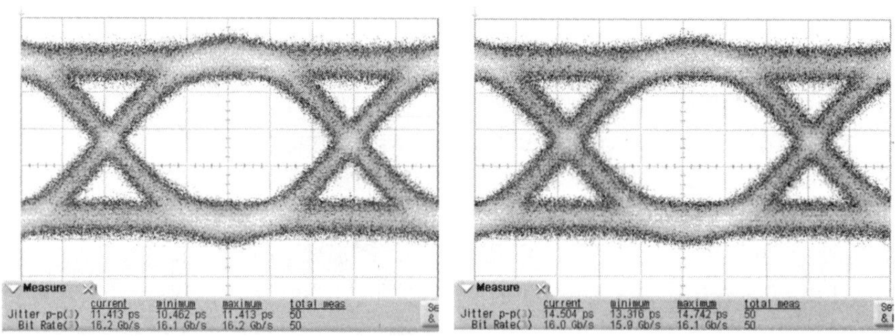

TRIGGER: ATE CHANNEL

TRIGGER: EXTERNAL RF SOURCE (10 MHZ LOCKED TO ATE)

Figure 7.34 Example of the trigger setup impact on the measured peak-to-peak jitter from an ATE channel.

cables use low loss dielectrics like Teflon. This means that for typical high-quality coaxial cables, the loss is dominated by the skin effect. To reduce the skin effect loss, it is necessary to increase the diameter of the cable core which implies an increase on the shield diameter to keep the characteristic impedance constant. It is a good rule of thumb to assume that thicker cables provide lower loss for the same length. Figure 7.36 shows a comparison of a data eye diagram measured with two high-quality cables with different lengths and different diameters (i.e., core diameters) and also a comparison of the measured insertion loss. Figure 7.37 shows a picture of the two cables.

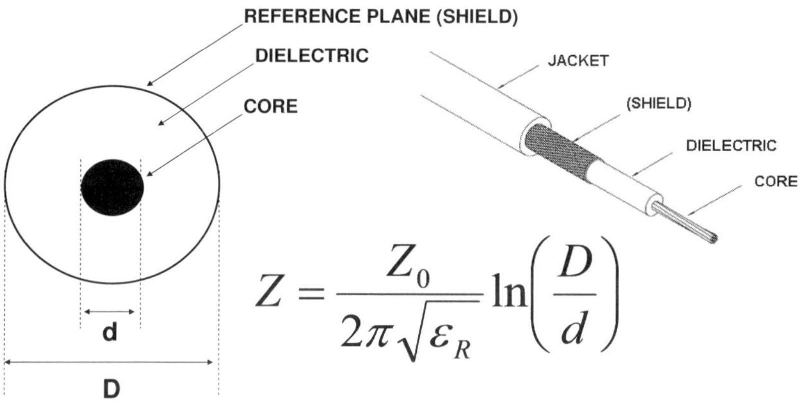

$$Z = \frac{Z_0}{2\pi\sqrt{\varepsilon_R}} \ln\left(\frac{D}{d}\right)$$

Figure 7.35 Geometry of a typical coaxial cable and its characteristic impedance equation.

Although one would expect the longer cable (80 cm) to have a worse performance than the shorter cable (61 cm), the larger diameter (10.6 mm compared to 6.5 mm) makes it a less lossy cable as shown by the improved rise time (36 ps versus 38 ps) and the measured insertion loss.

One point to note is that there is a drawback of using coaxial cables with large diameters. There is a relationship between the diameter of the cable and the frequency at which the cable starts to exhibit unwanted behaviors in the form of resonances. Figure 7.38 shows a comparison of the insertion loss of two cables with different diameters. One can observe that the larger diameter cable, although having a lower loss, does exhibit a resonant behavior after approximately 19 GHz.

This effect can also be observed in the time-domain by using a source with a very fast rise time. Figure 7.39 shows the output of a data source with 8 ps rise time and how that waveform looks like after the 80 cm length, 10.6 mm diameter coaxial cable in Figure 7.37.

The figure shows that the odd behavior of the cable at very high frequencies due to its diameter coupled with a pattern with a very fast rise

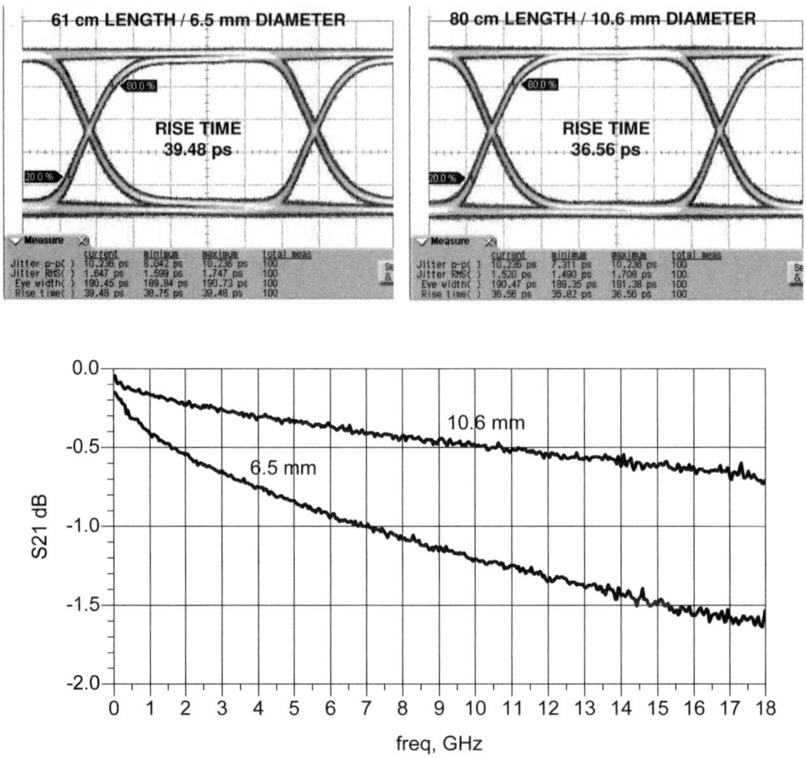

Figure 7.36 Data eye diagram comparison of two coaxial cables (top left: 61 cm length with 6.5 mm diameter; top right: 80 cm length with 10.6 mm diameter) at 5 Gbps with a PRBS $2^{31} - 1$ data pattern (top). Comparison of the insertion loss of the two cables (bottom).

time results in a waveform with a more pronounced oscillation as shown in Figure 7.39 (right).

It is also important to understand that even with a thin coaxial cable we can always have a lower loss than a PCB signal trace. Figure 7.40 shows this difference by comparing a microstrip in several geometries and different dielectrics to a coaxial cable with different diameters. This is further demonstrated in Figure 7.41.

Figure 7.42 shows an example of the specifications for a low loss coaxial cable family. For high-speed digital applications it is important to be able to map the application requirements into the coaxial cable specifications. The first important point to consider is if there are any mechanical restrictions that must be obeyed [22], for example:

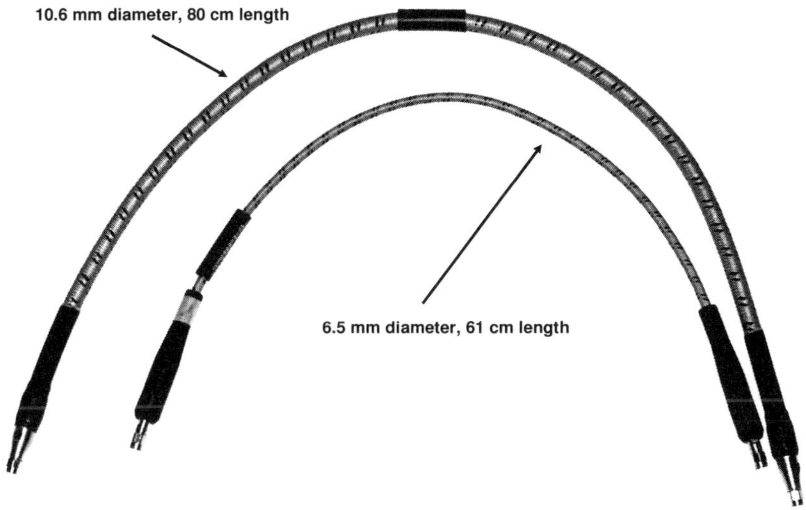

Figure 7.37 Picture of the two coaxial cables with different length and diameter.

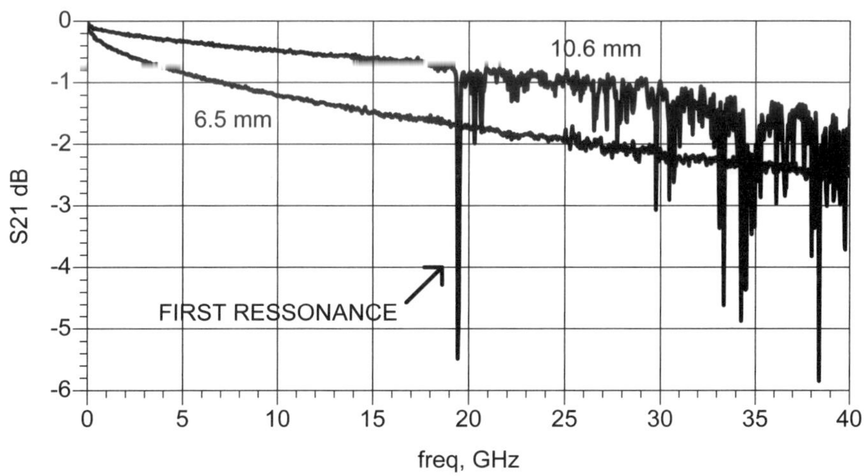

Figure 7.38 Insertion loss comparison of two cables with different core diameters showing the resonant behavior of the large diameter cable beyond 19 GHz.

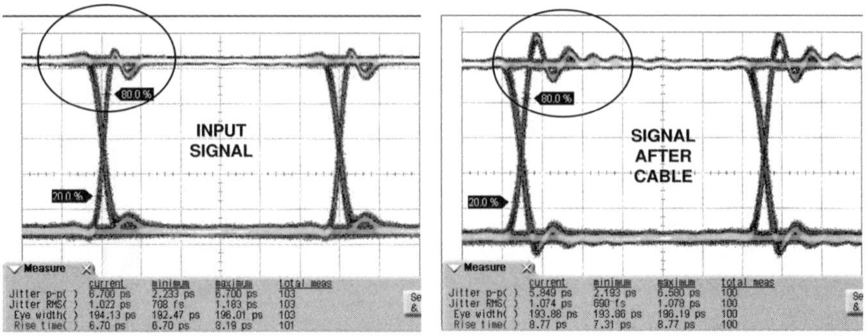

Figure 7.39 5 Gbps digital patten input signal with 6.7 ps (20/80) rise time (left) and data eye after a 80 cm length, 10.5 mm diameter coaxial cable (right).

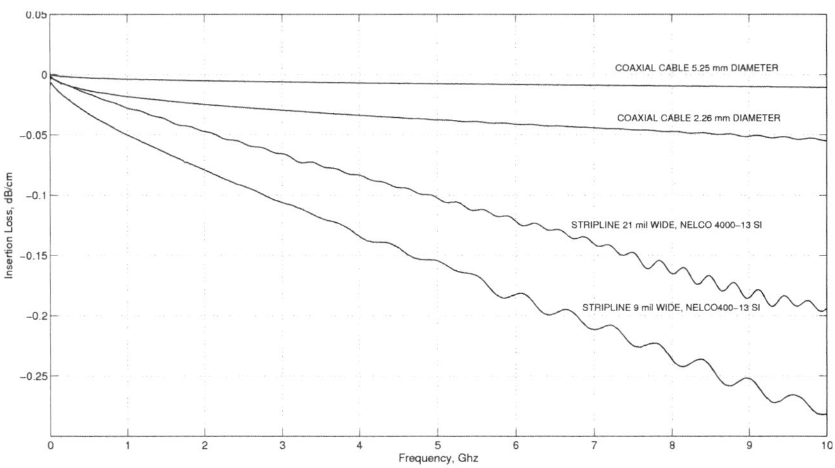

Figure 7.40 Comparison of the insertion loss between two coaxial cables with different diameters with two PCB striplines with different trace widths.

Cable Type: Is the cable of the flexible type, conformable (also known as hand-formable), or semi-rigid, as shown in Figure 7.43.

Cable Core: When the cable is of the flexible type and intended to be flexed multiple times, then one should consider the use of a stranded core in the cable (a stranded core will have more loss than a solid one) and also make sure that the maximum bending diameter is acceptable.

Diameter: Is there a maximum diameter for the cable? How bulky can the cable be?

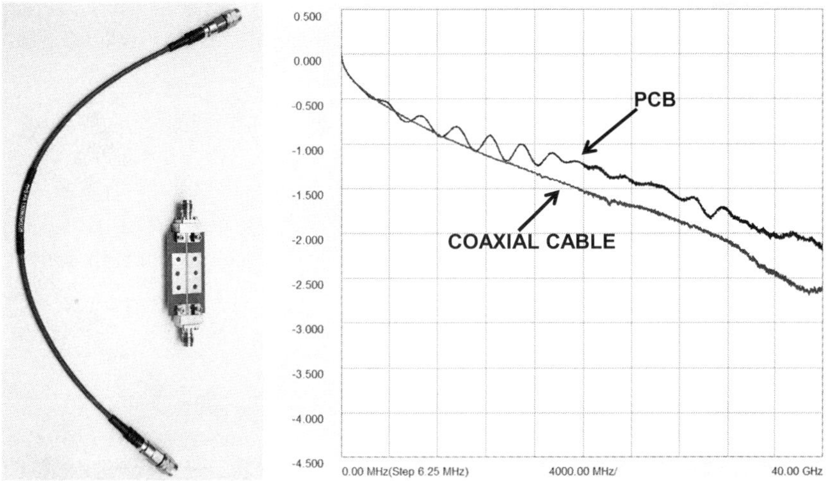

Figure 7.41 Comparison of the insertion loss between a 305 mm coaxial cable and a 40 mm microstrip trace implemented using rolled copper on a very low loss dielectric.

MECHANICAL CHARACTERISTICS							
UTiFLEX TYPE		UFB142C	UFB142A	UFB197C	UFB205A	UFB293C	UFB311A
Outer Diameter inch (mm)		0.142 (3.61)	0.142 (3.61)	0.197 (5.00)	0.205 (5.21)	0.293 (7.44)	0.311 (7.90)
Center Conductor Type		Stranded	Solid	Stranded	Solid	Stranded	Solid
Weight grams/ft (meter)		9 (29.5)	10 (32.8)	18 (59.1)	20 (65.6)	37 (121.4)	42 (137.8)
Minimum Bend Radius inch (mm)		0.38 (9.65)	0.38 (9.65)	0.50 (12.70)	0.50 (12.70)	0.75 (19.05)	1.25 (31.75)
Flexures		35,000	5,000	35,000	5,000	35,000	5,000
ELECTRICAL CHARACTERISTICS							
Impedance (ohms)		50	50	50	50	50	50
Frequency Range (GHz)		DC-18	DC-40	DC-26.5	DC-26.5	DC-18	DC-18
Velocity of Propagation		83%	83%	83%	83%	83%	83%
Capacitance pf/ft (meter)		24.5 (80.4)	24.5 (80.4)	24.5 (80.4)	24.5 (80.4)	24.5 (80.4)	24.5 (80.4)
Shielding Effectiveness (dB @ 1 GHz)		> 100	> 100	> 100	> 100	> 100	> 100
Maximum Insertion Loss dB/ft (meter)		See figure on next page					
	1 GHz	0.12 (0.39)	0.11 (0.36)	0.09 (0.30)	0.08 (0.26)	0.06 (0.20)	0.05 (0.16)
	10 GHz	0.38 (1.25)	0.33 (1.08)	0.30 (0.98)	0.23 (0.76)	0.19 (0.62)	0.15 (0.49)
	18 GHz	0.51 (1.67)	0.45 (1.48)	0.40 (1.31)	0.32 (1.05)	0.26 (0.85)	0.21 (0.69)
	26.5 GHz	–	0.55 (1.81)	0.49 (1.61)	0.39 (1.28)	–	–
	40 GHz	–	0.68 (2.23)	–	–	–	–
Phase Stability vs Flexure*	10 GHz	1°	2°	1°	1°	3°**	3°**
	18 GHz	1°	3°	1°	2°	5°**	5°**

Figure 7.42 Example of a low loss coaxial cable family spec sheet (Courtesy of Micro-Coax).

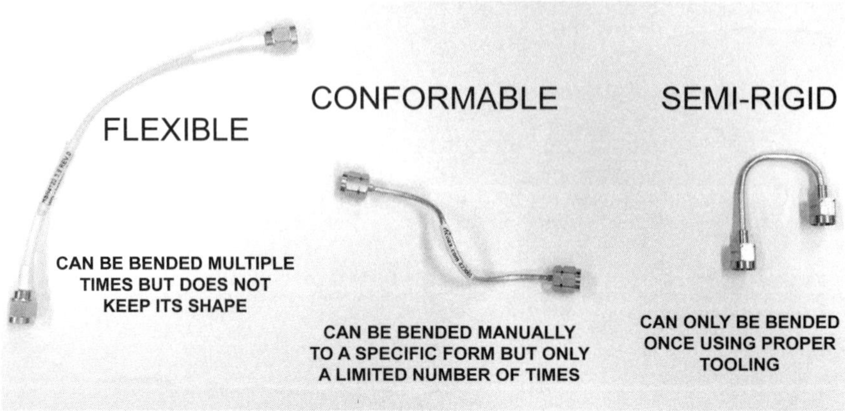

Figure 7.43 Examples of the three different types of coaxial cables: flexible type, conformable (or hand-formable), and semi-rigid.

After the mechanical requirements are evaluated, one should now address the cable bandwidth requirements. Note that the maximum frequency described on the cable specification refers to the maximum frequency for which it is guaranteed that the propagation through the coaxial cable is without any unwanted behaviors like resonances. For computing the cable bandwidth, one needs to take the loss at the frequencies specified and the cable length that one intends to use. Some cable manufacturers specify the cable loss through an equation that approximates the real loss, allowing the cable loss for a specific length to be easily evaluated.

7.16.2 Coaxial Connectors

Coaxial connectors are typically used on ATE applications in two ways: either as part of cable assemblies or as connectors soldered on the test fixture. It is also possible that for certain applications and ATE interface architectures, coaxial connectors are used on the interface between the test fixture and the ATE pin electronics.

Confronted with the huge variety of coaxial connectors, an inexperienced engineer can easily get lost on all the different names [23, 24]. In this subsection we will concentrate on the type of connectors that are most relevant for high-speed digital applications.

Figure 7.44 shows several examples of coaxial connectors for cable assemblies. When choosing a connector, the following items should be taken into consideration:

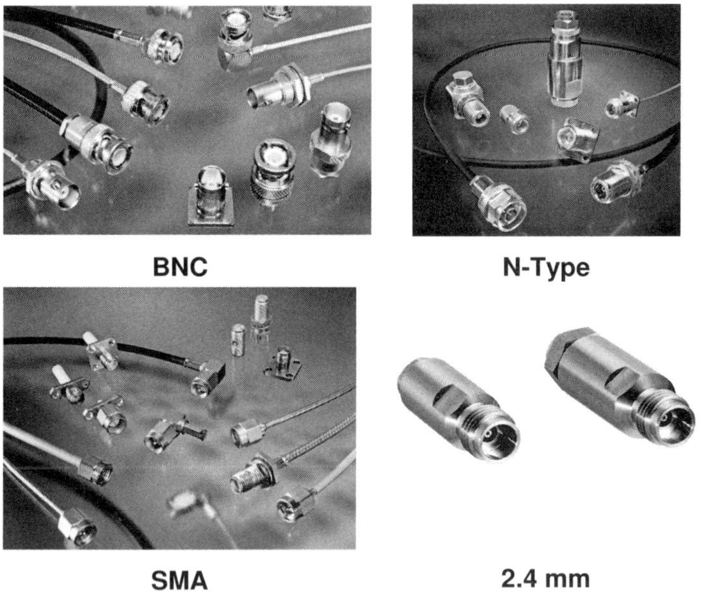

Figure 7.44 Example of coaxial connectors, top right: N-type; top left: BNC; bottom right: 2.4 mm; bottom left: SMA (Courtesy of Rosenberger).

Do I have to mate the connector to a specific instrument?

This might define which connector you must use.

What is the minimum bandwidth that I need for the connector?

There are 110 GHz bandwidth connectors available in the market if needed, but if the application only requires 10 GHz bandwidth, there is no need to go through the extra cost and effort. Other items like wear-out cycling and mechanical robustness might also be important depending on the application. Make sure you understand your application needs.

Table 7.1 shows some of the most important coaxial connector families [25]. Note that some connectors are sometimes further divided into categories like standard and precision, which typically imply different performances. One should always inspect the connector manufacturer specifications. Note also that although two different types of connectors might be mechanically compatible, it might not be appropriate to mate those different types since it can damage the higher precision connector type due to manufacturing tolerances.

It is always possible to move from one type of coaxial connector to another through the use of a coaxial connector adapter. Figure 7.45 shows

Table 7.1
Coaxial Connector Families

Type	BW	Compatible Mating Types	Notes
BNC	2 GHz	BNC	
Type N	18 GHz	Type N	
SMA	18 GHz	SMA; 3.5 mm; K-type	Typically Teflon dielectric core with some rated to 26.5 GHz
3.5 mm	26.5 GHz	3.5 mm; SMA; K-type	Air core with some rated at 34 GHz
K-type	40 GHz	K-type; SMA; 3.5 mm; 2.92 mm	2.92 mm connector version from Anritsu
2.92 mm	40 GHz	K-type; SMA; 3.5 mm; 2.92 mm	
2.4 mm	50 GHz	2.4 mm; 1.85 mm	
APC-7	18 GHz	APC-7	Sexless connector typically used in metrology
1.85 mm	67 GHz	1.85 mm; 2.4 mm	Also referred to as V-connector by Anritsu
1.0 mm	110 GHz	1.0 mm	Also referred to as W1-connector by Anritsu
0.8 mm	145 GHz	0.8 mm	Developed by Anritsu
SMP	40 GHz	SMP	Also referred to as GPO
MINI-SMP	65 GHz	MINI-SMP	Also referred to as GPPO or SMPM
SMPS	100 GHz	SMPS	Also referred to as G3PO
BMA	18 GHz	BMA	Blind mate connection
MMPX	67 GHZ	MMPX	Huber & Suhner coaxial-to-PCB connector

several different adapters. The drawback of using a coaxial adapter is that it will add cost, take space, and, of course, it degrades the signal, although the degradation can be minimal if an appropriate connector adapter is chosen.

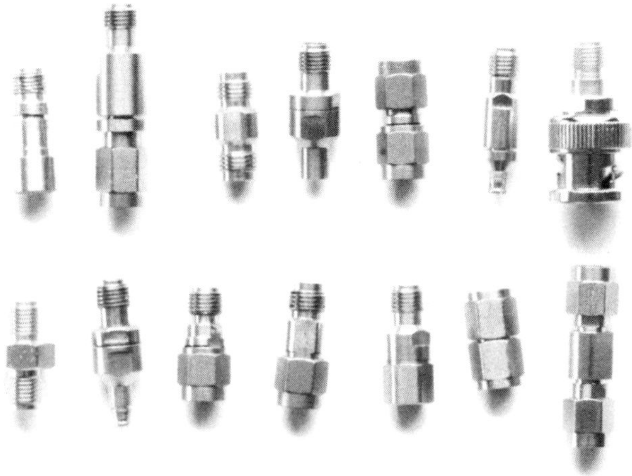

Figure 7.45 Example of several types of coaxial adapters.

Another important point regarding coaxial connectors on cable assemblies for high-speed digital applications is the importance of using an appropriate torque wrench as shown in Figure 7.46 (left) for mating the connectors. Note that each connector family might require a different torque value. Table 7.2 shows the parameters for the SMA, 3.5 mm and 2.4 mm connector types. Note that SMA and 3.5 mm connectors require different torque values.

Figure 7.46 Torque wrench for correctly connecting a connector (left) and connector measurement gauge (right).

Table 7.2
Torque Wrench Requirements for Typical Connector Families Used on High-Speed Digital Applications

Connector Type	Wrench Size (in)	Preset Torque (lb/in)	Preset Torque (N/cm)
3.5 mm, 2.4 mm, 1.85 mm	0.312 HEX	8	90
SMA	0.312 HEX	5	56

For some coaxial connector families where the connector cable assembly snaps into its mating connector (e.g., MMPX) there is no need for a torque wrench but usually there is the need to use a specialized disconnect tool to avoid damaging the connector when disconnecting the cable assembly. Figure 7.47 shows some examples of disconnecting tools.

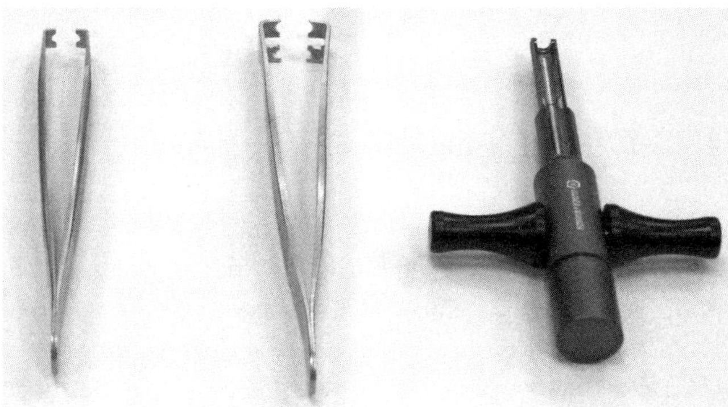

Figure 7.47 Examples of disconnect tools for mini-SMP (left), SMP (center) and MMPX (right).

Dirty connectors should be cleaned using a cotton swab and isopropyl alcohol (IPA) [23, 26]. In connectors with air dielectric, one should not clean the connector pin since it is easy to damage, and, in fact, any dirt on the inside of the connector will go to the back where you cannot reach it anymore. To avoid getting your connectors dirty, you should keep the plastic plugs on them when not used (that is why an expensive 3.5 mm adapter comes with plastic plugs for protection).

It is important to note that one damaged connector might damage any connectors that you mate it with. Meaning that if you have a damaged connector, you should simply throw it away. A more difficult connector damage to assess is if the height of the connector center pin is still within specifications. Changes due to damage might have performance impact on high-precision measurements or calibrations, and it is not easy to realize that a given connector has a problem. To address this issue, a connector gauge like the one shown in Figure 7.46 (right) should be used. With this instrument one can measure the height of the connector center pin and check if it is within the allowed range. As a final note on connector protection, remember that not all connectors are manufactured in the same way. There is a reason for the price difference between an SMA connector and a 3.5 mm connector. The 3.5 mm connector is a precise machined mechanical part and if you mate it with a cheap SMA connector, you might damage the expensive 3.5 mm connector. This is especially important with expensive bench equipment or calibration kits. In this case avoid connecting different connector types directly to the instruments or calibration standards and instead include an appropriate adapter to protect the most delicate connector.

The other type of coaxial connectors used on ATE applications are the ones that can be assembled into the test fixture. The number of designs and types of connectors for printed circuit board (PCB) assembly is even larger than for coaxial cable assemblies. PCB connectors for high-speed digital applications can be divided in the following groups:

- Edge mounted;
- Surface mounted;
- Through hole mounted;
- Hybrid through hole/surface mounted.

Figure 7.48 shows several type of connectors for PCB assembly. It is important to understand that the performance of a coaxial connector on a PCB assembly does not only depend on the performance of the connector itself, but also on the footprint used on the test fixture PCB for mounting the connector, and especially on how the transition is made from the signal trace to the connector [27]. Although some manufacturers provide default footprints for their connectors, for high-speed digital test and measurement applications it is sometimes necessary to develop a custom footprint and transition strategy for a specific connector. Figures 7.49 and 7.50 show examples of custom footprint and transition design developed using a 3D EM field solver (Appendix G) for an ATE test fixture.

7.17 Accessories

When connecting or integrating bench instrumentation to an ATE system, there are several accessories that might be needed. This section presents a brief description of some important accessories.

Figure 7.48 Example of coaxial connectors for PCB assembly.

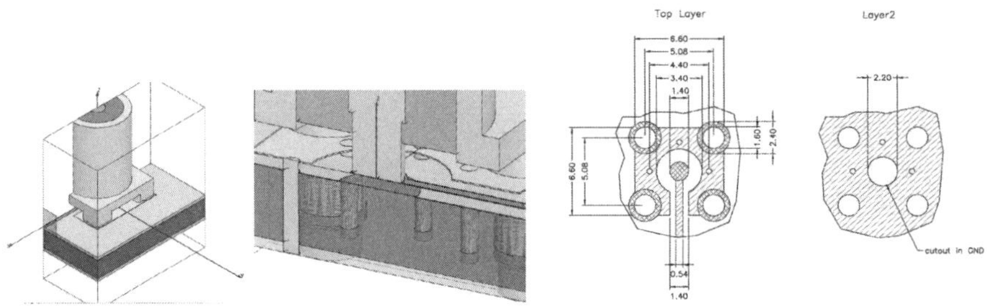

Figure 7.49 Example of an optimized footprint for a microstrip transition to a surface mounted SMA connector (Courtesy of Rosenberger).

7.17.1 Power Splitters and Power Dividers/Combiners

Power splitters and power dividers/combiners are two types of accessories that are frequently used for high-speed digital measurements. There is sometimes confusion regarding the difference between a power splitter and a power divider/combiner [13]. Figure 7.51 shows a circuit diagram of a power splitter and a power divider/combiner. Both the power splitter and the power divider allow the splitting of a signal in two identical signals, each one with half of the original power. The difference is that while on the power combiner all ports are backwards matched (i.e., any reflections that arrive on any port of the power combiner are absorbed), on the power splitter only port 1 is backward matched, meaning that any reflections that arrive on ports 2 and 3 of the power splitter will not be absorbed [28].

Sometimes it is very helpful to split a signal into more than two signals (e.g., four signals) but without resorting to cascading power splitters with the added loss. This is especially helpful for clock signals. For these cases there are other approaches to the design of power splitters that allows splitting a

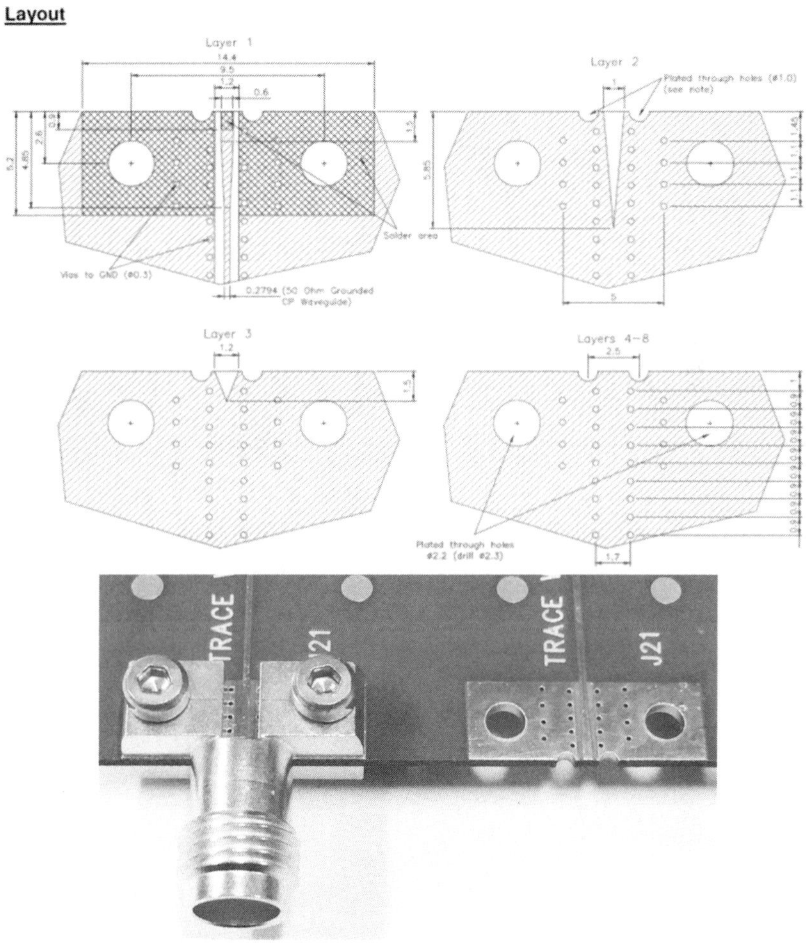

Figure 7.50 Example of an optimized footprint for a microstrip transition to an edge mounted 1.85 mm connector (Courtesy of Rosenberger).

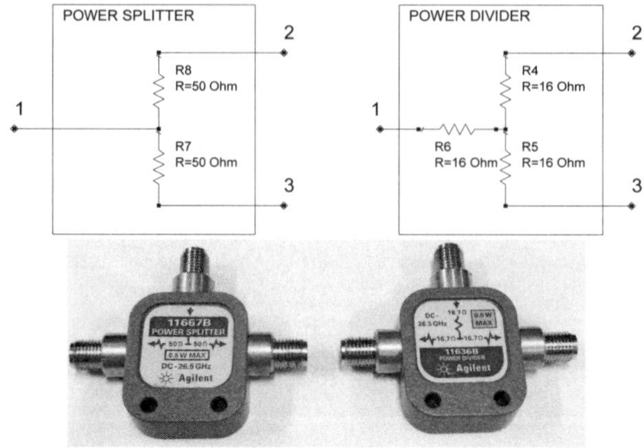

Figure 7.51 Block diagram and pictures of a resistive power splitter and a power divider/combiner.

signal into more than two signals without a major impact on the loss. The drawback is that these power splitters work only on a limited frequency band. Figure 7.52 shows an example of a one-to-six power splitter.

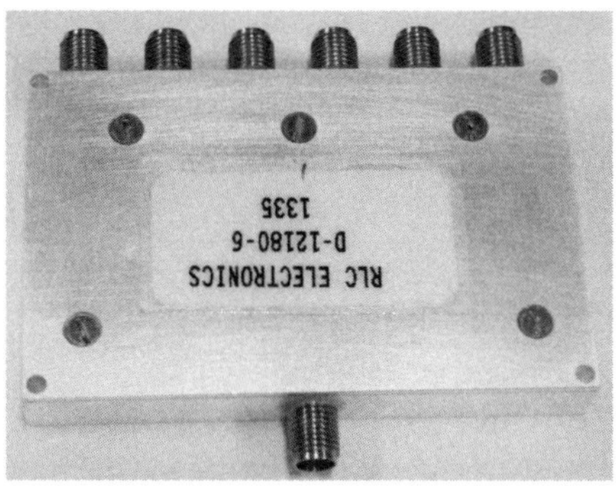

Figure 7.52 Example of a one-to-six power splitter with a frequency range of 12 to 18 GHz from RLC Electronics.

7.17.1.1 Skew Calibration of Two Oscilloscope Channels

In some applications it might be necessary to measure the skew between two signals, but usually one finds that measurement instruments already have an internal skew between the channels that is not fully calibrated. Assuming that we have a power splitter in which the skew between both paths is negligible for the application, the procedure described in Figure 7.53 can be used to deskew the measurement setup.

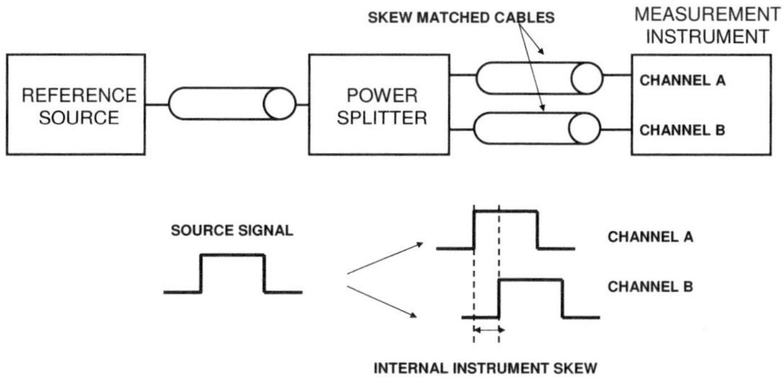

Figure 7.53 Calibrating the internal skew between two channels of a measurement instrument like an oscilloscope.

7.17.1.2 Providing a Termination Voltage When Using an Instrument with a Fixed Termination to Ground

Some measurement instruments only provide a fixed 50 Ω to ground termination that for some DUT outputs might be inappropriate, since they require a specific termination voltage. In these cases, one needs to add a small circuit to provide the 50 Ω to a specific termination voltage using power splitters and combiners before the measurement instrument input as shown in Figure 7.54 [4].

7.17.2 Attenuators, DC Blocking Capacitors, and Terminations

Attenuators, DC blocking capacitors (DC blocks), and coaxial terminations are three types of accessories that are very helpful when debugging an ATE application with bench instrumentation. Figure 7.55 shows a picture of these accessories.

The main application of DC blocking capacitors is to remove the DC component of a signal. In fact, it removes all the signal energy from DC to a certain frequency. This, in conjunction with the bandwidth of the DC blocking capacitor, defines its performance. Figure 7.56 shows the frequency response from 5 Hz to 3 GHz of two different DC blocking capacitors, showing they have different characteristics at very low frequencies.

Typical use cases are connecting a device output to an instrument with a fixed 50 Ω termination to ground or for protection of the measurement instrument (some measurement instruments inputs can be destroyed by a DC voltage above a certain threshold). When choosing a DC blocking capacitor, it is not only important to choose the DC blocking capacitor's maximum frequency according to the rise time of the signal to be measured but also

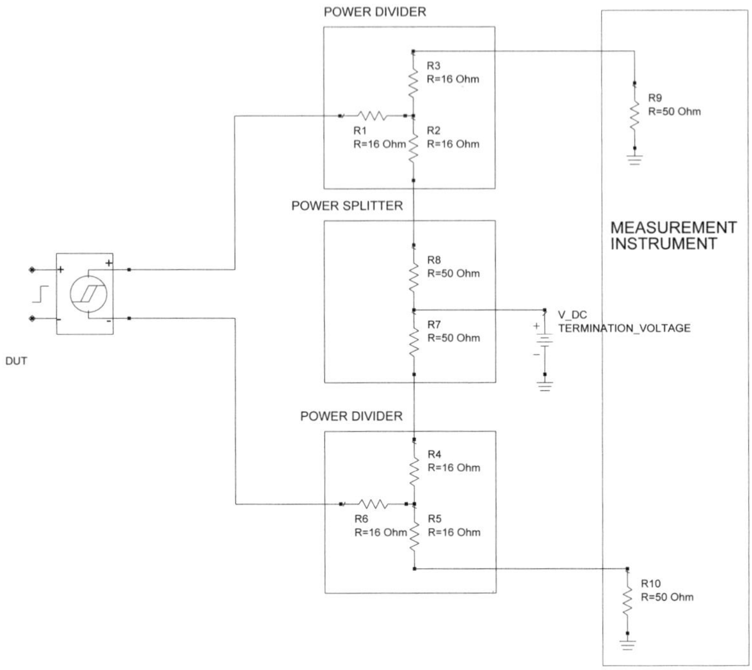

Figure 7.54 Possible circuit to provide a 50 Ω termination to a specific voltage before a measurement instrument with a fixed 50 Ω termination to ground.

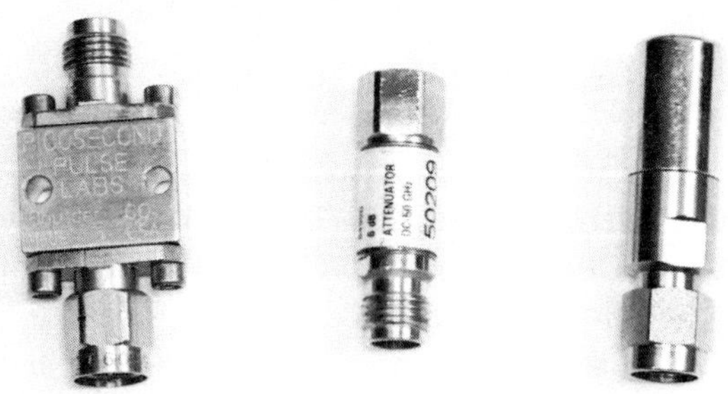

Figure 7.55 DC blocking capacitor or DC block (left), a 6 dB attenuator (center), and a 50 Ω coaxial termination (right).

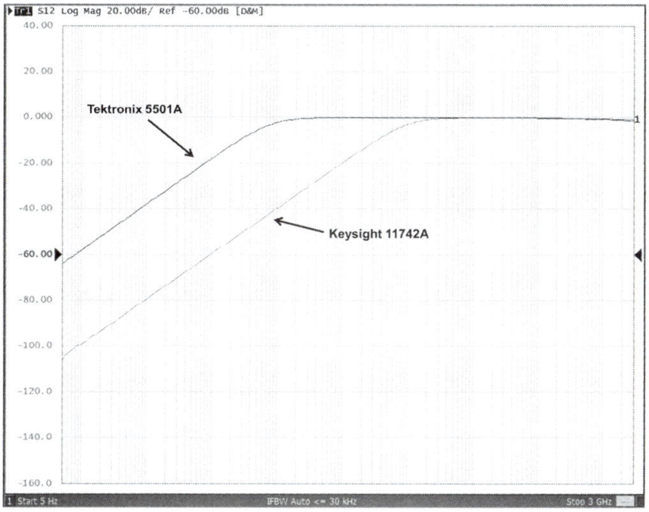

Figure 7.56 Low-frequency response of two different DC blocking capacitors (5 Hz to 3 GHz).

to choose the minimum frequency according to the pattern to be applied [29]. Table 7.3 shows an example of the lowest frequency component for a 10 Gbps signal for three different PRBS patterns.

Table 7.3
Lowest Frequency Component for a Digital Signal Running at 10 Gbps for Different PRBS Patterns

Pattern	Lowest Frequency Component
PRBS $2^7 - 1$	78.7 MHz
PRBS $2^{23} - 1$	1.2 kHz
PRBS $2^{31} - 1$	5 Hz

Attenuators as the name indicates are able to attenuate a signal. Attenuators are normally specified in decibels (dB); Table 7.4 provides a relation between dB and the attenuation factor. Typical applications are the attenuation of a signal before a measurement instrument input. The reason is that some measurement instruments have a maximum range for the signal amplitude they can measure and sometimes this range is smaller than the amplitude of the signal that is to be measured. Another application is to generate a low amplitude signal from a digital source by attenuation.

Coaxial terminations can be used to terminate nonused ports to prevent reflections. For example, when performing a measurement of a differential

Table 7.4
Voltage Attenuation in Decibels (dB) to Attenuation Factor Conversion Table

Attenuation (dB)	Attenuation Factor
3 dB	0.71
6 dB	0.5
10 dB	0.32
15 dB	0.18
20 dB	0.1
40 dB	0.01

signal using a single-ended instrument, in most situations it is necessary to properly terminate the nonused leg of the differential pair with a 50 Ω termination. Another important application is the use of the termination as a 50 Ω reference to calibrate a measurement (e.g., a TDR). It is important to note that coaxial terminations can vary significantly in performance and price and it is important to choose the correct one depending on the measurement needs.

7.17.3 Bias Network

A bias network[2] as the name indicates is an accessory that provides a bias voltage. Figure 7.57 shows a high-level diagram of a typical bias network. One critical application of a bias network is in conjunction with instruments that require AC-coupling or only provide a 50 Ω termination to ground when measuring a DUT that needs a nonzero termination voltage for proper operation.

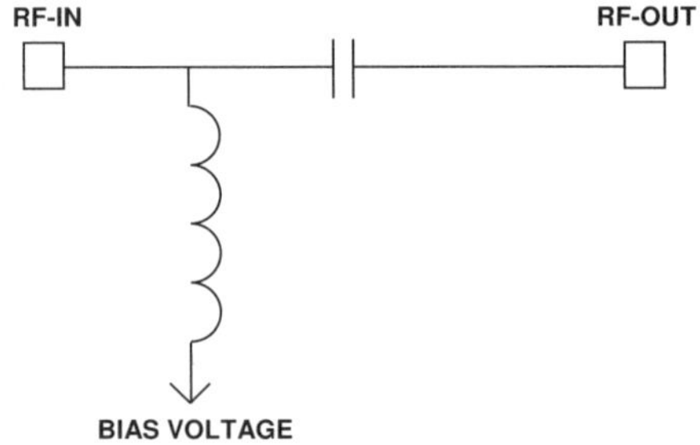

Figure 7.57 High-level diagram of a bias network.

Figure 7.58 shows a picture of a bias networks connected to an equivalent-time oscilloscope remote sampling head that only provides a 50 Ω

[2]Sometimes referred to as a Bias Tee.

termination to ground. The critical parameters of a bias network is not only its upper frequency bandwidth but like in the case of a DC blocking capacitor also its lower frequency. In the example presented in Figure 7.58, although the bias network bandwidth is 65 GHz, the lowest frequency that goes through the bias network is 50 MHz.

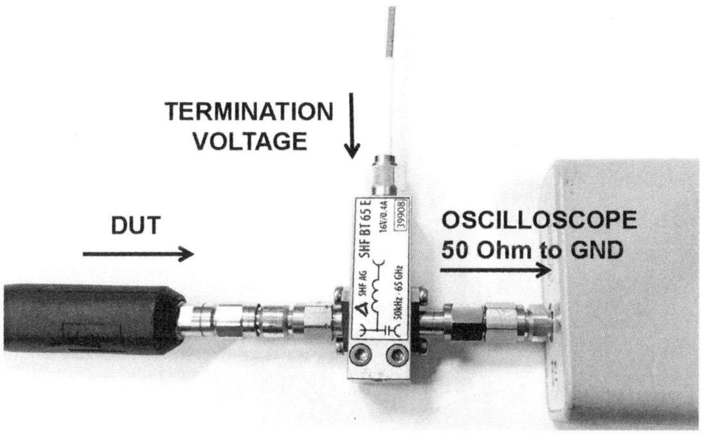

Figure 7.58 Picture of a bias network connected to an oscilloscope input.

Another application of a bias network is to provide DC access in a AC-coupled link as shown in Figure 7.59. In this case for functional testing the link is AC-coupled, but for DC testing another channel is used through the bias network coil. In some cases the link is already AC-coupled so a bias network would add a second capacitor to the signal path. In this cases another option is to use a bias network without the capacitor as shown in Figure 7.59. This kind of bias network is usually referred to as a DC feeder.

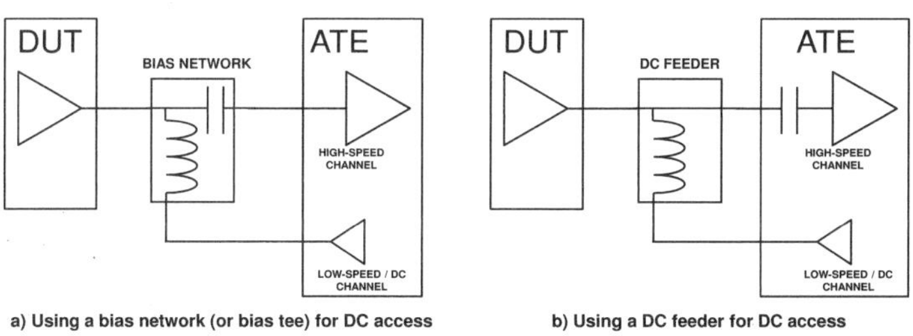

Figure 7.59 Using a bias network or a DC feeder in an AC-coupled link for DC access to the DUT I/O.

7.17.4 Pickoff Tee

The pickoff tee can be a helpful accessory to debug a high-speed digital application on an ATE system. It allows the user to probe the signal on a coaxial signal path with minimal disturbance of the signal. The drawback is that the probed signal has a significantly reduced amplitude compared to the signal that goes through the pickoff tee. Figure 7.60 shows a picture of two pickoff tees from different manufacturers and a circuit implementation example [30]. Note that the Marki Microwave pickoff tee has two equivalent pickoff ports while the Tektronix pickoff tee has only one.

As the diagram shows, one possible use of a pickoff tee is to monitor the signals between the ATE and DUT. Of course, the usage of this accessory for ATE applications requires some preconditions (i.e., the availability of appropriate connection points). Figure 7.60 also shows the insertion loss of the through path and of the pickoff path on both examples. There is approximately a 15 dB signal attenuation at the pickoff port. Figure 7.61 shows an example result with a 10 Gbps data signal for the Tektronix pickoff tee example. Although the data eye probed by the pickoff tee shows some degradation when compared to the through signal (which shows minimal degradation), it allows, for example, to easily monitor if the through data pattern is correct.

In Section H.1.3 we discuss another approach to monitor the signals between the ATE pin electronics and the DUT (and vice versa) using a PCB interposer between the ATE test fixture PCB and the DUT socket.

7.17.5 ESD/Overload Protection

Most measurement instruments are sensitive to electrostatic discharge (ESD) or overvoltage events [19] and usually also have a maximum voltage rating. Any of these events can damage the instrument measurement input stage. One way to avoid this kind of damage is to add an extra protection circuit to the instrument input. Figure 7.62 shows an example of a connectorized overload/ESD protection circuit that can be easily added to any instrument that uses a female 3.5 mm connector at its input and its performance in terms of insertion loss (S21) and return loss (S11). It also shows one example of how this circuit is able to reduce an overvoltage event.

The reason why these strong protection circuits are not included by default on the instrument input stage is that they also degrade the instrument performance. The specifics depend on the original instrument performance and the protection circuit design, but as a rule an extra protection circuit implies some performance degradation. One good use of protection circuits is in teaching classes or when engineers with limited experience are using

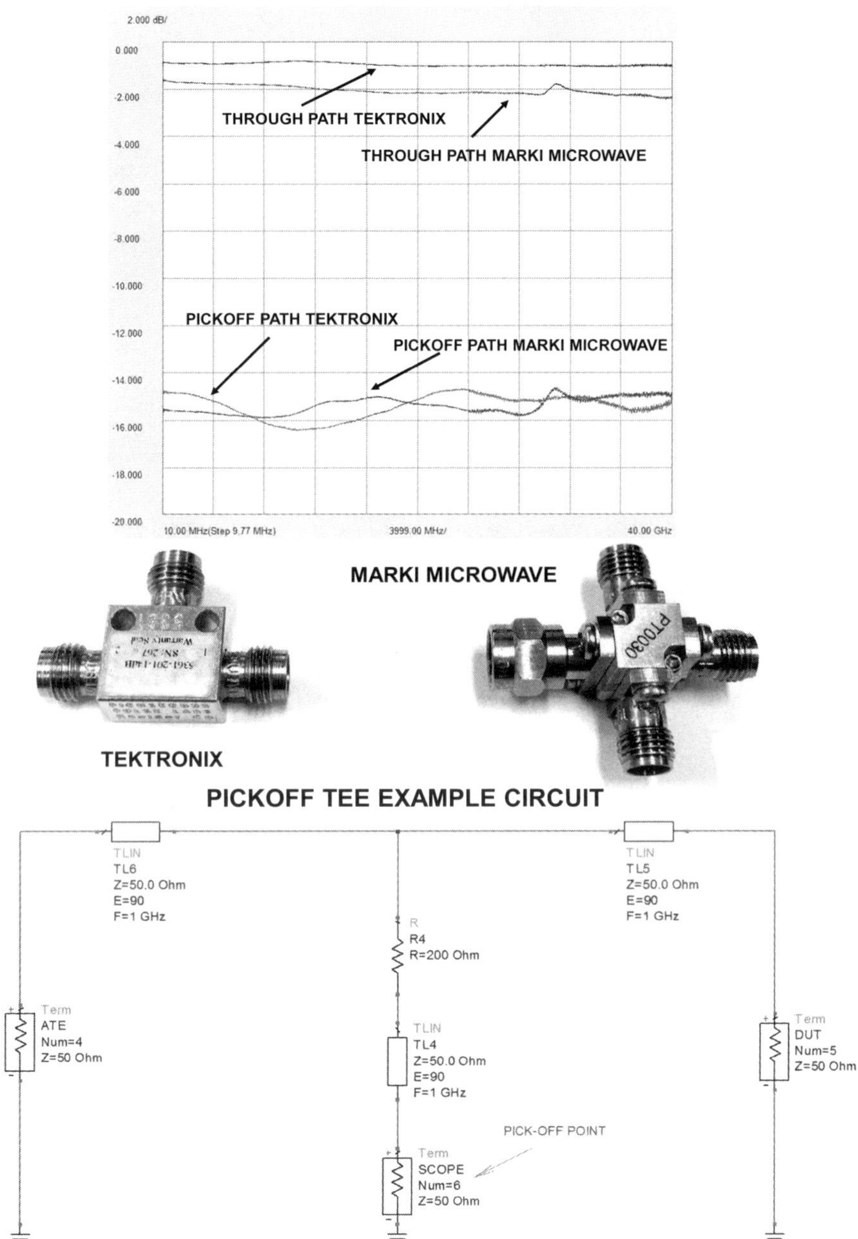

Figure 7.60 Picture of two pickoff tees from different manufacturers (Tektronix PSLP5361 and Marki Microwave PT0030), circuit implementation example and comparison of the through path and pickoff path insertion loss.

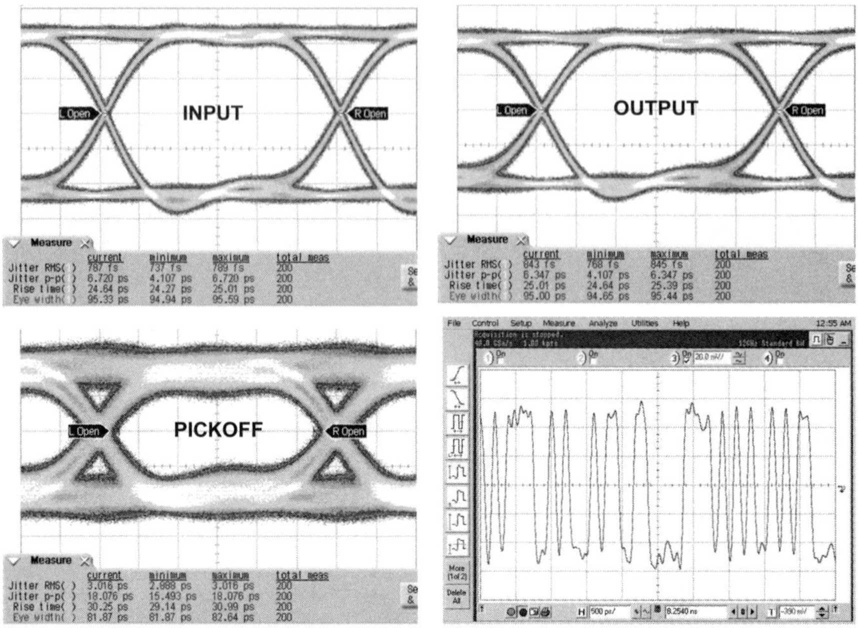

Figure 7.61 Example of the waveforms when using the pickoff tee (Tektronix PSPL5361) with 10 Gbps waveform and a PRBS $2^{31} - 1$ data pattern (top left: original signal; top right: signal after the pickoff tee; bottom left: probed signal measured with an equivalent-time oscilloscope; bottom right: probed signal measured with a real-time oscilloscope).

expensive instrumentation. The protection circuits allows the engineers to gain experience with the instrument without the fear of damaging it.

7.17.6 Delay Lines

A delay line is another accessory that might be helpful in some situations [13]. The type of delay lines in which we are mostly interested are of the passive type, also known as trombone delay lines. Examples for these types of delay line are shown in Figure 7.63. Typical usage for this accessory is during the validation of a measurement setup with external instrumentation. In some situations it is necessary to skew a defined signal relative to others and this type of accessory provides a simple solution. It also provides the high bandwidth that is needed for high-speed digital signals.

Another type of delay line is shown in Figure 7.64. This is an active delay line where the delay is determined by the voltage applied on the delay line voltage input. Although the usage of a voltage to control the delay value is very useful, this type of device is usually limited to a certain frequency band

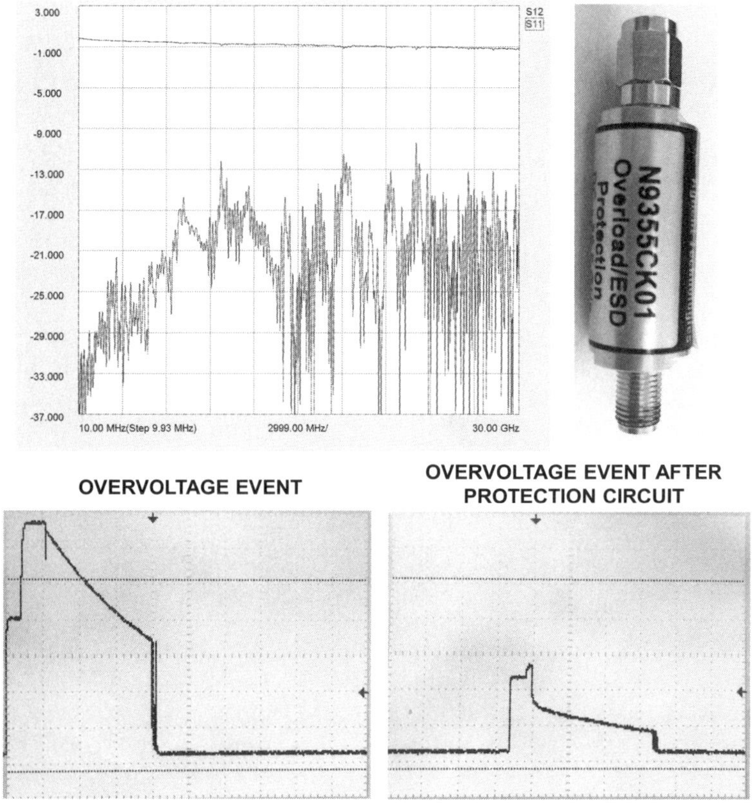

Figure 7.62 Example of a overload/ESD protection circuit in a coaxial package from Keysight Technologies.

Figure 7.63 Passive delay line examples. The delay line is adjusted by mechanically changing its length.

that limits its usage to narrow band signals like clocks. The delay lines shown in Figure 7.63 do not have this drawback and can be used for data signals.

Figure 7.64 Voltage controlled delay line example (Hittite HMC-C010).

7.17.7 Filters

Filters are also sometimes important tools on high-speed digital applications. For example, a band-pass filter might be needed to filter out unwanted harmonics in a clock signal. There are several excellent textbooks dedicated to filters and filter design (e.g., [31]). In this section we will just quickly review filter options from a test engineer perspective, where usually one either purchases the desired filters or has them custom designed and manufactured. Figure 7.65 shows some filter examples using commonly available technologies in a connectorized format. Note that the filters presented in Figure 7.65 are just a subset of the available technologies since there are others, such as tubular filters and notch filters.

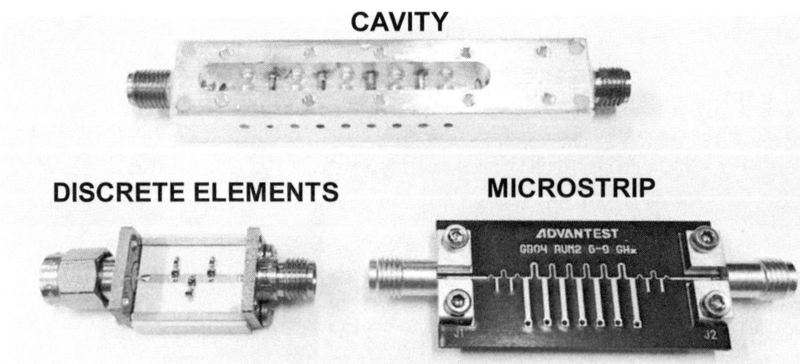

Figure 7.65 Examples of some filter technologies.

7.17.7.1 Rise Time Filters

Another type of filter commonly used on high-speed digital applications is the rise time filter or transition time converter (TTC). Typically the rise time of a stimulus source like a pattern generator is fixed to the best value the instrument manufacturer is able to achieve since this is considered an important specification (i.e., the smaller the rise time, the better the instrument driver performance). In some situations, the fast rise time of an instrument might present a problem to test a certain application. In this case the user needs to slow down the rise time of the instrument without a major impact on the other driver parameters like jitter. This can be achieved by using a rise time filter as shown in Figure 7.66. A rise time filter is basically a low-pass filter but it needs to have a constant group delay across the entire frequency range of interest for the high-speed digital. This is needed because high-speed digital signals are broadband by definition and frequency-dependent changes on the group delay are translated into added jitter. This limits the types of filter designs that can be used (e.g., Bessel filter), resulting in a cutoff frequency that is not as sharp as on a typical low-pass filter as shown in Figure 7.67. This is why standard low-pass filters should not be used for reducing the rise time of a driver as shown in Figure 7.68.

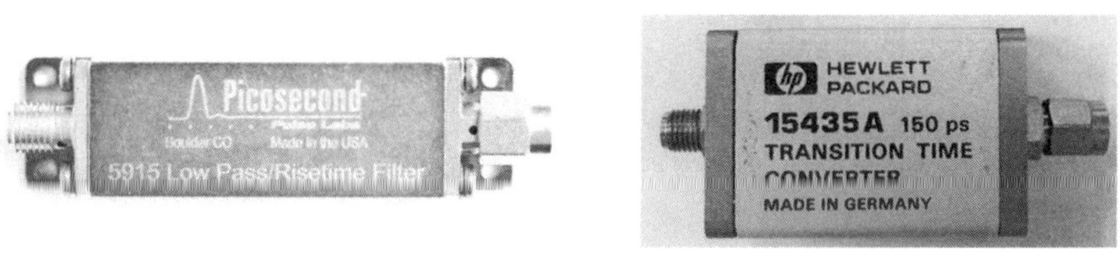

Figure 7.66 Examples of rise time filters (also referred to as transition time converters).

A rise time filter is designed to provide a specific rise time, assuming that the stimulus source has a faster rise time. The user needs to choose the correct rise time filter for the application. Please also note that rise time filters are typically specified with 10%–90% rise times. To obtain the 20%–80% rise time, the following formula is usually used assuming a linear rise time in the 10%–90% region:

$$T_R(10\% - 90\%) \times 0.75 = T_R(20\% - 80\%) \qquad (7.2)$$

7.17.7.2 Equalizers

Another type of filter that is of critical importance for high-speed digital applications is the continuous time linear equalization filter (CTLE) [32]. A

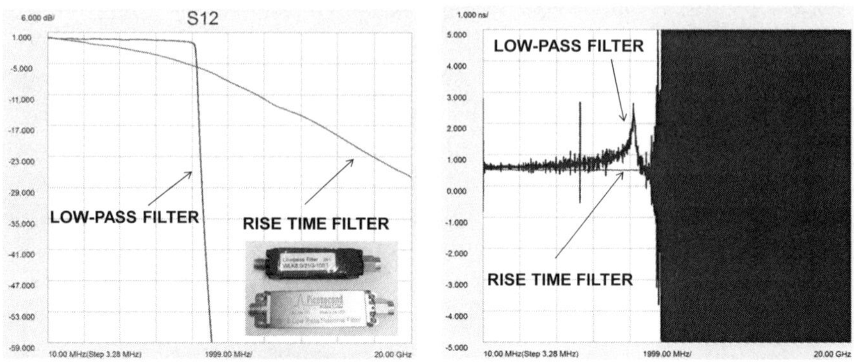

Figure 7.67 Comparison of a rise time filter with a typical low-pass filter (left: insertion loss, right: group delay).

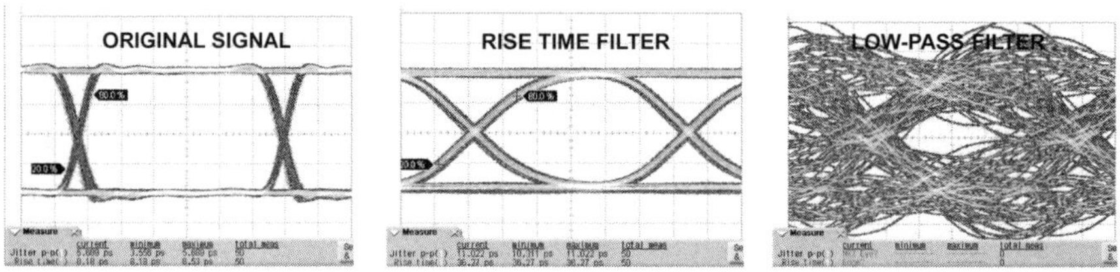

Figure 7.68 Example of using a rise time converter versus a standard low-pass filter for a stimulus source with 8 ps (20/80) rise time running at a data rate of 10 Gbps using a PRBS7 data pattern.

continuous time linear equalization filter in the context of high-speed digital applications is intended to equalize (or flatten) the frequency response of a system (usually in our context it is the test fixture). This is achieved by using filter that has the inverse frequency response of the test fixture [33].

Figure 7.69 shows one example of an equalization filter in a coaxial package and its corresponding insertion loss. Section 9.9.3 discusses equalization in detail for compensation of the test fixture loss.

7.17.8 Probes

Measurement probes are probably the most common instrumentation accessory among test engineers. There is a large amount of documentation material available on bench instrumentation probes. References [3, 4, 34] are good starting points. The Web sites of probe and bench instrumentation manufacturers are also good sources for information. In this section we will only provide a high-level overview on measurement probes for high-speed

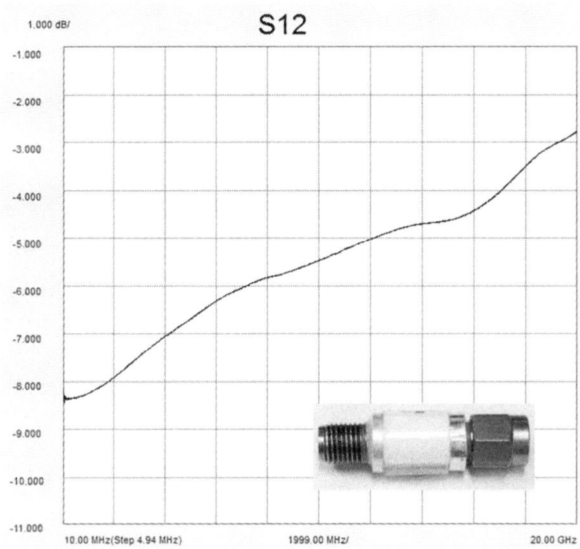

Figure 7.69 Example of a continuous time linear equalization filter in a coaxial package and its insertion loss.

digital signals. Measurement probes can be divided into two major families: passive probes and active proves.

7.17.8.1 Passive Probes

Figure 7.70 shows a picture of different commercially available passive probes. The name "passive probe" indicates that there are no active elements inside (like transistors). For high-speed digital signals we are interested in a probe that provides a connection to the measurement point with minimum amount of distortion. This means for the typical high-speed application where 50 Ω PCB signal traces are used, the probe design should create the minimum possible discontinuity at the probing point. The large passive probe on Figure 7.70 (left) connects to a coaxial cable and provides a signal and a ground tip to attach to the ground reference with a variable pitch. This type of probe provides significant flexibility but at the cost of signal performance. The critical geometry factor in a probe is the loop size between the signal and ground pins, and in this probe the flexibility to address any signal to ground pitches (e.g., the distance between the signal BGA pad and the closest ground BGA pad) creates a large loop reducing the probe bandwidth. However, the probe in Figure 7.70 (center) is also passive, can be hand-used, and provides a much higher frequency at the expense of a smaller pitch and less flexibility.

Finally the coplanar microcoaxial probe in Figure 7.70 (right) is also passive but the signal to ground pitch is fixed to a value of 400 micrometers in this example. This allows the highest bandwidth of the three probe examples but the least amount of flexibility.

Figure 7.70 Passive probe example. Left: 50 Ω low-frequency single-ended resistive divider passive probe (Keysight Technologies 54006A); center: high-frequency differential passive probe (Gigaprobe from DVT Solutions); right: microcoaxial 50 Ω passive coplanar probe (Picoprobe from GGB Industries).

It is also possible to construct a passive 50 Ω probe from a coaxial cable as shown in Figure 7.71. This type of self-made probe can be very useful and provides a high bandwidth if the signal ground leads are kept very short.

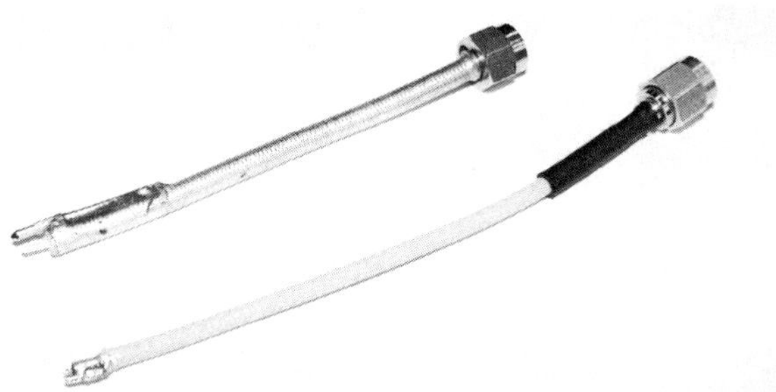

Figure 7.71 Examples of hand-made 50 Ω passive probes from a coaxial cable.

7.17.8.2 Active Probes

Figure 7.72 shows a picture of different active probes. An active probe tries to solve the challenge of how to probe a signal in situations where one does not want to disturb the probed signal (e.g., when probing a test point on a microstrip between the ATE and the DUT). Because of this, active probes are more sophisticated and contain electronic elements like a transistor. For active probes, the same concerns regarding the loop size between the signal and ground pins apply as with passive probes. With differential signaling becoming more prevalent, there are also differential active probes available on the market.

Figure 7.72 Examples of active probes. Left: differential active probe from Keysight Technologies; right: single-ended active probe (Courtesy of GGB Industries).

Although an active probe is usually applied in the context of ATE applications to monitor signals between the ATE pin electronics and the DUT, in some situations an active probe is needed to measure the performance of the ATE system at the DUT socket. One example is to access the performance of the ATE system for an application where the DUT receiver input is a high impedance type input (e.g., a DDR address input). One approach to perform this measurement is to measure the performance at the DUT socket using an interposer and an active high impedance probe (interposers are discussed in detail in Appendix H). Figure 7.73 shows an example of this measurement. In this case a capacitor was also soldered between the DUT receiver input and ground pads to simulate the input capacitance of the DUT receiver since this can have a significant influence on the waveform shape [35].

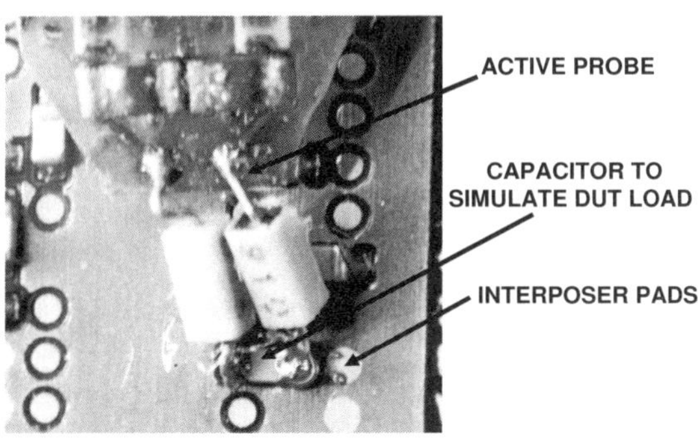

Figure 7.73 Probing a single-ended signal with a high impedance active probe (note that a capacitor was soldered to simulate the input capacitance at the DUT).

7.17.8.3 Probe Positioners

Although it is possible to use a probe manually to measure a signal, in some situations the test engineer needs the probe to be continuously connected to the probing point or the probe does not allow a simple manual positioning approach. In these cases a probe positioner like the ones shown in Figure 7.74 can be used. Another possible option is to use solderable probes, although this might not be possible in some cases because it damages the measurement point (e.g., the socket pads).

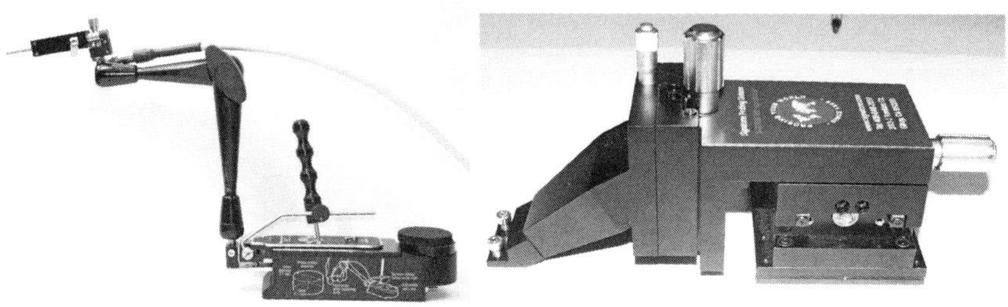

Figure 7.74 Examples of probe positioners (left: adjustable probe arm; right: micro-probe positioner with 3D adjustment).

7.17.9 Balun

Sometimes test engineers face the challenge of measuring a differential signal with single-ended measurement instrumentation or providing a differential stimulus signal with a single-ended source. A typical approach is to terminate one side of the differential pair and measure the other pair or use one single-ended instrument for each leg of the differential pair. Another option is to use a balun (BALanced UNbalanced transformer) [36]. Figure 7.75 shows a typical high-level schematic of a balun. Reference [37] describes a balun for high-speed digital applications. In Figure 7.75 the turns ratio of the transformer is one since it is assumed that both sides of the transformer have the same impedance. The input is the unbalanced (single-ended) input with one terminal of the transformer connected to ground. The output (secondary winding) is not connected to ground and in this way it is "floating" with respect to ground. This is a very simple example of balun construction and more complex designs are available for high-speed digital applications. Figure 7.76 shows a photograph of a ultrawideband balun. Another interesting application of a balun in the context of high-speed applications is to perform differential measurements with a standard two-port VNA (e.g., differential return loss) by using a pair of baluns as described in [38].

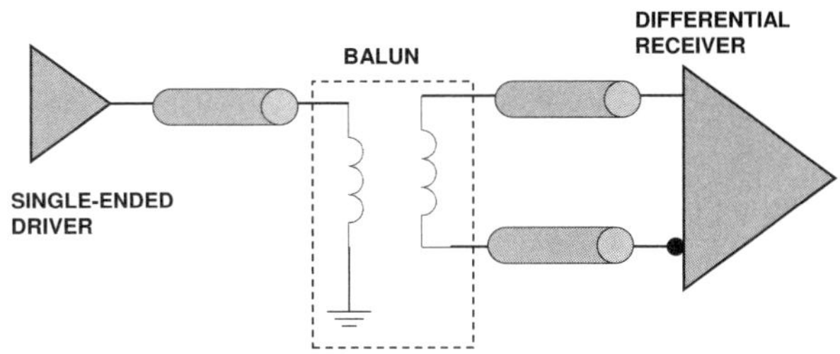

Figure 7.75 Typical high-level schematic of a balun.

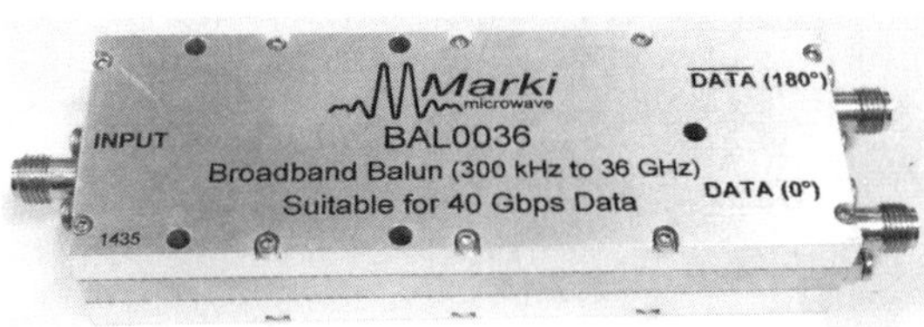

Figure 7.76 Picture of a wideband balun from Marki Microwave.

7.17.10 Frequency Doublers

In some situations is might be required to double or quadruple the frequency of a clock. This is possible by using a nonlinear passive device called frequency doubler [13, 39] together with an appropriate bandpass filter and a low phase noise amplifier. This can be very useful in some applications like the one shown in Figure 7.77 where using an ATE system digital channels with a maximum data rate of 12.8 Gbps, the user would like to develop a solution for delivering 25 Gbps using a retimed muxing solution on the test fixture with a 4 to 1 mux. The mux requires a 12.5 GHz clock (25 Gbps bit clock) to retime the data since it uses both edges of the latching clock. This would not be possible with this ATE system unless we can double the frequency of the ATE clock signal used for latching the data. This can be achieved with a frequency doubler [40, 41].

The block diagram in Figure 7.77 shows an amplifier before the frequency doubler. This is required to increase the amplitude of the signal from the ATE digital channel due to the conversion loss of the frequency doubler (e.g., 10 dB). The output from the frequency doubler will include not only the doubled frequency but also the corresponding harmonics which

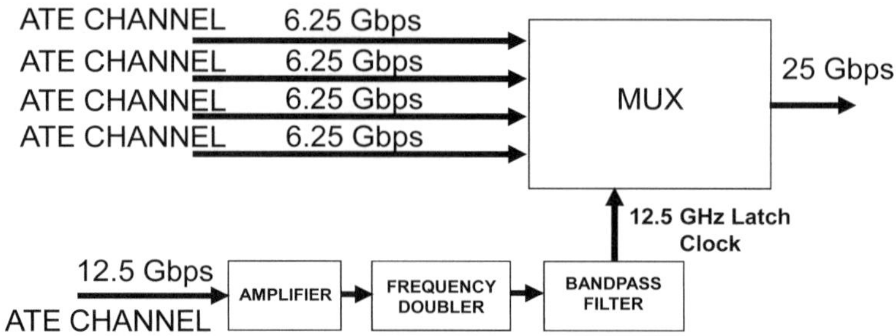

Figure 7.77 Using a 4:1 mux together with a frequency doubler to reach 25 Gbps on an ATE system with a maximum data rate of 12.8 Gbps.

degrade the input signal as shown in Figure 7.78 where a 6.25 GHz bit clock from a digital channel is doubled to 12.5 GHz. To address this problem an appropriate bandpass filter should be used after the clock doubler to remove these harmonics.

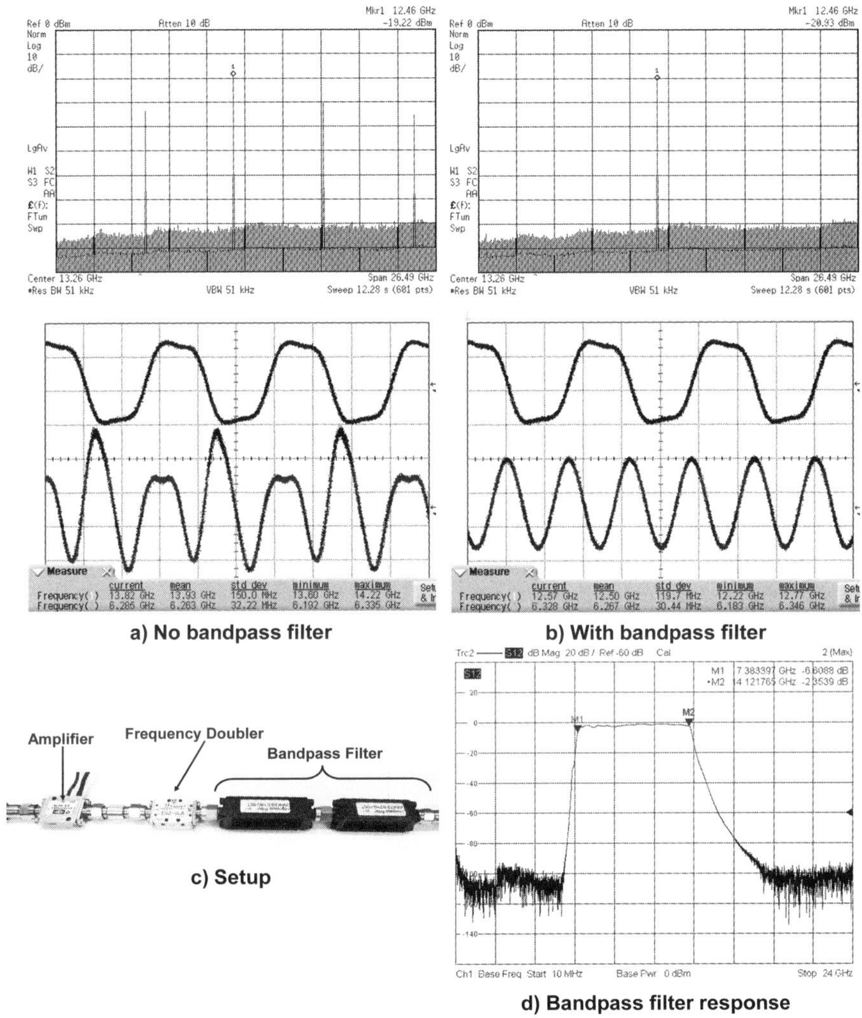

Figure 7.78 (a—d) Output from the clock doubler without any bandpass filter (a) and with bandpass filter (b). The input frequency was 6.25 GHz resulting in an output frequency of 12.5 GHz.

7.17.11 Frequency Dividers

Frequency dividers are active devices that can be used to divide the frequency of a clock signal [42]. Like the frequency doubler, this component can be very useful in some high-speed digital applications. Figure 7.79 shows an example of a frequency divider in operation.

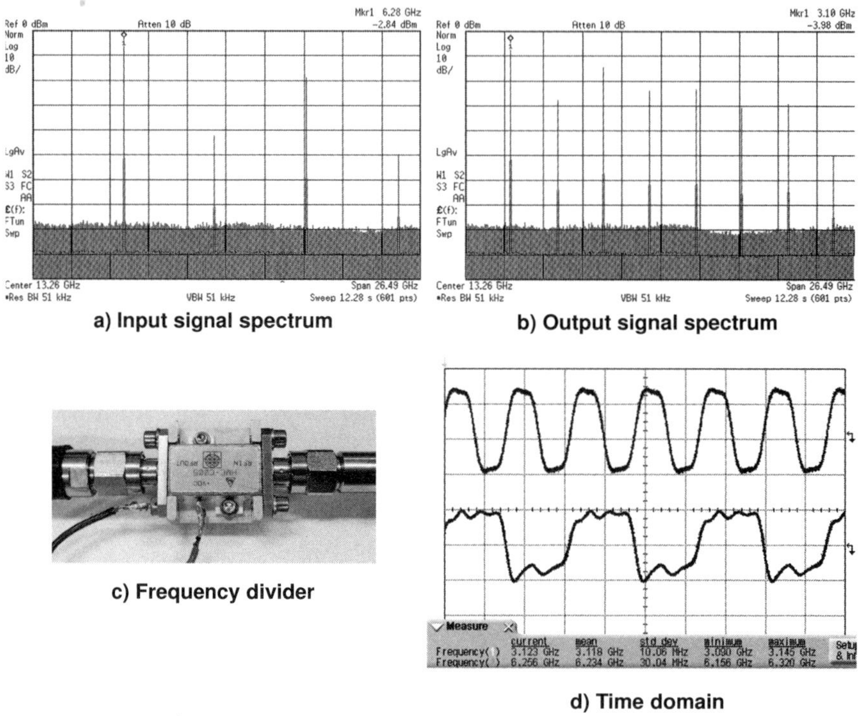

Figure 7.79 (a—d) Example of a frequency divider in operation.

References

[1] N. Kularatna, *Digital and Analog Instrumentation: Testing and Measurements.* New York: The Institution of Electrical Engineers, 2003.

[2] D. Derickson and M. Mueller, *Digital Communications Test and Measurement: High-Speed Physical Layer Characterization.* Upper Saddle River, NJ: Prentice-Hall, 2007.

[3] D. I. Geoff Lawday and G. Edlund, *A Signal Integrity Engineer's Companion.* Upper Saddle River, NJ: Prentice-Hall, 2008.

[4] W. Maichen, *Digital Timing Measurements.* New York: Springer, 2006.

[5] Agilent Technologies, "Fundamentals of Time Interval Measurements," *Application Note 200-3*, 1997.

[6] T. Lyons and E. Sang, "Zero Re-Arm Time Measurements," *Test & Measurement World*, Dec. 2010.

[7] R. A. White, *Spectrum & Network Measurements, 2nd Edition*. Scitech Publishing, 2014.

[8] C. Rauscher, *Fundamentals of Spectrum Analysis*. Rhode & Schwarz, 2004.

[9] M. Hiebel, *Fundamentals of Vector Network Analysis*. Rhode & Schwarz, 2007.

[10] S. M. Sandler, *Power Intergrity: Measuring, Optimizing and Troubleshooting Power Related Parameters in Electronic Systems*. New York: McGraw Hill, 2014.

[11] R. Stephens and R. Muro, "Characterization of Gaussian Noise Sources," *IEC DesignCon*, 2008.

[12] Tektronix, "Controlled Jitter Generation for Jitter Tolerance and Jitter Transfer Testing," *Application Note*, 2005.

[13] L. A. Belov, S. M. Smolskiy, and V. N. Kochemasov, *Handbook of RF, Microwave and Milimeter-Wave Components*. Norwood, MA: Artech House, 2012.

[14] H. W. Ott, *Electromagnetic Compatibility Engineering*. New York: John Wiley & Sons, 2009.

[15] C. Tursky, R. Gordon, and S. Cowie, *Test System Design*. Upper Saddle River, NJ: Prentice-Hall, 2001.

[16] Toroid Corporation. http://www.toroid.com/.

[17] J. Moreira, B. Sadler, and D. Ferguson, "Parametric Testing of a 10Gbs I/O Cell in Production Through a Parametric Loopback Approach with Retiming," *IEEE International Design and Test Workshop*, 2006.

[18] J. Moreira, G. Haensel, and F. Koban, "Addressing the Challenges of Implementing an At-Speed Production Test-Cell for 10Gb/s Wafer Probing," *DesignCon*, 2005.

[19] S. H. Voldman, *ESD Basics: From Semiconductor Manufacturing to Product Use*. New York: John Wiley & Sons, 2012.

[20] E. B. Joffe and K.-S. Lock, *Grounds for Grounding: A Circuit-to-System Handbook*. New York: John Wiley & Sons, 2010.

[21] G. Haensel and K. Stieglbauer, "Optimized Method for Pattern Matching of High Speed Devices on ATE," *IEEE International Test Conference*, 2005.

[22] D. Vye, "2012 RF Connector and Cable Assembly Outlook," *Cables and Connector Supplement of the Microwave Journal*, Mar. 2012.

[23] R. Collier and D. Skinner, *Microwave Measurements*. Institution of Engineering and Technology, 2007.

[24] P. Pino, "The Important of Knowing your Cable Constrains," *Cables and Connector Supplement of the Microwave Journal*, Mar. 2012.

[25] SHF Communication Technologies AG, "Microwave Connectors," *Tutorial #2*, 2009.

[26] Keysight Technologies, "Principles of Microwave Connector Care (For Higher Reliability and Better Measurements)," *HP Application Note 326*, 1996.

[27] E. Holzman, *Essentials of RF and Microwave Grounding*. Norwood, MA: Artech House, 2006.

[28] R. A. Johnson, "Understanding Microwave Power Splitters," *Microwave Journal*, Dec. 1975.

[29] G. Foster, "D.C. Blocks, a Trap for the Unwary When Using Long Patterns," *Synthesis Research SR-TN088 Measurement Brief*, 2009.

[30] Picosecond Pulse Labs, "Model 5361 5x Pickoff Tee," *Product Specification*, 2009.

[31] G. Bianchi and R. Sorrentino, *Electronic Filter Simulation and Design*. New York: McGraw-Hill, 2007.

[32] W. Humann, "Compensation of Transmission Line Loss for Gbit/s Test on ATEs," *IEEE International Test Conference*, 2002.

[33] B. Roth and J. Moreira, "Equalization in the HSM6800 Pin-Electronics," *Verigy VOICE Users Conference*, 2010.

[34] A. Heltborg, I. Pollock, and R. V. Epps, "High-Speed Probe InterConnect Techniques," *IEC DesignCon*, 2008.

[35] M. Moessinger, "Probing Inside the Socket," *Burn-In and Test Strategies (BITS) Workshop*, 2011.

[36] Doug Jorgesen and Christopher Mark, "Balun Basics Primer: A Tutorial on Baluns, Balun Transformers, Magic-Ts, and 180 °C Hybrids," *Marki Microwave*, 2014.

[37] J. R. Andrews, "Ultra-Wideband Differential Measurements Using PSPL Baluns," *Picosecond Pulse Labs Application Note AN-8*, 1999.

[38] J. R. Andrews, "Differential VNA Measurements Using Single-Ended, Two-Port Instruments and BALUNs," *Picosecond Labs Application Note AN-21*, 2008.

[39] S. A. Maas, *Nonlinear Microwave and RF Circuits 2nd Edition*. Norwood, MA: Artech House, 2003.

[40] J. Moreira, "Using Frequency Doublers in High-Speed Digital Applications," *Advantest GoSemi Newsletter*, Dec. 2012.

[41] J. Moreira, B. Roth, and C. McCowan, "An Active Test Fixture Approach for Testing 28 Gbps Applications Using a Lower Data Rate ATE System," *IEEE Asian Test Symposium*, 2012.

[42] U. Alvarado, G. Bistue, and I. Adin, *Lower Power RF Circuit Design in Standard CMOS Technology*. New York: Springer, 2011.

8

Test Fixture Design[1]

The test fixture interface between the ATE system and the DUT, also known as DUT board, loadboard, device interface board (DIB), or device interface (DI), is a critical element in the performance of high-speed I/O characterization and production testing. In the past, books dealing with ATE testing (e.g., [1]) spent a significant amount of time on transmission line termination issues and techniques. This is no longer the major topic of concern since modern I/O interfaces and ATE pin electronics incorporate transmission line terminations on the DUT and the ATE pin electronics. The main challenges with multigigabit data rates are the transmission line losses across a test fixture and the effects of discontinuities like vias, relays, and the DUT socket. Several excellent books exist on printed circuit board[2] (PCB) design for high-speed digital applications (e.g., [2–7]). For those interested in the PCB manufacturing process itself and its challenges, [8–11] are a good starting point.

The ATE engineer must keep in mind that most books on high-speed signal integrity are written for a generic audience that includes engineers working on high-volume consumer applications where minimizing PCB layer counts and routing distances are critical for reducing manufacturing costs. A different set of trade-offs on the ATE test fixture can result in a completely different physical topology from that of the DUT application board. These differences can be seen in Figure 8.1 where a high-volume desktop computer motherboard is compared with an ATE test fixture. The cost-sensitive consumer motherboard application has a high density of components and short routing distances as compared with the performance-driven ATE test fixture

[1]This chapter was written in collaboration with Heidi Barnes.

[2]Sometimes the name printed wiring board (PWB) is also used.

with only a single IC (the DUT) and long routing distances to the ATE pin electronics.

Rules of thumb or design guidelines are always made for a specific application and must be reviewed for accuracy when leveraged to a new design. When discussing high-speed PCB design guidelines with a colleague from a different area, it is important not to assume that his requirements and constraints are the same as the ones for an ATE test fixture. Remember that a test engineer's objective is to measure as accurately and cost-effectively as possible the performance of the DUT while conforming to the ATE interface. However, designers of a multicomponent PCB for an end user system, only need to guarantee functionality with the lowest cost and smallest form factor.

(a) (b)

Figure 8.1 Comparison of a (a) high-speed digital test fixture PCB design compared with (b) traditional PCB design.

The signal path on an ATE system is composed of a chain of four basic parts as shown in Figure 8.2. Starting with the ATE pin electronics, the signal travels through the ATE to test fixture interconnect (e.g., the pogo pin assembly) to the PCB test fixture and ends at the DUT socket. A simple DC continuity check to validate the netlist of the ATE to DUT connections along this signal path is no longer sufficient. The signal integrity of each of these parts is important and any one of them can break the high-speed digital connection.

The ATE pin electronics card and the ATE to test fixture interconnect are the responsibility of the ATE manufacturer. The situation is different for the test fixture and DUT socket performance where the ATE user must define the requirements for interfacing with their product. The ATE user will typically design this interface with the expectation that the design will work the first time and that no prototyping is needed. This requires close collaboration with the ATE manufacturer, the test fixture layout and fabricators, and socket manufacturers to leverage past experience and guarantee that the test fixture and socket have the required performance. The next sections of this chapter

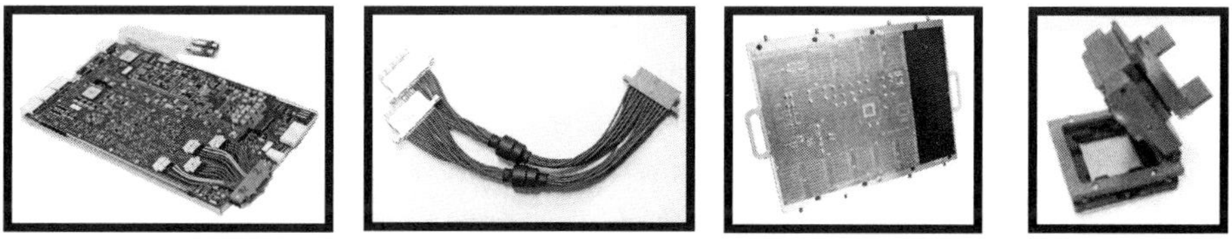

Figure 8.2 The four parts of the ATE signal integrity chain [the ATE pin electronics card, the ATE to test fixture interconnect (e.g., pogo pin assembly), the test fixture, and the DUT socket].

will present some of the important topics associated with the design of test fixtures for high-speed digital applications.

8.1 Test Fixtures

Test fixtures can be significantly different depending on the ATE system model, manufacturer, and the intended application (RF, high-speed digital, wafer test, package test). Figure 8.3 shows an example of two high-speed digital application test fixtures from two different ATE platforms. One of the test fixtures demonstrates a single socket interface, which is typical for higher performance characterization and design verification, while the other test fixture utilizes four sockets to increase the throughput and lower the production test cost of a large digital device such as a microprocessor. Each of these design examples depends upon the ability of a multilayer PCB to route the interconnections between the ATE system and the DUT. The larger size and high layer count of this type of ATE test fixture is not common within the PCB industry and often requires highly specialized PCB fabricators as well as specialized design techniques, which will be addressed throughout this chapter.

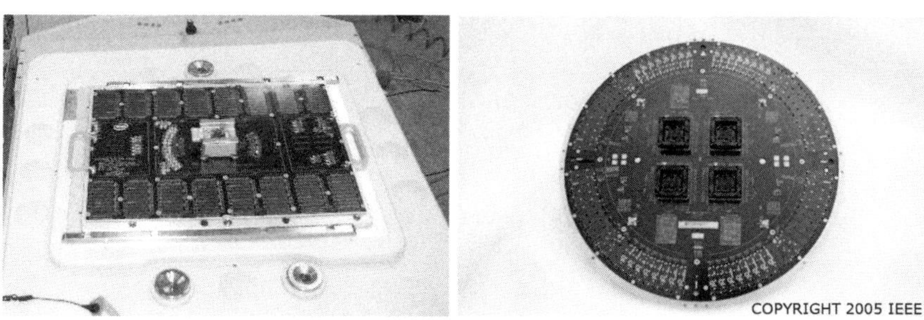

Figure 8.3 Examples of ATE test fixtures for high-speed digital applications (left: Advantest V93000; right: Advantest T2000 [12]).

Figure 8.4 shows a single-site test fixture with a mechanical plate on top for docking to a handler for production testing [13]. Figure 8.5 shows a multi-site test fixture PCB with 16 sites. Figure 8.6 shows another example of an ATE test fixture for IC characterization designed for usage on an ATE system and also on a pure bench measurement setup [14].

Figure 8.4 Example of an ATE volume production test fixture with the handler guiding plate on top (reprinted with permission from [13]).

The ATE test fixture also includes a mechanical metal frame to simplify the installation and removal (also known as docking and undocking) of a test fixture from an ATE system. This mechanical frame is often called a stiffener since it ensures that the PCB is flat and mechanically supported to handle the pressure from hundreds or even thousands of connecting pins on the ATE side and the insertion force of a DUT by an automated handler on the other side. Figure 8.7 shows an example of two types of these mechanical frames. The mechanical frame creates constraints on the test fixture PCB design in terms of mounting holes, outer layer clearances, and board thicknesses and will vary according to the ATE manufacturer and system model.

Figure 8.5 Example of an ATE volume production multi-site test fixture PCB with 16 sites (Courtesy of Advantest).

Figure 8.6 Pictures of a single test fixture designed for dual usage in the bench and ATE (reprinted with permission from [14]).

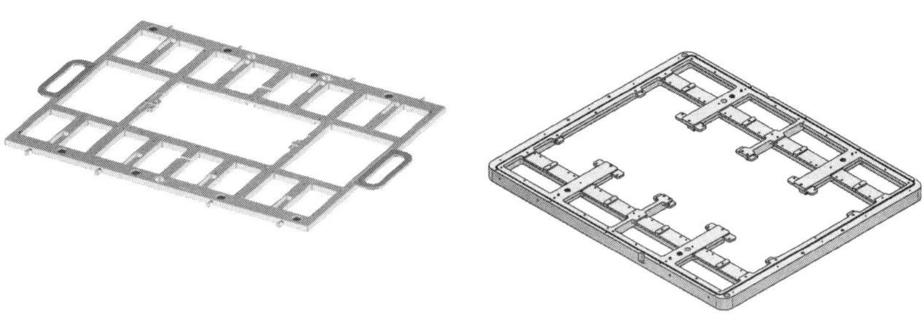

Figure 8.7 Pictures of a mechanical frames (also known as stiffener) for test fixture PCBs in the Advantest V93000 platform (left) and the Advantest T2000 platform (right) (Courtesy of Advantest).

8.1.1 Test Fixture to ATE Interconnect

There are two major approaches for ATE to test fixture PCB interconnect for high-speed digital applications: pogo pin and coaxial connector. Figure 8.8 shows an example of the two options implemented on the bottom side of an ATE test fixture. In Chapter 4 the two approaches are shown from the ATE side. In both approaches it is critical to use optimized PCB footprints as will be shown in Section 8.6 for coaxial connectors and in Section 8.5.2 for pogo pin connectors. Each approach has its disadvantages and advantages, and in most cases backwards compatibility reasons require that an ATE platform stays with only one of the approaches in most cases.

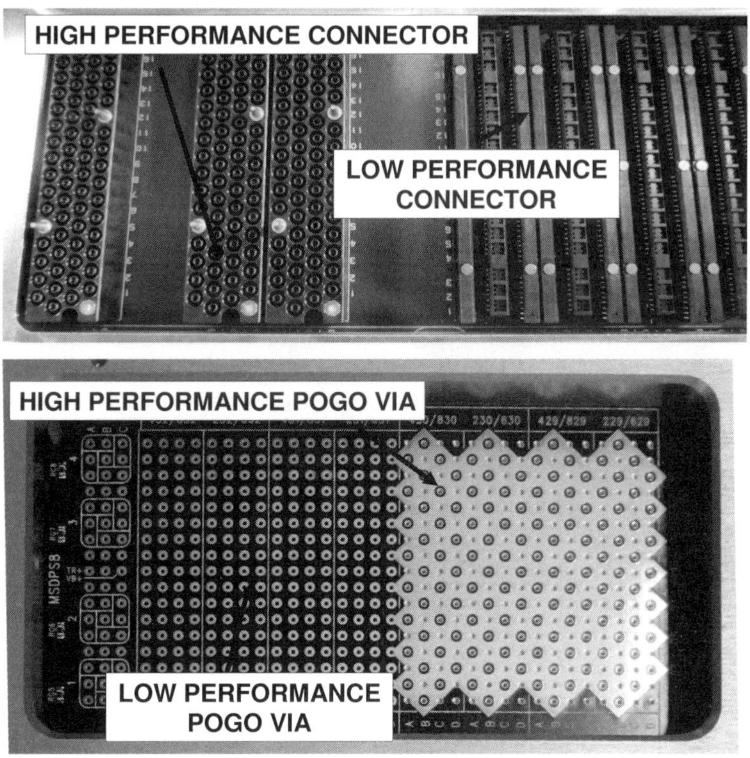

Figure 8.8 The two main approached for ATE to test fixture interconnect as seen from the test fixture bottom side (top: coaxial connector, bottom: pogo pin via).

8.2 High-Speed Design Effects

One of the difficulties in high-speed design is in understanding the multiple effects causing the degradation of a signal as the data rate increases and the rise time decreases. Signal degradation can be caused by impedance mismatches, material losses, conductor losses, radiation, and proximity effects as shown in Figure 8.9. One of the most confusing aspects of high-speed design is that the design rule trade-offs for minimizing these sources of signal degradation can be different depending upon the application and the PCB fabrication details. Since these effects are frequency-dependent, it is important to look at each one and understand when they start to degrade a high-speed I/O signal.

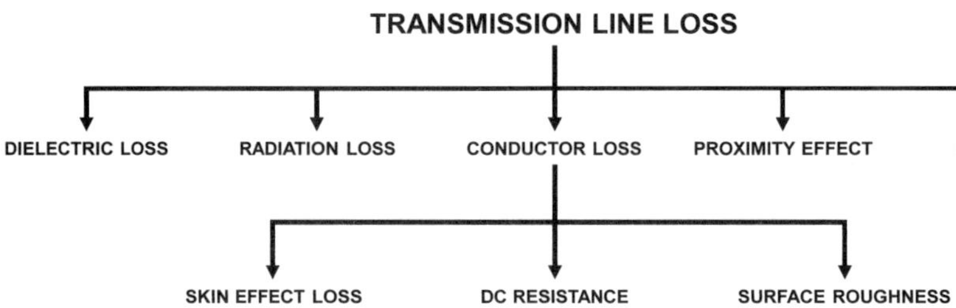

Figure 8.9 Loss factors for a printed circuit board signal trace.

8.2.1 Reflections Due to Impedance Mismatches

The problem of reflections due to impedance mismatches was first addressed by Oliver Heaviside in the late 1800s to describe the propagation of telegraph communication signals over long distances. He discovered that by breaking up the signal path into discrete sections with the incremental values of series inductance and parallel capacitance to ground and low resistive and conductive losses, one could describe the transmission line as having a characteristic impedance with the following unique properties:

$$\begin{aligned} &\text{Propagation Velocity: } \nu = \frac{1}{\sqrt{LC}} \\ &\text{Characteristic Impedance: } Z_o = \sqrt{\frac{L}{C}} \\ &\text{Reflection Coefficient: } \Gamma = \frac{Z_L - Z_S}{Z_L + Z_S} \end{aligned} \tag{8.1}$$

where Z_L is the impedance of the load and Z_S is the impedance of the source.

The important discovery was the relationship between the size of the reflected signal and a change of the impedance along a transmission line. If there are impedance mismatches in the transmission line, then some portion of the signal will be reflected and the transmitted signal at the end of the transmission line will be reduced in amplitude. A good analogy is to imagine a very long transmission line with constant impedance where a voltage step will result in a certain current flow determined by the transmission line characteristic impedance. Assuming that the transmission line is eventually terminated in the same impedance, then the rate of current flow never changes and no voltage reflections occur. If the line is left open at the end ($Z_L = \infty$), then the current traveling down the line must go somewhere and is reflected back resulting in a positive voltage amplitude reflecting back along the transmission line. If the end of the line is a short ($Z_L = 0$), then the voltage must go to zero and a negative voltage amplitude is reflected back along the transmission line.

The next consideration is the finite rise time or finite slew rate of a transmitted voltage step. Assume a transmission line that is much shorter than the distance for the propagation of either a rising or falling edge of a voltage step. Before such a slope reaches its full voltage, some of the reflected signal from a nonideal load is already returning to the source and in this way supports the maintenance of the same voltage at the source and load. The assumption for a lumped circuit design is that the voltage at one end of a connecting trace is the same as at the other end, and this approximation is valid when the rising edge of a signal transition is on the order of 6 to 10 times the length of the connecting trace.

The rise time of the step edges for a high-speed digital signal are typically given in terms of the time to go from 10% to 90% or 20% to 80% of the voltage amplitude. This can be converted to physical distance by rewriting the Telegrapher Equation for the propagation velocity of a signal in terms of the dielectric constant:

$$\text{Propagation Velocity: } \nu = \frac{c}{\sqrt{\varepsilon_r}} \tag{8.2}$$

where c is the speed of light in vacuum and ε_r is the dielectric constant of the material surrounding the signal trace. For FR4 as the PCB dielectric material with $\varepsilon_r = 4.4$, the velocity on an inner layer stripline is approximately 15 centimeters per nanosecond.[3] This shows that for a multigigabit signal with a subnanosecond rise time the signal will transition from a low to a high in less than 15 cm of distance. If an ATE signal path is on the order of 15 cm, then the signal at one end of the test fixture is not the same as at the other end

[3] 6 inches per nanosecond.

and it can no longer be considered lumped element design. Designers often want the distance required for the rising edge to be 6 to 10 times the signal path for a true lumped approximation and in this case it clearly requires a distributed transmission line theory approach with matched impedances for optimum signal integrity.

When multiple impedance discontinuities happen in the middle of the signal path, then one has to consider the double reflection that occurs when the reflection at one impedance discontinuity travels back towards the source and encounters another discontinuity so that the resulting double reflection is now traveling in the same direction as the original signal but is delayed and reduced in amplitude. The forward traveling reflection will then add and subtract from the original signal creating amplitude variations or ripple as shown in Figure 8.10. Again, if the impedance discontinuities are closer together than 6 to 10 times the length of the rising edge, then the voltage differences between them are small and the amplitude of the ripple can be neglected. A 10 Gbps signal typically has a rise time below 50 ps, which is approximately equivalent to a length of 7.6 mm (300 mil[4]) for a signal path on FR4 PCB material. A via on a high layer count ATE test fixture is close to this same length and can easily result in an impedance discontinuity at each end causing signal degradation due to this type of double reflection.

As multiple impedance discontinuities occur along a signal path, one quickly finds that the interactions are data rate and rise time dependent and require simulations in order to predict the signal path losses. Certain combinations of closely spaced discontinuities can act like a low-pass filter and quickly attenuate the higher frequencies. The challenge for the ATE test fixture designer is to ensure that each section of a transmission line including the transition structures are matched in impedance.

8.2.2 Conductor Losses

The simple resistivity of a conducting material can describe the conductor losses at DC, but is no longer valid for higher frequencies. The conductor losses shown in Figure 8.9 include the frequency-dependent skin effect and surface roughness losses in addition to the DC resistance.

8.2.2.1 Skin Effect Loss

The skin effect loss, as the name indicates, is based on the fact that for increasing frequencies, the magnetic fields set up by the rapidly changing

[4] mil is the standard unit used for the width of a PCB signal trace and corresponds to 1/1,000 of an inch or 0.0254 mm.

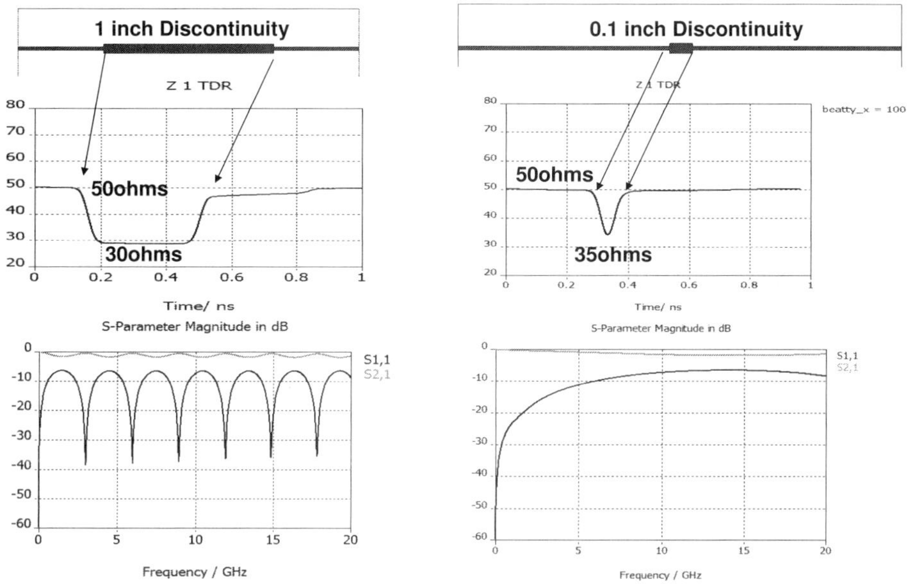

Figure 8.10 Comparison of the signal degradation in the time and frequency-domains versus the length of an impedance discontinuity for a given signal with a 30 ps rise time.

currents flowing in the conductor will force the electrical current to flow on the edge or "skin" of the conductor where the inductance is lowest. Figure 8.11 shows how this current distribution in a conductor changes with frequency and highlights that there is only a very small depth into a conductor where the high-frequency currents are actually flowing [15].

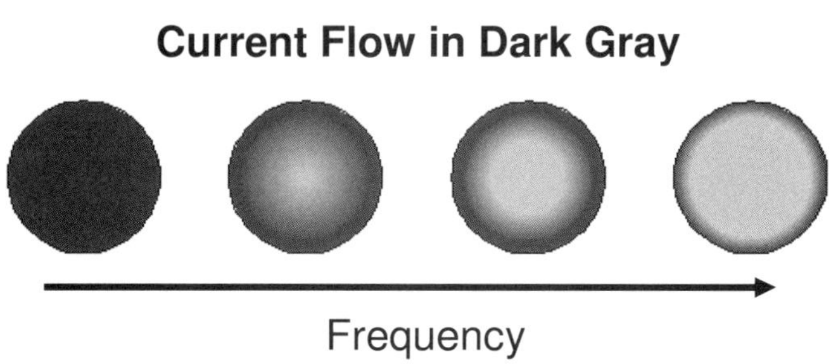

Figure 8.11 Skin effect on the center conductor of a coaxial cable.

The skin depth can be computed by (8.3) with ρ being the bulk resistivity of the core material (in Ω-m, e.g., for copper it is 1.673×10^{-4} Ω-m), in which f is the frequency, μ_0 is the permeability constant ($4\pi \times 10^{-7}\ N/A^2$), and μ_R is the relative permeability (usually 1 in most application cases).

$$\text{Skin Depth}(\delta_S) = \sqrt{\frac{\rho}{\pi\ f\ \mu_o\ \mu_r}} \tag{8.3}$$

The density of current flow is falling off exponentially with distance into the conductor, and the skin depth represents the thickness of the outer skin that is carrying $1/e$ or 37% of the total current flow. Conductor losses that include skin effect will depend on the trace geometry and a simple approximation for a low loss transmission line with trace thickness much larger than the skin depth is [7]:

$$\begin{aligned} &\text{Loss due to skin effect} = 4.34\ \frac{R_L}{Z_o}\ \text{dB/m} \\ &\text{Series resistance per length of stripline, } R_L \approx \frac{\rho}{2\text{w}\delta_S} \end{aligned} \tag{8.4}$$

where w is the trace width in meters.

The skin depth losses are proportional to $1/\delta_S$ and thus increase with the $\sqrt{f}$ making it difficult to predict the significance for a given digital signal. The skin depth can be increased and losses reduced by using a low resistivity conductor and avoiding magnetic materials. An ATE test fixture typically uses copper, which is one of the metals with the lowest resistivity, so little improvement can be made with the material selection; however, (8.4) shows that a shorter length conductor with a wider trace width will have lower losses.

8.2.2.2 Surface Roughness

Figure 8.12 shows a scanning electron microscope (SEM) picture of a microstrip cross section showing the surface roughness that is present on the bottom edge of a microstrip.

The skin depth for copper at 5 GHz is only 0.9 μm, which can easily be less than the height of surface roughness features on the conductor. This thin surface layer of current must follow the longer and higher resistance signal path of the conductor amorphous surface contours. Standard PCB fabrication processes often include surface treatment steps to roughen up the surface of the copper foil layers for improved mechanical adhesion [15–17] and at multigigabit data rates it becomes important to quantify the loss properties for a given lamination process [18]. The PCB industry mainly uses electro-deposited copper (ED), although rolled copper is another option for obtaining a very low surface roughness [19]. Table 8.1 presents a list of the typical

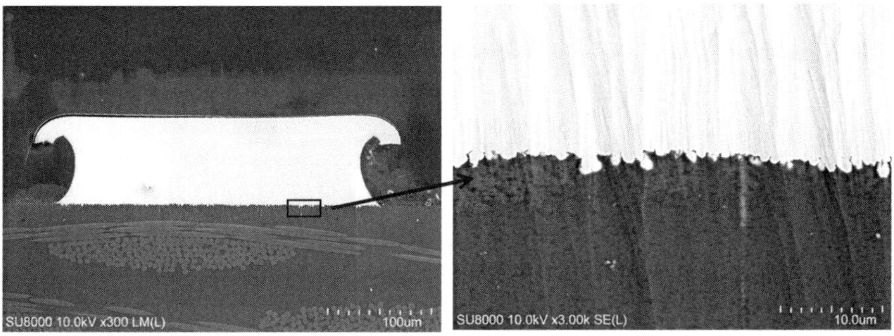

Figure 8.12 SEM cross section of a microstrip surface roughness.

Table 8.1
Copper Foil Surface Roughness Treatment Options

Treatment Options	Comments
STD (standard high tensile elongation)	Rough surface profile results in increased signal attenuation and delay due to increased propagation distance
RTF (reverse treated foil)	Reverse treatment of copper clad laminate allows for improved etching capabilities resulting in smaller variation in Zo
VLP (very low profile)	Smooth surface profile improves signal quality at higher frequencies where skin depth becomes a limiting factor to signal propagation
SVLP/HVLP (super/hyper very low profile)	An even smoother surface than the VLP option
Rolled copper	Best surface roughness, hard to use in PCBs due to its peel strength

options for copper foil surface roughness treatment. Some manufacturers might use other terms not shown in the table like ULP (ultra low profile).

Figure 8.13 shows a comparison of the loss for different copper foil surface roughness treatments. Note that a surface roughness factor is defined for each type of copper foil treatment. Modeling and measuring surface roughness are far from trivial and a lot of research is still going on for the best way to do it [17, 20–22].

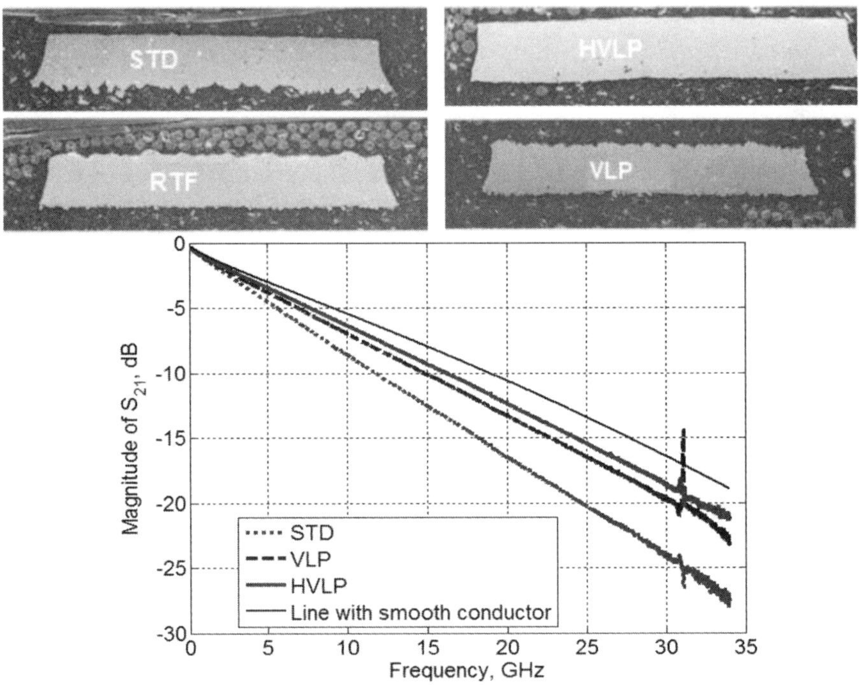

Figure 8.13 Loss comparison of different copper foil surface roughness treatments (reprinted with permission from [21]).

For applications where signal trace loss is very critical, it is important to pay close attention to the type of copper foil treatment used. However this needs to always be discussed with the PCB fabricator to avoid any mechanical issues with the PCB that can result in reduced PCB yield, increased cost, or reduced reliability.

8.2.3 Dielectric Losses

The dielectric loss occurs when electromagnetic fields lose energy to the molecular dipoles found in the dielectric material between the signal and reference conductors in a transmission line. The amount of dielectric loss depends on the type of material and the frequency or rate at which the magnetic fields are changing. Figure 8.14 shows this effect. Dielectric loss, unlike skin effect loss, is independent of the trace or cable geometry being used and only depends on the dielectric material properties. Equation (8.5) shows a closed formula approximation for the dielectric loss. The $\tan \delta$ is the loss tangent or dissipation factor of the dielectric material, which is a frequency-dependent property that is specified along with the dielectric constant to determine the amount of attenuation or loss per meter for a given material. Note that this is only an approximation since in reality even environmental factors like temperature and humidity have an impact on the dielectric loss [23].

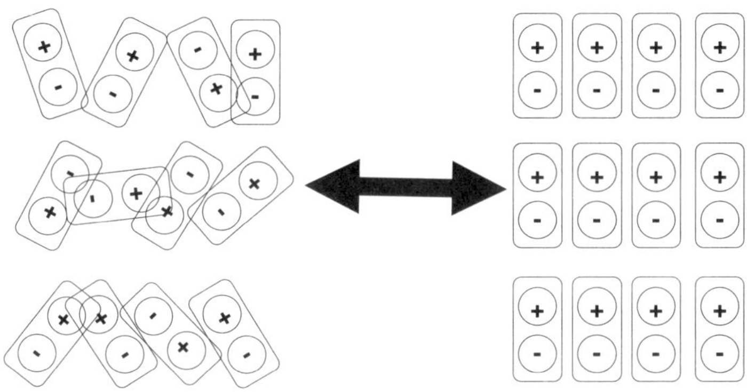

Figure 8.14 The polarization effect of the dielectric material molecular dipoles that generates the dielectric loss effect.

$$\text{Dielectric Loss} = 92.0216\, f\, \sqrt{\varepsilon_r}\, \tan \delta \text{ dB/m} \tag{8.5}$$

Test engineers looking for the lowest dielectric and skin effect losses on a transmission line topology have found that coaxial cables with their lower loss dielectrics and larger signal conductors are superior to the geometries of a PCB trace. Two different implementations of coaxial cable routing on an ATE test fixture are presented in Figure 8.15. This type of approach is common in test fixtures for device characterization with low I/O pin counts, but becomes less efficient as the pin count increases and connections must be spread further from the DUT.

Although a coaxial cable has a lower loss than a PCB trace (as shown in Section 7.16), this technique requires the addition of connectors to transition from the PCB to the low loss coaxial cable. Even though the connector PCB transitions are small relative to the total length of a cable, they can still

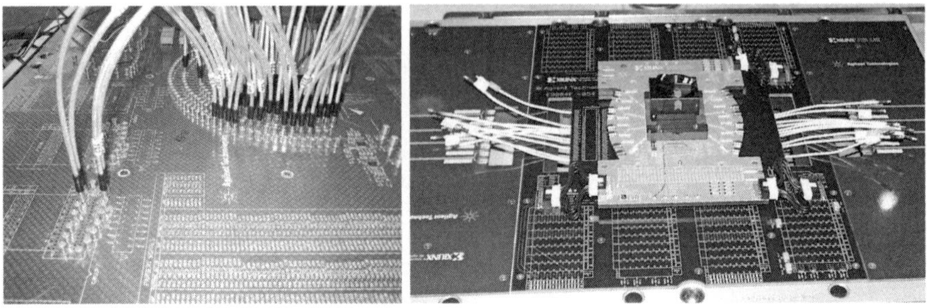

(a) (b)

Figure 8.15 Test fixtures for the Advantest V93000 ATE system using coaxial cables to address the PCB signal trace loss: (a) a single board approach and (b) a mother/daughter card approach.

have significant high-frequency losses if they are not impedance matched. The signal degradation from a poorly designed transition can quickly degrade any benefit of the low loss coaxial cable topology and one finds that coaxial cables often increase the complexity of an ATE test fixture with no measurable benefit. In [24] a more advanced approach is presented where a semi-rigid coaxial able is embedded in the test fixture PCB bypassing the need for a coaxial connector.

Optimizing the material selection and PCB trace geometries on an ATE test fixture can significantly reduce the high-frequency losses and make it quite attractive for at-speed, high density I/O testing. A PCB test fixture needs to have some type of dielectric material to separate the different copper layers that contain the signal traces and their respective return current path or ground. Ideally this would be air with the lowest possible dielectric loss (with the exception of a vacuum); however, this is not something that can be manufactured and one must search for the best balance of mechanical properties versus high-frequency electrical performance. The selection of the dielectric material and the separation that it provides between layers will also define the impedance of a given trace geometry.

The trade-offs between mechanical performance, electrical performance, and ease of fabrication have led to a wide selection of available PCB materials. The PCB dielectric materials can be divided in two groups: the ones that use a fiberglass mesh filled with resin and the ones that use no fiberglass. Table 8.2 provides a list of common dielectric materials used for high-speed digital test fixtures [25–27].

Figure 8.16 compares the insertion loss of different dielectric materials for a 482.6 μm (19 mil) wide inner stripline PCB trace using measured

Table 8.2
Some Dielectric Materials Typically Used for High-Speed Digital Test Fixtures PCB (FR4 is Included for Comparison)

Material	ε_r	tanδ (1 GHz)	tanδ (10 GHz)	Relative Cost
FR4	4.4	0.018	N/A	1
NELCO 4000-13 SI	3.4	0.008	0.008	1.5
MEGTRON 6	3.4	0.002	0.004	1.5
METEROWAVE 2000	3.4	0.004	0.003	1.5
ARLON 25FR	3.38	N/A	0.0038	1.75
ROGERS 4003	3.58	0.0027	0.0027	2
ROGERS 4350	3.5	0.0031	0.0037	2
SPEEDBOAD C	2.6	0.004	0.004	2
EZ–IO	2.8	N/A	0.0012	2
FASTRISE 27	2.7	<0.002	<0.002	2

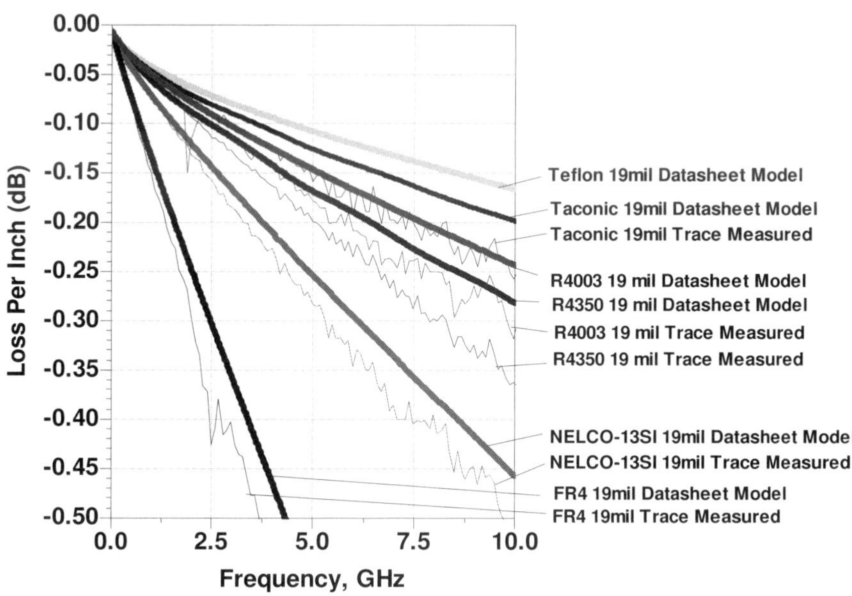

Figure 8.16 Simulation and some measurements for the insertion loss of different dielectric materials using the same trace width of 482.6 µm (19 mil).

data from ATE test fixtures and simulated data from manufacturer material specifications. One immediate observation is that the simulated results are more optimistic than the measured ones. This is expected since the simulation model is based on the material manufacturer's data sheet, which does not include the final "as-fabricated" properties such as surface roughness, moisture content, and lamination effects. The manufacturer's data sheet also relies on a simple IPC standard for measuring bulk dielectric constant and loss tangent in the z-axis direction, but PCB materials with glass weave are not homogeneous and signal traces running in the x- and y-axis directions can see different material properties as shown in Figure 8.17 with an impact on the signal trace loss and skew [28–30]. The best way to improve the accuracy of high-speed test fixture design and optimization is to obtain measured data of the material properties and ensure that all of the loss factors are included in the simulations.

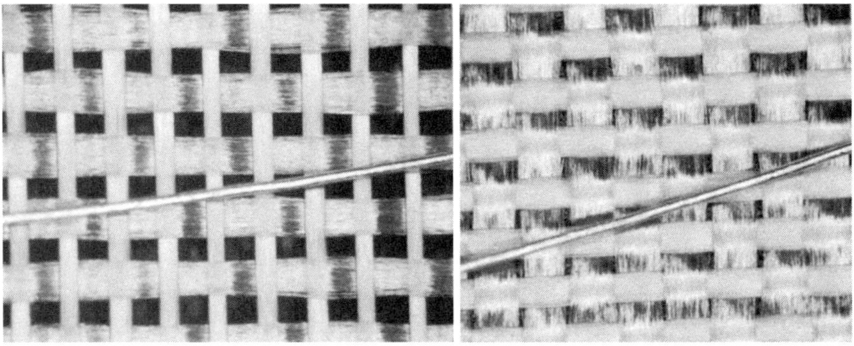

Figure 8.17 Photograph of two different types of fiberglass meshes with a copper wire on top. The glass mesh will change the local dielectric material properties, which impact the impedance of the signal trace and the propagation velocity. Tighter glass meshes mitigate this problem but at the expense of a higher dielectric loss due to the fiberglass.

Knowledge of the loss per length versus frequency for the different PCB materials is valuable for running design simulations to optimize the selection of dielectric material and trace geometries, but for comparison of high-speed digital performance it is also helpful as shown in Table 8.3 to view the digital signal performance in the time-domain using, for example, the data eye diagram. The data eye diagrams for the different materials shown in Figures 8.18, 8.19, and 8.20 demonstrate that for increasing data rates the data eye diagram for standard low cost FR4 PCB material degrades quite rapidly. Performance differences between the higher end materials are less significant and often become a balance of understanding the priorities of performance, cost, and manufacturability.

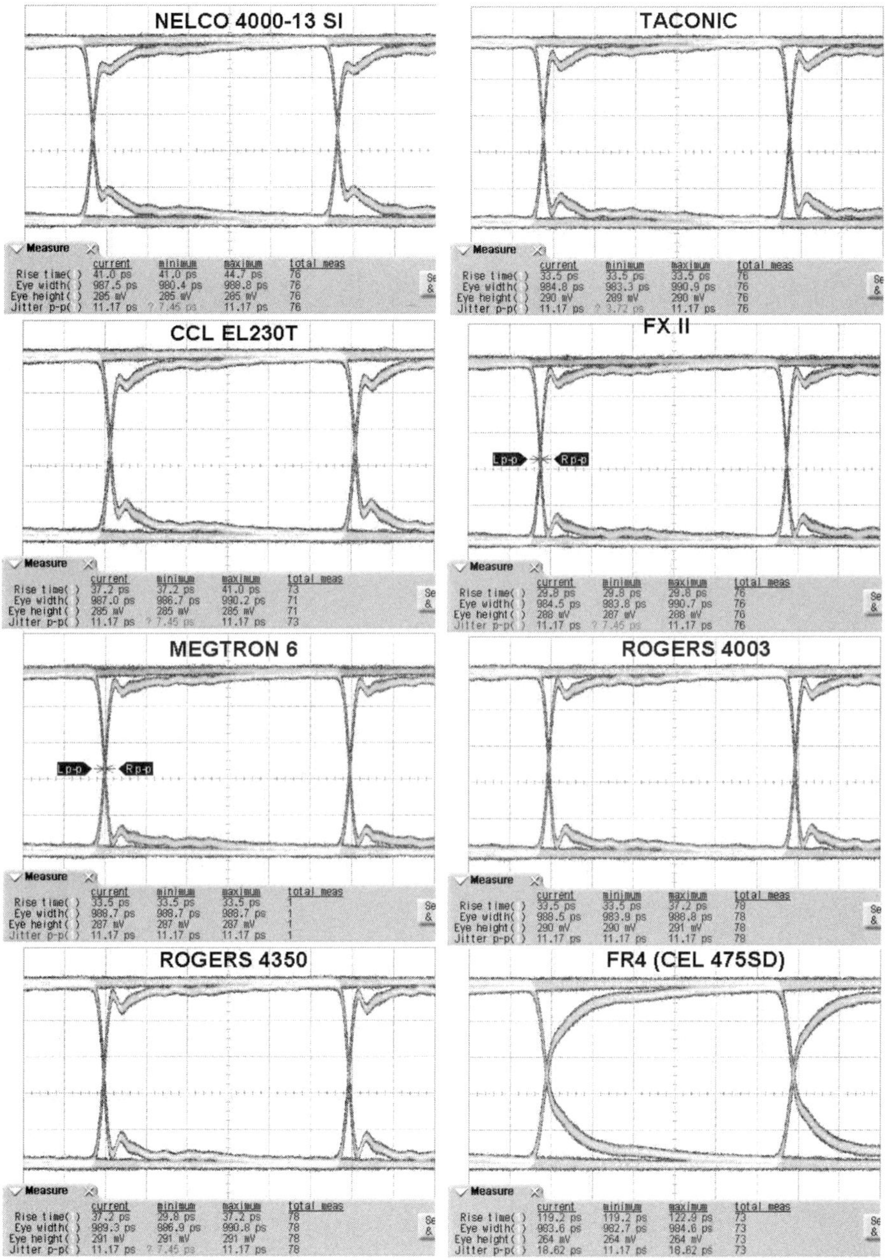

Figure 8.18 Data eye diagram comparison using different dielectric materials at 1 Gbps.

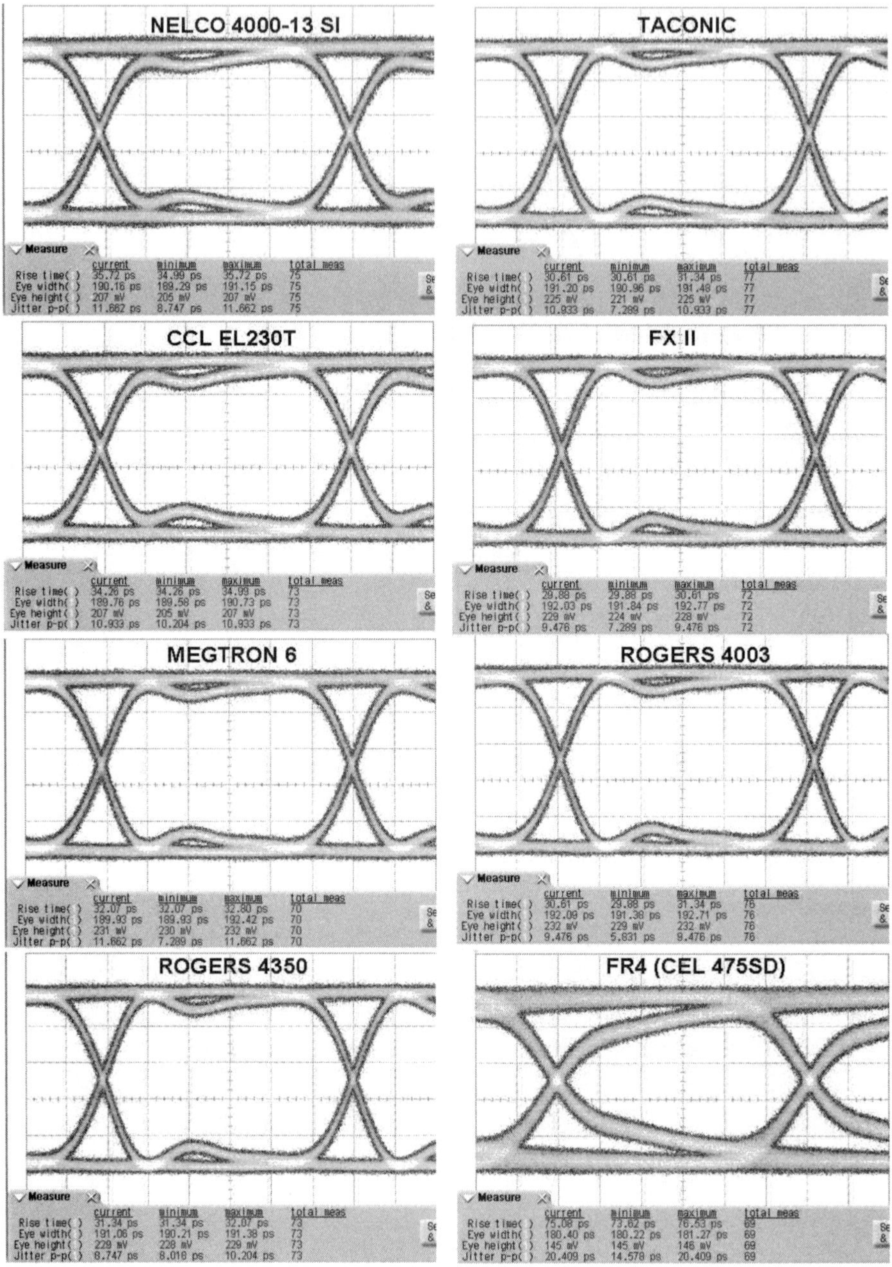

Figure 8.19 Data eye diagram comparison using different dielectric materials at 5 Gbps.

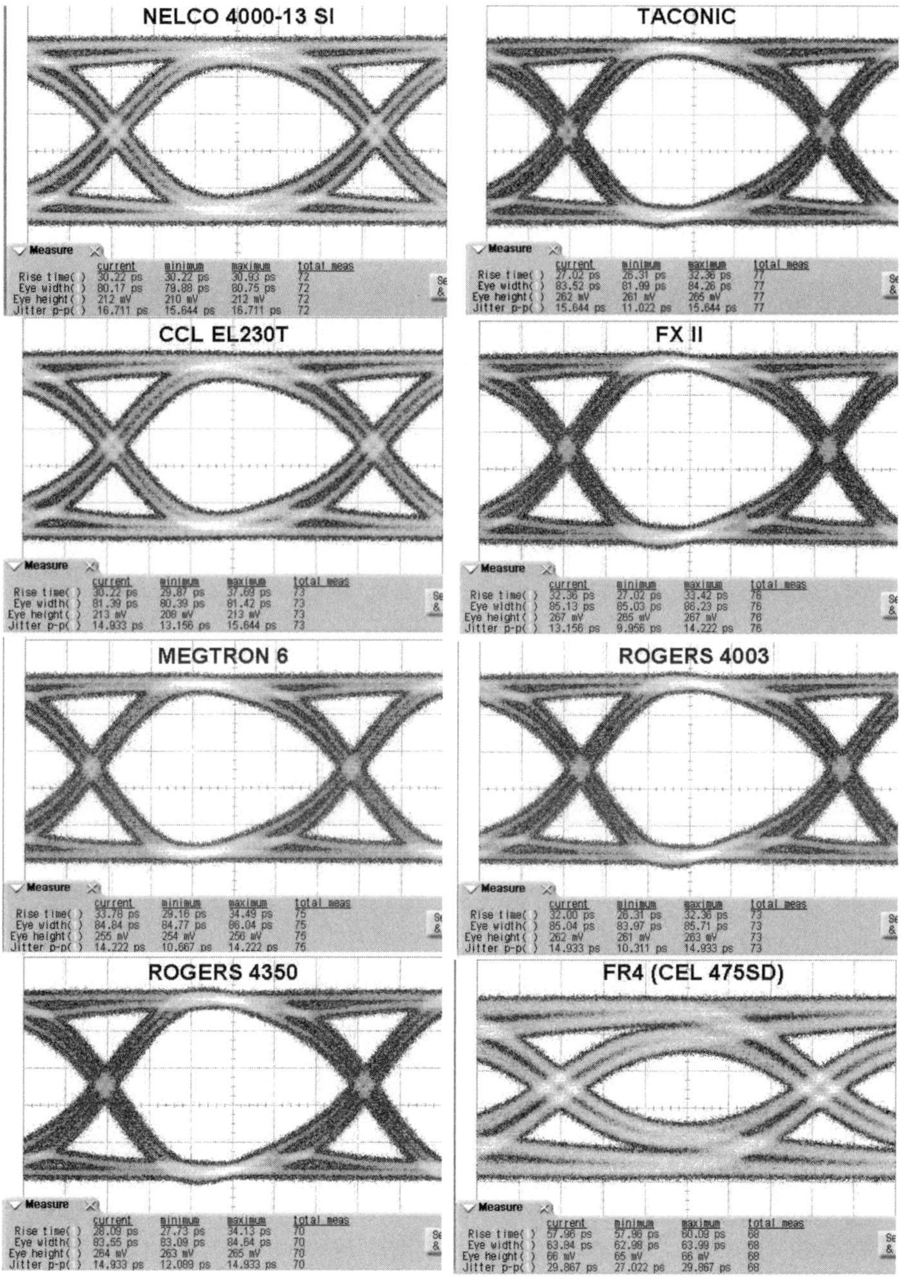

Figure 8.20 Data eye diagram comparison using different dielectric materials at 10 Gbps.

Table 8.3

Comparison of the Measured Values for an ATE Test Fixture with a 25 cm (9.8 in) Length, 482.6 μm (19 mil) Wide Trace in Different Dielectrics (FR4 Is Included for Comparison); Tr is the Measured Rise Time (20/80), Jpp is the Peak-to-Peak Jitter Value Measured at a Certain Data Rate

Material	Tr 10 Gbps	Jpp 1 Gbps	Jpp 5 Gbps	Jpp 10 Gbps
Taconic FastRise	27.0 ps	11.2 ps	10.9 ps	15.6 ps
ROGERS 4350	28.1 ps	11.2 ps	8.7 ps	14.9 ps
Nelco 4000-13SI	30.2 ps	11.2 ps	11.7 ps	16.7 ps
CCL EL230	30.2 ps	11.2 ps	10.9 ps	14.9 ps
ROGERS 4003	32.0 ps	11.2 ps	9.5 ps	14.9 ps
Hitachi FX-II	32.4 ps	11.2 ps	9.5 ps	13.2 ps
MEGTRON 6	33.8 ps	11.2 ps	11.7 ps	14.2 ps
FR4 (CEL 475SD)	57.9 ps	18.6 ps	20.4 ps	29.9 ps

8.2.4 Crosstalk

The final source of signal loss that cannot be ignored is the capacitive and magnetic interaction or crosstalk between adjacent signal traces when using fast switching signals. The amount of crosstalk between an aggressor signal and a victim signal will be proportional to the aggressor voltage amplitude, its rise time, and the physical proximity between the aggressor and victim structures (e.g., the distance between two PCB signal traces on the test fixture). Figure 8.21 shows a diagram representing the possible crosstalk points on a typical ATE test fixture.

The higher pin counts of complex SOC devices along with faster signal rise times has increased the amount of crosstalk for high-speed digital applications. The amount of crosstalk depends on the routing topology of the ATE DUT test fixture, which will differ from that of the end-user DUT application. A good introduction to the types of crosstalk between electrical signals, such as near-end crosstalk (NEXT) and far-end crosstalk (FEXT), can be found in the signal integrity references [4, 7]. Applications with only one DUT IC on the test fixture have some design flexibility when minimizing crosstalk effects while denser multi-site test fixtures can become quite challenging.

The best way to minimize both NEXT and FEXT is to simply space the signals farther apart, and one finds that with just five times the trace width between microstrip traces on a PCB, the NEXT can be reduced to less than 1%. Figure 8.22 shows this crosstalk reduction with gap spacing on the left

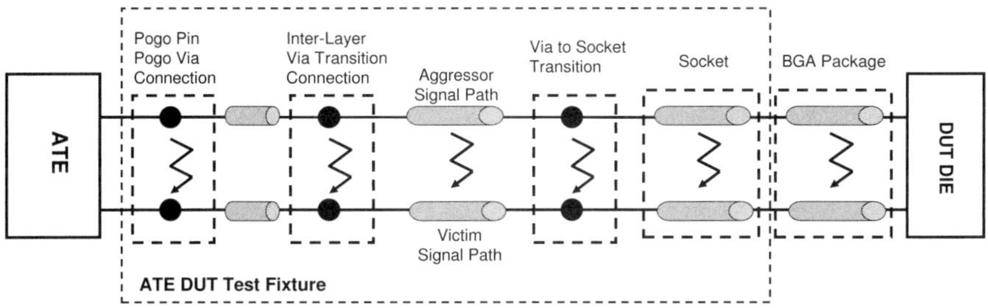

Figure 8.21 Diagram describing some of the areas where crosstalk might arise on an ATE test fixture.

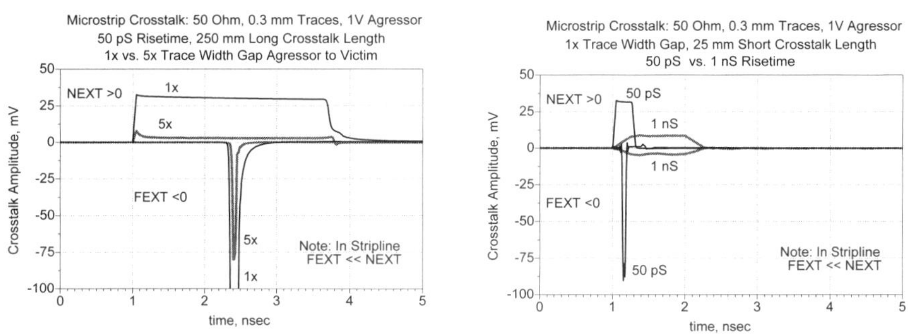

Figure 8.22 Simulation of adjacent stripline traces to show crosstalk reduction when increasing gap spacing from 1x to 5x on the left, and the change in crosstalk when the rise time is increased from 50 ps to 1 ns on the right.

for a long 250 mm routing distance and for decreasing rise time on a short 25 mm routing distance. The maximum amount of NEXT happens when the coupling length or distance of interaction is greater than half the length of the full step edge. The NEXT is the sum of the capacitive and inductive coupled currents, while FEXT is the difference. In the special case of a homogenous dielectric around the conductor, like in stripline, the capacitive and inductive couplings are about equal and essentially cancel the FEXT. This cancelation of FEXT makes stripline the best routing method for high-speed signals, since even 25 mm of microstrip routing with 5 times trace width spacing will have greater than 5% crosstalk for a signal with 50 ps rise time. Stripline routing is already quite common on high density digital applications where multiple signal layers are required for routing into the DUT BGA via field.

Interlayer vias that transition between different PCB layers or the pogo vias that connect the test fixture to the ATE pogo pin assembly [31] do not have well-behaved transmission line EM fields and can be a significant source of unwanted crosstalk. Via transitions at low speeds are significantly shorter than the distance occupied by a rising or falling edge and crosstalk is not a major concern, but as frequencies increase even the short distance across a via can end up with a maximum amount of crosstalk. Crosstalk at vias is usually caused by an increase in the magnetic loop inductance of the ground path interacting with neighboring signals and the crosstalk is reduced by ensuring adjacent return path ground vias. The magnitude of the crosstalk is difficult to predict without the use of a 3D EM field solver, which also has the benefit of providing visualization of how the crosstalk is occurring.

There will be some cases where crosstalk exists that is outside of the control of the test engineer or test fixture designer. For example, if the package signal and ground ballout design are not optimized, then one can get significant crosstalk on the vias and socket pins leading to the DUT. The ATE engineer cannot change the device ballout and is left with trying to shorten the distance of interaction by shortening the socket pogo pins and the PCB via length or by adding additional ground shielding or return paths. To a certain extent, the crosstalk in the ATE test fixture PCB via field for the DUT footprint is similar to what the application board will also see for a given ballout topology. The ATE system typically has longer distances through the via and socket to the DUT so the crosstalk will be worse than in the end-user application, but still a good indication of the crosstalk problems that the end-user application can encounter.

8.3 Impedance Controlled Routing

8.3.1 Microstrip and Striplines

An ATE test fixture is a multilayer PCB where the signals are routed through copper lines etched into the copper layers of the board. There are several geometries that one can use to transmit signals on a PCB. They include microstrip on the outer layers, stripline on the inner layers, and coplanar microstrip with ground and signal on the same layer to name a few of the more common ones. ATE applications typically use microstrip and stripline geometries as shown in Figure 8.23. The more complicated coplanar structure is often used just for the short distances when transitioning between transmission lines such as on an edge launch electrical connector footprint to improve the impedance matching.

Figure 8.24 shows a cross section picture of a PCB microstrip and stripline. The trace is not an ideal rectangle as one might expect, but it has a modified edge profile due to the etching process that etches sideways as well as down through the copper. The outer layer microstrip shown on the top with gold (Au) plating used as the mask for etching ends up with a combination

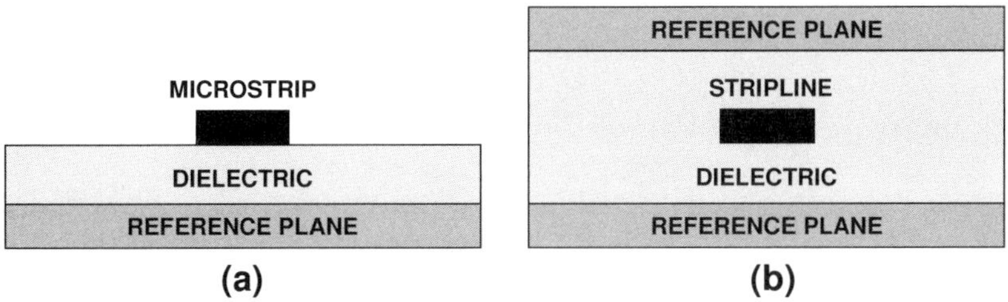

Figure 8.23 (a) Microstrip and (b) stripline geometries.

of trapezoidal with a thin cap layer that overhangs the edges of the trace. The gold plating is typically required on ATE test fixtures to provide a robust nonoxidizing contact for the ATE pogo pins and DUT socket pin contacts and to prevent any exposed traces from oxidizing. Typical simulation tools assume rectangular-shaped traces and will not always capture the additional high-frequency losses due to this nonuniform etching at the edges and it is helpful to obtain measurements to validate the losses for a given fabrication process.

The selection of a passivation layer (also referred to as surface finish or plating [32]) to protect the outer layer copper microstrip traces impacts the signal performance due to the interaction with the skin effect and dielectric properties at the edge of the conductor. Table 8.4 presents several outer layer plating options for optimizing the electrical performance for a given microstrip design [33–36]. At high frequencies the skin effect losses require the use of a low loss conductor for the outside plating with silver (Ag) providing the best performance. Hard gold is a good second choice for passivation, but requires an additional nickel (Ni) barrier to keep the gold from diffusing into the copper. Since nickel is not a good conductor and is ferromagnetic (relative magnetic permeability can be as high as 100 at 1 GHz), which greatly increases the skin effect losses, it is important to use greater than 1.9 μm of gold for the outer plating on microstrip traces.

Figure 8.25 shows a comparison of the insertion loss for different plating approaches on a 533 μm (21 mil) wide microstrip. The results show that, as expected, silver plating is the lowest loss option but also the most expensive. The thick gold over nickel (without solder mask) has almost the same loss as silver plating at a lower cost. The very thin Au plating can actually be worse then putting soldermask on the thick Au plating option and clearly demonstrates the need for greater than 1.9 μm of Au plating on any outer layer microstrip traces. Even though the data shows that soldermask adds additional losses, it is still used in cases where ESD protection is more important than the additional losses.

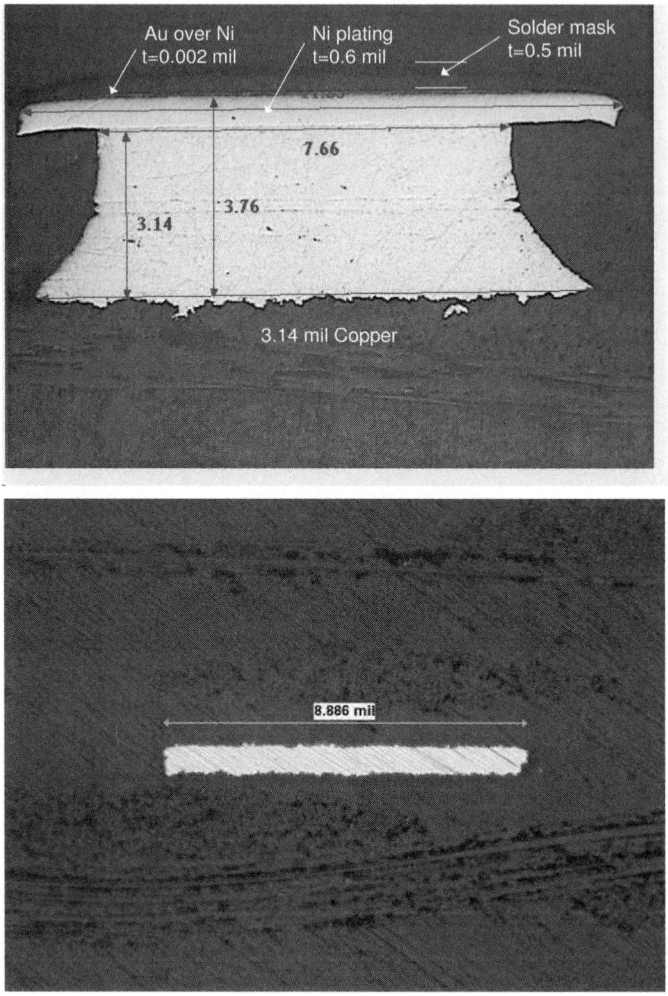

Figure 8.24 Cross section of a microstrip (top) and a stripline (bottom).

Table 8.4
Typical Surface Finish Options for a Microstrip Design

Plating Options	Advantages	Disadvantages
Thin Au: A thin <0.3 μm layer of gold (Au) is plated on top of a 5 μm nickel (Ni) barrier to the copper (Cu) trace.	Lower Au content improves the soldering of SMT components.	Higher skin effect losses than thick Au plating.
Thick Au: A layer of >1.9 μm is plated on top of a 5 μm Ni barrier to the copper trace.	Low skin-effect losses, almost as good as silver plating.	Higher Au content in the solder joint can reduce SMT reliability.
OSP: Organic surface protective finish for improved soldering of tight pitch components.	Used on SMT pads, traces still require solder mask.	Loss is higher due to solder mask.
Silver: A layer of silver (Ag) is plated directly on top of the copper.	Lowest loss since no nickel barrier or solder mask is required.	Popular choice for lead-free SMT soldering.
Solder Mask (HASL): Typically a layer of 10 μm goes on top of the Cu trace that may or may not be plated with Ni-Au.	Protects against electro-migration, ESD, and Cu oxidation.	Increases the loss of the microstrip trace.
Electroless Ni Immersion Au (ENIG): A layer of 0.05 to 0.1 μm immersion gold on top of 3 to 6 μm electroless nickel.	Lower Au content improves the soldering of SMT components, very flat surface, good for fine pitch components.	Higher skin effect losses due to thin Au layer, not typically used for ATE test fixtures.

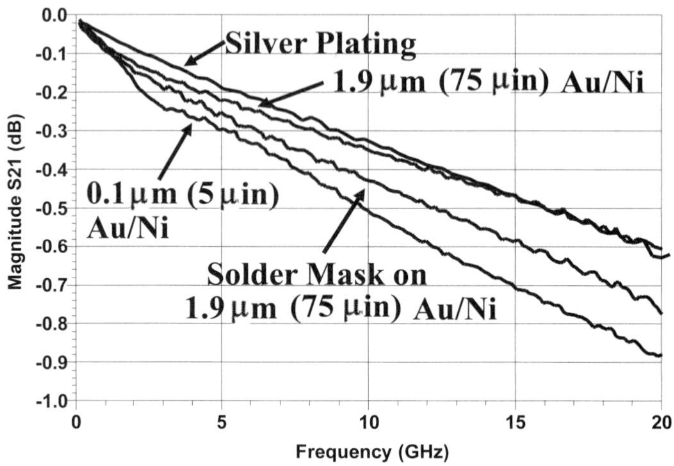

Figure 8.25 Comparison of the loss of different plating techniques for a microstrip trace.

8.3.2 Differential Routing

Differential signaling, as the name implies, uses the difference of two signals to determine the digital signal of interest. This type of signaling benefits from a constant di/dt current demand on the power supply and common-mode noise rejection; however, it doubles the number of I/O signal traces which increases the packaging and PCB costs. The difficulty with differential signaling is in understanding how the single-ended impedance and even mode signals interact with the differential impedance and the odd mode signals. A complete discussion of differential signaling and line impedances can be found in [7]. Figure 8.26 shows a diagram of an edge-coupled differential stripline pair with the key geometric values that determine the differential pair impedance. To compute the impedance of a given geometric configuration, it is necessary to use a 2D field solver or an impedance computation tool as shown in Appendix G since an accurate closed formula for the impedance of an edge-couple differential stripline is not available.

A 100 Ω differential signal trace can be composed of two single-ended 50 Ω signal traces as long as they are length-matched and separated by enough distance to prevent any strong coupling between the single-ended traces. Routing the two signals with a gap distance smaller than the trace width takes advantage of increased EM coupling between the traces to reduce the signal sensitivity to ground discontinuities and common-mode noise. However, the stronger the coupling the narrower the trace width needs to be designed to maintain 100 Ω differential impedance. The coupled routing approach is very popular on end user application boards where space is limited, the environment is noisy, and ground discontinuities are likely. The longer distances used on an ATE test fixture drive a different set of trade-offs when routing differential signals.

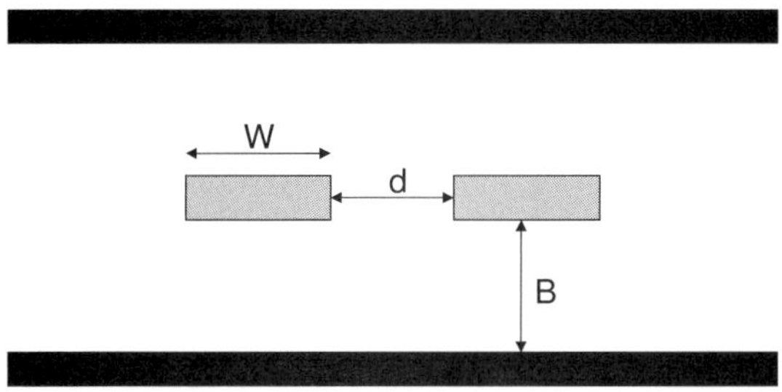

Figure 8.26 Diagram with key geometric values for determining the differential impedance of a differential PCB stripline.

The longer routing on an ATE application with the bends and turns can make it difficult to maintain a minimal skew between the coupled traces along the signal path. If the differences becomes too large, the hazard of common-mode to differential conversion grows. Furthermore the coupling of traces requires a reduced trace width, which generates increased losses for long routing distances. Single-site DUT characterization test fixtures often have dedicated internal signal layers with excellent noise immunity. This allows the test fixture designer to design a low loss differential pair by using a large trace width and using a distance between the single-ended signal traces of the differential pair that is at least three times the trace width. In terms of signal integrity, the timing skew of a differential pair can be understood as a degradation on the differential insertion loss as discussed in [37, 38] and shown in Figure 8.27.

8.4 Stack-Up

The stack-up is one of the key aspects of a test fixture PCB and needs to be documented in an appropriate way as shown in Figure 8.28. The stack-up describes how the test fixture layers are assembled and which dielectric materials are used and in the example of Figure 8.28, also shows the drill sizes that are needed for the pogo and interlayer vias. This information is typically found on the "fabrication document" that is included with the CAD files sent to the PCB fabricator.

On signal layers the dielectric height together with the trace width will define the impedance for that layer. For layers corresponding to power planes

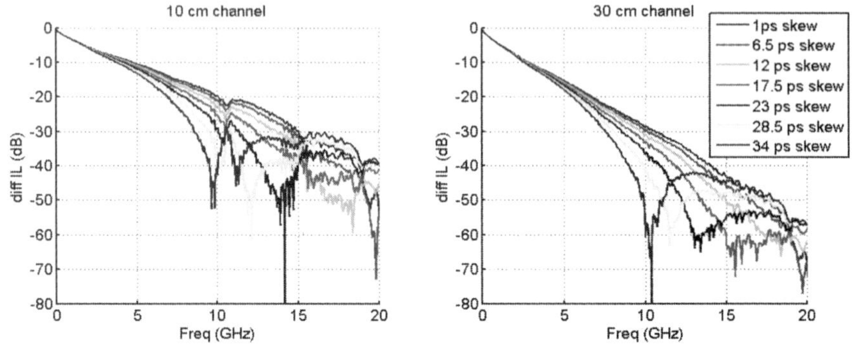

Figure 8.27 Impact of the differential skew on the differential insertion loss for two different signal trace lengths (reprinted with permission from [38]).

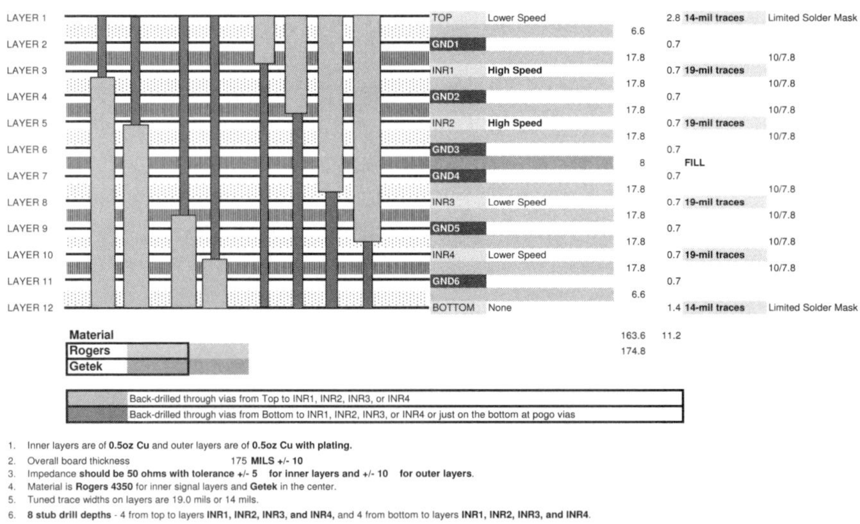

Figure 8.28 Example of a stack-up for a high-speed digital test fixture (Courtesy of R&D Altanova).

the dielectric height will define the capacitance and the loop inductance between the power plane(s) and the ground plane(s). One challenge is that, due to manufacturing constraints, the maximum stack-up height of a multilayer PCB is limited to around 8.128 mm (320 mil) with the exact value dependent on the manufacturer and yield issues. This creates restrictions on the combination of number of layers and the respective height of each layer, which, in turn, impacts the signal trace loss. Note that some PCB manufacturers can go above this stack-up height limit by processing the PCB as two subassemblies in which that area is then bonded together with prepreg/conductive paste. In this case stack-up heights of 12.7 mm (500 mil) and above are possible.

In some cases it is possible to use the power planes also as a reference for the stripline routing layers to reduce the number of required layers. In this case careful design of the power decoupling network in needed to minimize the impact on the signal performance on that stripline layer [39].

On the stack-up of Figure 8.28, the copper thickness is described in mil units but sometimes it is described as copper weight in ounces (oz) [27]. Table 8.5 provides a mapping from copper weight in oz to copper thickness in inch and μm.

Table 8.5
Copper Thickness Mapping from oz to μm and inch

Copper Thickness (oz)	inch	μm
0.5	0.0007	17.78 μm
1.0	0.0014	35.56 μm
2.0	0.0028	71.12 μm

One question that might arise when designing and manufacturing a multilayer test fixture PCB is if for a given multilayer board with several identical layers the signal trace performance in each layer will be identical [40]. Figure 8.29(a) shows an example of a high-speed digital test fixture with identical signal traces in each layer and where the insertion loss of each layer was measured with the result of almost identical graphs of the insertion loss versus frequency for every layer. Although this does not provide a guarantee for any other test fixture PCB, it is the typical result that one can expect if using a good PCB manufacturer. Another, more complicated question is if the signal trace performance of the exact same PCB design manufactured by two different companies might change. Figure 8.29(b) shows one example where a test fixture was manufactured by two different companies and although the trace loss is very similar, due to a difference in performance on the backdrilling (i.e., remaining via stub length) from one of the fabricators, the performance at frequencies higher than 23 GHz is very different. This example shows how important it is to reverify a design when changing PCB vendors for a test fixture PCB even if nothing else has changed [41]. To verify the test fixture performance, detailed measurements must be made on the manufactured test fixture. Appendix H provides some discussion on different techniques.

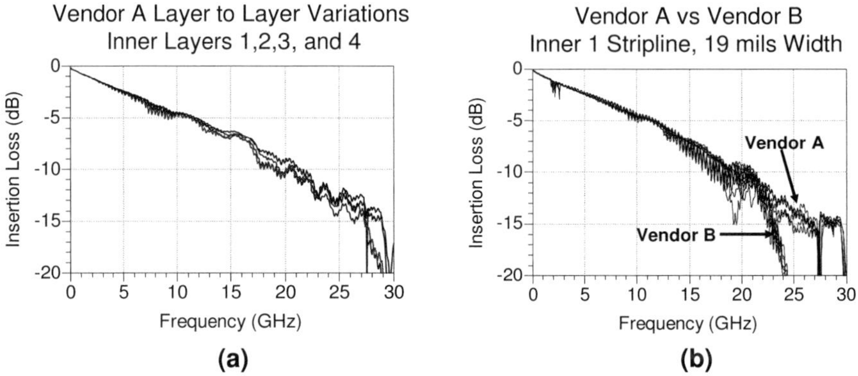

Figure 8.29 Comparison of (a) the insertion loss of identical traces in different layers of the test fixture and (b) the insertion loss comparison of identical traces and layers in two DUT test fixtures manufactured by different PCB fabrication houses.

8.5 Via Transitions

Vias are responsible for moving a signal between different layers of the PCB on an ATE test fixture as shown in Figure 8.30. At low speeds or DC levels the via topology is selected for robust manufacturing, high reliability, and available routing space. Looking back at Figure 8.9, it shows that at higher frequencies the impedance discontinuities from a change in the signal path topology will create reflections that can dominate the signal losses. The design of impedance controlled vias becomes critical for minimizing these reflections and applications like an ATE test fixture via design will depend on the required application performance and the available PCB topology [33, 42]. An ATE test fixture can easily be 6 mm thick, which is roughly 50 ps in length for the via transition and a significant portion of the rise time for a 10 Gbps signal.

The impedance of the via is determined by the relation of the signal via to the neighboring ground via structures. The parameters used to describe the via geometry are shown in Figure 8.31. Default combinations of these features defined by the PCB fabrication process may not have the required

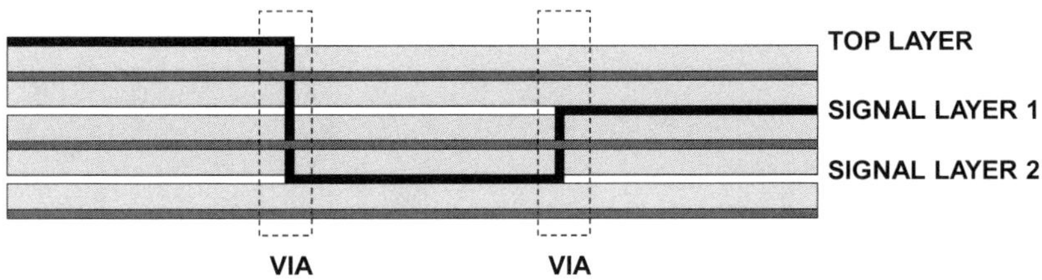

Figure 8.30 Vias on an ATE test fixture PCB.

performance, and as speeds increase, it is important that they are specified by the test fixture designer so that the appropriate engineering trade-offs can be made. The following parameters are important for defining a high-speed via:

- **Plane Clearance:** Plane clearance[5] defines the diameter of a circle that must be cleared of copper surrounding the via so that isolation is maintained when going through a power or ground plane layer or through a copper pour on a signal layer. The default size that guarantees no unintentional shorts due to manufacturing tolerances between the via and other signals may be too capacitive for a high-speed via. A 3D EM simulation provides the best method for optimizing the plane clearance for high-speed performance.
- **Pad Size:** Pad size refers to a disk of copper that is required on a given layer to ensure that even for the minimum and maximum drill location tolerances a connection will be made with the via. The pad is required on the top and bottom of the PCB as well as any internal connecting signal layer. Sometimes a pad is added on all layers that the signal via crosses as a default in the layout tool even if there is no signal trace to be connected. These types of pads are called nonfunctional pads and should be removed for high-speed designs [43] to increase performance; however, it is best to confirm with the PCB fabrication house to make sure that there are no manufacturing trade-offs for a given stack-up.
- **Drilled Hole Size:** This corresponds to the diameter of the drill used to make the via hole, prior to plating. Due to the skin effect, high-speed signals will see this larger outside diameter of the via hole for determining the impedance.
- **Finished Hole Size:** This corresponds to the inside diameter of the via hole after the plating process. Finished hole size (FHS) has been an industry standard since it is easy to measure with a feeler gauge at final inspection for PCB acceptance.

[5]Sometimes called antipad by the PCB layout tools.

- **Return vias[6]:** When a signal follows a via between layers, its return current signal must also find a way to change ground layers. In high-speed digital applications, return vias connecting the reference planes on each layer are added close to the signal via to provide this return current path and set the desired impedance. The number of these vias and their location can be optimized though 3D EM simulation.
- **Backdrilling:** Backdrilling allows the removal of any stub that remains at the signal via after it connects to the intended signal layer. This controlled depth drilling process is done after lamination along with any other unplated drills. If the via stub is significantly less than the rise time of the high-speed signal, then backdrilling may not be required.
- **Blind Via:** This is a via that stops at the connecting signal layer without the creation of any stub on the signal path. Blind vias on ATE test fixtures are typically done with sequential lamination steps.

A cross section of the PCB can be used to verify the via structure dimensions as shown in Figure 8.32 for the case of a backdrilled signal via with a return via on either side. It is important to understand that the transitioning of the return currents on the ground layer is just as important as the signal via at high frequencies and it is the relation between the signal and return vias that determines the high-frequency performance. A detailed discussion on the problems with discontinuities in the ground current path can be found in the reference literature [7].

An incorrect via design can significantly degrade a high-speed digital signal. Figure 8.33 shows two different via designs on an evaluation ATE test fixture. One of the via designs is a standard through via with no adjacent return vias while the other via design has been optimized for the PCB stack-up and includes four return current ground vias to maintain a 50 Ω impedance topology. To evaluate each via, a stimulus data signal is sent through four consecutive vias with the same design to look at a worst-case signal path with multiple vias.

Figure 8.34 shows the frequency-domain response of each of the via designs and the data eye diagrams obtained at 3.25 Gbps. The data eye clearly shows a significant degradation for the signal going through the four nonoptimized vias and in the frequency-domain it shows that above 1.5 GHz the signal has trouble getting through just a single via to transition between layers. Another parameter that easily might degrade the high-frequency performance is the remaining length of the via stub after a connection to the

[6]Sometimes they are called shorting vias or ground vias to indicate the shorting connection to the ground plane layers.

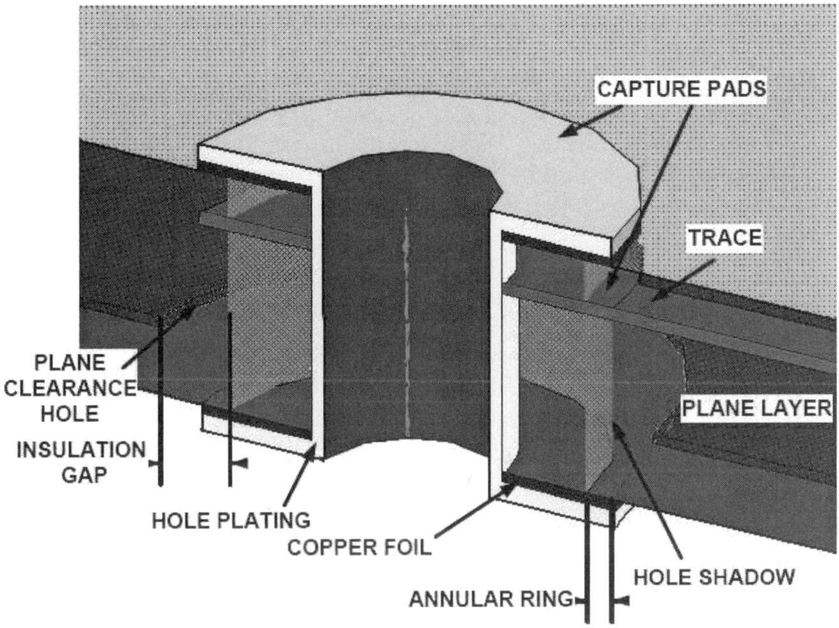

Figure 8.31 Diagram of a typical via with the respective important parameters. (From: [3]. ©2003 Lee W. Ritchey. Reprinted with permission.)

Figure 8.32 Micro photograph of a PCB cross section showing a thru-hole signal via with adjacent return vias (left), a backdrilled via (center), and a controlled depth (blind) via on the right (Courtesy of R&D Altanova).

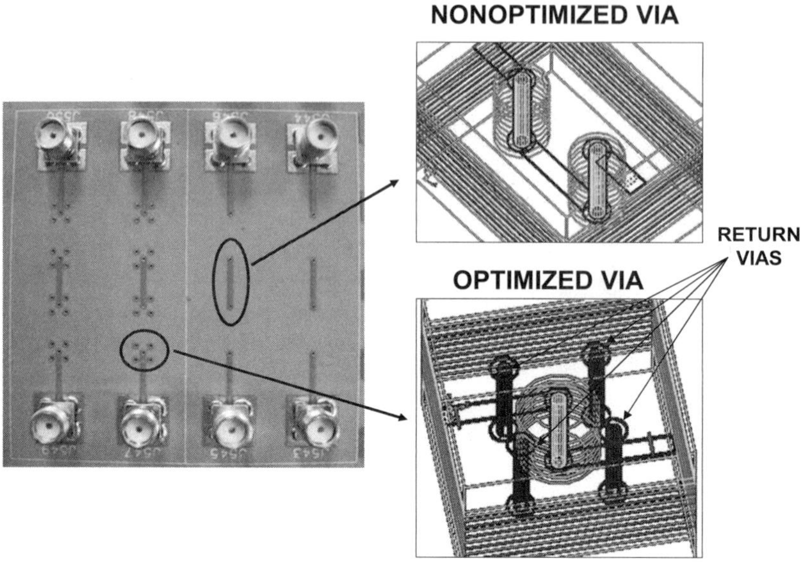

Figure 8.33 Test fixture for evaluation and demonstration of two different via designs (optimized and nonoptimized).

signal layer. The physical length of this via stub can resonate at multiples of the quarter-wavelength to create significant signal degradation.

It is important to understand the effect of a via stub at high data rates [44]. Figure 8.35 shows a cross section of two vias with different stub lengths. The measured S-parameters show that the short via with long stub starts to resonate in the frequency-domain already at 4 GHz. The long via with short stub performs significantly better. The quarter-wavelength resonance is a definite limit for the via stub length, but already for smaller values the added capacitance of the stub and its relation to traces, ground, and power planes on other layers can easily degrade the high-frequency performance. The 3D EM simulations are useful in understanding the impact of the via stub on the impedance profile as well as any probable radiation coupling into other layers of the PCB.

It is also important to understand the relative impact of the via design optimization and its backdrilling since sometimes a large amount of time is spent with 3D EM optimization of the via design but little attention is spent on the backdrilling accuracy. Remember that for backdrilling what matters is the worst-case stub length and not the average stub length. Figure 8.36 shows the differential insertion loss for two consecutive vias (and the corresponding signal traces) with combinations of optimized and nonoptimized vias and backdrilled and nonbackdrilled vias.

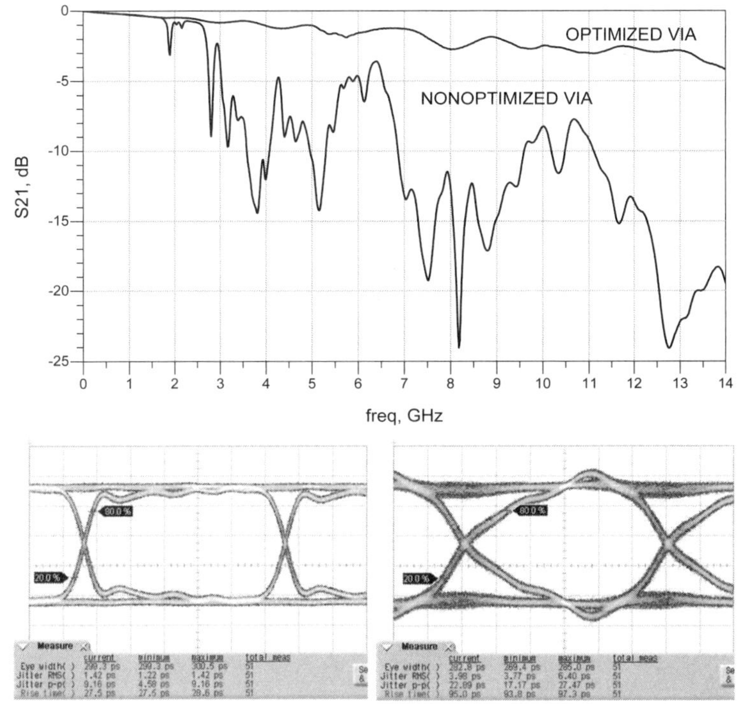

Figure 8.34 Performance comparison of a series of four via transitions for an optimized and nonoptimized via design in the frequency-domain (top) and on the time-domain (bottom) using the data eye diagram (PRBS $2^{31} - 1$ data pattern at 3.25 Gbps with an input rise time of 12 ps).

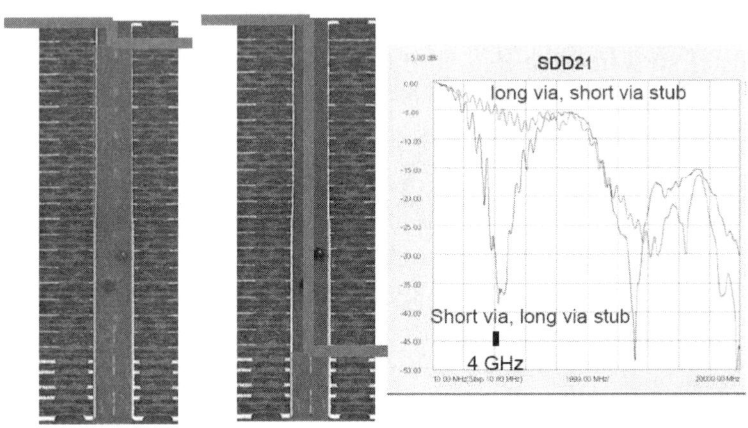

Figure 8.35 Cross section of two vias, one with a long stub (left) and one with a short stub (right), with their measured differential insertion loss. (From: [45]. ©2009 Mike Resso. Reprinted with permission.)

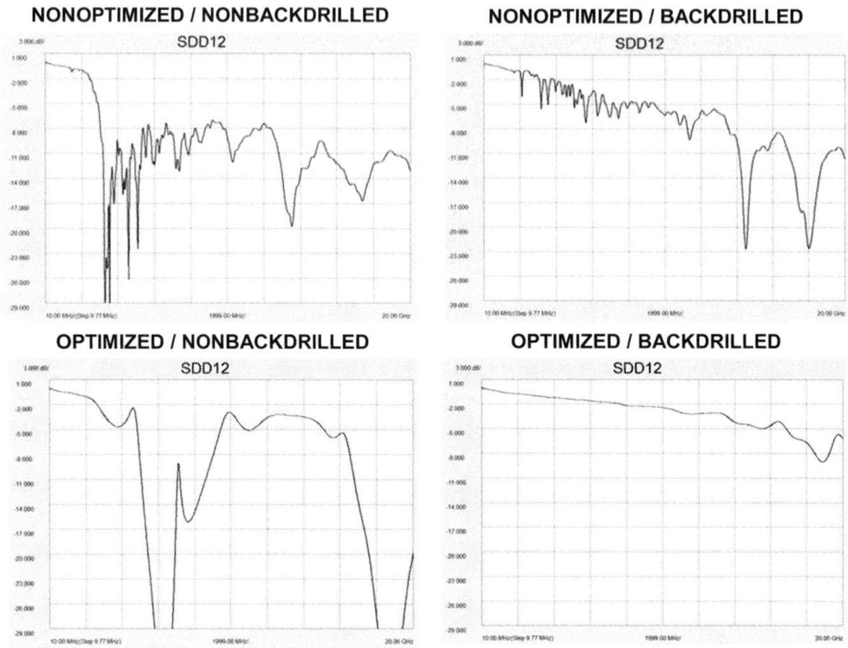

Figure 8.36 Comparison of the differential insertion loss for a series of two consecutive vias with different combinations of backdrilled/nonbackdrilled and optimizes/nonoptimized vias.

Figure 8.37 shows also the impact of the different combinations but using a loopback setup with an ATE driver and comparator measuring a 10 Gbps waveform going through the same via combinations. The results shown that in this case the impact of the backdrilling is higher than that from the via optimization. This shows how important it is not only to think about the optimization of a via transition but also make sure the backdrilling is done correctly by the PCB fabricator.

8.5.1 Interlayer Vias

The design of an interlayer via transition should not be blindly leveraged from one design to the next. Efforts to minimize the number of vias in a PCB for cost or routing densities may result in a single ground return via per signal via, which may work for one application but in another the crosstalk or signal losses may be too high. Designs using four surrounding ground vias can have a significant advantage when leveraging from one design to another. Designing with two return vias on either side of the signal path entering a via transition and then two more as the signal exits help to contain the fields as they make

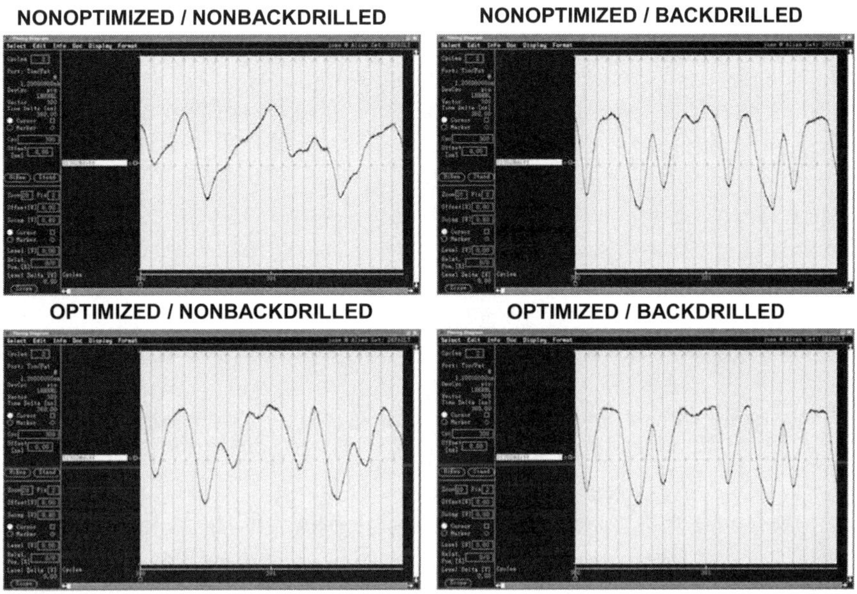

Figure 8.37 Comparison of the measured waveform by the ATE comparator for the different combinations of via optimization and backdrilling for a 10 Gbps waveform.

the right angle via transitions and limit the amount of signal loss caused by coupling to other layers of the PCB. This type of via is shown in Figure 8.38. By limiting the interaction with neighboring layers, the via is less sensitive to variations in the stack-up from one design to the next. The shielding provided by the return via structure reduces crosstalk between adjacent signals. The surrounding four return current vias also have the advantage of forcing the layout designer to keep other signal traces at a safe distance from the signal via, further minimizing crosstalk problems. The four ground via topology is also symmetrical and the connection to the via therefore well defined.

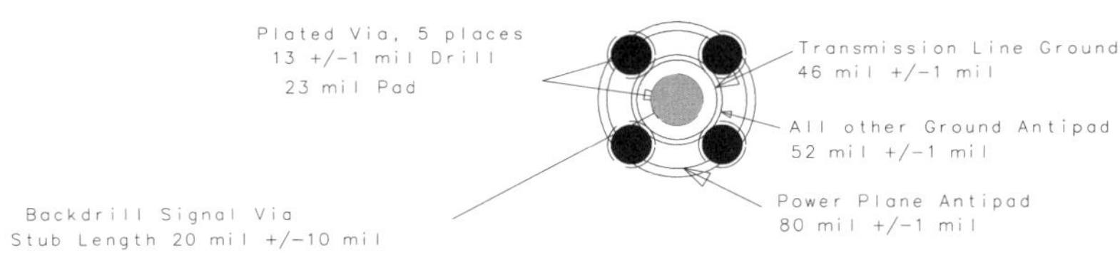

Figure 8.38 Example of an optimized interlayer via design using four surrounding ground vias.

8.5.2 Pogo Pin Vias

ATE systems that use an array of pogo pins, like the assembly shown in Figure 8.39, will require a specialized "pogo via" to provide the best signal integrity when transitioning from the ATE coaxial cables to the PCB transmission line on the DUT test fixture. Pogo pin assemblies can handle very high data rates when properly designed, but the pogo pin is only half of the transition and the mating via and transition design on the test fixture can easily be a source of significant signal degradation. This creates a challenge when specifying "at-the-pin" performance of an ATE system since the test fixture is under the control of the test fixture designer and not the ATE manufacturer. To achieve the best possible performance, the test engineer can utilize 3D EM modeling to optimize the transition design as shown in Figure 8.40 or rely on the generalized transitions provided by the ATE manufacturer. The ATE manufacturer has a significant interest in designing a test fixture to measure the highest quality "at-the-pin" performance and is less concerned with complexity and fabrication costs. The test engineer, however, must balance performance with cost as well as design risk when deviating from a verified topology.

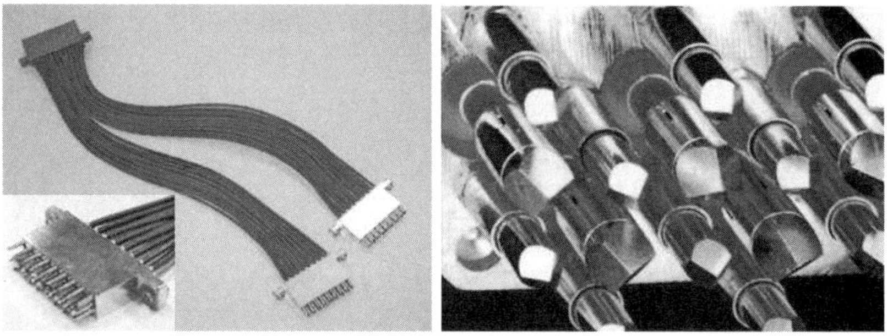

Figure 8.39 Example of a pogo pin assembly for an ATE system (Courtesy of Advantest).

The optimization of the pogo via transition can leverage various PCB via drilling processes to maximize performance. Utilization of controlled depth drilling for the pogo via contacts ensures a robust multipoint contact while at the same time freeing up the location of the through ground vias and the drill diameter of the signal via so that performance can be optimized. Figure 8.41 shows a picture and cross section of an ATE DUT test fixture pogo via that was the result of optimizing the various drilling features and ground clearances using 3D EM simulations that are verified and further optimized with measured data. More details on this type of approach for pogo via design can be obtained in [33, 46]. The ATE manufacturer may have a

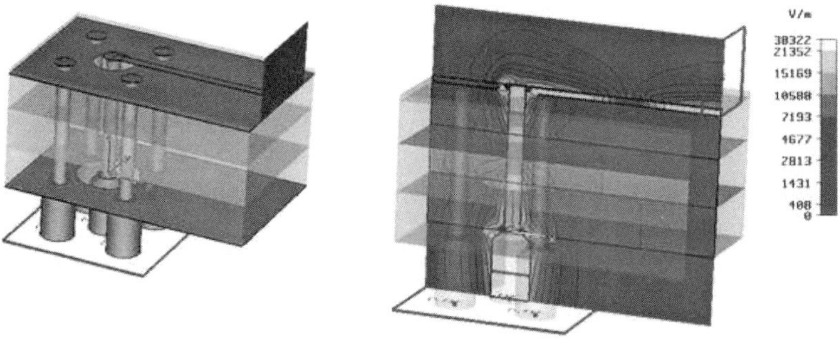

Figure 8.40 The 3D model of a pogo pin and pogo via for EM simulation.

variety of optimized pogo via transition designs depending on the interfacing pin electronics card and the performance range. This clearly shows that the test engineer must work closely with the ATE manufacturer to select the best pogo via transition for a given application. In some situations it is possible to use a more complex pogo via design to compensate for limitations on the pogo pin block design as shown in [47].

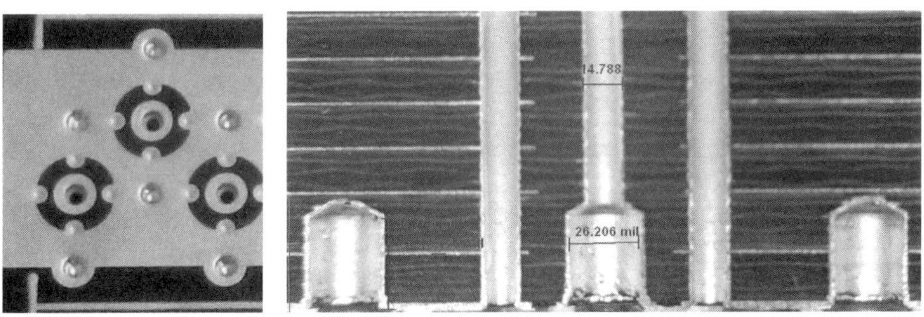

Figure 8.41 Picture and cross section of a manufactured ATE DUT test fixture pogo via for high-speed digital applications (Courtesy of Advantest).

There are also some important mechanical points to take into account when designing a pogo pin mating via. The via pad diameter needs to be large enough to take into account the worst-case misalignment between the pogo pin and the pogo via. This requirement might be contrary to the signal integrity requirements that might prefer a smaller pogo via pad. The plating is also an important consideration. Given the multiple dockings that a test fixture goes through in its lifetime, it is necessary to use an appropriate plating like a thick NiAu plating.

8.6 Coaxial Connector Footprint Design

Coaxial connectors are important in ATE test fixtures not only as possible ATE to test fixture interconnect solution as described in Section 8.1.1 but also for providing access points on the ATE test fixture, especially on a DUT characterization environment as shown in Figures 8.42 and 8.43. Coaxial connectors are also discussed in Section 7.16.2.

When using coaxial connectors in a test fixture it is not only important to choose a connector with the desired performance but also to design and optimize the entire connector footprint and transition from the signal trace to the connector using 3D EM simulation [48, 49]. Figure 8.44 shows the 3D EM simulation of a coaxial connector that is used as interconnect to the ATE system in the test fixture.

Figure 8.42 Example of an ATE characterization board with several SMA connectors (Courtesy of Advantest).

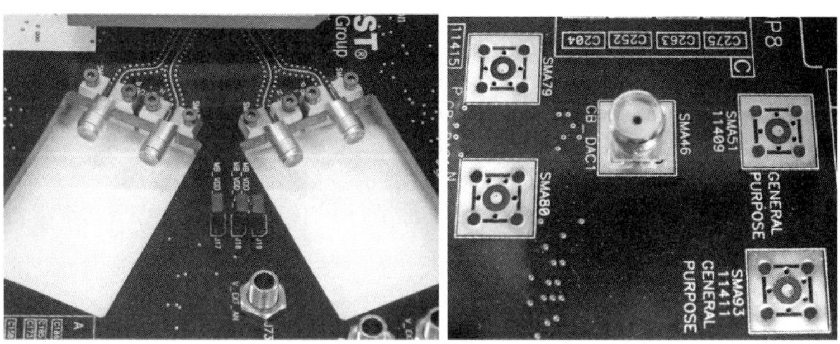

Figure 8.43 Example of a SMA edge mounted connector on an ATE test fixture cutout (left) and a SMA surface mounted connector (right).

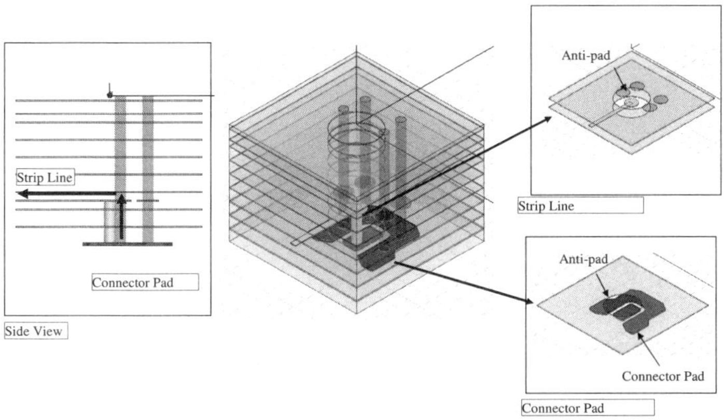

Figure 8.44 Optimization of a connector footprint through 3D EM simulation of the PCB footprint with connector model (Courtesy of Advantest).

8.7 DUT BGA Ballout

The DUT pin connections or ballout of a ball grid array (BGA) device are defined by the DUT package. The ATE test fixture cannot change the ballout of a device and must instead work with the topology of the BGA power, ground, and signal pin locations to provide the best performance possible. One of the first challenges is in routing signals through the BGA via field to reach the desired signal pad. The designer will find that for BGA pitches of 1 mm and smaller the allowable trace width can quickly push one to the limits of the PCB technology with 0.125 mm trace widths or less. To guarantee the specified transmission line impedance all the way to where the trace connects to the BGA via requires that this small trace width is used for the entire signal path between the ATE pogo via and the BGA via. However, this small trace width will have higher resistive and skin effect losses that must be considered when routing over the long distances found on ATE test fixtures and it will be more sensitive to etching tolerances that cause impedance variations. The test engineer must decide whether a wider trace and an impedance discontinuity when necking down into the BGA via field, as shown in Figure 8.45, has higher performance than a single narrow trace routed into the BGA via field.

As usual, the right answer will depend on the geometry of the specific application (i.e., how long is the signal trace, how large is the impedance discontinuity, and how long is the routing in the BGA via field). These parameters are shown in Figure 8.46 where L_T is the length of the signal trace before entering the BGA array area, W_T is the trace width that guarantees the 50 Ω impedance, L_N is the trace length where the signal trace width is

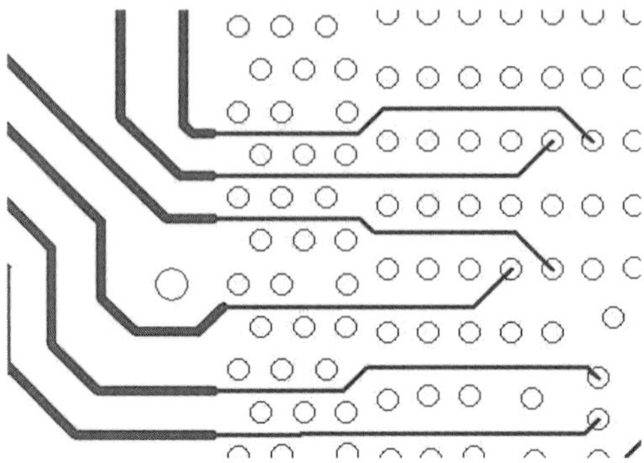

Figure 8.45 Addressing the routing challenges on a tight-pitch BGA through the reduction of the trace width (necking) on the BGA area.

reduced (necked), and L_B is the trace length with the reduced width of W_B until the BGA pad.

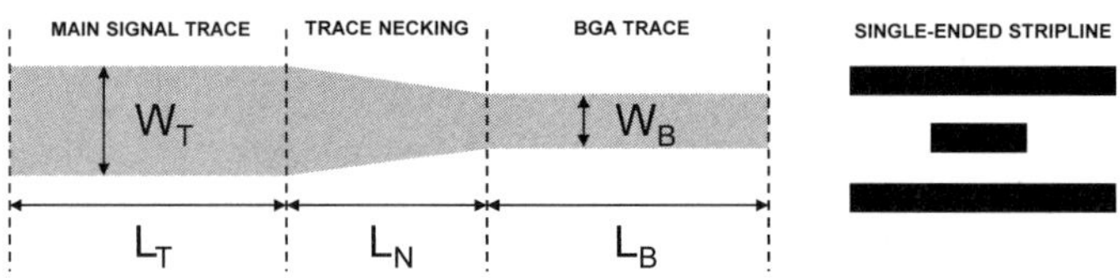

Figure 8.46 Signal trace necking parameters.

The challenge is to find the optimal value of W_T and W_B since the values of L_T and L_B are fixed from the layout requirements. L_N allows for some optimization but it is not a major factor. Figure 8.47 shows the data eye diagrams from one example of a signal trace with L_T=26 cm (10.2 in), L_B=15 mm (600 mil), and L_N=2 mm (80 mil) comparing a geometry of W_T=381 µm (15 mil) and W_B=101.6 µm (4 mil) with a geometry of W_T=215.9 µm (8.5 mil) and W_B=101.6 µm (4 mil). The data shows that for an ATE test fixture with long routing distances one should consider wider trace widths with trace necking. Measured trace loss data and a transmission line simulator make it possible to quantify the losses for different layout topologies as shown in Figure 8.48.

Another PCB layout item related to the BGA ballout is the implementation of the via from the stripline where the signal arrives from the ATE pin electronics to the pad on the top of the PCB test fixture. Two options are possible. One is to make the via very close to the pad position and then connect the via to the pad forming what is called a dog bone connection (see Figure 8.49). Another option is to have the via directly in the pad. This approach shown also in Figure 8.49 requires that the via is filled with epoxy so that it is

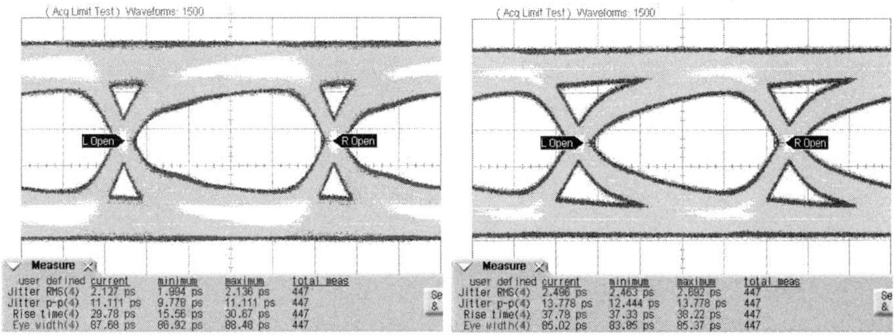

Figure 8.47 Example of a trace necking optimization with the data eye on the left obtained with a trace necking from 381 μm (15 mil) to 101.6 μm (4 mil) and on the right from 216 μm (8.5 mil) to 101.6 μm (4 mil) on a 26 cm (10.2 in) trace. (From: [33]. ©2006 Jose Moreira. Reprinted with permission.)

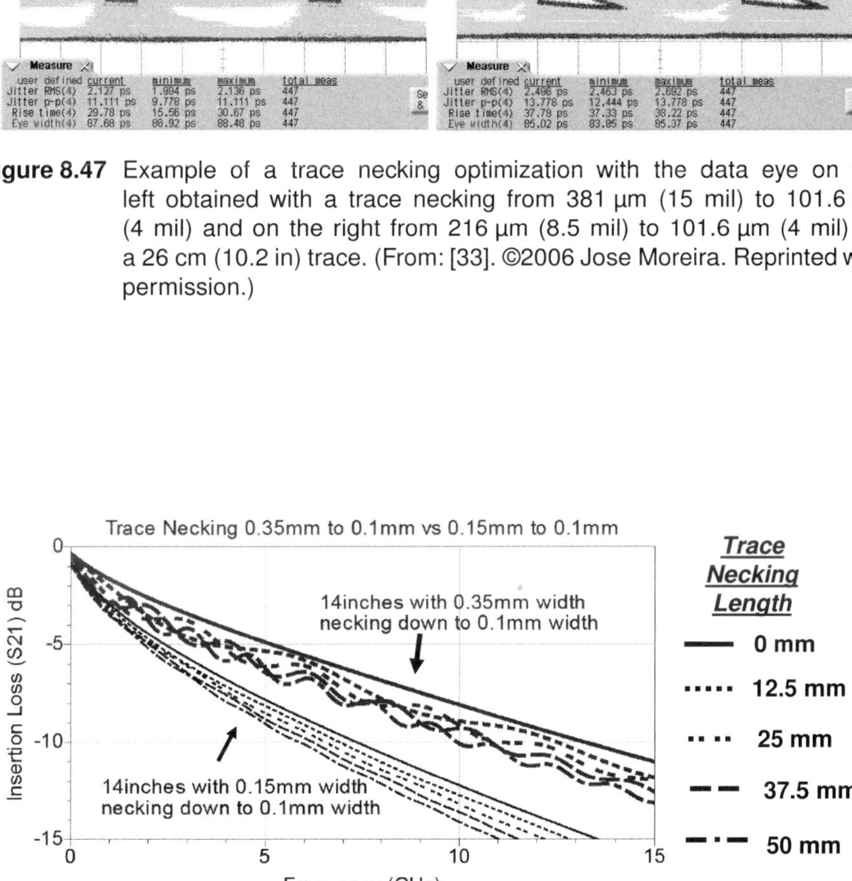

Figure 8.48 Loss versus frequency data for decreasing trace necking impedance discontinuity. The 50 Ω trace width is varied from 381 μm (15 mil) down to 152.4 μm (6 mil), and the routing distance of the 100 μm (4 mil) is varied from 0 to 50 mm (2 in).

possible to have a flat solid pad on top of the via. Although more expensive, this is the recommended solution for high-speed digital applications since it has a straighter signal path and less potential for crosstalk.

Note that it is not necessary to use a conductive epoxy like a silver filled one since due to the skin effect we know that at higher frequencies the signal will only flow on the edge of the via. By keeping the via core nonconductive, we are already providing some minimal equalization by forcing the signal at low frequencies to also flow on the edge of the via and have similar resistive losses. The other concern might be that for power distribution the vias need conductive silver filling, but it is important to understand that a conductive epoxy uses silver "flakes" to create a conductive path and is not solid metal, while the copper plating of the via barrel is solid and provides the lowest resistance path. Although conductive filling does not hurt, in the case of large ATE test fixtures, any increase in plating thickness and copper planes for heat spreading is more effective for thermal cooling than a conductive via fill.

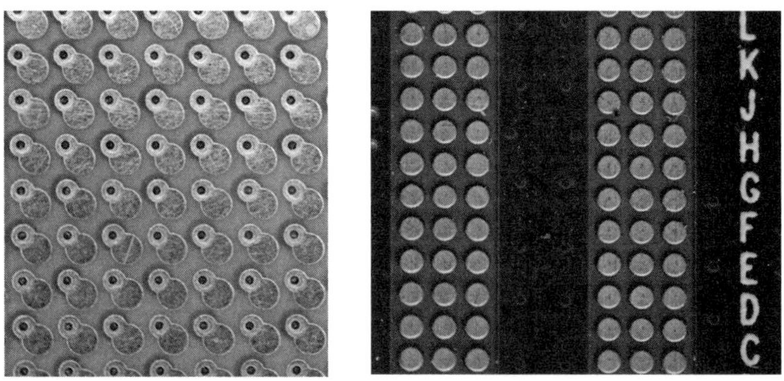

Dog Bone Connection **Via-in-Pad**

Figure 8.49 Comparison of a via in the pad approach with the "dog bone" type approach.

The physical dimensions of the DUT package ballout are not the only constraints for the performance of the test fixture design; it is also important to note that the performance of a given I/O connection will depend on crosstalk from the neighboring signals and the location of the reference pads (ground and power). The distribution of signal, power, and ground pins may work well for a DUT on a thin application PCB without a socket, but the longer path of the ATE via and socket could exacerbate any performance issues of the I/O due to a nonideal distribution of surrounding reference pads.

8.8 Relays

Relays are probably the most used component on test fixtures to extend the capabilities of an ATE system [50] as shown in the example of Figure 8.50. One very typical application is to use relays to multiplex the ATE resources by allowing several DUT channels to be tested by a single ATE channel as shown in Figure 8.51.

Figure 8.50 Example of a ATE test fixture with multiple relays on its bottom side.

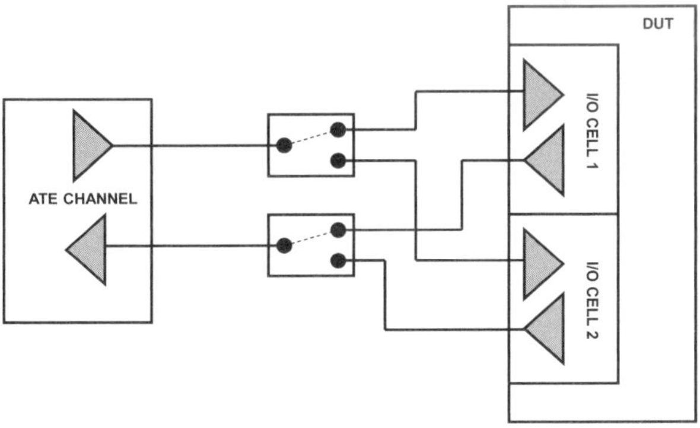

Figure 8.51 Using a relay to duplicate the number of DUT channels being measured by a single ATE channel.

Of course, there is a penalty in test time and also on signal integrity, but depending on the application requirements (e.g., test equipment cost is a major factor), it might be a good alternative. Another usage could be the ability to change between an ATE measurement setup and a loopback test setup as shown in Figure 8.52 and discussed in Chapter 6. Given the importance of this configuration, some relay manufacturers have developed special relay packages to handle a loopback setup as shown in Figure 8.53 [51].

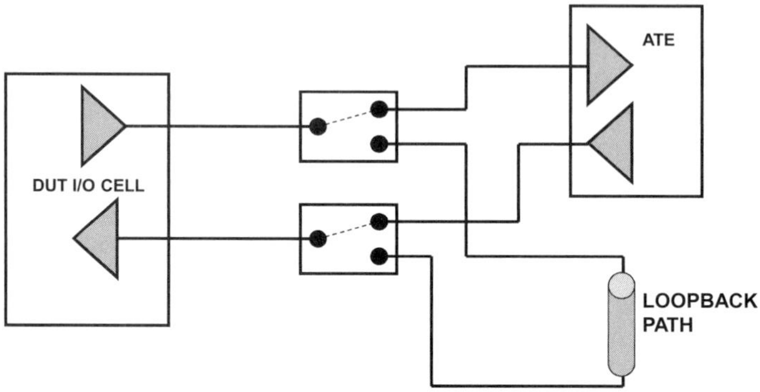

Figure 8.52 Using a relay to switch between an ATE measurement setup and a loopback test setup.

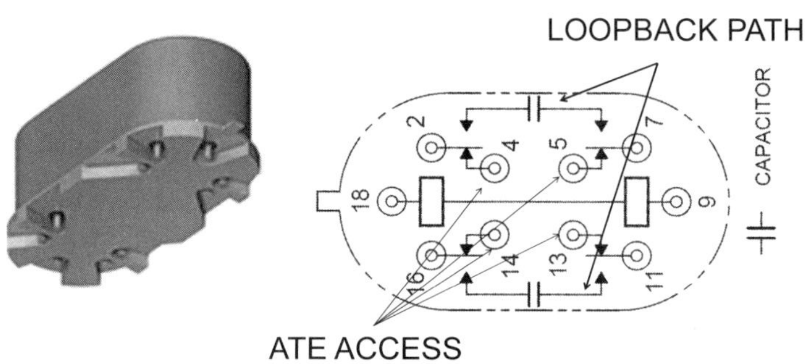

Figure 8.53 Example of a loopback relay (Teledyne GLB363).

Figure 8.54 shows pictures of different types of relays and Table 8.6 shows a comparison of various relay technologies. Electromechanical relays have historically provided a robust low loss signal path with stable performance over temperature and time, but can be expensive and larger in size and have a lower reliability than other families such as solid-state switches. Solid-state switches [52] have several disadvantages for high-speed digital applications. They have higher losses, temperature dependencies, broadband nonlinearities, and a very limited DC common-mode range. REED relays from a performance point of view are typically worse than mechanical relays due to the stub of the open path and the capacitance over the

Figure 8.54 Pictures of different relay families.

open contact. Their advantages are price, reliability (statistically better than mechanical but can have a higher percentage of early failures), and typically a smaller footprint than mechanical relays. MEMS-based [53] relays have received much attention lately and show great promise [54–56], but this relay family is still struggling with reliable long-term contacts. MEMS-based relays can be fragile and highly susceptible to contamination and face some challenges regarding hot switching, which is a significant issue when being considered for test and measurement applications. Another challenge of MEMS relays is the need for large control voltages that are usually not available on the ATE environment, but solutions are available in a small form factor [57]. The potential for extremely small size and high-frequency performance of the MEMS relay technology continues to drive significant research and investigation. There are four main types of approaches for designing MEMS-based relays: piezoelectric, thermal, electrostatic, and electromagnetic. Each one has its own advantages and disadvantages.

Relays are also classified by the number of ports and their functionality. Figure 8.55 shows the commonly used nomenclature for the description of the relay functionality and the associated schematic. For REED type relays a different nomenclature is sometimes used as shown in Figure 8.56.

One important question when selecting relays for DUT test fixtures is the influence of the relay on the overall measurement performance. Although proper choice of the relay model/family is important (i.e., choosing the relay with the required bandwidth for the intended data rate), the PCB footprint

Table 8.6
Relay Families

Relay Type	Advantages	Disadvantages
Nonhermetic Electromechanical: Switching by mechanical means	Repeatability, bandwidth	Size, cost, reliability, life expectancy
Hermetic Electromechanical: Switching by mechanical means	Repeatability, bandwidth, reliability	Size, cost, life expectancy
Solid-State: Switching by a FET switch or PIN diode	Reliability, size, cost	Trade off between DC common-mode range and performance
Mechanical REED: Switching by mechanical means in a glass casing	Long life, size, cost	Bandwidth with open stub capacitance in a SPDT configuration, reliability
MEMS: Switching by electrostatic mechanical means	Size, bandwidth	Fragile, hot switching concerns, bias circuitry

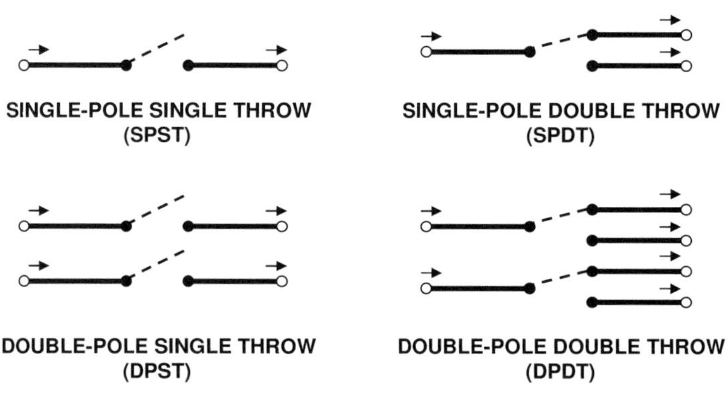

Figure 8.55 Nomenclature and schematics used to describe the different types of relay configurations.

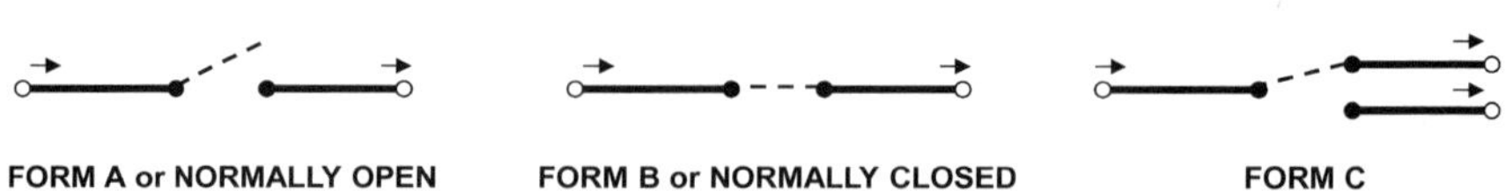

Figure 8.56 Nomenclature and schematics used for REED type relays.

is also of critical importance [50, 58]. Typically relay vendors will provide a suggested relay footprint, but unfortunately this footprint is in most cases optimized for an RF type application (i.e., requires high isolation between ports) and uses microstrip trace design. Usually in test fixture design for high-speed digital applications, stripline routing is used and high isolation is not needed and it is preferred to utilize a minimum number of ground vias for improved routing space. Figure 8.57 shows one example of an optimized PCB footprint for a mechanical relay where the optimized transition vias from the stripline to the relay and back are also included in the footprint.

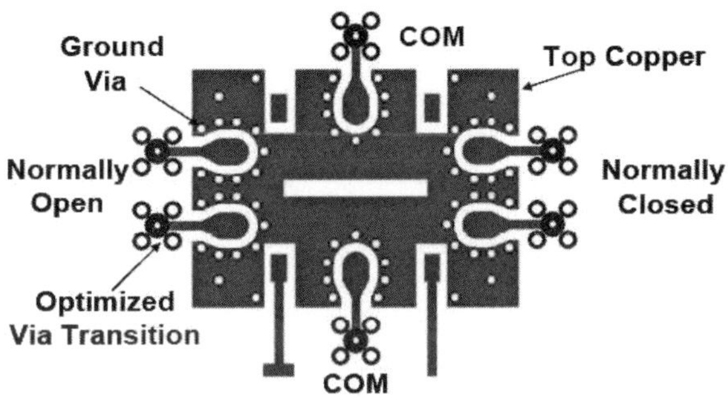

Figure 8.57 Footprint example for a mechanical relay.

Figure 8.58 shows the measured insertion loss comparison of a test fixture signal path with and without a relay for a 10 Gbps application. Although a state-of-the-art relay was chosen with an optimized footprint, there is a clear effect on the insertion loss.

The influence of the relay was also measured in the time-domain. Figure 8.59 shows a comparison of the measured data eye diagram of the signal path with and without the relay. One can observe a degradation on the rise time, but the jitter is very similar, showing that in this example the relay choice together with an optimized footprint provided an acceptable solution for the application.

In conclusion, it is possible to use relays on DUT test fixtures for high-speed digital applications, however it is important that in conjunction with the appropriate relay choice special attention is given to the optimization of the PCB footprint for the relay.

8.9 Bidirectional Layout

Bidirectional interfaces can require special layout techniques depending on the capabilities of the ATE pin electronics. If the pin electronics have inherent

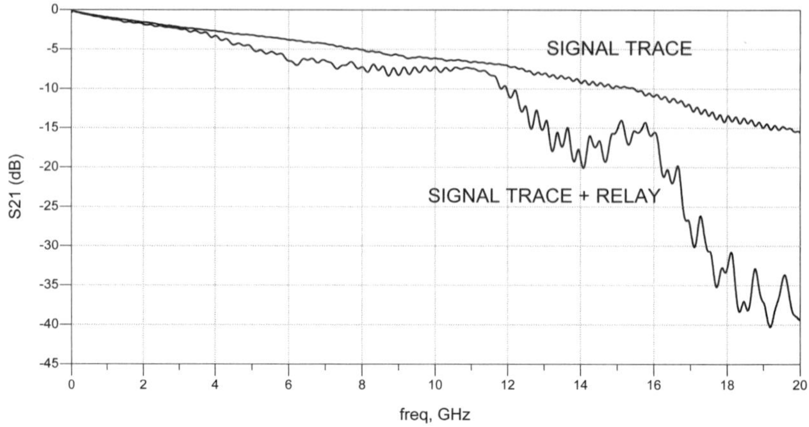

Figure 8.58 Insertion loss comparison of a signal path with and without a high-frequency relay for a 482.6 μm (19 mil) wide stripline and a length of 34 cm (13.4 in) in Rogers 4350 with a Matsushita NAIS ARJ relay).

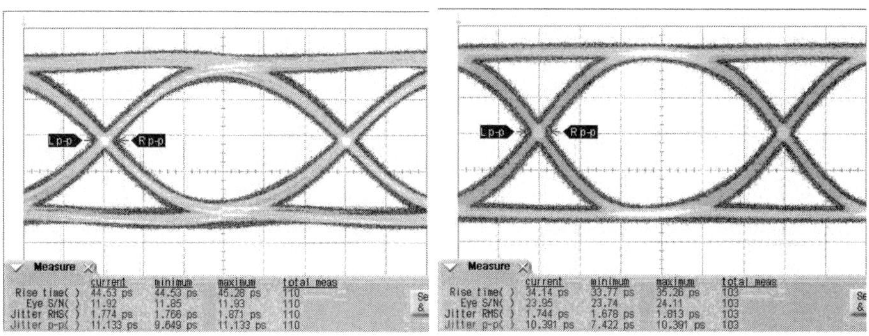

Figure 8.59 Data eye comparison of a signal path with (left) and without (right) a high-frequency relay [34 cm (13.4 in) long, 482.6 μm (19 mil) wide stripline in Rogers 4350 with a Matsushita NAIS ARJ relay].

support for bidirectional interfaces, then the layout for a bidirectional interface presents no additional challenges compared to a unidirectional interface. In the case where the pin electronics do not have an inherent support for bidirectional interfaces, it is necessary to use special techniques for their test [59]. For the test fixture layout, these techniques might imply the need to connect the ATE driver and receiver to the DUT I/O. The simplest option is the dual transmission line or fly-by approach where the ATE driver and receiver are connected to the DUT at a point very close to the DUT as shown in Figure 8.60.

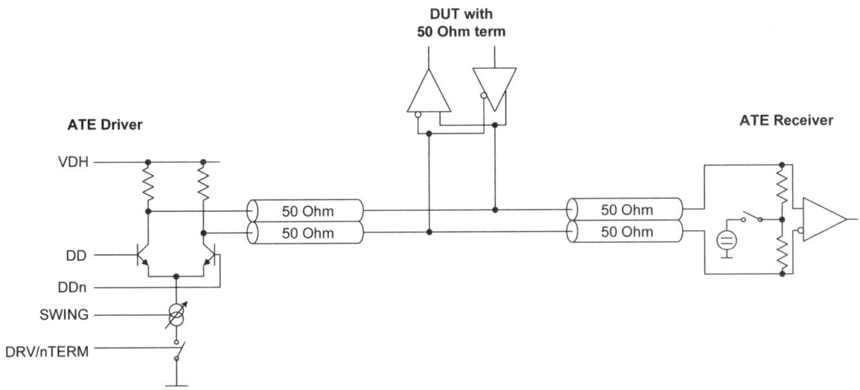

Figure 8.60 Bidirectional interface to unidirectional ATE channels connection diagram using a dual transmission line (fly-by) approach.

This approach introduces a discontinuity at the point where the ATE driver and receiver are connected together. To reduce the length of this discontinuity to a minimum, one possibility is to connect the ATE driver to the receiver at the socket via that goes to the DUT. This can be achieved with minimal coupling between driver and receiver signals if the ATE driver and receiver traces are on different signal layers as shown in Figure 8.61.

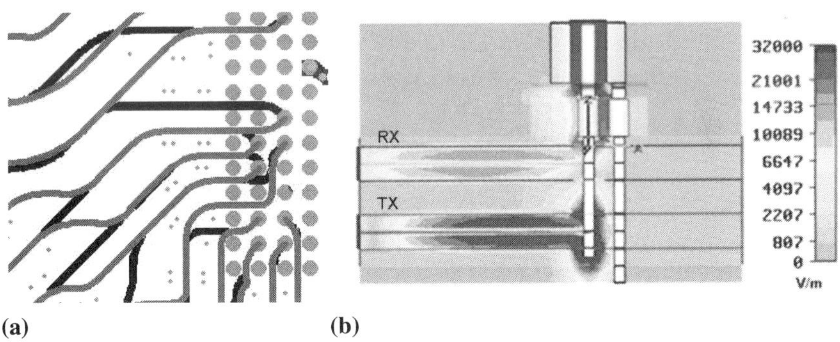

Figure 8.61 (a) Example of a bidirectional interface layout (the driver and receiver lines are on different layers and they join on the via below the I/O package pad) and (b) a 2D cross section of a 3D EM simulation.

Another option to address the discontinuity of connecting the ATE driver to the receiver is to use a power divider (Section 7.17.1) to maintain a 50 Ω impedance through the entire signal as shown in Figure 8.62.

Although this approach solves the discontinuity problem, it raises other challenges due to the need to add a surface mount component (the power

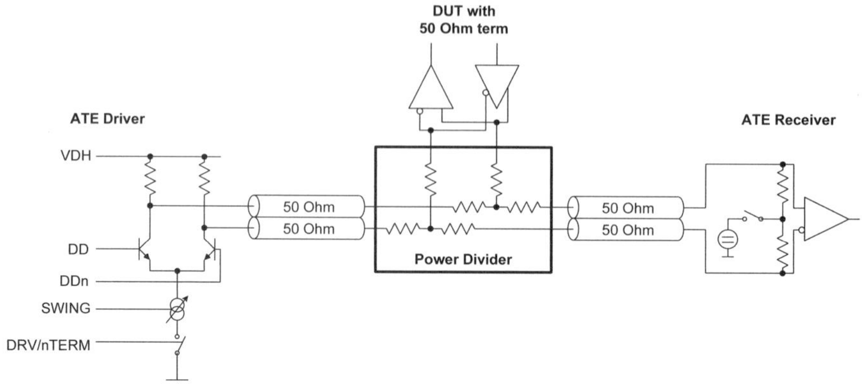

Figure 8.62 Bidirectional interface to unidirectional ATE channels connection diagram using a dual transmission line with a power divider.

combiner), which adds additional transitions and by design reduces the signal power by half. In memory applications, one could argue that when the DUT is receiving, then the fly-by connection is really just a long transmission line with a high impedance DUT receiver tapping into the center with minimal impact on the signal integrity. When the memory device is transmitting, the transmitter impedance is actually lower then 50 Ω and may find it better matched with the two fly-by transmission lines in parallel having a total of 25 Ω impedance as seen from the DUT. Additional comments regarding the application side of this approach are discussed in Section 9.4.

8.10 Sockets

The test fixture socket corresponds to the "last mile" of the signal path from the pin electronics to the packaged DUT and is no less important. Historically the key design goal of a socket was mechanical reliability in the form of repeatable low resistance DC connections and compliance for DUT package variations. At high-speeds one must also consider the impedance discontinuity determined by the socket materials, contactor dimensions, and if the high-frequency signal path of the contactor in the compressed position is repeatable [60, 61]. Production test socket requirements are often at odds with high-speed performance since one of the simplest ways to improve high-speed performance is to make the socket as thin as possible, which leaves little room for mechanical compliance. References [62, 63] provide an introduction to IC sockets and Figures 8.63 and 8.64 show examples of different socket types.

The socket is usually the most stressed element of a test fixture. The ATE test fixture might be docked to an ATE test head two to ten times

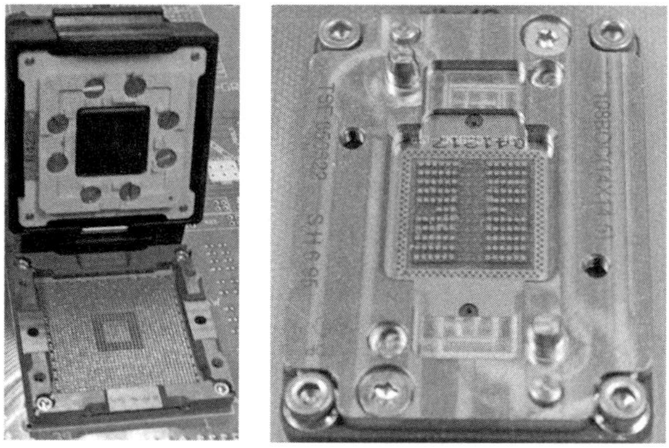

SPRING PIN CONTACT SOCKET FOR LARGE SIZE BGA

ELASTOMER CONTACT SOCKET FOR MEMORY APPLICATIONS

Figure 8.63 Example of a spring pin (pogo pin) and an elastomer sockets for BGA packages.

COAXIAL SOCKET

PACKAGE ON PACKAGE SOCKET

Figure 8.64 Examples of coaxial and package-on-package (POP) DUT sockets (Courtesy of WinWay®).

per day, while in one day a socket might have dozens of IC insertions for characterization/design verification or several thousands of IC insertions for production testing. Although it is expected that a socket should be repairable or replaceable, any reduction in the lifetime or reliability of the socket directly impacts cost of test [64].

A socket for a test and measurement application can be divided in three parts: the socket housing, the contactor, and the socket lid as shown in Figure 8.65. The socket housing is typically built from a polymer material like polyethylene terephthalate (PET) selected for mechanical stability and precision machining properties [65]. The socket housing provides a way to correctly push the DUT package into the socket contacts and keep it there. The socket lid will usually contain a heat sink and sometimes a fan for temperature control of the DUT, although in some cases more complex cooling approaches are necessary as discussed in Section 8.10.5. However, for applications requiring automatic DUT package insertion in the socket through a robotic handler, the socket will not have a lid and in this case uses an open top structure with the DUT package being "pushed" into the socket by the handler plunger. For BGA type packages that correspond to the majority of high-speed digital applications, Table 8.7 presents a list of socket contact technologies [62, 63]. Note that this list is not comprehensive since there are multiple socket contact technologies available and it is not easy to classify all of them into simple groups. Also, although a given contact technology might have a given disadvantage described in the table, there might be a socket vendor that developed a version of that contact technology with added features that addresses that disadvantage.

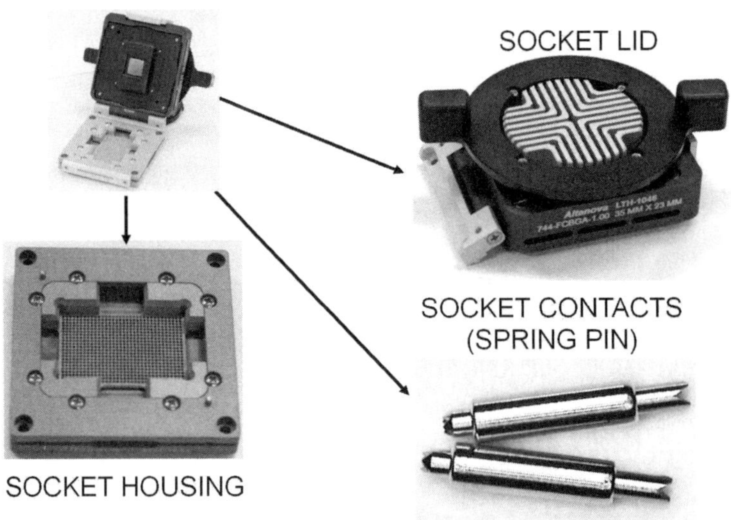

Figure 8.65 Breakup of a typical socket into its individual parts (Courtesy of R&D Altanova).

Spring pin or pogo pin type contact continues to be the dominant approach in sockets for high-speed digital applications. Figure 8.66 shows a spring pin (pogo pin) example. Spring pin based sockets because of the inherent mechanical advantage of a spring can have a very large compliance which is important for very large BGA packages. However, on the other side spring pins cannot be made arbitrary small both in terms of length and diameter. The compressed length of the spring pin is a critical factor for signal and power integrity. The spring pin diameter also has an impact on the signal

Table 8.7
Comparison of Contact Technologies for BGA Sockets

Contact Technology	Advantages	Disadvantages
Spring pin contact	High-bandwidth, high compliance	Spring contact high-frequency path length, no wipe
Elastomer contact	Fine pitch capability, minimal socket thickness (very low inductance)	No single contact replacement, elastomeric creep or stress relaxation, solder ball bottom contact, no wipe
Coaxial contact	High bandwidth even for long spring pin lengths	Spring pin length still impacts power integrity, high cost
Direct spring contact	Replaceable contact, short electrical path, high-frequency application. Examples are the fuzz button [66] and the BeCe contact [67]	No wipe action due to straight compression, no penetration action
Cantilever contact	Conventional technology, low force relaxation, replaceable contact	Limited pitch capability, long contact path, contacting solder ball bottom, no penetration action
Rocking or rolling contact	High bandwidth, wipe action, short signal path [68]	PCB pad offset, rocking board wear, Elastomer fatigue, difficult to replace contacts

integrity as will be discussed later in this section, but it is also a factor when designing sockets for DUT packages with a very small pitch.

From a signal integrity point of view the design of a spring pin socket requires optimizing the spring pin diameter with the socket pitch to achieve the target impedance as shown in Figure 8.67. However, in most cases the optimal design for signal integrity does not correspond to the best mechanical design, which means a compromise must be reached in the socket design [60].

Another important point when designing a contactor for a socket like a spring pin is the shape of the tip what will touch (probe) the solder ball. The objective is to have good contact to the DUT package solder balls, which might require breaking any layer that has formed around the solder ball to obtain a good contact. However, on the other side it cannot deform the package

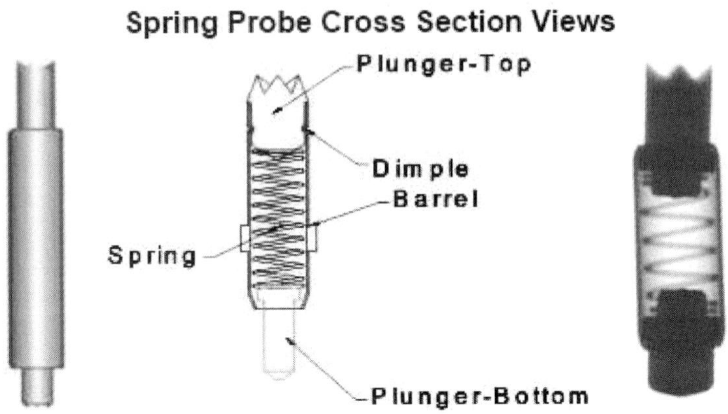

Figure 8.66 Spring pin (pogo pin) diagram (Courtesy of WinWay®).

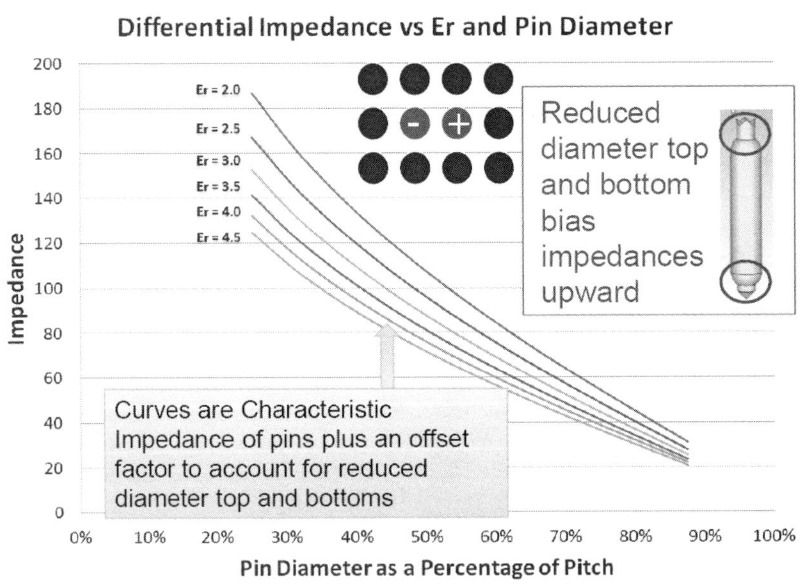

Figure 8.67 Impact of the socket pitch and spring pin diameter into the socket differential impedance (reprinted with permission from [60]).

solder balls in a way that can impact its soldering or reliability later. Figure 8.68 shows an example of different types of tips for a spring pin.

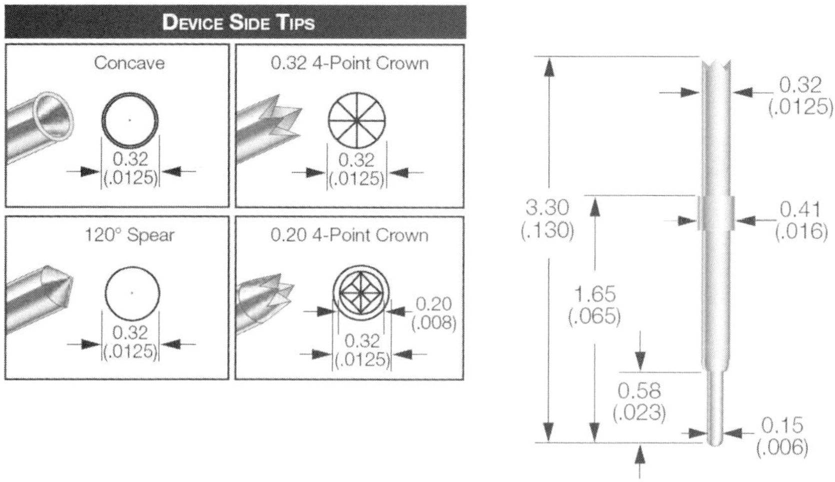

Figure 8.68 Example of a spring pin and the different options for the tip (Courtesy of Smiths Connectors).

Elastomer type sockets have been gaining traction due to their signal integrity advantages although compliance limitations especially for large BGA packages is a major challenge. Different technologies can be used to create an elastomeric socket which is usually composed of two parts: A metallic element and an elastomer [69]. The metal element can be in a powder form, particles, or wires. The elastomer element can be a flexible adhesive binder, silicon, epoxy or a synthetic rubber. Figure 8.69 provides a high level view of how an elastomer socket works.

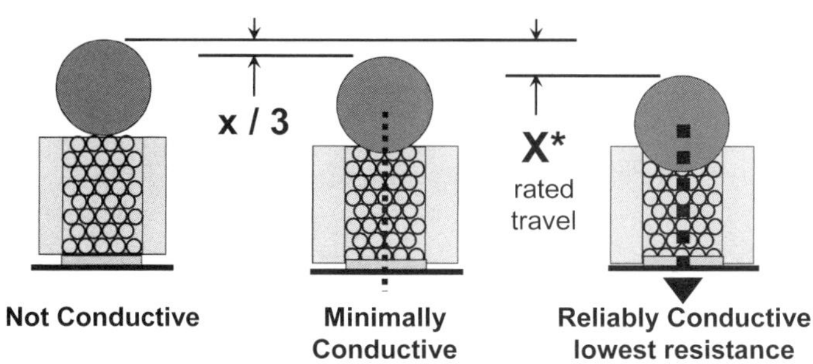

Figure 8.69 Diagram of how an elastomer socket works (reprinted with permission from [69]).

Another challenge of elastomer type sockets is the contact with the package BGA balls since there is no tip like on a spring pin contact. This is especially important in a production environment where a large number of parts are tested. Socket manufacturers have addressed this challenge by using a hybrid approach where a separable top layer consisting of contact pins is assembled to the top of the elastomer for a better contact as shown in Figure 8.70.

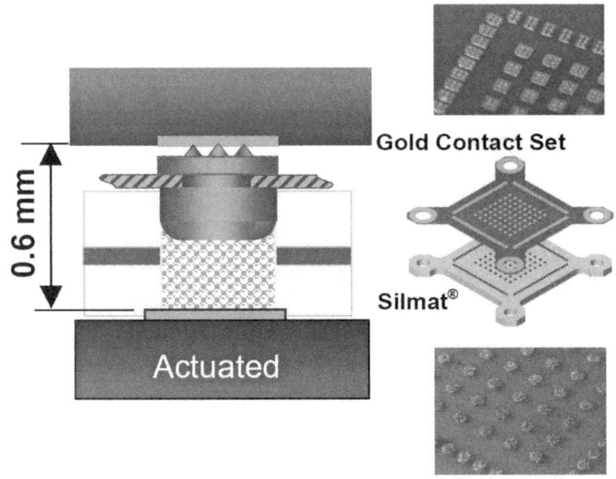

Figure 8.70 Hybrid elastomer socket with a separable top copper tip layer for better contact to the DUT (Courtesy of Phoenix Test Arrays).

Coaxial sockets [70, 71] try to address the signal integrity challenges that arise from having a DUT BGA ballout that is not optimized (e.g., nonideal BGA ground pin assignment) and the fact that the socket pitch and spring pin diameter have an impact on the socket impedance. This impact increases for longer spring pins which might be required for compliance in a large DUT BGA. A coaxial socket addresses this problem by creating a true coaxial structure for each signal pin as shown in Figure 8.71. The drawback of this approach is of course the higher cost that a coaxial socket entails.

8.10.1 Sockets for Package-on-Package (POP) Applications

Package-on-package (POP) applications [73, 74] present some new challenges for socket design [75, 76]. The challenge is that a POP application consists of two packages that are assembled together as shown in Figure 8.72.

The challenge for the test engineer is the testing of the bottom packaged IC. The top packaged IC can tested using a standard socket but the bottom package cannot use a standard socket because the pads on the top of the

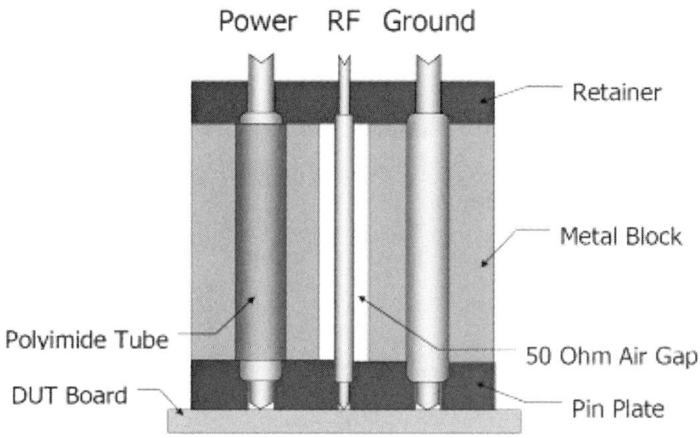

Figure 8.71 Coaxial socket diagram (reprinted with permission from [72]).

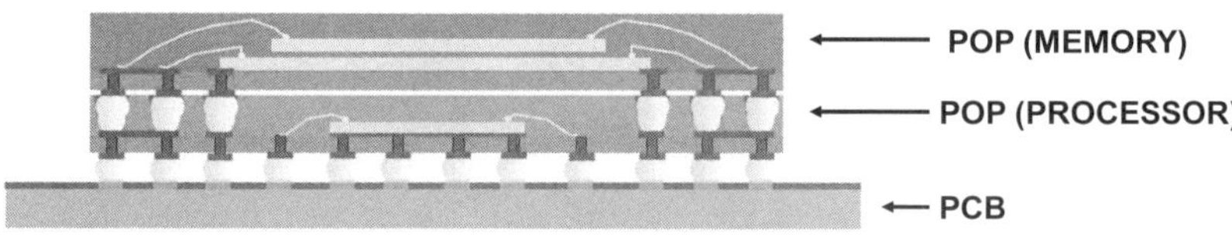

Figure 8.72 Diagram of a package on package (POP) application (picture courtesy of Bob Willis [77]).

package also need to be tested. Two types of approach are possible as shown in Figure 8.73. One is to use a socket that also allows the probing of the top package pads. The other approach is to use a socket where the top packaged IC (e.g., a LPDDR memory) is included on the socket so that bottom IC is tested with the top packaged IC also connected to it. The challenge with the second approach is that there is no access to the top pads of the package since the memory package will be there. On the other side, although the first approach provides access to the top pads of the package to the measurement system, there will be a signal integrity challenge due to the long signal path length to those pads.

8.10.2 Socket Electrical Characterization

Evaluating the electrical performance of a socket is a complex task [79–81]. Socket specifications are defined in a very specific framework that usually does not correspond to the environment where the socket is used on a test fixture. Table 8.8 shows one example of a manufacturer's specification for a socket.

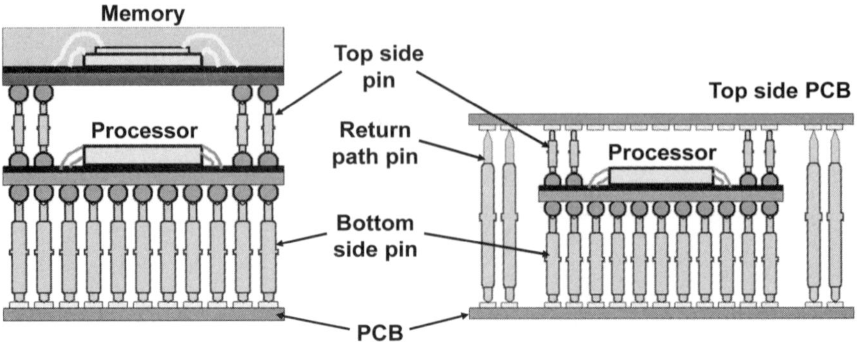

Figure 8.73 Example of a package on package (POP) socket options (reprinted with permission of Smiths Connectors [78]).

Table 8.8
Example of a Manufacturer Specification for Socket with a Spring Type Contactor

Electrical Specifications	Value
Contact resistance	18.62 mΩ
Self inductance	0.6 nH
Insertion loss	< 1 dB to 10 GHz
Pin to pin capacitance	0.075 pF
Current carrying capability	3 A

The first item described in the specification is the contact resistance. This is an important parameter especially for DC type tests. The challenge with contact resistance is that one desires it to be very low and not changing even after a large number of insertions. This usually conflicts with the requirement to use very low resistance materials (especially in the plating of spring pin type sockets) since low resistance is typically found in higher purity softer metals that do not have the mechanical properties to allow the socket to handle a large number of insertions. However, there are alloys that can withstand a very large number of insertions but have a higher contact resistance value. It is then important when looking into the contact resistance value for a specific socket to understand how it will change with the number of insertions.

Self-inductance is also an important value since it represents the physical dimensions of the socket and provides important data to understand how the socket will impact items like the power distribution to the DUT (see Section 8.11.4). It is important to note that self-inductance is usually measured in a very specific configuration (e.g., four ground pins surrounding the socket pin being measured) that might not correspond to the pin configuration in the final application.

The pin-to-pin capacitance specification allows a judgment of the capacitive coupling between the pins in the socket, which can be directly related to the possible crosstalk between the pins. However, a 3D EM field solver is needed to accurately predict both the capacitive and inductive coupling in the socket for a given DUT ballout.

The insertion loss specification provides an idea of the performance of the socket in terms of its frequency range. Note again that this specification might be based on a very specific pin configuration that may not represent the final application BGA ballout. The maximum current specification indicates the upper limit of the continuous current the socket can handle [82].

8.10.3 Socket Cleaning

Socket cleaning is an important topic, especially in a production environment. For example, the continuous contact of the spring pin tip with the BGA solder balls will contaminate the spring pin tip increasing its contact resistance until it is too high for properly testing the DUT [83]. Multiple cleaning technologies and strategies exist but an in-depth discussion on them is outside the scope of this book.

8.10.4 Space Transformers for ATE Test Fixtures

Space transformers are becoming critical components especially for low-pitch devices. A space transformer allows the test engineer to use a larger pitch on the ATE test fixture BGA area than the DUT pitch [84]. This can be very helpful since low pitch devices (e.g., 0.4 mm) can make ATE test fixtures more expensive or harder to manufacture given the typical thickness of ATE test fixtures. A space transformer can also be used not only to change the pitch but also to add additional components like power planes or embedded decoupling capacitors. Figure 8.74 shows an example of a space transformer.

8.10.5 Socket Temperature Control

Another critical item when designing a socket is to decide what type of temperature control (if any) is needed for the DUT package. Some devices like microprocessors consume a significant amount of current and can reach very high temperatures. Also some test modes like scan might have a power consumption that is higher than the device would have in normal operation and in this way generate a larger amount of heat. For some devices it is enough to add a heat sink on top of the socket that contacts the DUT package. The heat sink size and shape must be designed taking into account the heat dissipation

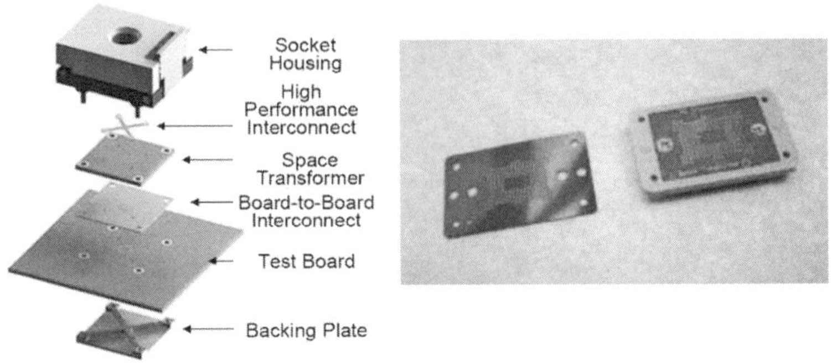

Figure 8.74 Example of a space transformer for an ATE test fixture (reprinted with permission from [84]).

needs [85]. In some instances it might be necessary to add an electrical fan on top of the heat sink to help to dissipate the heat as shown in Figure 8.75 (left).

Figure 8.75 Example of a socket with a heat sink and a electrical fan (left) and with water cooling (right).

However, in some cases this might not be enough and more active measures are needed to control the DUT package temperature. One option is to use water cooling on the socket. In this case water pipes are embedded in the heat sink and water flows through the socket heat sink to a chiller that keeps the water temperature at a specific value as shown in Figure 8.75 (right). Another approach is to use a peltier element to control the package temperature [86, 87]. Figure 8.76 shows one example of this type of approach.

Another option is to use a thermal airstream system as shown in Figure 8.77. This is an external instrument that connects to the socket and forces a certain temperature through an airstream. The peltier and airstream-based approaches are especially critical if the DUT needs to be tested at extreme temperatures.

Figure 8.76 Example of a socket with a temperature control through a peltier element from Silicon Thermal [86].

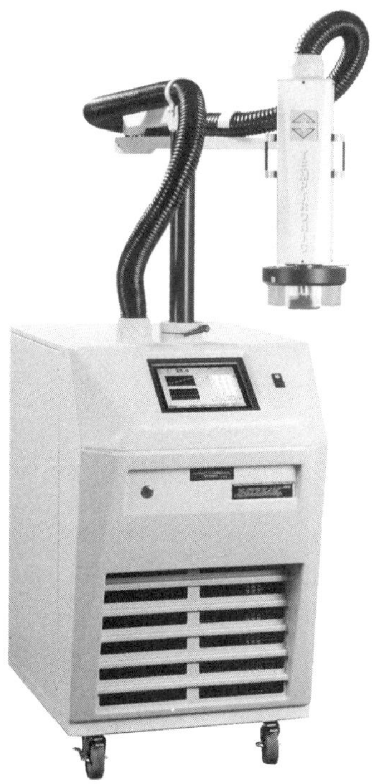

Figure 8.77 Example of an airstream system for temperature control (Courtesy of Temptronic).

8.11 Power Distribution Network Design

Power distribution on an ATE test fixture is a complex topic for which a complete detailed treatment is outside the scope of this book. Proper design of the power distribution network[7] (PDN) in the ATE test fixture is critical for the performance of high-speed digital devices. Applications like a microprocessor might require peak currents in the hundreds of amperes, creating significant challenges for the ATE power supplies and the test fixture PDN design [88, 89]. The references [3, 90–97] provide good starting points for a general treatment on this topic.

The PDN design challenges are directly related to the transistor density in the DUT, I/O pin count, data rate, and clock frequency as shown in Figure 8.78. This means that the challenges will continue to increase with the expected move to higher data rates and I/O pin counts together with lower supply voltages in the future.

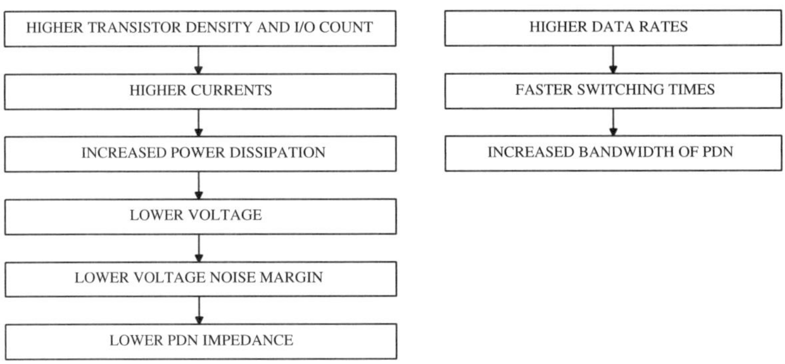

Figure 8.78 The PDN design challenge with the increased integration and data rates.

Noise in the PDN can have a direct impact on the DUT performance by increasing the maximum delay and the delay uncertainty inside the DUT, increasing the DUT jitter [98, 99], degrading the noise margin, and also degrading the gate oxide reliability [100]. A low noise PDN on a digital application PCB is also important to reduce EMI emission, although this is usually not an issue on test fixtures for ATE.

One important point to note is that the power distribution requirements of a DUT while being tested or characterized can be significantly different from the power requirements when the DUT is being used on its target application [i.e., when one designs a PCB board that will contain this device (and normally several others) to perform a certain task]. The reason is that when a DUT is being tested or characterized on an ATE system, the

[7] Also sometimes referred as the power distribution system (PDS).

switching behavior might be different from its mission mode. Typically, the switching behavior is higher, meaning that there are higher current requirements from the DUT (e.g., during scan test the number of flip-flops switching simultaneously is very high). However, when testing the DUT PDN performance, there is the question of whether the test fixture PDN should be the same as that on the PCB where the DUT will reside for its final application (e.g., a desktop computer motherboard). Some engineers advocate that the PDN performance on the test fixture should be the best possible independent of the PDN performance of the final application PCB. Note that, in some cases, the performance of the test fixture PDN might actually be worse than that of the final PCB (system board) even when using the best possible PDN design as shown in Figure 8.79. One of the reasons can be seen in Figure 8.80 where the thickness of an ATE test fixture is compared to the thickness of the PCB where the DUT will reside for its final application. The increased thickness in the ATE test fixture will result in increased inductance for the decoupling capacitors as will be shown in Section 8.11.2. The socket were the DUT will reside in the ATE test fixture also exacerbates this problem.

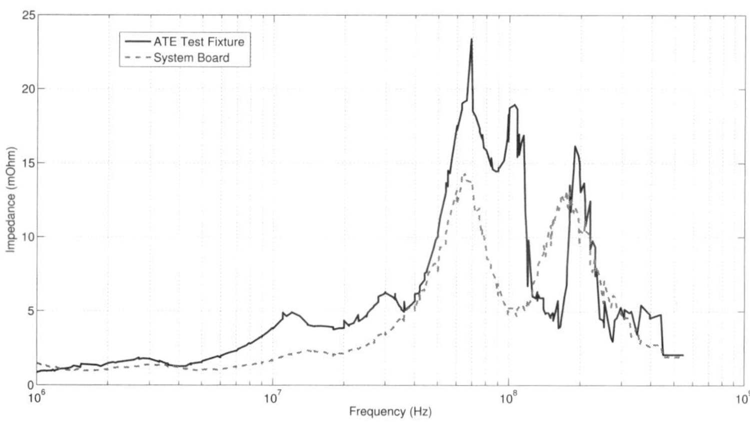

Figure 8.79 Example of the self-impedance difference between the application PCB board the DUT is intended to be used on and the ATE test fixture (reprinted with permission from [101]).

For ATE manufacturers, the challenge is not only to develop device power supplies (DPS) with the required current/voltage capabilities but also to be able to integrate a sufficient number of these power supplies into an ATE system, since ICs can require multiple different power supplies.

The test fixture PDN design challenge is presented in Figure 8.81. A PDN is not a simple power transmission line from the ATE DPS to the DUT, but it is a complex hierarchical system where power is stored and delivered

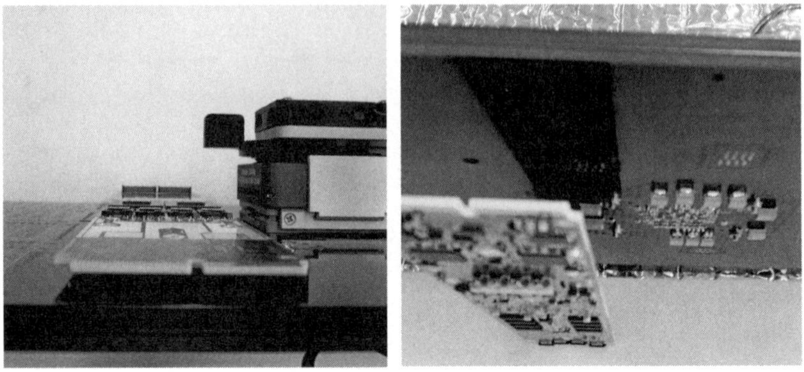

Figure 8.80 Comparison of the PCB thickness of an ATE test fixture to the PCB board where the DUT will be used in its final application.

by different components as shown in Figure 8.82. Each component of the PDN is responsible for providing power to the DUT in a different frequency range. The ATE DPS is responsible for supplying the bulk power, but its response time is very slow, which means that it cannot compensate for the power requirements of the DUT when it is switching its transistors at GHz frequencies. The decoupling capacitors and the power planes are responsible for storing and delivering power to the DUT for its needs in the hundreds of MHz range. Above these frequencies, the power delivery is the responsibility of the package and the on-die capacitance since the vias from the DUT socket to the PCB power planes and the DUT package will limit the maximum frequency of the PDN on the test fixture. All these components must work together to provide the proper power distribution to the DUT.

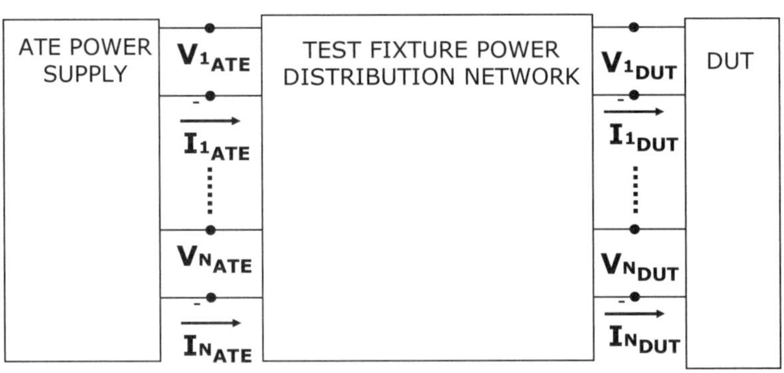

Figure 8.81 Diagram of the power distribution network (PDN) design challenge.

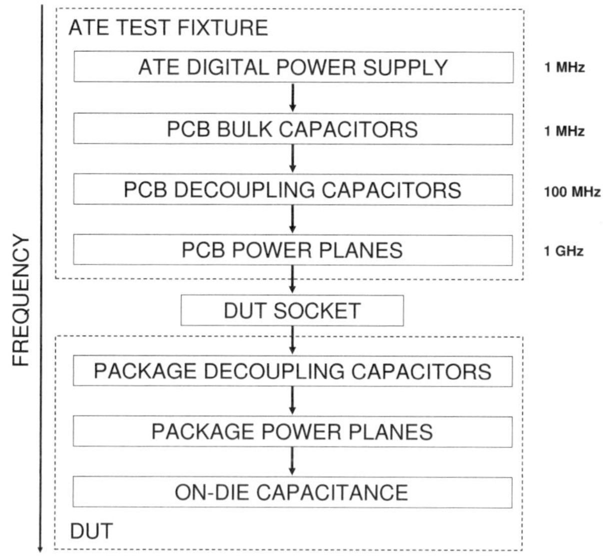

Figure 8.82 Hierarchical description of the components of a power distribution network.

When designing a power distribution system for a test fixture, the objective is to keep the voltage change at the DUT power supply inputs as small as possible when the DUT forces a change in the supply current (dI/dt) due to transistor switching. Typically, this requirement is defined in terms of the maximum ripple allowed on the power supply at the DUT power input package pads [102]. Assuming a maximum possible transient current for the DUT, it is possible to compute the needed impedance value using Ohm's law:

$$Z_{\text{TARGET}} = \frac{\text{Power Supply Voltage} \times \%\text{Allowed Voltage Ripple}}{\text{Transient Current}} \tag{8.6}$$

This value will correspond to the maximum allowed impedance that guarantees that the voltage ripple will be below the required threshold. Equation (8.7) shows an impedance computation example using this methodology:

$$\begin{aligned} &\text{Power Supply Voltage} = 1.8\ \text{V} \\ &\text{Allowed Voltage Ripple} = 5\% \\ &\text{Transient Current} = 1\ \text{A} \\ &Z_{\text{Target}} = \frac{1.8 \times 0.05}{1} = 0.09\ \Omega \end{aligned} \tag{8.7}$$

Unfortunately, a frequency independent value for the PDN is not possible. This is because each component of the PDN (power planes, decoupling capacitors, vias) has an impedance behavior that is not constant with frequency, as will be shown later. This means that one needs to define also a frequency range where the impedance of the PDN must be below the target impedance as shown in Figure 8.83. This frequency range will depend on the rise/fall time of the transient current to the DUT. While the magnitude of the transient current determines the target impedance, the rise/fall time determines the upper corner frequency where the target impedance should be met.

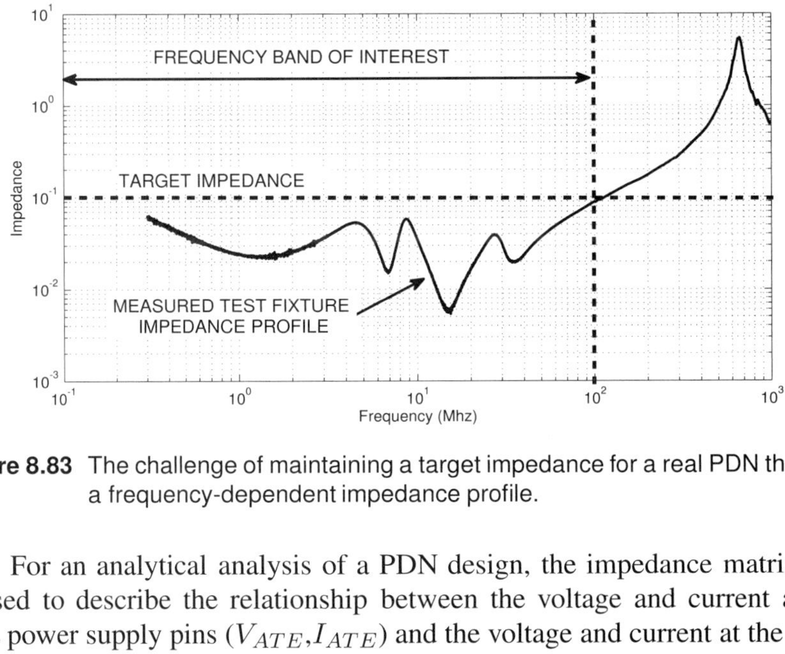

Figure 8.83 The challenge of maintaining a target impedance for a real PDN that has a frequency-dependent impedance profile.

For an analytical analysis of a PDN design, the impedance matrix (Z) is used to describe the relationship between the voltage and current at the ATE power supply pins (V_{ATE},I_{ATE}) and the voltage and current at the DUT package power supply (V_{DUT},I_{DUT}) as shown in Figure 8.81 and described in (8.8).

$$\begin{bmatrix} V_{ATE} \\ V_{DUT} \end{bmatrix} = \begin{bmatrix} Z_{11}(f) & Z_{12}(f) \\ Z_{21}(f) & Z_{22}(f) \end{bmatrix} \begin{bmatrix} I_{ATE} \\ I_{DUT} \end{bmatrix} \tag{8.8}$$

Z_{11} is the self-impedance seen from the ATE power supply pins side and Z_{22} is the self-impedance seen at the DUT power supply pads side. Z_{12} and Z_{21} are the transfer impedance values from the ATE power supply pins to the DUT power supply pads and vice versa. Since we assume the PDN network to be reciprocal [103], then $Z_{12} = Z_{21}$. Figure 8.84 shows one example of the measured impedance matrix Z magnitude from a real test fixture (Section H.2 discusses in detail the measurement of the PDN impedance in a test fixture).

The most important result is the self-impedance at the DUT (Z_{22}). By analyzing the impedance profile, it is then possible to validate if the PDN impedance is below the target impedance in the frequency range of interest. The self-impedance at the DPS (Z_{11}) and the transfer impedance (Z_{12},Z_{21}) are of less importance, although Z_{11} is sometimes of interest to evaluate possible issues of the PDN design in regards to the ATE DPS (e.g., the DPS feedback loop stability). For time-domain analysis, (8.8) can be converted into the circuit shown in Figure 8.85 [104]. The next sections will discuss in more detail the different components of the test fixture PDN.

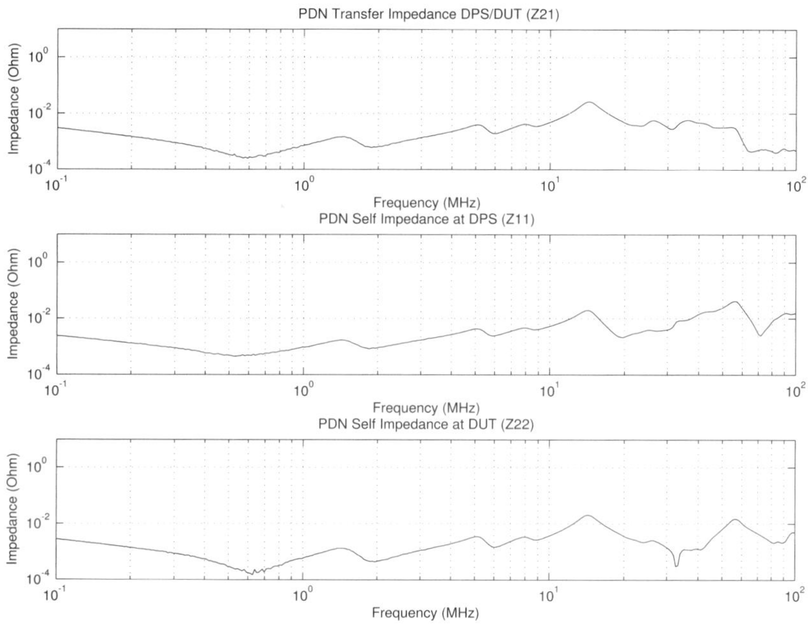

Figure 8.84 Impedance magnitude values for the power distribution network of a test fixture: Z_{21} transfer impedance from the DPS to the DUT (top), Z_{11} self-impedance at the DPS side (center), and the Z_{22} self-impedance at the DUT side (bottom). The Z_{12} transfer impedance is equivalent to Z_{21}.

It is important to understand that the PDN performance will have an impact on the DUT performance. Figure 8.86 shows a comparison of three different test fixture PDN designs for the VDDQ power supply of a DDR3 memory application. The figure shows the measured data eye at one DQ pin for each PDN design. The DQ pin was running at 1.6 Gbps. Note that in one of the PDN examples there are no decoupling capacitors mounted on the test fixture, just the capacitance of the power planes is available to the DUT. The results show the impact that the PDN design can have on the data eye in the

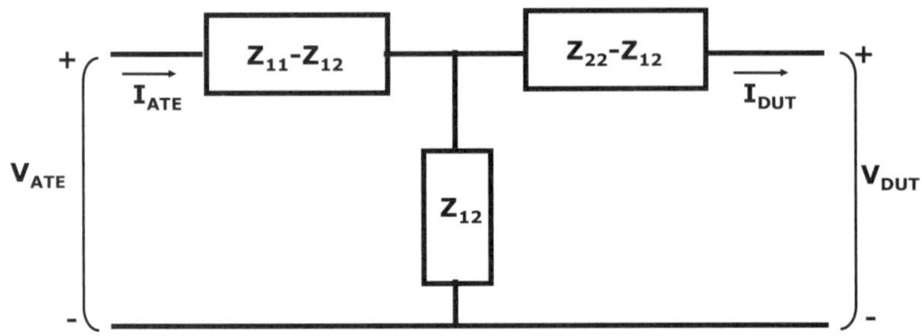

Figure 8.85 Equivalent circuit for time-domain analysis of a PDN based on the impedance matrix Z.

form of a reduced data eye width, especially in the case of single-ended I/O interfaces like DDR3. In this case the PDN design with the lowest impedance profile provides the best data eye performance in terms of the data eye width.

It is also interesting to note that even in the case where there were no decoupling capacitors assembled on the test fixture, the DUT was still able to function showing the importance of the power planes for the PDN performance as will be discussed in the next section. One of the reasons is that when a high-speed digital DUT is switching, the requirements on the power supply are across the frequency range and not at a specific frequency as shown in Figure 8.87, which implies the need to design the PDN taking into account its frequency response.

8.11.1 Power Planes

The task of designing a power distribution network starts with appropriate choices for the power and ground planes for the ATE test fixture PCB stack-up. The power and ground planes serve as a storage capacitor that is able to store and provide small amounts of charge at very high frequencies when compared to the ATE DPS or the decoupling capacitors. Its ability to store charge is characterized by the power plane capacitance which can be approximately computed by the following equation [4]:

$$C = \frac{0.0885\ \varepsilon_r\ A}{d}\ \text{pF} \tag{8.9}$$

where ε_r is the relative dielectric permittivity of the dielectric material used for the power plane, A is the area of the power plane in cm^2, and d is the height of the dielectric in cm between the power and ground planes as shown in Figure 8.88.

Since the objective is to minimize the loop inductance between the power plane and ground plane pair, the power plane should be made with

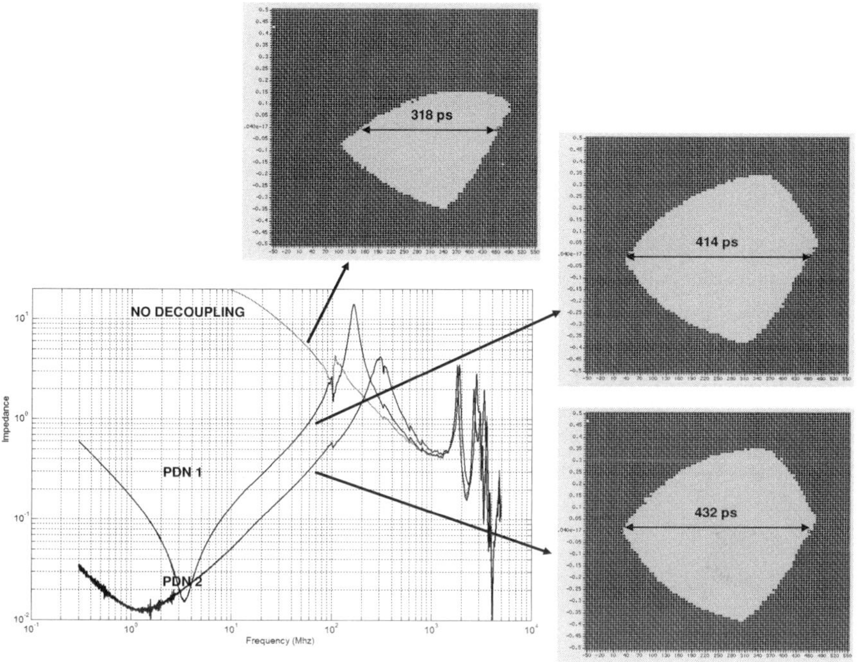

Figure 8.86 Comparison of the effect of different test fixture PDN designs on the data eye from a DQ pin of a DDR3 memory running at 1.6 Gbps.

the smallest dielectric height possible, typically 50.8 to 101.6 μm (2 to 4 mil) using a dielectric material with a large ε_r value. This is not the same as the low loss requirement for high-speed digital signal routing where a low ε_R and loss tangent are preferred. There are dielectric materials that have been developed specifically to be used in power planes, with a very small thickness and a large ε_r value. This type of dielectric material can have significant advantages as shown in [105].

Figure 8.89 shows an example of a stack-up for a power intensive application (a microprocessor). Note that each power supply has a separate power plane. This stack-up also shows one of the main challenges associated with the layer choice for the power planes in a multilayer test fixture PCB. The power planes should be as close as possible to the top of the test fixture PCB to reduce the inductance of the vias connecting the socket power pins to the power planes. However, to maintain stack-up symmetry for minimal PCB warpage, manufacturers prefer to place an equal number of power planes on the other side of the PCB, or place them all together in the middle. In the example of Figure 8.89, the power planes were placed at the middle of the

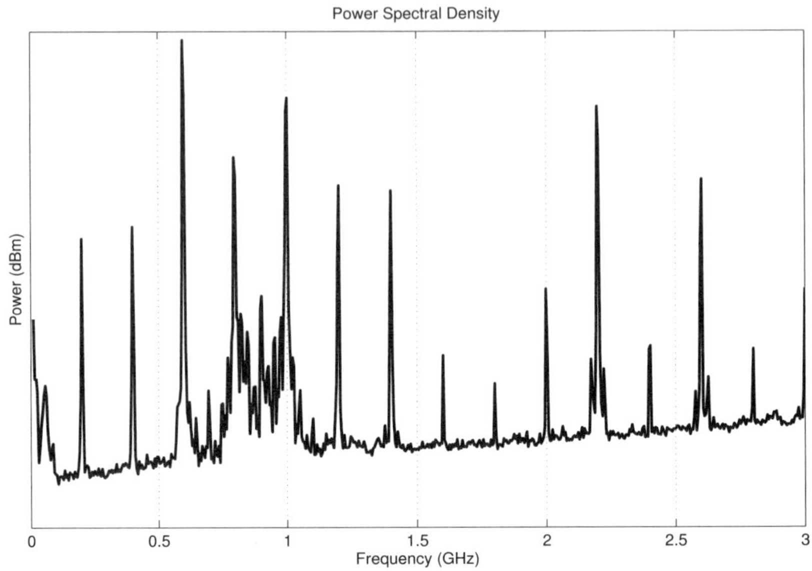

Figure 8.87 Power spectral density of the noise present on the power plane of an ATE test fixture for a high-speed memory running at 1.6 Gbps. The noise spectrum was measured by connecting a spectrum analyzer to a test point on the ATE test fixture power plane with a 50 Ω probe (high impedance compared to the PDN) while the DUT was running a pattern that generated a large number of logic transitions.

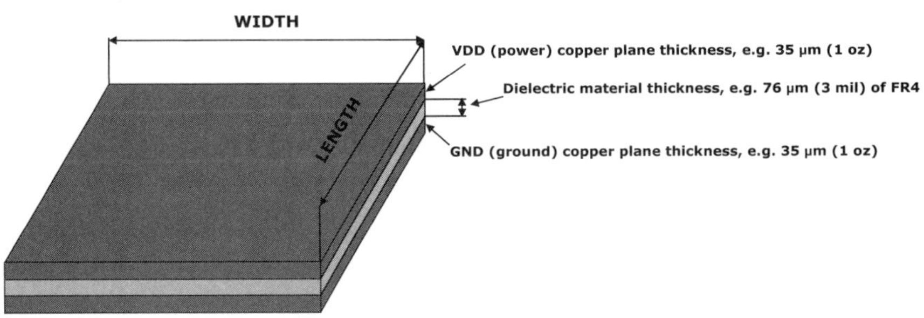

Figure 8.88 Geometry of a power plane.

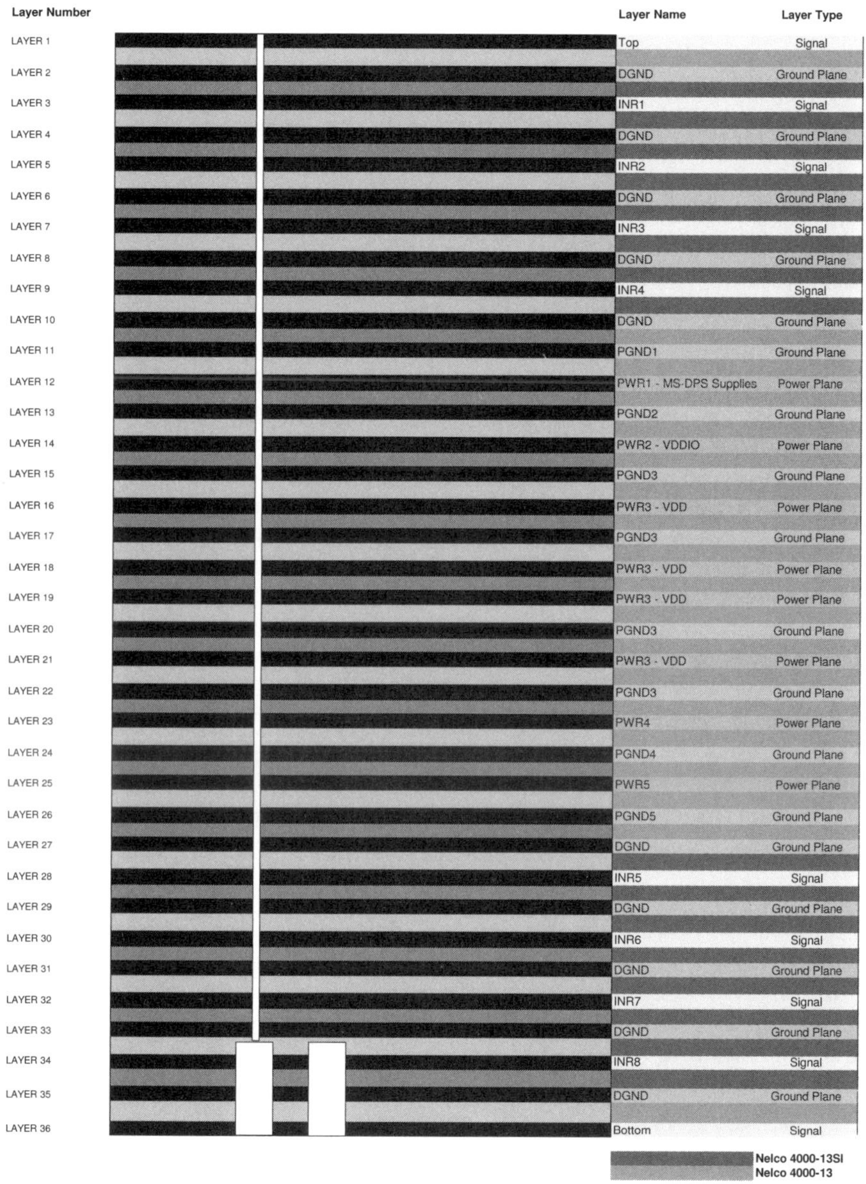

Figure 8.89 Example of the stack-up for a power intensive DUT (Courtesy of R&D Altanova).

stack-up due to manufacturing reasons, which implies additional inductance for the vias connecting the power planes to the DUT socket.

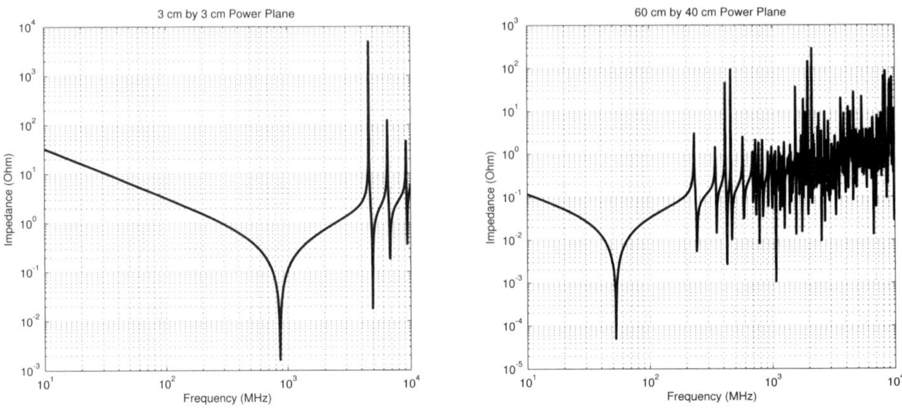

Figure 8.90 Example of the self-impedance profile for two different power planes at the geometric center. A 3 by 3 cm (1.2 by 1.2 in) power plane (left) and a 40 by 60 cm (15.8 by 23.6 in) power plane (right). Both power planes have a 76.2 μm (3 mil) dielectric thickness using a FR4 type dielectric [106].

It is important to understand that the impedance response of a power plane in regards to frequency will depend significantly on the geometry of the power plane and the specific locations where the DUT power pins reside. Figure 8.90 shows an impedance plot versus frequency of two different power planes with different areas but the same dielectric height. The impedance is measured at the geometric center of each power plane. The figure shows that the large power plane, although having a higher capacitance, has its resonant frequency around 50 MHz. On the smaller power plane, the resonant frequency is around 900 MHz.

ATE test fixtures are usually large in size, which means that their resonant frequencies will be very low if the power plane covers the entire PCB area. Figure 8.91 shows a picture of an investigation board created to measure the resonant frequency in an ATE test fixture. The board size is 58.2 by 42.9 cm (22.9 by 16.9 in) corresponding to a full-size test fixture for a particular ATE platform. Two power planes were implemented using an FR4 dielectric with a thickness of 63.5 μm (2.5 mil) and 190.5 μm (7.5 mil) and another power plane was implemented with an 11 μm (0.43 mil) 3M embedded capacitance dielectric material. The measured results shown in Figure 8.92 show that the resonant frequency is below 100 MHz for the full-size and half-size test fixture. Note also the results for the thicker power plane where it is possible to observe the reduced capacitance when compared to the thinner power plane.

Figure 8.91 Picture of a printed circuit board designed to investigate the power plane characteristics of a typical ATE test fixture (Courtesy of R&D Altanova and Advantest).

The resonant frequencies can have an impact on the PDN performance since they can interact with the decoupling capacitor resonant frequencies. One approach developed to smooth the resonant frequencies peaks at higher frequencies is to use a lossy decoupling or edge termination around the power planes [107, 108]. Notice that this is opposite to the need for lower loss dielectric materials on high-speed signal layers and often results in a mixture of laminate materials for a high-performance PCB test fixture. Note also that the impedance profile of a power plane not only varies with frequency but also with the geometric location where the impedance is being measured or calculated especially at high frequencies as shown in Figure 8.93 [91].

8.11.2 Decoupling Capacitors

Decoupling capacitors, also known as bypass capacitors, are used to provide a low impedance source of energy to middle and high-frequency changes in the DUT load current. One can imagine that they are like an energy reservoir for fast changes on the load current that cannot be compensated fast enough by the ATE power supply. The challenge is that capacitors are not ideal elements, which forces the power distribution network design to use a mixture of different decoupling capacitor types and multiple numbers of the same capacitor to reduce the parasitic inductance. Figure 8.94 shows a picture of

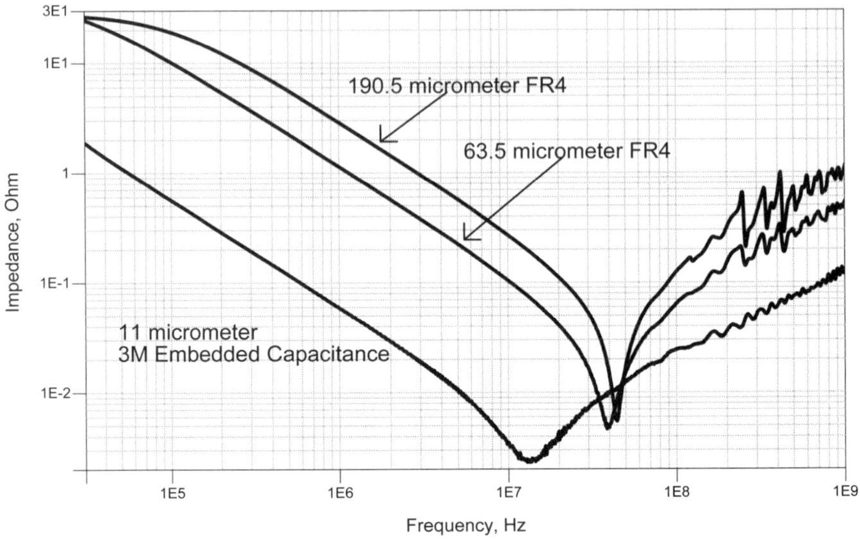

Figure 8.92 Impedance results at the power plane center for a full-size (58 by 43 cm) ATE test fixture with 63.5 μm (2.5 mil) and 190.5 μm (7.5 mil) FR4 dielectric thickness and an 11 μm (0.43 mil) 3M embedded capacitance dielectric material.

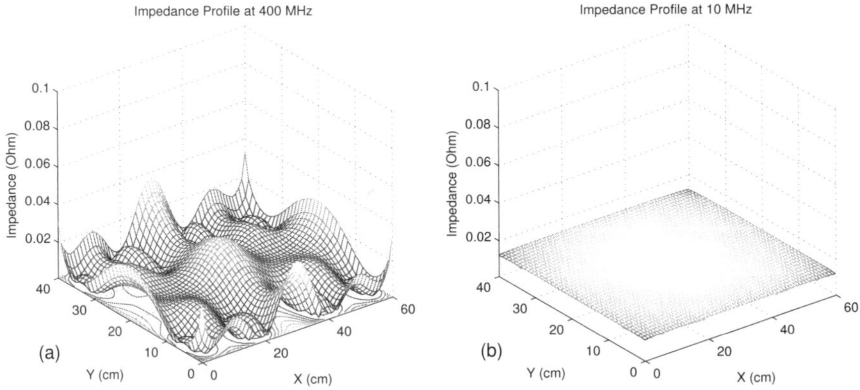

Figure 8.93 Example of (a) the impedance distribution at 400 MHz and (b) 10 MHz for a 40 by 60 cm (15.8 by 23.6 in) power plane with a 76.2 μm (3 mil) dielectric thickness using an FR4 type dielectric [106].

the decoupling capacitors on the bottom of an ATE test fixture under the DUT socket.

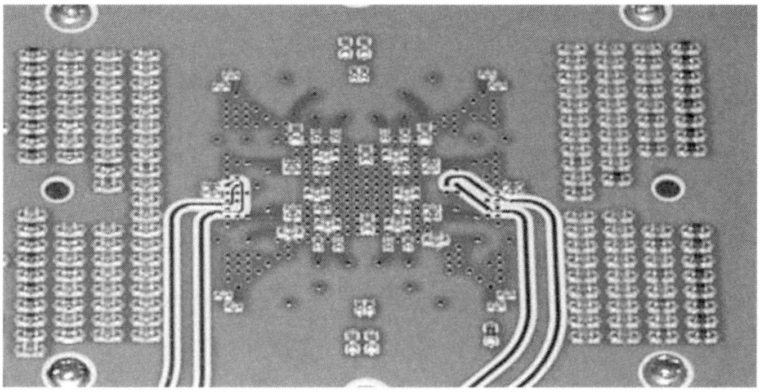

Figure 8.94 Example of the decoupling capacitors on the bottom of the test fixture below the DUT socket and also surrounding it.

A real capacitor does not behave as an ideal capacitor in the entire frequency range and is typically modeled by a capacitance value together with an equivalent series resistance (ESR) and an equivalent series inductance (ESL) as shown in Figure 8.95. Since a real capacitor is an RLC circuit, the ESR and ESL values will determine the resonant frequency for the capacitor. Note that this is still only an approximation since the ESR and ESL values of a capacitor will vary with frequency [91].

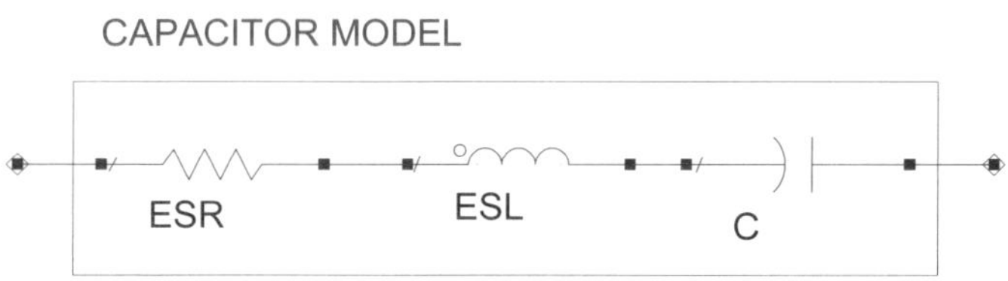

Figure 8.95 Typical model for a surface mounted decoupling capacitor.

To better understand the effect that the ESR and ESL have on a capacitor impedance, Figure 8.96 shows a plot of the impedance versus frequency for an ideal 1 μF capacitor and a 0603[8] type 1 μF capacitor with an ESR of 15 mΩ and an ESL of 2 nH.

The figure shows that there is a resonance due the capacitor reactance being equal to the inductor reactance [2]. After this resonance, the capacitor no longer behaves like a capacitor but instead like an inductor. It is important to note that the ESL value will depend on the PCB footprint and also on

[8]0603 refers to the physical size of the capacitor, which is standardized.

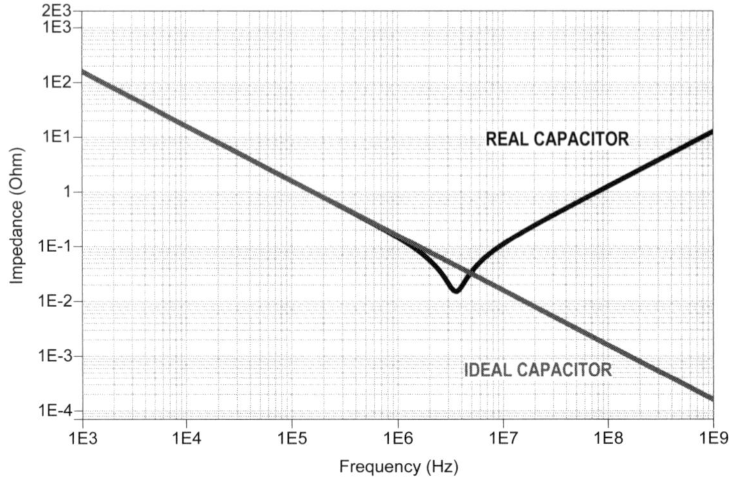

Figure 8.96 Comparison of an ideal 1 μF capacitor with a 0603 1 μF capacitor with an ESR of 15 mΩ and an ESL of 2 nH.

the length of the via connecting the decoupling capacitor to the respective power plane. Because of this, an appropriate footprint should be used for the decoupling capacitors to keep the inductance to a minimum as shown in Figure 8.97.

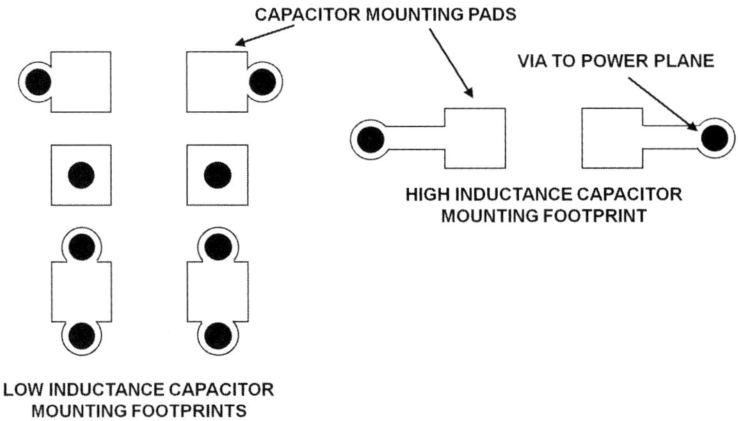

Figure 8.97 Capacitor mounting footprint influence on the capacitor ESL value.

To understand the effect of increasing or decreasing the ESR and ESL value of a decoupling capacitor on its frequency-domain response, Figure 8.98 shows a comparison of an ideal capacitor with that of a more realistic

ESR/ESL capacitor model. The left figure shows the effect of increasing the ESR value of the capacitor while keeping the ESL value constant. In this case the resonance of the capacitor is smoothed and the capacitor is no longer able to achieve the same low impedance value at the resonance frequency as before. This effect will be important when we discuss the combination of multiple different decoupling capacitors later. Figure 8.98 (right) shows the effect of reducing the ESL value of the capacitor while keeping the ESR value constant. In this case, the capacitor frequency response shows that the capacitor resonance happens at a higher frequency meaning that it is able to behave as a capacitor to a higher frequency. These results show that ideally one should try to use capacitors with the lowest possible ESL value.

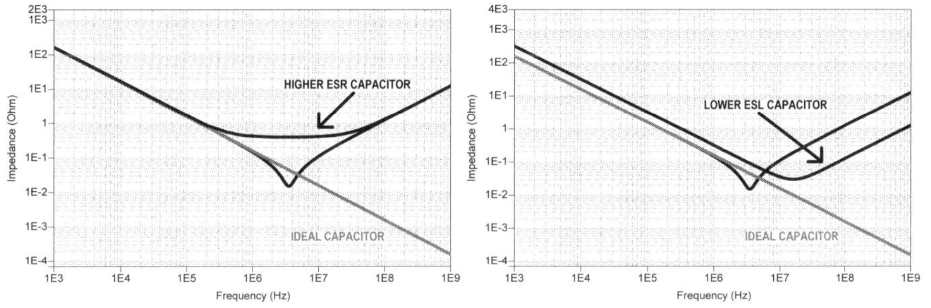

Figure 8.98 Comparison of the effects on the capacitor frequency response of an increase on the ESR value or a decrease on the ESL value.

Figure 8.99 shows how the vias connecting the decoupling capacitor to the power plane add to the decoupling capacitor parasitic inductance value and in this way increase its ESL value. To minimize this value, the power planes should be as close as possible to the decoupling capacitor although this is not always possible. The common approach to the capacitor parasitic inductance challenge is to reduce the inductance by using multiple capacitors to achieve a given capacitance and parasitic inductance value. For example, if the PDN design requires a 4 μF capacitor but the associated ESL and parasitic inductance are too high, it is possible to use four 1 μF capacitors instead. If the ESL value of each is the same, the parasitic inductance will be reduced by a factor of four when compared to using a single 4 μF capacitor. This effect is shown in Figure 8.100. The drawback of this approach is, of course, the layout space these capacitors will take while their additional cost is only a minor factor for an ATE test fixture.

Apart from the physical sizes, a capacitor type also depends on its underlying manufacturing technology. Table 8.9 describes some of the capacitor technologies available for use in a PDN for high-speed digital applications [2, 109].

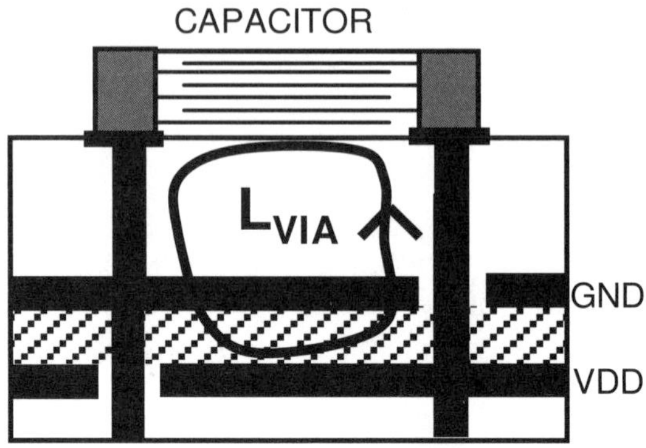

Figure 8.99 Graphic demonstration of the parasitic inductance created by the connection of a surface mounted decoupling capacitor to the power plane through vias.

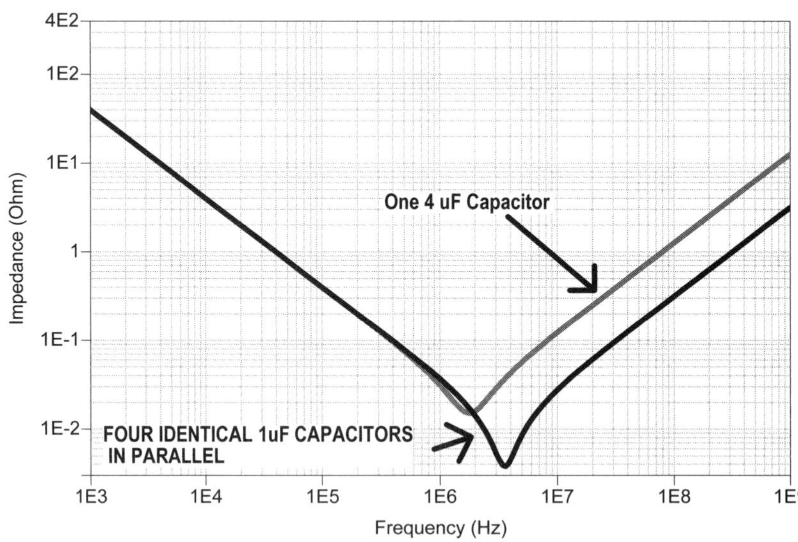

Figure 8.100 Comparison of the effect of adding four identical capacitors in parallel.

Table 8.9
Comparison of Capacitor Types

Capacitor Type	Comments
Tantalum	1 to 1,000 μF, medium physical size, low voltage, low ESR versions available (polymer tantalum and aluminum polymer). Usually considered as a low-frequency bulk capacitor. Avoid tantalum-MnO2
Ceramic	Small capacitance, small physical size, high and low voltage, very low ESR, highest reliability
Arrays	Ceramic capacitors with an array of contacts for very low ESL
Low inductance chip capacitor (LICC)	Sometimes referred to also as a reverse geometry capacitor (RGC), it has a very low ESL but with a nonstandard footprint
X2Y	Four-port decoupling capacitor, very low ESL, sizes 0602 to 2220
Interdigitated capacitor (IDC)	Very low ESL
Controlled ESR	Controlled ESR value for better resonance control (1 μF and 10 μF) [110]
OS-CON	Wound polymer aluminum capacitor, low ESR, usually considered as a low-frequency bulk capacitor
Aluminum electrolytic	High voltage, used as low-frequency bulk capacitor

One important topic when discussing decoupling capacitors is what happens when assembling multiple different decoupling capacitors in parallel. Since each decoupling capacitor has an ESR and an ESL value, this parallel circuit can have a complex behavior. To better understand this, Figure 8.101 shows two different circuits of two parallel decoupling capacitors where all the capacitors have the same ESR and ESL value but their capacitance value separation is different (one circuit contains a 1 μF/100 nF capacitor combination and the other a 1 μF/10 nF combination). The frequency-domain self-impedance results for both circuits show that after the self-resonance of the first capacitor occurs, another resonance between the two capacitors' self-resonances exists. The problem with this resonance is that its peak corresponds to a high impedance value, which is bad for the power distribution performance. The results also show that in the case where the capacitors are further apart in capacitance value (10 nF compared with 100 nF), the peak of the resonance between the two capacitors is higher. The conclusion to be taken from this small experiment is that when using multiple capacitor values in a PDN, it is important to use multiple capacitance values that are not very far apart in their capacitance value. This effect can also be reduced by using capacitors with higher ESR values, as will be discussed next.

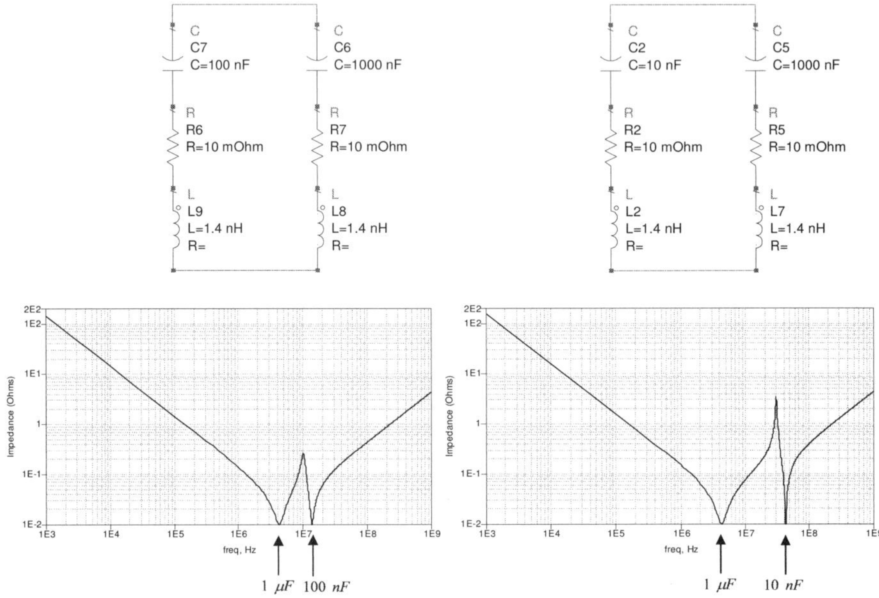

Figure 8.101 Example of the resonances of a two-capacitor decoupling configuration.

One new category of decoupling capacitor that has become available is the controlled ESR capacitor [110]. The ESR value of a capacitor can have a significant impact on the shape of the resonance associated with the decoupling capacitor by smoothing its resonance as already shown in Figure 8.98 (left). The important point is that in a configuration with multiple different capacitors, a higher ESR value generates smaller resonances between the different capacitors as shown in Figure 8.102. The figure shows the impedance profile of two circuits each composed of a 1 μF and a 10 nF capacitor. The ESL of each capacitor is 1.4 nH and the ESR of each capacitor is 10 mΩ, with the exception that in the second circuit the ESR value for the 1 μF capacitor is increased to 100 mΩ. This increase on the ESR is enough to bring the peak of the resonance below 1 Ω.

In the cases where layout space is at a premium (e.g., multi-site applications) and given the fact that a significant number of capacitors might be needed to achieve the desired power distribution network, one option to

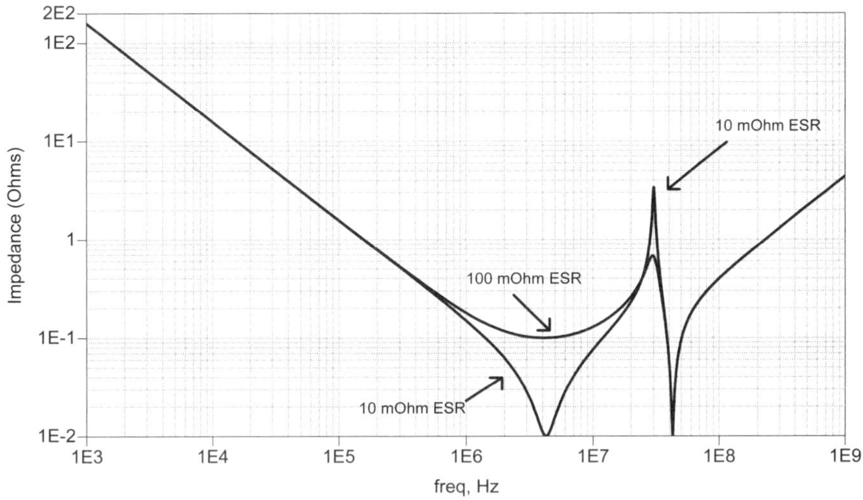

Figure 8.102 Comparison of the effects on the capacitor frequency response with an increase in the ESR value.

address this challenge is to stack the capacitors as shown in Figure 8.103. The drawback is that the higher the capacitor is on its stack, the larger will be its loop inductance. Because of this, if a stacked configuration is used, it is important to have the lowest value capacitors on the bottom of the stack.

Figure 8.103 Stacked capacitors in a test fixture.

In some situations, the test engineer designing the test fixture PDN might need to add some decoupling capacitors, especially on the low-frequency range after the DUT loadboard is manufactured to tune the PDN based on measured data. One approach to prepare for this kind of rework is to use a footprint consisting of long copper bars (bus bar design) as shown in Figure 8.104 where different types of decoupling capacitors can be easily soldered after the DUT loadboard is manufactured without needing to worry about a specific footprint [102]. Note that it is critical to include enough vias under the copper bars to keep the inductance to the power planes low.

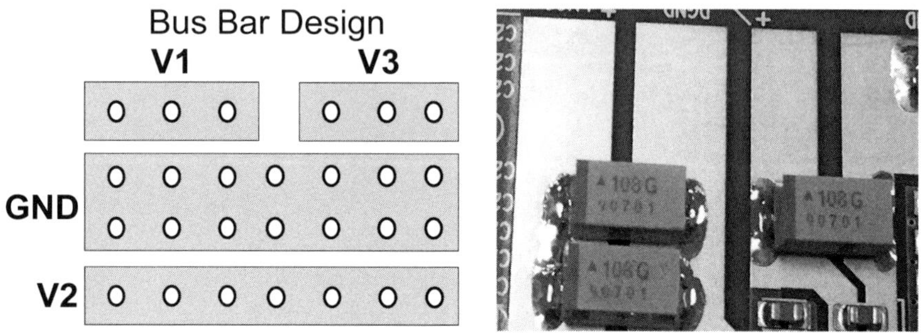

Figure 8.104 Generic footprint for adding decoupling capacitors to a DUT test fixture.

Another technique that has started to gain acceptance is to embed the capacitors in the test fixture PCB [111]. This approach allows the decoupling capacitors to be mounted much closer to the DUT BGA power pin so that the parasitic inductance of any mounting vias or connection traces can be reduced. This approach has been further extended to integrating the decoupling capacitors in the DUT socket [101, 112–114].

8.11.3 Inductors and Ferrite Beads

Inductors especially in the form of ferrite beads are commonly used to isolate power distribution network noise especially in noise sensitive applications like mixed-signal but also sometimes in high-speed digital applications. It is important to use them in an appropriate way, since if they are misused, they can have a significant impact on performance [115].

8.11.4 Socket Inductance

The need to use a socket for the DUT on an ATE application raises the issue of the inductance of the socket power pins since the socket pins will have a certain length. This issue becomes more problematic if in its final application the DUT is soldered to the PCB without any socket as shown in Figure 8.105.

As already discussed in Section 8.11.1, this problem can be exacerbated if the power planes cannot be set close to the top of the test fixture PCB due to

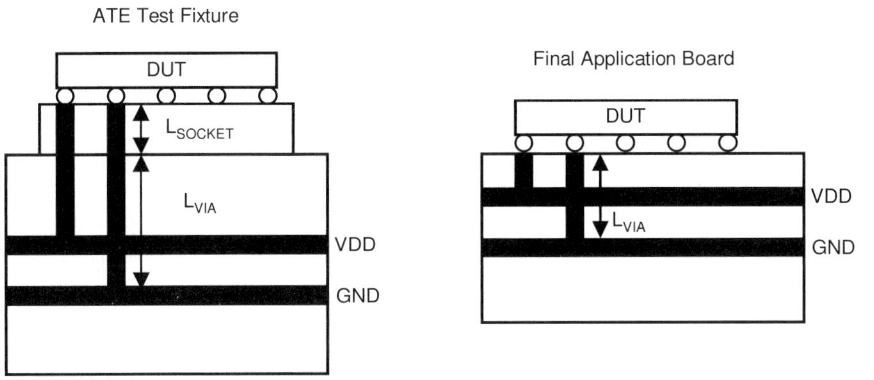

Figure 8.105 Comparison of the inductance of the DUT connection to the power planes for an ATE test fixture and a final application PCB where the DUT is soldered without any socket.

stack-up restrictions. This would make the length of the vias from the socket to the power plane longer than in the final application, thus increasing the inductance.

There are not too many alternatives available to address this issue. One option is to reduce the socket inductance by making the socket spring pins shorter or to use some type of elastomer socket. Reducing the socket pin length has mechanical limits and using a elastomer type contactor instead of a socket also raises other problems (reduced mechanical compliance being one of them). Another option is to add power decoupling into the socket itself [101, 113, 114]. This approach is especially suited to applications where the DUT BGA contains a central area that is dominated by power pins (like on a high-performance microprocessor). The drawbacks are the additional complexity and cost and also the fact that it still cannot guarantee that the performance would be equal to the case of the DUT BGA being directly soldered into the PCB. A similar technique is to add a power plane in the socket using a high-performance dielectric material [116]. Another approach is to evaluate if the added inductance from the socket (and the via to the power plane) is compensated enough by the package substrate and on-die power decoupling. The problem with this approach is that it requires a good knowledge and model of the DUT PDN.

8.11.5 Power Distribution Network Design

Designing a power distribution network for a test fixture consists first of identifying the power requirements of the DUT and its impact on the PDN impedance requirements as described in (8.6). The next step is to choose the stack-up for the power planes taking into account the number of power supplies in the DUT and the discussion on power planes in Section 8.11.1 followed by the choice of the needed decoupling capacitors. Note that choosing the decoupling capacitors entails not only choosing their values but also their type and location as discussed in Section 8.11.2.

Power distribution design is sometimes seen as a "black-magic" art, with several design philosophies available in the technical literature that are sometimes contradictory [117]. Some of the available design methodologies are:

- Frequency-domain target impedance method (FDTIM);
- Multipole (MP);
- Capacitors-by-the-decade (CBD);
- Big "V";
- Distributed matching bypassing.

Reference [93] provides more detailed discussion into some of the above methods. In most PDN designs the guiding objective is not only achieving the target impedance across the PDN frequency band of interest but also to make the impedance response as flat as possible across the frequency band of interest.

8.11.6 Power Distribution Network Simulation

It is important to be able to simulate a DUT loadboard PDN before physically building the test fixture to evaluate the choice of the decoupling capacitors and verify if the PDN fulfills the impedance requirements. Usually one starts with an initial idea consisting of the power planes and the number and types of decoupling capacitors. This could be based in the previous test fixture design for a similar application. The next step is to model each item using the ESR and ESL values for the capacitor and power planes. This is the tough part since sometimes these values are not readily available and one might need to make an educated guess, although some manufacturers provide design kits with capacitor models. It is also critical that the inductance value of the vias from the decoupling capacitors to the power planes is added to the ESL value of the decoupling capacitors. This value will be dependent on the stack-up and must be computed based on the via length (see [91] for some guidelines on computing the via inductance).

This type of simulation can be easily performed on a SPICE-like simulator as shown in Figure 8.106. The impedance profile is obtained by using a 1 A amplitude sinusoidal current source and measuring the voltage. The values will correspond to the impedance value since $Z = V/I = V/(1\ \mathrm{A}) = V\ (\Omega)$.

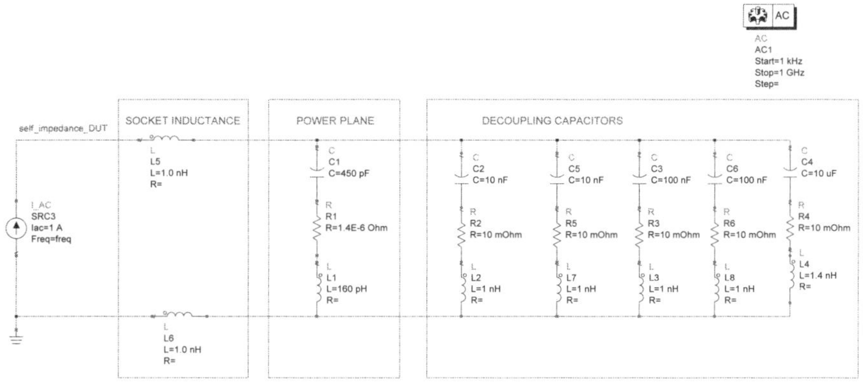

Figure 8.106 Example of a simplified test fixture power distribution network simulation example using an RLC model for the decoupling capacitors, socket via, and power planes.

Figure 8.107 shows the simulation results from the setup in Figure 8.106. In the simulated impedance profile it is easy to identify the resonance peaks due to the different decoupling capacitors. This allows the designer to identify where to try to improve the impedance profile by adding more decoupling capacitors with different capacitance values or changing to a different type of capacitor like a controlled ESR one. Because this type of PDN model treats the PDN as a lumped circuit (i.e., neglects the distributed nature of the design), one should not expect a high degree of accuracy. For example, such a model will overestimate the severity of resonances (and thus cause a conservative design in this respect), but also overestimate the high-frequency performance (where propagation times become important for the PDN's reaction to fast changes).

Apart from SPICE, several more complex tools exist that are specifically designed for power distribution network simulation. Some of these tools are discussed in Appendix G.

8.11.7 Disconnecting Bulk Capacitance for Faster DC Measurements

One important point when designing ATE test fixtures for high-current applications is that the need to have large values of bulk capacitance also means that for certain DC measurements a large settling time is required. This is due to the fact that the time constant associated with these bulk capacitors

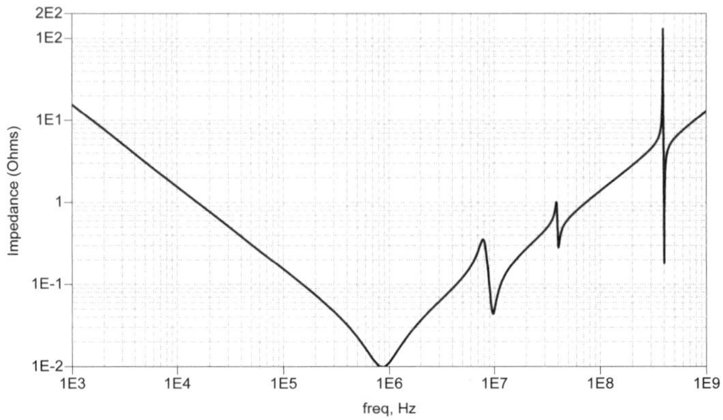

Figure 8.107 Simulated impedance results from the simulation setup of Figure 8.106.

is very large meaning that, for example, in an IDD measurement one needs to wait a long time for the measured value to settle.

One approach to circumvent this problem is to disconnect the bulk capacitance from the DUT PDN when performing DC measurements. This can be done by using a relay or a transistor as shown in Figure 8.108.

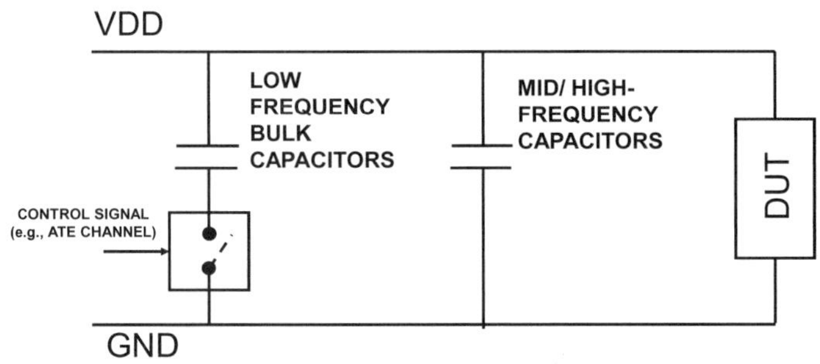

Figure 8.108 Using a relay or MOSFET transistor to connect/disconnect the bulk capacitors in an ATE test fixture PDN.

8.11.8 Stability of the ATE DUT Power Supply

When designing the PDN for an ATE test fixture, it is important to make sure that the PDN design will guarantee a stable power supply operation. Most ATE power supplies specifications will include a minimum of capacitance that is required on the ATE test fixture for the ATE DUT power supply (DPS) to properly operate. A maximum capacitance value is usually also specified. However, it usually does not specify how this capacitance should be implemented. The problem is that an ATE DPS with a very fast response time can have an unstable behavior if the test fixture PDN is not properly implemented as shown in [118]. The stability of the ATE DPS feedback loop will usually be determined by the PDN impedance response on the hundreds of kHz range where the bulk capacitors dominate. It is the choice of these bulk capacitors that is critical for the ATE DPS stability [102, 118]. Some ATE DPS modules address this challenge by having a selectable DPS control loop bandwidth. In this case, the user can reduce the loop bandwidth to guarantee the stability of the PDN with the drawback that the response time of the ATE DPS will also be slower.

8.12 HIFIX

A special style of ATE test fixtures that can be found in the production testing of high-speed memory devices where multiple sites are required on a single ATE test fixture. In this case a combination of high density coaxial cables and connectors with an array of small DUT socket board PCB assemblies is utilized. This type of test fixture is also sometimes referred to as a HIFIX.[9] Figure 8.109 shows an example of a HIFIX test fixture for a 256-site production testing of a DDR3 application [119]. Note that this type of test fixture is significantly different from the test fixtures shown on Figure 8.3. Figure 8.110 shows a diagram with the different components that are usually present on a HIFIX for production testing of memory ICs like DDR [119].

Figure 8.109 A 256-site test fixture for high-speed memory applications (Courtesy of Advantest).

[9]High-Fidelity Test Access Fixture.

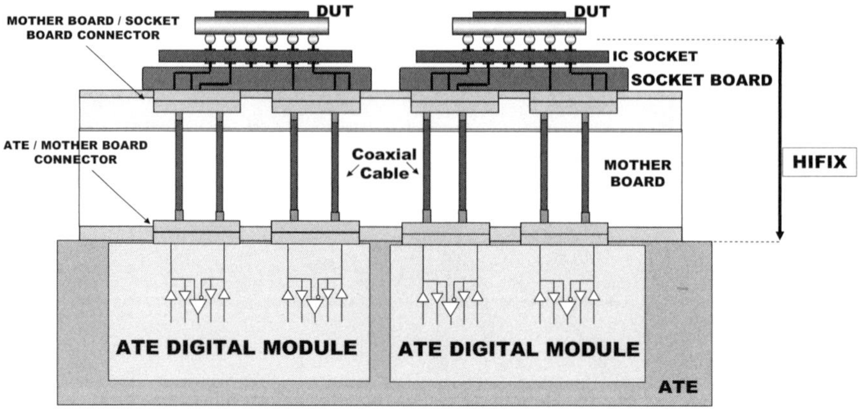

Figure 8.110 Block diagram showing the different components of a HIFIX intended for production testing of memory ICs like DDR.

The use of multiple small PCB socket board assemblies with coaxial cabling to the ATE interface presents additional challenges in the design of the connecting cable transitions and the limited space for layout and support circuitry. Figure 8.111 shows a HIFIX test fixture where the cover was removed to show the cable connections and a picture of a socket board.

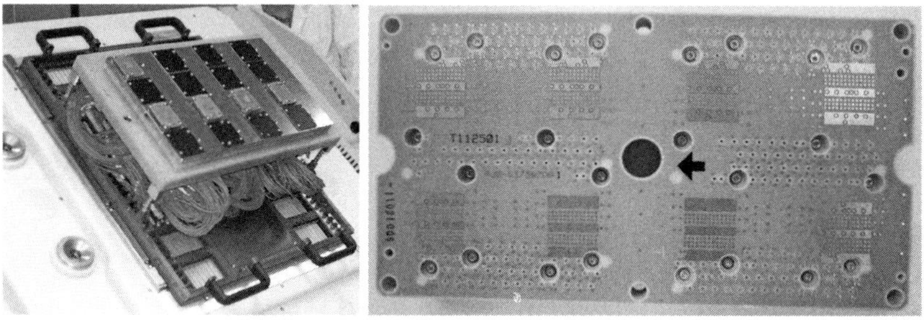

(a) (b)

Figure 8.111 Example of (a) a multi-site high-speed memory testing test fixture without the cover showing the cabling to the socket boards and (b) a socket board for four DUT sockets (Courtesy of Advantest).

8.13 Wafer Probing[10]

Wafer probing is such a large and important topic that it cannot be properly addressed in this short section. One good source for up-to-date information

[10]The term wafer testing is also used.

is the yearly IEEE Semiconductor Wafer Test Workshop [120], which is dedicated to semiconductor wafer testing. Wafer probing is part of almost every production test-flow [121], but in this section we will only address wafer probing for at-speed digital testing.

At-speed wafer probing has been in use for some time in the ATE industry even at data rates of 10 Gbps, although usually only for devices with limited pin count, in the wired communication area. An example is shown in Figure 8.112. The increased cost of packages and the need to provide known good die (KGD) for multichip module (MCM) and through silicon via (TSV) integration has raised the need for KGD testing. In KGD testing the die must be fully tested for failures, which usually implies not only the standard DC type tests but also at-speed functional testing [122].

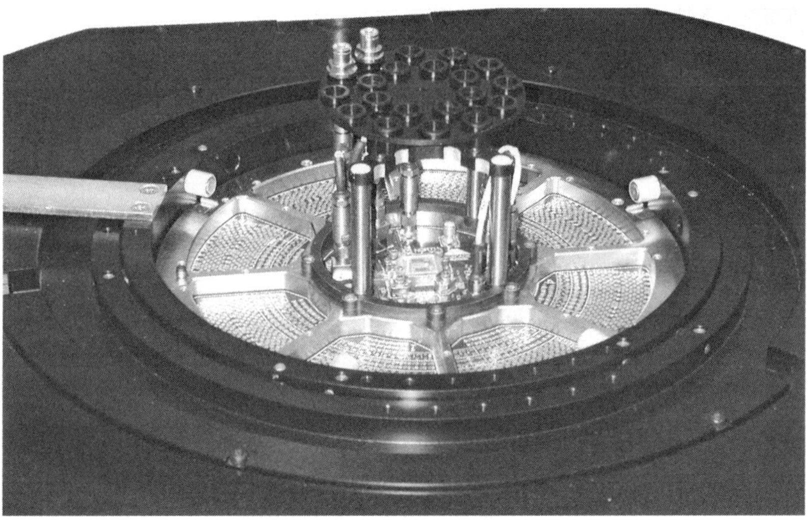

Figure 8.112 Probe card connected to a wafer prober for 10 Gbps at-speed wafer testing. (From: [123]. ©2005 Jose Moreira. Reprinted with permission.)

A typical test fixture for wafer probing is composed of three elements as shown in Figure 8.113. A wafer probe interface PCB connects the ATE pin electronics to the probe pogo tower. The pogo tower connects the wafer probe interface to the probe card. The probe card is a PCB board that contains the probes, which connect the DUT on the center of the probing interface. This type of configuration can typically accommodate a new application or device with just a redesign of the probe card. The wafer probe interface and the pogo tower are normally application independent and designed by the ATE manufacturer since they are only dependent on the specifics of the ATE platform. Figure 8.114 shows an example of a vertical probe card for a high-speed digital application.

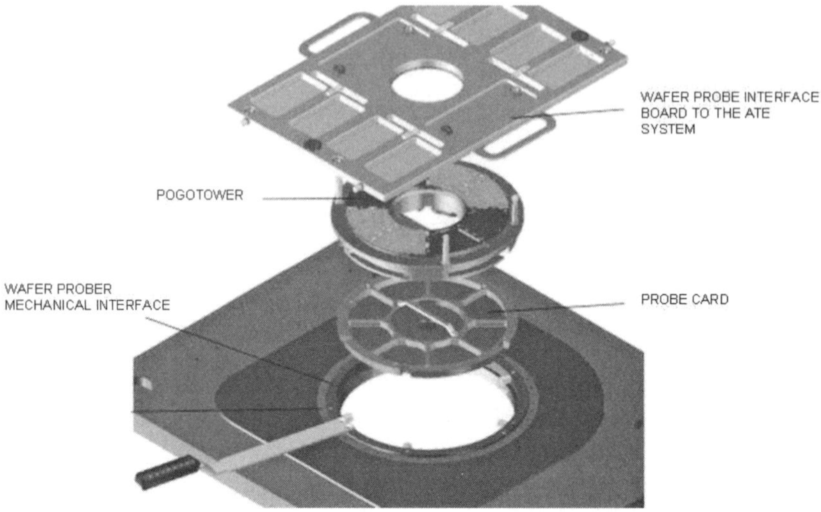

Figure 8.113 A typical standard wafer probing test fixture setup (Courtesy of Advantest).

Figure 8.114 Vertical probe card (Accumax probe card courtesy of Wentworth Laboratories, Inc.).

An alternative solution for the wafer probing test fixture design is to connect the probe interface directly to the test fixture as shown in Figure 8.115. The three separate interface parts (probe interface PCB, pogo tower, and the probe card) are substituted by a single PCB test fixture. The advantages of this approach are increased space available for items like relay or support circuitry, improved bandwidth, and the overall setup is closer to the final test fixture design for a packaged DUT [124–126]. Figure 8.116 shows one example of an ATE probe card using this approach.

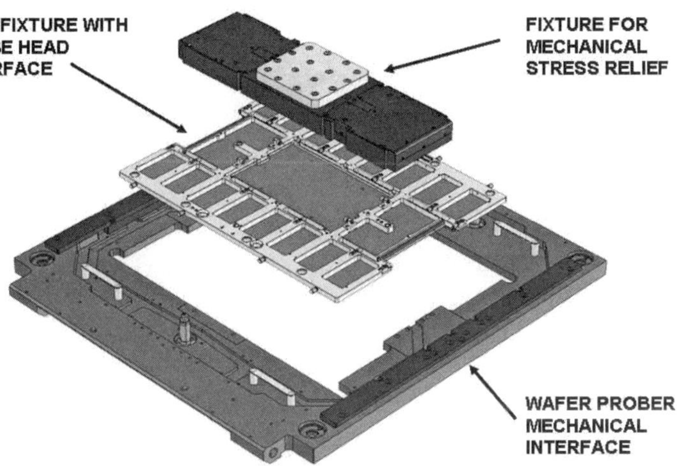

Figure 8.115 A wafer probing test fixture with direct docking of the wafer prober to the ATE interface PCB to eliminate the need for a pogo tower and probe card (Courtesy of Advantest).

For the die probing interface itself, there exist different probing technologies that can be used [127, 128]. Table 8.10 describes some of the main wafer probing technologies and Figure 8.117 shows pictures of the wafer probes in those technologies. Note that there are several other technologies like blade, pellicle, and elastomeric hybrids that are not described in this section.

Another important topic related to wafer probing is probe tip cleaning. This is a complex topic that is dependent on the probing technology used [129]. The lack of proper probe tip cleaning will result in reduced measurement performance and will cause yield loss in a production setup. However, improper cleaning technique might also reduce measurement performance by degrading the probing tips and reduce the probe interface lifetime [127].

When using wafer probing for at-speed digital testing applications, there is also the possibility of applying the loopback techniques described in Section

Figure 8.116 Advantest V93000 direct probe probe card with vertical probe array [124].

Table 8.10
Types of Wafer Probing Technologies for High-Speed Digital Applications

Type	Description	BW	Number of Pins
Cantilever	Needle soldered to probe card and held in place through an epoxy ring	< 400 MHz	Several hundred
Vertical probe	Vertical needle that delivers a tangential force at the top of the solder pads	< 5 GHz	Many thousand
Membrane	Photolithography defined metal deposited on a flexible membrane	< 20 GHz	Several hundred
MEMS	Photolithography defined with portions or all of the supporting substrate etched away, leaving the structure free	< 5 GHz	Many thousand

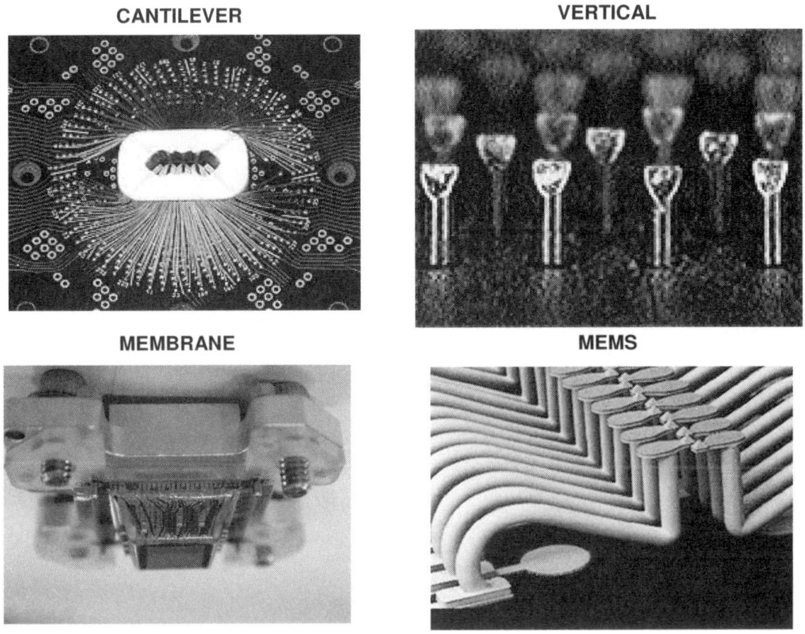

Figure 8.117 Picture of different types of wafer probing technologies (top left: cantilever; top right: vertical; bottom right: membrane; bottom left: MEMS) (FormFactor MicroSpring, photo used with permission of FormFactor, Inc. (c)2009).

6.4.3.4 on the probe card and in this way use a low-speed ATE system for at-speed I/O testing. The challenge is that for high pin count devices, the space needed for implementing the loopback approach might not be enough, especially if DC testing and at-speed testing are to be done with the same probe card.

References

[1] A. Mayer and R. Plitschka, *High Speed Testing: Introduction and Reference Guide to Testing High Speed Digital ICs*. Hewlett-Packard, 1989.

[2] L. W. Ritchey, *Right the First Time: A Practical Handbook on High Speed PCB and System Design, Volume One*. Speeding Edge, 2003.

[3] L. W. Ritchey, *Right the First Time: A Practical Handbook on High Speed PCB and System Design, Volume Two*. Speeding Edge, 2006.

[4] H. Johnson and M. Graham, *High-Speed Digital Design*. Addison-Wesley, 1993.

[5] H. Johnson and M. Graham, *High-Speed Signal Propagation*. Addison-Wesley, 2003.

[6] S. H. Hall, G. W. Hall, and J. A. McCall, *High-Speed Digital System Design: A Handbook of Interconnect Theory and Design Practices*. Wiley-IEEE Press, 2000.

[7] E. Bogatin, *Signal and Power Integrity Simplified*. Upper Saddle River, NJ: Prentice-Hall, 2010.

[8] D. Beaulieu, *Printed Circuit Board Basics*. Tulsa, OK: UP Media Group, 2003.

[9] C. Coombs, *Printed Circuits Handbook*. New York: McGraw-Hill, 2007.

[10] M. W. Jawitz, *Printed Circuit Board Materials Handbook*. New York: McGraw-Hill, 1997.

[11] T. Bresnan, "PCB Design, Fabrication and Assembly," *Burn-In and Test Strategies (BITS) Workshop*, 2010.

[12] R. Rajsuman, N. Masuda, and K. Yamashita, "Architecture and Design of an Open ATE to Incubate the Development of Third-Party Instruments," *IEEE Transactions on Instrumentation and Measurement*, vol. 54, Oct. 2005.

[13] J. Moreira, "Development of an ATE Test Cell for At-Speed Characterization and Production Testing," *IEEE International Test Conference*, 2011.

[14] D. Kather, J. Moreira, W.-J. Song, R. Nettles, W.-M. Zhang, and L. Ge, "A Unified Approach to Bench/ATE Testing with the V93000 Protocol Editor," *Advantest VOICE Users Group Meeting*, 2012.

[15] S. H. Hall and H. L. Heck, *Advanced Signal Integrity for High-Speed Digital Designs*. New York: John Wiley & Sons, 2009.

[16] G. Brist, S. Hall, S. Clouser, and T. Liang, "Non-Classical Conductor Losses Due to Copper Foil Roughness and Treatment," *ECWC 10 Conference at IPC Printed Circuits Expo*, 2005.

[17] Y. Shlepnev and C. Nwachukwu, "Practical Methodology for Analyzing the Effect of Conductor Roughness on Signal Losses and Dispersion in Interconnects," *DesignCon*, 2012.

[18] A. F. H. III, J. W. Reynolds, and J. C. Rautio, "Conductor Profile Effects on the Propagation Constant of Microstrip Transmission Lines," *IEEE International Microwave Symposium*, 2010.

[19] R. Corporation, "Copper Foils for High Frequency Materials," *Publication 92-243*, 2015.

[20] S. De, A. Gafarov, M. Y. Koledintseva, R. J. Stanley, J. L. Drewniak, and S. Hinaga, "Semi-Automatic Copper Foil Surface Roughness Detection from PCB Microsection Images," *IEEE International Symposium on Electromagnetic Compatibility*, 2012.

[21] M. Koledintseva, O. Kashurkin, T. Vincent, and S. Hinaga, "Effective Roughness Dielectric to Represent Copper Foil Roughness in Printed Circuit Boards," *DesignCon*, 2015.

[22] A. F. H. III, P. A. LaFrance, C. J. Caisse, J. P. Coonrod, and B. B. Fitts, "Effect of Conductor Profile Structure on Propagation in Transmission Lines," *DesignCon*, 2016.

[23] J. R. Miller, Y. Li, K. Hinckley, G. Blando, B. Guenin, I. Novak, A. Dengi, A. Rebelo, and S. McMorrow, "Temperature and Moisture Dependence of PCB and Package Traces and the Impact on Signal Performance," *DesignCon*, 2012.

[24] T. P. Warwick and D. Turpuseema, "The Quest for the Perfect Loadboard: Understanding the Issues and Technology Advances in Device-to-Tester Interface Assemblies," *Silicon Valley Test Workshop*, 2010.

[25] H. Barnes, J. Moreira, T. McCarthy, W. Burns, C. Gutierrez, and M. Resso, "ATE Interconnect Performance to 43 Gbps Using Advanced PCB Materials," *IEC DesignCon*, 2008.

[26] M. W. Jawitz and M. J. Jawitz, *Materials for Rigid and Flexible Printed Wiring Boards*. New York: John Wiley & Sons, 2007.

[27] T. S. Laverghetta, *Microwave Materials and Fabrication Techniques*. Norwood, MA: Artech House, 2000.

[28] J. R. Miller, G. J. Blando, and I. Novak, "Additional Trace Loss Due to Glass-Weave Periodic Loading," *DesignCon*, 2010.

[29] S. McMorrow and C. Heard, "The Impact of PCB Laminate Weave on the Electrical Properties of Differential Signaling at Multi-Gigabit Data Rates," *IEC DesignCon*, 2005.

[30] A. Morgan, "Developments in Glass Yarns and Fabric Constructions," *The PCB Magazine*, Mar. 2014.

[31] B. B. Szendrenyi, H. Barnes, J. Moreira, M. Wollitzer, T. Schmid, and M. Tsai, "Addressing the Broadband Crosstalk Challenges of Pogo Pin Type Interfaces for High-Density High-Speed Digital Applications," *International Microwave Symposium*, 2007.

[32] A. Wright, "The Pros and Cons of PCB Surface Finishes," *The PCB Magazine*, Jan. 2014.

[33] J. Moreira, M. Tsai, J. Kenton, H. Barnes, and D. Faller, "PCB Loadboard Design Challenges for Multi-Gigabit Devices in Automated Test Applications," *DesignCon*, 2006.

[34] G. Brist, D. Cullen, D. Luttrull, and J. Martin, "Reduction of High-Frequency Signal Loss Through the Control of Conductor Geometry and Surface Metallization," *International Conference of the Surface Mount Technology Association (SMTA)*, 2002.

[35] D. Staiculescu, J. Laskar, J. Mendelsohn, E. Sweetman, D. Rudy, and I. Artaki, "Ni-Au Surface Finish Effects on RF Performance," *IEEE MTT-S Digest*, 1999.

[36] X. Wu, D. Cullen, G. Brist, and O. M. Ramahi, "Surface Finish Effects on High-Speed Signal Degradation," *IEEE Transactions on Advanced Packaging*, vol. 31, Feb. 2008.

[37] M. Ahmad, M. Bugg, G. Fitzgerald, and M. Rost, "Skew in Twin-axial Cables and its Significance in Next Generation Differential Signaling," *DesignCon*, 2013.

[38] S. Farrahi, V. Kunda, Y. Li, X. Zhang, G. Blando, and I. Novak, "Does Skew Really Degrade SERDES Performance?," *DesignCon*, 2015.

[39] E. Acar and T. Swettlen, "Using Ground-Signal-Power Stack-Up for Striplines in ATE Load Boards," *Burn-In and Test Strategies (BITS) Workshop*, 2010.

[40] T. P. Warwick, "25GB/s Socket and Loadboard Test Issues," *Silicon Valley Test Workshop*, 2011.

[41] B. O. McCoy, R. Techentin, B. Buhrow, K. Buchs, H. Lin, B. K. Gilbert, and E. S. Daniel, "PWB Manufacturing Variability Effects on High-Speed SerDes Links: Statistical Insights from Thousands of 4-Port S-Parameter Measurements," *DesignCon*, 2010.

[42] M. Kikuchi and R. Yongho, "Via Structure Analysis for 10Gbps Signal Transmission on PCB," *Semicon Japan*, 2008.

[43] A. C. Segura, "Signal Integrity Analysis of a 26 Layers Board with Emphasis on the Effect of Non-Functional Pads," *IEEE International Symposium on Electromagnetic Compatibility*, 2008.

[44] A. Ippich, "Influence of Via Stub Length and Antipad Size on the Insertion Loss Profile," *The PCB Design Magazine*, Jan. 2014.

[45] E. Bogatin, L. Simonovich, S. Gupta, and M. Resso, "Practical Analysis of Backplane Vias," *IEC DesignCon*, 2009.

[46] H. Barnes, J. Moreira, H. Ossoining, M. Wollitzer, T. Schmid, and M. Tsai, "Development of a Pogo Pin Assembly and Via Design for Multi-Gigabit Interfaces on Automated Test Equipment," *Asia-Pacific Microwave Conference*, 2006.

[47] J. Moreira, H. Barnes, and V. Poisson, "PCB Via Field with Embedded Pitch Transformation for ATE Pogo Pin Blocks," *IEEE Workshop on Signal and Power Integrity*, 2015.

[48] D. Dunham, J. Lee, S. McMorrow, and Y. Shlepnev, "Design and Optimization of a Novel 2.4 mm Coaxial Field Replaceable Connector Suitable for 25 Gbps System and Material Characterization up to 50 GHz," *DesignCon*, 2011.

[49] R. Satrom, "Using Simulation to Optimize an Interface for 10GHz," *Silicon Valley Test Workshop*, 2010.

[50] J. Moreira, H. Barnes, and G. Hoersch, "Analyzing and Addressing the Impact of Test Fixture Relays for Multi-Gigabit ATE I/O Characterization Applications," *IEEE International Test Conference*, 2007.

[51] Teledyne Relay, "Series LB363/GLB363 Loopback Relay," 2011.

[52] Keysight Technologies, "Understanding RF/Microwave Solid State Switches and Their Applications," *Application Note 5989-7618EN*, 2014.

[53] G. M. Rebeiz, *RF MEMS Theory, Design, and Technology*. New York: John Wiley & Sons, 2003.

[54] B. Schoenlinner, "RF-MEMS Switches - The Status Compared to Its Solid-State Competitor Technologies," *42nd European Microwave Conference*, 2012.

[55] A.-Q. Liu, *RF Mems Switches and Integrated Switching Circuits Design Fabrication and Test*. New York: Springer, 2010.

[56] J. Bouchard and B. Knoblich, "RF MEMS Switches Deliver on Early Promise," *Sensor & Transducers Journal*, Dec. 2007.

[57] J. Rosales, "Low Power Boost Regulator with Dual Half-Bridge in 3mm x 2mm DFN Drives MEMS and Piezo Actuators," *Linear Technology Magazine*, Sept. 2009.

[58] H. Barnes and J. Moreira, "ATE Signal Integrity Challenges for Multi-Gigabit Communication Interfaces," *SouthWest DFT Workshop*, 2008.

[59] D. C. Keezer and Q. Zhou, "Alternative Interface Methods for Testing High-Speed Bidirectional Signals," *IEEE International Test Conference*, 1998.

[60] D. Thompson, "High Bandwidth Sockets for SERDES Applications on ATE Load Boards," *Burn-In and Test Strategies (BITS) Workshop*, 2014.

[61] D. Thompson and M. A. Salas, "High Speed BGA Sockets from a System Perspective," *Burn-In and Test Strategies (BITS) Workshop*, 2016.

[62] W. Liu and M. Pecht, *IC Component Sockets*. New York: John Wiley & Sons, 2004.

[63] J. Diller, "Sockets: A Comprehensive, Comprehensible Guide to Contact Technology," *Burn-In and Test Strategies (BITS) Workshop*, 2011.

[64] H.-C. Teng, M.-K. Chen, C.-H. Yeh, Y.-J. Huang, and S.-L. Fu, "Study of Contact Degradation in Final Testing of BGA Sockets," *IEEE International Conference on Electronic Materials and Packaging*, 2007.

[65] R. W. Campbell, "Materials For Test Sockets," *Burn-In and Test Socket (BITS) Workshop*, 2003.

[66] D. Carter, "'Fuzz Button' Interconnects at Microwave and mm-Wave Frequencies," *Packaging and Interconnects at Microwave and mm-Wave Frequencies IEE Seminar*, 2000.

[67] C.-Y. Li, "Braided Electrical Contact Element (BeCe)," *Burn-In and Test Socket (BITS) Workshop*, 2007.

[68] Johnstech, "PAD ROL 200 for QFN, DFN, and Other Pad-Style Applications," 2015.

[69] S. Nicholas Langston, F. Bumb, and J. Pereschuk, "Elastomeric Interconnects - Reliable Enough for Production," *Burn-In and Test Socket Workshop*, 2007.

[70] D. Mahoney, S. Halm, and N. L. Jr., "A Review of a Universal Coaxial Test SocketŠs Performance," *SiliconăValleyăTestăConference*, 2010.

[71] T. Watanabe, "Complete Z0=50 Ohm Coaxial Spring probe IC Socket," *Burn-In and Test Socket Workshop*, 2008.

[72] T. Yoshida, C. Hudson, T. Nagata, A. Sato, and S. Kakegawa, "High Frequency Wafer Level Test Approach with Coaxial Socket," *Burn-In and Test Strategies (BITS) Workshop*, 2008.

[73] R. R. Tummala and M. Swaminathan, *Introduction to System-on-Package (SOP): Miniturization of the Entire System*. New York: McGraw Hill, 2008.

[74] P. Viswanadham, *Essentials of Electronic Packaging, A Multidisciplinary Approach*. ASME Press, 2011.

[75] J. Zhou and S. Soh, "PoP Test Socket - Challenges and Approaches," *Chip Scale Review*, May 2012.

[76] S. Soh, F. Zhou, J. Diller, J. Spooner, and K. Elmadbouly, "Special Designs and Applications for PoP Device Testing," *Burn-In and Test Strategies (BITS) Workshop*, 2013.

[77] B. Willis, *Package on Package Assembly Inspection & Quality Control*. http://www.packageonpackagebook.com, 2013.

[78] J. Zhou, S. Soh, D. B. Bogardus, and B. L. Hahn, "PoP Contactor - Challenges & Solutions," *Burn-In and Test Socket (BITS) Workshop*, 2009.

[79] J. Brandes, "Specmanship," *Burn-In and Test Workshop*, 2012.

[80] E. Bogatin, K. DeFord, and M. Nagappan, "Practical Techniques for Measuring High Bandwidth Electrical Models of Test Sockets," *Burn-In and Test Socket Workshot*, 2005.

[81] H. Kilicaslan, B. Tunaboylu, and D. McDevitt, "Accuracy Improvements in Microwave Measurements by Double-Sided Probing," *Burn-In and Test Socket Workshop*, 2006.

[82] V. Treibergs, "Comparison of Different Methods in Determining Current Carrying Capacity of Semiconductor Test Contacts," *Burn-In and Test Strategies (BITS) Workshop*, 2015.

[83] N. L. Jr., "An Examination of the Causes of Cres Degradation Which Affect the Life of a Test Socket," *Burn-In and Test Socket (BITS) Workshop*, 2008.

[84] D. Young and T. Bresnan, "Super-Sockets: Integration of Technology from Test Board into Socket Assembly," *Burn-In and Test Socket (BITS) Workshop*, 2008.

[85] A. D. Kraus and A. Bar-Cohen, *Design and Analysis of Heat Sinks*. New York: John Wiley & Sons, 1995.

[86] Silicon Thermal. http://www.siliconthermal.com/.

[87] Sensor Controls Co. Ltd. http://www.scnt.co.jp/.

[88] G. H. Johnson, "Challenges of High Supply Currents During VLSI Test," *IEEE International Test Conference*, 2000.

[89] J.-P. Mallet, "High Current DPS Architecture for Sort Test Challenge," *IEEE International Test Conference*, 2002.

[90] I. Novak, "A New Dawn in R&D," *IEEE Microwave Magazine*, Mar. 2011.

[91] I. Novak and J. R. Miller, *Frequency-Domain Characterization of Power Distribution Networks*. Norwood, MA: Artech House, 2007.

[92] M. Swaminatan and A. E. Engin, *Power Integrity Modeling and Design for Semiconductors and Systems*. Upper Saddle River, NJ: Prentice-Hall, 2007.

[93] I. Novak, *Power Distribution Network Design Methodologies*. IEC, 2008.

[94] J. L. Knighten, B. Archambeault, J. Fan, G. Selli, S. Connor, and J. L. Drewniak, "PDN Design Strategies: I. Ceramic SMT Decoupling Capacitors - What Values Should I Choose?," *IEEE EMC Society Newsletter*, 2005.

[95] J. L. Knighten, B. Archambeault, J. Fan, G. Selli, S. Connor, and J. L. Drewniak, "PDN Design Strategies: II. Ceramic SMT Decoupling Capacitors - Does Location Matter?," *IEEE EMC Society Newsletter*, 2006.

[96] J. L. Knighten, B. Archambeault, J. Fan, G. Selli, S. Connor, and J. L. Drewniak, "PDN Design Strategies: III. Planes and Materials - Are They Important Factors in Power Bus Design?," *IEEE EMC Society Newsletter*, 2007.

[97] J. L. Knighten, B. Archambeault, J. Fan, G. Selli, S. Connor, and J. L. Drewniak, "PDN Design Strategies: IV. Sources of PDN Noise," *IEEE EMC Society Newsletter*, 2007.

[98] H. Lan, R. Schmitt, and C. Yuan, "Prediction and Measurement of Supply Noise Induced Jitter in High-Speed I/O Interfaces," *IEC DesignCon*, 2009.

[99] K. S. Oh and X. Yuan, *High-Speed Signaling: Jitter Modeling, Analysis, and Budgeting*. Upper Saddle River, NJ: Prentice-Hall, 2012.

[100] R. Jakushokas, M. Popovich, A. V. Mezhiba, S. Kose, and E. G. Friedman, *Power Distribution Networks with On-Chip Decoupling Capacitors*. New York: Springer, 2011.

[101] O. Vikinski, S. Lupo, G. Sizikov, and C. Y. Chung, "Embedded Power Delivery Decoupling in Small Form Factor Test Sockets," *IEEE International Test Conference*, 2008.

[102] H. Barnes, H. Nuessle, and K. Eichler, "V93000 DUT Board Ultra-High Current Power Distribution Network Design," *Advantest VOICE Users Group Meeting*, 2012.

[103] W. C. Johnson, *Transmission Lines and Networks*. New York: McGraw-Hill, 1950.

[104] D. M. Pozar, *Microwave Engineering*. New York: John Wiley & Sons, 2005.

[105] J. Peiffer, B. Greenlee, and I. Novak, "Electrical Performance Advantages of Ultra-Thin Dielectric Materials Used for Power-Ground Cores in High Speed, Multilayer Printed Circuit Boards," *IPC Expo*, 2003.

[106] I. Novak, "Electrical Integrity: Signal and Power Integrity," 2009. http://www.electrical-integrity.com.

[107] X. Wu, "Matched Lossy Decoupling Termination for Power Plane Noise Mitigation," *International Symposium on Electromagnetic Compatibility*, 2005.

[108] I. Novak, "Reducing Simultaneous Switching Noise and EMI on Ground/Power Planes by Dissipative Edge Termination," *IEEE Transactions on Advanced Packaging*, vol. 22, Aug. 1999.

[109] D. L. Sanders, J. P. Muccioli, T. M. North, and K. P. Slattery, "The Quantitative Measurement of the Effectiveness of Decoupling Capacitors in Controlling Switching Transients from Microprocessors," *25th Annual Passive Components Conference (CARTS)*, 2015.

[110] H. Ishida, J. Prymak, C. Burket, and I. Novak, "Controlled ESR Capacitors Have Arrived," *IEC DesignCon TecForum*, 2007.

[111] T. P. Warwick, "Mitigating Test Interconnect Issues for the Next Generation of High-Speed, High Power Devices," *Silicon Valley Test Workshop*, 2012.

[112] N. Langston, J. Zhou, and H. Yao, "Minimizing Socket & Board Inductance Using a Novel Decoupling Interposer," *Burn-In and Test Strategies (BITS) Workshop*, 2007.

[113] A. Detofsky, O. Vikinski, S. Lupo, and T. Swettlen, "Power Integrity Ingenuity at Test," *Burn-In and Test Socket Workshop*, 2009.

[114] S. Lupo, O. Vikinski, D. Bogardus, K. Elmadbouly, and C. Jacob, "Next Generation CiS (Capacitor in Socket) Featuring Discrete Capacitors and Elastomer Hybrid Schemes," *Burn-In and Test Socket Workshop*, 2010.

[115] S. Weir, "PDN Application of Ferrite Beads," *DesignCon*, 2011.

[116] M. Giesler, A. Barr, Y. Kawate, and Y. Tsubaki, "Impact of Embedded Capacitance on Test Socket and Test Board Performance," *Silicon Valley Test Conference*, 2011.

[117] I. Novak, "Comparison of Power Distribution Network Design Methods: Bypass Capacitor Selection Based on Time Domain and Frequency Domain Performances," *IEC DesignCon TecForum*, 2006.

[118] J. Moreira, H. Nuessle, and H. Barnes, "PDN Design Challenges for ATE Test Fixtures," *IEC DesignCon*, 2011.

[119] J. Moreira, M. Moessinger, K. Sasaki, and T. Nakamura, "Driver Sharing Challenges for DDR4 High-Volume Testing with ATE," *IEEE International Test Conference*, 2012.

[120] IEEE, "Southwest Test Workshop." http://www.swtest.org.

[121] W. R. Mann, F. L. Taber, P. W. Seitzer, and J. J. Broz, "The Leading Edge of Production Wafer Probe Test Technology," *IEEE International Test Conference*, 2004.

[122] T. Ueno and Y. Kondoh, "Membrane Probe Technology for MCM Known-Good-Die," *IEEE International Test Workshop*, 1994.

[123] J. Moreira, G. Haensel, and F. Koban, "Addressing the Challenges of Implementing an At-Speed Production Test-Cell for 10Gb/s Wafer Probing," *DesignCon*, 2005.

[124] L. Dibattista and D. Lam, "Verigy V93000 Direct-Probe Evolution of the Verigy V93000 SOC Tester in Wafer Probing," *Advantest*.

[125] D. Lefeve and R. Sinsheimer, "A New Tester-Prober Interface Paradigm: Direct Docking," *SW Test Workshop*, 1998.

[126] T. J. Sheng and E. Valderama, "Establish WLCSP Testing at Tri-Temp for RF and non-RF Products," *Burn-In and Test Strategies (BITS) Workshop*, 2016.

[127] J. Broz and B. Mann, "Wafer Probe Test Technology Tutorial," *IEEE International Test Conference*, 2008.

[128] I. Feldman, "Wafer Probe Technology & Application Overview," *Silicon Valley Test Conference*, 2010.

[129] J. Broz, G. Humphrey, and W. Fitzgerald, "Probe Card Cleaning: A Short Tutotial," *IEEE SW Test Workshop*, 2007.

9

Advanced ATE Topics

This chapter presents some advanced topics that are of interest for the test of high-speed digital I/O interfaces with automated test equipment.

9.1 ATE Specifications and Calibration

Understanding ATE specifications and calibration can be challenging for the test engineer. This is due to the fact that ATE manufacturers, although using the same nomenclature for a given specification parameter, might use different definitions or measurement methodologies to guarantee that parameter. This makes it more difficult to compare specifications across different ATE platforms. Also, unlike with bench instrumentation, the large number of channels associated with an ATE system presents challenges for defining and measuring specifications for the ATE system and its calibration. The next subsections discuss some of these topics in detail.

9.1.1 Accuracy and Resolution

It is important to understand the difference between resolution and accuracy. Resolution is the smallest change that can be forced or detected for an electrical or timing parameter (e.g., 1 mV for the receiver threshold voltage or 1 ps for an edge placement of a low to high transition on a tester pin driver). Note that although there is an inherent resolution in the hardware (e.g., quantization of the digital-to-analog converter that sets the threshold voltage value), the instrument software can implement any type of resolution. When discussing the resolution of a tester module, we should be aware to address the hardware resolution and not a theoretical software resolution without correspondence to the test hardware. However, it is also important to note that for some cases resolution can be improved by the software through the use of averaging.

The accuracy of a tester module is the degree of deviation of a measured parameter from its true or reference value (e.g., +/− 10 mV accuracy for the receiver threshold). The precision of an ATE instrument, also expressed as reproducibility or repeatability, is the degree to which repeated measurements under identical conditions show the same results as illustrated in Figure 9.1 [1]. In the ATE world the terms "accuracy" and "precision" are normally summarized as accuracy (e.g., edge placement accuracy in the following section). This means that for a specified accuracy, like the threshold voltage from above, we expect its validity for all tester channels on all test systems. We should never forget the statistical nature of measurement events and focus on a single measurement, which could accidentally lead to perfect accuracy with zero deviation.

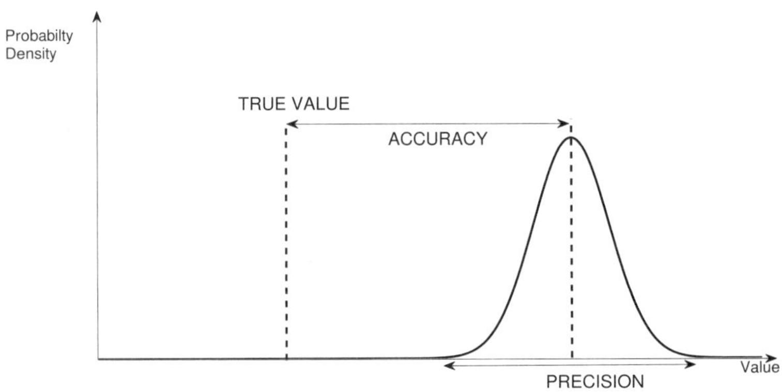

Figure 9.1 Comparing the definition of accuracy and precision for a measured value in regards to its true reference value.

9.1.2 Understanding OTA and EPA

Overall time accuracy (OTA) and edge placement accuracy (EPA) are two important specifications for an ATE system. OTA refers to the worst-case error between two drive or receive actions (or edges) in all the drive and receive channels of the ATE system. However, the EPA refers to the worst-case error that a single drive or receive action can have in any drive or receive channel of the ATE system. The definition of OTA and EPA is shown graphically in Figure 9.2 where all ATE channels have one single drive or receive edge that was programmed to occur at the same time instant but due to the timing accuracy of the ATE system, each drive and receive edge will have a certain timing error. In the figure the channel 1 drive edge is used as the reference. Note that any channel can be used as a reference without any impact on the measured EPA value.

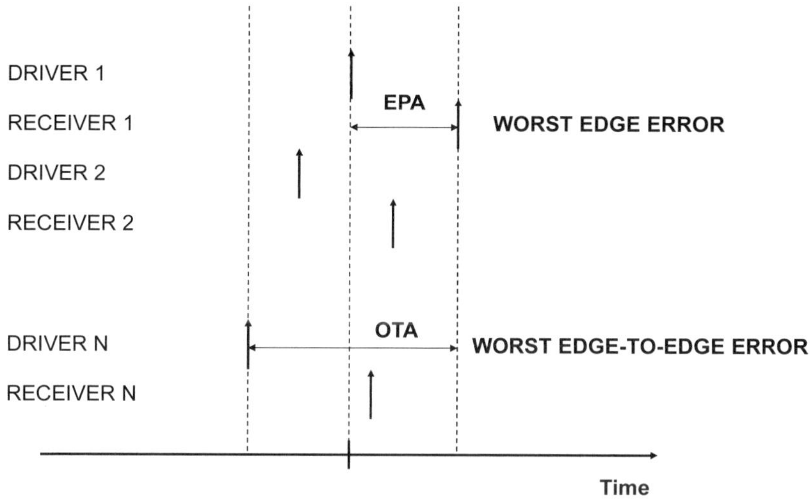

Figure 9.2 Diagram showing the OTA and EPA definitions.

Usually, the EPA value is defined as a signed value (e.g., $+/-$ 60 ps) because typically ATE manufacturers will define the maximum achievable edge error to be the same for either a positive or negative difference to the programmed value. If one assumes that the EPA of the driver edges is equivalent to the EPA of the receiver edges, which by itself is a significant assumption, the OTA value can be computed as twice the EPA value (i.e., $OTA = 2 \times EPA$).

Measuring OTA based on its definition can be very time consuming for ATE systems with a large number of channels, especially if the large number of different parameters that can have an influence on the OTA of an ATE system is taken into account. Reference [2] is an ANSI standard that presents a procedure for measuring the OTA of an ATE system. To address the OTA measurement challenge for a large amount of channels, the standard describes some time saving strategies like using a loopback between the driver and receiver of each channel for most of the measurements instead of measuring one ATE channel against all others.

9.1.3 Linearity and Edge Placement Accuracy

Another important difference that needs to be highlighted is the difference between linearity and edge placement accuracy. To understand the difference between these parameters, Figure 9.3 describes how the timing of a drive or receive edge of an ATE pin electronics channel is programmed at a very high level.

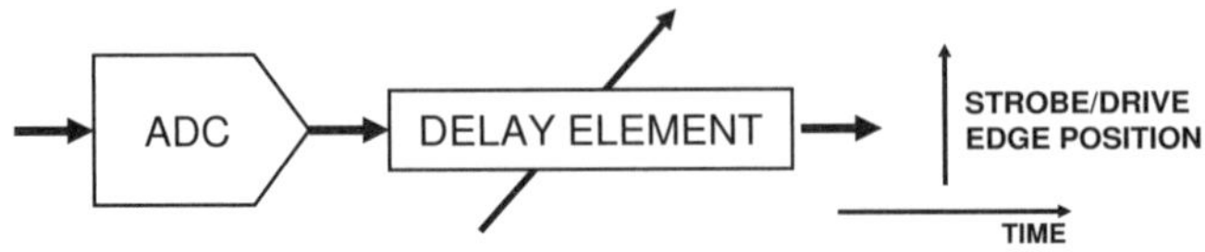

Figure 9.3 Conceptual block diagram of a driving/strobing edge delay programming.

The value to be programmed for an edge timing is transferred to an analog voltage provided by an ADC that controls a delay line, which sets the timing of the edge. Any offset error (i.e., an error that is independent of the programmed values) will have an effect on the EPA, which is important for measurements like the skew between two channels. For measurements like a jitter histogram, however, a fixed error is negligible since it only has influence on the mean value of the distribution but not on its peak-to-peak or variance value. For such self-related measurements, only the errors between programmed values are crucial as shown in Figure 9.4. In the presented architecture this kind of inaccuracy is caused by nonlinearities of the ADC (gain errors) or the delay element.

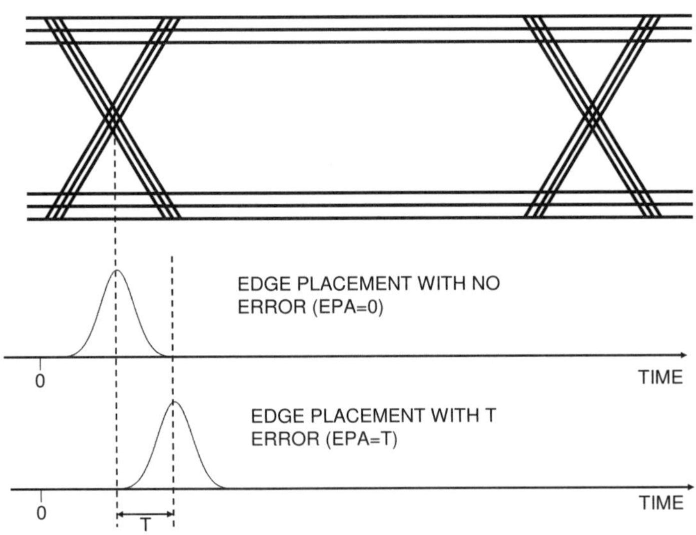

Figure 9.4 Comparison of the edge placement accuracy influence in a jitter histogram measurement.

One way to visualize the linearity of a drive/receive channel is to execute a shmoo (see Section 5.3) of the drive and compare timing for each ATE channel with pairs of an ATE driver and receiver connected together (ATE driver to receiver loopback) as shown in Figure 9.5. If the linearity of the ATE pin electronics is perfect, then one would expect a straight 45° transition boundary from the pass to the fail region in the shmoo plot. Reality, however, shows that this transition is not perfectly linear and sharp and reveals the nonlinearity of the driver and receiver timing.

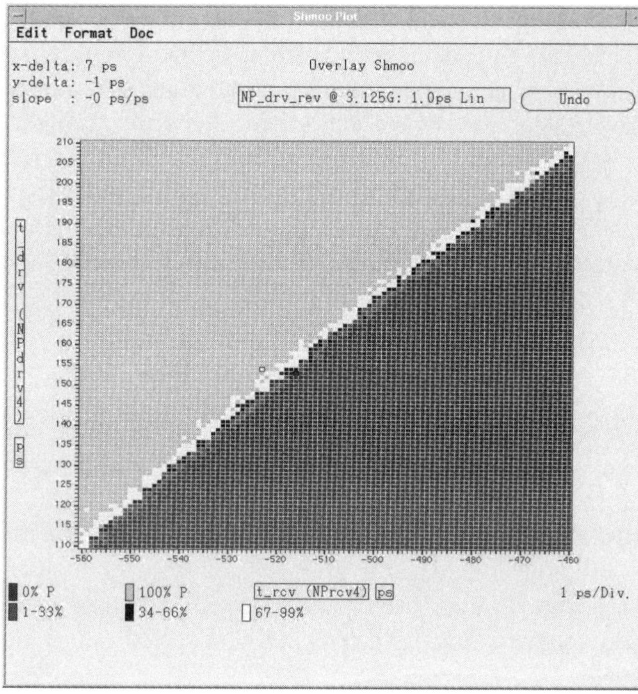

Figure 9.5 Example of a shmoo plot of a driver to receiver loopback for evaluating the driver/receiver edge linearity (Courtesy of Advantest).

9.1.4 Calibration

If we discuss calibration in an ATE-based test solution environment, the calibration of the ATE system itself is only one aspect. The other aspect is the application-specific part of the overall solution that changes with each device to be tested. The most critical component of this application-specific part that has influence on test accuracy certainly is the test fixture and potential external instrumentation. ATE manufacturers acknowledge this fact and typically structure their calibration tooling in a way that the system calibration which guarantees the performance up to the test fixture interface is separated from the test fixture-specific calibration tasks.

9.1.4.1 System Calibration

Calibration of an ATE system is by itself a large and complex topic that is very dependent on the ATE platform and pin electronics specifics. One challenge for ATE calibration is to determine the right compromise between the number and accuracy of the calibration steps and the time required to calibrate an ATE system. This is especially important in modern ATE systems with several thousands of pins that need to be calibrated. As a further complication of the situation, there are always calibration steps that cannot be done in parallel on all pins because they might require a unique external reference or measurement unit to be connected to each channel. There are currently two main approaches for system calibration: switch matrix and calibration robot. The other calibration technique described in this section is the compound dot technique.

Switch Matrix

The most straightforward and fastest way to calibrate the pin electronics channels of an ATE system is to connect each one to a reference channel through a matrix of relays (or switches) as shown in Figure 9.6. This allows, for example, to calibrate the timing of all the channels using a single channel as reference. Figure 9.7 (left) shows an example of a switch matrix based ATE calibration unit. In this case to keep the unit to a manageable size, the unit needs to be moved manually across the ATE system.

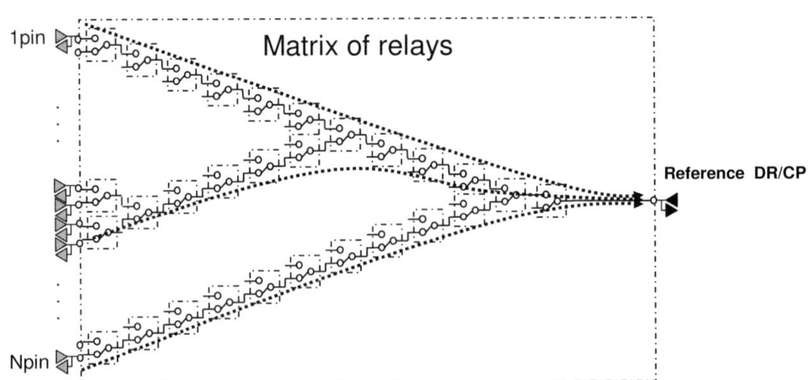

Figure 9.6 Diagram of a switch matrix for calibration of an ATE system pin electronics.

Although there will be a different delay for different paths in the switch matrix, these delays can be measured a priori and this calibration data can be saved on the calibration test fixture that contains the switch matrix. Although this approach allows for a very fast calibration time, it presents signal integrity challenges for high data rates due to the large number of relays that are necessary for a high density ATE system.

Compound Dot Technique

Another interesting calibration approach is the compound dot technique where all ATE channels are connected in a single point in the calibration test fixture [3]. This approach can be further extended to a hybrid setup using the compound dot technique together with a matrix of relays [4].

Calibration Robot

One approach to avoid the drawbacks of the switch matrix approach is to use a calibration robot where each channel is calibrated serially, one channel at a time. Since the robot physically moves the probing point, there is no need for relays. Figure 9.7 shows an example of a calibration robot for an ATE system. The main drawback of the calibration robot is the larger calibration time that is required in comparison with the switch matrix approach.

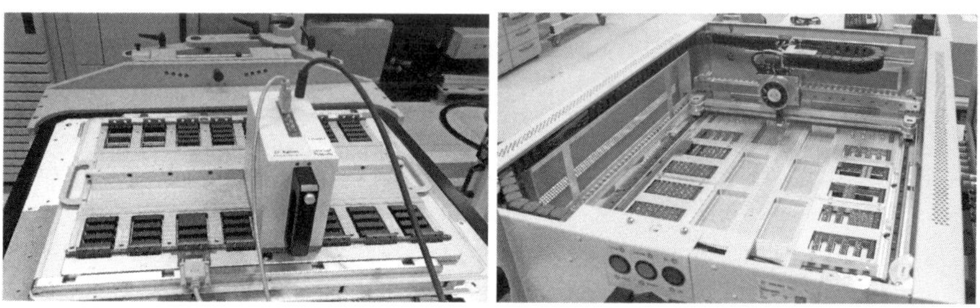

Figure 9.7 Picture of switch matrix based ATE calibration unit (left) and a robotic calibration unit docked to an ATE system with the top cover removed, showing the calibration robot (right) (Courtesy of Advantest).

To reduce the calibration time of an ATE system, parameters that need to be calibrated but are considered to be stable over time might simply be calibrated only at manufacturing time. The resulting calibration data is stored closely linked to the hardware it corresponds to (e.g., in a nonvolatile memory on the pin electronics ATE card). This approach avoids the need to calibrate those parameters on a regular basis and eliminates the corresponding calibration time from maintenance calibrations. This procedure is referred to by some ATE manufacturers as trimming. Another approach to reduce calibration time is to calibrate only the channels that really require calibration. This type of calibration approach is sometimes referred to as "smart calibration" and requires a methodology to first check with channels require calibration (e.g., using a diagnostic or channel loopback test fixture).

An important distinction exists between the standard calibration of an ATE system and its performance verification (PV). Typically the ATE manufacturer will perform a PV of the ATE system after manufacturing and calibration by measuring the parameters that are part of the ATE system specifications. The measurement allows the assessment of the real performance of the ATE system, especially how much margin it has with regard to its specifications. This type of data across multiple manufactured ATE systems provides the ATE manufacturer with an insight into whether the ATE specifications are too loose or whether additional margin is required. Some ATE manufacturers allow the purchaser of an ATE system to get access to the PV data of the delivered system.

The system calibration that needs to be executed in regular time intervals to keep the ATE system within the specified accuracy boundaries is typically divided into a DC and an AC calibration. The DC calibration ensures that the specified voltage levels and current drive capabilities for all ATE resources are valid. The main task during this calibration step is to measure the characteristics of analog-to-digital (ADC) and digital-to-analog (DAC) converters together with the other components (e.g., buffers). The result of these measurements is used to determine the required compensation parameters that allow adjustment of the DACs and ADCs to their target behavior. It is important to note that DC calibration not only covers the typical DC resources of an ATE such as device power supplies (DPS) and parametric measurement units (PMU) but also the level generators for digital drivers and receivers.

The AC calibration takes care of the adjustable timing parameters that are specified for the respective ATE system like specified flavors of skew values or timing linearities. Different ATE system architectures require the measurement of very different resources to achieve the specified timing values. Also a substantial subset of the timing values is typically guaranteed during PV measurements and does not require regular recalibration due to the stability of these parameters over time. The compensation of the timing errors determined during the AC calibration typically also involves DACs that are used to control timing delay lines in the ATE pin electronics.

9.1.4.2 Fixture Calibration

As discussed in Chapter 8, the test fixture is a critical part of the overall ATE system performance for high-speed digital applications. This means that for high-speed digital applications in some cases the standard ATE calibration is not sufficient and additional application-specific calibration steps are required to compensate the accuracy impact of the test fixture. These calibration steps can be separated into components that are the same for all test fixtures and into components that need to be adapted to the operating conditions of a DUT. The latter of these typically are handled by focus calibration procedures as discussed in Section 9.3.

The most prominent test fixture-specific calibration step that is independent of the DUT to a large extent is the fixture delay calibration. This calibration determines the electrical length for each of the different signal paths of the test fixture. With the knowledge of these values, the ATE system can compensate the fixture delays for each signal path. This allows the referencing of the timing system of the ATE to the pins of the DUT, which is not possible if there are unknown differences on the electrical lengths of the test fixture signal paths. While the fixture delay compensation mechanism is a standard part of ATE systems, the way to obtain the fixture delay values might vary from ATE system to ATE system. For ATE systems that do not have integrated support for fixture delay measurements, either external equipment is required for this measurement, the test fixture has to be designed with identical signal path lengths for all connections, or the electrical delays are determined by simulation. Especially for high-speed applications, it is difficult to achieve the required matching accuracy by design over a large amount of signals and the accuracy of the simulation approach always will depend on the quality of the simulation models.

Thus, the extraction of the fixture delay values by measurements typically is mandatory for high-speed I/O applications. Fixture delay values can be derived from skew measurements because the ATE channels are typically deskewed up to the test fixture already by the system calibration. This, however, will require signal probing in the test socket as for a focus calibration. Since automated probing in the socket is difficult to implement and will result in expensive instrument additions (e.g., a probing robot), fixture delay values typically are measured via TDR (see Appendix E). This method is implemented by contacting the test fixture at its interface to the ATE with a single contact point per measurement. Another advantage of the TDR methodology is that, in addition to being executed using dedicated TDR equipment, is can also be done with bidirectional ATE pin electronics. The driver of the pin electronics is used to launch a voltage step into the test fixture trace and the receiver of the bidirectional front-end scans for the response to that step stimulus. The ATE receiver typically uses the methodology described in Section 4.2.9 to retrieve that response. The accuracy of the TDR result measured with an ATE pin, like for dedicated TDR equipment, depends on the rise time that the driver can achieve and the receiver bandwidth. Due to the way the response is measured on the ATE receiver, timing linearity and level threshold accuracy also have an impact on the resulting accuracy.

In general, the ATE-based TDR measurement is possible into an open or a shorted socket. For dual transmission lines, the implementation of the dual transmission line (DTL) on the fixture determines whether a TDR into open or a TDR into short can be used. If a TDR into open is to be performed, an impedance adaptation is required at the DTL joint so that each of the three connections is loaded with an effective 50 Ω impedance. The drawback of this implementation is that the level swing of the response to the measurement step is reduced, which has an impact on the accuracy for the measurement of the response at the ATE receiver. Thus, a DTL topology with a direct DTL connection that requires a TDR into a short as shown in Figure 9.8 typically results in more accurate fixture delay values.

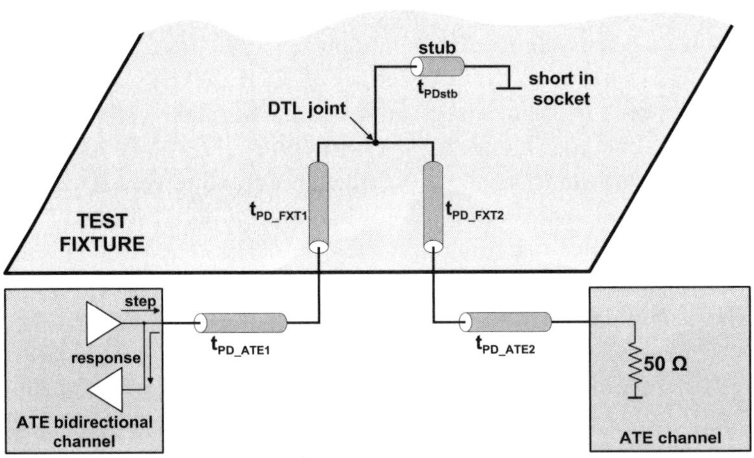

Figure 9.8 Dual transmission line structure with shorted socket pin.

A critical parameter for this kind of DTL topology is the electrical length of the stub from the DTL joint to the socket tip. This is due to the fact that the joint represents a discontinuity and reflections go back and forth on the stub between the socket tip and the DTL joint when the TDR measurement is done. The shorter the stub is, the less accuracy impact these reflections have on the measured fixture delay result.

Another parameter that influences the level of reflections on the stub and thus the accuracy of the shorted TDR measurement is the quality of the short in the socket. The lower impedance this short has, the more accurate measurement results are possible. Good low impedance shorts in a device socket are usually achieved by using gold plated device package dummies in the socket during the shorted TDR measurement as shown in Figure 9.9.

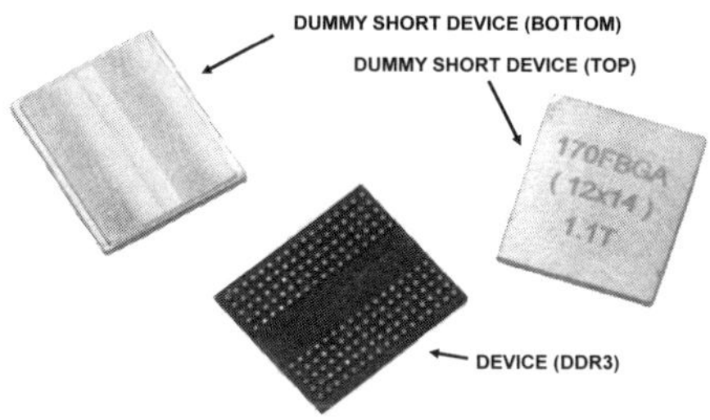

Figure 9.9 Device with corresponding gold plated package dummies (short device).

9.2 Multiplexing of ATE Channels

One typical approach to achieve higher data rates on ATE is to multiplex several low-speed ATE channels into a single high-speed digital signal [5]. This approach can be very appealing from a cost point of view since it makes use of low-speed ATE equipment to test higher data rate I/Os. For example, this could allow the use of an already available ATE system to test an application above its maximum data rate without buying a new ATE system or higher performance ATE cards.

The challenge with this approach is to design the multiplexing circuitry on the DUT test fixture, especially for multigigabit digital applications [6, 7]. ATE channel multiplexing can be divided into two major categories, standard and retimed channel multiplexing.

Standard Channel Multiplexing In this case a logic gate (e.g., a XOR) is used to generate a high-speed signal derived from the logical combination of several low-speed ATE channels. The ATE channels are skewed in time so that at the output of the logic gate, the waveform is running at the speed of the ATE multiplied by the number of multiplexed channels as shown in Figure 9.10.

Retimed Channel Multiplexing In this approach the signal derived from the logic gate multiplexer is retimed using a clean retiming clock that usually runs at half-rate (e.g., 16 GHz for 32 Gbps output data rate). Another approach is to use a dedicated multiplexing IC that already retimes the output signal. Retiming removes jitter from the multiplexing operation, and the jitter of the output signal is only defined by the jitter of the retiming clock and retiming circuitry [8]. Another advantage of the retiming clock is that it provides a simple way to add some amount of jitter to the output by injecting the jitter on the retiming clock.

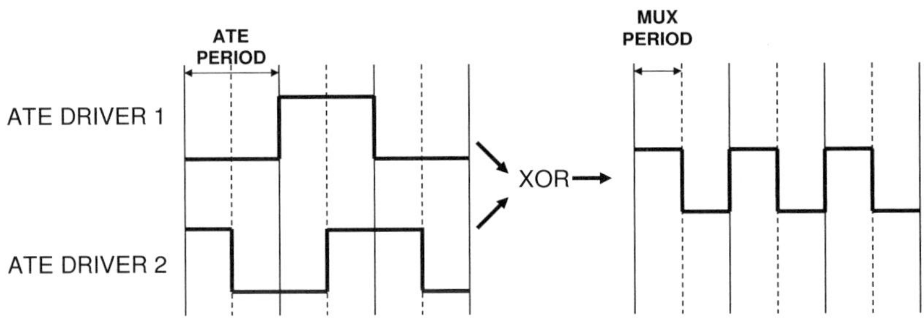

Figure 9.10 Multiplexing of two ATE driver channels using a XOR logic gate.

Figure 9.11 shows a comparison between a off the shelf MUX (Hittite HMC954) IC and a XOR IC (Hittite HMC844) where two ATE channels running at 16 Gbps are used to generate a 32 Gbps signal. On the MUX case a retiming clock signal is also needed. This retiming clock signal was generated using another ATE channel running a bit clock signal at 16 Gbps (8 GHz) followed by a passive frequency doubler and a bandpass filter (see Section 7.17.10) to generate the needed 16 GHz retiming clock for the MUX IC. The results show as expected that the MUX option provides a lower jitter output due to the retiming.

Another challenge with channel multiplexing is related to the compare side. One option is to use an analog instrument like a sampler with the required bandwidth although with this approach no at-speed functional testing would be possible. Another option is to use a demultiplexer IC to perform the data comparison using lower-speed ATE channels. In this case the demultiplexer retiming clock needs to strobe the input signal at its data eye center. By changing the retiming clock phase, it is then possible to perform BER bathtub measurements or jitter measurements. Also, with proper circuitry on the demultiplexer IC, it is possible to perform amplitude measurements like data eye height [10, 11].

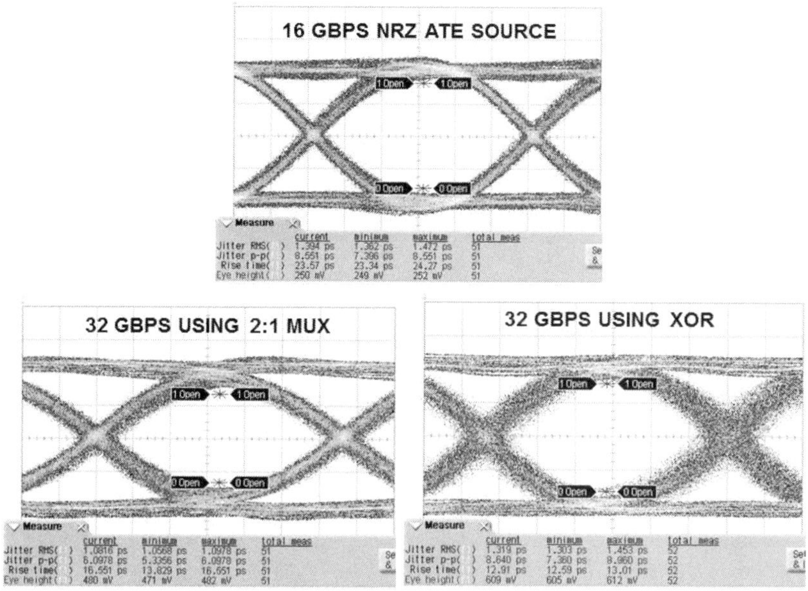

Figure 9.11 Comparison of generating a 32 Gbps NRZ stimulus signal from two ATE 16 Gbps NRZ channels using a XOR gate and a MUX IC [9].

9.3 Focus Calibration

A commercial ATE system needs to serve a broad range of different applications, but it is defined by a restricted set of specifications. This creates the situation that for some high-speed digital applications, the ATE specifications guaranteed by the standard calibration process are not sufficient. One possibility to address this challenge is to use the knowledge of the specific requirements of the application for which one intends to use the ATE equipment and improve on the ATE specifications by the use of a focus calibration procedure.

Focus calibration can be applied to address several issues [12]. One possibility is to verify whether an ATE system is able to address a set of specifications that are outside the guaranteed set for the ATE system. For example, ATE manufacturers must define the edge placement accuracy for the entire ATE system taking the maximum possible configuration, which corresponds to the worst-case scenario, into account. If for a specific application only the EPA of a set of eight pins that stimulate an I/O bus of the DUT is important, and if those eight pins belong to the same ATE pin electronics module, it is likely that the EPA of those eight pins is significantly better than the specified EPA for the entire ATE system, which is of little meaning for the application.

In this scenario, since the "improved" specifications cannot be guaranteed by the ATE manufacturer, it is important to measure the specification with the specific application setup for those eight pins. That is why although no focus calibration is required in this example, a focus measurement setup might be necessary for verification purposes.

In other situations, the ATE or measurement instrumentation manufacturers do not provide the necessary calibration for a given application measurement. This is especially true for high-speed digital applications where the DUT test fixture can have a significant influence that is not considered in the standard ATE calibration. This section presents three types of focus calibration items that are common for high-speed digital applications. These focus calibration items are skew calibration, data eye height calibration, and jitter injection calibration. The concept of data eye profile that provides a methodology to analyze the influence of the data rate on the data eye diagram parameters will also be discussed.

9.3.1 Skew Calibration

One very important test for some high-speed interfaces is the measurement of the minimum setup/hold time (Section 5.6.1). The accuracy of this measurement depends on the ability to guarantee the relative timing between the different ATE pins. Not only is there an inherent error from the pin electronics, but there is also an error induced by the different electrical lengths on the test fixture. One can minimize the effect of the electrical lengths in the test fixture by matching the signal trace physical lengths, but this does not guarantee a perfect timing skew matching.

9.3.1.1 External Instrumentation

One approach to address this issue would be to measure the specific timing of each ATE pin used on the test application including the test fixture. This can be accomplished by using external instrumentation with the required accuracy to measure the ATE timing [13, 14]. The idea would be to correct the ATE timing with correction factors until the timing between all channels required for the application are within the applicable specifications (note that these correction factors might depend on variables like the programmed data rate). This might even be an interactive process where calibration values are measured and programmed on the ATE system and measured again in order to obtain more accurate calibration factors. Figure 9.12 shows a high-level diagram of the procedure. This approach could also be used to verify the OTA/EPA of an entire ATE system.

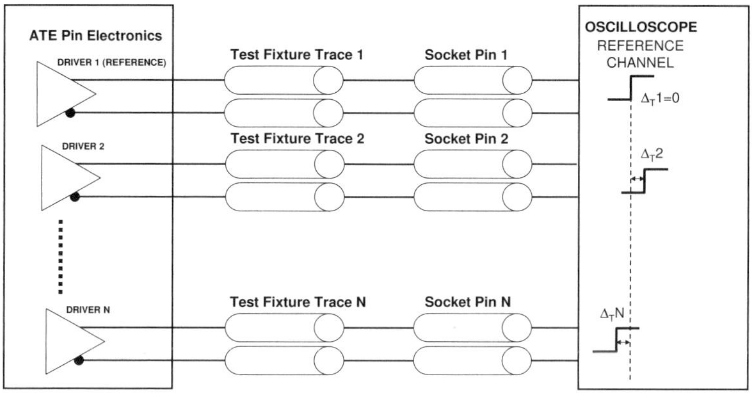

Figure 9.12 High-level diagram of a focus skew calibration measurement setup.

9.3.1.2 Loopback Measurements

Another approach to achieve minimum skew between ATE channels on a test fixture is to use loopback (ATE pin-to-pin) connections between ATE channels in different connection configurations and derive the single skew values per channel from measurements on these connection combinations. The underlying concept of this approach is to enable enough independent measurements between pins via the loopback connections to obtain an equation system that can be solved for the unknown skew values. The loopback measurement approach first assigns an even number of pins to skew groups. For the pins in such a skew group, all possible pin-to-pin connections are implemented either on a set of interposers that are mounted on the test fixture instead of the test socket for skew calibration (see Appendix H) or package dummies that can be seated inside the socket instead of the DUT. The way these loopback paths are implemented has to be customized to the actual DUT contacting environment. The electrical length (propagation delay) of the loopback paths has to be known in order to do the calculation of the single pin skew values and calculation is simplified if the same electrical length is kept for all loopback implementations on a single interposer or dummy device. The set of loopback implementations for a skew group then is used to perform all possible measurements of the position of a driver transition on one pin with the receiver of the connected pin. For each skew group this results in a set of measurements that depend on the skews of the pins of that skew group. The resulting equation system then is solved for the skews of the single pins. In order to get a unique solution for the equation system, one skew value of one pin has to be set to 0. This pin is considered the reference for all pins of the skew group and the resulting skew values within a skew group always relate to the skew component of that reference pin.

With this first step, the pins of a skew group can be corrected to have fully aligned timing edges. It is important to note that the number of pins in one skew group is not determined by the number or overall pins that have to be deskewed but by the amount of connection combinations that is required to solve the equation system derived from the measurements for the number of unknown skew values. For example, if one pin only has one unknown skew value, the number of pins in a skew group is lower than if one pin has separate skew values for its drive and receive edge. If the overall pins to be deskewed extend the number of pins in one skew group, then an additional interposer or dummy device is used that implements connections between pins of different skew groups. Only one connection and measurement between two different skew groups is required to retrieve skew correction values for all pins of the two skew groups because it is valid to apply the relative skew between the two connected pins to all pins of one of the two skew groups. The connections between the skew groups are fanned out and if required cascaded to cover all pins that have to be deskewed versus each other with the implemented connection network. Thus, in the end there will be one reference pin with one skew component that is zero. The skew values of all other pins that are covered by the combined interposer connection network are then related to that one skew component of this reference pin. It is important to note that independent connection networks that are implemented with a set of interposers or dummy devices are deskewed within themselves, but not between each other. This, however, is typically not problematic because the amount of pins that have to have minimum skew are limited to single at-cycle source synchronous groups that contain not more than only two or three dozen pins. Another point to note is that for cascaded connections between skew groups, the inherent error of the measured skew values accumulates along the cascaded path. Thus, it is best practice to design the skew group connections starting from one skew group as root of a tree and implement a connection tree with the maximum width possible with the amount of pins per skew group. This results in a tree with minimum depth and thus, with the minimum depth with minimum error propagation.

Example

Let's assume we have an interface consisting of 20 bidirectional ATE pins that need to be deskewed for better accuracy than what the standard ATE calibration provides. Each ATE pin has a drive and receive skew component that needs to be considered when deskewing the pins of the interface. So for each pin there are two unknown values that need to be determined from the equation system derived from the measurements performed during skew focus calibration. In order to obtain enough measurements to resolve the equation system, the 20 pins have to be subdivided into skew groups that consist of four pins as shown in Figure 9.13.

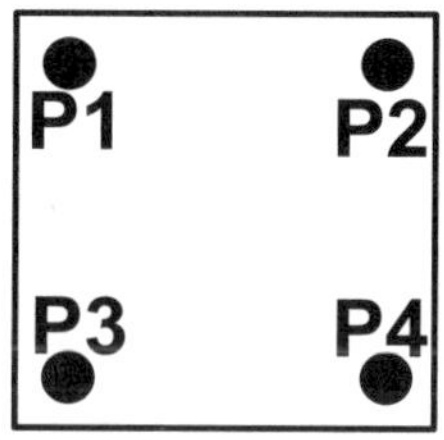

Figure 9.13 Cross-connect deskew skew group.

For each of the five skew groups, the measurement interposers or dummy devices have to implement cross-connections as shown in Figure 9.14. For each of these interposers, there will be two measurement equations per pin pair. One equation for the forward direction [e.g., P1 driving, P2 receiving for interposer (a)], and one equation for the reverse direction [e.g., P2 driving and P1 receiving for interposer (a)]. Thus, for the eight unknowns of each skew group there are 12 measurements if all cross-connects are measured. This is more than sufficient to solve the equation system for the single skew components per pin.

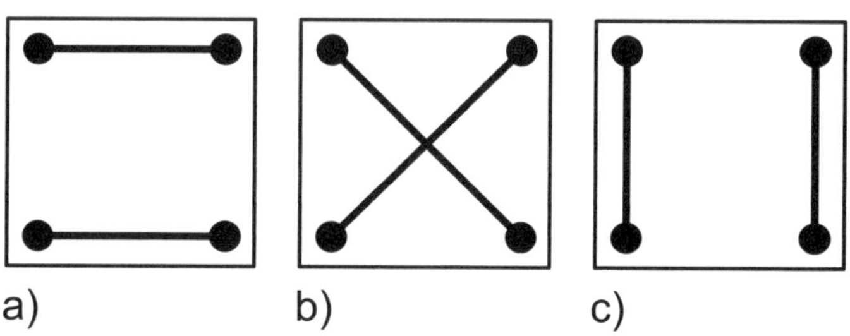

Figure 9.14 Cross-connect deskew interposer structure. Three interposers with connections according to (a—c) for each skew group.

Once the skew values for each skew group are derived, a fourth interposer or dummy device with the connections according to Figure 9.15 is measured. Here only one measurement per connection is required to determine the relative skew between the skew groups. The root skew group of the connection tree is considered as reference and all pins in this reference skew group keep their skew values as derived from the measurements on the first three interposers. For the pins of all other skew groups, the relative skew measured between the pin of the reference skew group (SKWGRP1) and the representative pin of each dependent skew group is added to all pins of that dependent skew group. After this step, the single skew values relative to one single reference pin in the reference skew group are known for all pins and can be used as correction values to deskew all 20 pins of the interface that has to be deskewed.

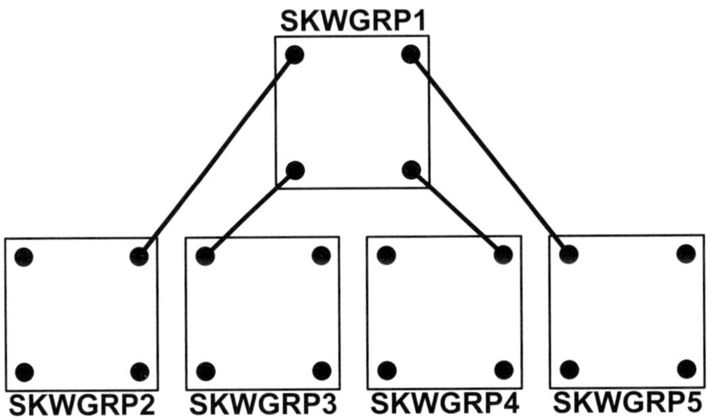

Figure 9.15 Cross-connect deskew skewgroup tree structure.

9.3.2 Data Eye Height Calibration

Another important specification that might lead to a focus calibration procedure is the data eye height of the ATE driver or receiver at the DUT [15]. This specification is very important in tests like the DUT receiver sensitivity or a data eye height measurement. The problem is that due to the frequency-dependent loss inherent to any test fixture, the data eye height at the DUT will not correspond to the expected value (the one programmed in the ATE software) due to the signal path attenuation that results from added level jitter in the case of a data pattern. To make this problem even more complex, the resulting data eye height will depend on the data rate and pattern as shown in Figure 9.16. The same reasoning applies to measuring the data eye from the DUT with an ATE receiver.

This problem can be significantly reduced by using techniques like equalization (Section 9.9.3). Another option is to simply overdrive the levels of the stimulus signal to obtain the desired data eye height at the DUT. Figure 9.17(a) shows a possible setup for an ATE driver eye height focus calibration

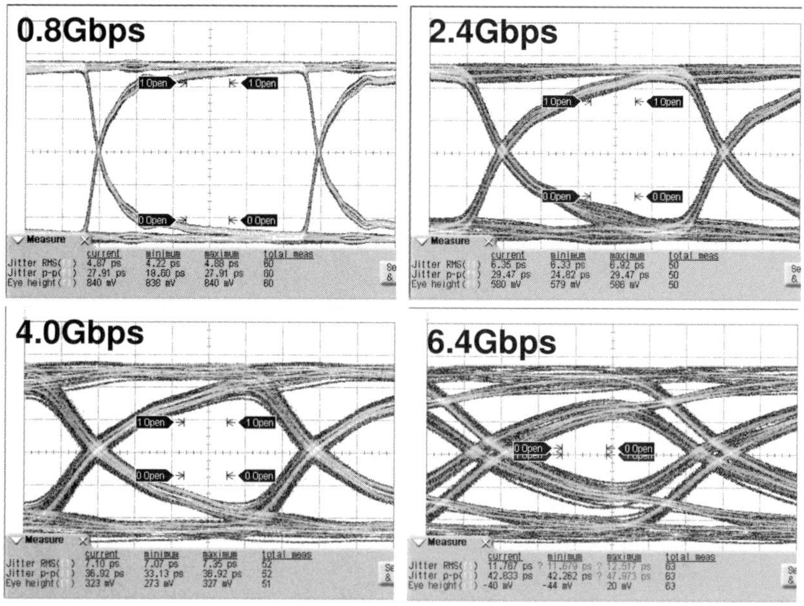

Figure 9.16 Example of the degradation of the data eye with increased data rate for a standard calibrated ATE system at the end of the test fixture.

measurement using an external oscilloscope and Figure 9.17(b) shows the setup for an ATE receiver eye height calibration where a pattern generator is used to provide a reference data eye to be measured by the ATE receiver.

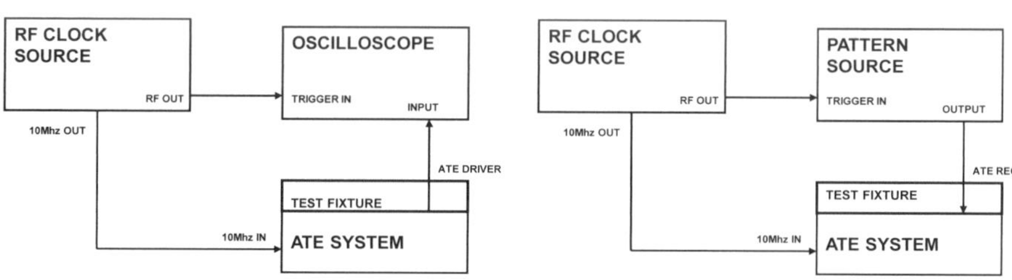

(a) (b)

Figure 9.17 Possible approach (a) for an ATE driver data eye height focus calibration setup and (b) for an ATE receiver data eye height focus calibration setup.

Since the obtained calibration factor will be data rate and pattern-dependent, the measurements have to be executed at the different data rates and with the patterns required by the application as shown in Figure 9.18 where an example for a lossy test fixture is shown. It is important to notice that this graph also can provide detailed data on the test fixture performance when comparing the focus calibration results with the measurements done without a test fixture. For the data eye height case, the calibration factors are obtained from the inverse of the measured data eye height curve as shown in Figure 9.19.

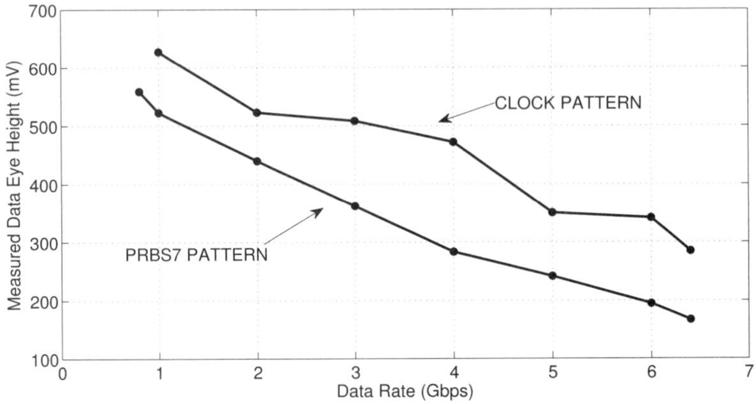

Figure 9.18 Example of the data eye height profile for the driver data eye height of an ATE system including a lossy test fixture. (From: [12]. ©2009 Jose Moreira. Reprinted with permission.)

9.3.3 Jitter Injection

Jitter injection is another important parameter that benefits from a focus calibration approach to compensate for the test fixture effects, especially the data-dependent jitter (DDJ) that is added by the test fixture due to its loss. In a receiver jitter tolerance test, it is typical to inject a certain amount of deterministic jitter (DJ) in the form of periodic sinusoidal jitter on the stimulus waveform to close the data eye by a certain amount. Even if the programmed amount of injected jitter is calibrated at the ATE pogo pin interface, the added DDJ from the test fixture will further reduce the date eye opening and increase the amount of DJ to values beyond the test limits for the receiver jitter tolerance. One solution is to calibrate the injected jitter value by measuring it at the test fixture DUT socket. Figure 9.20 shows one example of a measured focus calibration curve for jitter injection on an ATE system with a lossy test fixture. This curve can then be used for injecting a calibrated amount of jitter at the DUT by only injecting the sinusoidal jitter amplitude that is required to achieve the desired peak-to-peak jitter amplitude at the DUT. For example, to achieve 85 ps of peak-to-peak jitter at the DUT, only 40 ps of sinusoidal jitter should be injected by the ATE pin electronics on the driver signal. The remaining jitter will be added by the test fixture loss.

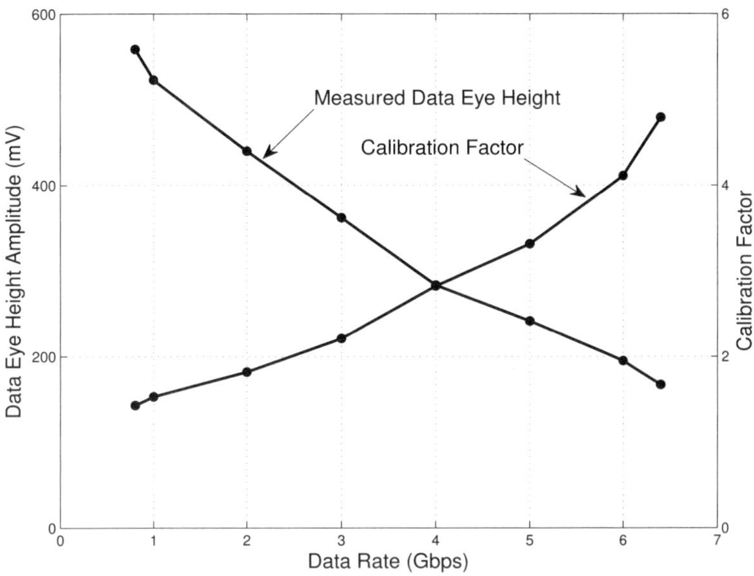

Figure 9.19 Example of the resulting focus calibration plot for the measured ATE driver data eye height at different data rates and the corresponding calibration factor curve to achieve an 800 mV target data eye height at the DUT.

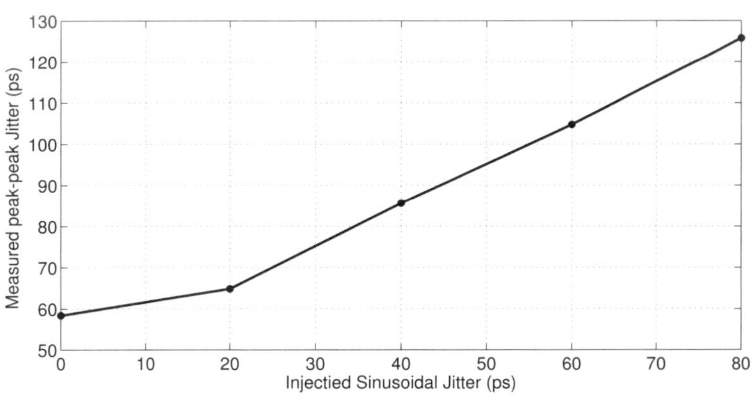

Figure 9.20 Example of the resulting focus calibration plot for the measured ATE driver jitter injection calibration when using 1 MHz injected sinusoidal periodic jitter for a PRBS $2^7 - 1$ data pattern at a data rate of 4 Gbps.

9.3.4 Data Eye Profile[1]

The concept of the data eye profile [12] is tightly connected to the characterization and focus calibration of an ATE system. As discussed in the previous sections, the parameters that characterize the ATE channel performance, including the test fixture, will change depending on the data rate at which the ATE system is running. This means that to properly characterize the performance of an ATE system at the DUT, including the test fixture, some of the characterization measurements will be data rate dependent. There are multiple parameters that can be defined to characterize the high-speed performance of an ATE system at the DUT. In this section we will use the following measurements:

Effective data eye height: The vertical data eye height at the middle of the data eye.

Nominal data eye height: The vertical data eye height 0.5 UI away from the time zero calibrated point (note that in systems with multiple channels, this time zero is the same for every ATE channel).

Maximum data eye height: The maximum eye height one can achieve at any point on the data eye.

Timing error (max and min): This value corresponds to the worst-case timing offset for a given bit transition which is before the calibrated time zero for the entire system (timing error max), or after the time zero (timing error min). The difference between these two values will correspond to the peak-to-peak deterministic jitter value. For an ATE system this value is directly related to the edge placement accuracy.

Timing shift: This value corresponds to the difference between the calibrated time zero and the average of the maximum and minimum timing error defined on the previous bullet.

Jitter: This is the peak-to-peak deterministic jitter value that corresponds to the maximum timing error variation.

Figure 9.21 shows one example of a data eye profile for the measurements defined above that are plotted in relation to the ATE system data rate. These graphs provide a profile of the ATE driver data eye height, width, and jitter at the DUT with regard to the data rate [16].

[1] This section was written in collaboration with Bernhard Roth.

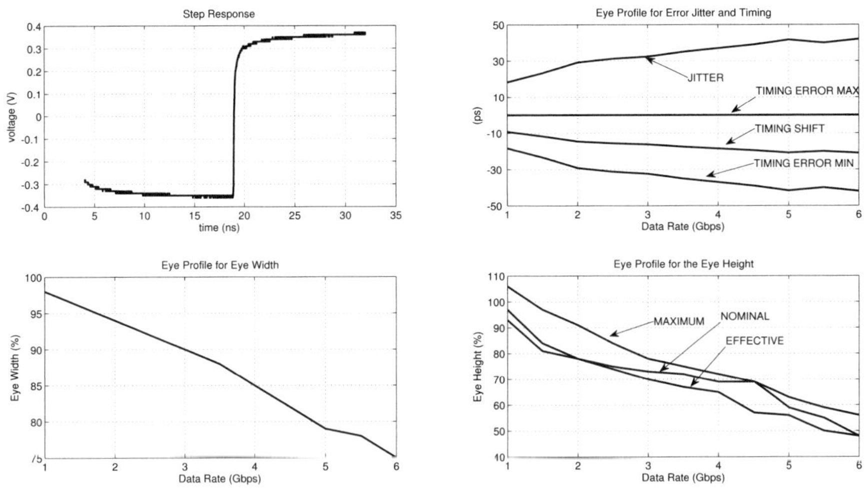

Figure 9.21 Example of a data eye profile obtained through simulation using a measured step response.

Direct measurement of the data eye profile parameters is not the only possible approach. The data eye profile can also be derived from simulation based on an ATE and test fixture model or on a single measurement point like a step response. Several options are available to simulate the data eye profile depending on the types of models that are available. One approach that allows for fast simulation time with minimum requirements on the model is to use the measured step response of the ATE system and then generate the data eye profile using techniques like double-edge response (DER) [17, 18]. This was the approach taken in Figure 9.21. Another possible approach is to use the system pulse response as described in [19].

The data eye profile presented in Figure 9.21 takes into account only the DDJ jitter added by the test fixture with the increased data rate based on the step response. It does not include the timing jitter and amplitude noise that is present on the ATE driver. It is possible to include this timing jitter and amplitude noise on the data eye profile generation. This is shown in Figure 9.22, which compares the generated data eye profiles with and without any timing jitter and amplitude noise.

Another extension of the data eye profile methodology is the concept of data eye profile correlation in a multi-site application. In this case a data eye profile is generated for each site based on the measured or simulated step response of each site and the difference between the two sites with the most extreme values is computed. In an ideal multi-site setup, the difference should be zero for all data eye profile measures. Figure 9.23 shows the step responses

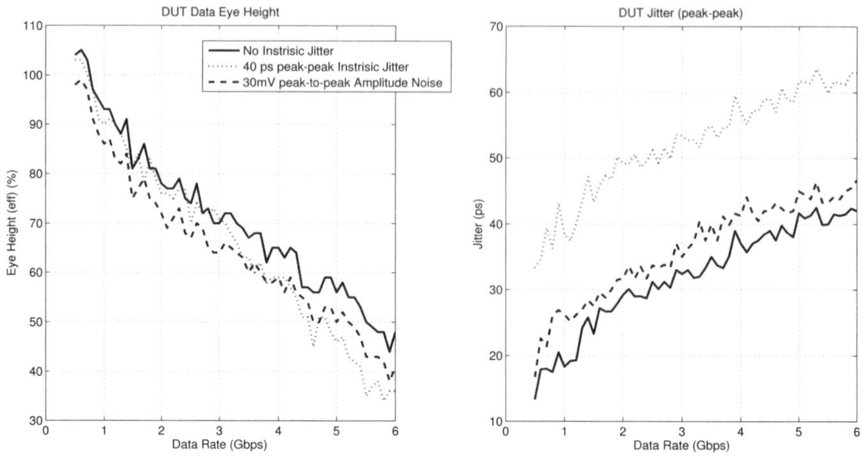

Figure 9.22 Example of a data eye profile computation including the ATE driver timing jitter and amplitude noise for the data eye height (left) and the peak-to-peak jitter (right).

and the computed data eye height profiles for four DUT positions in a multi-site test fixture. This data is used to generate the data eye height correlation plot for these four sites as shown in Figure 9.24. In this example there is a difference in the data eye height between the sites showing that there is some headroom in obtaining a better correlation between sites (e.g., by improving the test fixture).

9.4 Testing of High-Speed Bidirectional Interfaces with a Dual Transmission Line Approach

To measure bidirectional interfaces with semiduplex data transmission, one requires that the ATE pin electronics is capable of handling semiduplex signals. If this is not the case, one needs to connect the ATE driver and receiver on the test fixture to create a virtual semiduplex interface on the ATE side. The use of separate drive and receive channels in such a configuration is known as dual transmission line or fly-by configuration [20].

The use of DTL configurations is even sometimes required if the pin electronics has native handling of semiduplex signals. The requirement that might drive such a configuration is the test of the minimum bus turnaround time. The bus turnaround time defines the time a device takes to switch the data transmission direction on its bidirectional pin(s). Different connection configurations might be required for the usage of a device in its final system environment and in its ATE test environment because the connection length

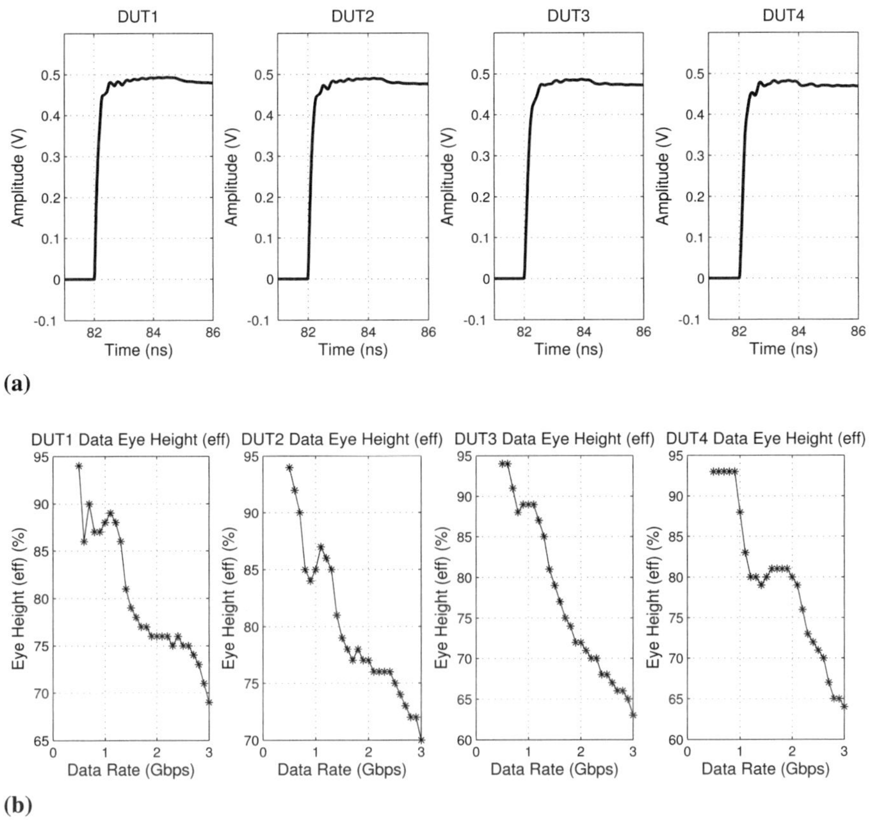

Figure 9.23 Example of (a) the measured step response for a signal pin in each site of a four-site test fixture and (b) the corresponding data eye profile for the data eye height.

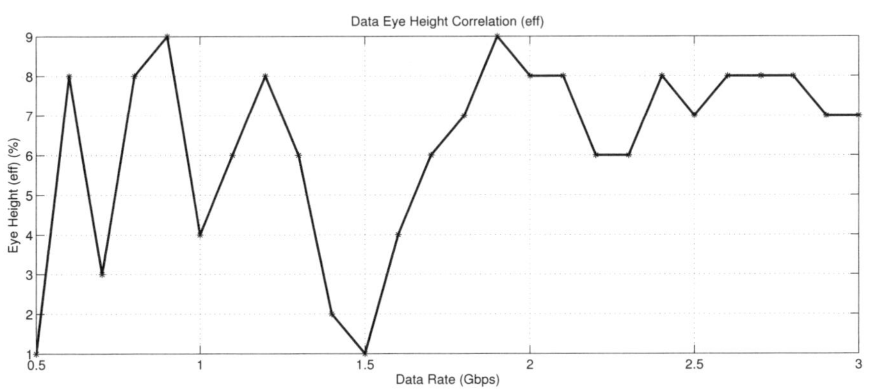

Figure 9.24 Data eye height profile correlation result for a multi-site test fixture using the data eye height profiles from Figure 9.23.

between the ATE pin electronics front-end and the DUT pins is significantly longer than the connections between two devices in the final system. The electrical length or propagation delay of this connection, however, defines the minimum bus turnaround time that can be applied without causing data collisions on the data connection.

The critical case on the ATE in this regard is to test the bus turnaround time of a DUT from its drive to its receive state as shown in Figure 9.25. Let's assume we have a device (device B) in its system environment with a propagation delay of $t_{PDS} = 1.0$ ns to its partner device (device A). The partner device has a receive-to-drive bus turnaround time of $t_{TARD} = 1.0$ ns. In such a configuration, the delay t_D between driving the last bit of data to the partner device and receiving the first data bit from the partner device is $t_D = 2\,t_{PDS} + t_{TARD} = 3.0$ ns because the drive data has to travel the connection completely before the partner device starts turning around its bus. After bus turnaround completion on the partner device, the data stemming from the partner device also has to travel the connection before the receiving device sees that data. Thus, the minimum drive-to-receive bus turnaround time t_{TADR} the device has to support needs to be less than t_D.

As already mentioned, the propagation delay t_{PDL} typically is significantly larger in an ATE environment than in the system environment. ATE systems usually compensate this propagation delay by generating drive data earlier and compare receive data later by the propagation delay between the pin electronics front-end and DUT pin. With this compensation, ATE systems can reference the timing values programmed by the ATE user to the DUT pins which is required for all timing measurements but which also makes it less obvious to the user that such a propagation delay exists.

For the measurement of the drive-to-receive turnaround time of a semiduplex DUT pin, of course the propagation delays of the ATE environment contribute to the overall turnaround delay time t_D that can be achieved at the DUT pin in the same way as in the system environment. With the longer propagation delay in the ATE environment, the achievable t_D also becomes significantly larger than in the system environment and in most cases exceeds the minimum bus turnaround time at the DUT pins that has to be ensured by the test. If the minimum bus turnaround time for a DUT is programmed on the ATE and this minimum bus turnaround time is less than two times the propagation delay, the ATE driver already will start sending data while there is still data to be received by the ATE on the transmission line between pin electronic front-end and DUT. Thus, the ATE receiver will see a signal that is generated by an overlay of the ATE drive data and the data stemming from the DUT as shown in the comparison between the system environment and the ATE environment in Figure 9.25.

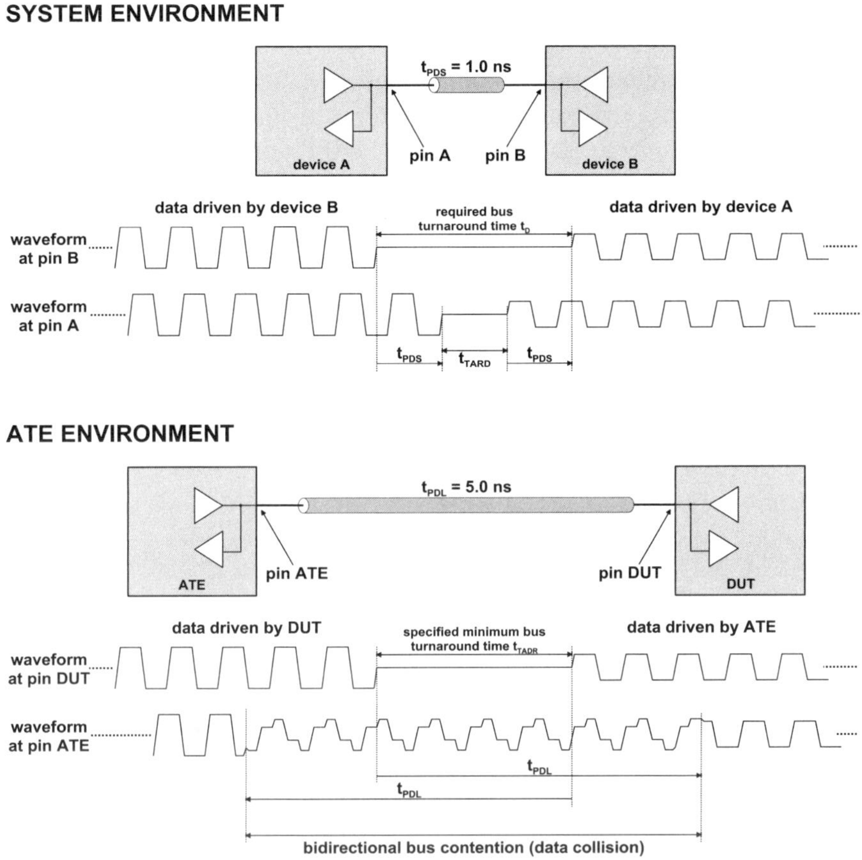

Figure 9.25 Data collision when trying to test a bidirectional application with a long test fixture [21].

Traditional ATE receivers cannot extract the data sent by the DUT from this overlayed signal and thus cannot do a correct comparison to expected data on the data part that is affected by such a data collision. One solution to such a data collision problem would be to consider the ATE propagation delays during test pattern generation and insert dead zones that are long enough between the affected signal direction changes. This, however, automatically would prevent the test of minimum bus turnaround times. Another solution is the use of a DTL configuration where the ATE driver and the ATE receiver are connected to the DUT pin via physically separated transmission lines that only have a connection at the DUT pin. This configuration prevents data collisions on the transmission lines that affect the compare capability of the ATE receiver.

Figure 9.26 shows a DTL setup using two ATE channels. On the receiving channel, the channel termination to a termination voltage is enabled to prevent signal reflections to travel back to the DUT. Also the driving channel provides a termination voltage on its driver during pattern sections that contain ATE compare data to absorb the DUT signals that travel down the transmission line connecting the ATE driver to the DUT. Section 8.9 discusses in more detail the impedance discontinuity that arises from connecting two 50 Ω transmission lines.

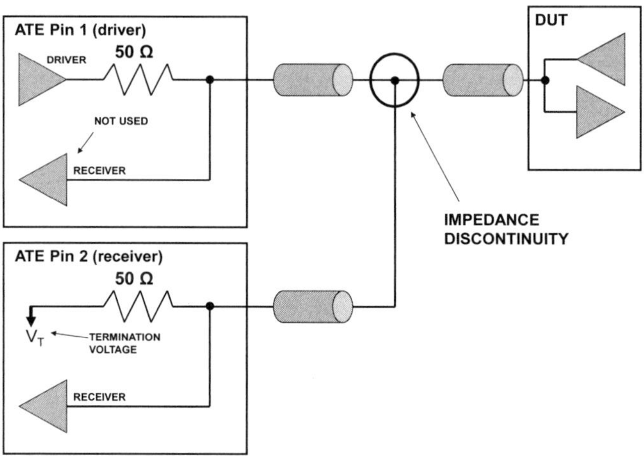

Figure 9.26 Bidirectional application being tested with a DTL (fly-by) configuration.

Clearly, a major drawback of the DTL configuration is the need to use two ATE channels to test a single DUT I/O channel. The other drawback is that due to the parallel 50 Ω terminations on the driver and the receiver during compare actions, the signal amplitude at the ATE receiver is reduced. The same is true for the signal amplitude at the DUT during ATE drive actions due to the parallel 50 Ω terminations in the DUT and the ATE receiver during ATE drive actions. This is shown in the fly-by simulation in Figure 9.27. This is not a problem if the ATE driver has enough margin to compensate for the loss during drive actions and the ATE receiver has sufficient accuracy to also recognize the lower amplitude signal during ATE receive actions correctly. However, in some situations this might cause test problems. The challenges for test fixture design with a DTL type layout are discussed in Section 8.9.

Note that simultaneous bidirectional (SBD) pin electronics architectures (see Chapter 4) do not have this drawback and can recover the correct data sent by the DUT on their receivers even in the data collision case. Thus, semiduplex signals can be tested using single transmission line configurations

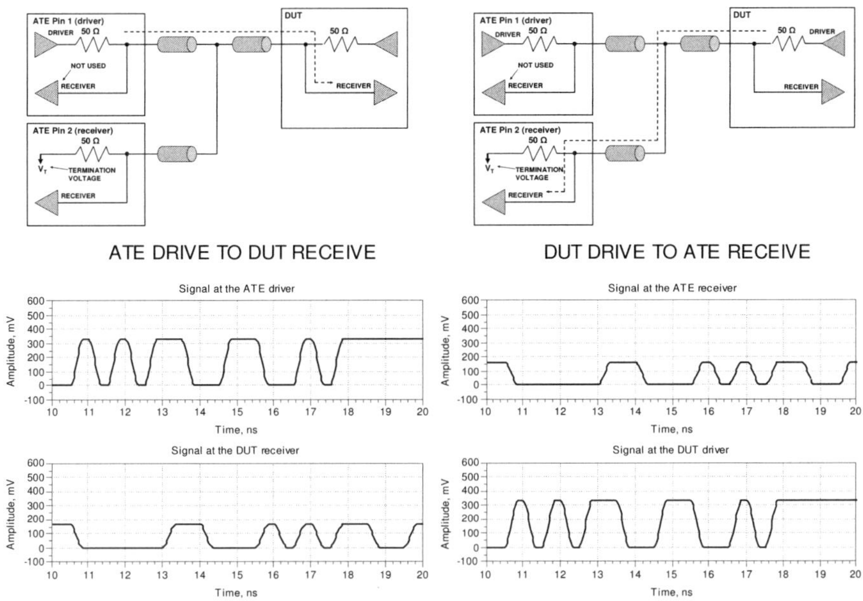

Figure 9.27 Simulation of the loss of the ATE driver signal at the DUT for a bidirectional test setup using fly-by wiring.

even with long test fixture trace lengths if the pin electronics support SBD operation modes.

9.5 Including the DUT Receiver Data Recovery in Driver Tests

The jitter generation measurements presented in Section 5.5 have the limitation that they do not take into account how the receiver in the link partner works. For some high-speed interfaces, the link partner receiver typically has either a complete clock data recovery (CDR) unit or at least a PLL in its sampling clock path. The jitter seen by the receiver after the CDR or PLL can be different from the jitter sent by the DUT driver as shown in Figure 9.28. A more detailed discussion on the background of this behavior can be found in Section 2.6.3.

Some standards (e.g., PCI Express) require that jitter tests (e.g., jitter generation) are performed using a reference or golden CDR or PLL that is defined by the standard. This allows the jitter generation specification to take into account the capabilities of the link partner receiver CDR/PLL to attenuate certain frequency ranges of the jitter magnitude.

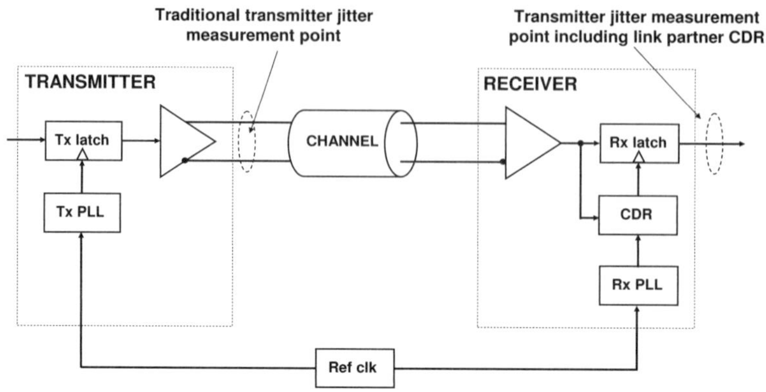

Figure 9.28 Difference of measuring the generated jitter by a DUT driver at the driver output and at the output of the link partner receiver CDR.

Example Let us assume that a DUT test requires the data eye total jitter (at a given BER) to be less than 0.5 UI. Let's also suppose that the DUT driver has a design problem that creates a periodic jitter component (e.g., sinusoidal) with 1 UI amplitude at a very low jitter frequency (e.g., 10 kHz). On a standard jitter test without any CDR unit, this part would fail the jitter generation test. Since a typical receiver CDR, however, can handle 10 kHz of sinusoidal jitter easily, this design problem would not have an effect on the real application. Thus, from the IC manufacturer's point of view, it makes sense to use a CDR for the jitter measurement with a defined specification so that this DUT would pass the jitter generation test.

Figure 9.29 shows the CDR requirements from the PCI Express standard [22] for any jitter measurements. The figure shows that any jitter components below 10 kHz should be attenuated by a factor of 10^{-3} since these jitter components will be tracked by the CDR.

If a CDR is really required for a jitter measurement, there are two ways to address this requirement. One option is to use a real-time sampling oscilloscope to measure the jitter. Since a real-time oscilloscope can acquire the entire waveform in real-time, one could perform a software-based CDR on the waveform and then compute the transmitted jitter value of the resulting waveform after the CDR operation. Although this is a very elegant solution because it allows the user to define the shape of the CDR response in a software postprocessing step, real-time sampling capabilities are typically not available in standard ATE pin electronics cards.

The second option is to implement the CDR on the measurement equipment hardware. This is currently the typical approach followed for ATE pin electronics cards. Although there is bench test and measurement

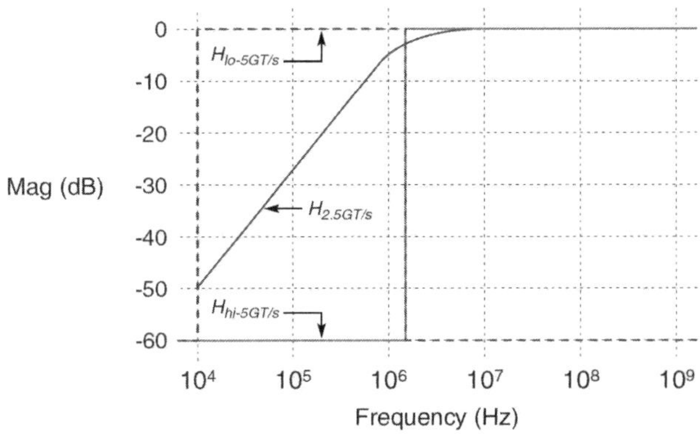

$$H_{2.5GT/s} = \frac{s}{s + w_c} \qquad w_c = 2\pi f_T$$

$$H_{hi\text{-}5GT/s} = if(f >= f_T)\ then\ 1.0\ else\ 10^{-3}$$

$$H_{lo\text{-}5GT/s} = if(f < f_{10kHz})\ then\ 10^{-3}$$

$$elseif\ (f < f_T)\ then\ 1.0$$

$$else\ 10^{-3}$$

$$f_T = 1.5\ MHz$$

Figure 9.29 CDR requirements for the transmitted jitter measurement as defined by the PCI Express standard. (From: [22]. ©2002–2009 PCI-SIG. Reprinted with permission.)

equipment that provides the capability to define the characteristic of the CDR being used for the jitter measurement, typically what one finds is a fixed CDR unit that is able to track the low-frequency jitter and in this way try to be compliant with most standard requirements.

9.6 DUT Reference Clock Jitter Attenuation Approaches

The DUT reference clock performance can be critical for properly testing a DUT. Even in a low cost test setup where the DUT high-speed I/Os are tested using an external wire loopback approach, the failure coverage might be impacted by the DUT reference clock performance. This can be a challenge since for the DUT reference clock usually one only needs a limited number of digital channels, but the need to have a low jitter (i.e. low phase noise) reference would require the ATE system to have at least one module with high-performance digital channels. This, in turn, increases the ATE system configuration cost especially if only a subset of the channels on that high-performance ATE digital channel module is used. One option to address the need for a low jitter DUT reference clock is to use an external RF clock source (see Section 7.10) together with a Balun (see Section 7.17.9) to generate a differential signal. Another option is to use a jitter attenuation IC that contains a integrated PLL to "attenuate" the jitter contents of the low cost digital channel as described in [23]. To provide DC access for parametric measurements, a set of relays can be used as shown in Figure 9.30. Figure 9.31 shows an example of the implementation of a jitter attenuator module and its use on a DUT test fixture.

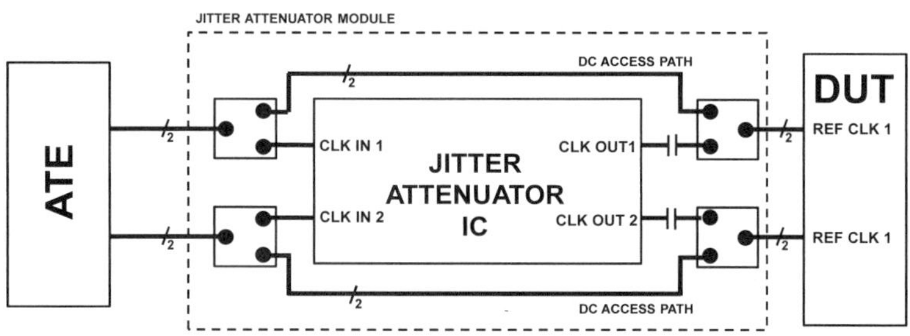

Figure 9.30 High level diagram of a jitter attenuator module implementation with a bypass path using relays for DC measurements.

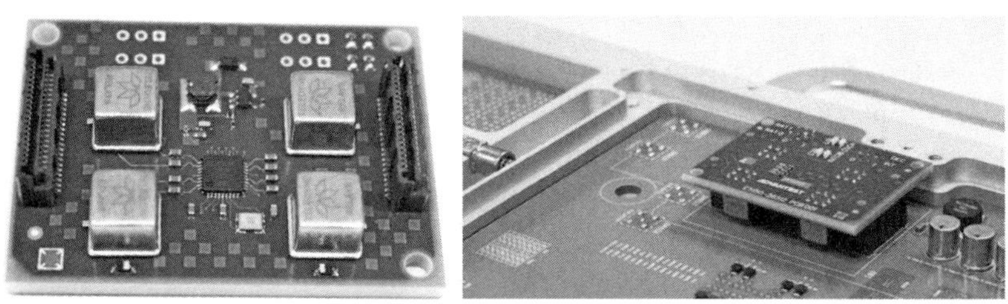

Figure 9.31 Example of jitter attenuator module and its use on a DUT test fixture [23].

Figure 9.32 shows one example of how a jitter attenuator IC can improve the phase noise of a low cost ATE digital channel. As a reference, the phase noise from a low phase noise external RF source is also included. There are of course some drawbacks from this approach. First, the output amplitude of the jitter attenuator is usually fixed to certain value. Second, the jitter attenuator needs to be programmed and locked to the input reference signal, which can take some time. Finally the jitter attenuator IC and any needed related components like relays will take space on the DUT test fixture, which can be problematic for a multi-site application. Another possible issue is the lock time of the jitter attenuator PLL, but this issue can be minimized by choosing an appropriate PLL bandwidth and also by hiding the PLL lock time on other tests like DC or scan.

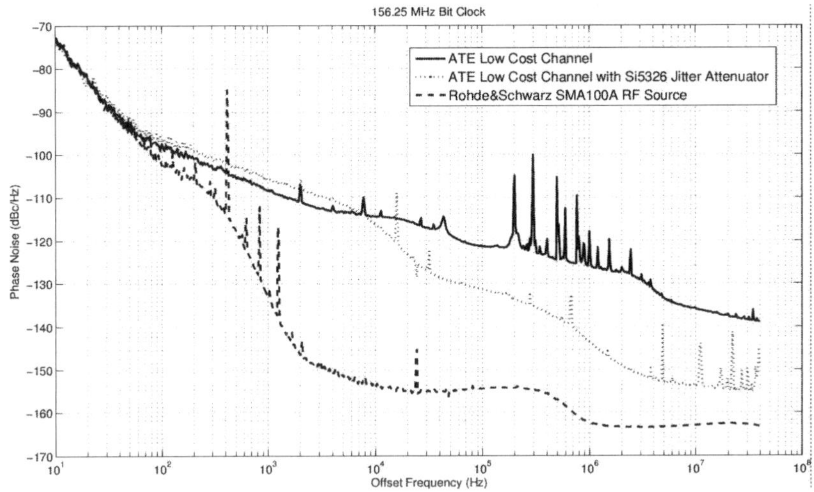

Figure 9.32 Jitter attenuator example for 156.25 MHz using a Silicon Labs Si5326 IC.

Figure 9.33 shows one example of how the reference clock phase noise can have a significant impact on the DUT I/O performance [23].

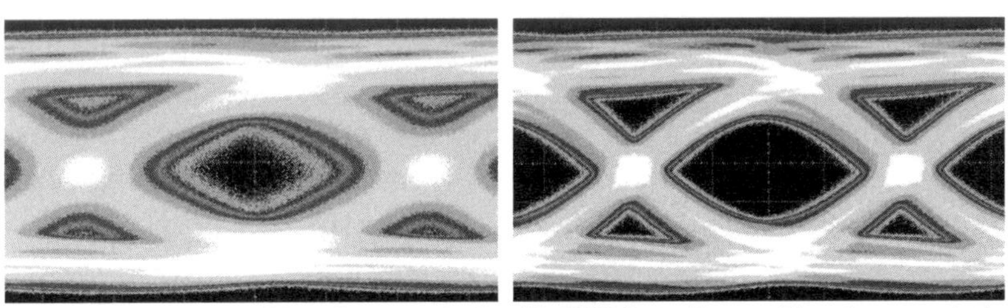

Figure 9.33 Impact of the jitter attenuator module on the DUT I/O data eye diagram performance [23].

9.7 Protocol-Awareness and Protocol-Based Testing

The typical ATE-based functional test of a device applies binary data to the inputs of the DUT and compares the response of the DUT to expected data as shown in Figure 9.34. Traditionally the binary stimulus and expected data is derived from device simulations and the ATE system does not interpret or modify the test data beyond the single bit logical states as, for example, protocol analyzers do.

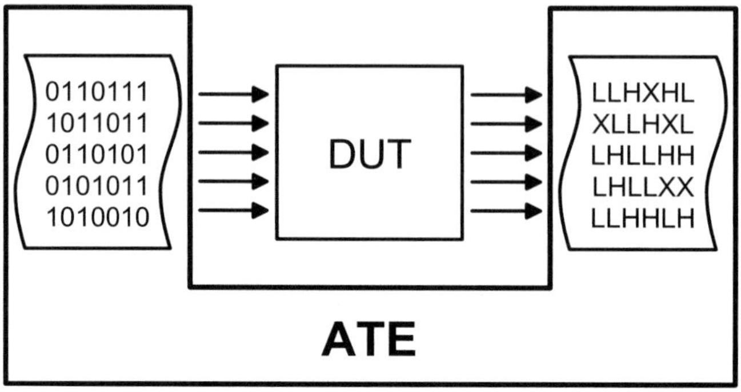

Figure 9.34 Classical view of an ATE functional pattern for a standard digital functional test.

Devices, however, operate based on protocols that define how data needs to be applied to a device and in which sequence this data can be applied. Of course, the devices also follow these protocol rules on the data that they generate on their output pins. For low-speed parallel buses or very simple serial protocols as, for example, I2C [24], it was a relatively simple task for a test engineer to interpret the binary data representation available on ATE systems to understand the functionality behind the test pattern. The ability to interpret this data is especially important during interactive device debugging to allow a test engineer to apply corrective actions without going through the complete simulation and pattern generation process for each debug step [25]. This is of increased importance for the protocols used to program and query the configuration registers of a device.

With the advent of high-speed interfaces and especially the increased application of serial and embedded clock data transmission for these devices, this situation changed significantly. Serial interfaces typically use packet-based protocols that embed their payload like address data or transfer data into complex packets that span over several serial bytes as shown in Figure 9.35 for an example of one of the transaction layer packets (TLP) defined for PCI Express.

Moreover, such complex high-level packets might be repackaged into several lower level packets that can be distributed over a potentially scalable number of parallel lanes on the physical layer of a serial link. Each of these parallel lanes that form the logical link transmits its data serially.

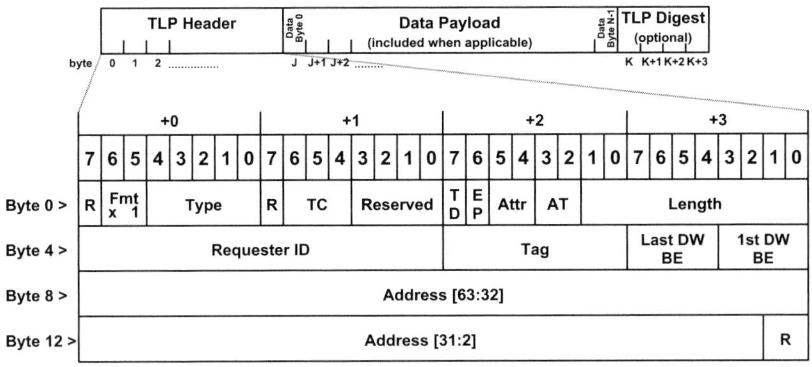

Figure 9.35 Example of a PCI Express transaction layer data packet (TLP) with TLP header for a 64-bit memory access [22].

In addition to the complex packet structure that makes interpretation of the data difficult for a test engineer, embedded clock interfaces usually apply coding, like for example, 8B/10B and/or data scrambling (see Appendix D). This makes the interpretation of the raw data during interactive debugging nearly impossible without additional support from the ATE system [26]. For production testing with debugged patterns, however, this does not pose a problem as long as the data coding and scrambling generates unique data. For some codes, this is sometimes not necessarily the case. For example, the data generated by 8B/10B coding depends on the starting disparity that is set by the device. If a DUT that uses 8B/10B coding does not set a specific known starting disparity during initialization, this can lead to varying data response codes of the device which can cause problems when the devices are tested on an ATE that only compares to predefined expected raw data bits.

Another issue for the debug scenario of high-speed I/O interfaces is caused by the nature of embedded clock interfaces. In order to keep the phase lock of a clock data recovery unit, the transmitted data stream needs to contain a sufficient number of data transitions as described in Section 2.6.3. This, however, means that in intervals with no transmitted data, a data connection cannot just be idled and therefore transitions need to be generated by the transmitter driving the transmission line. The relevant high-speed I/O standards usually define the exact data that needs to be transmitted in pauses where there is no payload data available (logical idle). For the SATA standard, this data, for example, is specified as SYNC primitive; for PCI Express, this data is called an "idle data symbol." The transmission of data-like patterns during these logical idle states makes it difficult for a test engineer to distinguish between payload and idle data during debugging.

The last protocol related test challenge to be mentioned here not only has consequences for device debugging on the ATE, but also for fully automated testing of devices based on predefined test patterns (e.g., in production test runs). With the clocking architectures used for systems that deploy high-speed I/O devices, it is possible that two communicating devices are not driven by one central system reference clock, but by two separate reference clocks. Although these separate clocks usually run at the same frequency, there typically is a frequency offset that might be only in the parts-per-million (ppm) range. Since the phase of the reference clocks of high-speed I/O devices directly translates to the phase of the high-speed I/Os, the same offsets are seen on the data connections between the two devices. Thus, devices connected in such an environment see a slow drift of the incoming data relative to their reference clocks. With the data tracking mechanisms of source-synchronous interfaces, this drift is not an issue for sampling the incoming data correctly. However, in the device, usually the clock domains of the device core and the high-speed I/O data section are decoupled via FIFO buffers. On the core side, data is read and written into the FIFO based on the reference clock phase. On the I/O side, the data is read from the FIFO also based on the phase of the reference clock, but the received data is written into the FIFO based on the phase information derived from the high-speed I/O data interface.

If there is a frequency offset between the reference clock and the clock derived from the data interface with the derived clock running at a slightly higher frequency, this decoupling FIFO will fill up more and more over time until it reaches its stall state and does not have any buffer space available to store received data. This will lead to data loss on the high-speed I/O connection. In order to prevent such a situation, especially high-speed I/O standards that are based on embedded clock data transmission, define skip data sequences that need to be transmitted regularly and that can be ignored (i.e., not transferred into the FIFO) on the receiving side of the data connection. Examples for skip data in high-speed I/O standards are skip ordered sets for PCI Express or ALIGN primitives for SATA. The intervals in which skip data needs to be transmitted are derived from the maximum frequency offsets that are allowed on the reference clocks by these standards.

The issue for ATE-based testing with this kind of frequency offset compensation is that the insertion intervals for skip data are not defined in a deterministic manner. This means devices usually generate skip data in a way that fulfills the specification, but which is not predictable. This means that it might not be possible to know a priori how the expected output bit stream from the DUT will look. This is exemplified in Figure 9.36.

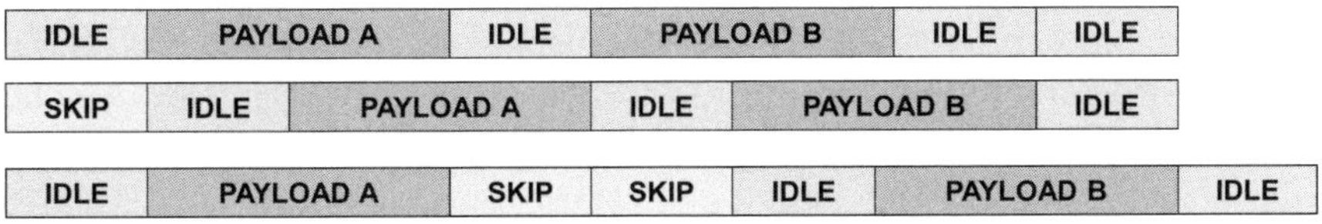

Figure 9.36 Example of three possible data patterns from a protocol-based I/O cell with the exact same payload.

In the figure there are three different cases where payload A and B are transmitted by the DUT together with command packets called "IDL" and "SKIP." The exact function of these control packets is irrelevant for this discussion since from the I/O test perspective we are only interested in verifying if the data content of the payload A and B packets is correct. This means that the three cases in Figure 9.36 would correspond to a functional pass as long as the data on the payload A and B packets are functionally correct. This means that simply defining a compare pattern would not be sufficient because it is not known during pattern generation how the DUT I/O protocol will generate the data packets A and B.

The lowest level of protocol-aware testing might simply be to allow the test engineer to look at the data pattern and program the expected pattern without any encoding, scrambling, or error correction required by the protocol (see Appendix D). In this case the ATE system (hardware or software) has to decode and encode the test patterns. The data presentation to the test engineer happens in a way that only the real data patterns that are used without any protocol requirements are visible. At the more complex level, protocol-aware ATE systems also might consider the packet structure of the DUT protocol and only present the payload data to the test engineer. Besides packet composition and decomposition, the ATE also has to take care of correct idle-data and skip data handling on the high-speed I/O.

There are two options to address the protocol-aware test challenge. Instead of performing a functional test by comparing the output data to an expected pattern, the first option is to capture the data from the DUT and post-process it. The post-processing operation uses the protocol knowledge to identify the various packets, extracts the relevant payload data from these packets, and just compares the extracted payload data to the expected payload data. This approach provides the maximum flexibility since any kind of protocol can be implemented with the post-processing software. However, it requires that the ATE system is able to do real-time data capture at the I/O data rate. Another issue is that the time spent for data transfer and post-processing might have a significant impact on the test time (compared to a standard functional test).

The second option is to integrate the protocol capability in the ATE pin electronics hardware (i.e., the pin electronics receiver is able to understand the protocol from the DUT and only compares the payload data). Typically, protocol-awareness of ATE systems is referring to such a kind of hardware implementation [27–29]. This means that the test engineer only needs to specify the expected data for the payload and the pin electronics takes care of the protocol specifics. This approach has the advantage that no overhead or at-speed capture capabilities are required because the pin electronics understands the protocol at-speed. The challenge of this option is that it adds complexity and cost to the pin electronics. Another difficulty is that one cannot expect that a hardware implementation is able to handle any possible protocol scheme (especially the ones that are not defined when the ATE system is designed). One possible approach to mitigate this challenge is to allow some type of programming flexibility on the ATE pin electronics protocol engine either by using an FPGA or other approaches.

In this discussion it is important to note that for characterizing an I/O cell that uses a complex protocol engine it is typically easier and overall more cost-efficient if precautions to simplify the test of such a device are already taken during the circuit design phase. As a minimum, it should be ensured that the device generates deterministic and predictable data. This means that, for example, devices that use encoded data offer the possibility to set the encoding circuit to a known state so that predictable codes are generated (e.g., controlled start disparity for 8B/10B code generators). For protocols that generate skip data, DfT circuitry that allows it to generate the skip data on a known timing grid helps to simplify the test of such devices without applying time consuming post-processing operations or dedicated protocol-aware ATE hardware. Very often the main target of ATE-based testing of high-speed I/Os is to ensure their analog specifications such as jitter, data eye height, and so on. In such cases DfT circuitry that helps to generate the required data streams in a way that is decoupled from the device functionality is helpful for the test on an ATE. Examples for such DfT circuitry are pattern generators (e.g., PCI Express compliance pattern generator, PRBS generators), pattern analyzers, or loopback capabilities (e.g., far-end loopback paths, near-end loopback paths) that are implemented for the high-speed I/Os of a DUT.

9.8 Testing Multilevel Interfaces with Standard Digital ATE Pin Electronics

Most digital pin electronics available for today's ATE systems are designed to test and characterize NRZ type digital interfaces with two voltage levels (see Section 2.1). Although in [30] an at-speed pin electronics design was proposed to test PAM type signal interfaces. When facing the challenge of testing a multilevel interface like a PAM-4 that uses four voltage levels, it is necessary to develop an approach to test this type of interface using a standard ATE system.

One possible approach is to use analog ATE instrumentation for testing this type of interface where an arbitrary waveform generator is used to provide the stimulus signal and a sampler/digitizer is used to characterize/test the DUT output waveform. This was the typical approach used to test LAN applications like 1000 BASE-T [31] where PAM-5 signaling is used.

Another possibility is to use multiple ATE digital channels to generate and receive a multilevel signal. Figure 9.37 shows a diagram of how the setup would look for generating a PAM-4 waveform using two standard ATE channels connected through a power combiner.

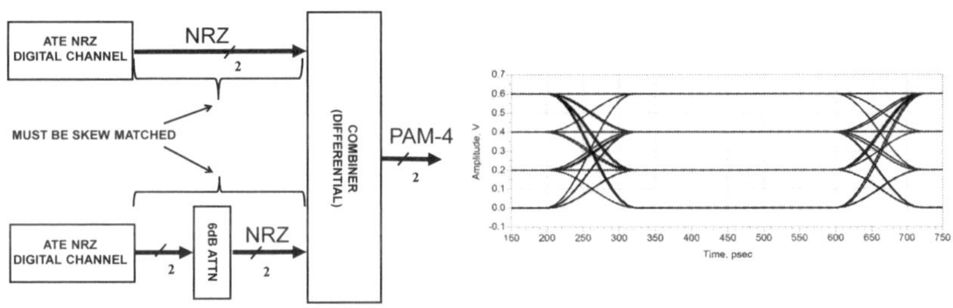

Figure 9.37 Example of generating a PAM-4 multilevel signal using standard ATE channels and a power combiner.

The two channels run standard NRZ data patterns with one of the channels amplitude reduced to half (using a 6 dB attenuator or programming the amplitude by half) as shown in Figure 9.38 [9, 32].

The data patterns in each channel needs to be set to obtain the desired PAM-4 data pattern. When using a PRBS pattern on the ATE NRZ digital channels, then it is necessary to select the right seeds in each ATE channel [32]. Figure 9.39 shows an example of a 32 Gbaud PAM-4 waveform obtained from two 32 Gbps NRZ channels. Because we are using a passive power combiner the performance of the output PAM-4 waveform depends on the performance (e.g., jitter) of the two NRZ sources that are combined. This means that when using the power combiner approach, it is critical that the two NRZ sources that are combined and the signal path to the combiner have a very good performance. Another challenge of using a power combiner approach to generate a PAM waveform is that it becomes more complex for a higher multilevel encoding like PAM-16.

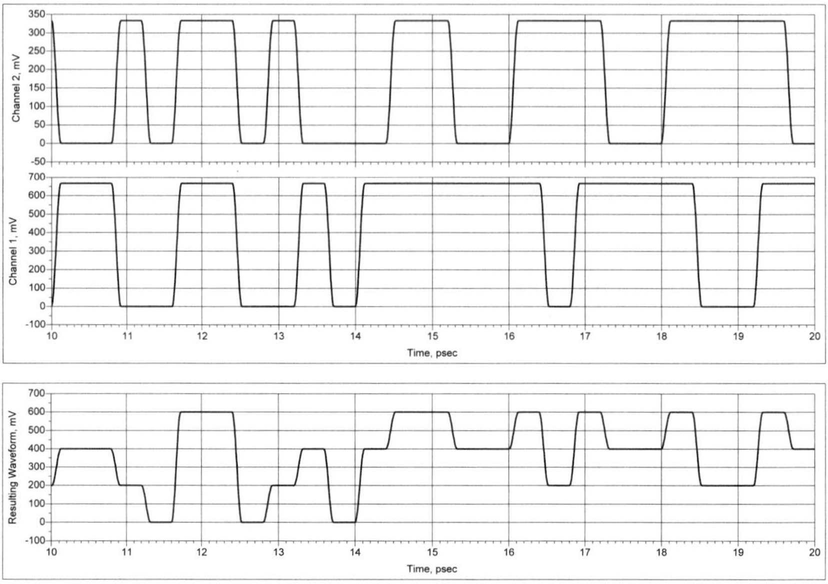

Figure 9.38 Individual ATE digital channel programmed NRZ waveforms (top) and combined waveform (bottom).

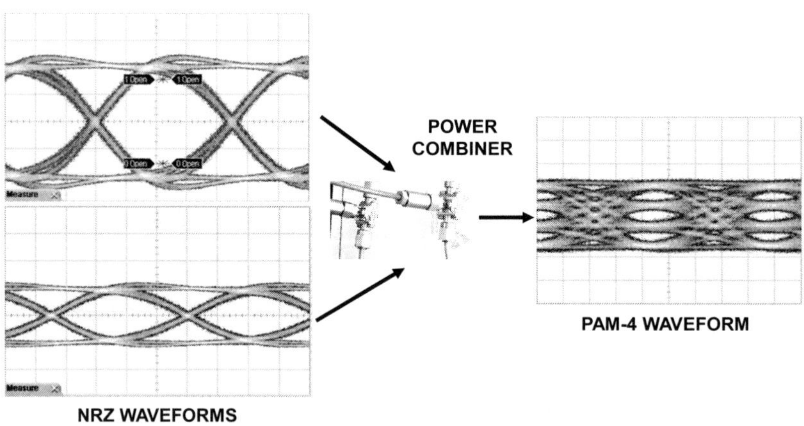

Figure 9.39 Example of generating a 32 Gbaud PAM-4 multilevel signal using standard ATE channels and a power combiner.

Another approach is to use a digital to analog converter (DAC) to generate the PAM waveform [9, 32]. For a PAM-4 waveform, we need a 2-bit DAC as shown in Figure 9.40.

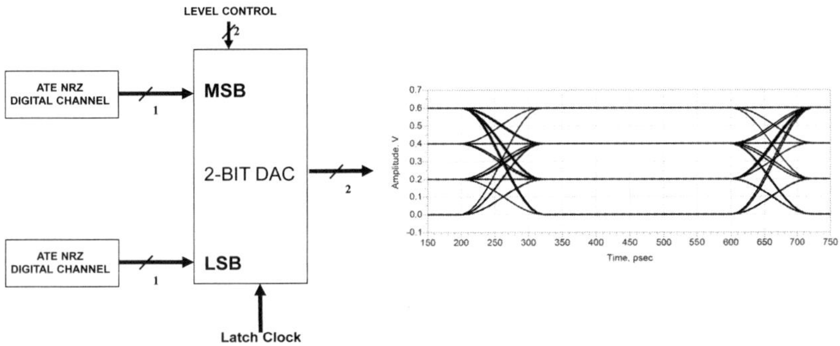

Figure 9.40 Example of generating a PAM-4 multilevel signal using standard ATE channels and a digital-to-analog converter (DAC).

The disadvantage of the DAC approach is that it is an active component requiring power and also a retiming clock. Also a DAC is more expensive than a power combiner.

On the compare side there are also two solution options using standard ATE NRZ digital channels. The first solution is to use a series of NRZ comparators, each one set for a different threshold [33]. A resistive network is needed to connect the different comparators while keeping a impedance matched environment, which results in additional attenuation to the signal at each comparator. This approach has the advantage that it is possible to do a full multilevel functional test in a single execution. The main disadvantages is the need to use three comparators (for the case of a PAM-4 signal) and the additional signal loss at each comparator.

The other option for performing a functional compare of a PAM waveform using a standard ATE digital channel is to use a single comparator and multiple functional tests with different patterns and compare thresholds as shown in Figure 9.41. The disadvantage of this approach is the increase on the test time by performing three functional tests instead of one. By performing a shmoo on the strobe time and compare threshold, it is also possible to generate a data eye diagram by combining the shmoo results for each functional test as shown in Figure 9.42.

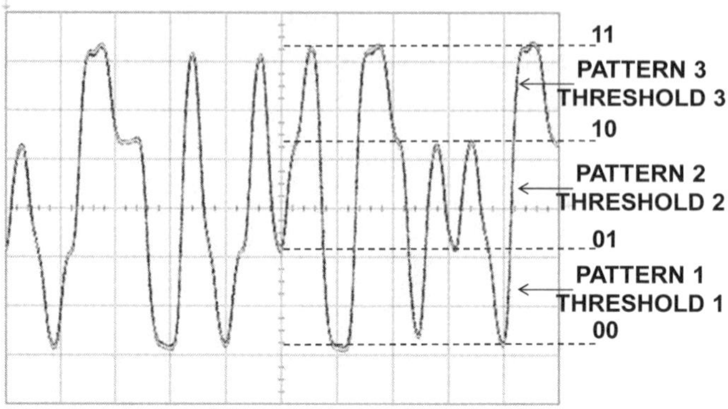

Figure 9.41 Performing a functional test on a PAM-4 waveform using a single comparator and three functional tests with different compare voltage thresholds.

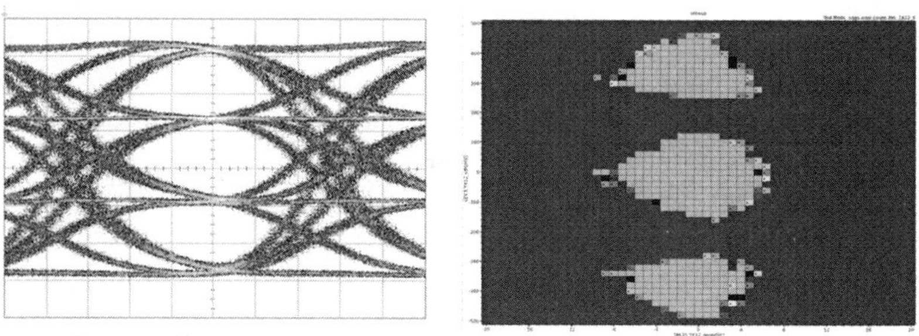

Figure 9.42 Example of a PAM-4 data eye measured with an equivalent-time oscilloscope and with an ATE integrated PAM-4 measurement setup [9].

9.9 Signal Path Characterization and Compensation

9.9.1 Signal Path Loss Compensation: De-Embedding

Time-domain de-embedding or time-domain deconvolution [34] corresponds to a post-measurement step where measurement data is processed to remove the effects of the test fixture loss or the bandwidth limitations of the pin electronics receiver (assuming that those effects are known and have been measured or simulated). The idea is to assume the test fixture and the pin electronics to be a linear time invariant (LTI) system (see, for example, [35] for a proper treatment of LTI systems). One simple model for the loss that a signal suffers when traveling from the DUT I/O package pad to the measurement instrument is shown in Figure 9.43 where the DUT output goes through a low-pass filter representing the test fixture (and the DUT socket) and then another one representing the bandwidth limitations of the ATE pin electronics.

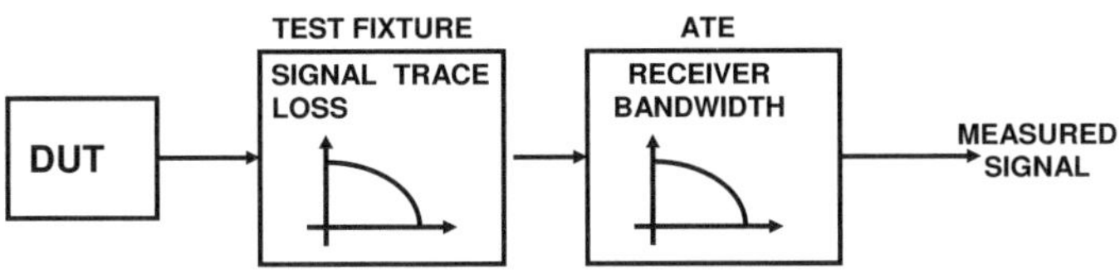

Figure 9.43 Modeling the measurement of the output of a DUT through a test fixture and a bandwidth limited ATE pin electronics.

One straightforward approach to measure the real output waveform of the DUT would be to compensate for the signal degradation by "inverting" the effect of the "low-pass filters" on the measurement setup by an appropriately designed "high-pass" filter as shown in Figure 9.44. This filter would be applied by means of a software algorithm on the measured data. In order to obtain this de-embedding filter, we need to formalize the problem by using a black box approach as shown in Figure 9.45.

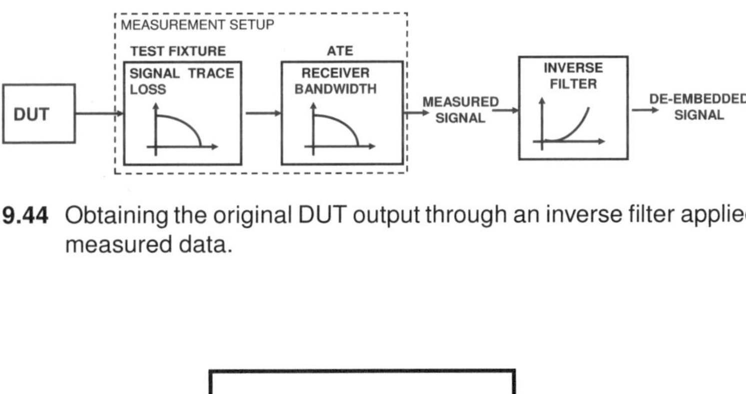

Figure 9.44 Obtaining the original DUT output through an inverse filter applied on the measured data.

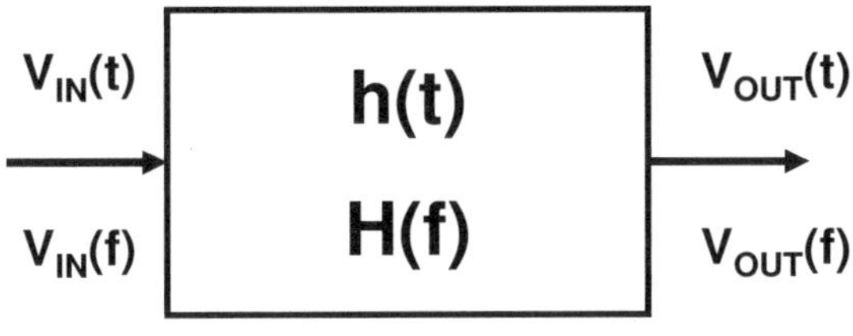

Figure 9.45 Black box model of the de-embedding problem: $h(t)$ is the time-domain impulse response of the measurement system and $H(f)$ is the frequency-domain transfer function.

The output function (V_{OUT}) can be computed by the following equations in the time and frequency-domains where $*$ represents the convolution operator[2]:

$$V_{OUT}(t) = h(t) * V_{IN}(t) \tag{9.1}$$

$$V_{OUT}(f) = H(f) \cdot V_{IN}(f) \tag{9.2}$$

The de-embedding problem consists of computing the original input waveform function V_{IN} by using the measured output waveform function V_{OUT} and the system transfer function H. In the frequency-domain this is represented by the following equation:

$$V_{IN}(f) = \frac{V_{OUT}(f)}{H(f)} \tag{9.3}$$

The frequency-domain is used since a convolution in the time-domain is simply a multiplication on the frequency-domain. The time-domain waveform can then be obtained from $V_{IN}(f)$ through an inverse Fourier transform.

One open question is how to obtain the system transfer function $H(f)$ since it is typically unknown. If the test fixture loss is the only item to be de-embedded, then one approach is to obtain the insertion loss of the test fixture. This can be achieved by using the test fixture measurement approaches discussed in Appendix H. The measured insertion loss (S21) will then correspond to $H(f)$.

To include the ATE pin electronics receiver in the model, it is typically necessary to obtain the $H(f)$ model first in the time-domain since this is how a high-speed digital ATE receiver works. One way to measure the transfer function $H(f)$ is to use a test signal like a step function and measure the resulting step function with the ATE receiver (i.e., using some of the techniques of Appendix H, a step signal is injected at the DUT socket). The $H(f)$ function can then be computed by dividing the Fourier transform of both step functions:

$$H(f) = \frac{V_{OUT}(f)}{V_{IN}(f)} = \frac{FFT\{V_{OUT}(t)\}}{FFT\{V_{IN}(t)\}} \tag{9.4}$$

The previous discussion shows the basic theory behind time-domain de-embedding or deconvolution, but it is important to realize that there are several technical details regarding the Fourier transform of step functions,

[2]The convolution operator in the time-domain between two functions is defined by the expression: $x(t) * y(t) = \int_{-\infty}^{+\infty} x(\delta).y(t - \delta)d\delta$.

causality, and stability of the numerical algorithms that are outside the scope of this book. Reference [36] provides an excellent discussion of time-domain deconvolution.

Figure 9.46 shows one example of a stimulus step response (generated with a bench pattern generator and measured with a high-bandwidth equivalent-time oscilloscope) and the step response measured by the ATE pin electronics, which is degraded by the test fixture loss and the pin electronics bandwidth limitations. Both measured step responses are used to obtain the frequency response of the measurement setup $H(f)$ through the Fourier transform of the step waveforms as shown in Figure 9.47.

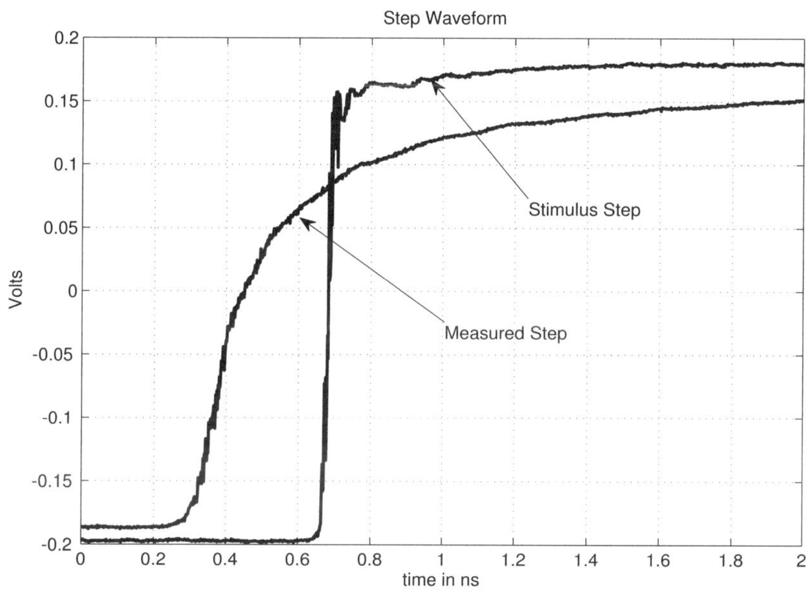

Figure 9.46 Measured step stimulus waveform and the step waveform measured by the ATE pin electronics after a lossy test fixture.

With the knowledge of the computed transfer function $H(f)$, it is now possible to de-embed a measured waveform of the DUT. Figure 9.48 shows a comparison of a de-embedded waveform obtained through post-processing of the measured waveform with the transfer function $H(f)$ as described in (9.3).

This example shows that time-domain de-embedding can improve the measured results, but it is also important to be aware of the limitations. The first one is that perfect de-embedding in a real application is not possible due to several factors like the random noise inherent to the pin electronics that might dominate in the frequency region where we still want to de-embed a measured signal or the computational issues associated with obtaining the

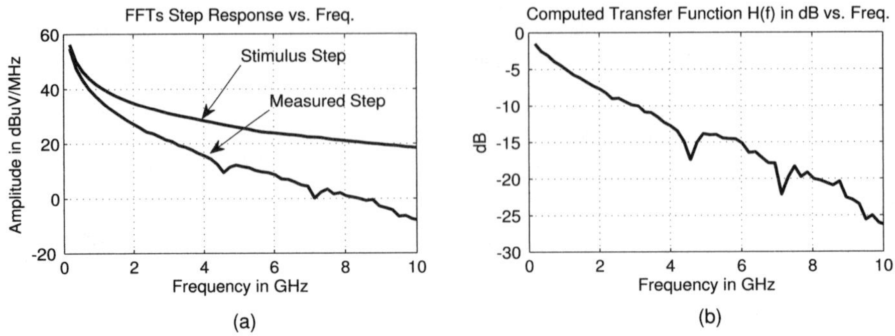

Figure 9.47 (a) Computed FFT of the input and output step responses and (b) computed system transfer function based on the measured step responses.

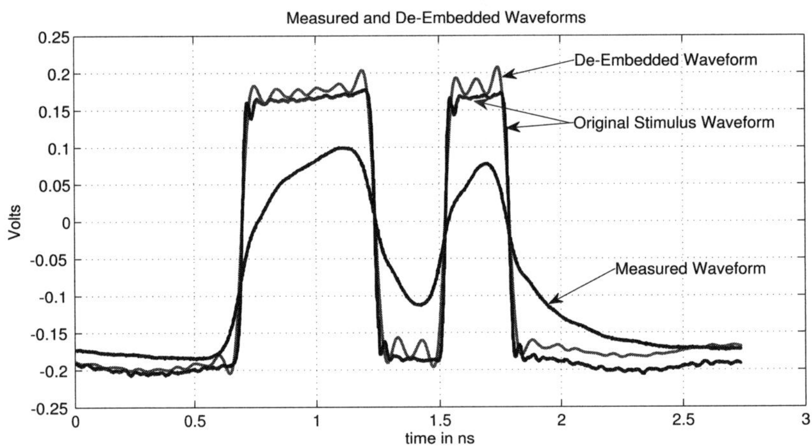

Figure 9.48 Results of applying the de-embedding procedure to a measured 5 Gbps waveform.

transfer function using a step response. Another challenge in ATE applications is the fact that only time-equivalent (undersampled) waveforms are measured either by a sampler ATE card or the digital pin electronics. In the pin electronics case there is the additional challenge that digitizing a waveform can be very time consuming. It is also important to note that this type of post-processing technique cannot be used to improve measurements that depend on at-speed functional testing like a bathtub curve measurement. To address these drawbacks, Section 9.9.3 discusses possible hardware-based techniques in the form of equalization.

9.9.2 Characterization in the Frequency-Domain

The previous discussion can also be used as an approach to measure the performance of an ATE receiver channel in the frequency-domain with or without the test fixture loss included or even for characterizing the test fixture performance separately from the ATE driver/receiver channel performance.

Figure 9.49 shows an example of measuring the ATE receiver performance in the frequency-domain with a reference test fixture using an external pattern generator that provides a step waveform that is measured by the ATE receiver.

The frequency response of the ATE receiver plus the reference test fixture can be computed from the Fourier transform of the measured step responses as described in (9.4) and shown in Figure 9.49.

If no reference test fixture is to be included and only the receiver performance is to be measured, then the pattern generator should be connected directly to the ATE receiver pogo pin.

If the objective is to measure the frequency response of the test fixture, this can be accomplished either for the drive or the receive channels. For the ATE driver channels, one needs to measure the step response at the ATE driver without the test fixture and at the DUT socket with the test fixture docked on the ATE system. The measurement of the driver step response can be done with an external instrument like an equivalent-time oscilloscope. The two measured step responses will then provide the frequency response of the test fixture through (9.4). For the receiver side the process is similar. One measures the step response with the ATE receiver when a pattern generator is connected directly to the ATE receiver pogo pin and when it is connected to the DUT socket on the test fixture. From these two step responses, the frequency response of the test fixture can be computed.

9.9.3 Signal Path Loss Compensation: Equalization

The objective of equalization is to substitute the inverse filter function in Figure 9.44 with a hardware component so that the compensation is done before the signal is sampled by the receiver. This means that, in contrast to the de-embedding technique presented in Section 9.9.1, it will apply also to at-speed tests like a BER bathtub curve measurement. The other advantage is that equalization can be applied to both the driver and receiver on the ATE pin electronics.

The term "equalization" is used in several different contexts and it is important to first describe how it will be used in the context of this section. By equalization, we mean a system that compensates for the degradation of the waveform due to the test fixture loss or the measurement instrumentation.

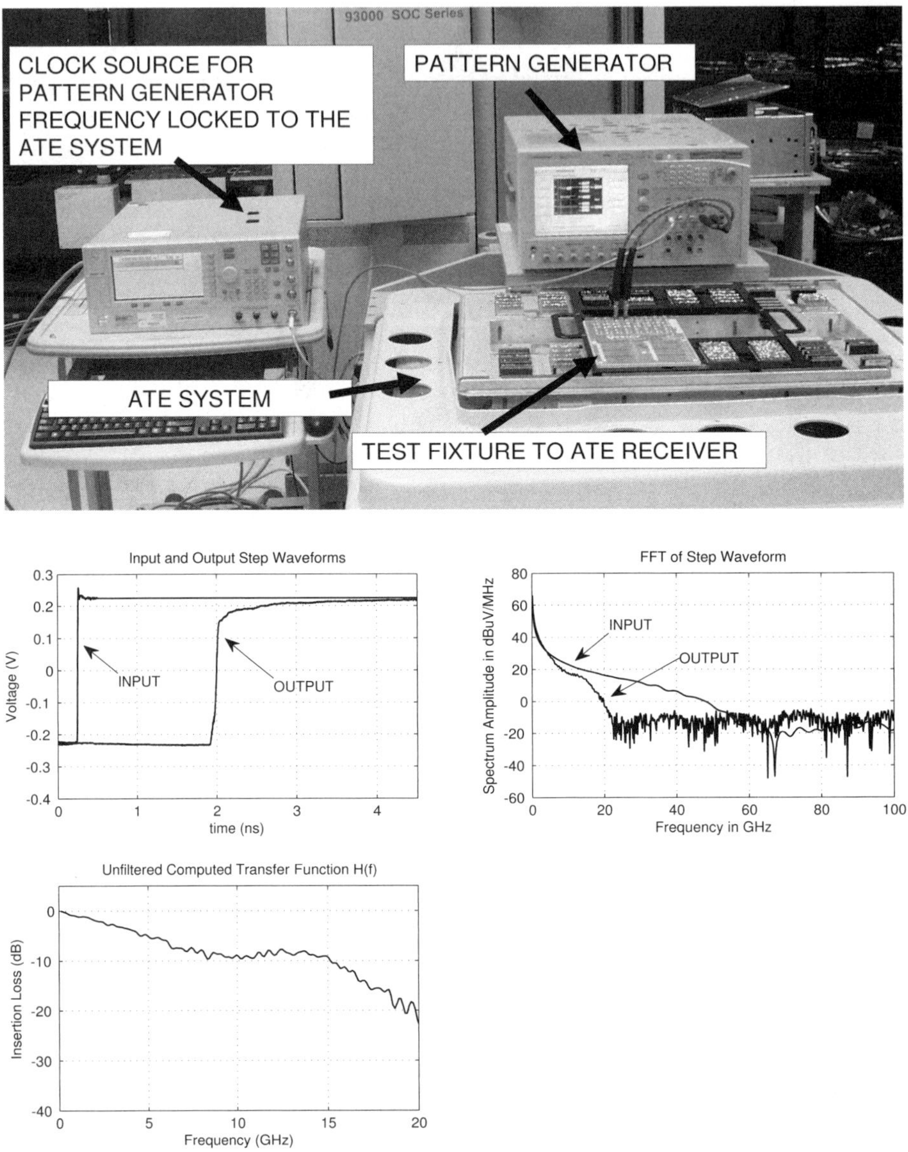

Figure 9.49 Example of the measurement setup for characterizing an ATE receiver in the frequency-domain using a step waveform (top) and the measured step response and computed frequency transfer function of the ATE receiver (bottom) (Courtesy of Advantest).

That is, in the case of a waveform from a DUT driver, equalization should ensure that the waveform at the ATE receiver input is exactly the same as at the DUT package pad (independent of whether the waveform is good or bad).

On the stimulus signal to the DUT (i.e., the ATE driver), the term "pre-emphasis" is commonly used, although we will also use the term "equalization" when referring to the ability of the ATE driver to compensate for the loss on the test fixture.

In a modern I/O cell receiver for high-speed digital applications, the objective is to correctly identify the bit stream logic values with the exact shape of the electrical waveform being of lower importance. In this context techniques like decision feedback equalization (DFE) [37] (see also Section 2.6.4) have shown significant advantages since they target the key objective, which is to figure out the logic values of the received bits and not on the waveform shape at the link partner driver. However, on a test and measurement application like ATE, the electrical waveform is the primary objective together with the bit sequence. Because of this, techniques like DFE are not discussed in the context of equalization on ATE although it is important to point out that testing the DFE circuitry on a DUT is an important topic for ATE (see Section 5.7.5). However, this is not related to the test fixture loss compensation.

The next sections will present the three main approaches for equalization on ATE: passive, active, and hybrid equalizations [38]. Note that in this section only continuous time linear equalization (CLTE) approaches for ATE pin electronics equalization are discussed [37].

9.9.3.1 Passive Equalization

Passive equalization is the longest established approach to compensate for the loss of a test fixture using a hardware approach [39]. The idea is to compensate the test fixture loss, which can be modeled as a low-pass filter, by a matching filter similar to a high-pass filter that flattens the frequency response. The idea is shown in Figure 9.50 [40–42].

From the figure it is also possible to notice that the combined response does imply a certain constant loss through the entire frequency spectrum of interest. This is the price to be paid for using a passive equalizer. Since the loss is constant over the frequency range of interest, it easily can be calibrated. Figure 9.51 shows one example of a single-ended equalizer topology and the simulation results in the frequency-domain of applying a specific implementation of this equalization filter to a lossy signal path. Note that although there is a 8 dB DC loss, the frequency-dependent behavior of the signal path is reduced significantly. The DC loss will reduce the data eye

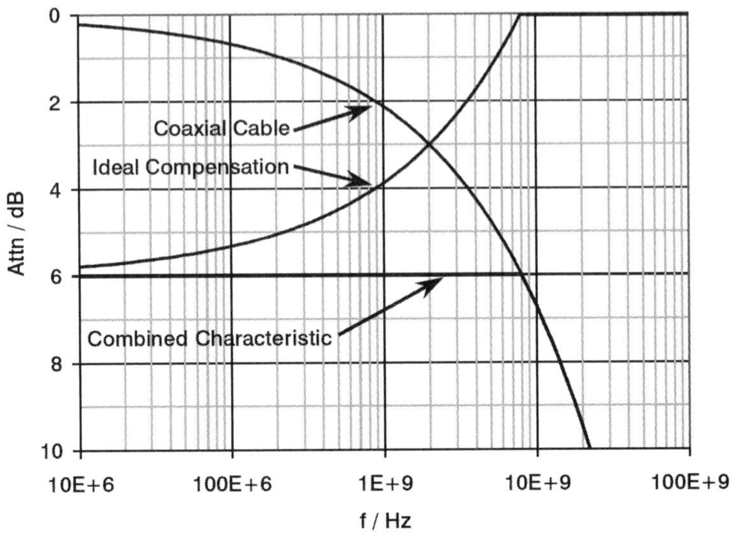

Figure 9.50 Compensating for a coaxial cable signal path loss through a passive equalization filter.

height amplitude and can be calibrated. The important fact is that in the time-domain, equalization will improve level and timing jitter significantly, as will be demonstrated later using a real example.

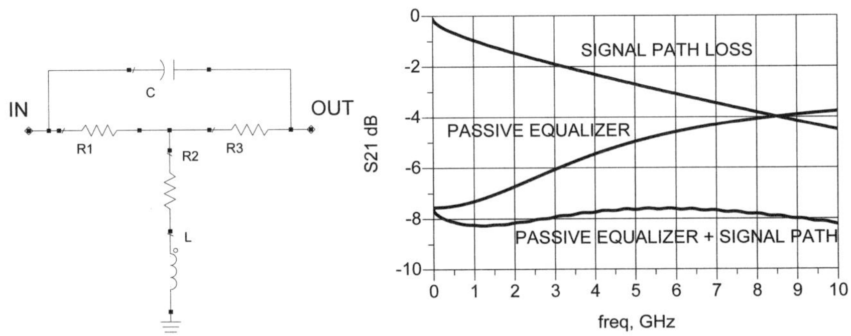

Figure 9.51 Example of a simple passive equalizer circuit and its compensation effect on the signal trace loss.

The challenge with implementing a passive equalizer for high-speed digital applications is to make sure that the manufactured equalizer provides the needed frequency response. Also, for high pin count applications, this might require a small form factor [43]. Figure 9.52 shows some examples of passive equalizers for coaxial cables and PCB test fixtures.

Figure 9.52 Examples of passive equalizers. Left: 0603 type surface mounted passive equalizers (Courtesy of Thin Film Technology Corporation [43]); right: passive equalization filter implemented in thin-film technology with wire bounding in a coaxial package [44].

Figures 9.53 and 9.54 show a real example of using a passive equalizer to compensate for the test fixture loss. In Figure 9.53 the frequency response of the test fixture without any compensation is shown together with the frequency response of the passive equalization filter that is used and the combined response of the test fixture loss with the passive equalization filter (the DC loss is compensated out on the figure). The figure shows that the bandwidth of the signal path between the DUT and the ATE pin electronics is improved from 2.8 GHz to 8.3 GHz, an improvement of 5.5 GHz in terms of bandwidth. Figure 9.54 shows this improvement in the time-domain using the data eye diagram. Although the signal amplitude has been reduced due to the passive equalization, the jitter and the data eye opening have been improved.

It is also possible to add a passive equalizer directly into the ATE pin electronics. This has the advantage that it allows the ATE manufacturer to already include all the required calibration factors into the system's standard calibration. The remaining question is if a single passive equalizer could handle all the different possibilities of loss for an ATE test fixture. This depends on the design of the passive equalizer and on the strategy for designing the DUT test fixtures. Figure 9.55 shows both options with an example of an ATE pin electronics channel that contains an integrated passive equalizer and a test fixture with an integrated passive equalizer [43].

9.9.3.2 Active Equalization[3]

Another approach to implement equalization on the ATE pin electronics is to use an active circuit. This circuit can be implemented as part of the driver/receiver ASIC on the ATE pin electronics. An active circuit

[3] This section was written in collaboration with Bernhard Roth.

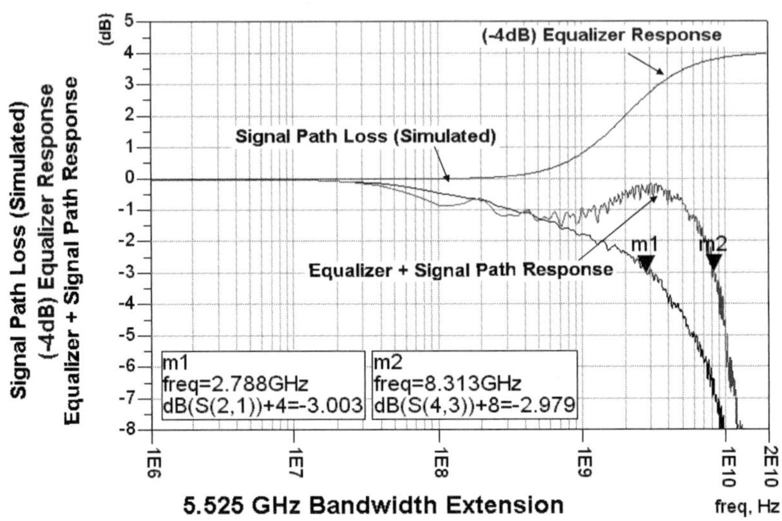

Figure 9.53 Example in the frequency-domain of the bandwidth extension through the use of passive equalization (the DC loss of the passive equalizer is calibrated out in the graph).

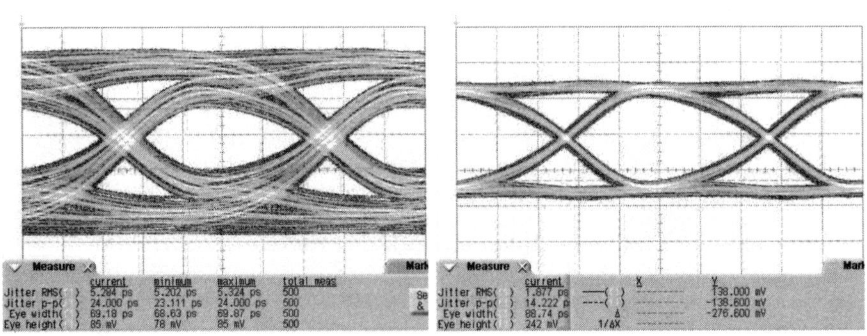

(a) (b)

Figure 9.54 Data eye diagram comparison after a lossy signal path (a) without equalization and (b) with equalization.

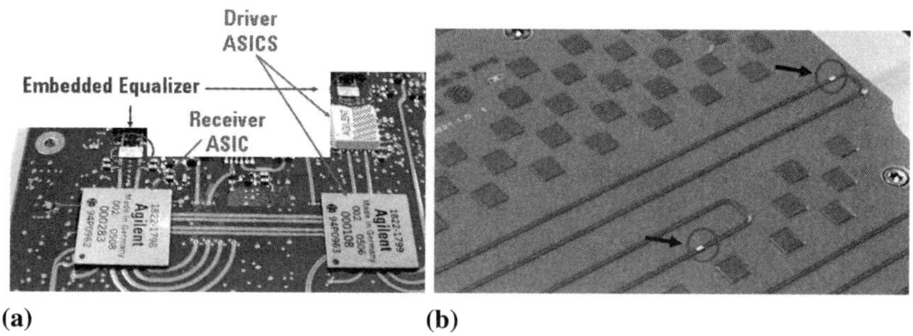

Figure 9.55 Picture of a passive equalizer (a) integrated on an ATE pin electronics card and (b) integrated on the test fixture. Reprinted with permission from [43].

implementation, unlike with passive equalization, allows it to have different types of equalization settings (e.g., for different test fixtures losses).

The important point when discussing an active compensation approach is that it must be a continuous time compensation [37, 41]; that is, the compensation must literally shape the waveform at the DUT and not simply create an open data eye. Figure 9.56 shows a simple implementation of an active equalizer where a copy of the signal is sent through a high-pass filter and then added to the original signal resulting in a frequency response that amplifies the higher frequencies and in this way compensates for the frequency-dependent loss of a signal path. Note that unlike passive equalization, there is no DC loss with an active equalizer, because the compensated AC loss gets added before the signal is launched into the transmission line.

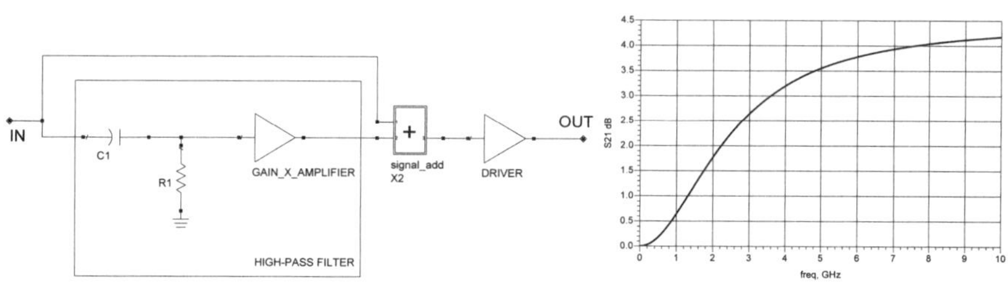

Figure 9.56 Example of a simple active equalizer topology and its frequency response.

The possibility of integrating the active equalizer in silicon also allows the option to increase the complexity of the active equalizer with multiple degrees of freedom. Figure 9.57 shows a more realistic approach for an active equalization circuit that would allow programmable equalization on the pin electronics driver. There are two high-pass type filters with different time constants that tap into the input signal and then are added to the input signal to generate the output waveform [45]. Each filter has a programmable gain K1 and K2. Figure 9.58 shows the time-domain step response of each stage of the equalizer circuit for a certain value of the K1 and K2 gains and the final combined step response. This approach can be extended by adding additional filter stages.

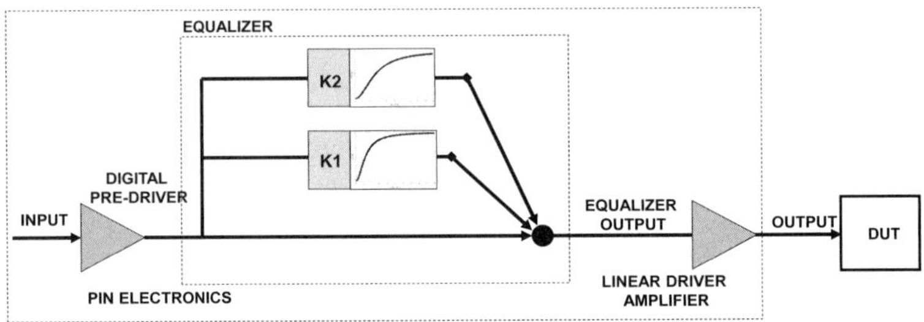

Figure 9.57 High-level architecture of an equalization approach for integration on the pin electronics driver with variable control.

Figure 9.59 shows the data eye diagram at the various stages of an ATE system with active equalization: at the ATE pin electronics before equalization, after the equalization, and at the DUT I/O input. It also presents the data eye at the DUT I/O without equalization. The figure shows how the waveform at the DUT is approximated to the original waveform at the ATE driver with the active equalization in comparison to the waveform at the DUT without equalization.

On the pin electronics receiver side the procedure is exactly the same, although in this case the input signal to the equalizer is the DUT output as shown in Figure 9.60.

9.9.3.3 Hybrid Equalization[4]

One drawback of active equalization compared to passive equalization is shown on Figure 9.61 [46]. The figure shows that on an active equalization setup, to achieve a 1 V amplitude waveform after a lossy signal path the ATE pin electronics has to be able to generate 1.5 V slopes with a 2 V dynamic range. On a passive equalization setup, it is only necessary for the ATE pin electronics to generate the 1.5 V slopes within a 1.5 V dynamic range. This is a critical point in pin electronics design since the use of faster silicon processes

[4]This section was written in collaboration with Bernhard Roth.

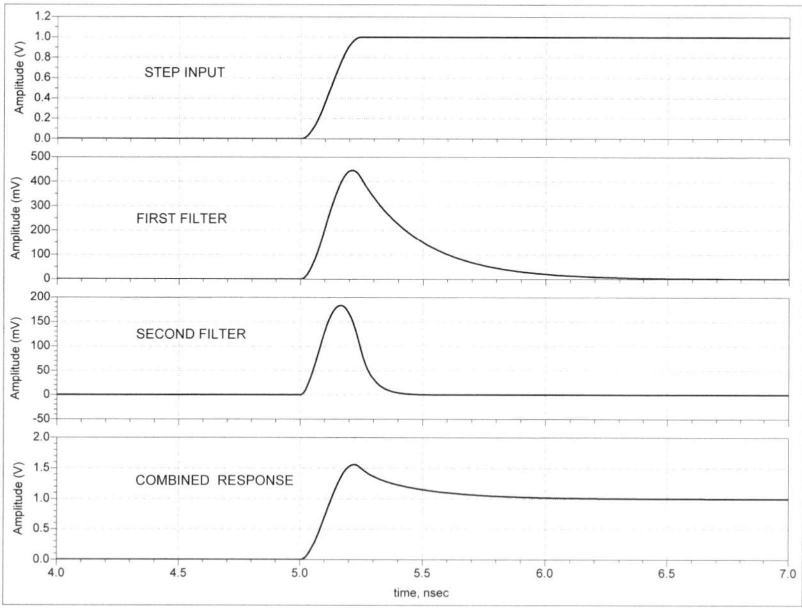

Figure 9.58 Comparison of the initial step response before the equalizer with the individual step responses of each filter on the equalizer and the combined step response of the equalizer.

to obtain faster data rates usually comes at the expense of a reduction on the transistor breakdown voltage.

One approach to address this challenge is to combine passive and active equalization into a hybrid equalization approach and in this way combine the advantages of both approaches. Figure 9.61 shows an example of a real implementation.

9.9.3.4 Pitfalls of Overequalization

Equalization is critical for testing high-speed digital applications where the loss of the test fixture can dominate the ATE system performance. However, equalization can have a negative impact on the ability to properly test a DUT if it is not applied correctly. The reason is that equalization should be used to compensate for any loss from the test fixture or pin electronics but not to improve the performance of the DUT (i.e., it should not compensate for any possible deficiencies on the DUT, for example, input capacitance).

Figure 9.62 shows how overequalization can mask a defect in a DUT I/O with a passing functional test when in fact the DUT should fail the functional test. In the figure the DUT driver waveform, which consists of a single pulse

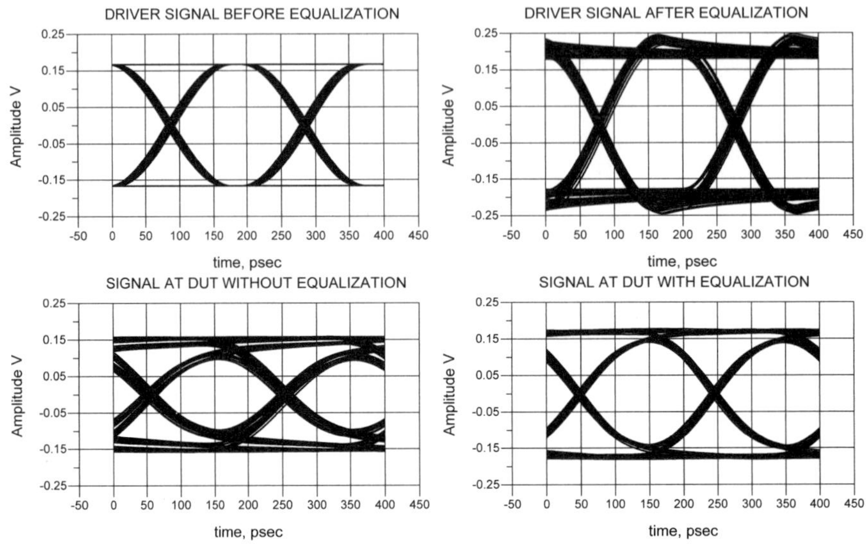

Figure 9.59 Demonstration of the loss compensation capability of the active equalization approach on improving the waveform at the DUT. Data eye diagram before equalization, after equalization, and at the DUT. The data eye diagram at the DUT without equalization is also included for comparison.

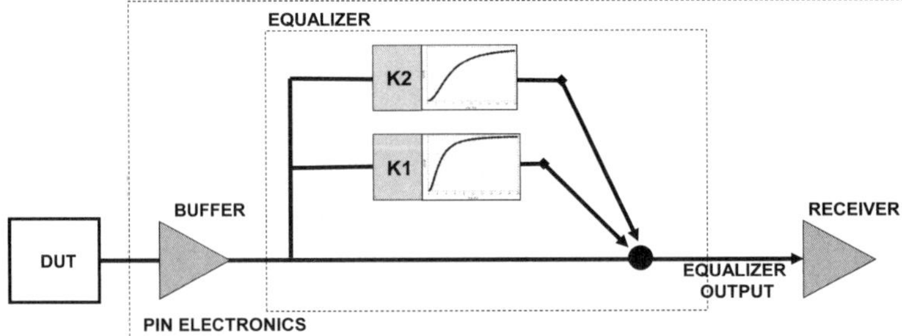

Figure 9.60 High-level architecture of an equalization approach for integration on the pin electronics receiver with variable control.

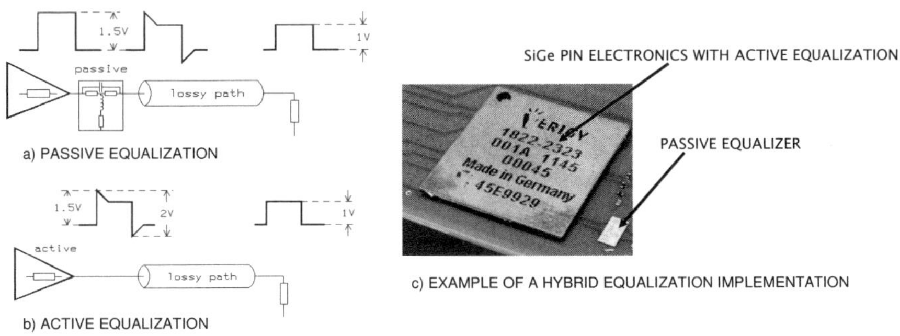

Figure 9.61 Comparison of the difference between passive and active equalization in regards to the required driver output amplitude and example of an hybrid equalization approach (Courtesy of Advantest).

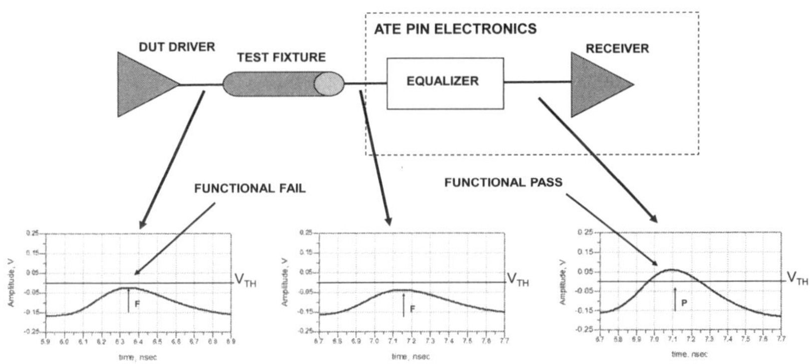

Figure 9.62 Example of how overequalization can mask the problems of a DUT I/O cell.

representing a differential logic high, is compared against a fixed threshold to check if the pulse corresponds to a logic high or low. The waveform at the DUT I/O driver is unable to reach this threshold, which would represent a bad driver and should result in a failing functional test. The waveform is further degraded by the test fixture during its transmission to the ATE pin electronics. In this case the equalization in the pin electronics was not properly set to compensate for the test-fixture loss only but was set to overcompensate the test fixture loss and in this way improve the waveform from the DUT (i.e., compensating for the possible deficiency in the DUT driver). The waveform after the equalizer is now able to achieve the threshold value and would be interpreted by the ATE receiver as a functional pass. The same case can be constructed for the DUT receiver.

The overequalization effect demonstrated in Figure 9.62 can also be shown in the frequency-domain. Figure 9.63 shows the insertion loss of the test fixture plus the ATE pin electronics equalizer. We would expect that the equalizer improves the bandwidth of the test fixture but what we observe is that it overcompensates the test fixture loss in the frequency range below 5 GHz. This overcompensation ends up masking the real performance of the DUT.

The danger of overequalization, especially in the case where there is an associated control "knob," is that the test engineer in the search for the maximum possible yield turns this knob to its maximum value. This strategy can in some situations result in an increased yield but at the expense of a reduced fault coverage (i.e., more field returns).

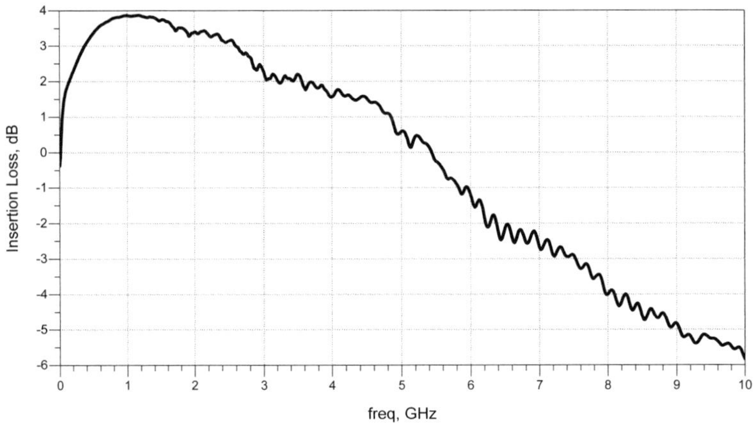

Figure 9.63 Test fixture plus ATE pin electronics insertion loss in the case of overequalization.

9.9.3.5 Using Active Equalization for DDJ Injection

The ability to have a programmable equalization on the ATE pin electronics driver using an approach like the one presented in Figure 9.57 opens the possibility of using the equalization for DDJ injection. This can be achieved by programming each filter stage of the equalizer to have a low-pass behavior instead of the expected high-pass behavior from a loss compensation equalizer as shown in Figure 9.64. This can also be observed on the simulated data eye diagram in Figure 9.65 [38].

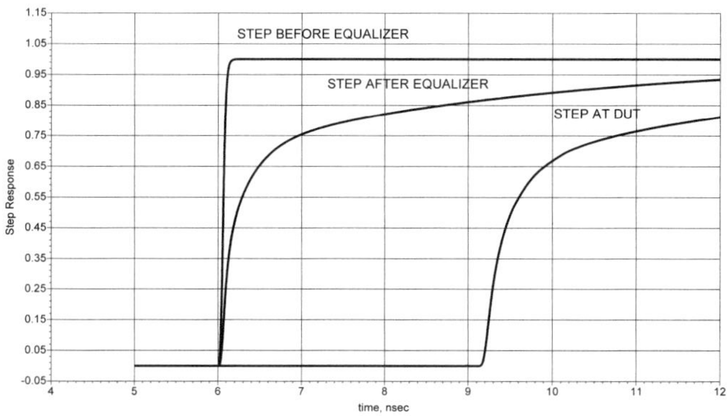

(a)

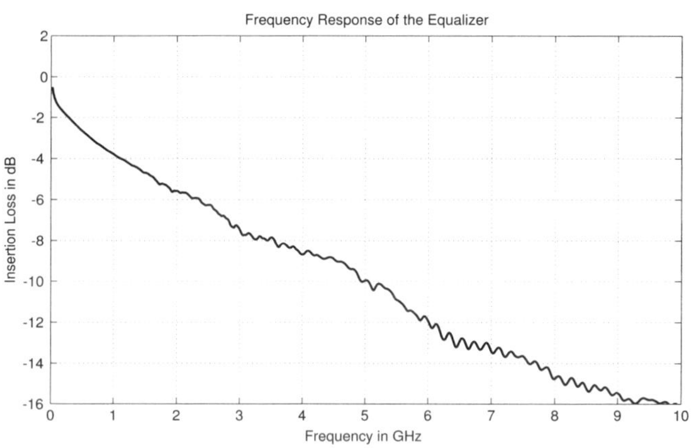

(b)

Figure 9.64 DDJ injection through an active equalizer on the ATE pin electronics. (a) Step response before and after the equalizer and at the DUT, and (b) frequency response of the equalizer.

9.10 Test Fixture and ATE Pin Electronics Co-Simulation

When simulating an ATE test fixture to analyze whether the performance is enough for a certain application, it is not only important to have good models of the DUT test fixture signal path, but it is also critical to have a good model of the ATE pin electronics. This is especially true for high-speed digital applications where the ATE pin electronics might include variable equalization. The reason is that simulating the test fixture without taking into account the ATE pin electronics equalization results in a very pessimistic result especially for lossy signal traces on the DUT test fixture [47]. Figure

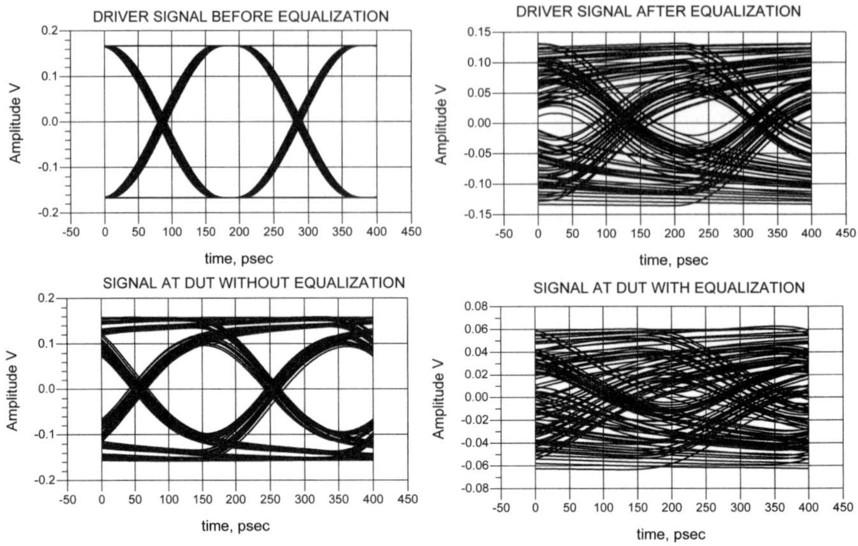

Figure 9.65 DDJ injection through an active equalizer on the ATE pin electronics. Data eye diagram before the equalizer, after the equalizer, and at the DUT. The data eye diagram at the DUT without equalization is also included for comparison.

9.66 shows a simulation setup comparing the results obtained using a simple driver model available on the used simulation tool that was tuned for the ATE pin electronics performance at the ATE to DUT test fixture connection point and a realistic model of the ATE pin electronics. This realistic model included the embedded variable equalization that is part of the ATE pin electronics that was intended to be used for this application. Four simulations were performed. The first two compare the performance at the ATE to DUT test fixture connection point and the other two at the DUT with a lossy signal trace but where the ATE pin electronics model is set with the proper equalization values to compensate for the signal trace loss. The DUT was modeled as a simple 50 Ω termination.

Figure 9.67 shows the time-domain simulation results in the form of data eye diagrams. It is possible to see that when comparing the data eye diagram at the ATE to DUT test fixture interconnect point (in this case the pogo pin via), the performance is very similar since we tune the simple driver model with the specifications of the ATE pin electronics at this point. However, when we perform the simulation at the DUT with a model of the DUT test fixture lossy signal trace, the difference is substantial because the realistic ATE pin electronics model includes the correct equalization setting for the

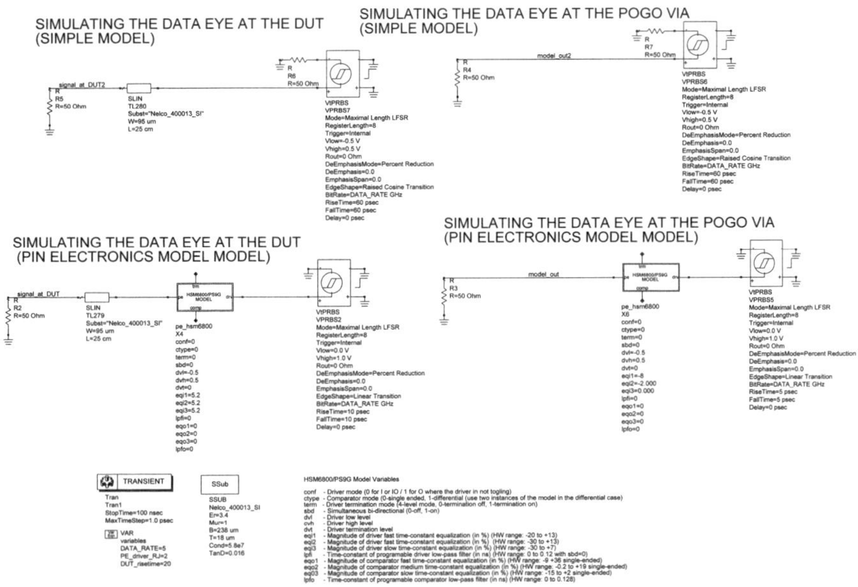

Figure 9.66 Example of a simulation setup that includes a simulation of a DUT test fixture signal trace with a simple model available on the simulation tool and with a realistic ATE pin electronics model that includes equalization.

signal trace loss [38] while the simple ATE pin electronics model does not have any equalization.

Because transistor level models of the ATE pin electronics are usually not available, one approach that ATE vendors use to model ATE pin electronics for high-speed digital applications is to use behavioral models or a SPICE model consisting of voltage controlled sources. One should discuss with the vendor of the used ATE platform about what type of pin electronic models are available.

There are different simulation tools that can be used for the co-simulation of ATE pin electronics with the DUT test fixture (see Appendix G). The important point is to make sure that the used models are realistic. It is also possible to improve on the simulation setup shown in Figure 9.66 by adding models for the DUT socket, relays (if any) and even a more realistic model of the DUT I/O (e.g., HSPICE model). Another option is to use a measurement-based model of the test fixture (e.g., S-parameters) using the techniques described in Appendix H, although there are usually more advantages in simulating the test fixture before manufacturing to correct any possible signal integrity problems.

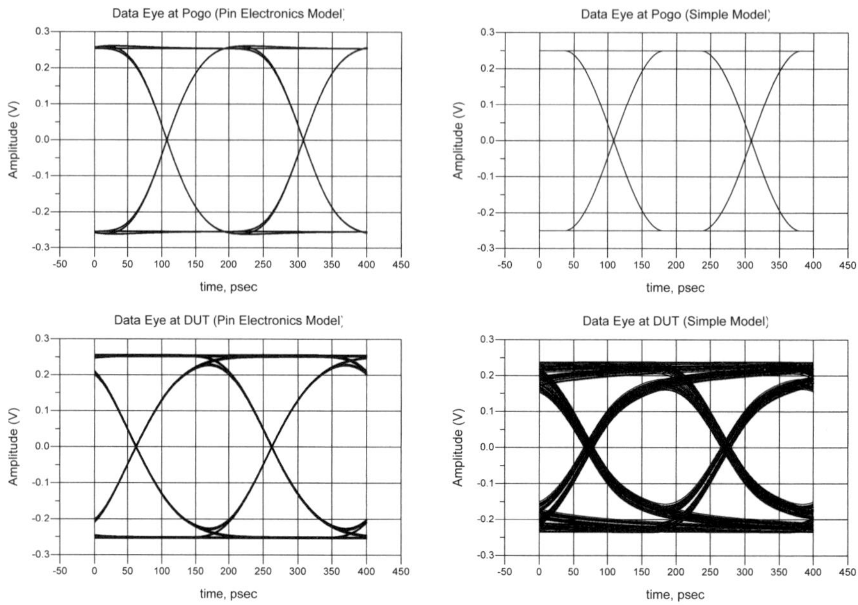

Figure 9.67 Simulation results comparing a simplistic pin electronics model with a realistic pin electronics model that includes equalization.

9.11 ATE DC Level Adjustments

ATE drivers and receivers designed for testing high-speed I/O circuits typically provide impedances that match the 50 Ω single-ended or 100 Ω differential impedances that are predominantly used for the high-speed I/Os of a DUT. In order to achieve the high data rates, proper terminations in the ATE pin electronics and the DUT I/Os are mandatory [48–50]. For a DUT that is to be deployed in a 50 Ω system environment, the programmed ATE level values can be directly applied from the respective datasheets with potential corrections due to level loss on the signal path as described in Section 9.3.2. Some high-speed I/O standards do not use a 50 Ω environment. However, ATE high-performance pin electronics cannot be implemented in a cost-effective way to support all possible impedance environments. If a device that is designed for a system environment impedance different than the 50 Ω/100 Ω mainstream of the ATE pin electronics needs to be tested according to its specifications, a DC level adjustment of the drive and compare levels used by the ATE is required. The reader should be aware that in this section we only consider the DC effects of such an impedance mismatch. If the impedance difference between ATE pin electronics and DUT I/Os becomes too large, also

AC effects as described in Section 8.2.1 have to be taken into consideration and have to be corrected by means of loadboard design.

The guiding principles for the correction of the DC levels for non-matching source-impedance and receiver-termination pairs is that for inputs, the DUT should be stimulated at its pin with the same levels as in an impedance matched environment. For output pins, the levels that are measured by the ATE on the DUT pins are based on different load currents I_{OL} and I_{OH} than is the case in a system environment. Thus, the voltages measured at the pins of the DUT need to be corrected to the corresponding values that would be valid in the system environment before comparing them against the specified levels. If the termination scheme in the system environment is symmetrical (i.e., uses identical source and load impedances), the compare threshold level used by the ATE receiver does not need any adjustment because the effects on the measured high and low levels also are symmetrical and thus do not have an effect on the threshold that usually is at the center of high and low levels. If the termination scheme, however, is not symmetrical (e.g., as for POD15; see Section 3.3.5), the threshold value that is used to distinguish between logical high and low levels also needs to be adjusted.

If one wants to apply the load currents as seen in a system environment also in the ATE environment, impedance adjustments by means of test fixture design need to be applied or appropriate adjustments of the termination voltages used by the ATE receivers need to be done. The difficulty with the latter of these two is that this would result in two different termination voltages that would need to be used for received high and low levels.

9.11.1 Correction of Force Levels for DUT Input Pins

The general configuration for an input pin of the DUT is shown in Figure 9.68. In this figure, V_{ATE} denotes the voltage that is programmed as force voltage of the ATE driver. V_{DTerm} is the termination voltage used by the DUT, R_{DTerm} is the termination impedance of the DUT receiver, and V_I denotes the resulting voltage at the input pin of the DUT.

The goal of our calculation is to determine V_{ATE} for a given V_{DTerm} and R_{DTerm}. In order to get there, we first calculate the drive current that the ATE driver has to provide according to (9.5) and the resulting voltage at the DUT input pin according to (9.10).

$$I_D = \frac{V_{ATE} - V_{DTerm}}{50\Omega + R_{DTerm}} \tag{9.5}$$

$$V_I = V_{DTerm} + I_D R_{DTerm} \tag{9.6}$$

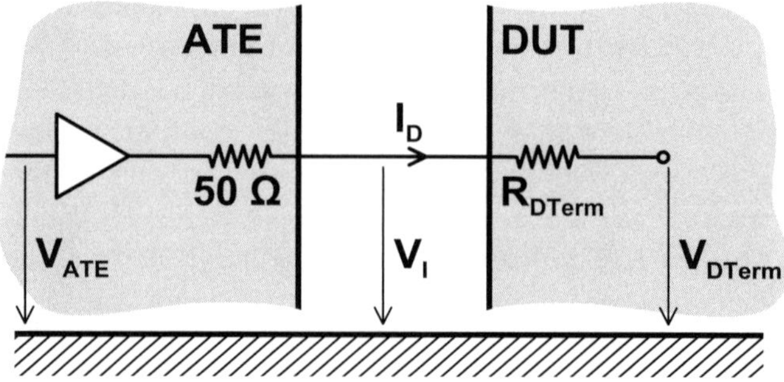

Figure 9.68 General configuration of an ATE to DUT connection for a DUT input pin.

Substituting I_D in (9.6) with (9.5) yields

$$V_I = V_{DTerm} + \frac{V_{ATE} - V_{DTerm}}{50\Omega + R_{DTerm}} R_{DTerm} \tag{9.7}$$

With this, V_{ATE} can be calculated as

$$V_{ATE} = \left(1 + \frac{50\Omega}{R_{DTerm}}\right) V_I - \frac{50\Omega}{R_{DTerm}} V_{DTerm} \tag{9.8}$$

9.11.2 Correction of Levels for DUT Output Pins

For the correction of the measured DUT output voltage, the source voltage V_{DSrc} used by the DUT to drive the source impedance R_{DSrc} on its outputs to achieve a certain device output voltage V_{OSYS} in the system environment is calculated in a first step. The basis for this calculation is the system configuration with the termination impedance R_{DTerm} and termination voltage V_{DTerm} of the receiving device as shown in Figure 9.69. From the figure we get

$$I_O = \frac{V_{DSrc} - V_{DTerm}}{R_{DSrc} + R_{DTerm}} \tag{9.9}$$

$$V_{DSrc} = V_{DTerm} + I_O(R_{DSrc} + R_{DTerm}) \tag{9.10}$$

We also can derive I_O dependent on the voltage V_{OSYS} at the device pin as

$$I_O = \frac{V_{DSrc} - V_{OSYS}}{R_{DSrc}} \tag{9.11}$$

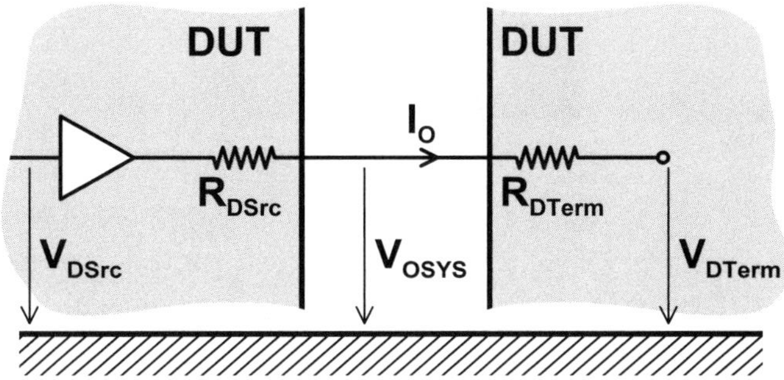

Figure 9.69 General configuration of a DUT system connection for a DUT output pin.

Equation (9.11) in (9.10) yields

$$V_{DSrc} = V_{DTerm} + \frac{V_{DSrc} - V_{OSYS}}{R_{DSrc}} (R_{DSrc} + R_{DTerm}) \qquad (9.12)$$

By solving this equation for V_{DSrc}, we get

$$V_{DSrc} = \left(1 + \frac{R_{DSrc}}{R_{DTerm}}\right) V_{OSYS} - \frac{R_{DSrc}}{R_{DTerm}} V_{DTerm} \qquad (9.13)$$

For the ATE environment as shown in Figure 9.70, we can derive the DUT source voltage required to obtain a certain device output voltage V_{OATE} in the ATE environment in a similar manner.

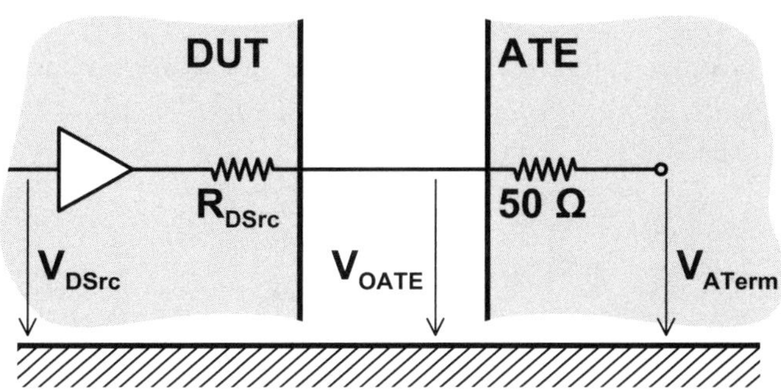

Figure 9.70 General configuration of an ATE to DUT connection for a DUT output pin.

From Figure 9.70 together with (9.13), we get

$$V_{DSrc} = \left(1 + \frac{R_{DSrc}}{50\Omega}\right) V_{OATE} - \frac{R_{DSrc}}{50\Omega} V_{ATerm} \tag{9.14}$$

The DUT source voltage V_{DSrc} stays the same, whether we use the device in the impedance matched system environment or in a potentially non-impedance matched ATE environment. Thus, we can link the DUT output voltage V_{OSYS} in a system environment to the DUT output voltage V_{OATE} in a nonimpedance matched ATE environment by equalizing (9.13) with (9.14). This yields

$$\left(1 + \frac{R_{DSrc}}{R_{DTerm}}\right) V_{OSYS} - \frac{R_{DSrc}}{R_{DTerm}} V_{DTerm} = \left(1 + \frac{R_{DSrc}}{50\Omega}\right) V_{OATE} - \frac{R_{DSrc}}{50\Omega} V_{ATerm} \tag{9.15}$$

Solving this equation for V_{OSYS} results in (9.16), which is used to calculate the device output voltage that would be present in a system environment from a measured output voltage value in a nonimpedance matched ATE environment.

$$V_{OSYS} = \frac{R_{DTerm}}{R_{DTerm} + R_{DSrc}} \left(\left(1 + \frac{R_{DSrc}}{50\Omega}\right) V_{OATE} - \frac{R_{DSrc}}{50\Omega} V_{ATerm} + \frac{R_{DSrc}}{R_{DTerm}} V_{DTerm} \right) \tag{9.16}$$

For nonsymmetrical termination schemes, this formula also can be used to calculate the compare threshold that has to be used for an ATE receiver to match the threshold given in the system specification. Equation (9.16) has to be solved for V_{OATE} in this case. With the resulting (9.17), the ATE threshold for a given system threshold V_{OSYS} can be determined.

$$V_{OATE} = \frac{50\Omega}{50\Omega + R_{DSrc}} \left(\left(1 + \frac{R_{DSrc}}{R_{DTerm}}\right) V_{OSYS} - \frac{R_{DSrc}}{R_{DTerm}} V_{DTerm} + \frac{R_{DSrc}}{50\Omega} V_{ATerm} \right) \tag{9.17}$$

Example POD15 signaling (see Section 3.3.4.2) as used by GDDR5 is an example for a nonsymmetrical termination scheme that does not use 50 Ω terminations. Instead, it uses a 40 Ω drive source impedance R_{DSrc} and a 60 Ω termination impedance R_{DTerm} on its receivers (for details, see Section 3.3.5). The termination voltage V_{DTerm} that is used by POD15 is the supply voltage VDDQ. With this boundary condition and (9.8), the resulting correction equation to calculate the ATE drive levels to be programmed into open for the GDDR5 drive levels is

$$V_{ATE} = \frac{11}{6} V_I - \frac{5}{6} VDDQ \tag{9.18}$$

For the ATE comparators, we can derive the correction rule from (9.16) as

$$V_{OSYS} = \frac{27}{25} V_{OATE} - \frac{2}{25} VDDQ \tag{9.19}$$

Since POD15 is nonsymmetrical, threshold voltages programmed on the ATE also need to be corrected due to the nonmatching impedance environment. For POD15, the standard threshold voltage is $V_{OSYS} = 0.7\, VDDQ$. With this and (9.17), we can calculate the corrected threshold voltage to be programmed for the ATE receivers to $V_{OATE} = (13/18)\, VDDQ$ if a termination voltage $V_{ATerm} = VDDQ$ is used by the ATE receiver.

References

[1] N. Kularatna, *Digital and Analog Instrumentation: Testing and Measurements.* New York: The Institution of Electrical Engineers, 2003.

[2] SEMI G80-0200, *Test Method for the Analysis of Overall Digital Timing Accuracy for Automated Test Equipment*, 2000.

[3] S. Kikuchi, Y. Hayashi, T. Suga, J. Saitou, M. Kaneko, T. Matsumoto, and R. Yoshino, "A Gate-Array-Based 666MHz VLSI Test System," *IEEE International Test Conference*, 1995.

[4] F. Z. W. Necoechea, P. Reiter, Y.-B. Kim, and F. Lombardi, "Load Board Design Using Compound Dot Technique and Phase Detector for Hierarchical ATE Calibrations," *IEEE Symposium on Defect and Fault-Tolerance in VLSI Systems*, 2006.

[5] D. C. Keezer, "A Multiplexing Test System Channels for Data Rates Above 1 Gb/s," *IEEE International Test Conference*, 1990.

[6] D. C. Keezer, D. Minier, and M.-C. Caron, "Multiplexing ATE Channels for Production Testing at 2.5 Gbps," *IEEE Design and Test of Computers*, 2004.

[7] D. Keezer, D. Minier, P. Ducharme, and A. Majid, "An Electronic Module for 12.8 Gbps Multiplexing and Loopback Test," *IEEE International Test Conference*, 2008.

[8] J. Moreira, B. Roth, and C. McCowan, "An Active Test Fixture Approach for Testing 28 Gbps Applications Using a Lower Data Rate ATE System," *IEEE Asian Test Symposium*, 2012.

[9] J. Moreira, H. Werkmann, V. Filsinger, and B. Roth, "At-Speed Testing of 32 Gbaud PAM-4 Interfaces Using Automated Test Equipment," *DesignCon*, 2015.

[10] J. Moreira, B. Roth, H. Werkmann, L. Klapproth, M. Howieson, M. Broman, W. Ouedraogo, and M. Lin, "An Active Test Fixture Approach for 40 Gbps and Above At-Speed Testing Using a Standard ATE System," *IEEE Asian Test Symposium*, 2013.

[11] J. Moreira, F. Pizza, C. Borelli, F. Corneo, H. Werkmann, S.-X. Yang, D. Lam, and B. Roth, "A Pragmatic Approach for At-Speed Characterization and Loopback Correlation at 28 Gbps," *Advantest VOICE Users Group Conference*, 2014.

[12] J. Moreira and B. Roth, "Characterization and Focus Calibration of ATE Systems for High-Speed Digital Applications," *IEC DesignCon*, 2009.

[13] M. Shimanouchi, "Timing Accuracy Enhancement by a New Calibration Scheme for Multi-Gbps ATE," *IEEE International Test Conference*, 2004.

[14] G. Haensel, "Beating V93000 Timing Specification with Focus Calibration," *Verigy VOICE Users Conference*, 2010.

[15] J. Moreira, C. McCowan, Z.-Y. Wen, Y. Zhan, Y. He, and W.-M. Zhang, "Calibration of the ATE Driver Data Eye Height for Receiver Sensitivity Measurements," *Verigy VOICE Users Conference*, 2010.

[16] B. Roth, "Eye-Profile: Generation and Display," *Verigy Technical Note*, 2009.

[17] J. Ren and K. S. Oh, "Multiple Edge Responses for Fast and Accurate System Simulations," *IEEE Transactions on Advanced Packaging*, vol. 31, Nov. 2008.

[18] D. Oh, J. Ren, and S. Chang, "Hybrid Statistical Link Simulation Technique," *IEEE Transactions on Components, Packaging and Manufacturing Technology*, vol. 1, May 2011.

[19] A. Sanders, M. Resso, and J. D'Ambrosia, "Channel Compliance Testing Utilizing Novel Statistical Eye Methodology," *DesignCon*, 2004.

[20] D. C. Keezer and Q. Zhou, "Alternative Interface Methods for Testing High-Speed Bidirectional Signals," *IEEE International Test Conference*, 1998.

[21] J. Moreira, H. Barnes, and H. Werkmann, "DUT Loadboard Layout for Multi-Gigabit Bi-Directional Interfaces," *Verigy Users Conference*, 2007.

[22] PCI-SIG, *PCI Express Base Specification Revision 2.1*, Mar. 2009.

[23] H. Werkmann, J. Moreira, and S.-X. Yang, "A DUT Reference Clock Jitter Attenuator Module for the V93000 ATE Platform," *Advantest VOICE Users Group Conference*, 2015.

[24] NXP Semiconductors, *I2C-Bus Specification and User Manual Rev. 3.0*, June 2007. UM10204.

[25] D. Kather, J. Moreira, W.-J. Song, R. Nettles, W.-M. Zhang, and L. Ge, "A Unified Approach to Bench/ATE Testing with the V93000 Protocol Editor," *Advantest VOICE Users Group Meeting*, 2012.

[26] J. Rivoir, "Protocol-Aware ATE Enables Cooperative Test Between DUT and ATE for Improved TTM and Test Quality," *IEEE International Test Conference*, 2007.

[27] E. Larson, "Protocol Aware ATE," *Beijing Advanced Semiconductor Technology Symposium*, 2008.

[28] A. C. Evans, "The New ATE: Protocol Aware," *IEEE International Test Conference*, 2007.

[29] T. Lyons, "Speeding SOC Test: Protocol-Aware ATE Hardware Boosts the Efficiency of Parallel-Site and Concurrent Tests for SIP, SOC, and MCM Designs," *Test & Measurement World*, Sept. 2011.

[30] M. Ishida, K. Ichiyama, D. Watanabe, M. Kawataba, and T. Okayasu, "Real-Time Testing Method for 16 Gbps 4-PAM Signal Interfaces," *IEEE International Test Conference*, 2011.

[31] Gigabit Ethernet Alliance, "Gigabit Ethernet 1000BASE-T Whitepaper," 1997.

[32] J. Moreira, H. Werkmann, I. Masahiro, B. Roth, V. Filsinger, and S.-X. Yang, "An ATE Based 32 Gbaud PAM-4 At-Speed Characterization and Testing Solution," *IEEE Asian Test Symposium*, 2014.

[33] Anritsu, "High-Speed PAM Signal Generation and BER Measurement," *Application Note*, 2015.

[34] S. M. Riad, "The Deconvolution Problem: An Overview," *Proceedings of the IEEE*, vol. 74, Jan. 1986.

[35] W. J. Rugh, *Linear System Theory*. Upper Saddle River, NJ: Prentice-Hall, 1996.

[36] J. R. Andrews, "Deconvolution of System Impulse Responses and Time Domain Waveforms," *Picosecond Labs Application Note AN-18*, 2004.

[37] B. Casper, P. Pupalaikis, and J. Zerbe, "Serial Data Equalization," *IEC DesignCon*, 2007.

[38] B. Roth and J. Moreira, "Equalization in the HSM6800 Pin-Electronics," *Verigy VOICE Users Conference*, 2010.

[39] H. W. Bode, "Attenuation Equalizer," *U.S. Patent 2,096,027,027*, 1936.

[40] W. Humann, "Compensation of Transmission Line Loss for Gbit/s Test on ATEs," *IEEE International Test Conference*, 2002.

[41] S. H. Hall and H. L. Heck, *Advanced Signal Integrity for High-Speed Digital Designs*. New York: John Wiley & Sons, 2009.

[42] L. Zhang, W. Yu, Y. Zhang, R. Wang, A. Deutsch, G. A. Katopis, D. M. Dreps, J. Buckwalter, E. S. Kuh, and C.-K. Cheng, "Analysis and Optimization of Low-Power Passive Equalizers for CPU-Memory Links," *IEEE Transactions on Components, Packaging and Manufacturing Technology*, vol. 1, Sept. 2011.

[43] J. Moreira, M. Howieson, M. Petersen, M. Broman, and J. Kenton, "Passive Equalization of Test Fixtures for High-Speed Digital Measurements with Automated Test Equipment," *IEEE International Design and Test Workshop*, 2006.

[44] J. Moreira, G. Haensel, and F. Koban, "Addressing the Challenges of Implementing an At-Speed Production Test-Cell for 10Gb/s Wafer Probing," *DesignCon*, 2005.

[45] H. Higashi, "5-6.4 Gbps 12 Channel Transceiver with Pre-Emphasis and Equalizer," *Symposium on VLSI Circuits*, 2004.

[46] J. Moreira, "High-Speed Digital I/O Characterization with ATE: Is There a Future?," *VDEC D2T Symposium*, 2012.

[47] B. Roth, J. Moreira, and R. Harjung, "ATE Pin Electronics Behavioural Modeling Using the Measured Step Response," *VOICE, Advantest Users Conference*, 2013.

[48] G. Esch and T. Chen, "Design of CMOS IO Drivers with Less Sensitivity to Process, Voltage, and Temperature Variations," *Electronic Design, Test and Applications, IEEE International Workshop on*, 2004.

[49] G. Esch and R. B. Manley, "Theory and Design of CMOS HSTL I/O Pads," *The Hewlett-Packard Journal*, Aug. 1998.

[50] G. Esch and T. Chen, "Near-Linear CMOS I/O Driver with Less Sensitivity to Process, Voltage, and Temperature Variations," *IEEE Trans. VLSI Syst.*, vol. 12, no. 11, 2004.

A

Introduction to the Gaussian Distribution and Analytical Computation of the BER

This appendix provides some basic concepts on the Gaussian (or normal) probability distribution that is important for the analysis of high-speed digital signals. It also shows how to apply those concepts to the analytical computation of the bit error rate (BER) of a system. Most of the variables that we measure on a typical ATE application exhibit some type of random behavior that normally is described as noise on the measurement. However, in other cases like a jitter measurement, the objective is to measure and characterize this random (noisy) behavior. When we have a system that exhibits this type of behavior, we can analyze it using the mathematical arsenal that is available in the form of probability theory, stochastic processes, and so on. For the interested reader [1, 2] provide a good starting point. For the topics addressed in this book, all random variables are modeled as Gaussian distributions. The choice of the Gaussian distribution is not by chance. There is a mathematical principle that brings some insight into this choice; this principle is known as the Central Limit Theorem.

Central Limit Theorem

> *Given N independent random variables X_i, we can add them in the following way: $X = X_1 + ... + X_N$*
> *This is a random variable with mean $\mu = \mu_1 + ... + \mu_N$ and variance $\sigma^2 = \sigma_1^2 + ... + \sigma_N^2$. The Central Limit Theorem states that under certain general conditions, the distribution $F(x)$ for X approaches a Gaussian distribution with the same mean and variance.*

The Central Limit Theorem tells us that if we have a system where the result of a given measurement is due to the contributions of a large number of independent events that are controlled by any kind of random distribution, then the distribution of the sum of these events will be a Gaussian.

As an example, take a crystal used for the reference clock of a system. The crystal is composed of several atoms in a crystalline structure where each atom will have the same response (with an unknown distribution) to external factors like temperature. When looking at the measurement value for this crystal, we are interested in the jitter from the crystal over a certain period of time; this jitter value will be the result of the behavior of all the atoms in the crystal. We can then assume that the resulting measured distribution will follow a Gaussian distribution because of the Central Limit Theorem.

The mathematics of a Gaussian distribution imply that the tails of the curve are infinite, which we know is not true because it would not be possible in the real physical world. However, one needs to understand that the tails of the Gaussian distribution reduce to low probability values very quickly which implies that the probabilities of events at the end of the tails will be computed in the "billion of years" type of numbers. This is so far outside our area of interest that it makes little difference to using the Gaussian model on analyzing the worst-case jitter scenario on a real data transmission system (e.g., how many failed bits in one day).

Another example on how to look at the Central Limit Theorem and the infinite tails of a Gaussian distribution is the following:

Take N dice, roll them, and sum up the result. The distribution converges towards a Gaussian distribution. For example, for N=10 it already starts to look Gaussian. But notice that it is bounded because we have only 10 dice (the minimum value one can obtain is 10 and the maximum is 60). To have an unbounded Gaussian distribution we would need an infinite number of dice which is the requirement of the Central Limit Theorem.

With random jitter we have a similar behavior. It looks Gaussian, but somehow it is still bounded because the number of contributors is limited.

A.1 The Gaussian Distribution

The Gaussian distribution probability density function[1] (PDF) is described by the following equation:

[1]The probability density function is also sometimes referred to as the probability density distribution.

$$p(t) = \frac{1}{\sqrt{2\pi}\sigma} e^{-\frac{(t-\mu)^2}{2\sigma^2}} \tag{A.1}$$

where μ is the mean value of the Gaussian distribution (also sometimes referred to as the average value, although incorrectly) and σ^2 its variance.

The square root of the variance (σ) is called the standard deviation. The estimation of the standard deviation value based on several measurements is called the root mean square value (RMS). Figure A.1 shows the Gaussian distribution for different σ and μ values. Notice the effect that the standard deviation σ has on the spreading of the Gaussian distribution. If the distribution represents the noise associated with a measurement, clearly we would like to have the smallest possible variance value.

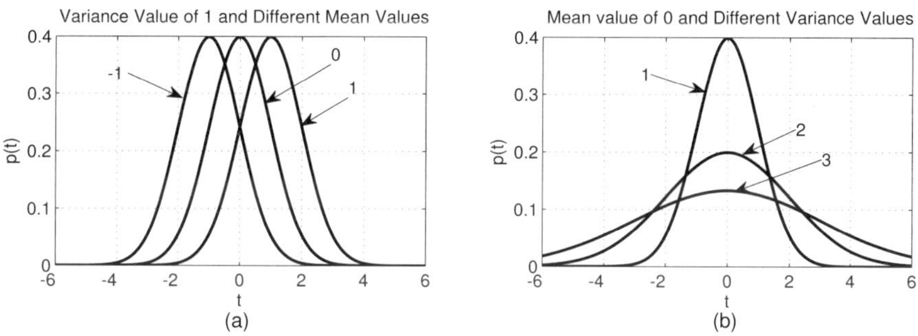

Figure A.1 (a) The Gaussian distribution with $\sigma = 1$ and different μ and (b) the Gaussian distribution with $\mu = 0$ and different standard deviation σ.

One important property of Gaussian distributions is that the sum of two Gaussian distributions X_1, X_2 will also be a Gaussian distribution with mean $\mu_1 + \mu_2$ and variance $\sigma_1^2 + \sigma_2^2$.

The probability density function is not of much help since typically one is interested in the probability of values smaller or higher than a certain t happening. For computing this probability value, one uses the cumulative probability density function (CDF). The CDF function is computed by integrating the PDF function in the interval of interest. For the Gaussian distribution, the CDF is computed using the following equation where the objective is to compute the probability of the measured value to be between $-\infty$ and x:

$$P(t) = \int_{-\infty}^{t} p(x)dx = \int_{-\infty}^{t} \frac{1}{\sqrt{2\pi}.\sigma} e^{-\frac{(x-\mu)^2}{2\sigma^2}} dx \tag{A.2}$$

In a real application, one typically starts with a set of measurements (sometimes called a time series) and from this set of measurements a normalized histogram can be computed that will correspond to an approximation of the probability density function of the random process that generated those measured values. Figure A.2 shows one example of this procedure.

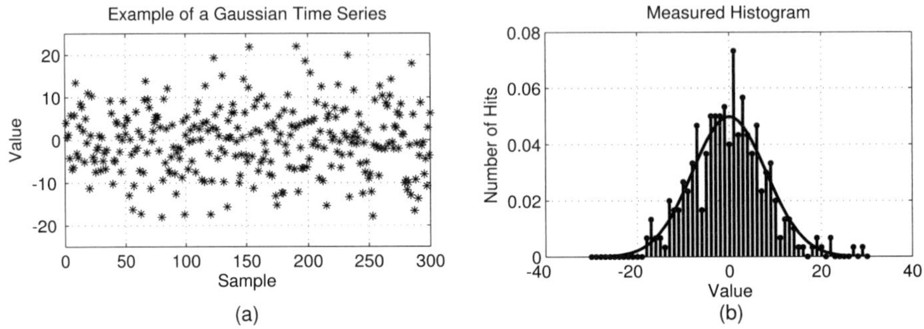

Figure A.2 Example of (a) a time series from a Gaussian distribution and (b) the computed normalized histogram (probability density function) from that time series with the theoretical Gaussian curve that would be obtained with an infinite series of values.

The best estimate for the mean value for the measured time series is:

$$\bar{\mu} = \frac{1}{N} \sum_{i=1}^{N} X_i \tag{A.3}$$

where $\bar{\mu}$ is the estimate of the mean value, X_i is the measured value i of the set of measure values, and N is the total number of samples measured. The best estimate for the standard deviation[2] is:

$$\bar{\sigma} = \sqrt{\frac{1}{N} \sum_{i=1}^{N} (X_i - \bar{\mu})^2} \tag{A.4}$$

The important question on the previous estimation equations is the value of the associated error. For the standard deviation estimation the following formula applies [3]:

[2]There is an alternative definition for the standard deviation which is:

$$\bar{\sigma} = \sqrt{\frac{1}{N-1} \sum_{i=1}^{N} (X_i - \bar{\mu})^2}$$

This definition is sometimes called population standard deviation or sample standard deviation and it has some advantages in regard to (A.4) [3].

$$(\text{fractional uncertainty in } \bar{\sigma}) = \frac{1}{\sqrt{2(N-1)}} \tag{A.5}$$

This result shows that multiple samples are needed to know with a good certainty the value of the standard deviation. For example, with three samples the uncertainty in the value of $\bar{\sigma}$ is 50%.

For the associated error to the estimate of the mean, the uncertainty is given by the standard deviation of the mean [3]:

$$\sigma_{\bar{\mu}} = \frac{\bar{\sigma}}{\sqrt{N}} \tag{A.6}$$

this means that the error associated with the estimation of the mean $\bar{\mu}$ will be:

$$\bar{\mu} \pm \frac{\bar{\sigma}}{\sqrt{N}} \tag{A.7}$$

larger sample sizes reduce the error associated with the estimation of the average.

Note that, as discussed in [4], if the mean value μ or the standard deviation value σ is calculated using a histogram, trade-offs between variance error and bias error, which are due to the width of bin interval, are inevitable.

A.2 Computation of the BER for a System with Only Gaussian Random Jitter

In this section we will calculate the BER function for a system where only Gaussian random jitter exists. Figure A.3 shows a graph of the jitter histogram for the right and left edges of the data eye. The horizontal axis corresponds to the time instant used to strobe the data eye (t_S). Ideally, to obtain the minimum BER, the strobing point should be on the time instant were the sum of the right and left jitter probability distributions is lowest, which in this case corresponds to the middle of the data eye.

The probability distributions for the jitter on the right and left edges of the data eye are described by the following Gaussian distributions:

$$p_{left}(t) = \frac{1}{\sqrt{2\pi}\sigma_{left}} e^{-\frac{(t-\mu_{left})^2}{2\sigma_{left}^2}} \tag{A.8}$$

$$p_{right}(t) = \frac{1}{\sqrt{2\pi}\sigma_{right}} e^{-\frac{(t-\mu_{right})^2}{2\sigma_{right}^2}} \tag{A.9}$$

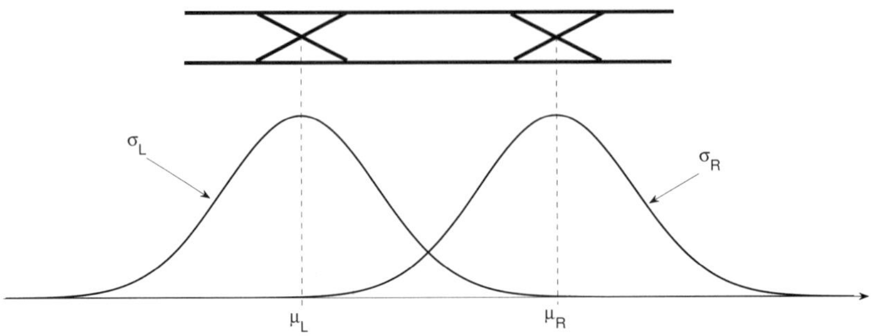

Figure A.3 Jitter probability distribution on the left and right transitions of the data eye.

The distance between the means of the two Gaussian distributions will correspond to the bit period. σ_{left} and σ_{right} are always assumed to be equal and in the remainder of this section we will then assume $\sigma_{left} = \sigma_{right} = \sigma$.

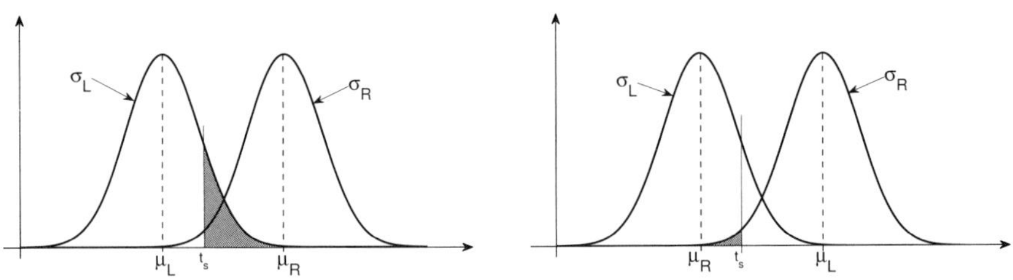

Figure A.4 Computation of the BER function.

To compute the BER value for a given strobe position, we need to integrate the contributions from the left and right Gaussian distributions (see Figure A.4) including also the probability that the measured bit is different from the expected bit (see Figure A.5). This factor is usually named transition density and it is important since although we might be strobing the wrong bit (timing-wise) due to the jitter, the bit value that is measured might still be identical to the expected bit value. On a truly random binary sequence, the transition density value is 0.5 and on a clock pattern it is 1.0. The BER value for a given strobing time t can then be computed by the following equation:

$$\begin{aligned} BER(t) = \; & p(b_N \neq b_{N-1}) \int_t^{+\infty} p_{left}(x)dx \\ & + p(b_N \neq b_{N+1}) \int_{-\infty}^{t} p_{right}(x)dx \end{aligned} \tag{A.10}$$

where p_{left} and p_{right} represent the probability distributions for the left and right edges of the data eye. $p(b_N \neq b_{N+1})$ and $p(b_N \neq b_{N-1})$ are the probabilities of the Nth bit being different from bit (N+1)th or bit (N-1)th (i.e., the transition density).

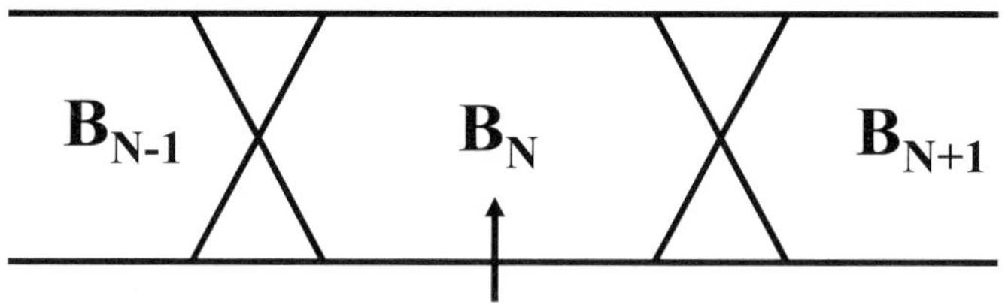

Clock Pattern: $P(b_N \neq b_{N+1}) = P(b_N \neq b_{N-1}) = 1$
True Random Pattern: $P(b_N \neq b_{N+1}) = P(b_N \neq b_{N-1}) = 0.5$

Figure A.5 Probability of adjacent bits being different.

If we suppose a Gaussian distribution for the random jitter, and assuming the same standard deviation for the right and left Gaussian ($\sigma = \sigma_{right} = \sigma_{left}$) and the same value for the transition density [$T_D = p(b_N \neq b_{N+1}) = p(b_N \neq b_{N-1})$], we obtain the following result:

$$\begin{aligned} BER(t) = {} & T_D \frac{1}{\sqrt{2\pi}\sigma} \int_t^{+\infty} e^{-\frac{(t-\mu_{left})^2}{2\sigma^2}} dt \\ & + T_D \frac{1}{\sqrt{2\pi}\sigma} \int_{-\infty}^{t} e^{-\frac{(t-\mu_{right})^2}{2\sigma^2}} dt \end{aligned} \tag{A.11}$$

we will now use the following variable change:

$$x_1 = \left(\frac{t-\mu_{left}}{\sqrt{2}\sigma}\right);\ x_2 = \left(\frac{t-\mu_{right}}{\sqrt{2}\sigma}\right) \tag{A.12}$$

to obtain the following expression for the BER:

$$\begin{aligned} BER(t) = {} & \frac{T_D}{\sqrt{\pi}} \int_{\frac{t-\mu_{left}}{2\sigma}}^{+\infty} e^{-x_1{}^2} dx_1 \\ & + \frac{T_D}{\sqrt{\pi}} \int_{\frac{\mu_{right}-t}{2\sigma}}^{+\infty} e^{-x_2{}^2} dx_2 \end{aligned} \tag{A.13}$$

Using the complementary error function (*erfc*) defined in Section A.4, the BER function can then be described by the following equation:

$$BER(t) = \frac{T_D}{2} erfc\left(\frac{t - \mu_{left}}{\sqrt{2}\sigma}\right) + \frac{T_D}{2} erfc\left(\frac{\mu_{right} - t}{\sqrt{2}\sigma}\right) \quad \text{(A.14)}$$

Figure A.6 shows the BER function, also known as the BER bathtub curve, for the following parameters: $\mu_{right} = 600$, $\mu_{left} = 200$, and $\sigma = 8$ using a clock pattern $(T_D = 1)$.

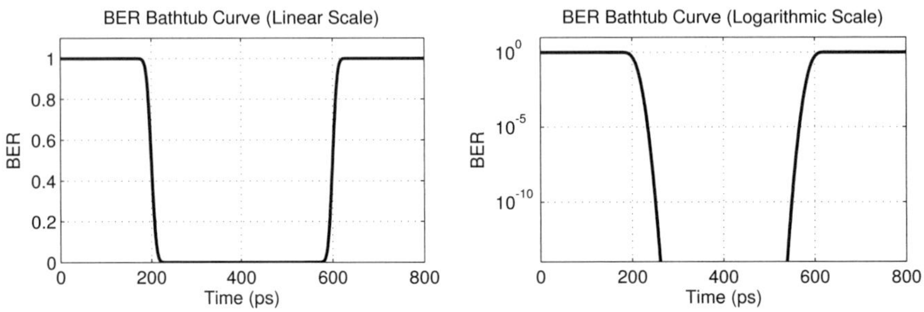

Figure A.6 BER bathtub curve for $\mu_{right} = 600$ ps, $\mu_{left} = 200$ ps, and $\sigma = 8$ ps using a clock pattern $(p(b_N \neq b_{N+1}) = p(b_N \neq b_{N-1}) = 1)$ plotted in a linear BER scale (left) and logarithmic BER scale (right).

A.3 Computation of the α(BER) Value

In a system with Gaussian random jitter and deterministic jitter, the total jitter value at a given BER is usually computed through the following equation [5, 6]:

$$TJ = DJ + \alpha(BER)\, RJ \quad \text{(A.15)}$$

where DJ is the peak-to-peak value of the deterministic jitter and RJ is the standard deviation (σ) or RMS jitter value of the random jitter Gaussian distribution.

The $\alpha(BER)$ factor is a value that is used to compute a peak-to-peak value of the random jitter for a given BER. This is exemplified in Figure A.7 where we define the peak-to-peak value for a Gaussian distribution as the interval $[-t, t]$ where $1 - BER$ of the jitter samples reside.

From Figure A.7 we can extract the following equation:

$$\frac{1}{2} BER = T_D \frac{1}{\sqrt{2\pi}.\sigma} \int_t^{+\infty} e^{-\frac{x^2}{2\sigma^2}} dx \quad \text{(A.16)}$$

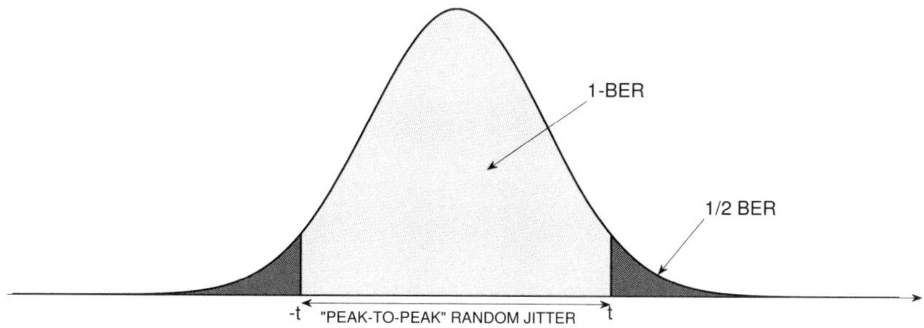

Figure A.7 Computing the $\alpha(BER)$ value.

Note that we need to include the transition density factor T_D which is the probability of the expected value of the bit under measurement, which may or may not be different from the previous or next bit. Equation (A.16) can be simplified as the following equation:

$$BER = T_D \, erfc\left(\frac{t}{\sqrt{2}\sigma}\right) \tag{A.17}$$

$$t = \sqrt{2}\, erfc^{-1}\left(\frac{BER}{T_D}\right)\sigma \tag{A.18}$$

the peak-to-peak value of the random jitter can now be computed:

$$RJ_{PK-PK} = 2\sqrt{2}\, erfc^{-1}\left(\frac{BER}{T_D}\right)\sigma \tag{A.19}$$

from (A.15) the $\alpha(BER)$ is:

$$\alpha(BER) = 2\sqrt{2}\, erfc^{-1}\left(\frac{BER}{T_D}\right) \tag{A.20}$$

For example, if we assume a true random pattern ($T_D = 0.5$), for a BER of 10^{-12} the $\alpha(BER)$ value is 14.069. Table A.1 shows some of the typical used values for $\alpha(BER)$ for a transition density of 0.5 and 1. In most applications a transition density value of 0.5 is explicitly assumed.

Figure A.8 shows the link between a specific desired BER value, the percentage of jitter events that need to be contained on the interval that guarantees that BER value, and the corresponding $\alpha(BER)$ factor that multiplied by the standard deviation σ provides the random jitter peak-to-peak value for a specific BER.

Table A.1
Some Typical Used Values for $\alpha(BER)$

BER	α(BER), $T_D = 0.5$	α(BER), $T_D = 1$
10^{-9}	11.9956	12.2188
10^{-12}	14.0690	14.2610
10^{-16}	16.4442	16.6096

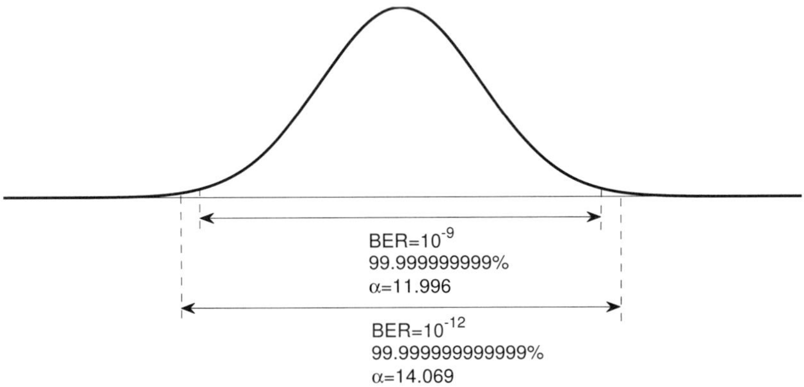

Figure A.8 Relationship between the BER value and the percentage of occurrences that need to be included for the computation of the $\alpha(BER)$.

A.4 Properties of the Error Function erf(x) and Complementary Error Function erfc(x)

The error function erf(x) is defined by the following equation:

$$erf(x) = \frac{2}{\sqrt{\pi}} \int_0^x e^{-y^2} dy \tag{A.21}$$

it has the following properties:

$$\begin{aligned} erf(0) &= 0 \\ erf(+\infty) &= 1 \\ erf(-\infty) &= -1 \end{aligned} \tag{A.22}$$

the complementary error function erfc(x) is defined by the following equation:

$$erfc(x) = \frac{2}{\sqrt{\pi}} \int_x^{+\infty} e^{-y^2} dy \tag{A.23}$$

and it has the following property:

$$\int_0^x erfc(y)dy = x\, erfc(x) + \frac{1}{\sqrt{\pi}}(1 - e^{-x^2}) \tag{A.24}$$

The error function and the complementary error function are related by the following expression:

$$erfc(x) = 1 - erf(x) \tag{A.25}$$

the inverse of the complementary error function is also important:

$$erfc^{-1}(erfc(x)) = x \tag{A.26}$$

For values of $x >> 1$, the complementary error function can be approximated by the following expression:

$$erfc(x) \approx \frac{2}{\sqrt{\pi}} \frac{e^{-x^2}}{x} \quad \text{for } x >> 1 \tag{A.27}$$

Note that the used definitions of the *erf* and *erfc* functions are the same as the ones used by software packages like MATLAB and Mathematica. Other authors might use slightly different definitions.

References

[1] A. Papoulis and S. U. Pillai, *Probability, Random Variables and Stochastic Processes.* New York: McGraw-Hill, 2002.

[2] S. M. Ross, *Introduction to Probability and Statistics for Engineers and Scientists.* New York: John Wiley & Sons, 1987.

[3] J. R. Taylor, *An Introduction to Error Analysis.* Sausalito, CA: University Science Books, 1997.

[4] J. S. Bendat and A. G. Piersol, *Random Data Analysis and Measurement Procedures.* New York: John Wiley & Sons, 4 ed., 2010.

[5] Maxim, "Converting Between RMS and Peak-to-Peak Jitter at a Specified BER," *Application Note HFAN-4.0.2*, 2000.

[6] INCITS, *Fiber Channel – Methodology for Jitter and Signal Quality Specification – MJSQ*, 2004. Technical Report for Information Technology.

B

The Dual Dirac Model and RJ/DJ Separation

This appendix presents in detail the dual Dirac jitter model and the random and deterministic jitter separation algorithm based on this model, which is used in several standards [1–3].

B.1 The Dual Dirac Jitter Model

The dual Dirac jitter model was developed to model the random and deterministic jitter components of a digital signal. The reasoning comes from the fact that the deterministic components of the jitter will be bounded. This means that the simplest model for deterministic jitter is a kind of ping-pong jitter that jumps between the two boundaries of the deterministic jitter range as shown in Figure B.1.

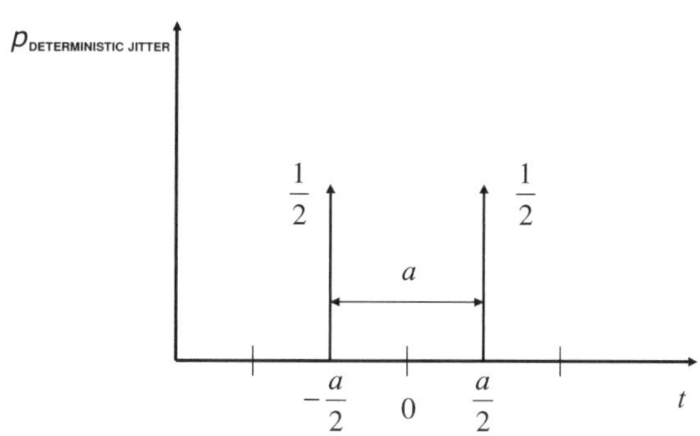

Figure B.1 The dual Dirac deterministic jitter model for a peak-to-peak jitter value of a.

This behavior can be modeled by the use of the Dirac delta function. A Dirac delta function is part of what are called generalized functions [4]. A Dirac delta function can be visualized as a zero width pulse at a specific point in time. Mathematically, the deterministic jitter model will have the following form:

$$DJ = \frac{1}{2}\delta\left(t - \frac{a}{2}\right) - \frac{1}{2}\delta\left(t + \frac{a}{2}\right) \tag{B.1}$$

where $\delta(t)$ is the Dirac delta function and a is the peak-to-peak value of the deterministic jitter. This model is referred to as the dual Dirac jitter model.

We can now complete the model by convolving the random jitter in the form of a Gaussian distribution as shown in Figure B.2.

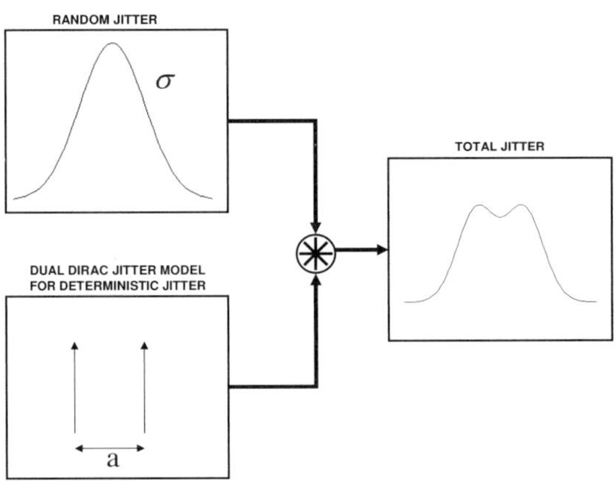

Figure B.2 The dual Dirac jitter model.

To obtain an analytical expression for this model, one needs to perform the convolution of the dual Dirac model for the deterministic jitter with the Gaussian distribution model for the random jitter (A.1) with a mean value of zero ($\mu = 0$). Mathematically, the model is now:

$$\begin{aligned} p_{\text{jitter}}(t) &= p_{\text{random}}(t) * p_{\text{dual\_dirac}}(t) \\ &= \int_{-\infty}^{+\infty} p_{\text{random}}(\tau) p_{\text{dual\_dirac}}(t-\tau) d\tau \end{aligned} \tag{B.2}$$

substituting the probability distributions, we obtain:

$$p(t) = \int_{-\infty}^{+\infty} \left[\frac{1}{2}\delta\left(\tau - \frac{a}{2}\right) \frac{1}{\sqrt{2\pi}\sigma} e^{-\frac{(t-\tau)^2}{2\sigma^2}} + \frac{1}{2}\delta\left(\tau + \frac{a}{2}\right) \frac{1}{\sqrt{2\pi}\sigma} e^{-\frac{(t-\tau)^2}{2\sigma^2}} \right] d\tau \tag{B.3}$$

solving this integral, we obtain the following expression for the total jitter distribution of the dual Dirac jitter model:

$$p(t) = \frac{1}{2\sqrt{2\pi}\sigma} \left[e^{-\frac{(t+\frac{a}{2})^2}{2\sigma^2}} + e^{-\frac{(t-\frac{a}{2})^2}{2\sigma^2}} \right] \tag{B.4}$$

Figure B.3 presents the total jitter probability distribution for the parameters $\sigma = 8$ and $a = 20$.

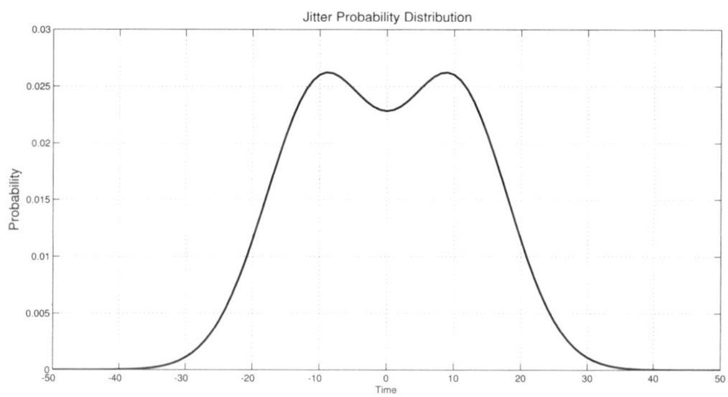

Figure B.3 Total jitter probability distribution for a dual Dirac jitter model ($\sigma =$ 8 and $a = 20$).

Using the computed distribution it is now possible to write the jitter distributions for the right and left edges of the data eye.

$$\begin{aligned} p_{right}(t) &= \frac{1}{2\sqrt{2\pi}\sigma} \left[e^{-\frac{(t+\frac{a}{2}-E_{right})^2}{2\sigma^2}} + e^{-\frac{(t-\frac{a}{2}-E_{right})^2}{2\sigma^2}} \right] \\ p_{left}(t) &= \frac{1}{2\sqrt{2\pi}\sigma} \left[e^{-\frac{(t+\frac{a}{2}-E_{left})^2}{2\sigma^2}} + e^{-\frac{(t-\frac{a}{2}-E_{left})^2}{2\sigma^2}} \right] \end{aligned} \tag{B.5}$$

where E_{left} and E_{right} correspond to the left and right crossing points of the data eye. $E_{right} - E_{left}$ should correspond to the bit period. The BER for a sampling point t is then computed by the following expression where T_D is the transition density (see Appendix A).

$$
\begin{aligned}
BER(t) = {} & T_D \int_{t}^{+\infty} p_{left}(t)dt \\
& + T_D \int_{-\infty}^{t} p_{right}(t)dt
\end{aligned}
\tag{B.6}
$$

substituting by the respective probability distributions, we obtain:

$$
\begin{aligned}
BER(t) = {} & \frac{T_D}{2\sqrt{2\pi}\sigma} \int_{t}^{+\infty} \left[e^{-\frac{(t-\frac{a}{2}-E_{left})^2}{2\sigma^2}} + e^{-\frac{(t+\frac{a}{2}-E_{left})^2}{2\sigma^2}} \right] dt \\
& + \frac{T_D}{2\sqrt{2\pi}\sigma} \int_{-\infty}^{t} \left[e^{-\frac{(t-\frac{a}{2}-E_{right})^2}{2\sigma^2}} + e^{-\frac{(t+\frac{a}{2}-E_{right})^2}{2\sigma^2}} \right] dt
\end{aligned}
\tag{B.7}
$$

applying the same techniques used in Section A.2, we then obtain:

$$
\begin{aligned}
BER(t) = {} & \frac{T_D}{4} \left[erfc\left(\frac{t - \frac{a}{2} - E_{left}}{\sqrt{2}\sigma} \right) + erfc\left(\frac{t + \frac{a}{2} - E_{left}}{\sqrt{2}\sigma} \right) \right] \\
& + \frac{T_D}{4} \left[erfc\left(\frac{E_{right} + \frac{a}{2} - t}{\sqrt{2}\sigma} \right) + erfc\left(\frac{E_{right} - \frac{a}{2} - t}{\sqrt{2}\sigma} \right) \right]
\end{aligned}
\tag{B.8}
$$

Figure B.4 shows the BER bathtub curve for the following parameters: E_{right}=600 ps, E_{left}=200 ps, σ=8 ps, and $a = 100$ ps using a clock data pattern $T_D = 1$. The bit period is 400 ps.

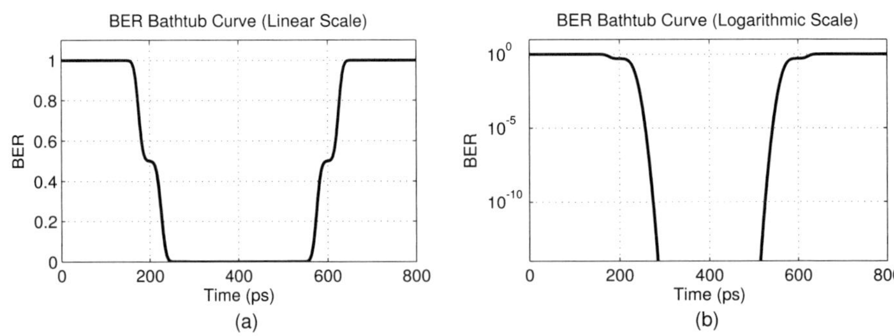

Figure B.4 BER bathtub curve for the dual Dirac jitter model in (a) a linear BER scale and (b) a logarithmic BER scale.

B.2 RJ/DJ Separation with the Q-Factor Algorithm

The idea behind RJ/DJ separation using the BER bathtub curve is that below a certain BER value, the BER will be dominated by the random jitter which follows a Gaussian distribution. In this region the BER can be approximated by a normalized complementary error function:

$$\begin{aligned} BER_{RJ}^{left}(t) &= N_{left} erfc\left(\frac{t-\mu_{left}}{\sqrt{2}\sigma_{left}}\right) \\ BER_{RJ}^{right}(t) &= N_{right} erfc\left(\frac{\mu_{right}-t}{\sqrt{2}\sigma_{right}}\right) \end{aligned} \tag{B.9}$$

We will now assume a normalization factor corresponding to the dual Dirac jitter model in the previous expressions ($N_{left} = T_D/4$ and $N_{right} = T_D/4$).

Let us now define the Q-factor[1] [5] for the right and left sides of the BER bathtub curve using the following equations:

$$Q_{right} = \frac{\mu_{right}-t}{\sigma_{right}}; Q_{left} = \frac{t-\mu_{left}}{\sigma_{left}} \tag{B.10}$$

we obtain the following relation between the BER and the Q-factor:

$$BER_{RJ}(t) = \frac{T_D}{4} erfc\left(\frac{Q}{\sqrt{2}}\right) \tag{B.11}$$

by rearranging the terms, we obtain:

$$\frac{4}{T_D} BER_{RJ} = 1 - erf\left(\frac{Q}{\sqrt{2}}\right) \Leftrightarrow Q = \sqrt{2}\, erf^{-1}\left(1 - \frac{4}{T_D} BER_{RJ}\right) \tag{B.12}$$

we can look at this expression as a mapping of the BER bathtub curve into the Q-factor space:

$$\begin{aligned} Q &= F(BER_{RJ}) \\ F(x) &= \sqrt{2}\, erf^{-1}\left(1 - \frac{4}{T_D} x\right) \end{aligned} \tag{B.13}$$

Figure B.5 presents a graphic example of this mapping. The important factor is that an *erfc* function (i.e., a Gaussian jitter distribution) is a straight line when represented in the Q-factor space (i.e., when we are on the

[1] Also referred to as the Q-scale.

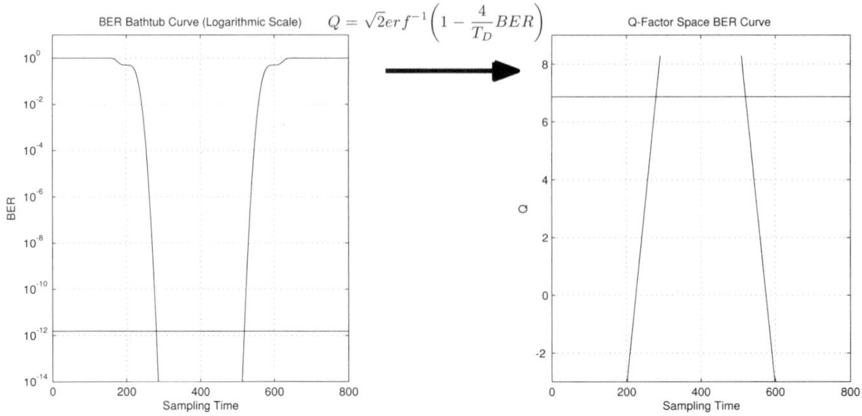

Figure B.5 Mapping the BER bathtub curve into the Q-factor space.

"Gaussian" part of the BER bathtub curve). Given the definition of the Q-factor, it is now easy to obtain the random and deterministic jitter values. First, it is necessary to perform a linear fitting to the left and right curve obtained after the transformation (see [6]), but only to the part where the random jitter is the dominant factor (below a certain BER). Let us now suppose we compute the fitting parameters ($A_{right}, A_{left}, B_{right}, B_{left}$) shown on (B.14).

$$\begin{aligned} Q_{right}(t) &= t\,A_{right} + B_{right} \\ Q_{left}(t) &= t\,A_{left} + B_{left} \end{aligned} \tag{B.14}$$

Given the previous definitions for the right and left Q-factor, it is then possible to create an equality between the fitting parameters and the Q-factor of each curve.

$$\begin{aligned} \frac{\mu_{right} - t}{\sigma_{right}} &= t\,A_{right} + B_{right} \\ -\frac{t}{\sigma_{right}} + \frac{\mu_{right}}{\sigma_{right}} &= t\,A_{right} + B_{right} \end{aligned} \tag{B.15}$$

$$\begin{aligned} \frac{t - \mu_{left}}{\sigma_{left}} &= t\,A_{left} + B_{left} \\ \frac{t}{\sigma_{left}} - \frac{\mu_{left}}{\sigma_{left}} &= t\,A_{left} + B_{left} \end{aligned} \tag{B.16}$$

From the above equation, it is possible to infer the values of the random jitter Gaussian distribution standard deviation and mean of each curve from the obtained fitting parameters:

$$
\begin{aligned}
\sigma_{right} &= -\frac{1}{A_{right}} \\
\sigma_{left} &= \frac{1}{A_{left}} \\
\mu_{right} &= B_{right}\,\sigma_{right} \\
\mu_{left} &= -B_{left}\,\sigma_{left}
\end{aligned} \tag{B.17}
$$

then the random jitter RMS value will correspond to the mean of the left and right standard deviations:

$$
RJ = \frac{\sigma_{right} + \sigma_{left}}{2} \tag{B.18}
$$

To compute the deterministic jitter, it is necessary to compute the spacing between the two Gaussian distributions. However, given that each Gaussian will reside on one of the sides of the BER bathtub curve, the following expression will provide the deterministic jitter value:

$$
DJ = 1.0\ \mathrm{UI} - (\mu_{right} - \mu_{left}) \tag{B.19}
$$

The advantage of this approach is that by transforming the Gaussian fitting into a linear fitting, it allows the computation of the random and deterministic jitter separation to be done in a very fast way given that there is an easy and fast solution for the linear fitting of a set of points [6].

To compute the total jitter (TJ) at a given BER, a different approach can be used instead of (A.15) by using the linear fitting parameters in (B.16). This can be accomplished using (B.20) and is shown graphically in Figure B.6.

$$
\begin{aligned}
TJ(BER) &= 1.0\ \mathrm{UI} - [t_{right}(BER) - t_{left}(BER)] \\
t_{right}(BER) &= \frac{Q(BER) - A_{right}}{B_{right}} \\
t_{left}(BER) &= \frac{Q(BER) - A_{left}}{B_{left}} \\
Q(BER) &= \sqrt{2}\, erf^{-1}\left(1 - \frac{4}{T_D} BER\right)
\end{aligned} \tag{B.20}
$$

As a final point, it is important to note that this jitter separation algorithm is exact as long as the model it is based on is exact. In a real application the double Dirac jitter model is an approximation that impacts the accuracy of the obtained jitter separation values [7–10].

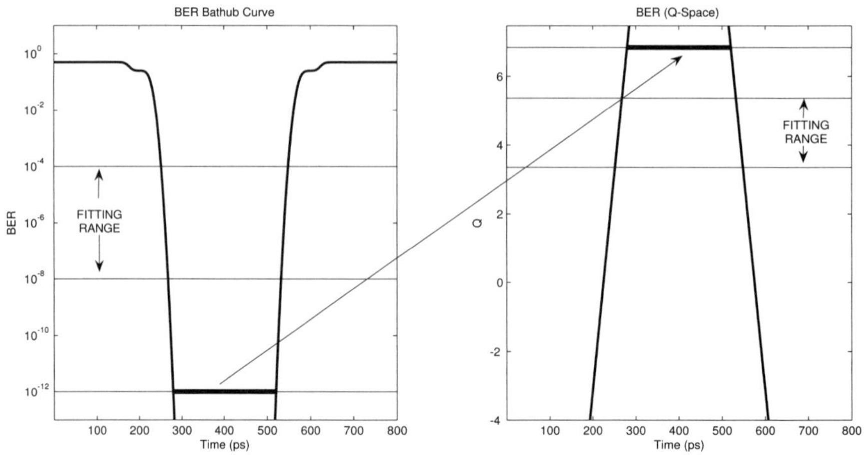

Figure B.6 Computing the total jitter value for a certain BER by using the results from the linear fitting on the right and left sides of the BER bathtub curve.

References

[1] M. P. Li, J. Wilstrup, R. Jessen, and D. Petrich, "A New Method for Jitter Decomposition Through Its Distribution Tail Fitting," *IEEE International Test Conference*, 1999.

[2] G. Haensel, K. Stieglbauer, G. Schulze, and J. Moreira, "Implementation of an Economic Jitter Compliance Test for Multi-Gigabit Device on ATE," *IEEE International Test Conference*, 2004.

[3] Agilent Technologies, "Jitter Analysis: The Dual-Dirac Model, RJ/DJ, and Q-Scale," *White Paper 5989-3206*, 2005.

[4] R. P. Kanwal, *Generalized Functions: Theory and Applications*. Birkhauser, 2004.

[5] N. S. Bergano, F. W. Kerfoot, and C. R. Davidson, "Margin Measurements in Optical Amplifier Systems," *IEEE Photonics Technology Letters*, vol. 5, no. 3, 1993.

[6] P. R. Bevington and D. K. Robinson, *Data Reduction and Error Analysis for the Physical Sciences*. New York: McGraw-Hill, 2003.

[7] R. Stephens, "What the Dual-Dirac Model Is and What It Is Not," *Tektronix 360 Knowledge Series*, 2006.

[8] M. Miller and M. Schnecker, "A Comparison of Methods for Estimating Total Jitter Concerning Precision, Accuracy and Robustness," *IEC DesignCon*, 2007.

[9] T. J. Yamaguchi, K. Ichiyama, H. X. Hou, and M. Ishida, "A Robust Method for Identifying a Deterministic Jitter Model in a Total Jitter Distribution," *IEEE International Test Conference*, 2009.

[10] T. J. Yamaguchi, H. X. Hou, K. Takayama, D. Armstrong, M. Ishida, and M. Soma, "An FFT-Based Jitter Separation Method for High-Frequency Jitter Testing with a 10X Reduction in Time," *IEEE International Test Conference*, 2007.

C

Pseudo-Random Bit Sequences and Other Data Patterns

This appendix provides a background on pseudo-random bit sequences (PRBS) and a few of the most common data patterns used for testing and characterization of high-speed digital interfaces.

C.1 Pseudo-Random Bit Sequences

A pseudo-random bit sequence data pattern is frequently used for the characterization of high-speed digital I/O interfaces. A PRBS data pattern generates all possible bit sequences for a given length. This means that for a given word with a set length of 7 bits, for example, the corresponding PRBS sequence will contain all the possible permutations of 1s and 0s for a 7-bit word. A PRBS sequence is easily implemented in hardware or software using a linear feedback shift register (LFSR) as shown in Figure C.1 [1–3]. Note that the figure only shows some of the typical PRBS data patterns. Reference [4] provides a more comprehensive list.

PRBS sequences are based on the mathematical theory of "irreducible polynomial of length N." Each PRBS $2^N - 1$ data sequence will contain all the N-bit long sequences (except $N \times 0$) and its cycle length will be $2^N - 1$. After this length the sequence will start again. Note that although the data word length is fixed for each N, the exact full length PRBS sequence will depend on the initial seed (initial values set in the storage elements of the LFSR) provided to the algorithm. Sometimes a PRBS $2^N - 1$ sequence is referred to as a PRBS N sequence for a compact notation. However, it is important not to mistake it for a PRBS 2^N data pattern which is different from

PRBS N	FB1	FB2	Pattern Length
3	1	3	7
5	2	5	31
7	1	7	127
11	2	11	2047
15	1	15	32767
21	2	21	2097151
31	3	31	2147483647

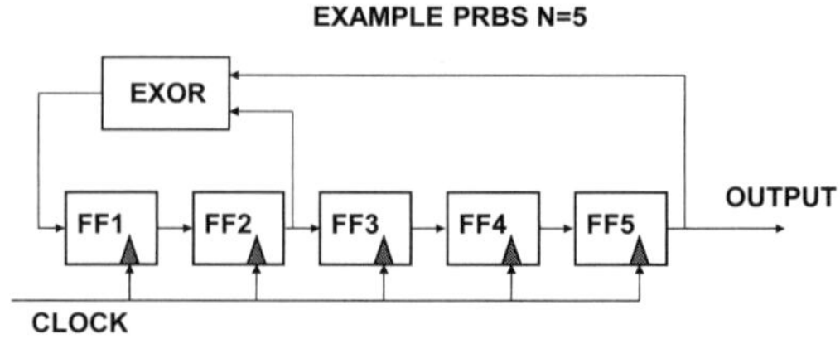

Figure C.1 Basic block diagram example of a PRBS implementation with a feedback shift register and the design parameters for different word length PRBS data patterns.

a PRBS $2^N - 1$ pattern. Figure C.2 shows an example of the bit sequence of a PRBS $2^7 - 1$ pattern.

0000001000000110000010100011110010001
011001110101001111101000001110001001
00110110101101111011000110100101110
1110011001010101111111

Figure C.2 Example of a PRBS $2^7 - 1$ bit sequence with a 0000001 seed.

Given its simple hardware implementation, PRBS sequences are used extensively in test and measurement equipment. However, it is important to understand their properties when choosing a PRBS sequence to stimulate a DUT. Figure C.3 shows a comparison of the power spectrum for two different PRBS sequences (PRBS $2^{31} - 1$ and PRBS $2^7 - 1$) at 5 Gbps. One notes that the spectrum for both PRBS sequences has the same bandwidth (the bandwidth is defined by the signal rise time, which is the same for both patterns). The spectrum may look similar for both patterns with the expected *sinc* function shape[1]; however, on closer inspection of the details, it is possible to observe that the PRBS $2^{31} - 1$ sequence has additional spectrum lines since it can generate a higher number of symbol combinations.

The fact that a PRBS $2^{31} - 1$ pattern has a higher number of spectral lines that are spread through the frequency band has consequences on the measured performance of a DUT or a signal path that is stimulated with a PRBS pattern. Figure C.4 shows the data eye diagrams obtained from a lossy signal path with a PRBS $2^{31} - 1$ and a PRBS $2^7 - 1$ data pattern. Although the stimulus driver performance is the same for both patterns (i.e., intrinsic jitter, rise time, and so forth), the PRBS $2^{31} - 1$ data pattern shows worse

[1]More details in Section 2.2.

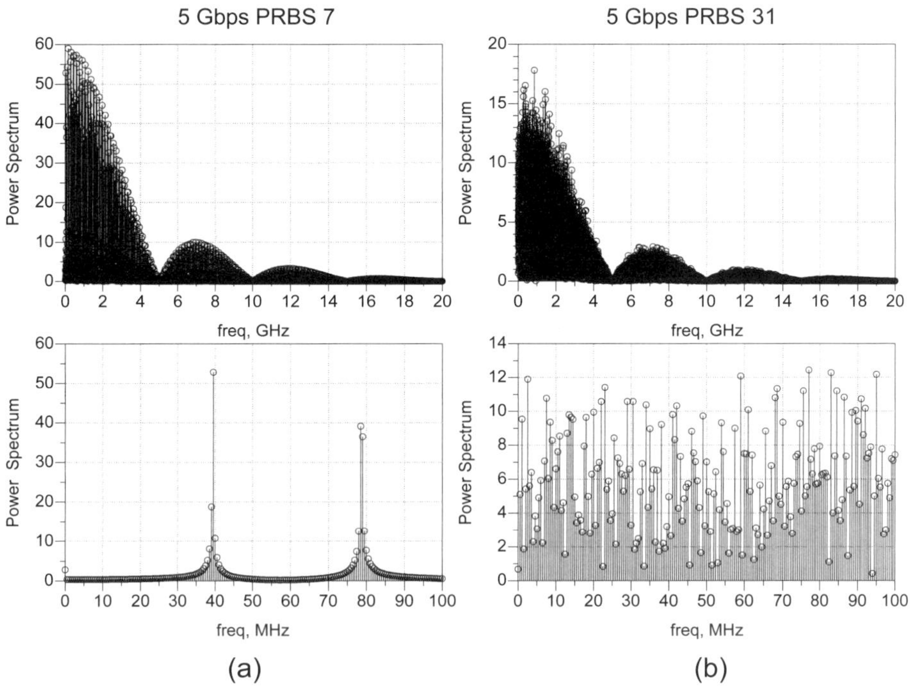

Figure C.3 Comparison of the power spectrum of (a) a PRBS $2^7 - 1$ pattern with (b) a PRBS $2^{31} - 1$ pattern.

performance than the PRBS $2^7 - 1$ pattern since its power spectrum "stresses" the DUT more.

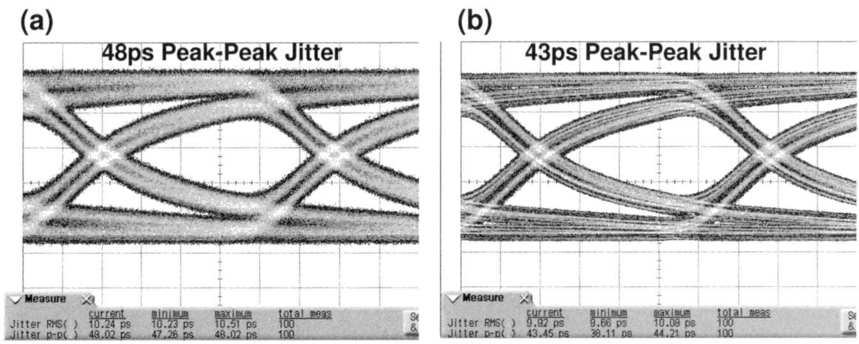

Figure C.4 Comparison of the measured data eye diagram of a lossy signal path with (a) a PRBS $2^{31} - 1$ and (b) a PRBS $2^7 - 1$ pattern.

C.1.1 Challenges of the PRBS31 Data Pattern

The PRBS31 data pattern although used in several standards presents several specific challenges due to the its large size (2,147,483,647 bits). Although easily generated using setups like the one described in Section C.1, it presents a very tough challenge for any memory-based measurement approach. Measurement instruments like real-time and equivalent-time oscilloscopes do not have enough memory to acquire the entire data pattern since a PRBS31 pattern at 10 Gbps has a time duration of 214.76 ms.

Another challenge of the PRBS31 data pattern is its low-frequency content. At 10 Gbps the lowest frequency component is at 5 Hz, which means AC-coupled components like a Bias-TEE can have an impact on the pattern since they can block the very low-frequency components present on the PRBS31 data pattern (see Section 7.17.2).

C.2 Pseudo-Random Word Sequences

Pseudo-random word sequences (PRWS) are a special application of PRBS sequences. One way to understand PRWS is to look at their typical usage which is the test of applications like a multiplexer. Figure C.5 shows the need to define a proper bit sequence for the 2-bit word that is an input to the multiplexer in a way that guarantees that the pattern at the output of the multiplexer is a PRBS sequence [5, 6].

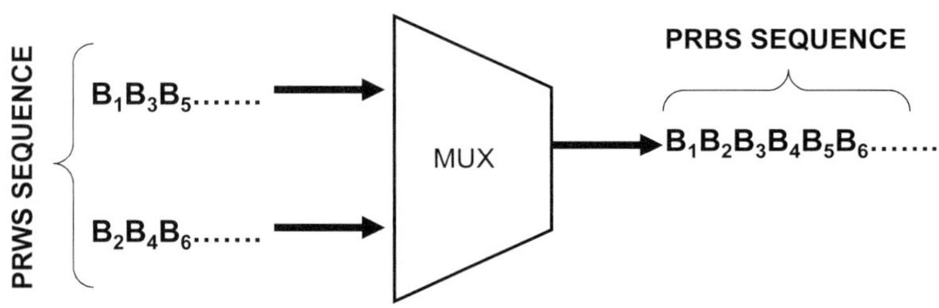

Figure C.5 The need to use a PRWS data pattern when testing a multiplexer.

Likewise for the PRBS case, PRWS generators are easily implemented into hardware through shift registers [7].

C.3 Other Important Patterns

Several patterns are important enough that their names are used instead of the corresponding bit sequence. One is the PCI Express compliance pattern shown in Figure C.6.

Figure C.6 The PCI Express compliance pattern (0011111010101010101010 1100000101010101010101) [8].

Another pattern is the K28.5 pattern shown in Figure C.7 that is generated by two consecutive K28.5 characters defined in the 8B/10B code space (for more details, see Appendix D).

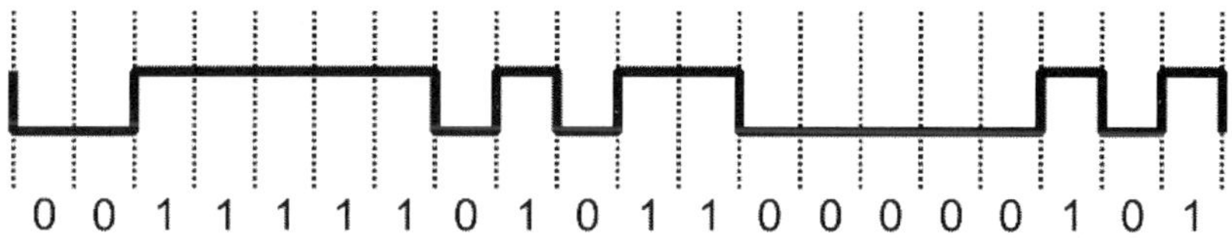

Figure C.7 The K28.5 pattern.

Other important patterns are the continuous random test pattern (CRPAT) and the continuous jitter test pattern (CJPAT). Both patterns are used in several standards (e.g., 10 Gigabit Ethernet). These patterns are not simple and their definitions can be found in [9]. The CRPAT has a broad and relatively flat spectral content that is targeted to provide high-frequency jitter components like ISI but with little content at low frequencies. The CJPAT in contrast was designed to have spectral content at low frequencies where the DUT PLL bandwidth will reside while also containing high-frequency components.

References

[1] P. H. Bardell, W. H. McAnney, and J. Savir, *Built-in Test for VLSI: Pseudorandom Techniques*. New York: John Wiley & Sons, 1987.

[2] S. W. Golomb, *Shift Register Sequences*. Aegean Park Press, 1982.

[3] D. Derickson and M. Mueller, *Digital Communications Test and Measurement: High-Speed Physical Layer Characterization*. Upper Saddle River, NJ: Prentice-Hall, 2007.

[4] Peter Alfke, "Efficient Shift Registers, LFSR Counters, and Long Pseudo-Random Sequence Generators," *Xilinx Application Note XAPP 052*, 1996.

[5] Agilent Technologies, "Agilent Technologies ParBERT 81250 Mux/Demux Application," *Application Note 5968-9695*, 2004.

[6] Centellax, "Generating 40G PRBS patterns," *Application Note AN13*, 2008.

[7] W. J. McFarland, K. H. Springer, and C.-S. Yen, "1-Gword/s Pseudorandom Word Generator," *IEEE Journal of Solid State Circuits*, vol. 24, June 1989.

[8] PCI-SIG, *PCI Express Base Specification Revision 2.1*, Mar. 2009.

[9] IEEE, *802.3 Part3: Carrier Sense Multiple Access with Collision Detection (CSMA/CD) Access Method and Physical Layer Specifications*, 2005.

D

Coding, Scrambling, Disparity, and CRC

Coding or scrambling of bit sequences has the purpose to adapt the properties of a data stream as much as possible to limitations and requirements of the signal path and the receiver that has to interpret the signal it receives. In real systems, signal paths have certain transmission characteristics that are mainly determined by the materials and geometries used to implement the signal path. Since system designers are usually not free in the selection of the materials and geometries that they use for signal path implementation, they are also limited in their influence on the transmission characteristic of the signal path. In such cases, coding and scrambling can help to modify the data stream properties in a way that the characteristic of the signal path has less influence on the transmitted signal.

One example of this is to spread out the spectral energy of a data stream over a wider frequency range using scrambling to minimize EMI caused by peaks in the spectral power distribution of a data stream due to repetitive data that is transmitted. Another example to minimize the influence of the signal path on the transmitted signal is coding to achieve DC-balance in the data stream. Each signal path has an inevitable capacitance that is charged and discharged depending on the data stream which is transmitted. The baseline voltage observed on the transmission path will be modulated with a low-pass filter characteristic according to the charging and discharging of the signal path.

If a pattern contains long sequences of charging bits (ones) and/or long sequences of discharging bits (zeros), there is the possibility that the baseline voltage of the signal path wanders away too far from the desired voltage level. This can cause errors in the data transmission because bits with a value other than the ones that caused this baseline wander are not recognized correctly anymore. One way to avoid this behavior is to code the transmitted data stream in a way that limits the number of consecutive bits having the same logical value and to make sure that the number of zeros and ones in the transmitted data stream are balanced over a certain range of bits. A measure for this balance over a certain bit range is the disparity, which will be explained in more detail in Section D.1. The amount to which subsequent bits of the same logical value are limited and the observation range for the disparity calculation a transmission path requires depends on its low-pass filter characteristics and thus on its capacitance.

As mentioned already, besides optimizing data stream properties to match signal path characteristics, coding and scrambling also are used to guarantee signal conditions that device receivers require to be able to correctly interpret the received signals. An example for this use scenario is embedded clock receivers that have to extract the clock phase information for the received signal from this signal itself. The phase information from the received data stream is used to align the phase of the device internal sampling clock to the actual bit positions in the received data.

If such a receiver cannot readjust its sampling clock phase regularly to the incoming data, there is the potential to miss incoming data bits or to sample more bits than there actually are in the data stream due to phase offsets between the incoming data stream and the device's sampling clock. In order to be able to extract the required clock phase information, embedded clock receivers rely on transitions in the received data stream. Thus, these receivers typically require that the data streams they receive contain transitions that are not too far apart from each other. Scrambling or coding usually is applied to ensure that there is a sufficient amount of transitions in the data stream and that the distance between the transitions is not exceeding the limits required by the receiver.

D.1 Disparity

Disparity is a term used to define the relation between the number of zeros and ones over a sequence of binary data (e.g., a byte or a data word). The disparity for a given bit sequence is calculated by increasing a counter for each one in the analyzed data and by decreasing this counter for each zero as shown in Figure D.1. With this calculation rule, the disparity number that is obtained represents a measure for how many more ones are contained in an analyzed data stream section than there are zeros.

If the disparity calculation as described above with a counter increment for each one and a counter decrement for each zero is applied to a continuously

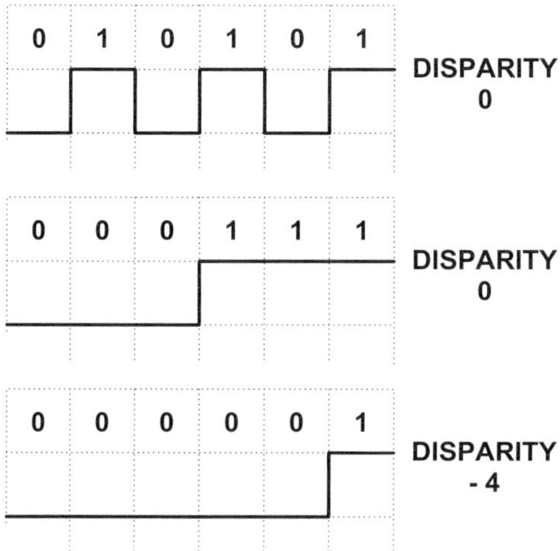

Figure D.1 Example of computing the disparity in a 6-bit symbol.

running data stream, a disparity number is obtained after each single bit. This disparity information is also referred to as running disparity. The running disparity indicates at each instance during transmission of a data stream the difference between transmitted ones and zeros over the whole data transmitted so far.

As already discussed before, it is important for some applications to keep the average numbers of zeros and ones over a transmitted data stream balanced. In order to keep such a balanced or neutral disparity, the easiest approach would be to use coding that transfers disparity-unbalanced data words into disparity-balanced codes. However, it is obvious that with such a truly balanced code space, the amount of bits required per code word would be significantly larger than the number of bits in the original data word because only code words having the same amount of zeros and ones would be allowed. Since the effects such as baseline wander that result from unbalanced disparity are usually found in the DC space and the data rates we are talking about are in the Gigabits per second range, it is typically not required that the running disparity is kept neutral for each and every transmitted bit to achieve DC-balance over the data stream. Usually it is sufficient to ensure by suitable code selection that the running disparity is kept at a minimum after transmission of a code word. One example for a code that follows this methodology to create a DC-balanced code space is the 8B/10B code, which was developed by IBM in 1983. In the following section we will give an overview of this 8B/10B code. More detailed information can be obtained in the original paper describing the 8B/10B coding [1].

D.2 8B/10B Coding

The 8B/10B is a line code that maps 8-bit symbols to 10-bit codes to achieve DC-balance and bounded disparity and provides enough transitions to allow for reasonable clock recovery. As a result of coding, any combination of two code words has neutral disparity and the longest sequence of bits with the same logical value (run length) is limited to 5 consecutive bits.

The 8B/10B coding scheme transmits 1 byte of data as a 10-bit code, which also is called symbol or character. The lower 5 bits of the transmitted byte are encoded into a 6-bit group and the upper 3 bits of the byte are encoded into a 4-bit group. Both of these code groups are concatenated together to form the 10-bit symbol that is transmitted. The data symbols are often referred to as Dxx.y where the number xx ranges from 0 to 31 and represents the value of the lower 5 bits before conversion. The number y is in the range from 0 to 7 and represents the value of the upper 3 bits of the transmitted byte before conversion. Thus, the original byte value can easily be determined from the symbol name.

Standards using the 8B/10B encoding also define special symbols (or control characters) that can be sent in place of a data symbol. They are often used to indicate end-of-frame, link idle, skip, and similar link-level conditions. These control characters are referred to as Kxx.y and have different encoding from any of the Dxx.y symbols. As a result of the differences in encoding data and control symbols, the maximum run length of five for 8B/10B codes can only occur within control symbols, whereas data symbols have a maximum run length of four. This distinct property of some control symbols allows easy identification of these in a data stream and typically is used for synchronization to the data stream by the standards that apply to 8B/10B coding.

One goal when developing the 8B/10B code was to achieve DC-balance for the coded data streams. This is achieved by tracking the running disparity for each code segment used and selecting one of up to two code alternatives for the next code segment based on the current status of the running disparity. Two code alternatives are usually available for all 3- or 5-bit portions of the original byte that translate into a code segment, which does not have neutral disparity or only contains a single bit transition. A single bit transition within the code automatically results in a code that contains only zeros in one half of the code word and only ones in the other half.

One characteristic of these two alternative code segments is that one is the inverse of the other. Since the disparity of any code segment is +2, 0, or −2, it is obvious that in all the cases where a choice exists between two code segments due to nonneutral disparity, one of these increases the running disparity by two, and the other one reduces it by two. In cases of alternative code segments with neutral disparity, there is the potential of a temporary increase or decrease of the running disparity within the code section by three because all zeros and ones in the code segment are cumulated. If there are alternative code segments, the code segment that keeps the running disparity closer to zero is selected based on the current value of the running disparity. Since 8B/10B coding requires setting the start value for the running disparity to +1 or to −1, the running disparity after each code segment also will be +1 (rd+) or −1 (rd−).

The coding tables for the 5B/6B coding (Table D.1) and the 3B/4B coding (Table D.2) that are the basis for 8B/10B codes show the mapping between an original byte HGFEDCBA to the 5B/6B code abcdei derived from EDCBA and the 3B/4B code fghj derived from HGF. In the header columns of the tables, the current status of the running disparity (rd+ or rd−) is listed to allow the selection of the required code segment for code words that offer this alternative. The column rd' in the tables indicates the status of the running disparity after the code segment is applied. The entry −rd in this column means that the running disparity will be reversed by the applied code segment, and the entry rd means that the running disparity will keep its state.

Table D.2 specifies a special handling for the coding of Dxx.7 symbols. There is a primary coding (Dxx.P7) and an alternative coding (Dxx.A7) for the 3B/4B code segment. The two alternatives are required to avoid a run length of five in combination with the previous 5B/6B code segment for the data symbols. Thus, the alternative Dxx.A7 code is used for D17, D18, and D20 if the running disparity is rd− and for D11, D13, and D14 if the running disparity is rd+.

Example Let's assume we want to transmit the data bytes 0xAA and 0x17 in this sequence. First, we separate these bytes into the 5B and 3B components that will then be mapped to their 5B/6B and 3B/4B codes. Since the bytes follow the bit sequence HGFEDCBA, the lower 5 bits, EDCBA, will be the source for the 5B/6B code section and the upper 3 bits, HGF, are the source for the 3B/4B code section. Thus, byte 0xAA = 10101010 is separated into the 5B section 01010 = 10 and the 3B section 101 = 5. Byte 0x17 = 00010111 is separated into 10111 = 23 and 000 = 0. With the binary representation of this separation, 0xAA will be coded by D10.5 and 0x17 will be coded by D23.0. Using Tables D.1 and D.2, these codes are translated into their 5B/6B and 3B/4B codes. Since the coding can vary depending on the running disparity, a start value for the running disparity has to be assumed. With a negative running disparity (rd−) as starting disparity, D10.y is coded to 010101 and the running disparity keeps its state. The code for Dxx.5 is 1010 independently of

Table D.1
5B/6B Data Character Coding

Inputs		abcdei outputs		rd'	Inputs		abcdei outputs		rd'
Dx	**EDCBA**	**rd+**	**rd−**	**rd'**	**Dx**	**EDCBA**	**rd+**	**rd−**	**rd'**
D0	00000	011000	100111	−rd	D16	10000	100100	011011	−rd
D1	00001	100010	011101	−rd	D17	10001	100011		rd
D2	00010	010010	101101	−rd	D18	10010	010011		rd
D3	00011	110001		rd	D19	10011	110010		rd
D4	00100	001010	110101	−rd	D20	10100	001011		rd
D5	00101	101001		rd	D21	10101	101010		rd
D6	00110	011001		rd	D22	10110	011010		rd
D7	00111	000111	111000	rd	D23	10111	000101	111010	−rd
D8	01000	000110	111001	−rd	D24	11000	001100	110011	−rd
D9	01001	100101		rd	D25	11001	100110		rd
D10	01010	010101		rd	D26	11010	010110		rd
D11	01011	110100		rd	D27	11011	001001	110110	−rd
D12	01100	001101		rd	D28	11100	001110		rd
D13	01101	101100		rd	D29	11101	010001	101110	−rd
D14	01110	011100		rd	D30	11110	100001	011110	−rd
D15	01111	101000	010111	−rd	D31	11111	010100	101011	−rd

Table D.2
3B/4B Data Character Coding

Inputs		fghj outputs		rd'
Dx	**HGF**	**rd+**	**rd−**	**rd'**
Dxx.0	000	0100	1011	−rd
Dxx.1	001	1001		rd
Dxx.2	010	0101		rd
Dxx.3	011	0011	1100	rd
Dxx.4	100	0010	1101	−rd
Dxx.5	101	1010		rd
Dxx.6	110	0110		rd
Dxx.P7	111	0001	1110	−rd
Dxx.A7	111	1000	0111	−rd

the running disparity and the disparity is kept. Thus, D10.5 with a starting negative running disparity is coded to 0101011010 and the running disparity is kept negative. This negative running disparity selects 111010 as coding for D23.y. The code for D23.y reverses the running disparity from negative to positive. Thus, Dxx.0 is coded with the rd+ representative to 0100 and the running disparity again is reversed from positive to negative. With this, the overall 10B code for D23.0 in our case is 1110100100 and the running disparity overall stays negative.

As mentioned previously, the 8B/10B code space contains distinct control characters Kxx.y that follow a different code mapping than the data characters. Overall, the 8B/10B coding scheme defines 12 control characters K28.0–K28.7, K23.7, K27.7, K29.7, and K30.7. Three of these special symbols (K28.1, K28.5, and K28.7) have a run length of five. This unique property makes them suitable for synchronization to the data stream because they can be identified easily. Characters that allow such an easy identification and synchronization are also called comma characters or comma symbols.

The use of the K28.7 comma character has to follow some special rules in order to avoid ambiguities during comma character identification. For example, it can only be transmitted isolated from other K28.7 characters and there are restrictions in regard to data character combinations because these could create comma characters consisting of two adjacent sections of two characters. For completeness, Tables D.3 and D.4 show the code mapping for all control characters defined in the 5B/6B and 3B/4B coding schemes.

Table D.3
5B/6B Command Character Coding

Inputs		abcdei outputs		rd’
Kx	**EDCBA**	**rd+**	**rd−**	**rd’**
K23	10111	000101	111010	−rd
K27	11011	001001	110110	−rd
K29	11101	010001	101110	−rd
K30	11110	100001	011110	−rd
K28	11100	001111	110000	−rd

D.3 128B/130B Coding

The 128B/130B encoding scheme was defined and introduced with the third generation of the PCI Express (PCIe) standard [2]. 128B/130B encoding is used for the 8 Gbps and higher data rate modes of PCIe. In fact, 128B/130B

Table D.4
3B/4B Command Character Coding

Inputs		fghj outputs		rd'
Dx	HGF	rd+	rd−	rd'
Kxx.0	000	0100	1011	−rd
Kxx.1	001	1001	0110	rd
Kxx.2	010	0101	1010	rd
Kxx.3	011	0011	1100	rd
Kxx.4	100	0010	1101	−rd
Kxx.5	101	1010	0101	rd
Kxx.6	110	0110	1001	rd
Kxx.7	111	1000	0111	−rd

is less of a typical encoding mechanism, but more of a packetization and scrambling definition that is applied on a per-lane basis for PCIe interfaces that can consist of multiple high-speed serial lanes. One 128B/130B packet starts with a 2-bit synchronization header (sync header) that is followed by payload data. The sync header content defines the payload type that follows the header. A sync header "10" indicates that the payload is a data block that consists of 16 bytes (symbols). A sync header "01" is used if the payload is an ordered set block. Ordered set blocks have a length of 128 bits with the exception of the skip ordered set block that can vary in length. Ordered set blocks provide a communication mechanism on the physical layer between link partners to implement low level functions like link training, lane-to-lane deskewing, frequency offset compensation, and others without the need to access the higher protocol layers.

The transmission of 128B/130B encoded data happens with the LSB per symbol being transmitted first. Also, the LSB of the sync header is transmitted before its MSB. For the 16 byte payload that consists of symbols S15 to S0 with S0 being the least significant symbol, S0 is transmitted first and S15 is transmitted last. If data is transmitted on a multilane implementation, the sync header has to occur on all lanes in parallel. The payload data then is spread per symbol sequentially over the lanes. Figure D.2 shows an example of how a data stream is transmitted over a single lane interface and a multilane interface.

D.3.1 128B/130B Ordered Set

As mentioned above, ordered sets implement a low level communication mechanism for the physical layer. In multilane configurations, identical ordered sets have to be transmitted on all lanes in parallel. Some of the

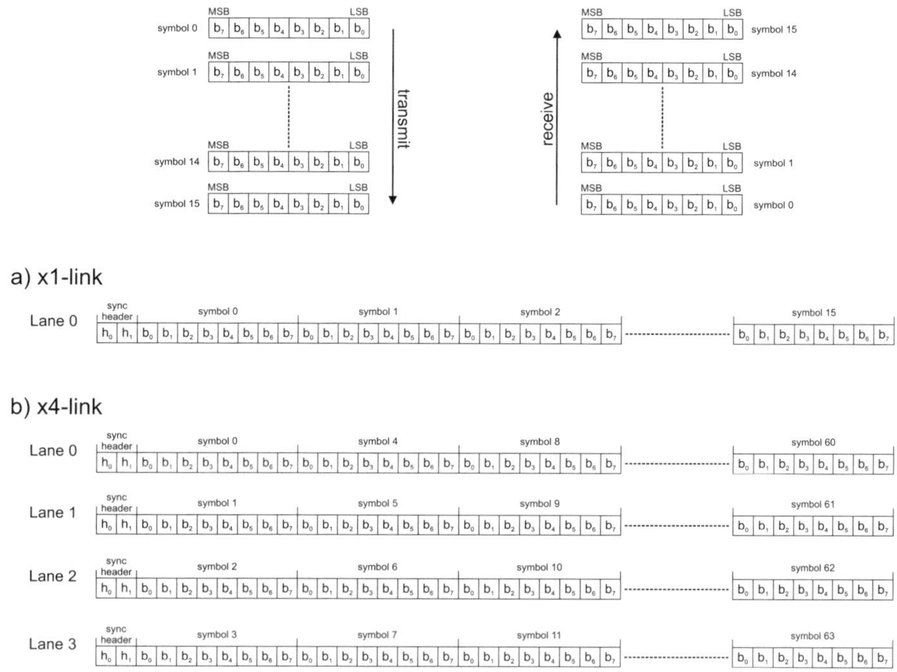

Figure D.2 Data transmission concept for 128B/130B encoding using (a) a single lane interface and (b) a multilane interface.

most basic functions that are implemented by ordered sets are 128B/130B block alignment between lanes, frequency offset compensation with skip ordered sets and control of the configuration and initialization of the links for normal operation. Of all ordered sets, the skip ordered set that is used for frequency offset compensation has the unique characteristic that a transmitter or a receiver can add or remove four symbols to adjust the transmitted data amount to a potential frequency offset between transmitter and receiver. Due to the addition or removal of four symbols per transmitter and receiver, the final length of a skip ordered set can vary between 8 and 24 symbols on a four symbol grid distance. The variable length of the skip ordered set has the consequence that a special skip symbol (SKP_END) has to identify the location of the end of the skip ordered set four symbols before the end of the skip ordered set.

D.3.2 128B/130B Data Block

The payload of a data block can consist of higher level protocol packets (Data Link Layer Packets [DLLP] or Transaction Layer Packets [TLP]) and framing tokens. Framing tokens indicate the start of higher level protocol packets, the end of a data stream as well as nullified TLPs and idle states of the link. The framing tokens that are defined with 128B/130B encoding are listed in Table D.5.

Table D.5
128B/130B Framing Tokens

Token	Description
IDL	Logical Idle - this 1-symbol token is transmitted when no other data is transmitted.
SDP	Start of DLLP - this 2-symbol token is followed by the Data Link Layer Packet (DLLP) data.
STP	Start of TLP - this 4-symbol token contains the 12-bit TLP number and the length of the TLP as well a framing CRC and farming parity bit. The token is followed by the Transaction Layer Packet (TLP) data.
EDB	End Bad token - this 4-symbol token confirms that the preceding TLP was nullified.
EDS	End of data stream - this 4-symbol token indicates that the next block will be an Ordered Set Block.

The token that indicates the start of a TLP (STP token) contains the length of the TLP as well as a 4-bit frame CRC (FCRC) and a frame parity bit (FP). The FCRC is generated based on the bits in the STP token that contain the length of the TLP to be transmitted. The parity bit is calculated based on the FCRC and also the bits that define the length of the TLP. FCRC and FP both are used to implement a low level error detection mechanism that can detect 3-bit errors on the physical layer.

D.3.3 128B/130B Scrambling

For 128B/130B coded data, each lane is scrambled and descrambled according to the polynomial of (D.1). The scrambling rules exempt the sync headers and most of the ordered sets from scrambling. Also, loopback slaves (devices configured in far end loopback mode) are not allowed to descramble and rescramble the data that is looped back.

$$G(x) = x^{23} + x^{21} + x^{16} + x^8 + x^5 + x^2 + 1 \tag{D.1}$$

D.4 Scrambling

Scrambling is a methodology to create a random distribution of zeros and ones in the transmitted data stream. When scrambling is applied, the transmitter performs a scrambling operation before the data is transmitted and the receiver reverses this operation in a descrambler before the data is handed over to the following operational device blocks for further processing. The purpose of such a randomization of the data stream is to reduce the likelihood for long run lengths to occur and to reduce repetitions of data sequences within the transmitted data stream.

With scrambling, average run length reduction can be achieved without generating data traffic overhead as it is the case with 8B/10B coding, for example, where a byte that is to be transmitted needs to be expanded to a 10-bit word causing a data transmission overhead of 25%. Scrambling has other challenges (for more detailed information, refer to [3]).

The reduction of data repetitions due to scrambling has the desired effect to spread out the spectrum of the energy transmitted with a data stream and thus reduces the radiated EMI energy at single distinct frequencies. One extreme example for this would be the unscrambled transmission of a data pattern that is a bit clock. This would create a peak in the frequency spectrum of the transmitted data at the frequency representing the data rate of the transmitted data. With scrambling of the clock pattern, such a peak is reduced and the energy originally contained in the peak is transferred to other frequency bins as shown in Figures D.3 and D.4.

The functionality of scramblers and descramblers can be modeled best by a linear feedback shift register (LFSR) as shown in Figure D.5 for a very simple scrambling/descrambling operation. It has to be noted that this block diagram is very basic and in a real system additional blocks are needed to avoid desynchronization.

The similarities of the scrambler/descrambler structure to an LFSR-based PRBS generator are obvious and should be no surprise to the reader since both circuits have the goal to generate random data. The only difference is that for the scrambler/descrambler, the generated data depends on data fed into the LFSR and the seed of the LFSR, while for a PRBS generator, the output data only depends on the seed.

Real scrambler/descrambler implementations usually do not use LFSRs since they would have to run at the speed of the transmitted data, which is a challenge for high-speed interfaces. Instead, the LFSR functionality is transferred into a parallel architecture consisting of a combinatorial network of exclusive OR (XOR) gates (operator symbol $\oplus$) [4].

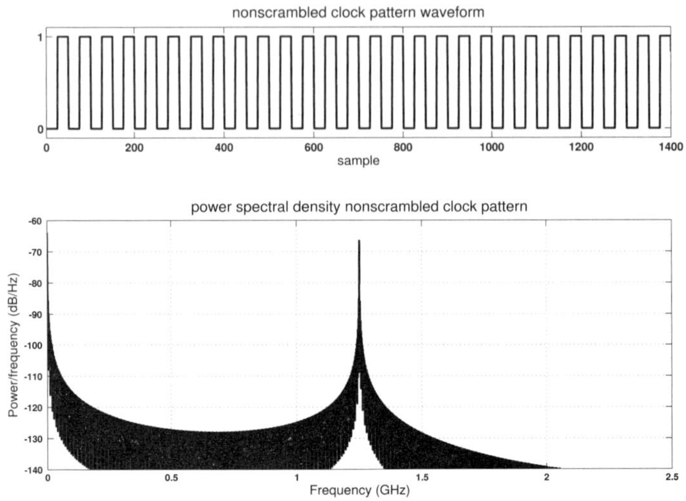

Figure D.3 Nonscrambled bit clock pattern waveform and associated power density spectrum.

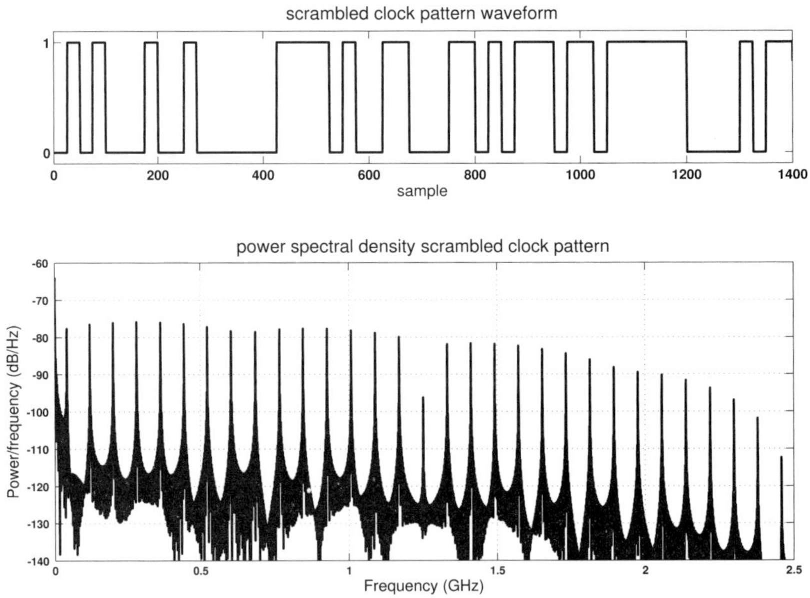

Figure D.4 Scrambled bit clock pattern waveform and associated power density spectrum.

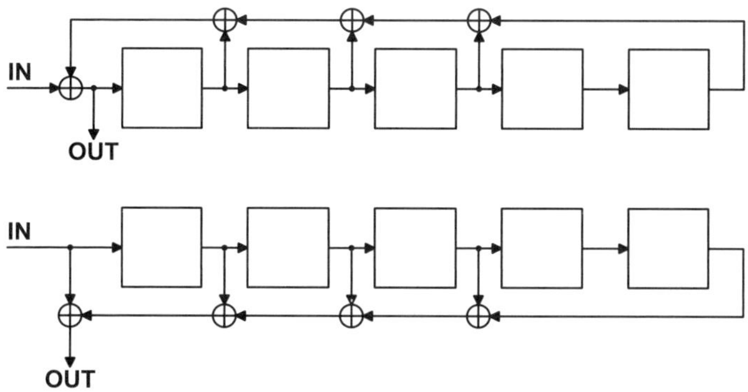

Figure D.5 Basic block diagram example of a scrambler (top) and descrambler (bottom) implementation.

D.5 Error Detection

With increasing data rates and shrinking parameter margins for the transmission of high-speed signals, methodologies to detect errors that occur during the data transmission start to gain traction in some of the high-speed I/O standards. For high-speed I/O interfaces that are designed for in-system or local communication, deploying error detection usually is sufficient because retransmission of corrupted data is affordable. For long-haul communication standards, error correction schemes are applied because the latencies caused by retransmissions are not tolerable. In this appendix we will restrict the discussion to error detection methodologies that are required to get a better understanding of the applications discussed in this book. If the reader is interested in more detailed information on error correction, it can be found in [5], for example.

D.5.1 Parity Bits

The most simple form of an error detection mechanism probably is the use of parity bits that are attached to each entity of bits on which they are based and are then transmitted together with this entity. The calculation of a parity bit is simply a bitwise recursive XOR operation on a set of bits of the data stream (e.g., one data word, a byte, and so forth) as shown in Figure D.6. Since the XOR function represents a modulo-2 operation on binary data, the result of this operation indicates whether the analyzed set of bits contains an even or odd number of ones. If the receiver of the data also builds a parity bit over the same set of bits including the received parity bit itself, it can detect an error if the result of that operation is not zero. With this parity bit based error detection, all errors that affect an odd number of bits can be detected and data retransmission can be requested. It is obvious that the error detection rate of parity bit based error detection is limited and that it does not work if an even number of bits are affected by errors. While this is sufficient for applications where single bit errors are expected as the main failure type (like in early times on PC main memory components), it is not necessarily suited for state-of-the-art, high-speed I/O interfaces where distortions often affect multiple bits due to the short bit times used. Thus, usually more sophisticated checksum-based error detection mechanisms are applied for this type of interface.

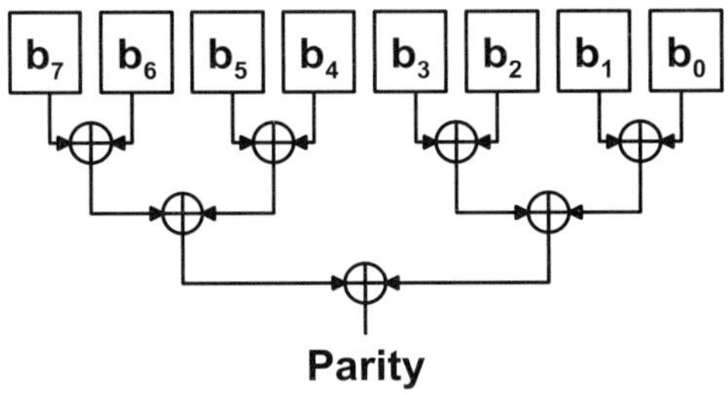

Figure D.6 Parity bit generation with recursive XOR gates.

D.5.2 Checksums

A logical step for a checksum-based error detection scheme is to extend the parity bit methodology to a parity word. An example of such an approach is the bit interleaved parity N (BIP-N) error detection scheme [6]. Here, a section (e.g., one packet) of the data stream is split into N-bit words and the parity bit calculation is done over corresponding bits within these words for all data words. This operation will result in a N-bit parity checksum.

Of course, the BIP-N error detection scheme has the same limitations as the standard parity bit approach regarding an even number of errors on the bits forming the basis for one parity bit within the parity checksum. However, the advantage of building one parity bit over several words instead of one parity bit for each word is that the bits that are used for the parity calculation are timely separated from each other. Thus, the likelihood that distortions during data transmission affect multiple bits that are used for the calculation of a single parity bit is lower with the BIP-N approach than with the classical parity bit error detection.

Besides the nonsensitivity for an even number of errors, parity-based approaches also have the issue that they are not sensitive to changes in the sequence of the data on which they calculate the parity. Thus, checksum error detection methodologies that are used in real applications predominantly make use of position-dependent checksum approaches that do not have the disadvantages of the previously discussed parity-based algorithms. While there are quite a few possible implementations of position-dependent error detection methodologies, we want to discuss the most prominent representative of this group, which is the cyclic redundancy check (CRC) [7–9].

The CRC calculates a checksum based on a polynomial division on subsets of the data to be transmitted by a generator polynomial. The remainder of this division serves as checksum that is transmitted together with the data subsets. The receiver also calculates the division remainder between received data and generator polynomial and compares its local remainder with the received CRC checksum. If these are not the same, an error is present on the received data or checksum. The identification of a transmission error by the receiver can be simplified for an n-bit CRC checksum by appending n zeros to the original dividend before the polynomial division by the sender. In this case, the receiver can append the received checksum to the received data before doing its division. If the remainder of the polynomial division in the receiver is not zero, there was a data transmission error.

The selection of the generator polynomial is a critical step for CRC error detection because the polynomial defines what types of errors can be detected. Applications that use CRC checksums usually have polynomials defined in their standards that are selected to best fit to the most likely errors that can occur in the respective environment of that application. An incomplete selection of CRC generator polynomials is shown in Table D.6. As a side note on this table, we want to highlight that the previously discussed parity bit error detection approach also can be implemented as a CRC checksum of the length one using the polynomial $x + 1$.

Table D.6
Selection of CRC Generator Polynomials

Name	Polynomial	Application
CRC-1	$x + 1$	Parity bit
CRC-5-USB	$x^5 + x^2 + 1$	USB
CRC-8-ATM	$x^8 + x^2 + x + 1$	ATM HEC, GDDR5
CRC-16-CCITT	$x^{16} + x^{12} + x^5 + 1$	CDMA, Bluetooth, IrDA, MMC, SD

In order to explain how a polynomial division can be done easily in a real implementation, we represent both the payload data as well as the generator polynomial in a binary form. The binary representation of a polynomial is done in a way that the single bits of the binary data form the coefficients of the polynomial according to the following example:

$$
\begin{aligned}
M &= 10100101 \\
\Leftrightarrow M &= 1x^7 + 0x^6 + 1x^5 + 0x^4 + 0x^3 + 1x^2 + 0x^1 + 1x^0 \\
&= x^7 + x^5 + x^2 + 1
\end{aligned}
$$

The division of the two numbers then works exactly as a polynomial division. An important fact to notice is that the modulo-2 subtraction we have to apply during the division process is a simple bitwise exclusive OR operation. An example for such a division is shown in Figure D.7. Since we are not interested in the division result, but only the remainder of the division, we do not keep track of the actual result in this example.

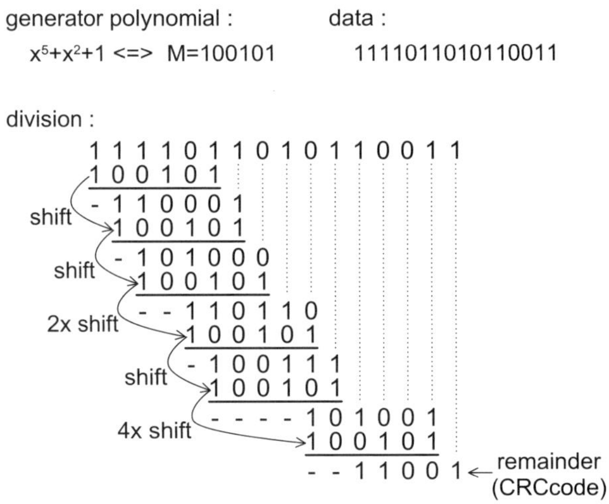

Figure D.7 Polynomial division example.

From this example it becomes obvious that the polynomial division just consists of XOR operations with shifts of the generator polynomial versus the data polynomial. From a hardware point of view this functionality can be implemented using a linear feedback shift register with a seed value of zero. This linear feedback shift register represents the CRC generator polynomial which is fed serially by the data for which the CRC checksum is to be built. After all data bits are consumed, the shift register contains the CRC checksum. The LFSR representation of the CRC generator polynomial places its XOR gates after the CRC bits that have a coefficient of one in the generator polynomial as shown for the CRC-8-ATM generator polynomial in Figure D.8.

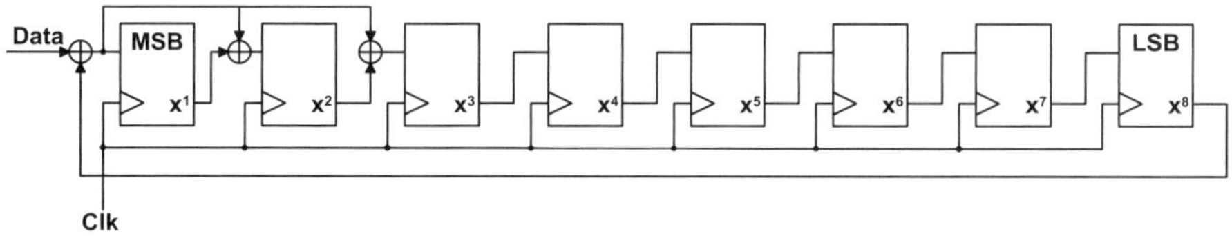

Figure D.8 LFSR representation of CRC-8-ATM polynomial.

As the reader can imagine, the LFSR implementation to calculate CRC checksums in hardware raises some significant challenges for high-speed interfaces due to the raw speed at which the shift operations and XOR operations between shift stages have to operate. If the LFSR operation is analyzed in detail, it becomes obvious that the CRC checksum is generated by repeated XOR operations on the data bits and the seed initially present in the shift register. Thus, the functionality of the LFSR can be mapped to a combinatorial XOR network to which the bits of the data word can be applied in parallel. Since the seed of the LFSR is set to zero and the rule $A \oplus 0 = A$ applies, the seed values do not have to be considered as inputs for the combinatorial network. A methodology to transfer the LFSR representation of a CRC generator polynomial to a combinatorial XOR network is given in [10]. The same transformation from an LFSR representation to pure XOR terms also can be applied for scrambling LFSRs.

References

[1] A. Widmer and P. Franaszek, "A DC-Balanced, Partitioned-Block, 8b/10B Transition Code," *IBM Journal of Research and Development*, vol. 17, Sept. 1983.

[2] PCI-SIG, *PCI Express Base Specification Revision 3.0*, Nov. 2010.

[3] Y. Takasaki, *Digital Transmission Design and Jitter Analysis*. Norwood, MA: Artech House, 1991.

[4] D. R. Stauffer, J. T. Mechler, M. A. Sorna, K. Dramstad, C. R. Ogilvie, A. Mohammad, and J. D. Rockrohr, *High Speed Serdes Devices and Applications*. New York: Springer, 2008.

[5] H. S. Warren, "Hacker's Delight – Error Correcting Codes," 2003. http://www.hackersdelight.org/ecc.pdf.

[6] American National Standards Institute, *American National Standard for Telecommunications - Digital Hierarchy Optical Interface Rates and Formats Specification*, 1988. ANSI T1.105-1988.

[7] W. W. Peterson and D. T. Brown, "Cyclic Codes for Error Detection," *Proceedings of the IRE*, Jan. 1961.

[8] A. S. Tanenbaum, *Computer Networks*. Simon & Schuster, 1983.

[9] H. S. Warren, "Hacker's Delight – Cyclic Redundancy Check," 2003. http://www.hackersdelight.org/crc.pdf.

[10] Cypress Semiconductor Corporation, *Parallel Cyclic Redundancy Check (CRC) for HOTLink*™, Mar. 1999. AN1089.

E

Time-Domain Reflectometry and Time-Domain Transmission (TDR/TDT)[1]

This appendix presents some important concepts regarding time-domain reflectometry(TDR) and time-domain transmission (TDT).

TDR/TDT is a tool that has been critical in electrical engineering from its beginning and continues to be of significant importance today [1]. Introductory references on TDR/TDT for electrical characterization are [2–7]. The evaluation of a printed circuit board (PCB) test fixture benefits from a number of TDR and TDT measurements to validate and determine performance with the following being the most common:

- Measuring the characteristic impedance of a PCB signal trace (TDR);
- Measuring and identifying discontinuities on a signal path (TDR);
- Measuring the step response of a signal path (TDT);
- Obtaining a model of a transmission line (TDR/TDT);
- Determining the location and magnitude of near-end and far-end (NEXT/FEXT) crosstalk (TDT).

Figure E.1 shows a very simple block diagram of a TDR/TDT instrument. The TDR/TDT timing resolution depends mainly on the step generator rise time and the sampler bandwidth. The TDR/TDT information can also be obtained with the conversion to the time-domain of the S-parameters measured with a vector network analyzer (VNA). A typical TDR instrument has the step generator and sampler connected together inside the instrument for ease of use; however, a high-performance instrument with very

[1]This appendix was written in collaboration with Heidi Barnes.

fast rise times may require separate specialized modules as shown in Figure E.2 [1].

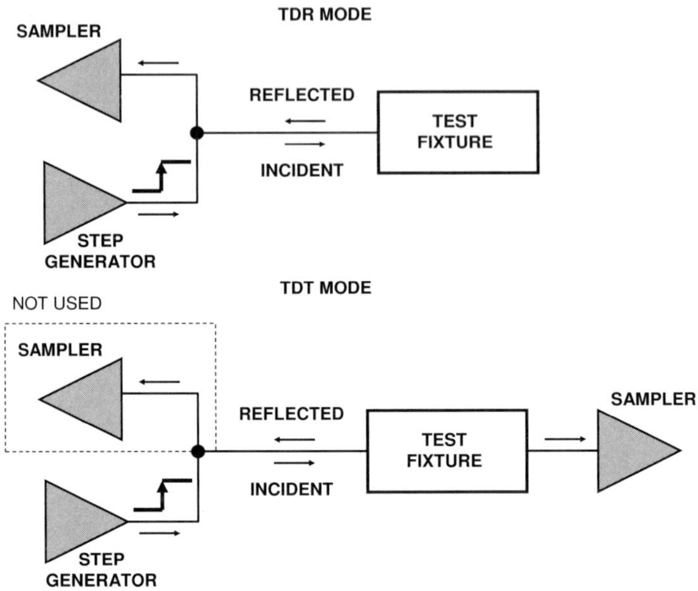

Figure E.1 Block diagram of a TDR/TDT instrument.

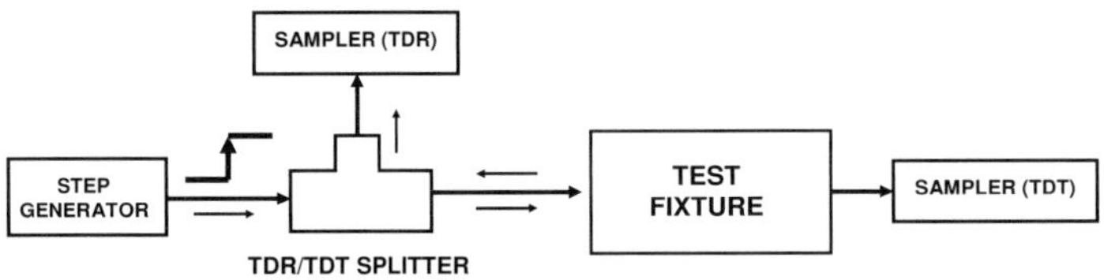

Figure E.2 Block diagram of a TDR/TDT instrument with independent step generator and sampler modules.

E.1 TDR

When performing a TDR measurement, a step waveform is sent into the transmission line path of interest, such as a test fixture. This step edge will travel the length of the path and reflect at the end if there is an open (e.g., at the empty DUT socket) or a short (DUT pins including ground pins shorted together), and return to the beginning of the path where the TDR sampler can then measure the reflected waveform. Reflections will also occur at any changes in impedance (e.g., a change on the PCB trace width or a discontinuity like a via or a ground plane discontinuity interrupting the return current path) as shown in Figure E.3.

The polarity of the reflection provides additional information on the type of discontinuity that was encountered. Higher series inductance than the characteristic impedance or an open provides a positive voltage reflection. A parallel capacitance or a short provides a negative voltage reflection as shown in Figure E.3. The equivalent excess inductance or capacitance of a discontinuity can be calculated from the step response by integrating the difference between the reflected response waveform and a reference waveform without the excess inductance or capacitance using (E.2) and (E.3). This approach is more accurate than comparing peak reflections [8]. This is not to be confused with the transmission line equation used for calculating total capacitance or inductance of a signal path as shown in (E.4) and (E.5).

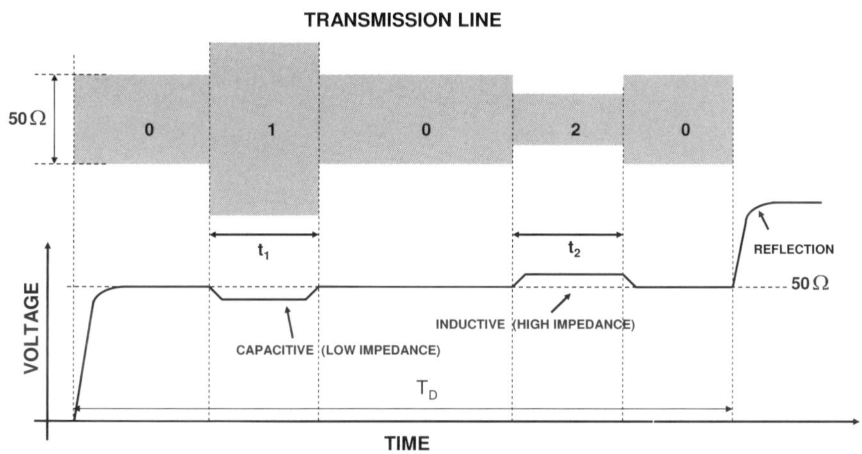

Figure E.3 Diagram showing the effect of discontinuities on the transmission line on the response to a step waveform (TDR).

E.1.1 Measuring the Impedance of a Trace with a TDR

One of the first measurements that should be done on the high-speed signal paths of a PCB test fixture is to verify the signal trace impedance with TDR [9]. It is important with this measurement to use the built-in calibration support of the TDR instrument to set the 50 Ω reference of the TDR measurement. Figure E.4 presents an example of an impedance measurement with a TDR. Transmission line losses and discontinuity reflections will mask the true impedance of the transmission line the further the TDR travels into a structure and a "peeling algorithm" may be required to analyze the impedance as the TDR reflection gets further from the input port [10]. However, unless the PCB transmission line is extremely low loss, empirical data suggests that the peeling algorithm has little effect on the impedance measurement itself. In fact for real and complex signal paths with significant loss and a mixture of PCB and cable signal paths, analyzing a TDR plot is far from trivial [11, 12].

Computing the Excess Capacitance or Inductance

Define τ to be the integral of the portion of the TDR reflection that contains the discontinuity of interest, and $\rho(t)$ is the reflected TDR voltage waveform:

$$\tau = \int_{t_0}^{t_1} \rho(t)dt \tag{E.1}$$

If $\tau > 0$, there is excess L:

$$L = 2Z_o\tau\frac{(Z_0 + Z_N)^2}{4Z_0 Z_N} \tag{E.2}$$

If $\tau < 0$, there is excess C:

$$C = -\frac{2\tau}{Z_0}\frac{(Z_0 + Z_N)^2}{4Z_0 Z_N} \tag{E.3}$$

Z_0 is the impedance of the non-50 Ω system where the integral starts and is used to normalize the integral to the height of the step incident on the discontinuity. Z_N is the nominal impedance of the TDR system (usually 50 Ω).

Computing the Total Capacitance and Inductance

The total capacitance and inductance for an ideal (lossless) transmission line can be found from the TDR measured delay and impedance (Z) using the following equations:

$$C(\text{Total}) = \frac{t(\text{delay})}{Z} = \frac{t(\text{TDR round-trip delay})}{2}\frac{1}{Z} \tag{E.4}$$

$$L(\text{Total}) = t(\text{delay})\, Z = \frac{t(\text{TDR round-trip delay})}{2} Z \tag{E.5}$$

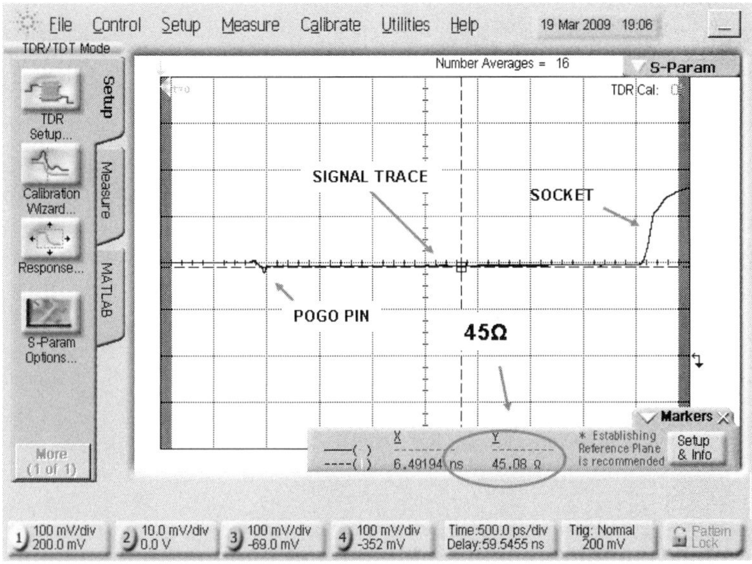

Figure E.4 Example of an impedance measurement on an ATE test fixture using a TDR.

E.1.2 Measuring the Round-Trip Delay of a Signal Trace

Another important measurement that can be done with a TDR on a test fixture is to determine the round-trip delay between the DUT socket and the test fixture pogo via connection. This measurement is especially important when evaluating the phase matching or time skew between different signal traces. Figure E.5 presents an example of a round-trip delay measurement with a TDR. It is important to remember that the TDR will always display the round-trip delay. A simple approximation for the signal path delay is to divide the round-trip delay by two. This approximation is only exact for ideal (lossless) signal paths [13].

E.1.3 Measuring Discontinuities on a Signal Path with a TDR

A TDR is extremely useful in evaluating the discontinuities on a test fixture and is one of the few methods that enables one to identify locations for signal integrity improvement. These discontinuities can be interlayer vias, relays, SMT footprints, and so on. The ability to see the details of a discontinuity is directly related to the rise time of the TDR pulse generator (the TDR step waveform is approximated by using a pulse waveform). Faster rise times provide finer time resolution, enabling the separation and measurement of

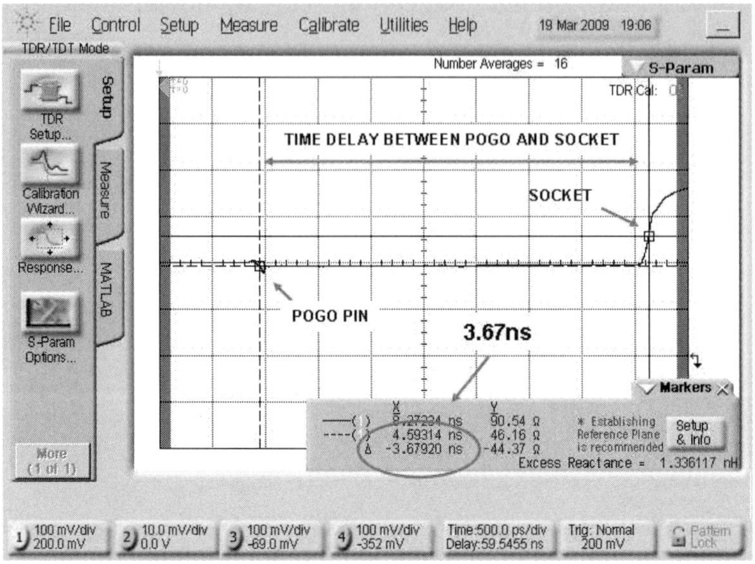

Figure E.5 Example of a round-trip delay measurement on an ATE test fixture with a TDR.

closely spaced impedance discontinuities. Figure E.6 shows an example of measuring test fixture discontinuities using a TDR.

E.1.4 Measuring the Return Loss with a TDR

With the appropriate calibration techniques and software, it is possible to obtain a frequency-domain return loss measurement (S11) through a TDR measurement [14–16]. The magnitude of the spectral content of a TDR step falls off as 1/f so these types of measurements often require a significant amount of averaging to improve the S/N ratio at the higher frequencies. Most TDR instruments have this algorithm integrated into the available software tools. References [6, 17] provide detailed discussions of the differences between a TDR and a VNA approach for signal path characterization measurements.

E.2 TDT

The TDR provides valuable information on the test fixture, but what really matters for signal integrity is the signal that is received or transmitted by the DUT, which can be investigated with a TDT measurement. When performing a TDT measurement, a step waveform is sent through the test fixture signal

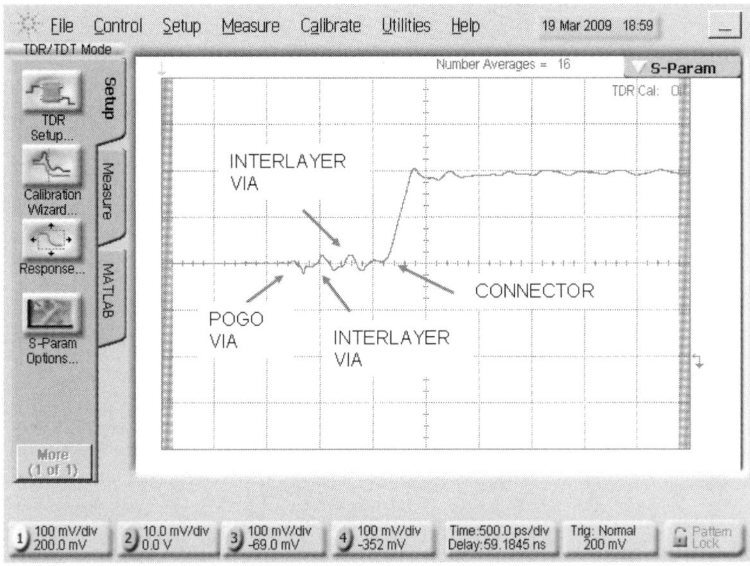

Figure E.6 Example of discontinuities measurements using a TDR.

trace and is measured at the other end of the signal trace with a sampler. The loss of the signal trace, impedance changes, and discontinuities will have an effect on the shape of the step after traveling through the signal trace as shown in Figure E.7. Note that unlike the TDR, for a TDT measurement it is necessary to have access to both ends of the PCB signal trace being measured, which is not always easy in the case of a test fixture. See Appendix H for an in-depth discussion of this topic.

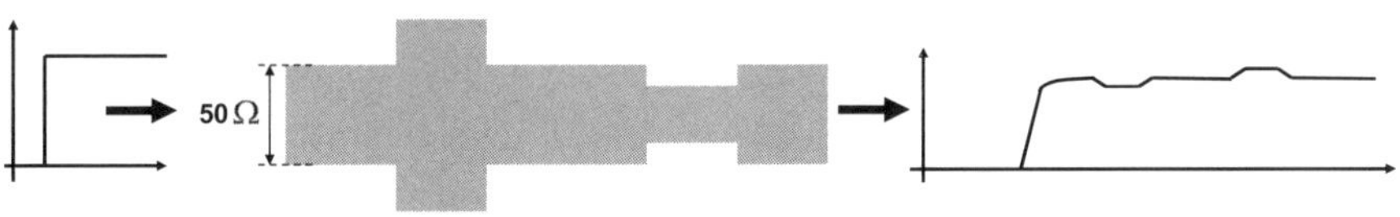

Figure E.7 Diagram showing the effect of discontinuities on the transmission line on the response to a step waveform (TDT).

E.2.1 Measuring the Step Response

The key result from a TDT measurement is the measured step response of the test fixture signal path. If one assumes the signal trace transmission line to be a time invariant linear system, then the system response to a known stimulus step will provide a complete description of the test fixture signal path performance [18]. Figure E.8 shows an example of the measured step response for a test fixture signal trace.

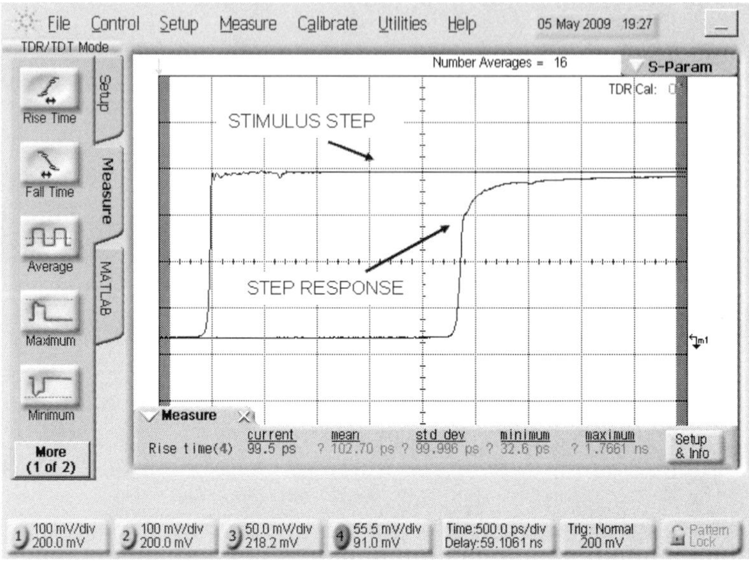

Figure E.8 Measuring the step response of a test fixture transmission line.

The rise time of the measured step response is related to the bandwidth of the test fixture signal trace being measured. Short rise times (faster slew rates) contain higher frequencies and a wider bandwidth. Accurate measurement of the step response relies on the intrinsic rise time of the TDR module step being much smaller than the expected rise time at the end of the signal. This ensures that the measured rise time value is dominated by the bandwidth limitations of the signal trace under measurement and not the input step.

It is also possible to take the step response and use it to obtain a "synthesized" data eye diagram as shown in Figure E.9 [19]; it is synthesized because it is not directly measured but instead it is generated from the measured step response or frequency-domain S-parameters.

E.2.2 Measuring the Insertion Loss with a TDT

Similar to obtaining frequency-dependent reflection information (S11) from the TDR measurement, it is also possible to obtain the insertion loss measurement (S21) through a TDT measurement with the appropriate calibration techniques and software [14–16]. Observing the performance in the frequency-domain can provide insight into signal path resonances and bandwidth limitations.

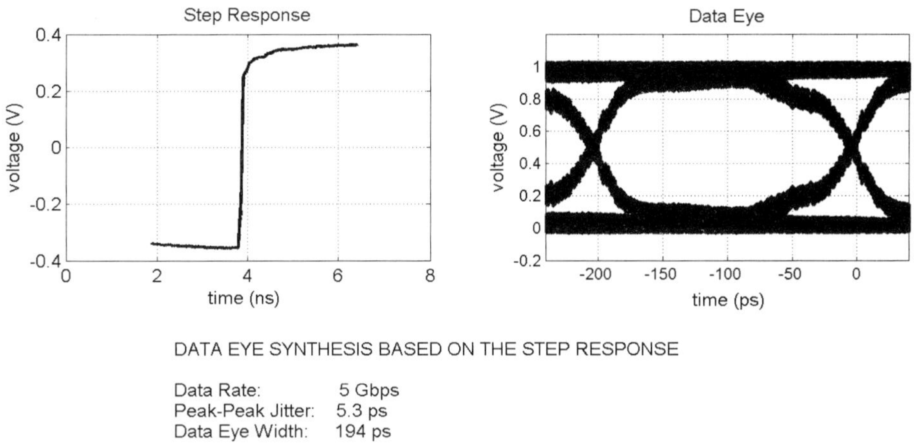

Figure E.9 Example of a synthesized data eye diagram from the step response of a test fixture measured with the ATE pin electronics (Courtesy of Advantest).

E.2.3 Measuring Crosstalk Using a TDT and an Extra Sampler

Another interesting use for TDT measurements is the ability to measure aggressor-to-victim crosstalk. Figure E.10 shows a block diagram of a measurement scenario that would require only an extra sampler to measure both the far-end crosstalk (FEXT) and the near-end crosstalk (NEXT) on a victim trace from an adjacent aggressor trace [2, 20].

Figure E.11 shows an example of a crosstalk measurement of a pogo via on an ATE system using the setup described in Figure E.10 [21].

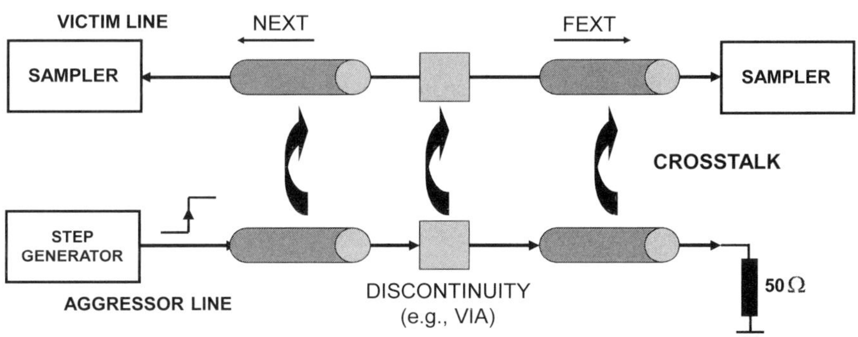

Figure E.10 Diagram showing a crosstalk measurement using a TDT signal source and two extra samplers.

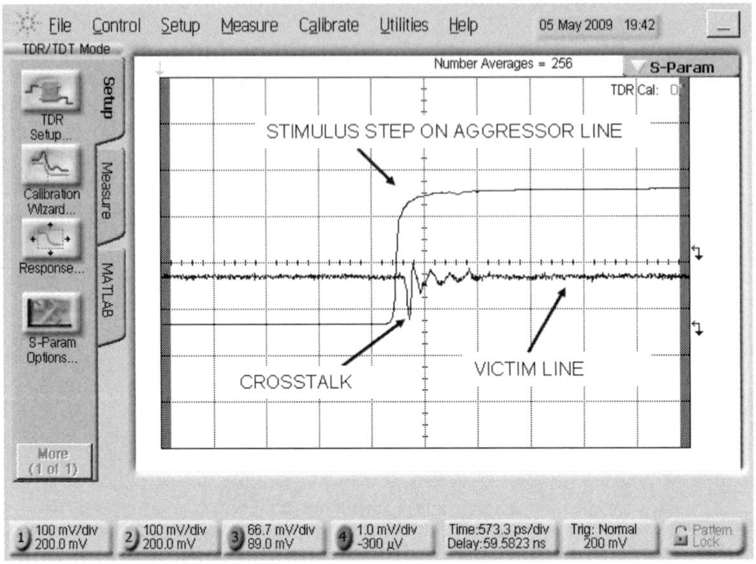

Figure E.11 Example of measuring the crosstalk between two adjacent pogo vias using a TDT instrument.

E.3 Differential TDR/TDT Measurements

As the name implies a differential TDR or TDT measurement requires the measurement of the difference of two signals and can be obtained by using two step sources to stimulate a given differential structure to determine its differential response. The differential stimulus is important for accurately capturing the effects of coupling throughout the signal path and the effects of common-mode to differential and differential to common-mode conversions. The TDR/TDT instrument as shown in Figure E.12 can drive a pair of traces in the even mode with a common signal that has the same polarity step on each trace or in the odd mode with a differential signal that has opposite polarity steps on each trace. This is needed for obtaining both the common and differential modes of the step response [6, 22].

Like a four-port S-parameter measurement (see Appendix F), a differential TDR/TDT measurement is also a four-port measurement that will result in 16 separate measurements [3] as shown in Figure E.13. Measurements in the frequency-domain with a VNA can also provide differential TDR and TDT information with the use of a multiport network analyzer.

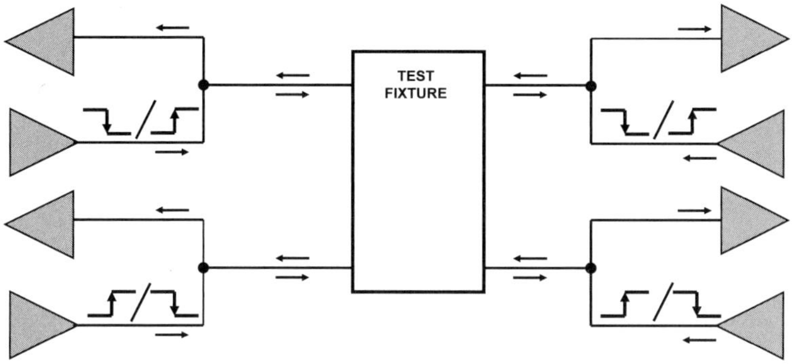

Figure E.12 A differential TDR/TDT measurement setup.

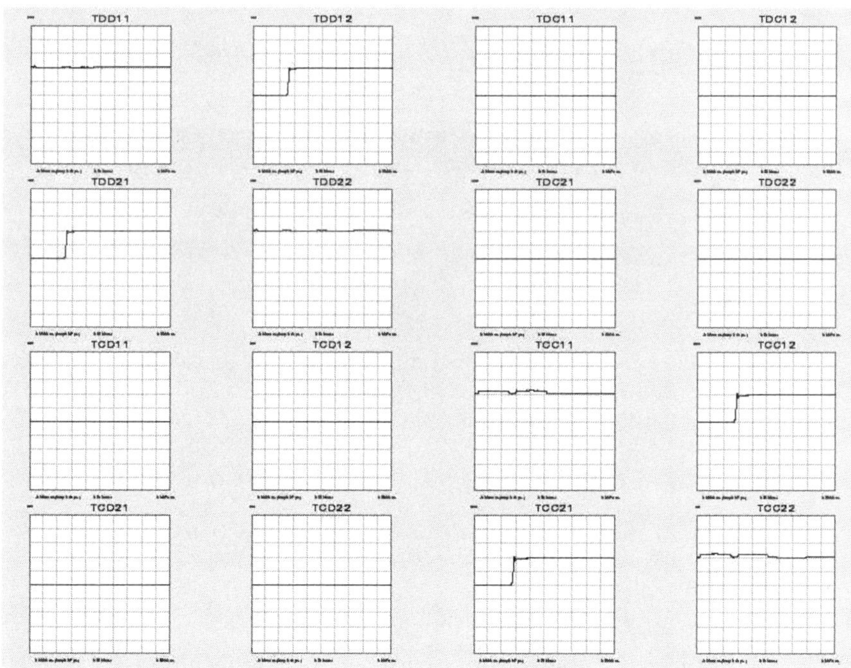

Figure E.13 Example of a four-port TDR/TDT measurement on a coupled differential line.

References

[1] D.-H. Han, B.-S. Xu, M. J. Choi, J. He, S. Gardiner, and C. Lee, "Realization of Ultra-Wideband High-Resolution TDR for Chip-Carrier Packages," *Proceedings of the ASME InterPACK*, 2005.

[2] Agilent Technologies, "Agilent Signal Integrity Analysis Series, Part 1: Single-Port TDR, TDR/TDT, and 2-Port TDR," *Application Note 5989-5763EN*, 2007.

[3] Agilent Technologies, "Agilent Signal Integrity Analysis Series, Part 2: 4-Port TDR/VNA/PLTS," *Application Note 5989-5764EN*, 2007.

[4] Agilent Technologies, "Time Domain Reflectometry Theory," *Application Note 1304-2*.

[5] D. I. Geoff Lawday and G. Edlund, *A Signal Integrity Engineer's Companion*. Upper Saddle River, NJ: Prentice-Hall, 2008.

[6] M. Resso and E. Bogatin, *Signal Integrity Characterization Techniques*. IEC, 2009.

[7] J. R. Andrews, "Time Domain Reflectometry (TDR) and Time Domain Transmission (TDT) Measurement Fundamentals," *Picosecond Labs Application Note AN-15*, 2004.

[8] D. J. Dascher, "Measuring Parasitic Capacitance and Inductance Using TDR," *Hewlett-Packard Journal*, 1996.

[9] Association Connecting Electronics Industries (IPC), "Characteristic Impedance of Lines on Printed Boards by TDR," *IPC–TM–650 2.5.5.7*, 2004.

[10] D. Miller, *Designing High-Speed Interconnect Circuits*. Intel Press, 2004.

[11] M. Steinberger, "TDR: Reading the Tea Leaves," *SiSoft, Advanced Signal Integrity Solutions*, 2012.

[12] I. Novak, Y. Li, E. Kunz, S. Paydavosi, L. Kocubinski, K. Hinckley, A. Nosovitski, N. Shannon, J. Miller, and G. Blando, "Determining PCB Trace Impedance by TDR: Challenges and Possible Solutions," *DesignCon*, 2013.

[13] W. Maichen, *Digital Timing Measurements*. New York: Springer, 2006.

[14] J. R. Andrews, "Time Domain Spectrum Analyzer and 'S' Parameter Vector Network Analyzer," *Picosecond Labs Application Note AN-16a*, 2004.

[15] Cherry Wakayama and Jeff Loyer, "Correlation between VNA and TDR/TDT Extracted S-Parameters up to 20 GHz," *White Paper*, 2005.

[16] V. Teppati, A. Ferrero, and M. Sayed, *Modern RF and Microwave Measurement Techniques*. Cambridge, UK: Cambridge University Press, 2013.

[17] Agilent Technologies, "Limitations and Accuracies of Time and Frequency Analysis of Physical Layer Devices," *White Paper 5989-2421*, 2005.

[18] J. R. Andrews, "Deconvolution of System Impulse Responses and Time Domain Waveforms," *Picosecond Labs Application Note AN-18*, 2004.

[19] J. Ren and K. S. Oh, "Multiple Edge Responses for Fast and Accurate System Simulations," *IEEE Transactions on Advanced Packaging*, vol. 31, Nov. 2008.

[20] Tektronix, "Time Domain Methods for Measuring Crosstalk for PCB Quality Verification," *Application Note*, 2003.

[21] B. B. Szendrenyi, H. Barnes, J. Moreira, M. Wollitzer, T. Schmid, and M. Tsai, "Addressing the Broadband Crosstalk Challenges of Pogo Pin Type Interfaces for High-Density High-Speed Digital Applications," *International Microwave Symposium*, 2007.

[22] M. M. McTigue and C. P. Duff, "Differential Time-Domain Reflectometry Module for a Digital Oscilloscope and Communications Analyzer," *Hewlett-Packard Journal*, 1996.

F

S-Parameters[1]

Scattering parameters (S-parameters) provide a powerful approach to analyze the performance of a test fixture and its individual components. S-parameters work by creating a black-box model that relates the voltage waves incident at each port of the model with the voltage waves that exit from the same ports. This appendix provides an overview of S-parameters, and for more detailed information on the definitions and conversions, [1–5] are good starting points for further reading.

For a two-port system like a single-ended transmission line, the S-parameters are described by the following equation:

$$\begin{bmatrix} V_1^-(f) \\ V_2^-(f) \end{bmatrix} = \begin{bmatrix} S_{11}(f) & S_{12}(f) \\ S_{21}(f) & S_{22}(f) \end{bmatrix} \begin{bmatrix} V_1^+(f) \\ V_2^+(f) \end{bmatrix} \tag{F.1}$$

where V_1^+ and V_2^+ are the waves traveling into the two-port network at ports 1 and 2 and V_1^- and V_2^- are the waves traveling out of the two-port network at ports 1 and 2. Note that the S-parameters matrix is defined at multiple frequency values, since in most applications there is a frequency band of interest and not a single value. That is the reason for the term (f) in the matrix.

The S-parameter equation is analyzed in more detail in the flow diagram in Figure F.1. The diagram represents the four different signal paths that a two-port system has and the corresponding S_{ij} terminology used to describe them. The "i" term is the network port being measured and the "j" term is where the input signal is located. The S_{21} insertion gain for a two-port network is the signal measured at port 2 when a voltage wave is input at port 1 and is also known as insertion loss for a passive system. The S_{11} term is the return loss at port 1, which corresponds to the signal reflected at port 1 when a voltage

[1]This appendix was written in collaboration with Heidi Barnes.

wave is input at port 1. The S-parameter measurement captures both phase and magnitude of the voltage waves to provide a complete electrical description of this two-port, black-box network for the mathematical case of a linear time invariant (LTI) system. This information can then be used to determine the performance of a system such as that shown in Figure F.1 where the ATE pin electronics connects to port 1, the DUT connects to port 2, and the two-port network would be the passive ATE test fixture with its LTI properties.

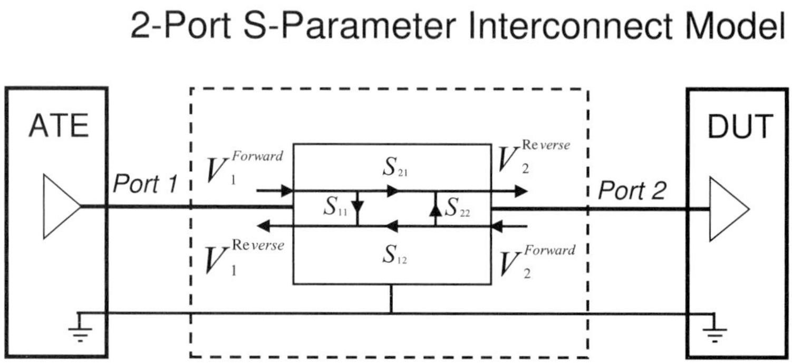

$$S_{21} = \frac{V_2^{Reverse}}{V_1^{Forward}} \Bigg|_{V_2^{Forward} = 0}$$ = ***Forward insertion gain with Port 2 terminated in matched Z_0***

$$S_{21} = \frac{V_1^{Reverse}}{V_2^{Forward}} \Bigg|_{V_1^{Forward} = 0}$$ = ***Reverse insertion gain with Port 1 terminated in matched Z_0***

$$S_{11} = \frac{V_1^{Reverse}}{V_1^{Forward}} \Bigg|_{V_2^{Forward} = 0}$$ = ***Input reflection coefficient with Port 2 terminated in matched Z_0***

$$S_{22} = \frac{V_2^{Reverse}}{V_2^{Forward}} \Bigg|_{V_1^{Forward} = 0}$$ = ***Output reflection coefficient with Port 1 terminated in matched Z_0***

Figure F.1 Block diagram of a two-port configuration that could correspond to a single-ended transmission line. Z_0 is the characteristic impedance of the measurement setup, which in most situations will correspond to 50 Ω.

It is possible to move from the frequency-domain representation of the S-parameters to a time-domain representation through Fourier transform techniques [6]. Figure F.2 is an example of a two-port S-parameter measurement for a PCB test fixture (see Appendix H on techniques to make this type of measurement). In the figure, the S-parameter data is also transformed to the time-domain for analysis of transition discontinuities, impedance, and step response. Note that the four S-parameter signal paths (S_{11},S_{12},S_{21},S_{22}) for the two-port network also have a corresponding time-domain path (T_{11},T_{12},T_{21},T_{22}). The S-parameter magnitude data is plotted here without the phase; however, the conversion to the time-domain is dependent on both the magnitude and phase information. With the ability to convert between the time and frequency-domains (and vice versa), it is also possible to use other techniques like gating for de-embedding [6].

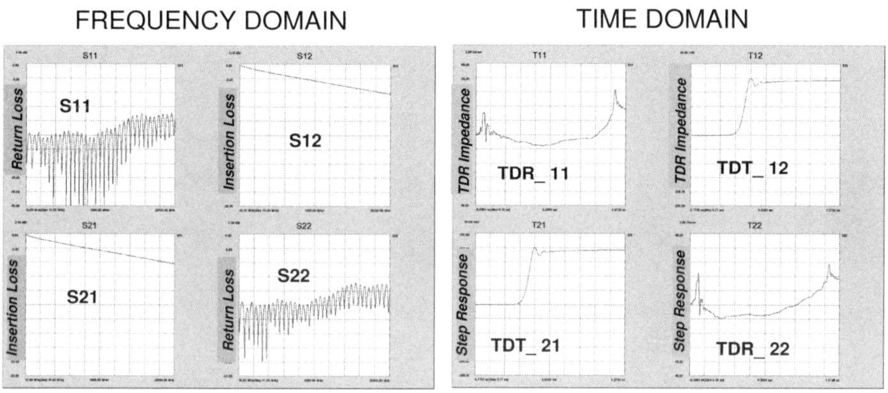

Figure F.2 Example of two-port S-parameter measurement and the corresponding conversion to time-domain information.

The simple two-port S-parameter network is a good starting point for measurements and analysis, but it is limited to a single mode of propagation. Most high-speed digital applications utilize differential signaling with multiple modes of propagation. The S-parameter network can easily be expanded to accommodate a four-port network as shown in Figure F.3. In this case there are 16 possible signal paths that show up in the four-port S-parameter matrix that is shown in (F.2). Figure F.4 shows an example of the measured S-parameters for a differential pair in an ATE test fixture. The same matrix would also describe the interaction between a victim and an aggressor signal path including the crosstalk between the two.

$$\begin{bmatrix} V_1^-(f) \\ V_2^-(f) \\ V_3^-(f) \\ V_4^-(f) \end{bmatrix} = \begin{pmatrix} S_{11}(f) & S_{12}(f) & S_{13}(f) & S_{14}(f) \\ S_{21}(f) & S_{22}(f) & S_{23}(f) & S_{24}(f) \\ S_{31}(f) & S_{32}(f) & S_{33}(f) & S_{34}(f) \\ S_{41}(f) & S_{42}(f) & S_{43}(f) & S_{44}(f) \end{pmatrix} \begin{bmatrix} V_1^+(f) \\ V_2^+(f) \\ V_3^+(f) \\ V_4^+(f) \end{bmatrix} \tag{F.2}$$

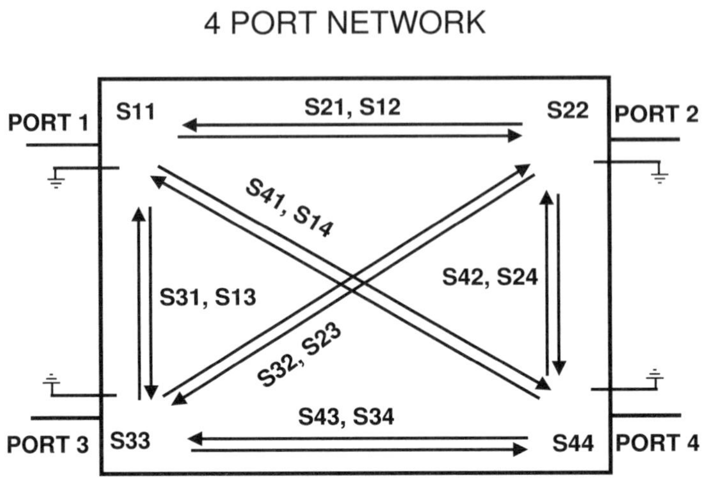

Figure F.3 Block diagram of a four-port system that could correspond to a differential transmission line.

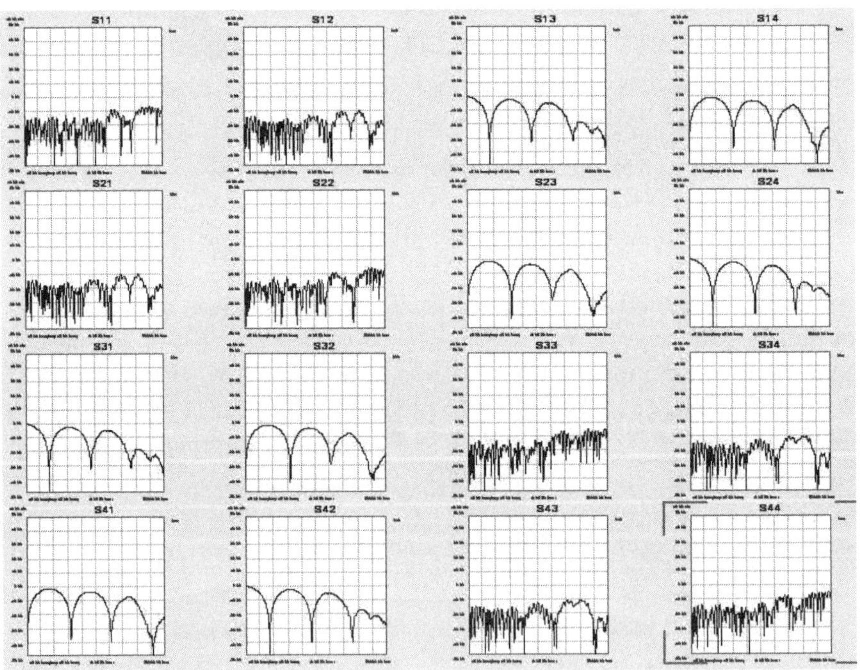

Figure F.4 Measured S-parameters from a four-port system (a differential pair on a test fixture).

If one groups the through paths so that they are in reference to each other, then one can rewrite the S-parameter matrix in terms of two-port networks representing the differential and common types of excitation for both the through path and the crosstalk paths. In this way the differential S-parameters can be regrouped in a way that allows quick analysis of the signal integrity of a serial I/O link and of any issues with crosstalk or even mode EMI (e.g., S_{13} is the near-end crosstalk between ports 1 and 3 and S_{14} is the far-end crosstalk between ports 1 and 4).

This mixed mode S-parameter data set is now in reference to the excitation as well as the ports and utilizes a new set of descriptors. The signal paths can be excited or measured as a differential signal (D) or a common signal (C). It is possible to identify on the four-port S-parameter matrix (F.3) the submatrices that correspond to the differential response to the differential signal (S_{DD}), the common-mode response to a differential signal (S_{CD}), the differential response to the common-mode signal (S_{DC}), and finally the common-mode response to a common-mode signal (S_{CC}). The port $1'$ notation refers to the signal across the single mode ports 1 and 3, and likewise port $2'$ refers to the signal across the single mode ports 2 and 4 using the convention that the through path always connects 1 to 2, 3 to 4, and so on, for commonality in port names as the number of ports is increased. Again, the output port is listed first and then the input port. We can then write the S-parameter matrix for the four-port system in the following form:

$$\begin{bmatrix} V_D 1'^- \\ V_D 2'^- \\ V_C 1'^- \\ V_C 2'^- \end{bmatrix} = \begin{pmatrix} S_{DD1'1'} & S_{DD1'2'} & S_{DC1'1'} & S_{DC1'2'} \\ S_{DD2'1'} & S_{DD2'2'} & S_{DC2'1'} & S_{DC2'2'} \\ S_{CD1'1'} & S_{CD1'2'} & S_{CC1'1'} & S_{CC1'2'} \\ S_{CD2'1'} & S_{CD2'2'} & S_{CC2'1'} & S_{CC2'2'} \end{pmatrix} \begin{bmatrix} V_D 1'^+ \\ V_D 2'^+ \\ V_C 1'^+ \\ V_C 2'^+ \end{bmatrix} \tag{F.3}$$

References [7, 8] provide a more detailed discussion on the four-port S-parameter representation and its properties. Figure F.5 shows the measured S-parameters in the format of (F.3) from a four-port VNA measurement of a differential signal trace of a test fixture.

Mathematically, it is easy to extend the S-parameter concept to any number of ports but realistically for a high-speed digital I/O the next step would be a 12-port network. This would allow the measurement of three adjacent differential pairs to capture not just the differential transmission of the center signal path, but also the crosstalk from the two adjacent differential transmission lines as shown in Figure F.6. An example of a 12-port network analyzer system is shown in Figure F.7. Note that calibration can become a significant challenge with so many ports if it is done manually.

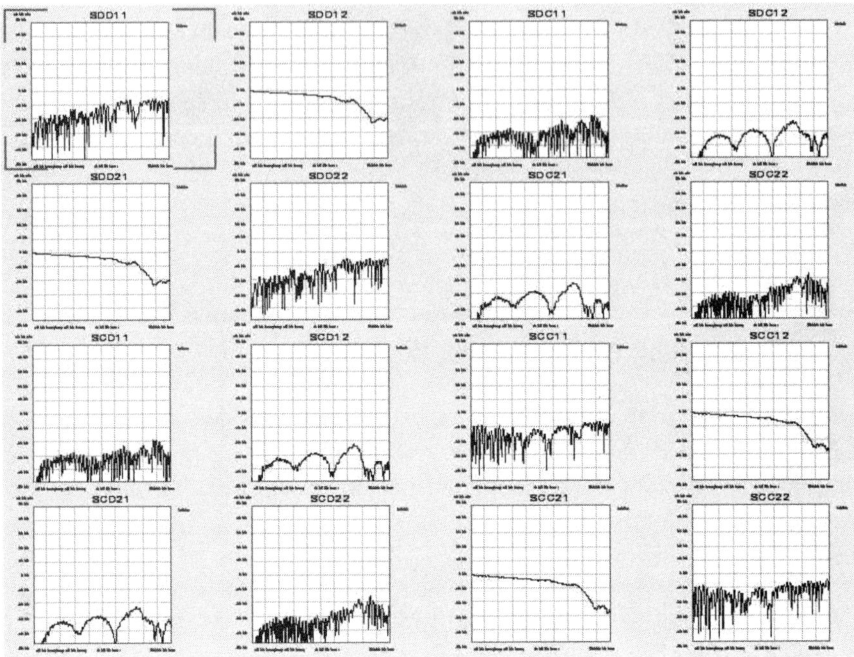

Figure F.5 Measured differential S-parameters from a four-port DUT using the format of (F.3) (a differential pair of a test fixture).

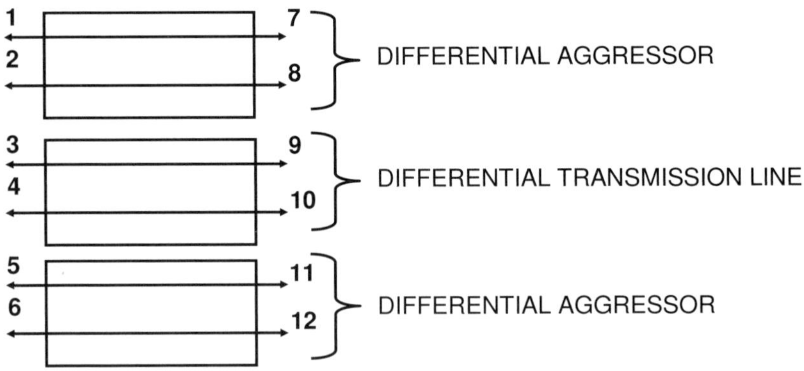

Figure F.6 Block diagram of a 12-port system.

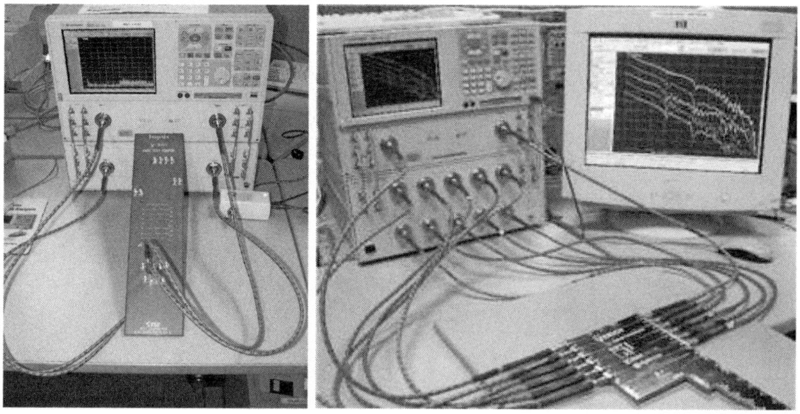

(a) (b)

Figure F.7 Picture of (a) a four-port VNA measurement setup and (b) a twelve-port VNA measurement setup.

F.1 Simulating and Synthesizing Time-Domain Responses from S-Parameters

As mentioned before, S-parameters are not only useful for visualizing the response of a system in the frequency-domain but also for analyzing the system in the time-domain. Several software packages can take the measured S-parameters usually using the Touchstone file format [9], and compute either the time-domain response in the form of a TDR/TDT or provide a data eye diagram for a given data rate and rise time as shown in Figure F.8.

The other important usage of S-parameters is in a simulation where the ultimate goal is to have the simulation predict the performance of a measurement. Just like with the measured S-parameters, the simulated S-parameters of a test fixture performance can be integrated into a more complex simulation that can include models of the ATE pin electronics and of the DUT (Figure F.9). Ideally, the transformation between time-domain and frequency-domain results in no loss of data or corruption of the signal; however, real data both simulated and measured can have nonreal effects at the edges or unwanted noise relative to the desired DUT electrical data. With measured data from a VNA the DC point is extrapolated to zero, and in some cases a poor calibration can even result in gain on a passive system. It is important when using band limited S-parameter data to understand the challenges and importance of using passive and causal data [10, 11] and one must always be watchful of data "features" caused by mathematical transformations on

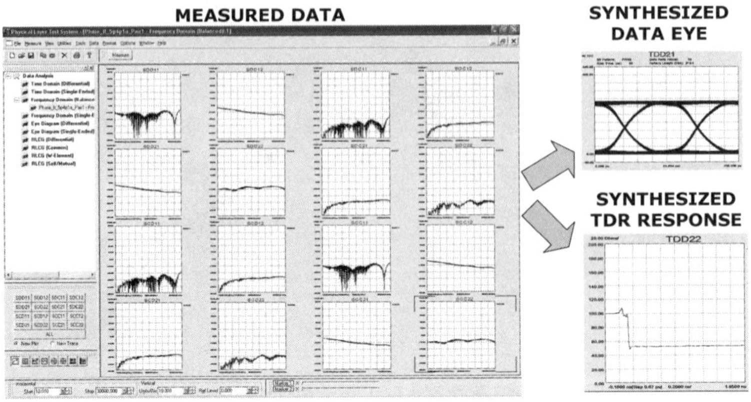

Figure F.8 Example of synthesizing the time-domain response from the S-parameters.

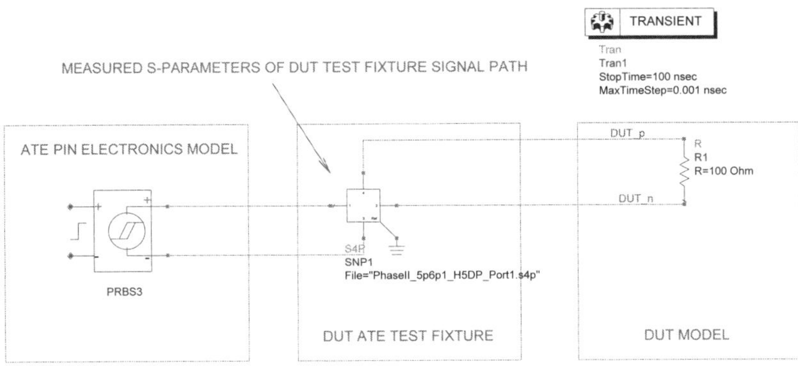

Figure F.9 Example of using S-parameters measured on a test fixture on a time-domain simulation of the ATE stimulus waveform at the DUT.

nonideal data sets [12, 13]. Note that some software packages (e.g., IdEM from IdemWorks and PLTS from Keysight Technologies) allow the evaluation of a S-parameter matrix for passivity and causality and if needed generate a new passive and causal S-parameter matrix.

It is also possible to cascade S-parameters to obtain the simulation of a signal path that consists of multiple elements (connectors, vias, signal traces), each described by an S-parameter data block. Although a CAD simulation tool will show this cascade of S-parameters, it should be noted that mathematically the S-parameters must be converted to a scattering transfer or T-matrix format to properly handle the reflections between two networks [4]. In other words, one cannot simply matrix-multiply the S-parameter matrices of the single path components together to get the total signal path response. The T-matrix has the property that:

$$\mathrm{T}_{Total}(f) = \mathrm{T}_A(f)\,\mathrm{T}_{DUT}(f)\,\mathrm{T}_B(f) \tag{F.4}$$

where:

$$[\mathrm{T}] = \begin{bmatrix} \mathrm{T}_{11}(f) & \mathrm{T}_{12}(f) \\ \mathrm{T}_{21}(f) & \mathrm{T}_{22}(f) \end{bmatrix} \tag{F.5}$$

and:

$$\begin{bmatrix} \mathrm{V}_1^{Reverse}(f) \\ \mathrm{V}_1^{Forward}(f) \end{bmatrix} = \begin{bmatrix} \mathrm{T}_{11}(f) & \mathrm{T}_{12}(f) \\ \mathrm{T}_{21}(f) & \mathrm{T}_{22}(f) \end{bmatrix} \begin{bmatrix} \mathrm{V}_2^{Forward}(f) \\ \mathrm{V}_2^{Reverse}(f) \end{bmatrix} \tag{F.6}$$

Once cascaded together, the T-matrix terms are converted back to the S-matrix for direct analysis of reflection and transmission performance. The T-matrix can also be used to go the other direction and actually de-embed a known network element connected to the input or output of a system [14]. In the case of a DUT with test fixture connections on either side:

$$\mathrm{T}_{DUT}(f) = \mathrm{T}_A(f)^{-1}\,\mathrm{T}_{Total}(f)\,\mathrm{T}_B(f)^{-1} \tag{F.7}$$

This frequency-domain technique of de-embedding the test fixture effects (also known as "test fixture adapter") from the measured performance of a DUT is an effective tool for obtaining a library of components for designing and simulating a full-path system. The accuracy of the de-embedded DUT performance will depend on the calibration techniques used to obtain the "test fixture adapter" data and typically requires the design of custom calibration structures in the form of transmission through lines, reflects, shorts, and loads [15].

F.2 S-Parameters of Coupled Differential Pairs and Structures

The significance of being able to look at differential S-parameters becomes apparent when one starts using differential transmission lines with increased coupling between the two single transmission lines that make up the differential pair. The impedance of the differential system is 100 Ω, or 50 Ω in series with 50 Ω when there is no coupling between the two transmission lines, but as the traces come together to utilize coupled routing for dense routing applications or to minimize sensitivity to ground discontinuities the issues increase in complexity. The single-ended performance with the changes in impedance due to coupling will no longer represent the same electrical performance as that of the differential path with the matched system impedance so it is important to display and simulate the S-parameters differentially [3, 16]. For example, Figure F.10 compares the single-ended and differential insertion loss of a coupled differential pair, while Figure F.11 shows the synthesized single-ended and differential data eye diagrams.

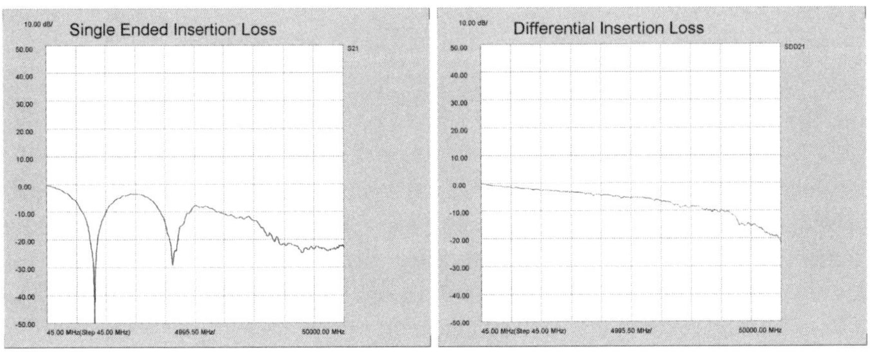

(a) (b)

Figure F.10 Comparing (a) the single-ended insertion loss with (b) the differential insertion loss.

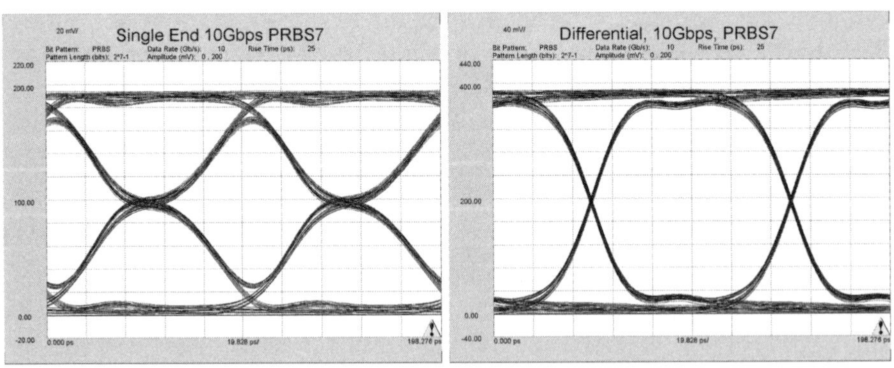

(a) (b)

Figure F.11 Comparison of (a) the synthesized single-ended data eye with (b) the synthesized differential data eye.

From the figures it is easy to see that, in this case, relying on the single-ended measurements to extrapolate the differential performance of the differential signal path with coupling would result in a significant error.

F.3 S-Parameters: Calibration and De-Embedding

S-parameters are usually measured with a VNA (see Section 7.7), although it is also possible to use a TDR/TDT instrument with the appropriate SW to measure S-parameters (see Appendix E). One critical point when measuring S-parameters of a DUT is the calibration and de-embedding of the test fixture. There are several VNA calibration algorithms [17, 18] with the Short, Open, Load and Thru (SLOT) calibration being the typical one used by test engineers [19]. Unfortunately, this type of calibration procedure only calibrates to the end of the measurement cables connectors (or to the tip of a coaxial probe), which in some situations might not even be compatible with the DUT connectors. In this case the DUT is referred to as a noninsertable DUT [20]. If the DUT is not connectorized and a PCB test fixture is necessary to measure it, then the problem becomes more complicated [21]. In this case more complex calibration algorithms like TRL can be used [15, 22–24].

Another option to de-embed a PCB test fixture is to use the automatic fixture removal algorithm [5, 25, 26]. If the signal path to the DUT is symmetric, then a single thru line is enough for de-embedding the test fixture. Figure F.12 shows an example of a PCB test fixture for a packaged DUT that includes a thru line for the usage of the automatic fixture removal algorithm. This algorithm has been improved recently to allow de-embedding using single port measurements [27].

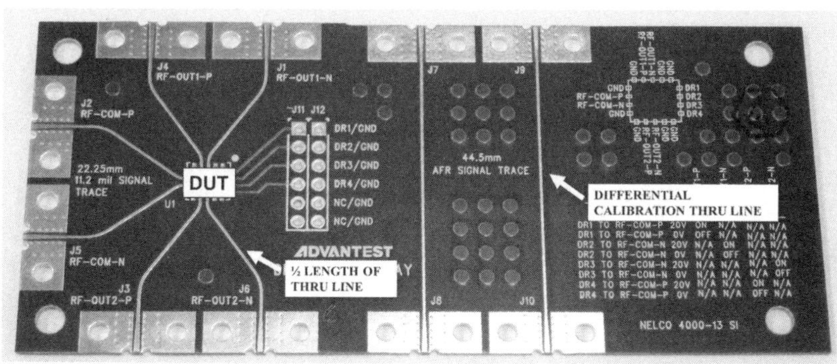

Figure F.12 Example of a PCB test fixture with a thru signal trace for the fixture removal calibration (Courtesy of Advantest).

Another approach is to compute the test fixture de-embedding parameters using simulation [28, 29]. This is not a trivial approach and requires good correlation of the simulation models to reality but in some situations where measurement of the test fixture is not possible, it is the only option available.

References

[1] D. M. Pozar, *Microwave Engineering*. New York: John Wiley & Sons, 2005.

[2] M. Resso and E. Bogatin, *Signal Integrity Characterization Techniques*. IEC, 2009.

[3] S. H. Hall and H. L. Heck, *Advanced Signal Integrity for High-Speed Digital Designs*. New York: John Wiley & Sons, 2009.

[4] R. Sorrentino and G. Bianchi, *Microwave and RF Engineering*. New York: John Wiley & Sons, 2010.

[5] J. P. Dunsmore, *Handbook of Microwave Component Measurements with Advanced VNA Techniques*. New York: John Wiley & Sons, 2012.

[6] Agilent Technologies, "Time Domain Analysis Using a Network Analyzer," *Application Note 1287-12*, 2012.

[7] D. Derickson and M. Mueller, *Digital Communications Test and Measurement: High-Speed Physical Layer Characterization*. Upper Saddle River, NJ: Prentice-Hall, 2007.

[8] L. Besser and R. Gilmore, *Practical RF Circuit Design for Modern Wireless Systems, Volume I: Passive Circuits and Systems*. Norwood, MA: Artech House, 2003.

[9] IBIS Open Forum, *Touchstone File Format Specification Version 2.0*, 2009.

[10] P. Trivero, S. Grivet-Talocia, M. S. Nakhla, F. G. Canavero, and R. Achar, "Stability, Causality, and Passivity in Electric Interconnect Models," *IEEE Transactions on Advanced Packaging*, vol. 30, Nov. 2007.

[11] S. G. Talocia, "Towards Real-Time S-Parameter Qualification and Macromodeling," *IEEE Workshop on Signal and Power Integrity*, 2012.

[12] D. Kaller, C. Schuster, Y. Kwark, D. Altabella, B. Truong, Z. Chen, A. Haridass, and E. Klink, "Using S-Parameters Successfully in Time Domain Link Simulations," *IEEE 14th Topical Meeting on Electrical Performance of Electronic Packaging*, 2005.

[13] F. Rao, C. Morgan, S. Gupta, and V. Borich, "The Need for Impulse Response Models and an Accurate Method for Impulse Generation from Band-Limited S-Parameters," *IEC DesignCon*, 2008.

[14] Agilent Technologies, "De-Embedding and Embedding S-Parameter Networks Using a Vector Network Analyzer," *Application Note 1364-1*, 2004.

[15] V. Duperron, D. Dunham, and M. Resso, "Practical Design and Implementation of Stripline TRL Calibration Fixtures for 10-Gigabit Interconnect Analysis," *IEC DesignCon*, 2006.

[16] E. Bogatin, *Signal and Power Integrity Simplified.* Upper Saddle River, NJ: Prentice-Hall, 2010.

[17] M. Hiebel, *Fundamentals of Vector Network Analysis.* Rhode & Schwarz, 2007.

[18] R. Collier and D. Skinner, *Microwave Measurements.* Institution of Engineering and Technology, 2007.

[19] S. A. Wartenberg, *RF Measurements of Die and Package.* Norwood, MA: Artech House, 2002.

[20] "Techniques for VNA Measurements of Non-insertable Devices," *Anritsu Technical Article*, 2005.

[21] Agilent Technologies, "Agilent Signal Integrity Analysis Series, Part 3: The ABCs of De-Embedding," *Application Note 5989-5765EN*, 2007.

[22] J. Fleury and O. Bernard, "Designing and Characterizing TRL Fixture Calibration Standards for Device Modeling," *Applied Microwave and Wireless.*

[23] "Agilent Network Analysis Applying the 8510 TRL Calibration for Non-Coaxial Measurements," *Product Note 8510-8A.*

[24] J. Ellison, S. Sercu, N. Altland, and S. Smith, "Strengths and Weaknesses of Various Calibration Techniques," *DesignCon*, 2014.

[25] R. Schaefer, "Comparison of Fixture Removal Techniques for Connector and Cable Measurements," *International Microwave Symposium*, 2010.

[26] H. Barnes, J. Moreira, M. Resso, and R. Schaefer, "Advances in ATE Fixture Performance and Socket Characterization for Multi-Gigabit Applications," *IEC DesignCon*, 2012.

[27] R. Schaefer, R. Oppelt, and J. Schuber, "A Simple Method to Characterize and Accurately Remove the Effects of Push-On Connectors," *DesignCon*, 2015.

[28] G. Antonini, A. C. Scogna, and A. Orlandi, "De-Embedding Procedure Based on Computed/Measured Data Set for PCB Structures Characterization," *IEEE Transactions on Advanced Packaging*, vol. 27, Nov. 2004.

[29] H. Barnes, A. Ciccomancini, M. Resso, and M. Tsai, "Differential PCB Structures using Measured TRL Calibration and Simulated Structure De-Embedding," *IEC DesignCon*, 2007.

G

Engineering CAD Tools[1]

This appendix presents additional detail on several engineering CAD tools that can be important to the test engineer working on high-speed digital applications with ATE. The tools can be divided into model-based circuit simulators, finite element electromagnetic (EM) field solvers, data transformation tools, and physical layout tools. There is a significant amount of overlap between the tools and the true art of simulation becomes one of knowing the strengths and weaknesses of the tools and which ones work the best for a given application. Additional references on these topics can be found in [1, 2].

G.1 Circuit Simulators

The primary simulation tool for electrical engineers has always been the circuit simulator, such as the well-known SPICE tool [3] with built-in models for each of the elements in a circuit. Originally, these tools focused on lumped element design forcing transmission line effects to be modeled with large quantities of lumped elements to accurately predict the phase and magnitude of a high-speed signal. Increasing the number of elements in a circuit simulator to handle all of the transmission line paths increases the simulation time and reduces the productivity of the tool. The significant increase in data rates for digital I/O on today's circuits has forced most of the circuit simulators to include more efficient transmission line models and some tools are even starting to provide integrated cosimulation with EM field solvers that work into the microwave frequencies. Examples of this type of circuit simulator are HSPICE from Synopsis, Hyperlinx from Mentor, and ADS from Keysight Technologies.

---

[1]This appendix was written in collaboration with Heidi Barnes.

Figure G.1 shows an example using ADS where the objective is to compare the loss versus frequency profile for two transmission lines of different lengths. The simulation tool provides a variety of transmission line models based on the physical topology of the transmission line such as microstrip routed on an external PCB layer or stripline routed on an internal layer. Each model also requires the user to correctly set up a substrate definition to define the properties of the materials being used, the dimensions of the signal trace, and the distances to the ground planes. Figure G.2 compares the simulated insertion loss versus frequency for the two different lengths of stripline that are being modeled. Figure G.3 shows the same model being simulated but now the excitation is in the time-domain using a pseudo-random bit sequence pattern source.

A simulation tool is just that, a simulated prediction of a circuit and not a real physical signal. This means that no simulation tool provides perfect results and there are always simplifications and model assumptions inherent to a given circuit simulation tool. The best way to understand the limitations of a given tool and find ways to verify and improve the accuracy is through the use of measurements. Utilizing measurement-based modeling provides the ultimate in cross-checking of simulation results so that a model can be optimized for a given application and provides the necessary insights for solving circuit design problems.

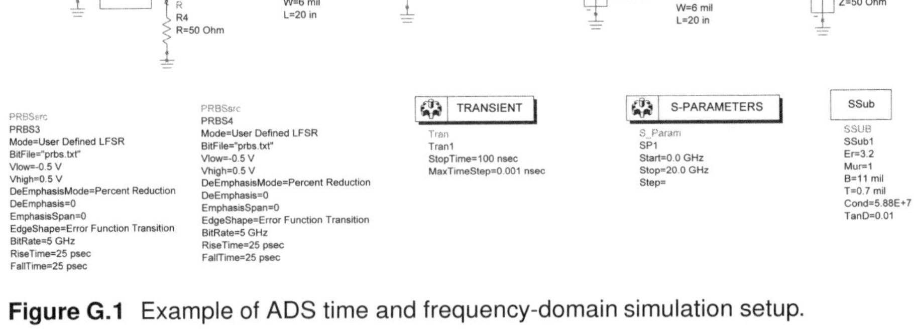

Figure G.1 Example of ADS time and frequency-domain simulation setup.

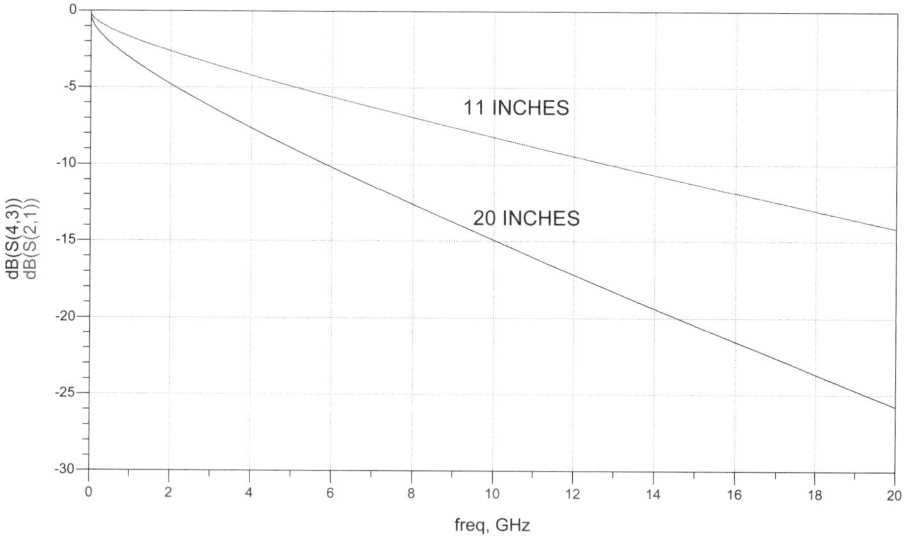

Figure G.2 Results from the ADS simulation in the frequency-domain.

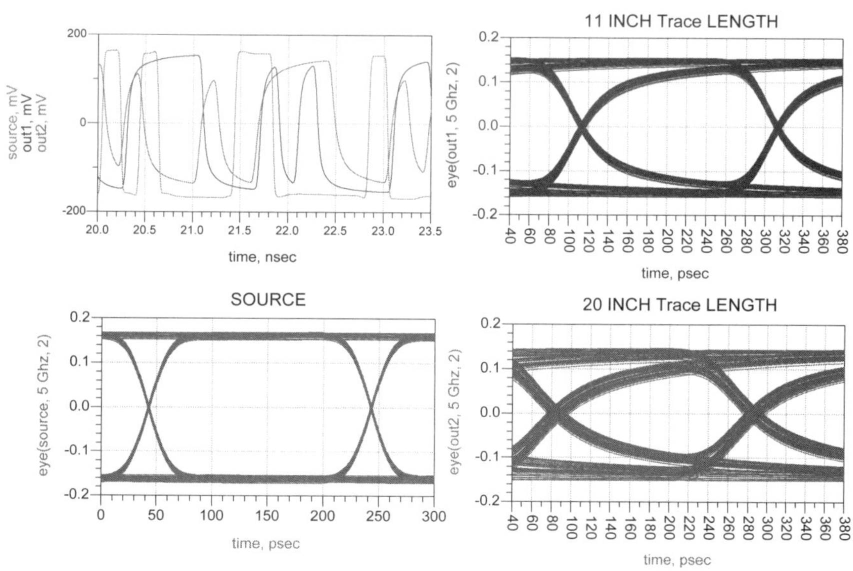

Figure G.3 Results from the ADS simulation in the time-domain.

G.2 3D EM Field Solvers

The 3D EM field solvers provide a very powerful way to analyze a circuit when simplified models are unavailable or no longer accurate enough. In a 3D EM simulation a 3D model of the problem is created with the appropriate material properties defined and then Maxwell's equations for the fundamental properties of electricity and magnetism are solved through a finite element process that divides the 3D structure into small units [4]. Some of the major differences in the available 3D EM field solvers are in the methodology of how the small units are created and whether Maxwell's equations are solved in the time-domain or in the frequency-domain. References [2, 5] provide a good overview of 3D EM simulation methodologies and Figure G.4 shows an example of a 3D EM field solver with a pogo via design problem.

In the context of test fixture design, 3D EM simulation is critical for the analysis and optimization of key structures like vias, pogo vias, transitions to connectors, transitions to relays, and so on. These structures can vary significantly from those used on the final DUT application board and the 3D EM field solver is the best way to quantify the differences and optimize the impedances for high-speed ATE performance. The 3D EM field solver is based on the fundamental properties of electromagnetic fields; however, it is still a simulator with some model assumptions and user-entered material properties that can still benefit from measurement-based modeling techniques to improve the accuracy [6].

G.3 2D Planar Field Solvers

Since the individual layers of a PCB are planar structures, it is possible to simulate them without resorting to a full 3D EM field solver. There is a special class of EM field solvers called 2D EM field solvers that are able to simulate planar structures in a faster way than a 3D EM field solver [2]. Figure G.5 shows one example of a commercial 2D EM field solver. Although 2D EM field solvers are faster than full 3D EM field solvers, they are not able to properly address problems that are truly 3D in nature like a connector, a via, or a DUT socket. However, via models are typically added so that they can be used for signals that route on more then one layer of a PCB. The faster simulation time makes it practical to simulate multiple signal traces on an ATE test fixture to look at signal performance and crosstalk issues.

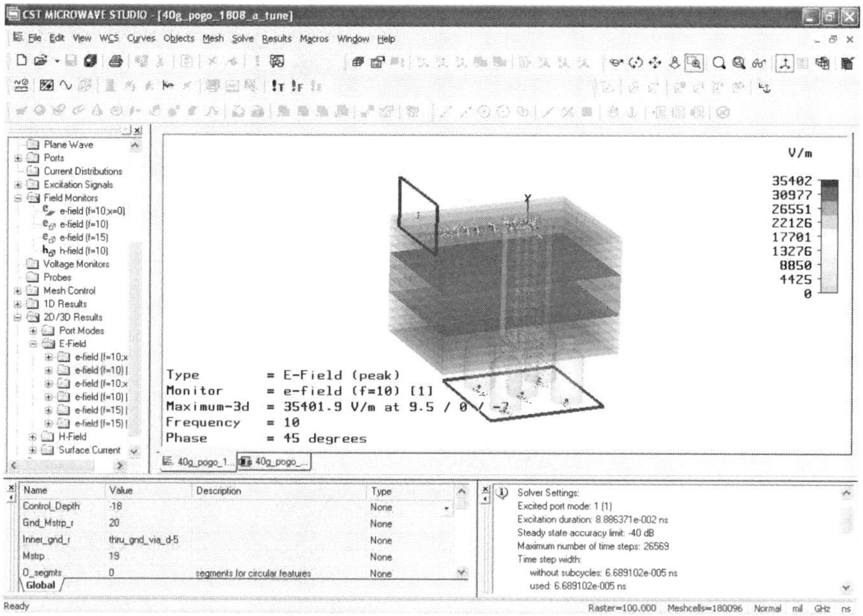

Figure G.4 Example of 3D EM simulation of a pogo via design using CST Microwave Studio.

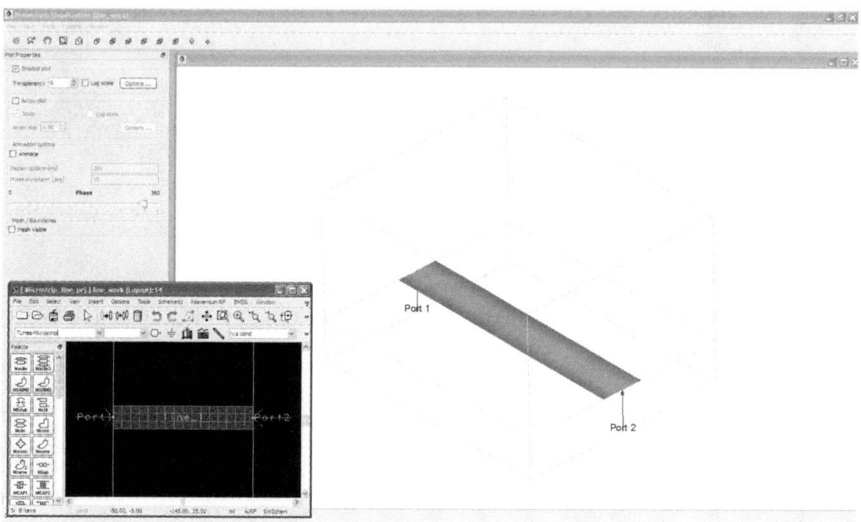

Figure G.5 Example of a microstrip line simulation using Keysight Technologies Momentum 2D EM field solver.

G.4 Power Integrity

Highly specialized tools can be found in the area of power integrity where field solvers and models are combined to specifically address the design of a PCB power distribution network (PDN) [7]. These tools simplify the user interface for modeling a PDN network, such as importing from a PCB layout tool, and entry and placement of decoupling capacitors.

The models and field solvers are also optimized for the large planar surfaces and hundreds of connecting ground vias so that simulation times remain reasonable. These tools allow the user to address some of the specific topics that are associated with a test fixture PDN design like optimum placement of decoupling capacitors, power plane resonances, unwanted coupling between PCB features, and so on. Figure G.6 shows an example of a power distribution network simulation software package.

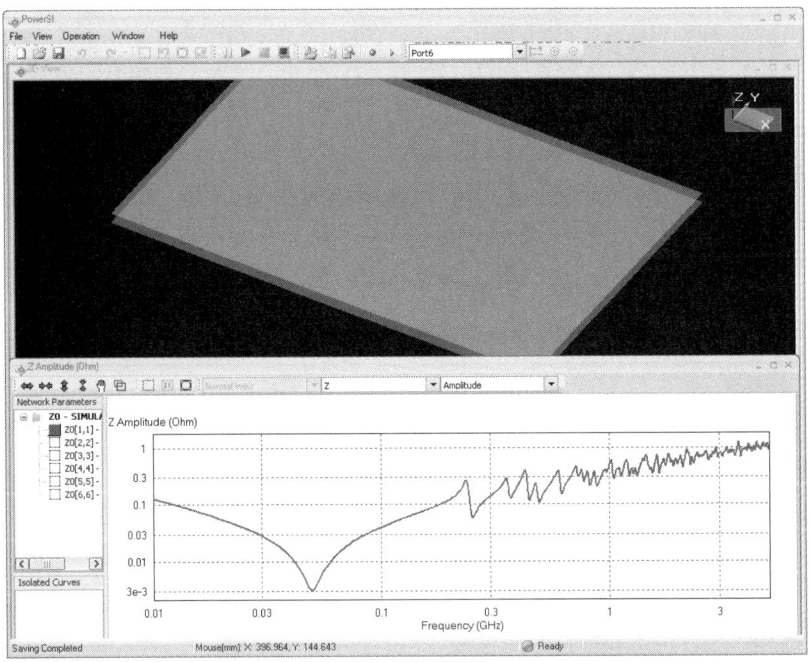

Figure G.6 Example of a PCB power distribution network analysis software modeling the impedance of a power plane (Courtesy of Sigrity).

G.5 Model Generation

Ideally, it is possible to use the different simulation tools previously described together to address an engineering project; for example a 3D EM simulator might be used for the connector and via design, a 2D EM simulator for the PCB board layout, and a SPICE model for the I/O cell. This approach can be quite challenging since tool integration is not always optimal and one may need to translate or develop a model of the data to import the data from one tool to another. Model generation is also quite useful for transforming measured data into a format that is compatible with a simulator tool. In some cases this model creation can be automated like, for example, generating a SPICE model from measured S-parameters. Another critical point is to make sure the generated model is passive and causal [8, 9]. A full-featured simulation package typically includes this type of model generation and verification capabilities, but stand-alone software packages are also available as shown in the example of Figure G.7.

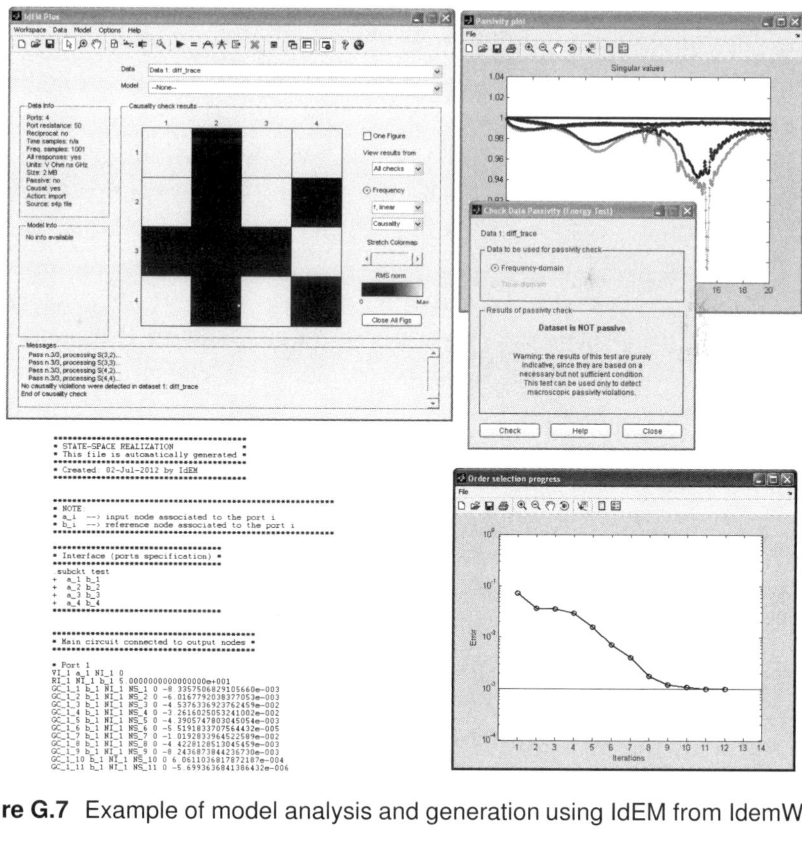

Figure G.7 Example of model analysis and generation using IdEM from IdemWorks.

G.6 Other Tools

One very important CAD tool in the design of high-speed digital I/Os is the layout tool for simulating how the fabricated PCB layers will look. The word simulate is used here to emphasize that there are manufacturing tolerances like etching and drilling that need to be considered as data rates move into the multigigabit domain, and one may need to work closely with the fabricator to understand any modifications that are made to the data. Layout tools are rather prolific, but few are specifically designed for high-speed digital layout and simple things like backdrilling to remove signal via stubs, trace necking into a BGA, and rounded trace bends may not be completely automated and therefore require a significant amount of manual checking. Figure G.8 shows an example of these types of tools.

Typically, when one only needs to visualize or check a PCB layout there exists a free layout viewer provided by the PCB layout software company. The layout viewer should be used with some caution since it is not always clear how the different positive and negative layers may be merged together to form a single PCB layer. To verify that the layout is being interpreted correctly, it is always important on high-speed designs to look at the final CAD layers that are generated at the end of layout during the Gerber generation process. This process converts the variety of layout layers to a single image layer for each PCB layer and saves it in a Gerber format that is compatible with the tools and equipment at the PCB fabricator.

The Gerber layers are far easier to overlay and turn on and off for visual checking of ground and power plane antipad features that are critical in achieving multigigabit performance. In some special cases where the fabricator requests to add additional metal in the empty spaces of a signal layer to improve etching uniformity, one should also request the modified Gerbers from the PCB fabricator. Figure G.9 shows one example of a Gerber visualization tool.

Finally, another important tool when planning or designing a test fixture PCB is an impedance calculator that allows the design of the correct stack-up and trace geometry for a given target impedance and dielectric material. On the planning phase of a PCB test fixture this type of tool is very helpful in analyzing possible trade-offs and limitations on the stack-up for complex multilayer PCB test fixtures. Figure G.10 shows an example of this type of tool. Once the stack-up is determined, it is important to request that the PCB fabricator provides the exact trace widths for a desired impedance since this could vary by 25 or 50 μm (1 or 2 mil) depending on the fabrication process.

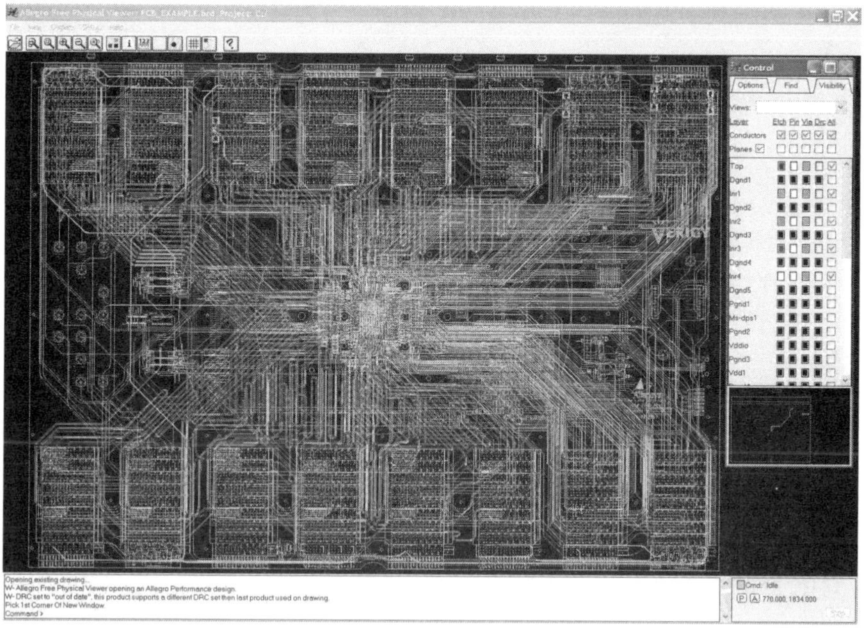

Figure G.8 Visualizing a test fixture layout using Cadence Allegro free physical viewer.

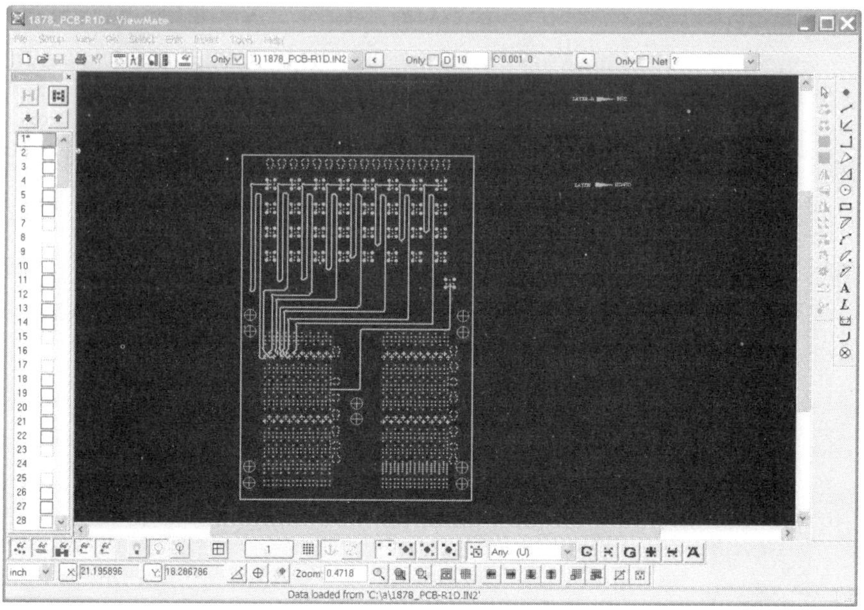

Figure G.9 Visualizing the Gerber files of a test fixture using Pentalogix ViewMate.

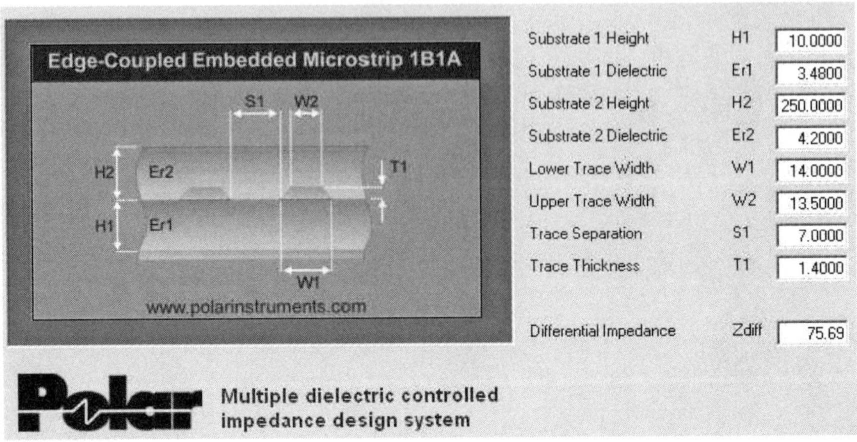

Figure G.10 Example of impedance computation for a microstrip using Polar Instruments impedance design system.

References

[1] G. Edlund, *Timing Analysis and Simulation for Signal Integrity Engineers*. Upper Saddle River, NJ: Prentice-Hall, 2008.

[2] D. G. Swanson and W. J. R. Hoefer, *Microwave Circuit Modeling Using Electromagnetic Field Simulation*. Norwood, MA: Artech House, 2003.

[3] G. W. Roberts and A. S. Sedra, *SPICE, 2nd Edition*. Oxford University Press, 1997.

[4] K. S. Yee, "Numerical Solution of Initial Boundary Value Problems Involving Maxwell's Equations in Isotropic Media," *IEEE Transactions on Antennas and Propagation*, 1966.

[5] T. Weiland, M. Timm, and I. Munteanu, "A Practical Guide for 3-D Simulation," *IEEE Microwave Magazine*, 2008.

[6] Y. Shlepnev, A. Neves, T. Dagostino, and S. McMorrow, "Practical Identification of Dispersive Dielectric Models with Generalized Modal S-Parameters for Analysis of Interconnects in 6-100 Gb/s Applications," *IEC DesignCon*, 2010.

[7] O. P. Mandhana and J. Zhao, "Comparative Study of Effectiveness of On-Chip, On-Package and PCD Decoupling for Core Noise Reduction by Using Broadband Power Delivery Network Models," *Electronic Components and Technology Conference*, 2005.

[8] P. Trivero, S. Grivet-Talocia, M. S. Nakhla, F. G. Canavero, and R. Achar, "Stability, Causality, and Passivity in Electric Interconnect Models," *IEEE Transactions on Advanced Packaging*, vol. 30, Nov. 2007.

[9] S. G. Talocia, "Towards Real-Time S-Parameter Qualification and Macromodeling," *IEEE Workshop on Signal and Power Integrity*, 2012.

H

Test Fixture Evaluation and Characterization[1]

H.1 Measuring the Test Fixture Signal Performance

Measuring the performance of a test fixture is an important and yet difficult task for high-speed digital applications. The ability to accurately measure the signal degradation at these higher data rates is necessary in order to fully characterize the test fixture and identify how it will impact the characterization and testing of a DUT. The measured performance of the test fixture also provides the ability to create full path simulations with models of the ATE pin electronics and DUT to further optimize future test fixture designs and evaluate methods of de-embedding the test fixture effects. To understand how critical it is to measure the performance of an ATE test fixture, one can ask the following questions:

- Is this the first time that the PCB design house and/or fabricator has built this test fixture?
- Is a previous high-speed design being leveraged for use at even higher data rates?
- Were there signal performance problems with the prior test fixture design/fabrication?
- Have any of the PCB materials or components in the high-speed paths changed?

If the answer is yes to any of these questions, then measuring the test fixture performance is critical before starting implementation of the

[1]This appendix was written in collaboration with Heidi Barnes.

application on the ATE system. Following through with this measurement step is not simple and is often skipped for one of the following reasons:

- **Lack of the needed equipment:** High-speed measurements with gigahertz bandwidths need specialized oscilloscopes, TDR/TDT modules, and vector network analyzers that are not always available in a standard ATE lab.
- **Lack of measurement expertise:** High-speed measurements can easily be degraded by the probes or techniques being used. Without the experience to understand when a measurement is being done poorly, it is very easy to end up with erroneous data. A lack of confidence in the measurement methods and equipment can then make it difficult to invest further in improving or fixing the measurement accuracy.
- **Lack of time:** This is the typical reason for skipping the measurement of a test fixture. Test engineers prefer not to delay the schedule for test fixture measurements and go straight to running the application on the ATE system in the hope that if everything works then no measurements are required. On high-speed designs this can often lead to longer delays with extremely long application debug times as one tries to sort between a test problem, a DUT silicon problem, and a test fixture problem.

The inability to accurately measure the test fixture performance makes it difficult to understand how the test fixture is influencing the DUT measurement and reduces the ability to correlate ATE measurements with bench instrumentation measurements of the DUT. One possible alternative is to subcontract the measurement of the test fixture. Figure H.1 shows an example setup of a probing station coupled with high-end bench instrumentation and micro-coaxial probes used to measure the performance of a test fixture PCB board. Of course, this task will entail a financial cost. One disadvantage is that the subcontractor evaluating your test fixture might not know your application in detail, so close cooperation may be required.

Since the accuracy of the test fixture measurements can depend on the techniques and probes being used, it is important to start planning for how the measurements will be made at the same time as the test fixture design. The addition of test coupons, interface boards for probing at the DUT socket, and calibration structures on a test fixture panel can have significant benefits in improving the ability to accurately characterize an ATE test fixture. The following sections go into additional detail on these methods for improving test fixture characterization.

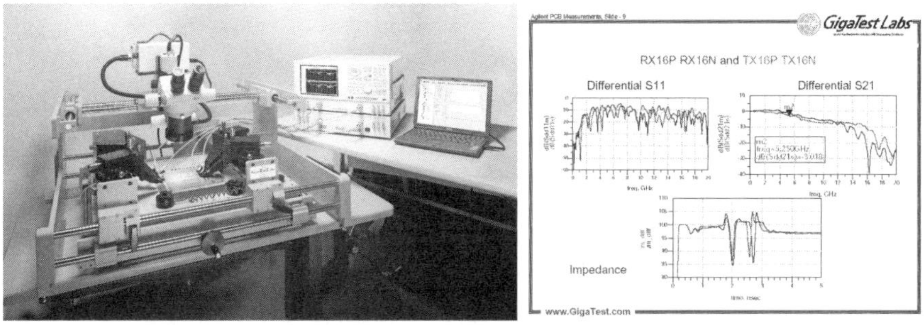

Figure H.1 GigaTest Labs probing station with measurement instrumentation and a report example (Courtesy of GigaTest Labs).

H.1.1 Test Coupons

Test coupons are an excellent method for obtaining a significant amount of information on a test fixture. A test coupon takes advantage of the empty space left on a test fixture PCB panel to build a separate board with test patterns. The test patterns replicate some of the signal paths from the main board to check losses and impedance matching on the different routing layers, via transitions, back drilling performance, relays, and so on. To simplify the measurement with bench instrumentation, the test coupon can make use of coaxial connectors to connect to the PCB. For example, instead of the DUT PCB footprint and socket, one could use a coaxial connector where the DUT socket pad would be.

An example of a test fixture and its associated test coupon is shown in Figure H.2. This example has several traces on the test coupon that correspond to the shortest and longest length traces on the test fixture with a connection on one end to a coaxial edge connector and at the other end to a pogo pin PCB via array. Trace bends and via transitions were also included for obtaining information on how an individual element of a design can affect the signal integrity.

This test coupon highlights the need to include the transition from the ATE pogo assembly to the test fixture via array since the overall performance can be significantly influenced by the PCB via topology. This can be achieved by using a modified pogo pin assembly with short cables and SMA connectors together with a mechanical bracket to perform the test coupon measurement as shown in Figure H.3 [1].

In the case of an ATE test fixture with coaxial type connectors to the ATE system (see Section 8.1.1), it is very easy to probe the ATE side of the test fixture or test coupon by using an appropriate mating adapter in a coaxial

Figure H.2 Example of a test fixture (top) and its corresponding test coupon (bottom) (Courtesy of Advantest).

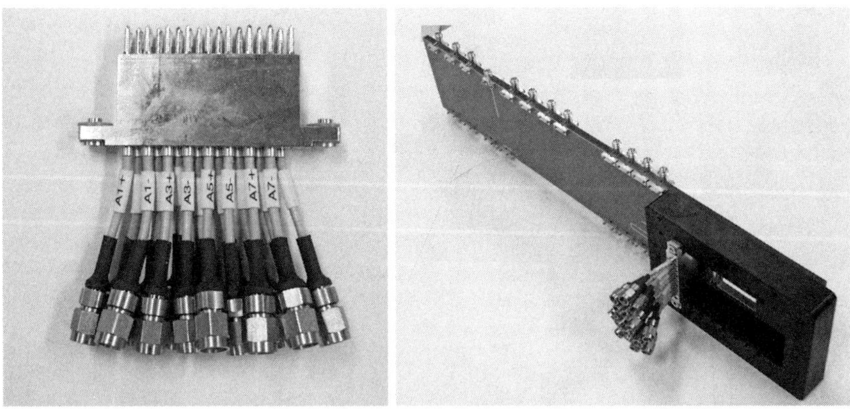

Figure H.3 An ATE bench pogo pin assembly (left) docked to the test coupon in a lab environment (right) (Courtesy of Advantest).

cable assembly with a coaxial connector on the other side for connecting to measurement instrumentation as shown in Figure H.4.

Figure H.4 Measuring a test coupon with a coaxial type interconnect to the ATE system using a single pin probing assembly with a mating connector, coaxial cable and SMA connector to connect to a bench instrument (left) and a probing assembly for the entire connector (right) (Courtesy of Advantest).

With this measurement setup, it is possible to measure the performance of the different traces on the test coupon including in this case the effect of the pogo via and pogo assembly as shown in Figure H.5. The measurements were done by connecting the ports on a VNA to the pogo assembly SMA connector and to the edge mounted connector at the end of the trace and measuring the insertion loss. Other types of measurements using different instruments are also possible with this approach.

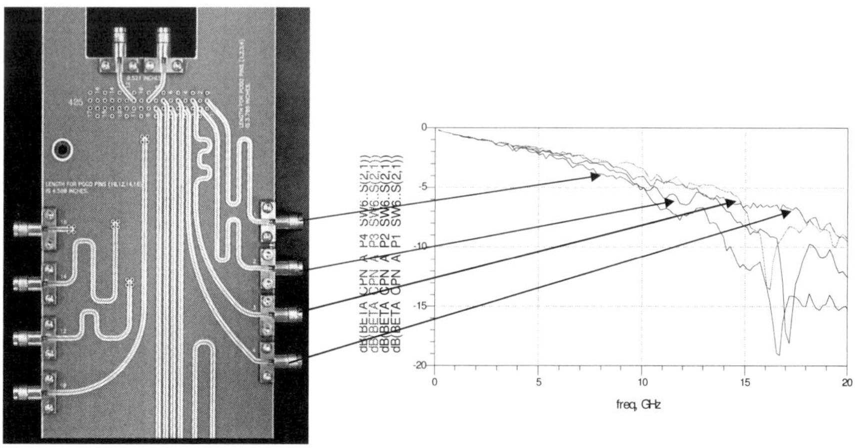

Figure H.5 Measurement results for some structures on the test coupon in Figure H.3.

The test coupon does a good job of characterizing the basic design of the test fixture and identifying any major design flaws, but it is not a total guarantee of the test fixture performance. Variations in etching and registration across a panel will result in differences between the test coupon at the edge of the panel and signal paths routed in the center. These small variations are typically ignored at data rates below 1 Gbps, but as data rates increase, so does the sensitivity to these PCB tolerances. Also, it is difficult to replicate the DUT footprint, power distribution network, and the routing density to accurately predict all of the crosstalk and signal losses at higher speeds on a test coupon. Test coupons can also be used to measure PCB parameters like loss that can then be used for tuning simulation of the test fixture PCB signal traces. The next two sections discuss two possible approaches: two transmission line segments and the Beatty resonant standard [2]. Other approaches have been proposed, especially for a production environment [3].

H.1.1.1 Two Transmission Line Segment

To calculate the loss tangent and the dielectric constant, one can use the two transmission line method shown in Figure H.6 that enables the removal of connector fixturing losses and impedance mismatches so that the total attenuation for a given length L of transmission line can be measured. At high frequencies, impedance reflections from the fixturing can be significant and must be removed in order to see the true loss of the transmission line [2, 4]. Since the transmission line is typically low loss, it is helpful to select L such that it is always longer (higher loss) then the connecting fixture and thus less sensitive to the fixture removal quality. This approach can be further extended by using multiple line segments using the multiline measurement technique [5].

H.1.1.2 Beatty Resonant Standard

To find the remaining as-fabricated variables including dielectric height, etched trace width, and copper thickness. The series resonant Beatty structure (Beatty standard) shown in Figure H.7 has the advantage of allowing additional data, which includes the delta change in impedance for a given delta change in physical trace width along with a resonant ripple in the insertion loss, which is affected by dielectric constant and transmission line losses. The Beatty standard is constructed by increasing the trace width of the transmission line for a specific distance that results in reflection resonances and broadband ripple on the insertion loss [2, 6].

Figure H.8 shows an example of manufactured PCB with a two transmission line segment and a Beatty standard structure.

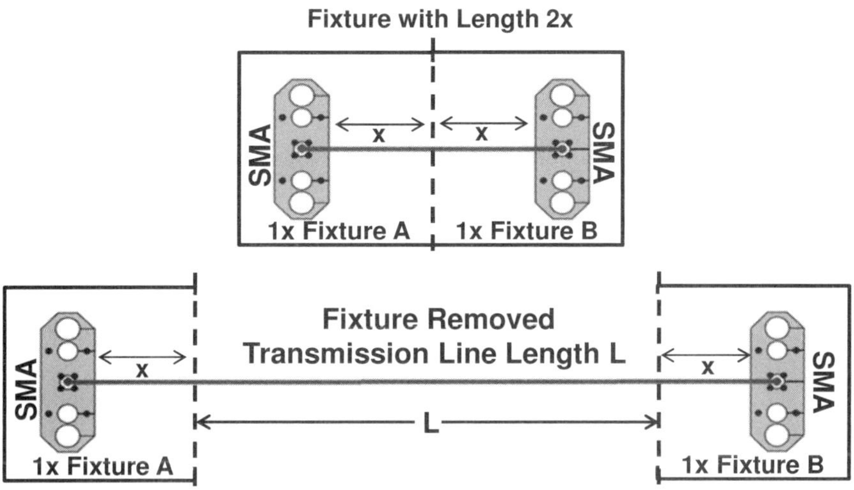

Figure H.6 Two transmission line test coupon structures for measuring the S-parameters of the length L of a transmission line with the connector fixturing removed.

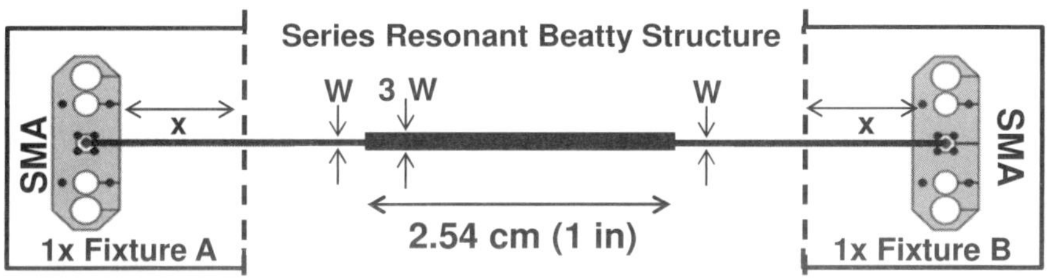

Figure H.7 The series resonant Beatty structure provides a delta impedance change for additional information in determining the as-fabricated dielectric height and trace width.

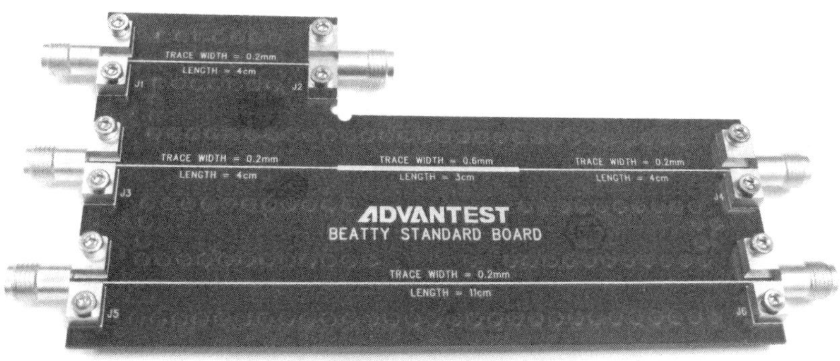

Figure H.8 Example of a PCB with a two transmission line segment and a Beatty standard (Courtesy of Advantest).

H.1.2 Test Fixture Socket and Socket Via Field Probing Interposer

Directly measuring the test fixture high-speed signal paths is the ideal characterization and verification method for a test fixture PCB. To perform this measurement the connection to the ATE pogo via arrays on one end of the test fixture can be done using the same modified pogo assembly that was used on the test coupon with the short cables and SMA connectors or in the case of an ATE test fixture with an ATE coaxial connector interface type using the appropriate connector adapter. Connecting to the other end with the DUT socket is not as simple. The challenge at the DUT socket is to provide a matched impedance probe connection to the DUT socket that replicates the signal and ground topology that the DUT will see. Variations in BGA pitch and ballout topology for signal and ground connections along with single-ended and differential connections increase the complexity of defining a method for accurate and repeatable measurements as shown in Figure H.9.

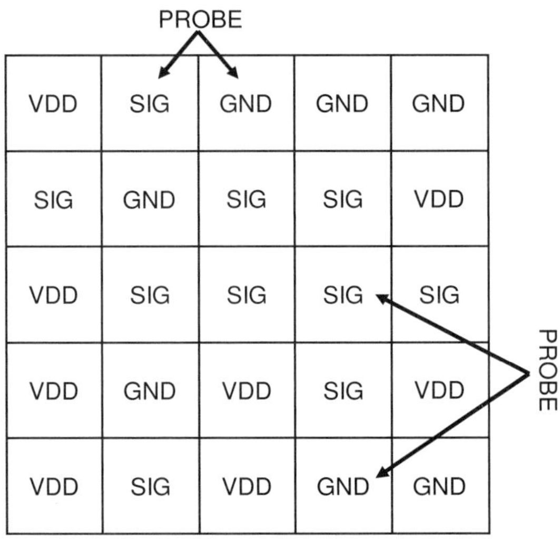

Figure H.9 The challenge of a probing a nonuniform BGA.

An understanding of the probe performance for a specific application may not be as easy as just looking at the data sheet from the probe vendor. Physical differences between the probe topology and the test fixture can lead to additional impedance mismatches that can significantly degrade a multigigabit signal. Common oscilloscope type probes with a side ground lead like those discussed in Section 7.17.8 suffer from the added inductance or higher impedance as this ground wire moves further from the signal path.

Commercially available impedance matched probes with extremely small submillimeter size probe tips have long been used in the wafer probing industry for RF/microwave applications and for package characterization [7] and have the ability to provide the best possible high-speed connection to the ATE test fixture. Leveraging this probing technology does require a special interface or "interposer" PCB [8, 9] to optimize the transition from the probe to the test fixture both mechanically and electrically. This probing interposer provides the mechanical strength to compress the DUT socket pogo pins and an optimized electrical transition from the ground signal spacing of the probe to that of the test fixture. The term surrogate package is also used instead of interposer by some authors [10].

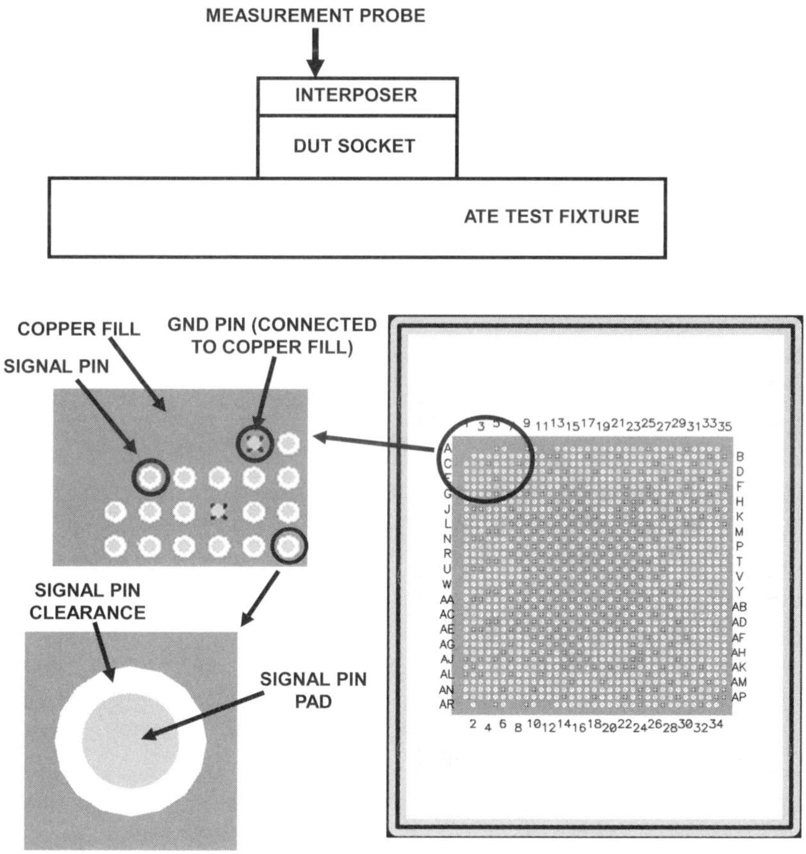

Figure H.10 Description of the layout details for one example of a DUT socket probing interposer.

The PCB probing interposer design is simply an array of vias that line up with the exact DUT BGA footprint on the bottom side. The interposer vias are filled and plated over to provide a flat surface for the socket pogo pin connection. On the top side a ground fill is added that connects all of the ground vias together and surrounds each signal and power via so that the via pad to ground fill spacing matches the ground to signal spacing of the micro-coaxial probe. Figure H.10 shows the details of the top ground fill for a probing interposer board. Figure H.11 shows how the probing interposer is attached to the test fixture socket with the mounting holes and alignment pins. The figure also shows the micropositioner used to correctly position the probe with the help of a microscope. Advanced interposer designs utilize 3D EM simulations to add internal ground layers and additional interstitial ground vias to reduce crosstalk and improve impedance matching between the test fixture and the micro-coaxial probes as shown in Figure H.12. The performance improvement can be easily see in Figure H.13 [11].

Note that to use a probing interposer with a socket it is necessary to mechanically modify the socket to obtain a flat surface where the probing interposer can attach as shown in Figure H.14. It is also possible to design an interposer that works on an unmodified socket as shown in Figure H.15. In this case it is necessary to design a special mechanical interface that attaches to the top of the socket (the socket lid needs usually to be removed) and compresses the interposer into the socket pogo pins while still keeping the top surface free for probing. Guaranteeing the alignment of the socket pogo pins with the interposer bottom pads is also not trivial.

The PCB probing interposer also enables the measurement of the test fixture without the DUT socket. The footprint on the bottom of the probing interposer matches directly with that of the test fixture PCB. With the addition of a thin vertical interconnect material compressed between the PCB probing interposer and the test fixture DUT PCB footprint a good electrical connection can be made. Figure H.16 shows an example of a thin <0.5 mm vertically conductive elastomer material that uses columns of silver plated nickel balls with high current carrying capacity to make multiple connections per signal path [12]. Another option is to add to the bottom of the probing interposer an elastomeric interconnect layer as shown in Figure H.17. This approach has the advantage that it is much easier to use but the elastomeric layer needs to be custom designed for the BGA ballout increasing the costs. The ability to measure the test fixture with and without the ATE socket provides additional information on the impact this transition has on the overall signal integrity of the signal path.

The selection of which micro-coaxial probes to use is less dependent on the manufacturer and more a function of selecting the best configuration and pitch for the ground and signal connections [8]. Micro-coaxial probes with two side ground connections (ground-signal-ground or GSG topology),

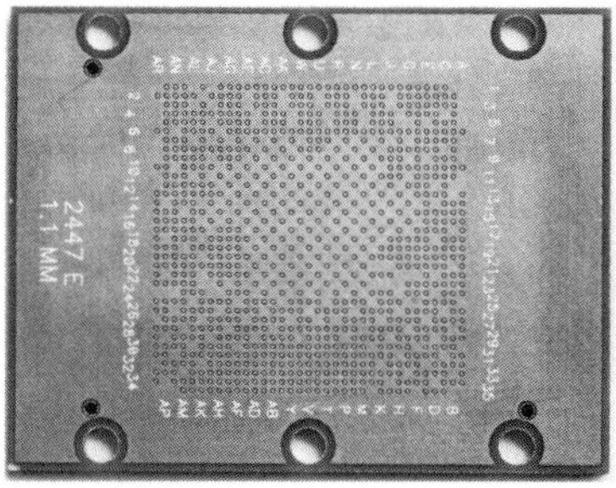

(a) Probing interposer (top). **(b)** Probing interposer (bottom).

(c) Using a probe together with a probing interposer to measure the test fixture performance including the DUT socket.

(d) Probing interposer connected directly to the test fixture without the socket using a elastomeric material sheet.

Figure H.11 (a–d) Example of using a probing interposer in test fixture characterization (Courtesy of Advantest).

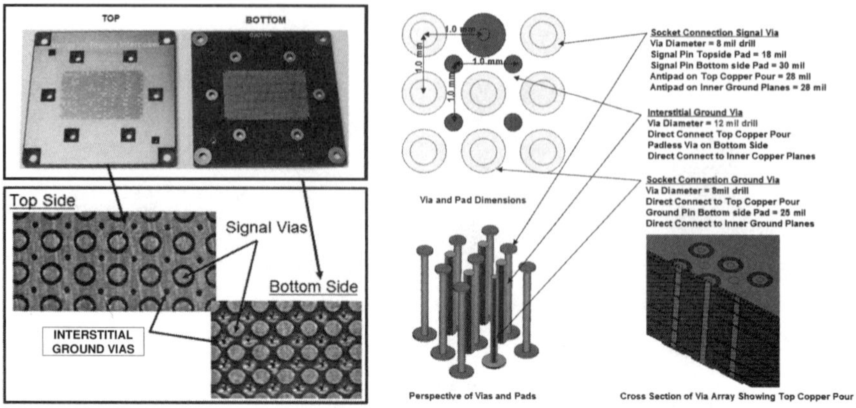

Figure H.12 Example of a probing interposer design with interstitial ground vias surrounding the signal vias used for probing.

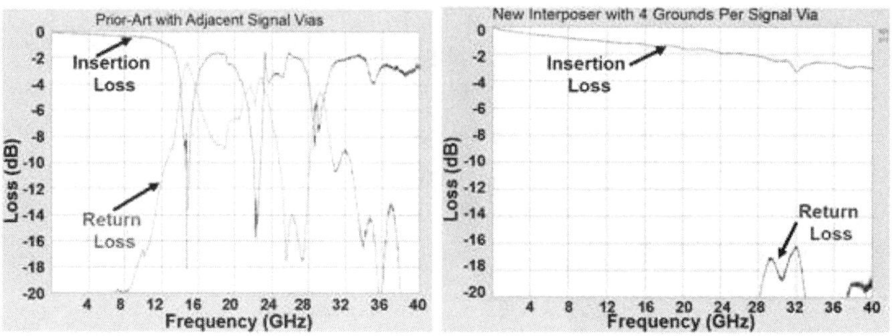

Figure H.13 Comparison of the probing interposer performance with and without the interstitial ground vias (reprinted with permission from [11]).

provide a low loss connection that is more tolerant of the variability in ground via locations on the probing interposer. A tighter spacing between the GSG leads will reduce crosstalk with adjacent probes and reduce the size of the transition discontinuity to a micro-coaxial probe. Figure H.18 shows two examples of micro-coaxial probes in use with the probing interposer to measure an ATE test fixture. Figure H.19 shows the complete view of an ATE test fixture with pogo assembly connections and micro-coaxial probe connections to a bench VNA for complete frequency characterization of a signal path using a customized support mechanical assembly. Figure H.20 shows another setup using a two-side probing station with micro-coaxial probes and a pogo via to coaxial cable adapter attached to a positioner.

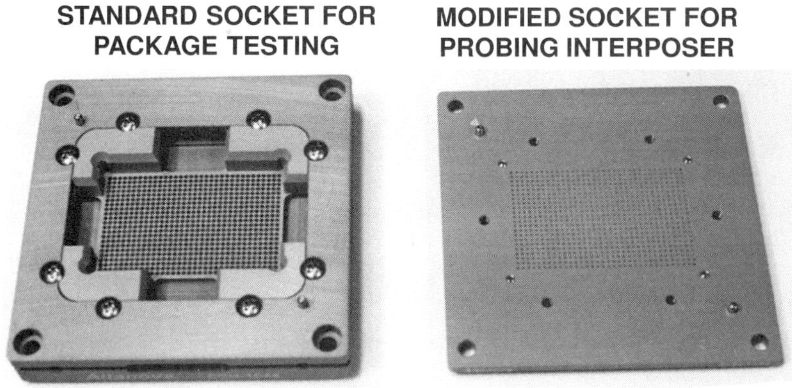

Figure H.14 Example of a modified socket for use with a probing interposer (Courtesy of Advantest).

Figure H.15 Example of a probing interposer for a unmodified socket (Courtesy of Advantest).

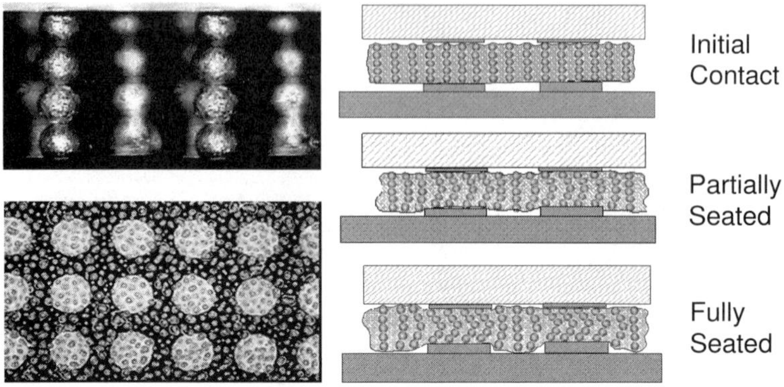

Figure H.16 Picture showing the Paricon material and how it works (Courtesy of Paricon Technologies).

Figure H.17 Picture of a probing interposer with an attached elastomeric interconnect layer in the bottom (Courtesy of Advantest).

It is also important to remember that when performing any measurement to characterize a test fixture that the calibration and de-embedding of the measurement setup is critical [13, 14].

Another approach for designing a probing interposer is to use coaxial connectors to avoid the use of micro-coaxial probes especially for multiport measurement with a large number of ports [9, 15, 16]. Figure H.21 shows one example of a probing interpose using mini-SMP type coaxial connectors attached to a DUT socket. A further possible addition to a DUT socket probing interposer is to use solder balls on the interposer bottom instead of copper pads [17] as shown in the example of Figure H.22. This approach allows including the solder ball impact on the measurement.

The ability to probe the DUT ATE test fixture at the DUT location can also be used for in situ measurements with the test fixture docked on the tester. The micro-coaxial probe makes it possible to send a signal from a

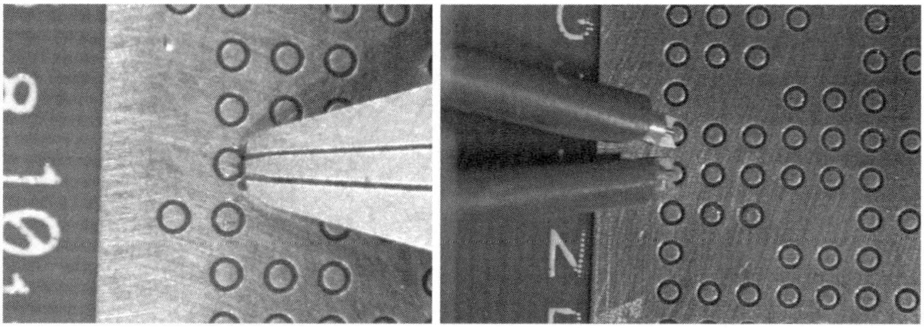

(a) **(b)**

Figure H.18 Probing the signal via at the probing interposer (a) using a SUSS single-ended (GSG) coaxial probe, and (b) using a GGB picoprobe dual probe (GSGGSG).

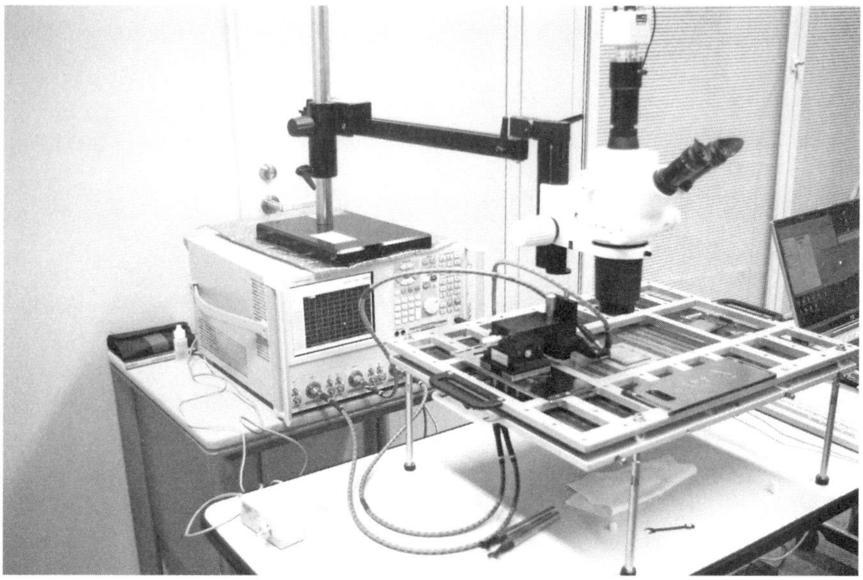

Figure H.19 Example of a customized bench characterization setup for test fixtures from a specific ATE platform (Courtesy of Advantest).

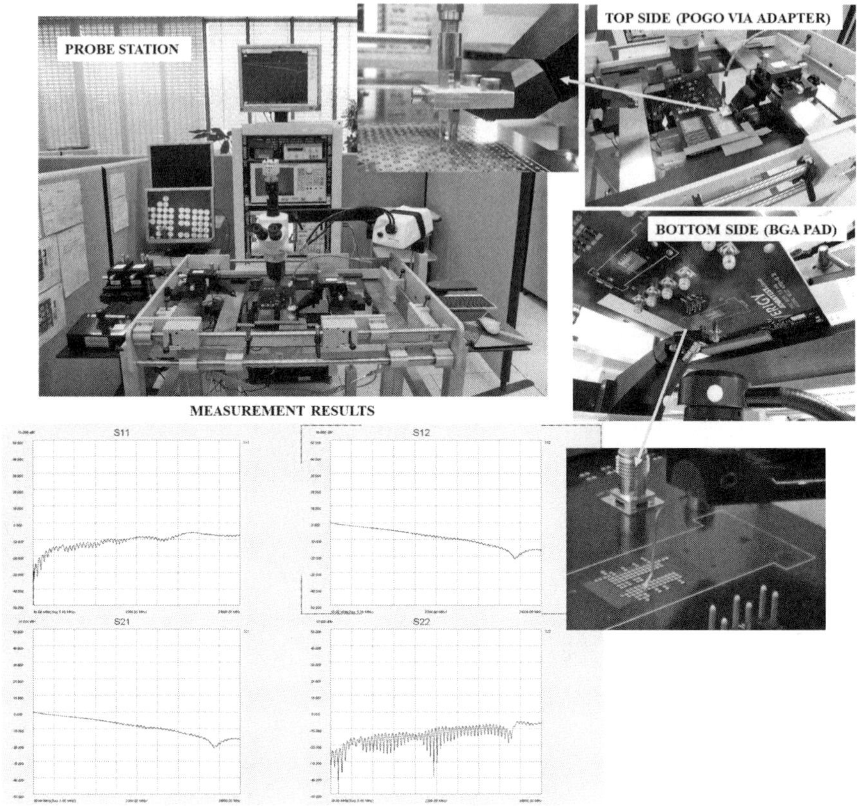

Figure H.20 Example of the characterization setup of an ATE test fixture using a two-side probe station and the appropriate adapter probe for the pogo via side (Courtesy of Advantest).

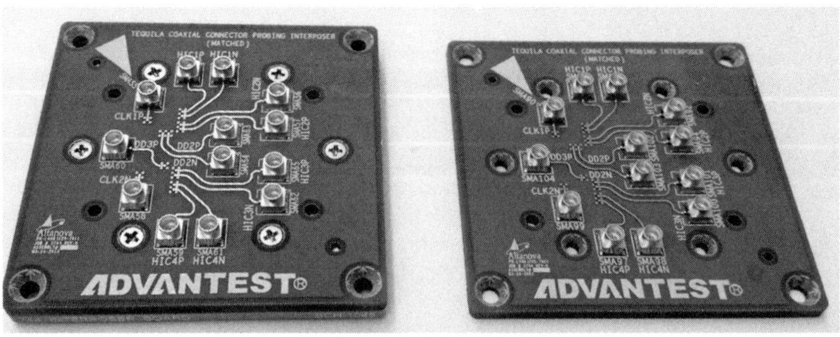

Figure H.21 Example of a probing interposer using mini-SMP type coaxial connectors with a modified socket (left) and without socket (right).

Figure H.22 Example of a coaxial connector socket interposer (or surrogate package) with solder balls on the bottom (Courtesy of R&D Altanova).

calibrated source into the test fixture for measurement by the ATE system or to utilize calibrated bench instrumentation to measure the performance of the ATE transmitted signal that reaches the DUT socket interface. This type of in situ measurement is discussed in Section 9.3 and [18] captures the full path performance of the test fixture and the ATE system, which can then be used for focus calibrating an ATE system for improving its accuracy. Another possible variation of the probing interposer is to use it to create a loopback between the ATE driver and the ATE receiver that includes the ATE test fixture and the DUT socket. Figure H.23 shows an example of a loopback interposer. This loopback interposer concept can also be used for skew calibration as described in Section 9.3.1.2. Figure H.24 shows an example of using a coaxial probing interposer and a loopback interposer in a test fixture docked to an ATE system [19].

H.1.3 Monitoring Interposer

In some applications the test engineer would like to observe the signals being exchanged between the ATE pin electronics and the DUT using an external measurement instrument (e.g., oscilloscope) as shown in Figure H.25. In a coaxial cable environment this can be easily implemented by using for example, a Pickoff Tee adapter as discussed in Section 7.17.4. However in an ATE test fixture without any connectors/cables to provide easy access, this approach is not feasible.

Figure H.23 Example of a set of loopback interposers for different I/O pairs in a single IC package.

Figure H.24 Using a coaxial probing interposer and a loopback interposer on a DUT test fixture socket [19].

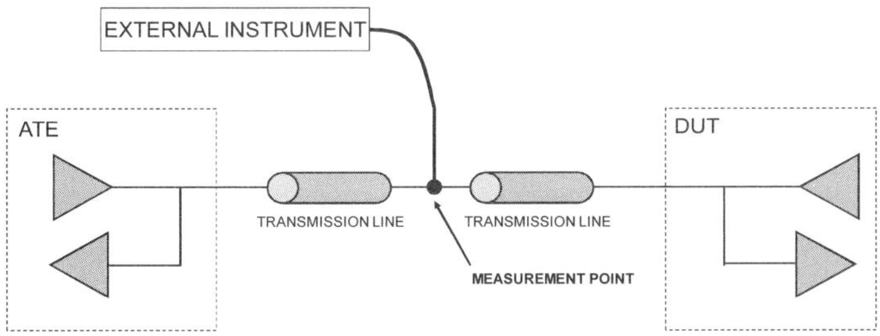

Figure H.25 Monitoring with an external instrument the signals being exchanged between the ATE pin electronics and the DUT.

Another approach is to keep all vias to the DUT BGA pads nonback-drilled so that it is possible to probe the vias on the backside pad with a high-impedance probe (this requires the DUT to be mounted on the back of the test fixture). This approach has the drawback that the stub to the BGA via can have a significant impact on the signal integrity. Another similar option is shown in Figure H.26 where a probing resistor is added to the signal path on the test fixture allowing for the DUT socket to be mounted in the standard position in the top of the test fixture but the signal integrity issues of the via stub still remain.

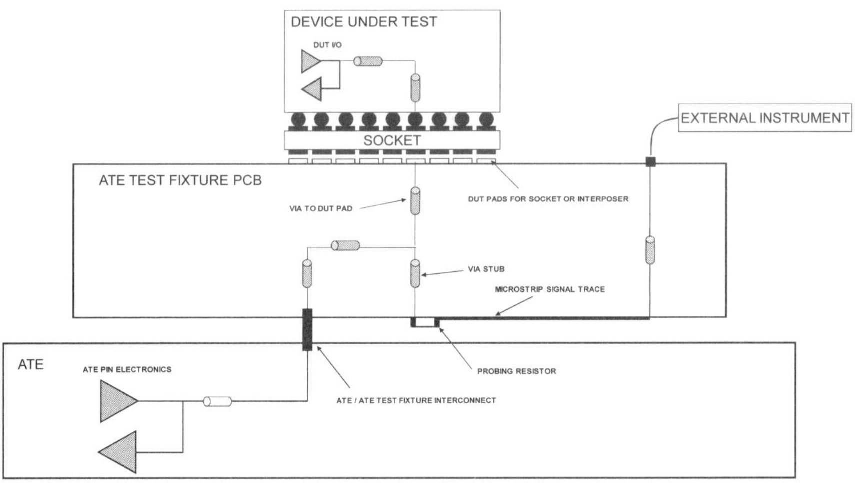

Figure H.26 High level diagram of a monitoring approach integrated on the ATE test fixture using a probing resistor.

One solution to address the via stub challenge is to use a monitoring interposer between the ATE test fixture PCB and the DUT socket as shown in Figure H.27 [20, 21]. The main idea is to use a PCB interposer with a probing circuit using embedded passive components to monitor the signal between the ATE pin electronics and the DUT. One simple probing circuit example would be a single resistor (e.g., 200 Ω). When using this type of approach to probe the signal between the DUT and the ATE pin electronics, it is important to understand the impact of the probing circuit and the measurement instrument termination on the DUT since it might require adjusting the ATE pin electronics levels and termination. Figure H.28 shows an example of a manufactured monitoring interposer for a DDR4 ATE test fixture application. In Figure H.28 a monitoring interposer example is shown using a coaxial probing pad that is probed using a micro-coaxial probe. Another option is to use a monitoring interposer with a coaxial connector as shown in Figure H.29.

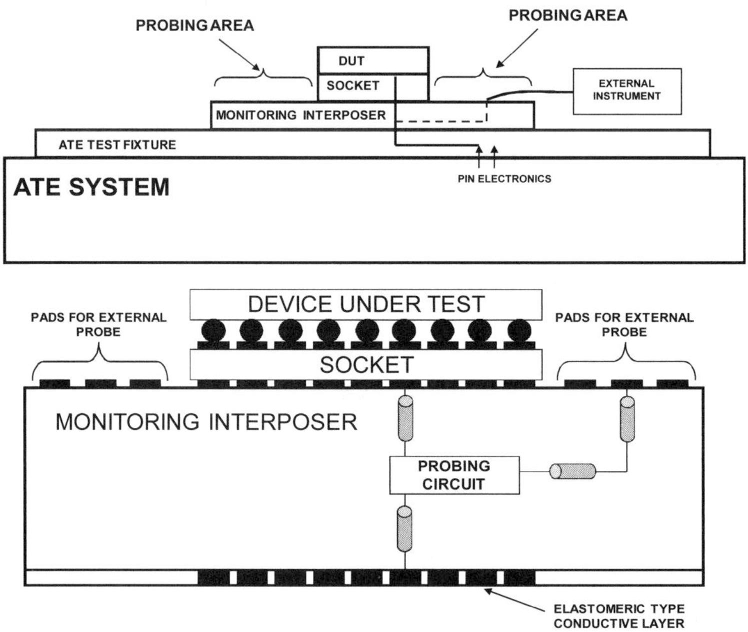

Figure H.27 High level diagram of the monitoring interposer concept.

Another possible extension of the interposer concept is to use it to measure the current going through one of the BGA power pins as shown in Figure H.30. This current monitoring interposer is positioned under the socket and contains a small sense resistor [22]. The value of the current sensing resistor should be selected based on the expected current values to be measured and the voltage noise floor of the measurement instrument to be used.

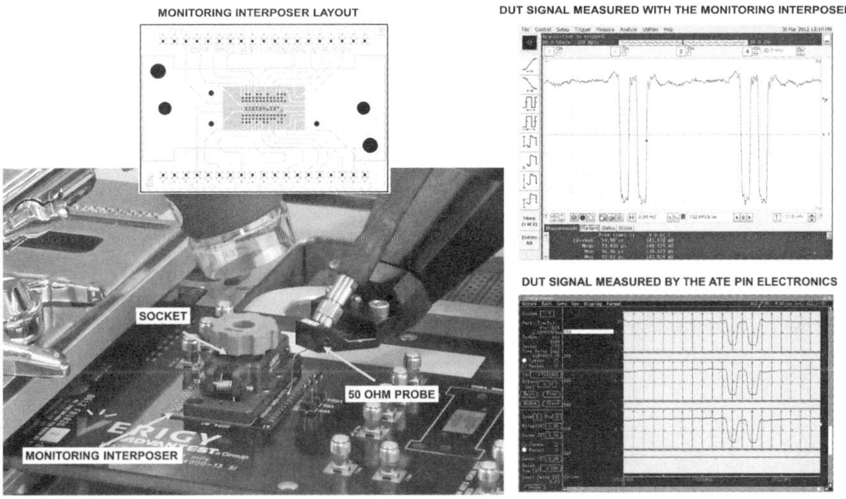

Figure H.28 Example of a monitoring interposer for a DDR4 application test fixture (Courtesy of Advantest).

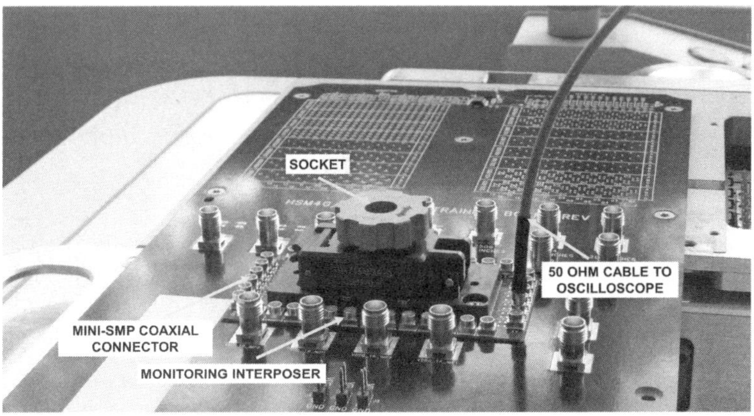

Figure H.29 Example of a monitoring interposer using a coaxial mini-SMP connector on the probing port for a DDR4 application test fixture (Courtesy of Advantest).

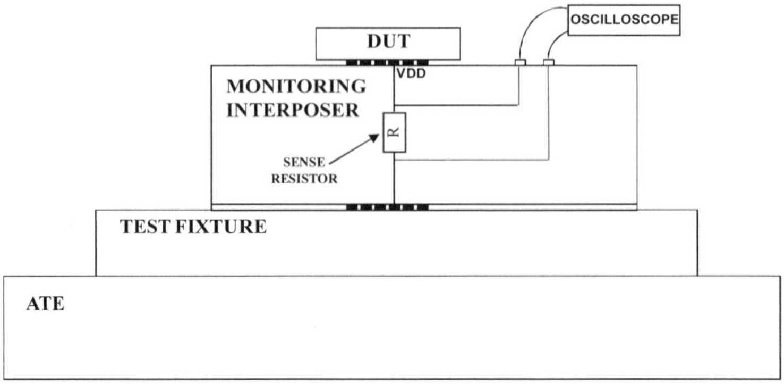

Figure H.30 Measuring the current flowing through one DUT BGA power pin using a sense resistor embedded in a monitoring interposer.

H.2 Measuring the Test Fixture Power Distribution Network

The power distribution network (PDN) of a test fixture is a critical part of a high-speed digital application, especially when high-current applications like microprocessors are involved. It is important to note that some test procedures like a scan test can have significantly different power requirements when compared to the nominal operation of the DUT [23].

Given the importance of the test fixture PDN, it becomes necessary to measure its performance on the final manufactured test fixture with all the decoupling capacitors already assembled. There are several techniques to measure the performance of a power distribution network. One typical approach is characterization on the frequency-domain by using a VNA. Figure H.31 shows a diagram of a setup for measuring the performance of a test fixture PDN with a VNA based on the techniques presented in [24–27]. The figure shows that to obtain a complete impedance measurement it is necessary to perform three measurement steps. The final values of the impedance matrix Z can then be computed using (H.1) [25]:

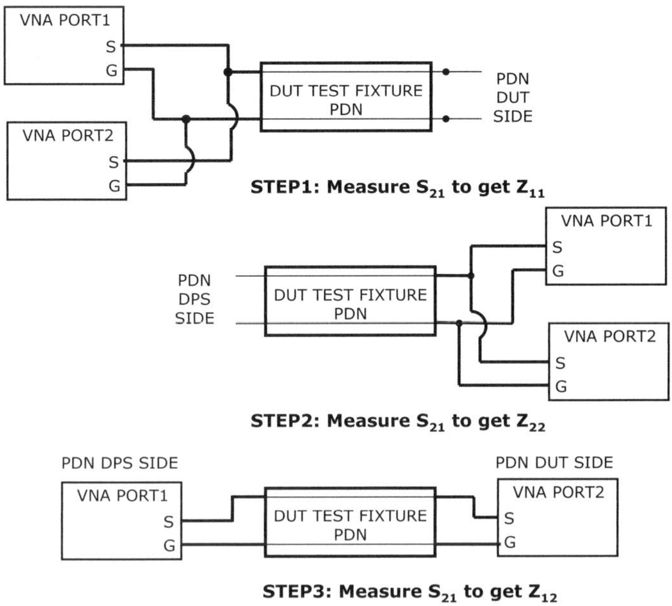

Figure H.31 Setup for a PDN measurement in the frequency-domain with a VNA.

$$
\begin{aligned}
Z_{11} &= \frac{Z_{\text{VNA}}}{2} \frac{S_{21}^{(1)}}{1 - S_{21}^{(1)}} \\
Z_{22} &= \frac{Z_{\text{VNA}}}{2} \frac{S_{21}^{(2)}}{1 - S_{21}^{(2)}} \\
Z_{12} = Z_{21} &= \frac{Z_{\text{VNA}}}{2} S_{21}^{(3)} \frac{1 + \frac{Z_{11}}{Z_{\text{VNA}}} + \frac{Z_{22}}{Z_{\text{VNA}}} + \frac{Z_{11}}{Z_{\text{VNA}}} \frac{Z_{22}}{Z_{\text{VNA}}}}{1 + S_{21}^{(3)} \frac{Z_{11}}{2Z_{\text{VNA}}}}
\end{aligned}
\tag{H.1}
$$

where $S_{21}^{(1)}$ is the S_{21} value measured in step 1 of Figure H.31, $S_{21}^{(2)}$ is the S_{21} value measured in step 2, and $S_{21}^{(3)}$ is the S_{21} value measured in step 3. Z_{VNA} is the VNA impedance value, which is typically 50 Ω. For low impedance values (which is the case of a properly designed PDN), (H.1) can be further simplified into (H.2).

$$
\begin{aligned}
Z_{11} &= 25\, S_{21}^{(1)} \\
Z_{22} &= 25\, S_{21}^{(2)} \\
Z_{12} &= 25\, S_{21}^{(3)}
\end{aligned}
\tag{H.2}
$$

In most cases, the test engineer is only interested in the performance of the power planes and decoupling capacitors in the immediate vicinity of the DUT. In this case it is enough to perform step 2 measurement of Figure H.31 to obtain the PDN self-impedance at the DUT. Another critical point when measuring a test fixture PDN is how the probes are connected to the test fixture. Note that since one usually needs to use a measurement point connected to a via to probe the PDN, the inductance of the via will dominate the measured results unless a proper probing configuration is used. This is shown in Figure H.32 where the optimum configuration requires that the PDN is probed simultaneously from both the top and the bottom of the PCB via [25].

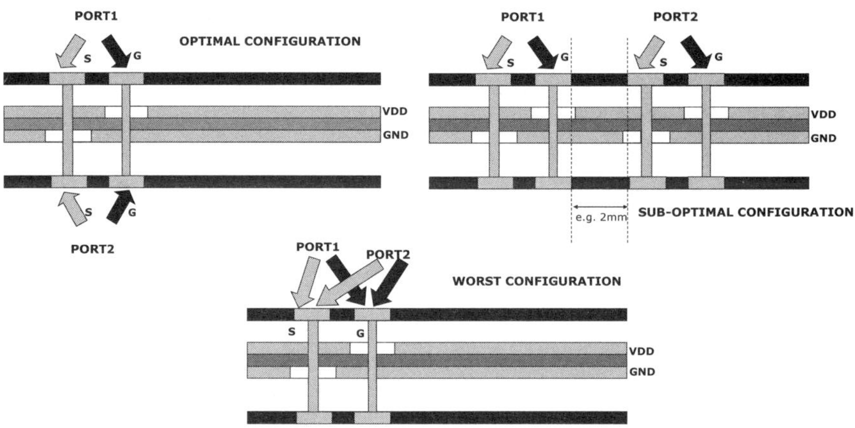

Figure H.32 Possible approaches to probe the test fixture PDN showing how to avoid that the via inductance masks the PDN measurement.

After computing the Z matrix, it is important to check if the matrix is passive.[2] Passivity of a Z impedance matrix can be verified by computing the eigenvalues of the $Z + Z'$ matrix at each frequency point where Z' is the conjugate transpose matrix. For the Z matrix to be passive, all the eigenvalues must be positive or zero [28, 29].

It is important to note that although, on a simulation setup, a S11 measurement is used to measure the PDN self impedance, in reality the dynamic range of a VNA does not allow the measurement of a low impedance PDN using a S11 measurement as discussed in [30].

Figure H.33 shows a picture of two measurement setup examples used to measure the power distribution network impedance of an ATE test fixture.

[2]A system is passive if it is unable to generate energy. Clearly this is a requirement for a PDN since the PDN itself cannot generate energy, only the power supply. Reference [28] provides a precise mathematical definition.

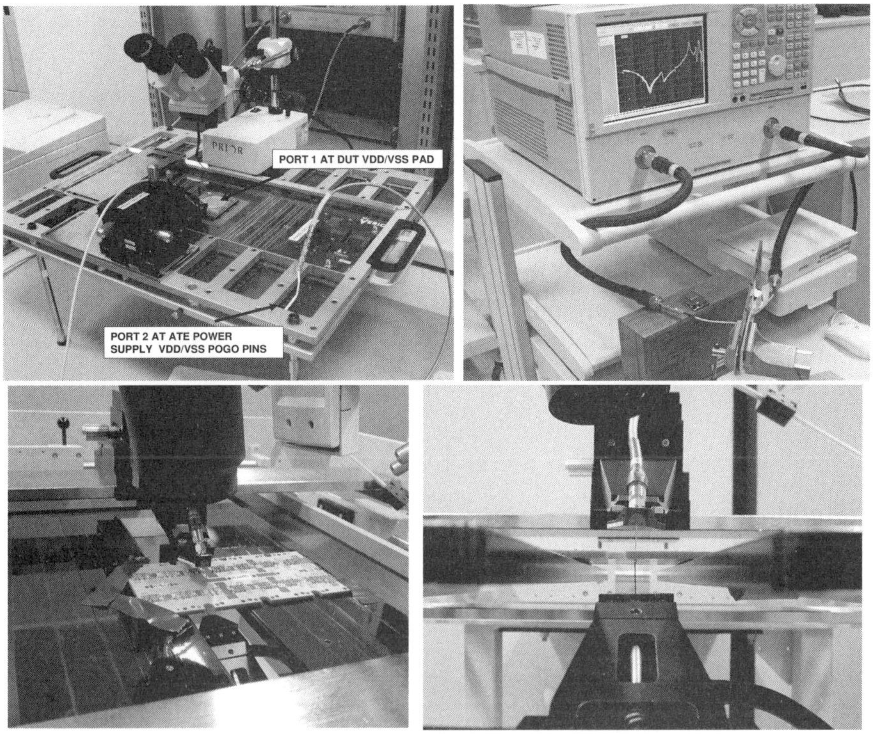

Figure H.33 Pictures of power distribution network measurement setups (top left: measuring Z_{12}; top right: measuring Z_{22}; bottom: measuring Z_{22} with a two-side probe station).

Figures H.34 and H.35 present the results of measuring the PDN of a real DUT ATE test fixture. Figure H.34 shows the measured insertion loss (S21) in each step of the measurement procedure shown in Figure H.31, and Figure H.35 shows the computed impedance values using (H.1).

These results can then be used for verifying the performance of the test fixture PDN by comparing with the expected results (e.g., from simulation) or used as the basis to develop a simulation model of the PDN to be used together with models of the ATE DUT power supply and the DUT power requirements.

As a final note, Figure H.36 shows the measured results for the optimal and suboptimal case PDN measurement methods as described in Figure H.32. Note that when using the worst-case configuration the via inductance will dominate the measured data after 200 kHz and in this way mask the real impedance of the test fixture PDN. In this example the test fixture height was about 5.8 mm (200 mil) with the power planes situated in the middle of the

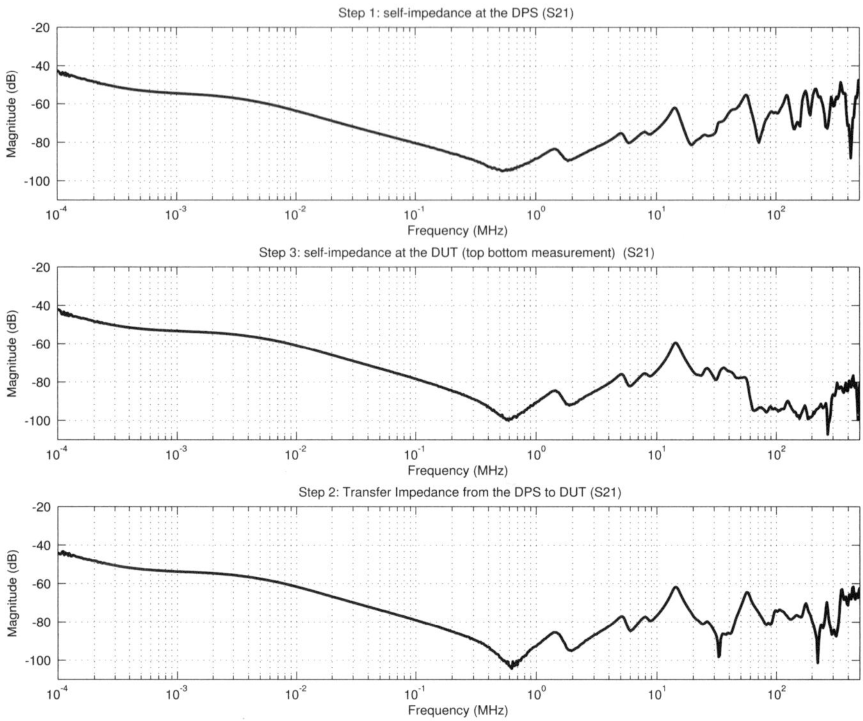

Figure H.34 Measured insertion loss (S21) for step 1, step 2, and step 3 of Figure H.31.

stack-up. This shows the importance of optimizing the probing approach for PDN measurements [25].

H.2.1 Measuring the PDN Voltage

Another measurement of interest when analyzing a test fixture PDN is the power spectrum of the voltage at the DUT power supply terminal. Since this measurement requires that the DUT is being stimulated with a certain pattern, it is necessary to add a monitor mechanism during the test fixture design. Figure H.37 shows one possible approach where a 50 Ω transmission line is connected to one DUT power supply vias and then routed to a probing point or a connector on the top of the test fixture. The 50 Ω transmission line functions as a high impedance probe in comparison with the low impedance of the test fixture PDN. Figure H.38 shows an example of the implementation of this concept to a DDR4 application ATE test fixture.

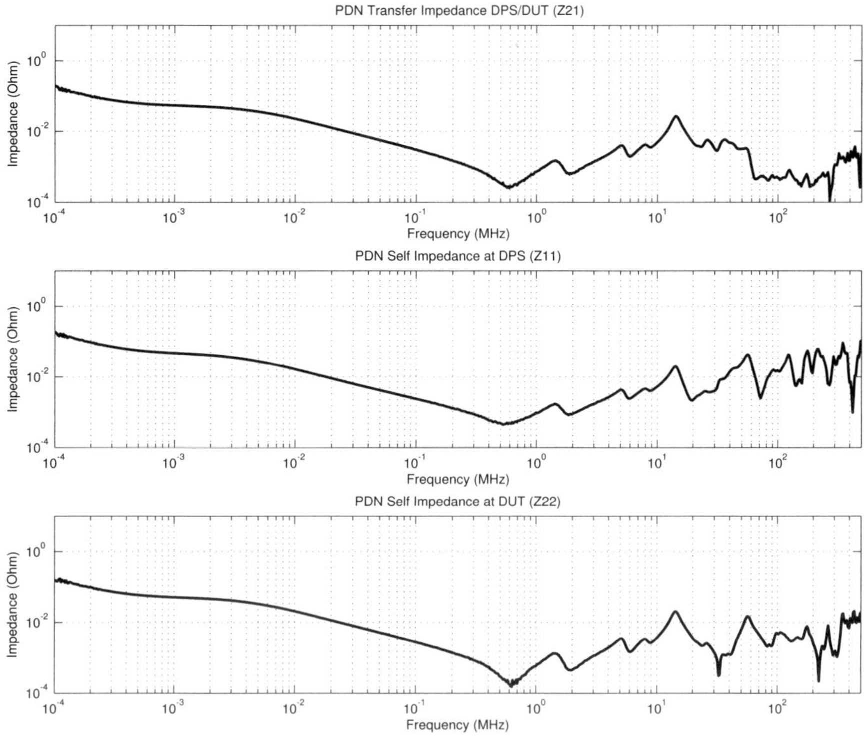

Figure H.35 Computed impedance values for the power distribution network of a test fixture: Z21 transfer impedance between the DPS and the DUT (top); Z11 self-impedance at the DPS (center); and Z22 self-impedance at the DUT (bottom).

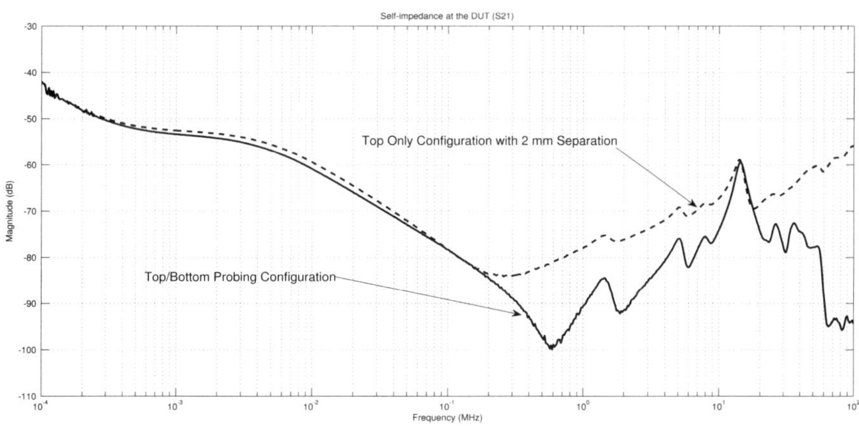

Figure H.36 Comparison of the measured S21 for a top/bottom probing configuration with a top only configuration with the two probes 2 mm apart.

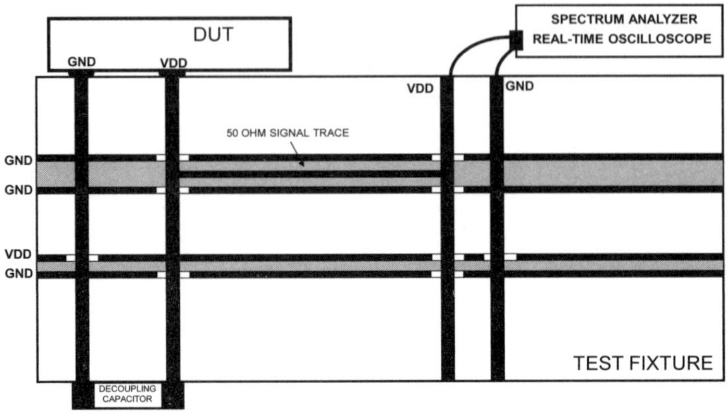

Figure H.37 Measuring the power spectrum of the voltage at the DUT power pad using a 50 Ω signal trace connected to the DUT power supply via.

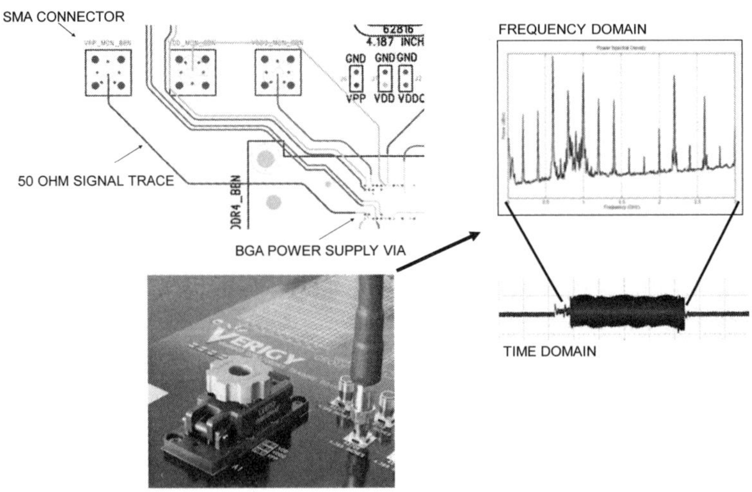

Figure H.38 Implementation example for a DDR4 application (Courtesy of Advantest).

References

[1] J. Moreira, H. Barnes, C. McCowan, and R. Winters, "A Time Domain Reflectometry Kit for ATE Test Fixtures," *Verigy Users Conference*, 2008.

[2] H. Barnes, R. Schaefer, and J. Moreira, "Analysis of Test Coupon Structures for the Extraction of High Frequency PCB Material Properties," *IEEE Workshop on Signal and Power Integrity*, 2013.

[3] Association Connecting Electronics Industries (IPC), "Test Methods to Determine the Amount of Signal Loss on Printed Boards," *IPC–TM–650 2.5.5.12*, 2012.

[4] J. P. Dunsmore, *Handbook of Microwave Component Measurements with Advanced VNA Techniques*. New York: John Wiley & Sons, 2012.

[5] D. D. Groot, P. Pupalaikis, and B. Shumaker, "Total Loss: How to Qualify Circuit Boards," *DesignCon*, 2011.

[6] C.-T. W. Lee and A. P. Neves, "Old School RF Structure Meets Modern Signal Integrity: The Beatty Standard," *DesignCon*, 2016.

[7] L. Martens, *High-Frequency Characterization of Electronic Packaging*. Boston, MA: Kluwer Academic Publishers, 1998.

[8] H. Barnes, J. Moreira, M. Comai, A. Islas, F. Tamayo-Broes, M. Resso, A. Ciccomancini, O. Bell, and M. Tsai, "Performance at the DUT: Techniques for Evaluating the Performance of an ATE System at the Device Under Test Socket," *IEC DesignCon*, 2008.

[9] J. Moreira, "Design of a High Bandwidth Interposer for Performance Evaluation of ATE Test Fixtures at the DUT Socket," *IEEE Asian Test Symposium*, 2012.

[10] J. Adley, E. Leung, and J. Sherry, "Electrical Modeling and Contactor Performance in a RF Test System," *Burn-In and Test Socket (BITS) Workshop*, 2002.

[11] H. Barnes, J. Moreira, M. Resso, and R. Schaefer, "Advances in ATE Fixture Performance and Socket Characterization for Multi-Gigabit Applications," *IEC DesignCon*, 2012.

[12] R. Weiss, "Interconnecting at 40 GHz and Beyond," *Burn-In and Test Socket Workshop*, 2002.

[13] M. Resso and E. Bogatin, *Signal Integrity Characterization Techniques*. IEC, 2009.

[14] H. Barnes, A. Ciccomancini, M. Resso, and M. Tsai, "Differential PCB Structures using Measured TRL Calibration and Simulated Structure De-Embedding," *IEC DesignCon*, 2007.

[15] Y. H. Kwark, M. Kotzev, C. Baks, X. Gu, and C. Schuster, "Novel Multiport Probing Fixture for High Frequency Measurements in Dense Via Arrays," *IEEE International Microwave Symposium*, 2011.

[16] M. Kotzev, Y. H. Kwark, C. Baks, X. Gu, and C. Schuster, "Electrical Performance of a Multiport Interposer for Measurements of Dense Via Arrays," *15th IEEE Workshop on Signal Propagation on Interconnects*, 2011.

[17] D. Thompson and J. Moreira, "Designing Sockets for Ludicrous Speed (80 GHz)," *Burn-In and Test Socket Workshop*, 2015.

[18] J. Moreira and B. Roth, "Characterization and Focus Calibration of ATE Systems for High-Speed Digital Applications," *IEC DesignCon*, 2009.

[19] J. Moreira, F. Pizza, C. Borelli, F. Corneo, H. Werkmann, S.-X. Yang, D. Lam, and B. Roth, "A Pragmatic Approach for At-Speed Characterization and Loopback Correlation at 28 Gbps," *Advantest VOICE Users Group Conference*, 2014.

[20] J. Moreira, M. Moessinger, D. Thompson, and M. Takahashi, "A Removable Signal Probing and Monitoring Solution for Gigabit Memory ATE Applications," *DesignCon*, 2013.

[21] J. Socha, J. Dandy, P. Pun, and P. Thota, "Designing High Performance Interposers with 3–Port and 6–Port S–Parameters," *DesignCon*, 2015.

[22] S. Lupo and O. Vikinski, "Pin Grid Array Current Sense Interposer Application Featuring Vertical Embedded Resistors," *Burn-In and Test Strategies (BITS) Workshop*, 2011.

[23] P. Girard, N. Nicolici, and X. Wen, *Power-Aware Testing and Test Strategies for Low Power Devices*. New York: Springer, 2009.

[24] I. Novak, "Frequency-Domain Power-Distribution Measurements – An Overview," *IEC DesignCon East*, 2003.

[25] I. Novak and J. R. Miller, *Frequency-Domain Characterization of Power Distribution Networks*. Norwood, MA: Artech House, 2007.

[26] I. Novak, *Power Distribution Network Design Methodologies*. IEC, 2008.

[27] I. Novak, Y. Mori, and M. Resso, "Accuracy Improvements of PDN Impedance Measurements in the Low to Middle Frequency Range," *IEC DesignCon*, 2010.

[28] P. Trivero, S. Grivet-Talocia, M. S. Nakhla, F. G. Canavero, and R. Achar, "Stability, Causality, and Passivity in Electric Interconnect Models," *IEEE Transactions on Advanced Packaging*, vol. 30, Nov. 2007.

[29] S. G. Talocia, "Towards Real-Time S-Parameter Qualification and Macromodeling," *IEEE Workshop on Signal and Power Integrity*, 2012.

[30] I. Novak, "Why S11 VNA Measurements Don't Work for PDN Measurements," *PCB Design 007 QuietPower Column*, Mar. 2010.

I

Jitter Injection Calibration

One of the most complicated challenges for a test engineer is to understand the error associated with a certain measurement setup and how to correlate the setup with other measurement instruments. For measurements like the frequency of a clock signal, it is possible to obtain an accurate estimation of the measurement uncertainty, since the measurement setup calibration uncertainty can be traced to an international recognized standard (e.g., a cesium clock).

As already discussed in Section 5.5.9, correlation of jitter measurements between different instruments and algorithms can be very challenging. One possible solution to this challenge is to analyze the jitter measurement of a given measurement setup or algorithm by using a source with a calibrated amount of jitter.

This appendix presents some techniques to address the challenge of calibrating a stimulus source (e.g., ATE driver) to deliver a specific amount of random and deterministic jitter [1]. The objective is to use calibration techniques that do not depend on a specific instrument (both hardware and proprietary algorithms), but use as much as possible standard measurement instrumentation that itself can be calibrated to known standards.

I.1 Sinusoidal Jitter Injection Calibration

Sinusoidal jitter injection is of key importance for I/O cell characterization, especially in tests like receiver jitter tolerance and transfer. Two methods for sinusoidal jitter injection calibration are presented in this section. The first method, called the J_1/J_0 Bessel approach, is based on an amplitude ratio measurement between the main carrier of the clock signal with sinusoidal jitter injected and its first phase modulation spectral line. This measurement is done with a spectrum analyzer and is an established procedure for sinusoidal jitter injection calibration [2, 3].

The second method, called the RJ Subtraction Approach, is based on the assumption that only random jitter is present on the driver when no sinusoidal jitter is injected. This method uses an oscilloscope for performing the calibration measurement in the time-domain by subtracting the random jitter value from the measured total jitter value.

I.1.1 The J_1/J_0 Bessel Approach

In this approach we assume that the DUT signal is a clock pattern, and if we ignore the harmonics due to the usual shape of a quasi-square wave bit clock (i.e., assuming that the bit clock is a pure sinusoidal signal), then injecting a sinusoidal jitter signal can be thought of as a simple phase modulating signal. The jittered (i.e., phase modulated) signal would then be described by:

$$V(t) = \sin\left(2\pi F_{BC}t - \pi J_{AMPL}\cos\left(2\pi f_J t\right)\right) \tag{I.1}$$

where F_{BC} is the frequency of the bit clock for a given data rate (e.g., 3.2 GHz for a 6.4 Gbps data rate). The sinusoidal jitter modulating signal has the expression $J_{AMPL}cos(2\pi f_J t)$ with f_J the value of the injected sinusoidal jitter frequency and J_{AMPL} the value of the injected sinusoidal jitter amplitude (the peak-to-peak jitter value will be twice the sinusoidal jitter amplitude, i.e., $J_{P_K-P_K} = 2J_{AMPL}$). One can now define a modulation index M_f that is related to the injected jitter amplitude by the following expression:

$$M_f = \pi J_{AMPL} \tag{I.2}$$

$$J_{AMPL} = \frac{M_f}{\pi} \tag{I.3}$$

The spectrum of the phase modulated clock will consist of a carrier and a theoretically infinite series of sideband pairs. The value of each sideband pair is computed by the Bessel function value for the modulation index (M_f) being used as shown in Figure I.1 for the carrier amplitude J_0 and the first pair of sidebands J_1.

One important point is that typically one does not have a sinusoidal clock when using a clock pattern but a bit clock (i.e., theoretically the waveform should be a square clock). Figure I.2 shows a comparison of the spectrum of a phase modulated sinusoidal clock with a bit clock. Notice that although the spectrum is different, around the carrier frequency it is easy to distinguish the sidebands due to the phase modulation. Although the presented approach

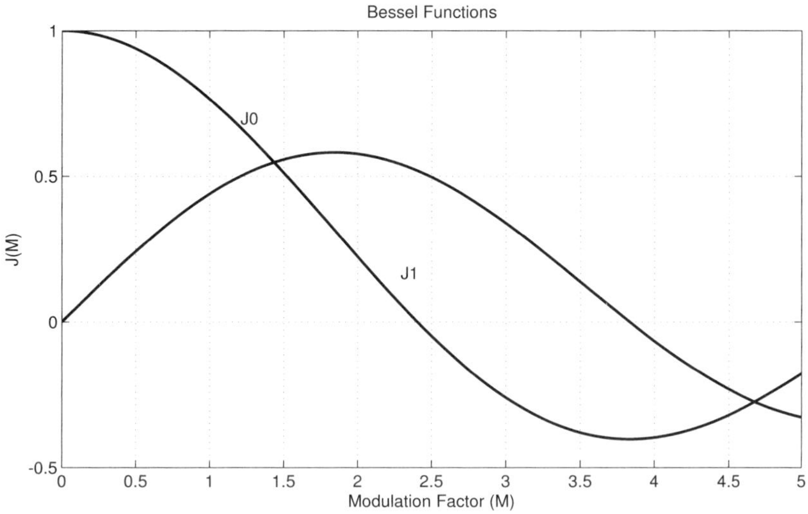

Figure I.1 J_0 and J_1 Bessel function plot.

assumes a sinusoidal clock, for most situations we can apply the method directly to a bit clock signal.

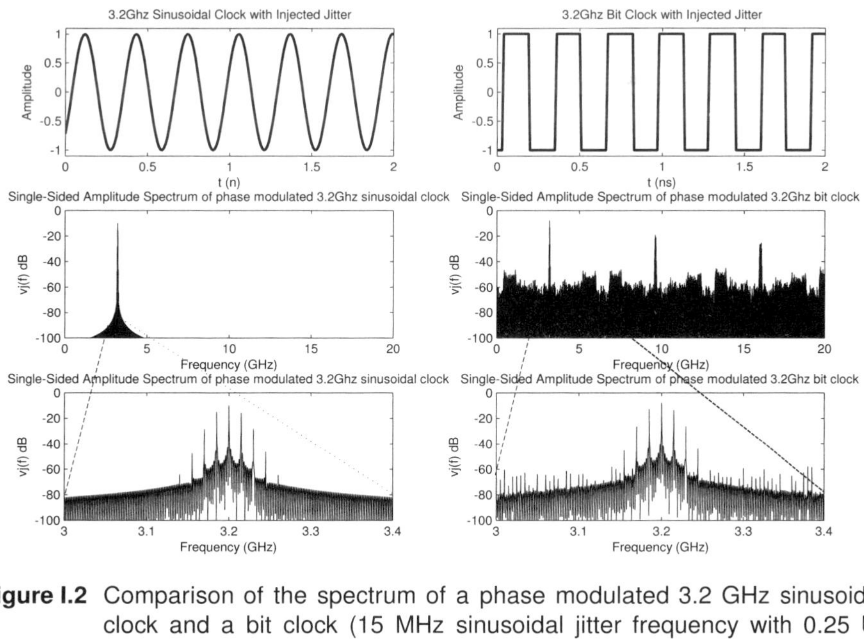

Figure I.2 Comparison of the spectrum of a phase modulated 3.2 GHz sinusoidal clock and a bit clock (15 MHz sinusoidal jitter frequency with 0.25 UI amplitude).

The methodology for calibrating the sinusoidal jitter injection amplitude is then to first compute the needed modulation index from (I.3) and then compute the values of the J_0 and J_1 Bessel functions for that modulation index. With these values we can compute the ratio J_1/J_0. When calibrating a stimulus source for a given injected sinusoidal jitter amplitude, one needs to change the driver parameters (e.g., the sinusoidal jitter source amplitude) until the desired J_1/J_0 ratio is obtained in the bit clock measured spectrum [2]. Using (I.3) now and a Bessel function table or a mathematical software package (e.g., MATLAB), it is possible to generate a table that provides a J_1/J_0 value for each value of sinusoidal jitter injection amplitude as shown in Table I.1.

Let us now take a real example. Let us suppose we would like to inject 0.25 UI sinusoidal jitter amplitude at 15 MHz on a 6.4 Gbps signal (0.5 UI peak-to-peak sinusoidal jitter). The bit clock frequency for this signal is 3.2 GHz and Table I.1 shows that for 0.25 UI of injected sinusoidal jitter amplitude, the J_1/J_0 ration needs to be 0.426. Figure I.3 shows the jittered bit clock signal measured with an equivalent-time oscilloscope and the signal spectrum measured with a spectrum analyzer where one can easily see the main carrier harmonics from the fact that we are measuring a quasi-square wave bit clock and not a sinusoidal clock.

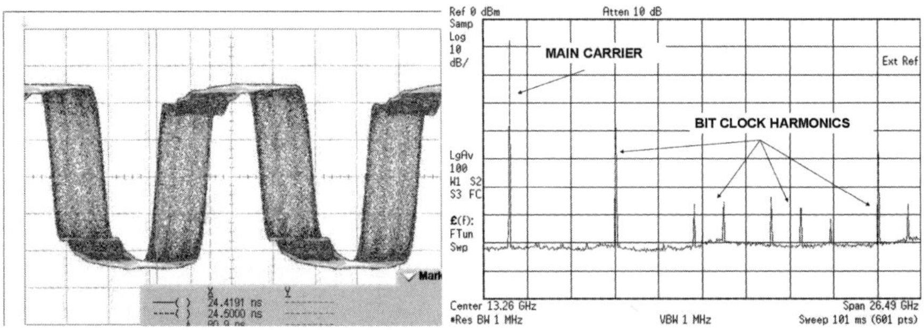

(a) (b)

Figure I.3 (a) Waveform of a jittered 6.4 Gbps clock pattern with 15 MHz sinusoidal jitter injection with 0.25 UI amplitude and (b) its spectrum.

Figure I.4 shows the main carrier and the first sidebands in a linear scale so that a measurement of the ratio J_1/J_0 can be made using the spectrum analyzer markers and the marker delta function. The measured value was 42.94% (0.4294), which is very close to the desired value (a ratio value of 0.428 corresponds to a jitter injection amplitude value of 0.251 UI by consulting Table I.1, which would be an approximately 156 fs (0.001 UI)

Table I.1

Table for Calibrating the Injected Sinusoidal Jitter Using the J_1/J_0 Method

INJECTED JITTER (UI)	MOD FACTOR	J0	J1	J1/J0	INJECTED JITTER (UI)	MOD FACTOR	J0	J1	J1/J0
0.01	0.0314	0.9998	0.0157	0.0157	0.51	1.6022	0.4541	0.5701	1.2554
0.02	0.0628	0.999	0.0314	0.0314	0.52	1.6336	0.4362	0.573	1.3137
0.03	0.0942	0.9978	0.0471	0.0472	0.53	1.665	0.4181	0.5755	1.3763
0.04	0.1257	0.9961	0.0627	0.063	0.54	1.6965	0.4	0.5776	1.4438
0.05	0.1571	0.9938	0.0783	0.0788	0.55	1.7279	0.3819	0.5792	1.5169
0.06	0.1885	0.9911	0.0938	0.0947	0.56	1.7593	0.3636	0.5805	1.5963
0.07	0.2199	0.9879	0.1093	0.1106	0.57	1.7907	0.3454	0.5813	1.6832
0.08	0.2513	0.9843	0.1247	0.1267	0.58	1.8221	0.3271	0.5818	1.7785
0.09	0.2827	0.9801	0.14	0.1428	0.59	1.8535	0.3088	0.5818	1.8839
0.1	0.3142	0.9755	0.1551	0.159	0.6	1.885	0.2906	0.5815	2.0012
0.11	0.3456	0.9704	0.1702	0.1754	0.61	1.9164	0.2723	0.5807	2.1325
0.12	0.377	0.9648	0.1852	0.1919	0.62	1.9478	0.2541	0.5795	2.2809
0.13	0.4084	0.9587	0.2	0.2086	0.63	1.9792	0.2359	0.578	2.4501
0.14	0.4398	0.9522	0.2146	0.2254	0.64	2.0106	0.2178	0.576	2.6451
0.15	0.4712	0.9452	0.2291	0.2424	0.65	2.042	0.1997	0.5737	2.8725
0.16	0.5027	0.9378	0.2435	0.2596	0.66	2.0735	0.1817	0.5709	3.1416
0.17	0.5341	0.93	0.2576	0.277	0.67	2.1049	0.1638	0.5678	3.4654
0.18	0.5655	0.9216	0.2716	0.2947	0.68	2.1363	0.1461	0.5643	3.8632
0.19	0.5969	0.9129	0.2854	0.3126	0.69	2.1677	0.1284	0.5604	4.3645
0.2	0.6283	0.9037	0.2989	0.3308	0.7	2.1991	0.1109	0.5561	5.0164
0.21	0.6597	0.8941	0.3122	0.3492	0.71	2.2305	0.0935	0.5514	5.9005
0.22	0.6912	0.8841	0.3253	0.368	0.72	2.2619	0.0762	0.5464	7.17
0.23	0.7226	0.8737	0.3382	0.3871	0.73	2.2934	0.0591	0.5411	9.1507
0.24	0.754	0.8628	0.3508	0.4066	0.74	2.3248	0.0422	0.5353	12.6798
0.25	0.7854	0.8516	0.3632	0.4265	0.75	2.3562	0.0255	0.5292	20.7583
0.26	0.8168	0.84	0.3753	0.4467	0.76	2.3876	0.009	0.5228	58.2918
0.27	0.8482	0.8281	0.3871	0.4675	0.77	2.419	-0.0074	0.516	-70.207
0.28	0.8796	0.8157	0.3986	0.4887	0.78	2.4504	-0.0235	0.5089	-21.7019
0.29	0.9111	0.803	0.4099	0.5104	0.79	2.4819	-0.0393	0.5015	-12.7533
0.3	0.9425	0.79	0.4208	0.5327	0.8	2.5133	-0.055	0.4938	-8.9844
0.31	0.9739	0.7766	0.4315	0.5556	0.81	2.5447	-0.0703	0.4857	-6.9048
0.32	1.0053	0.7629	0.4418	0.5791	0.82	2.5761	-0.0855	0.4774	-5.5848
0.33	1.0367	0.7488	0.4518	0.6033	0.83	2.6075	-0.1003	0.4687	-4.6714
0.34	1.0681	0.7345	0.4614	0.6283	0.84	2.6389	-0.1149	0.4598	-4.0007
0.35	1.0996	0.7198	0.4708	0.654	0.85	2.6704	-0.1292	0.4506	-3.4866
0.36	1.131	0.7049	0.4798	0.6806	0.86	2.7018	-0.1432	0.4411	-3.0794
0.37	1.1624	0.6897	0.4884	0.7082	0.87	2.7332	-0.1569	0.4313	-2.7483
0.38	1.1938	0.6742	0.4967	0.7367	0.88	2.7646	-0.1703	0.4213	-2.4734
0.39	1.2252	0.6585	0.5046	0.7663	0.89	2.796	-0.1834	0.411	-2.2411
0.4	1.2566	0.6425	0.5122	0.7972	0.9	2.8274	-0.1962	0.4005	-2.042
0.41	1.2881	0.6263	0.5194	0.8293	0.91	2.8588	-0.2086	0.3898	-1.869
0.42	1.3195	0.6099	0.5262	0.8628	0.92	2.8903	-0.2206	0.3789	-1.7171
0.43	1.3509	0.5932	0.5326	0.8978	0.93	2.9217	-0.2324	0.3677	-1.5824
0.44	1.3823	0.5764	0.5387	0.9346	0.94	2.9531	-0.2437	0.3564	-1.462
0.45	1.4137	0.5594	0.5444	0.9731	0.95	2.9845	-0.2548	0.3448	-1.3535
0.46	1.4451	0.5422	0.5497	1.0137	0.96	3.0159	-0.2654	0.3331	-1.255
0.47	1.4765	0.5249	0.5545	1.0565	0.97	3.0473	-0.2757	0.3212	-1.1651
0.48	1.508	0.5074	0.559	1.1018	0.98	3.0788	-0.2856	0.3092	-1.0825
0.49	1.5394	0.4898	0.5631	1.1498	0.99	3.1102	-0.2951	0.297	-1.0063
0.5	1.5708	0.472	0.5668	1.2009	1	3.1416	-0.3042	0.2846	-0.9355

amplitude error compared to the desired 0.25 UI at 6.4 Gbps). Also note that we have not discussed the accuracy of the spectrum measurement since the spectrum analyzer noise floor is expected to be below the needed measurement accuracy for a jitter injection calibration application.

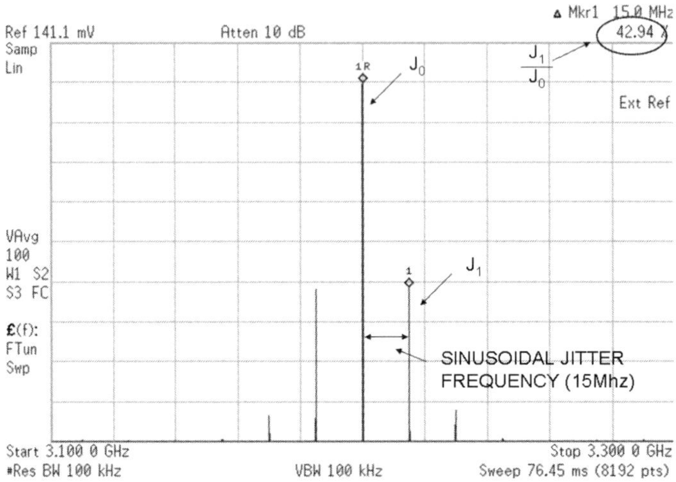

Figure I.4 Spectrum of a jittered 6.4 Gbps clock pattern with 16 MHz sinusoidal jitter injection with 0.25 UI amplitude (0.5 UI peak-to-peak) zoomed around the 3.2 GHz carrier frequency.

The method presented in this subsection is appropriate for low amplitudes of sinusoidal jitter injection (e.g., 1 to 5 UI). For higher amplitudes (e.g., 50 UI) similar methodologies are available, like the null Bessel method, which are more appropriate [2, 3].

I.1.2 The RJ Subtraction Approach

In this approach the main idea is to assume that all the jitter present in the stimulus driver is random for a bit clock pattern. In this case any injected sinusoidal jitter will add on top of the random jitter and since they should be independent processes it is possible to add their variances [4].

Sinusoidal jitter with a peak-to-peak jitter value $J_{P_K-P_K}$ is described by the following probability distribution [5]:

$$p_{\text{sinusoidal}}(x) = \begin{cases} \dfrac{1}{\pi\sqrt{\left(\frac{J_{P_K-P_K}}{2}\right)^2 - x^2}} & |x| < \dfrac{J_{P_K-P_K}}{2} \\ 0 & \text{Otherwise} \end{cases} \tag{I.4}$$

Note that the probability distribution is independent of the frequency of the sinusoidal jitter. The variance of the distribution is computed by the following equation:

$$\sigma^2_{SINUSOIDAL} = \frac{\left(\frac{J_{P_K-P_K}}{2}\right)^2}{2} \tag{I.5}$$

Figure I.5 presents a graph of the probability distribution for a signal with a sinusoidal jitter peak-to-peak value of 4 ps. Note that the sinusoidal jitter amplitude value will be half of the peak-to-peak value for sinusoidal jitter.

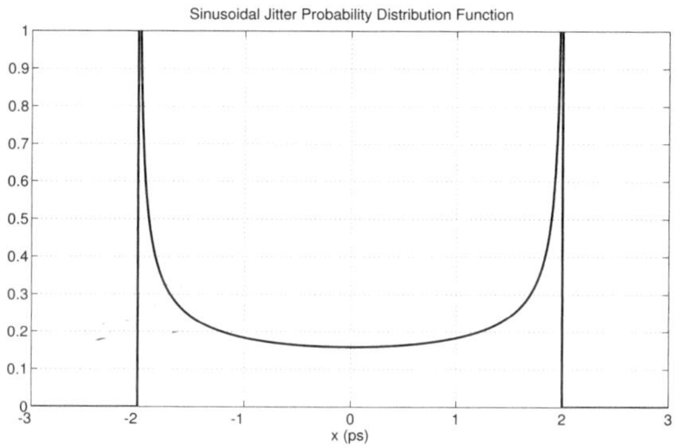

Figure I.5 Sinusoidal jitter probability distribution function ($J_{P_K-P_K} = 4$ ps).

Assuming that there is only random jitter and the injected sinusoidal jitter on the data signal from the stimulus driver, since one typically can consider the random jitter process independent (not correlated) to the sinusoidal jitter injection process, the total variance can then be computed by the square sum of the random and sinusoidal jitter variances.

$$\sigma^2_{TOTAL} = \sigma^2_{SINUSOIDAL} + \sigma^2_{RANDOM} \tag{I.6}$$

Substituting (I.5) into (I.6) and solving it for the sinusoidal peak-to-peak jitter value, we obtain:

$$J_{P_K-P_K} = 2\sqrt{2}\sqrt{(\sigma^2_{TOTAL} - \sigma^2_{RANDOM})} \tag{I.7}$$

This means that to compute the value of the injected peak-to-peak sinusoidal jitter amplitude it is necessary to measure the variance of the random jitter and the variance of the total jitter. Figure I.6 shows an example on how (I.7) can be used for sinusoidal jitter injection calibration with a time-domain measurement instrument (in this case an equivalent-time oscilloscope). Since the methodology assumes that when no sinusoidal jitter is injected only random jitter exists on driver output, one needs to use a bit clock pattern to prevent any jitter due to ISI. The driver can have other jitter forms like DCD (duty-cycle distortion) or other forms of periodic jitter. Since test and measurement equipment drivers are typically well designed it is acceptable to assume that the other types of deterministic jitter are very small, although one should always verify this assumption. Figure I.6 (left) shows the measured variance using the histogram function of the oscilloscope when no sinusoidal jitter is injected, and on the right the measured variance when the sinusoidal jitter is injected. In this example 78.125 ps of sinusoidal peak-to-peak jitter at 6.4 Gbps (0.5 UI) was injected at 15 MHz. The injected jitter was calibrated with the J_1/J_0 Bessel approach of Section I.1.1.

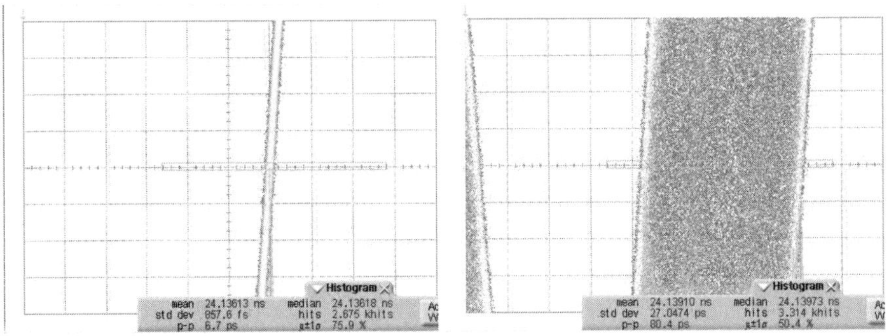

Figure I.6 Measured histogram of the driver output without jitter injection (left) and with sinusoidal jitter injected (right).

From the measurements and using (I.7) the peak-to-peak amplitude of the sinusoidal jitter is:

$$J_{PK-PK} = 2\sqrt{2}\sqrt{(27.05)^2 - (0.86)^2} = 76.47 \text{ ps} = 0.4894 \text{ UI} \qquad \text{(I.8)}$$

This means that, in this specific example, this methodology showed an error of approximately 1.6 ps (78.125 ps – 76.47 ps) for the injected peak-to-peak jitter amplitude compared with the J_1/J_0 Bessel methodology of Section I.1.1.

On the measurements shown in Figure I.6, the measured peak-to-peak total jitter value was 80.4 ps, which corresponds to an error of 2.27 ps if the measured peak-to-peak total jitter value was used as the peak-to-peak value of the sinusoidal jitter injected. Note that the error on the total jitter measurement peak-to-peak value is low (2.27 ps). This is because in the presented case, the random jitter of the driver was very low, which means that the measured total peak-to-peak jitter is very close to the injected sinusoidal peak-to-peak jitter value. Figure I.7 shows another example where the driver has significantly more random jitter and in this case the measured peak-to-peak total jitter value is 105.6 ps (we calibrated a 0.5 UI sinusoidal peak-to-peak jitter injection using the J_1/J_0 Bessel method) which means a 27.5 ps error, while if one uses (I.7), the value is $2\sqrt{2}\sqrt{27.75^2 - 6.3^2} = 76.4$ ps. An error of 1.7 ps is computed from the measurement.

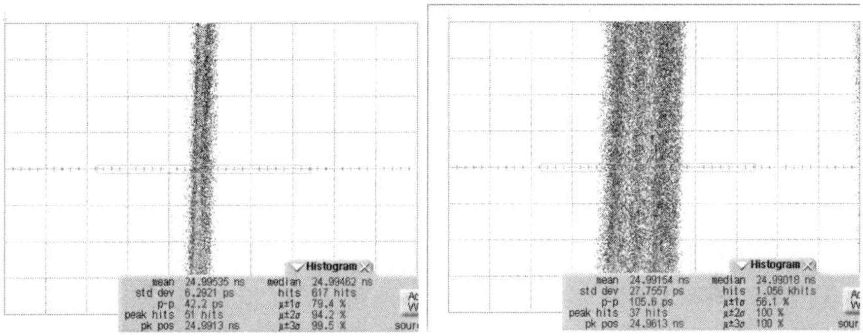

Figure I.7 Measured histogram of the driver output without jitter injection (left) and with sinusoidal jitter injected (right) for a driver with a significant amount of random jitter.

I.2 Random Jitter Injection Calibration

Random jitter injection calibration typically requires the calibration of the standard deviation σ (also known as RMS jitter) of a noise source that injects random jitter into a data signal. It is important to note that in some high-speed standards like PCI Express [6], the standard deviation for random jitter injection is specified for specific frequency bands as shown in Figure I.8.

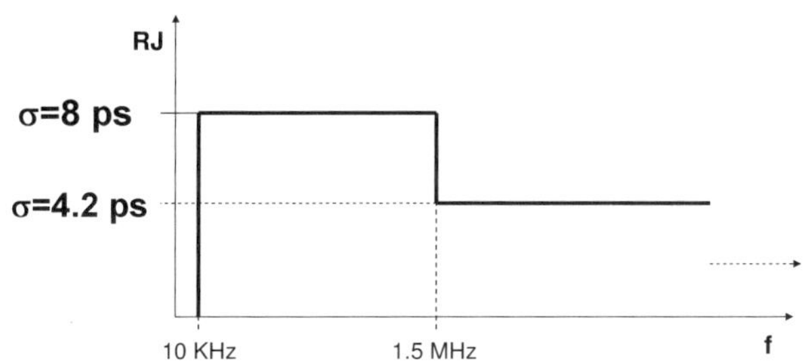

Figure I.8 PCI Express random jitter tolerance requirements [6].

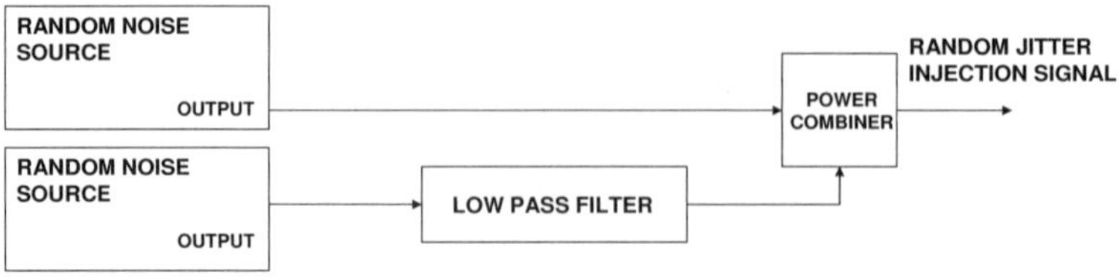

Figure I.9 Creating the random jitter injection profile of Figure I.8 using two random noise sources, a low-pass filter, and a power combiner.

This implies the usage of multiple random noise sources with appropriate band limiting filters as shown in Figure I.9. It is important to note that limiting the bandwidth of a Gaussian noise source can have consequences on its properties [7]. Figure I.10 shows the frequency profile of an example implementation of the methodology described in Figure I.9.

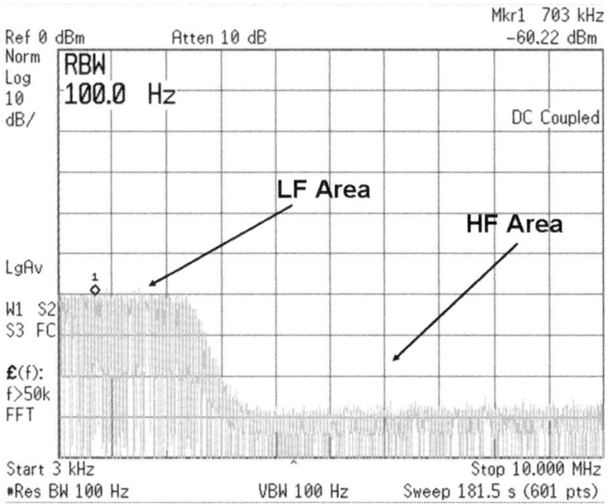

Figure I.10 Example of a measured frequency profile using the random jitter injection methodology of Figure I.9.

Section 7.8 discusses noise sources in more detail. In this section we will assume that one needs to calibrate a random jitter source that has a predefined noise bandwidth to a given standard deviation value (random jitter RMS value). One option is to measure the standard deviation of the data waveform with the random jitter injected using a time-domain measurement instrument like an oscilloscope as shown in Figure I.11. In this example a clock pattern was used at the data rate of 6.4 Gbps. The usage of a clock pattern is important since with this approach we can ensure that no data-dependent jitter is present in the signal. The measured standard deviation (RMS jitter) is approximately

5.3 ps. Note that in this example the histogram shows some asymmetries due to the fact that the delay line used for the jitter injection is not completely linear.

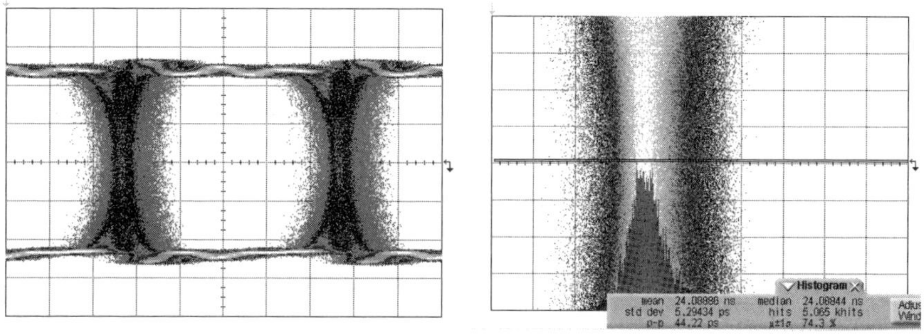

Figure I.11 Measuring the standard deviation (RMS value) of a bit clock signal with random jitter injected.

With the approach described above, the user would tune the amplitude of the noise source until the measured standard deviation corresponds to the desired value of random jitter. The problem with this approach is that a distribution with a given standard deviation does not mean that it corresponds to a Gaussian distribution with that standard deviation value. This is the case when the driver delay line linearity is not able to translate a Gaussian noise distribution perfectly into Gaussian random jitter in the signal. Since noise sources and stimulus sources can vary significantly in terms of quality, this is important (see Section 7.8). Another option would be to perform a fit of the histogram data to a Gaussian distribution to infer the standard deviation value. Figure I.12 shows an example of this procedure using the histogram data measured in Figure I.11. The standard deviation of the fitted Gaussian is now approximately 4 ps (compared to 5.3 ps of simply computing the standard deviation of the histogram). Note that verifying that a random distribution is truly Gaussian or which methodology to use for fitting a Gaussian curve to a measured distribution is not trivial [8].

One point that was not discussed is the intrinsic random jitter of the measurement instrument used for measuring the jitter histogram. For an accurate calibration of the injected random jitter in a stimulus driver, this intrinsic jitter should be calibrated out. This is easy if one assumes that the intrinsic jitter present on the measurement instrument is also random with a Gaussian distribution (a reasonable assumption for well-designed measurement instrumentation). In this case the intrinsic jitter from the measurement instrument can be removed from the measured random jitter standard deviation by the following equation:

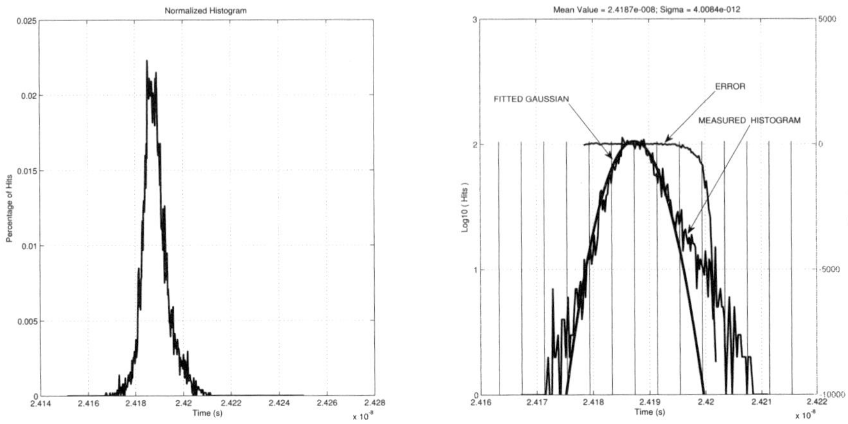

Figure I.12 Fitting a histogram to a Gaussian distribution. Note that a logarithmic scale is used on the right figure (based on the algorithm presented in [9]).

$$\sigma_{RJ} = \sqrt{\sigma_{MEASURED}{}^2 - \sigma_{INTRINSIC}{}^2} \tag{I.9}$$

This still leaves the question of how one should measure the intrinsic jitter of a measurement instrument. One option is to use the measurement setup described in Figure I.13. In this setup a very low phase noise RF source is provided to both the trigger input and the measurement input of the instrument. In this case any jitter measured by the instrument is due to the jitter of the RF source which should be very low compared to the expected instrument jitter, and the instrument intrinsic jitter. Figure I.14 presents an example of this approach using the Agilent Technologies 86100C equivalent-time oscilloscope.

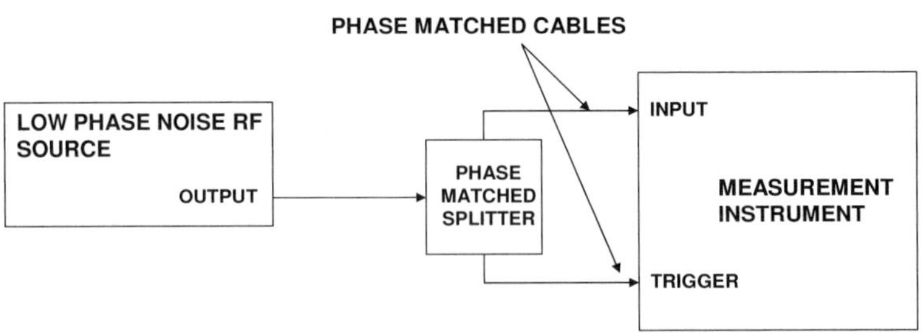

Figure I.13 Measuring the intrinsic jitter of a time-domain measurement instrument (e.g., equivalent-time oscilloscope).

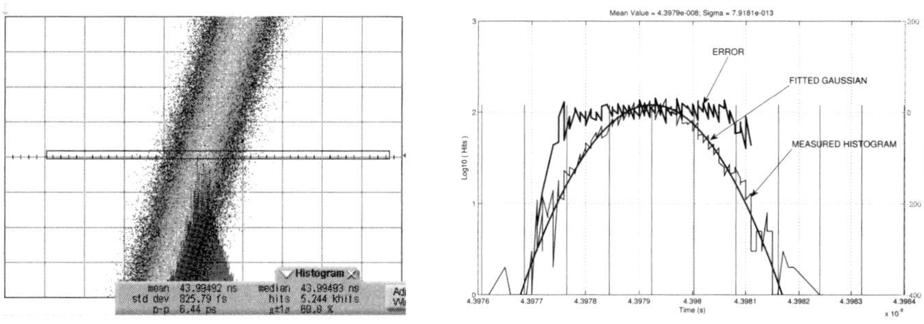

Figure I.14 Measuring the intrinsic jitter of an equivalent-time oscilloscope using a RF source.

I.3 ISI Jitter Injection Calibration

Calibrating data-dependent jitter injection in the form of ISI presents significant challenges. ISI can be defined via a filter specification, a lossy PCB signal trace [e.g., 50.8 cm (20 in), 127 μm (5 mil) trace on FR4], or sometimes by a simple peak-to-peak jitter value.

In standards like PCI Express [6], DDJ injection is defined by a PCB signal trace with a certain geometry on an FR4 type dielectric material. This type of definition creates significant difficulties since it does not provide an exact description of the amount of DDJ that is injected. This can be observed in the example shown in Figure I.15 where the PCB board includes a 127 μm (5 mil) wide, 50.8 cm (20 in) long microstrip and stripline trace. The measured insertion loss shows a difference between both traces with the microstrip trace showing the worst performance. Although the performance of the microstrip should be better since half of its dielectric is expected to be air, in reality the microstrip plating and solder mask can significantly increase the microstrip loss as shown in Section 8.3.1.

If we now synthesize the data eyes for a 2.5 Gbps PRBS $2^7 - 1$ data pattern for each signal trace using the measured S-parameter data, we obtain the data eyes shown in Figure I.16. The data eyes show that there is a significant difference on the measured injected jitter. Note that on the simulation no random jitter or any other forms of deterministic jitter are included, only ISI jitter from the signal trace loss. This means that although following the guideline of a particular standard, the way that the PCB is implemented can have a significant impact on the obtained ISI value. Although this example concentrates on the differences between a microstrip or stripline implementation, similar results could be obtained for differences between the specific FR4 material used. Such issues become more significant when moving to higher data rates.

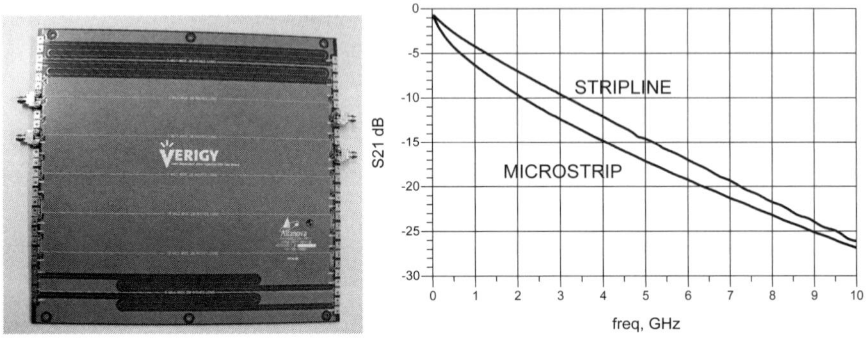

Figure I.15 Comparison of the insertion loss (S21) of a 127 μm (5 mil) wide, 50.8 cm (20 in) long trace implemented as microstrip and stripline on a FR4 dielectric.

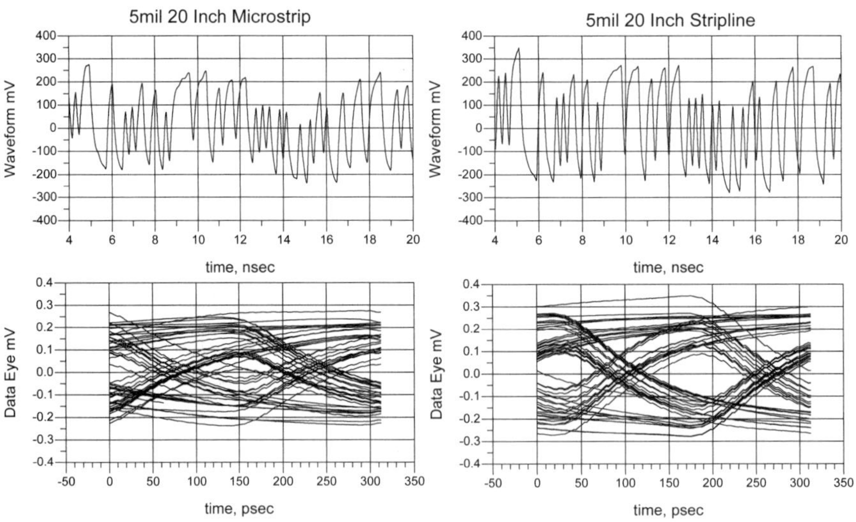

Figure I.16 Comparison of the synthesized data eye at 6.4 Gbps for a 127 μm (5 mil) wide, 50.8 cm (20 in) long microstrip in FR4 (left) and a stripline (right).

This means that for ISI jitter injection, the appropriate way to define and calibrate the amount of jitter injected is to define the response of the lossy signal channel (either a filter or a PCB trace) through the full S-parameter response of the channel (or the time-domain step response). From that data it is then possible to compute the corresponding peak-to-peak jitter value for the ISI jitter through simulation.

It is very important to remember that although the injected ISI jitter is defined by a lossy channel, its peak-to-peak value will depend on the spectral properties of the used pattern that are determined, for example, by the PRBS sequence type as shown in Figure I.17.

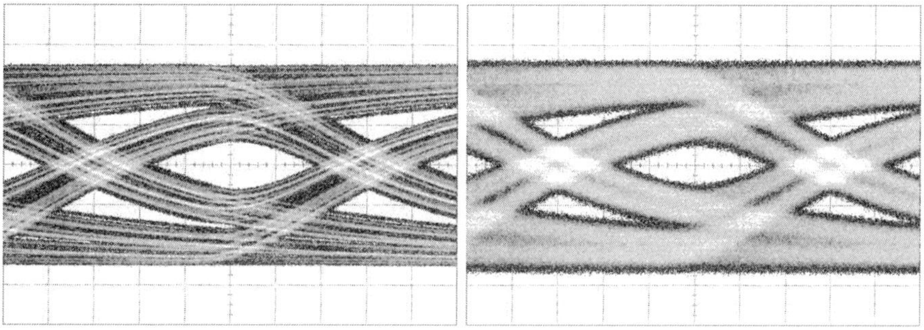

Figure I.17 Influence of the used pattern on the amount of injected DDJ (127 μm (5 mil) wide, 50.8 cm (20 in) long trace on an FR4 dielectric), left: a PRBS $2^7 - 1$ pattern; right: a PRBS $2^{31} - 1$ pattern.

In the case the DDJ is defined as a peak-to-peak value, one approach to generate a PCB signal trace definition is to use a simulation tool to optimize the PCB trace geometry to obtain the desired peak-to-peak DDJ value for a given data pattern. When following this approach, it is critical that the simulation models for the PCB trace loss have been verified and optimized with test boards.

References

[1] J. Stimple and R. Stephens, "Precision Jitter Transmitter," *IEC DesignCon*, 2005.

[2] Agilent Technologies, "Understanding Jitter and Wander Measurements and Standards," *Application Note 5988-6254EN*, 2003.

[3] D. Derickson and M. Mueller, *Digital Communications Test and Measurement: High-Speed Physical Layer Characterization*. Upper Saddle River, NJ: Prentice-Hall, 2007.

[4] A. Papoulis and S. U. Pillai, *Probability, Random Variables and Stochastic Processes*. New York: McGraw-Hill, 2002.

[5] W. Maichen, *Digital Timing Measurements.* New York: Springer, 2006.

[6] PCI-SIG, *PCI Express Base Specification Revision 2.1*, Mar. 2009.

[7] R. Stephens and M. Mueller, "Analysis of Random Noise and the Effect of Band-Limited Noise on Stressed-Eye Receiver Tolerance Tests," *IEC DesignCon*, 2009.

[8] R. Stephens and R. Muro, "Characterization of Gaussian Noise Sources," *IEC DesignCon*, 2008.

[9] Tektronix, "Controlled Jitter Generation for Jitter Tolerance and Jitter Transfer Testing," *Application Note*, 2005.

J

Phase Noise, RMS Jitter, and Random Jitter

In some cases the jitter of a DUT, especially in the case of clock oscillators, PLLs, or any other output in the form of a bit clock pattern is specified in terms of phase noise. It is important to understand the relationship between phase noise, RMS jitter, and random jitter. Phase noise is measured in the frequency-domain [1, 2]. Figure J.1 (left) shows the output RF power spectrum around the fundamental frequency of a 1 Gbps bit clock signal (500 MHz fundamental frequency).

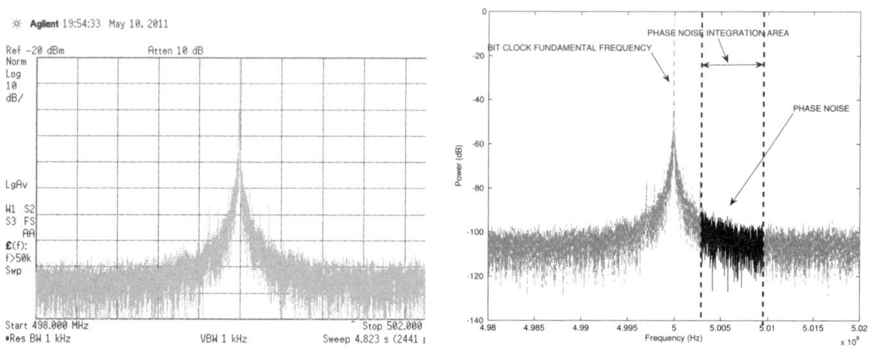

Figure J.1 Frequency spectrum of a 1 Gbps bit clock sinal (left) and the definition of phase noise (right).

Phase noise is computed by integrating a specified area on one sideband of the main carrier of the bit clock signal as shown in Figure J.1 (right). The phase noise (RMS phase jitter) can then be computed using (J.1) [3], where

f_0 is the carrier frequency (500 MHz in the case of a 1 Gbps bit clock) and $\int_{f_1}^{f_2} L(f)df$ is the noise integration area from frequency f_1 to f_2.

$$PhaseJitter_{RMS} = \frac{\sqrt{2. \int_{f_1}^{f_2} L(f)df}}{2\pi f_0} \tag{J.1}$$

It is important to take into account the units for the integration. The phase noise spectrum is usually represented in frequency (Hz) versus power (dBc/Hz) as show in the simple example of Figure J.2. Note that in the graph, the zero frequency point corresponds to the carrier frequency and only one side of the spectrum around the carrier frequency is shown. This is the typical way that phase noise plots are shown especially in specialized phase noise measurement instruments (see Section 7.6) or software.

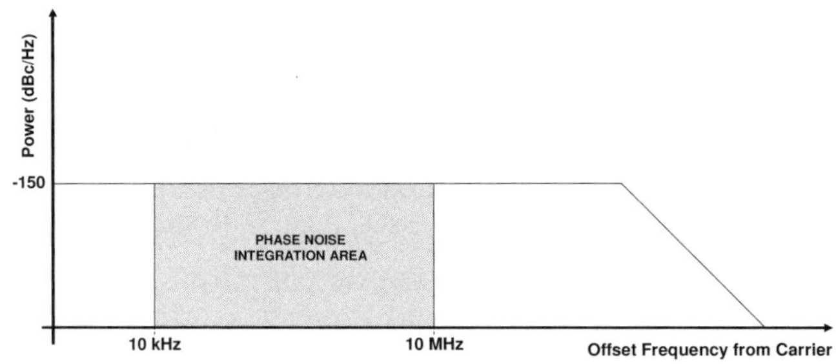

Figure J.2 Simple phase noise example.

The phase noise jitter in the example of Figure J.2 can be easily computed as shown in (J.2) assuming that we are measuring a 500 MHz clock. The term $10^{\frac{dBc/Hz}{10}}$ is necessary when using dBc/Hz units.

$$PhaseJitter_{RMS} = \frac{\sqrt{2 \times (10 \times 10^6 - 10 \times 10^3) \times 10^{\frac{-150}{10}}}}{2.\pi.500 \times 10^6} \tag{J.2}$$

The critical point with phase noise is that as seen in (J.1) it will depend on the integrated area. The proper choice of this area depends on the application requirements. For example, if the bit clock signal to be measured will be fed to a clock and data recovery (CDR) unit in its final application, that is able to track jitter to a certain frequency, then it makes sense to compute the phase noise using that frequency as the low value of the integration range.

This critical dependence on the integrated phase noise area is demonstrated in Figure J.3 where a 2.6 ps phase noise RMS jitter is measured with

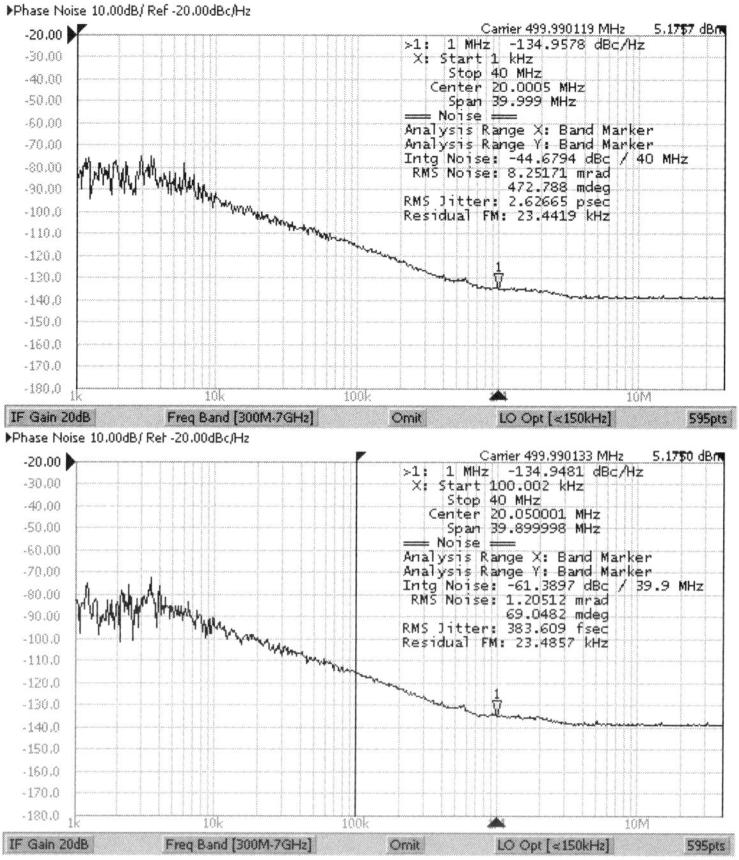

Figure J.3 Example of the phase noise RMS jitter value dependency on the integrated frequency range. Integrated from 1 kHz to 40 MHz gives 2.6 ps RMS (top) and from 100 kHz to 40 MHz gives 0.384 ps RMS (bottom).

a 1 kHz to 40 MHz phase noise integration frequency range, but this value drops to 384 fs if the frequency range is reduced to 100 kHz to 40 MHz.

This also raises an issue when phase noise RMS jitter measurements have to be correlated to jitter measurements performed using other instruments like an oscilloscope as shown in the example of Figure J.4.

In the figure a jitter histogram is measured and the standard deviation of the jitter histogram is computed which corresponds to the jitter RMS value assuming the measured signal is mainly dominated by random jitter. In this example a large difference between the measured RMS jitter value on the real-time oscilloscope (1.28 ps) and the measured phase noise RMS jitter value from the signal source analyzer (0.271 ps) is observed. One of the

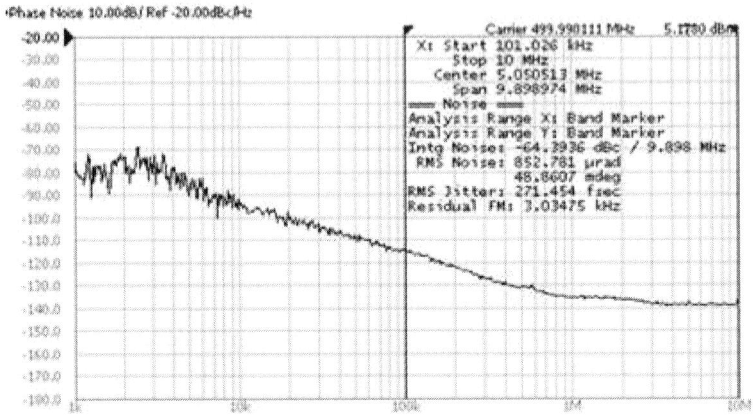

a) SIGNAL SOURCE ANALYZER, RMS JITTER: 0.271 ps

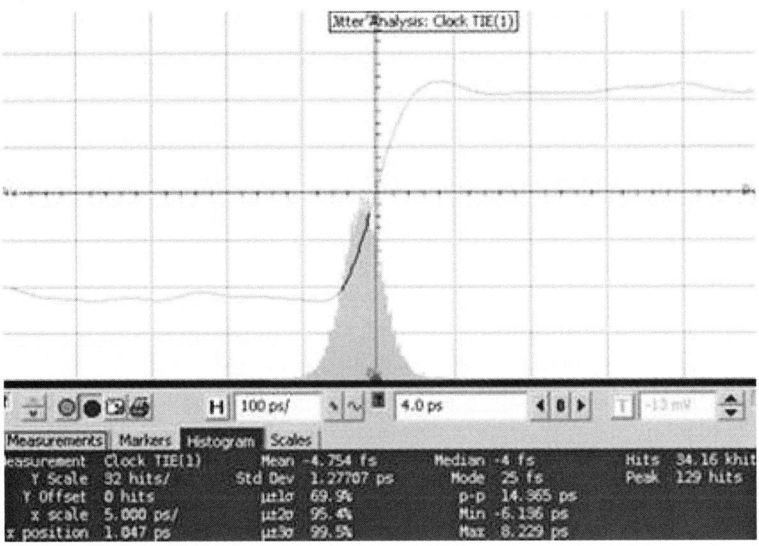

b) REAL-TIME SAMPLING OSCILLOSCOPE, RMS JITTER: 1.277 ps

Figure J.4 RMS jitter from a 1 Gbps bit clock measured with an Keysight Technologies signal source analyzer integrated from 100 kHz to 10 MHz (a) and with an Keysight Technologies real-time sampling oscilloscope (b).

reasons for this difference is that in the signal source analyzer the phase noise integration area starts at 100 kHz and goes to 10 MHz while on the real-time sampling oscilloscope there is no frequency range choice for the RMS jitter measurement. It will measure the entire frequency range. It might be possible to improve this correlation by using a software CDR with the appropriate frequency transfer characteristic on the real-time sampled waveform before computing the jitter histogram. In the case of an equivalent-time oscilloscope or a BER tester, it is possible to use a hardware-based CDR with the correct loop filter bandwidth. Note that adding a CDR does not guarantee a perfect correlation with a phase noise measurement. It only improves the correlation in some cases.

The previous example shows how critical it is to understand how a phase noise measurement was performed when trying to correlate to other time-domain instruments (e.g., real-time sampling oscilloscope, equivalent-time sampling oscilloscope, time interval analyzer, bit error rate tester, and ATE digital pin electronics), but one needs to understand that this correlation becomes even harder when other jitter components exist apart from the random jitter.

A phase noise measurement usually only measures the random jitter component on the signal being measured. This means that all deterministic jitter components or spurs need to be removed from the signal before the phase noise measurement since it can have an impact on the measured value as shown in Figure J.5. This sometimes requires that the phase noise measurement instrument contains the required software to remove the deterministic jitter components. In the example of Figure J.5 the phase noise with the deterministic jitter components removed is 2.47 ps compared to 4.68 ps if the deterministic jitter components are not removed.

This discussion also applies to the measurement instruments to which one is trying to correlate the phase noise measurement. For example, the RMS jitter value obtained from a real-time oscilloscope jitter histogram has a better correlation with the phase noise value if the signal being measured has no deterministic jitter components. If the signal contains other jitter components apart from random jitter, then it is necessary to use a jitter separation algorithm to obtain the RJ value from the histogram. Note that this challenge is on top of the already discussed phase noise integration frequency region and how that translates to correlation challenges with other instruments.

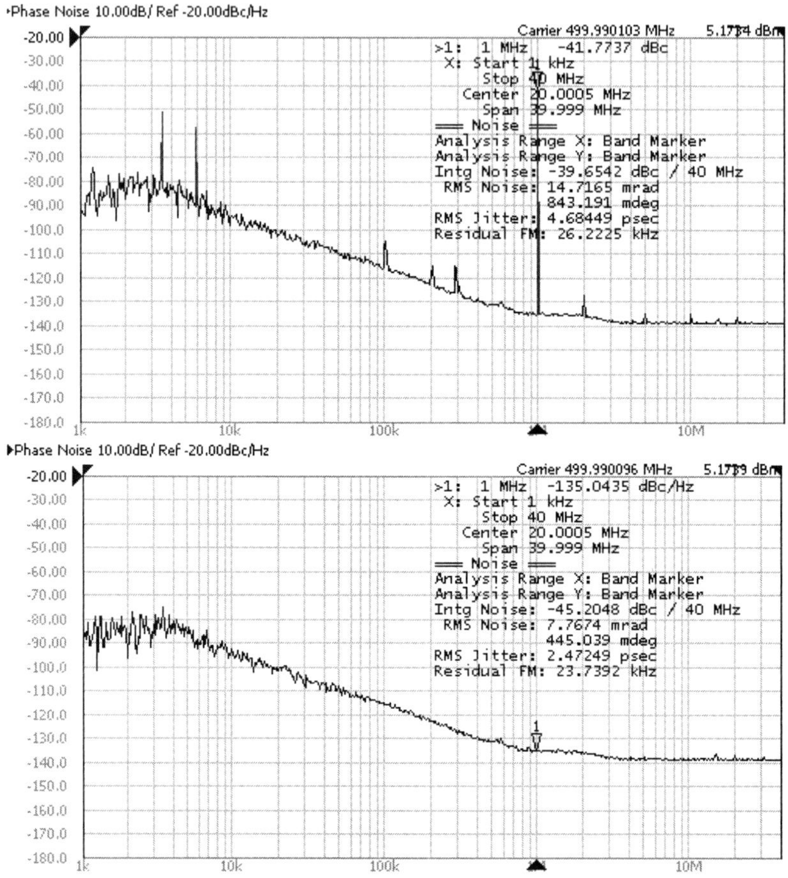

Figure J.5 Phase noise spectrum with deterministic jitter components in the measured signal (top) and with deterministic jitter components removed (bottom).

References

[1] C. Rauscher, *Fundamentals of Spectrum Analysis*. Rhode & Schwarz, 2004.

[2] R. A. Witte, *Spectrum and Network Measurements*. Noble Publishing Corporation, 2001.

[3] W. P. Robins, *Phase Noise in Signal Sources*. The Institution for Engineering and Technology, 2007.

About the Authors

José Moreira has worked for the last 15 years in the ATE industry where he has held several positions at Agilent Technologies, Verigy and Advantest. He has supported multiple customers and developed customized hardware and software solutions for ATE characterization and production testing of high-speed digital interfaces. He has also submitted and been awarded several patents and published multiple papers in the areas of ATE high-speed digital testing, jitter testing, test fixture design, and focus calibration. He is also working in the areas of mmwave test fixture design and silicon photonics testing. Mr. Moreira received a master of science degree in electrical and computer engineering from the Instituto Superior Técnico of the Technical University of Lisbon and he is currently a senior staff engineer at Advantest R&D in Boeblingen, Germany. He is a senior member of the IEEE.

Hubert Werkmann is currently employed as consulting director at Advantest Europe GmbH, where he works on ATE test solution development for high-speed I/O interfaces and high-speed memory devices in combination with the deployment of these test solutions to leading semiconductor manufacturers. His most recent work areas are ATE-based test implementations and test strategies for 100 Gigabit Ethernet devices, PAM4 and silicon photonics devices. He has over 25 years of experience in the semiconductor design and test industry with various positions at IMS Chips, Agilent Technologies, Verigy and Advantest. Dr. Werkmann has multiple published papers in the areas of scan-based design for testability and ATE high-speed testing of computation and memory devices. Dr. Werkmann received a Diploma degree in computer science from the University Stuttgart, Germany, and holds a Ph.D. in engineering from the University Stuttgart.

Index

Artech House Microwave Library

Behavioral Modeling and Linearization of RF Power Amplifiers, John Wood

Chipless RFID Reader Architecture, Nemai Chandra Karmakar, Prasanna Kalansuriya, Randika Koswatta, and Rubayet E-Azim

Control Components Using Si, GaAs, and GaN Technologies, Inder J. Bahl

Design of Linear RF Outphasing Power Amplifiers, Xuejun Zhang, Lawrence E. Larson, and Peter M. Asbeck

Design Methodology for RF CMOS Phase Locked Loops, Carlos Quemada, Guillermo Bistué, and Iñigo Adin

Design of CMOS Operational Amplifiers, Rasoul Dehghani

Design of RF and Microwave Amplifiers and Oscillators, Second Edition, Pieter L. D. Abrie

Digital Filter Design Solutions, Jolyon M. De Freitas

Discrete Oscillator Design Linear, Nonlinear, Transient, and Noise Domains, Randall W. Rhea

Distortion in RF Power Amplifiers, Joel Vuolevi and Timo Rahkonen

Distributed Power Amplifiers for RF and Microwave Communications, Narendra Kumar and Andrei Grebennikov

Electronics for Microwave Backhaul, Vittorio Camarchia, Roberto Quaglia, and Marco Pirola, editors

EMPLAN: Electromagnetic Analysis of Printed Structures in Planarly Layered Media, Software and User's Manual, Noyan Kinayman and M. I. Aksun

An Engineer's Guide to Automated Testing of High-Speed Interfaces, Second Edition, José Moreira and Hubert Werkmann

Envelope Tracking Power Amplifiers for Wireless Communications, Zhancang Wang

Essentials of RF and Microwave Grounding, Eric Holzman

FAST: Fast Amplifier Synthesis Tool—Software and User's Guide, Dale D. Henkes

Feedforward Linear Power Amplifiers, Nick Pothecary

Filter Synthesis Using Genesys S/Filter, Randall W. Rhea

Foundations of Oscillator Circuit Design, Guillermo Gonzalez

Frequency Synthesizers: Concept to Product, Alexander Chenakin

Fundamentals of Nonlinear Behavioral Modeling for RF and Microwave Design, John Wood and David E. Root, editors

Generalized Filter Design by Computer Optimization, Djuradj Budimir

Handbook of Dielectric and Thermal Properties of Materials at Microwave Frequencies, Vyacheslav V. Komarov

Handbook of RF, Microwave, and Millimeter-Wave Components, Leonid A. Belov, Sergey M. Smolskiy, and Victor N. Kochemasov

High-Linearity RF Amplifier Design, Peter B. Kenington

High-Speed Circuit Board Signal Integrity, Stephen C. Thierauf

Integrated Microwave Front-Ends with Avionics Applications, Leo G. Maloratsky

Intermodulation Distortion in Microwave and Wireless Circuits, José Carlos Pedro and Nuno Borges Carvalho

Introduction to Modeling HBTs, Matthias Rudolph

Introduction to RF Design Using EM Simulators, Hiroaki Kogure, Yoshie Kogure, and James C. Rautio

Introduction to RF and Microwave Passive Components, Richard Wallace and Krister Andreasson

Klystrons, Traveling Wave Tubes, Magnetrons, Crossed-Field Amplifiers, and Gyrotrons, A. S. Gilmour, Jr.

Lumped Elements for RF and Microwave Circuits, Inder Bahl

Lumped Element Quadrature Hybrids, David Andrews

Microstrip Lines and Slotlines, Third Edition, Ramesh Garg, Inder Bahl, and Maurizio Bozzi

Microwave Circuit Modeling Using Electromagnetic Field Simulation, Daniel G. Swanson, Jr. and Wolfgang J. R. Hoefer

Microwave Component Mechanics, Harri Eskelinen and Pekka Eskelinen

Microwave Differential Circuit Design Using Mixed-Mode S-Parameters, William R. Eisenstadt, Robert Stengel, and Bruce M. Thompson

Microwave Engineers' Handbook, Two Volumes, Theodore Saad, editor

Microwave Filters, Impedance-Matching Networks, and Coupling Structures, George L. Matthaei, Leo Young, and E. M. T. Jones

Microwave Materials and Fabrication Techniques, Second Edition, Thomas S. Laverghetta

Microwave Materials for Wireless Applications, David B. Cruickshank

Microwave Mixer Technology and Applications, Bert Henderson and Edmar Camargo

Microwave Mixers, Second Edition, Stephen A. Maas

Microwave Network Design Using the Scattering Matrix, Janusz A. Dobrowolski

Microwave Radio Transmission Design Guide, Second Edition, Trevor Manning

Microwave and RF Semiconductor Control Device Modeling, Robert H. Caverly

Microwave Transmission Line Circuits, William T. Joines, W. Devereux Palmer, and Jennifer T. Bernhard

Microwaves and Wireless Simplified, Third Edition, Thomas S. Laverghetta

Modern Microwave Circuits, Noyan Kinayman and M. I. Aksun

Modern Microwave Measurements and Techniques, Second Edition, Thomas S. Laverghetta

Neural Networks for RF and Microwave Design, Q. J. Zhang and K. C. Gupta

Noise in Linear and Nonlinear Circuits, Stephen A. Maas

Nonlinear Microwave and RF Circuits, Second Edition, Stephen A. Maas

Q Factor Measurements Using MATLAB®, Darko Kajfez

QMATCH: Lumped-Element Impedance Matching, Software and User's Guide, Pieter L. D. Abrie

Passive RF Component Technology: Materials, Techniques, and Applications, Guoan Wang and Bo Pan, editors

Practical Analog and Digital Filter Design, Les Thede

Practical Microstrip Design and Applications, Günter Kompa

Practical Microwave Circuits, Stephen Maas

Practical RF Circuit Design for Modern Wireless Systems, Volume I: Passive Circuits and Systems, Les Besser and Rowan Gilmore

Practical RF Circuit Design for Modern Wireless Systems, Volume II: Active Circuits and Systems, Rowan Gilmore and Les Besser

Production Testing of RF and System-on-a-Chip Devices for Wireless Communications, Keith B. Schaub and Joe Kelly

Radio Frequency Integrated Circuit Design, Second Edition, John W. M. Rogers and Calvin Plett

RF Bulk Acoustic Wave Filters for Communications, Ken-ya Hashimoto

RF Design Guide: Systems, Circuits, and Equations, Peter Vizmuller

RF Linear Accelerators for Medical and Industrial Applications, Samy Hanna

RF Measurements of Die and Packages, Scott A. Wartenberg

The RF and Microwave Circuit Design Handbook, Stephen A. Maas

RF and Microwave Coupled-Line Circuits, Rajesh Mongia, Inder Bahl, and Prakash Bhartia

RF and Microwave Oscillator Design, Michal Odyniec, editor

RF Power Amplifiers for Wireless Communications, Second Edition, Steve C. Cripps

RF Systems, Components, and Circuits Handbook, Ferril A. Losee

The Six-Port Technique with Microwave and Wireless Applications, Fadhel M. Ghannouchi and Abbas Mohammadi

Solid-State Microwave High-Power Amplifiers, Franco Sechi and Marina Bujatti

Stability Analysis of Nonlinear Microwave Circuits, Almudena Suárez and Raymond Quéré

Substrate Noise Coupling in Analog/RF Circuits, Stephane Bronckers, Geert Van der Plas, Gerd Vandersteen, and Yves Rolain

System-in-Package RF Design and Applications, Michael P. Gaynor

Terahertz Metrology, Mira Naftaly, editor

TRAVIS 2.0: Transmission Line Visualization Software and User's Guide, Version 2.0, Robert G. Kaires and Barton T. Hickman

Understanding Microwave Heating Cavities, Tse V. Chow Ting Chan and Howard C. Reader

Understanding Quartz Crystals and Oscillators, Ramón M. Cerda

For further information on these and other Artech House titles, including previously considered out-of-print books now available through our In-Print-Forever® (IPF®) program, contact:

Artech House Publishers
685 Canton Street
Norwood, MA 02062
Phone: 781-769-9750
Fax: 781-769-6334
e-mail: artech@artechhouse.com

Artech House Books
16 Sussex Street
London SW1V 4RW UK
Phone: +44 (0)20 7596 8750
Fax: +44 (0)20 7630 0166
e-mail: artech-uk@artechhouse.com

Find us on the World Wide Web at: www.artechhouse.com